Data Structures and Algorithms in C++

THIRD EDITION

Adam Drozdek

Andover • Melbourne • Mexico City • Stamford, CT • Toronto • Hong Kong • New Delhi • Seoul • Singapore • Tokyo

Data Structures and Algorithms in C++, 3e
Adam Drozdek

© 2006 by Course Technology, a part of Cengage Learning

This edition is reprinted with license from Course Technology, a part of Cengage Learning, for sale in India, Pakistan, Bangladesh, Nepal and Sri Lanka.

ALL RIGHTS RESERVED. No part of this work covered by the copyright herein may be reproduced, transmitted, stored or used in any form or by any means graphic, electronic, or mechanical, including but not limited to photocopying, recording, scanning, digitizing, taping, Web distribution, information networks, or information storage and retrieval systems, except as permitted under Section 107 or 108 of the 1976 United States Copyright Act, without the prior written permission of the publisher

ISBN-13: 978-81-315-0115-3
ISBN-10: 81-315-0115-9

Cengage Learning India Private Limited
418, F.I.E., Patparganj
Delhi 110092
India

Tel: 91-11-43641111
Fax: 91-11-43641100
Email: asia.infoindia@cengage.com

Cengage Learning is a leading provider of customized learning solutions with office locations around the globe, including Andover, Melbourne, Mexico City, Stamford (CT), Toronto, Hong Kong, New Delhi, Seoul, Singapore, and Tokyo. Locate your local office at: www.cengage.com/global

Cengage Learning Products are represented in Canada by Nelson Education, Ltd.

For product information, visit our website at www.cengage.co.in

Printed in India
Eighth Indian Reprint 2012

TO MY DAUGHTERS, JUSTYNA AND KASIA

Contents

1 OBJECT-ORIENTED PROGRAMMING USING C++   1
   1.1   Abstract Data Types   1
   1.2   Encapsulation   1
   1.3   Inheritance   6
   1.4   Pointers   9
       1.4.1   Pointers and Arrays   12
       1.4.2   Pointers and Copy Constructors   14
       1.4.3   Pointers and Destructors   16
       1.4.4   Pointers and Reference Variables   17
       1.4.5   Pointers to Functions   19
   1.5   Polymorphism   21
   1.6   C++ and Object-Oriented Programming   23
   1.7   The Standard Template Library   24
       1.7.1   Containers   24
       1.7.2   Iterators   24
       1.7.3   Algorithms   25
       1.7.4   Function Objects   26
   1.8   Vectors in the Standard Template Library   27
   1.9   Data Structures and Object-Oriented Programming   35
   1.10   Case Study: Random Access File   36
   1.11   Exercises   47
   1.12   Programming Assignments   49
       Bibliography   51

2 Complexity Analysis — 52

- 2.1 Computational and Asymptotic Complexity 52
- 2.2 Big-O Notation 53
- 2.3 Properties of Big-O Notation 55
- 2.4 Ω and Θ Notations 57
- 2.5 Possible Problems 58
- 2.6 Examples of Complexities 59
- 2.7 Finding Asymptotic Complexity: Examples 60
- 2.8 The Best, Average, and Worst Cases 62
- 2.9 Amortized Complexity 65
- 2.10 NP-Completeness 69
- 2.11 Exercises 72
- Bibliography 75

3 Linked Lists — 76

- 3.1 Singly Linked Lists 76
  - 3.1.1 Insertion 82
  - 3.1.2 Deletion 84
  - 3.1.3 Search 90
- 3.2 Doubly Linked Lists 91
- 3.3 Circular Lists 95
- 3.4 Skip Lists 97
- 3.5 Self-Organizing Lists 102
- 3.6 Sparse Tables 107
- 3.7 Lists in the Standard Template Library 110
- 3.8 Deques in the Standard Template Library 114
- 3.9 Concluding Remarks 119
- 3.10 Case Study: A Library 120
- 3.11 Exercises 131
- 3.12 Programming Assignments 133
- Bibliography 136

4 STACKS AND QUEUES 137

- 4.1 Stacks 137
- 4.2 Queues 145
- 4.3 Priority Queues 154
- 4.4 Stacks in the Standard Template Library 155
- 4.5 Queues in the Standard Template Library 155
- 4.6 Priority Queues in the Standard Template Library 157
- 4.7 Case Study: Exiting a Maze 159
- 4.8 Exercises 166
- 4.9 Programming Assignments 167
  Bibliography 169

5 RECURSION 170

- 5.1 Recursive Definitions 170
- 5.2 Function Calls and Recursion Implementation 173
- 5.3 Anatomy of a Recursive Call 175
- 5.4 Tail Recursion 178
- 5.5 Nontail Recursion 179
- 5.6 Indirect Recursion 185
- 5.7 Nested Recursion 187
- 5.8 Excessive Recursion 187
- 5.9 Backtracking 191
- 5.10 Concluding Remarks 198
- 5.11 Case Study: A Recursive Descent Interpreter 199
- 5.12 Exercises 208
- 5.13 Programming Assignments 211
  Bibliography 214

6 BINARY TREES 215

- 6.1 Trees, Binary Trees, and Binary Search Trees 215
- 6.2 Implementing Binary Trees 220
- 6.3 Searching a Binary Search Tree 223

- 6.4 Tree Traversal 225
  - 6.4.1 Breadth-First Traversal 226
  - 6.4.2 Depth-First Traversal 227
  - 6.4.3 Stackless Depth-First Traversal 234
- 6.5 Insertion 241
- 6.6 Deletion 244
  - 6.6.1 Deletion by Merging 245
  - 6.6.2 Deletion by Copying 247
- 6.7 Balancing a Tree 251
  - 6.7.1 The DSW Algorithm 254
  - 6.7.2 AVL Trees 257
- 6.8 Self-Adjusting Trees 262
  - 6.8.1 Self-Restructuring Trees 263
  - 6.8.2 Splaying 264
- 6.9 Heaps 269
  - 6.9.1 Heaps as Priority Queues 271
  - 6.9.2 Organizing Arrays as Heaps 272
- 6.10 Polish Notation and Expression Trees 277
  - 6.10.1 Operations on Expression Trees 279
- 6.11 Case Study: Computing Word Frequencies 282
- 6.12 Exercises 290
- 6.13 Programming Assignments 293
  - Bibliography 297

7 MULTIWAY TREES 300

- 7.1 The Family of B-Trees 301
  - 7.1.1 B-Trees 302
  - 7.1.2 B*-Trees 312
  - 7.1.3 B$^+$-Trees 314
  - 7.1.4 Prefix B$^+$-Trees 317
  - 7.1.5 Bit-Trees 318
  - 7.1.6 R-Trees 321
  - 7.1.7 2–4 Trees 322
  - 7.1.8 Sets and Multisets in the Standard Template Library 338
  - 7.1.9 Maps and Multimaps in the Standard Template Library 344
- 7.2 Tries 349
- 7.3 Concluding Remarks 358
- 7.4 Case Study: Spell Checker 358

	7.5	Exercises 369
	7.6	Programming Assignments 370
		Bibliography 374

8 GRAPHS   376

	8.1	Graph Representation 378
	8.2	Graph Traversals 380
	8.3	Shortest Paths 383
		8.3.1 All-to-All Shortest Path Problem 390
	8.4	Cycle Detection 393
		8.4.1 Union-Find Problem 394
	8.5	Spanning Trees 396
	8.6	Connectivity 400
		8.6.1 Connectivity in Undirected Graphs 400
		8.6.2 Connectivity in Directed Graphs 403
	8.7	Topological Sort 406
	8.8	Networks 408
		8.8.1 Maximum Flows 408
		8.8.2 Maximum Flows of Minimum Cost 418
	8.9	Matching 423
		8.9.1 Stable Matching Problem 427
		8.9.2 Assignment Problem 430
		8.9.3 Matching in Nonbipartite Graphs 432
	8.10	Eulerian and Hamiltonian Graphs 434
		8.10.1 Eulerian Graphs 434
		8.10.2 Hamiltonian Graphs 438
	8.11	Graph Coloring 444
	8.12	NP-Complete Problems in Graph Theory 447
		8.12.1 The Clique Problem 447
		8.12.2 The 3-Colorability Problem 448
		8.12.3 The Vertex Cover Problem 450
		8.12.4 The Hamiltonian Cycle Problem 451
	8.13	Case Study: Distinct Representatives 452
	8.14	Exercises 465
	8.15	Programming Assignments 470
		Bibliography 471

9 Sorting — 474

- **9.1 Elementary Sorting Algorithms** 475
  - 9.1.1 Insertion Sort 475
  - 9.1.2 Selection Sort 478
  - 9.1.3 Bubble Sort 480
- **9.2 Decision Trees** 482
- **9.3 Efficient Sorting Algorithms** 486
  - 9.3.1 Shell Sort 486
  - 9.3.2 Heap Sort 489
  - 9.3.3 Quicksort 493
  - 9.3.4 Mergesort 499
  - 9.3.5 Radix Sort 502
- **9.4 Sorting in the Standard Template Library** 506
- **9.5 Concluding Remarks** 510
- **9.6 Case Study: Adding Polynomials** 512
- **9.7 Exercises** 520
- **9.8 Programming Assignments** 521
- **Bibliography** 523

10 Hashing — 525

- **10.1 Hash Functions** 526
  - 10.1.1 Division 526
  - 10.1.2 Folding 526
  - 10.1.3 Mid-Square Function 527
  - 10.1.4 Extraction 527
  - 10.1.5 Radix Transformation 528
- **10.2 Collision Resolution** 528
  - 10.2.1 Open Addressing 528
  - 10.2.2 Chaining 534
  - 10.2.3 Bucket Addressing 536
- **10.3 Deletion** 537
- **10.4 Perfect Hash Functions** 538
  - 10.4.1 Cichelli's Method 539
  - 10.4.2 The FHCD Algorithm 542
- **10.5 Hash Functions for Extendible Files** 544
  - 10.5.1 Extendible Hashing 545
  - 10.5.2 Linear Hashing 547

10.6	Case Study: Hashing with Buckets	550
10.7	Exercises	560
10.8	Programming Assignments	561
	Bibliography	562

11 Data Compression   564

11.1	Conditions for Data Compression	564
11.2	Huffman Coding	566
	11.2.1 Adaptive Huffman Coding	575
11.3	Run-Length Encoding	580
11.4	Ziv-Lempel Code	581
11.5	Case Study: Huffman Method with Run-Length Encoding	584
11.6	Exercises	596
11.7	Programming Assignments	596
	Bibliography	598

12 Memory Management   599

12.1	The Sequential-Fit Methods	600
12.2	The Nonsequential-Fit Methods	601
	12.2.1 Buddy Systems	603
12.3	Garbage Collection	610
	12.3.1 Mark-and-Sweep	611
	12.3.2 Copying Methods	618
	12.3.3 Incremental Garbage Collection	620
12.4	Concluding Remarks	628
12.5	Case Study: An In-Place Garbage Collector	629
12.6	Exercises	638
12.7	Programming Assignments	639
	Bibliography	642

13 String Matching   644

13.1	Exact String Matching	644
	13.1.1 Straightforward Algorithms	644
	13.1.2 The Knuth-Morris-Pratt Algorithm	647
	13.1.3 The Boyer-Moore Algorithm	655

		13.1.4	Multiple Searches 665

13.1.4 Multiple Searches   665
13.1.5 Bit-Oriented Approach   667
13.1.6 Matching Sets of Words   670
13.1.7 Regular Expression Matching   677
13.1.8 Suffix Tries and Trees   681
13.1.9 Suffix Arrays   687

13.2 **Approximate String Matching**   689
  13.2.1 String Similarity   690
  13.2.2 String Matching with k Errors   696

13.3 **Case Study: Longest Common Substring**   699

13.4 **Exercises**   708

13.5 **Programming Assignments**   710

Bibliography   711

APPENDIXES

A **Computing Big-O**   713
  A.1 Harmonic Series   713
  A.2 Approximation of the Function $\lg(n!)$ 713
  A.3 Big-O for Average Case of Quicksort   715
  A.4 Average Path Length in a Random Binary Tree   717
  A.5 The Number of Nodes in an AVL Tree   718

B **Algorithms in the Standard Template Library**   719
  B.1 Standard Algorithms   719

C **NP-Completeness**   728
  C.1 Cook's Theorem   728

Name Index   741
Subject Index   745

Preface

The study of data structures, a fundamental component of a computer science education, serves as the foundation upon which many other computer science fields are built. Some knowledge of data structures is a must for students who wish to do work in design, implementation, testing, or maintenance of virtually any software system. The scope and presentation of material in *Data Structures and Algorithms in C++* provide students with the necessary knowledge to perform such work.

This book highlights three important aspects of data structures. First, a very strong emphasis is placed on the connection between data structures and their algorithms, including analyzing algorithms' complexity. Second, data structures are presented in an object-oriented setting in accordance with the current design and implementation paradigm. In particular, the information-hiding principle to advance encapsulation and decomposition is stressed. Finally, an important component of the book is data structure implementation, which leads to the choice of C++ as the programming language.

The C++ language, an object-oriented descendant of C, is widespread in industry and academia as an excellent programming language. It is also useful and natural for introducing data structures. Therefore, because of the wide use of C++ in application programming and the object-oriented characteristics of the language, using C++ to teach a data structures and algorithms course, even on the introductory level, is well justified.

This book provides the material for a course that includes the topics listed under CS2 and CS7 of the old ACM curriculum. It also meets the requirements for most of the courses C_A 202, C_D 202, and C_F 204 of the new ACM curriculum.

Most chapters include a case study that illustrates a complete context in which certain algorithms and data structures can be used. These case studies were chosen from different areas of computer science such as interpreters, symbolic computation, and file processing, to indicate the wide range of applications to which topics under discussion may apply.

Brief examples of C++ code are included throughout the book to illustrate the practical importance of data structures. However, theoretical analysis is equally important, so presentations of algorithms are integrated with analyses of efficiency.

Great care is taken in the presentation of recursion because even advanced students have problems with it. Experience has shown that recursion can be explained best if the run-time stack is taken into consideration. Changes to the stack are shown when tracing a recursive function not only in the chapter on recursion, but also in other chapters. For example, a surprisingly short function for tree traversal may remain a mystery if work done by the system on the run-time stack is not included in the explanation. Standing aloof from the system and retaining only a purely theoretical perspective when discussing data structures and algorithms are not necessarily helpful.

The thrust of this book is data structures, and other topics are treated here only as much as necessary to ensure a proper understanding of this subject. Algorithms are discussed from the perspective of data structures, so the reader will not find a comprehensive discussion of different kinds of algorithms and all the facets that a full presentation of algorithms requires. However, as mentioned, recursion is covered in depth. In addition, complexity analysis of algorithms is presented in some detail.

Chapters 1 and 3–8 present a number of different data structures and the algorithms that operate on them. The efficiency of each algorithm is analyzed, and improvements to the algorithm are suggested.

- Chapter 1 presents the basic principles of object-oriented programming, an introduction to dynamic memory allocation and the use of pointers, and a rudimentary presentation of the Standard Template Library (STL).
- Chapter 2 describes some methods used to assess the efficiency of algorithms.
- Chapter 3 presents different types of linked lists with an emphasis on their implementation with pointers.
- Chapter 4 presents stacks and queues and their applications.
- Chapter 5 contains a detailed discussion of recursion. Different types of recursion are discussed, and a recursive call is dissected.
- Chapter 6 discusses binary trees, including implementation, traversal, and search. Balanced trees are also included in this chapter.
- Chapter 7 details more generalized trees such as tries, 2–4 trees, and B-trees.
- Chapter 8 presents graphs.

Chapters 9–12 show different applications of data structures introduced in the previous chapters. They emphasize the data structure aspects of each topic under consideration.

- Chapter 9 analyzes sorting in detail, and several elementary and nonelementary methods are presented.

- Chapter 10 discusses hashing, one of the most important areas in searching. Various techniques are presented with an emphasis on the utilization of data structures.
- Chapter 11 discusses data compression algorithms and data structures.
- Chapter 12 presents various techniques and data structures for memory management.
- Chapter 13 discusses many algorithms for exact and approximate string matching.
- Appendix A discusses in greater detail big-O notation, introduced in Chapter 2.
- Appendix B presents standard algorithms in the Standard Template Library.
- Appendix C gives a proof of Cook's theorem and illustrates it with an extended example.

Each chapter contains a discussion of the material illustrated with appropriate diagrams and tables. Except for Chapter 2, all chapters include a case study, which is an extended example using the features discussed in that chapter. All case studies have been tested using the Visual C++ compiler on a PC and the g++ compiler under Unix except the von Koch snowflake, which runs on a PC under Visual C++. At the end of each chapter is a set of exercises of varying degrees of difficulty. Except for Chapter 2, all chapters also include programming assignments and an up-to-date bibliography of relevant literature.

Chapters 1–6 (excluding Sections 2.9, 3.4, 6.4.3, 6.7, and 6.8) contain the core material that forms the basis of any data structures course. These chapters should be studied in sequence. The remaining six chapters can be read in any order. A one-semester course could include Chapters 1–6, 9, and Sections 10.1 and 10.2. The entire book could also be part of a two-semester sequence.

Teaching Tools

Electronic Instructor's Manual. The Instructor's Manual that accompanies this textbook includes complete solutions to all text exercises.

Electronic Figure Files. All images from the text are available in bitmap format for use in classroom presentations.

Source Code. The source code for the text example programs is available via the author's Web site at http://www.mathcs.duq.edu/drozdek/DSinCpp.

It is also available for student download at course.com. All teaching tools, outlined above, are available in the Instructor's Resources section of course.com.

CHANGES IN THE THIRD EDITION

The new edition primarily extends the old edition by including material on new topics that are currently not covered. The additions include

- Pattern matching algorithms in the new Chapter 13
- A discussion of NP-completeness in the form of a general introduction (Section 2.10), examples of NP-complete problems (Section 8.12), and an outline of Cook's theorem (Appendix C)
- New material on graphs (Sections 8.9.1, 8.10.1.1, 8.10.2.1, and 8.11)
- A discussion of a deletion algorithm for vh-trees (Section 7.1.7)

There are also many small modifications and additions throughout the book.

ACKNOWLEDGMENTS

I would like to thank the following reviewers, whose comments and advice helped me to improve this book:

> James Ball, Indiana State University
> Kevan Croteau, Francis Marion University
> Shiev Hong Lin, Biola University
> William Thacker, Winthrop University

However, the ultimate content is my responsibility, and I would appreciate hearing from readers about any shortcomings or strengths. My email address is drozdek@duq.edu.

Adam Drozdek

Object-Oriented Programming Using C++

1.1 ABSTRACT DATA TYPES

Before a program is written, we should have a fairly good idea of how to accomplish the task being implemented by this program. Hence, an outline of the program containing its requirements should precede the coding process. The larger and more complex the project, the more detailed the outline phase should be. The implementation details should be delayed to the later stages of the project. In particular, the details of the particular data structures to be used in the implementation should not be specified at the beginning.

From the start, it is important to specify each task in terms of input and output. At the beginning stages, we should be more concerned with what the program should do, not how it should or could be done. Behavior of the program is more important than the gears of the mechanism accomplishing it. For example, if an item is needed to accomplish some tasks, the item is specified in terms of operations performed on it rather than in terms of its inner structure. These operations may act upon this item, for example, modifying it, searching for some details in it, or storing something in it. After these operations are precisely specified, the implementation of the program may start. The implementation decides which data structure should be used to make execution most efficient in terms of time and space. An item specified in terms of operations is called an *abstract data type*. An abstract data type is not a part of a program, because a program written in a programming language requires the definition of a data structure, not just the operations on the data structure. However, an object-oriented language (OOL) such as C++ has a direct link to abstract data types by implementing them as a class.

1.2 ENCAPSULATION

Object-oriented programming (OOP) revolves around the concept of an object. Objects, however, are created using a class definition. A *class* is a template in accordance to which objects are created. A class is a piece of software that includes a data specification and functions operating on these data and possibly on the data belonging to other class instances. Functions defined in a class are called *methods, member functions,* or *function members,* and variables used in a class are called *data members* (more properly, they should be called datum members). This combining of the data and related operations is called *data encapsulation.* An *object* is an instance of a class, an entity created using a class definition.

In contradistinction to functions in languages that are not object-oriented, objects make the connection between data and member functions much tighter and more meaningful. In languages that are not object-oriented, declarations of data and definitions of functions can be interspersed through the entire program, and only the program documentation indicates that there is a connection between them. In OOLs, a connection is established right at the outset; in fact, the program is based on this connection. An object is defined by related data and operations, and because there may be many objects used in the same program, the objects communicate by exchanging messages that reveal to each other as little detail about their internal structure as necessary for adequate communication. Structuring programs in terms of objects allows us to accomplish several goals.

First, this strong coupling of data and operations can be used much better in modeling a fragment of the world, which is emphasized especially by software engineering. Not surprisingly, OOP has its roots in simulation; that is, in modeling real-world events. The first OOL was called Simula, and it was developed in the 1960s in Norway.

Second, objects allow for easier error finding because operations are localized to the confines of their objects. Even if side effects occur, they are easier to trace.

Third, objects allow us to conceal certain details of their operations from other objects so that these operations may not be adversely affected by other objects. This is known as the *information-hiding principle.* In languages that are not object-oriented, this principle can be found to some extent in the case of local variables, or as in Pascal, in local functions or procedures, which can be used and accessed only by the function defining them. This is, however, a very tight hiding or no hiding at all. Sometimes we may need to use (again, as in Pascal) a function $f2$ defined in $f1$ outside of $f1$, but we cannot. Sometimes we may need to access some data local to $f1$ without exactly knowing the structure of these data, but we cannot. Hence, some modification is needed, and it is accomplished in OOLs.

An object in OOL is like a watch. As users, we are interested in what the hands show, but not in the inner workings of the watch. We are aware that there are gears and springs inside the watch, but because we usually know very little about why all these parts are in a particular configuration, we should not have access to this mechanism so that we do not damage it, inadvertently or on purpose. This mechanism is hidden from us, we have no immediate access to it, and the watch is protected and works better than when its mechanism is open for everyone to see.

An object is like a black box whose behavior is very well defined, and we use the object because we know what it does, not because we have an insight into how it does it. This opacity of objects is extremely useful for maintaining them independently of each other. If communication channels between the objects are well defined, then changes made inside an object can affect other objects only as much as these changes affect the communication channels. Knowing the kind of information sent out and received by an object, the object can be replaced more easily by another object more suitable in a particular situation: A new object can perform the same task differently but more quickly in a certain hardware environment. An object discloses only as much as is needed for the user to utilize it. It has a public part that can be accessed by any user when the user sends a message matching any of the member function names revealed by the object. In this public part, the object displays to the user buttons that can be pushed to invoke the object's operations. The user knows only the names of these operations and the expected behavior.

Information hiding tends to blur the dividing line between data and operations. In Pascal-like languages, the distinction between data and functions or procedures is clear and rigid. They are defined differently and their roles are very distinct. OOLs put data and methods together, and to the user of the object, this distinction is much less noticeable. To some extent, this incorporates the features of functional languages. LISP, one of the earliest programming languages, allows the user to treat data and functions similarly, because the structure of both is the same.

We have already made a distinction between particular objects and object types or classes. We write functions to be used with different variables, and by analogy, we do not like to be forced to write as many object declarations as the number of objects required by the program. Certain objects are of the same type and we would like only to use a reference to a general object specification. For single variables, we make a distinction between type declaration and variable declaration. In the case of objects, we have a class declaration and object instantiation. For instance, in the following class declaration, C is a class and object1 through object3 are objects.

```
class C {
public:
    C(char *s = "", int i = 0, double d = 1) {
        strcpy(dataMember1,s);
        dataMember2 = i;
        dataMember3 = d;
    }
    void memberFunction1() {
        cout << dataMember1 << ' ' << dataMember2 << ' '
             << dataMember3 << endl;
    }
    void memberFunction2(int i, char *s = "unknown") {
        dataMember2 = i;
        cout << i << " received from " << s << endl;
    }
protected:
    char dataMember1[20];
    int dataMember2;
    double dataMember3;
};

C object1("object1",100,2000), object2("object2"), object3;
```

Message passing is equivalent to a function call in traditional languages. However, to stress the fact that in OOLs the member functions are relative to objects, this new term is used. For example, the call to public `memberFunction1()` in `object1`,

```
object1.memberFunction1();
```

is seen as message `memberFunction1()` sent to `object1`. Upon receiving the message, the object invokes its member function and displays all relevant information. Messages can include parameters so that

```
object1.memberFunction2(123);
```

is the message `memberFunction2()` with parameter 123 received by `object1`.

The lines containing these messages are either in the main program, in a function, or in a member function of another object. Therefore, the receiver of the message is identifiable, but not necessarily the sender. If `object1` receives the message `memberFunction1()`, it does not know where the message originated. It only responds to it by displaying the information `memberFunction1()` encapsulates. The same goes for `memberFunction2()`. Therefore, the sender may prefer sending a message that also includes its identification, as follows:

```
object1.memberFunction2(123, "object1");
```

A powerful feature of C++ is the possibility of declaring generic classes by using type parameters in the class declaration. For example, if we need to declare a class that uses an array for storing some items, then we may declare this class as

```
class intClass {
    int storage[50];
    ..................
};
```

However, in this way, we limit the usability of this class to integers only; if we need a class that performs the same operations as `intClass` except that it operates on float numbers, then a new declaration is needed, such as

```
class floatClass {
    float storage[50];
    ..................
};
```

If `storage` is to hold structures, or pointers to characters, then two more classes must be declared. It is much better to declare a generic class and decide what type of items the object is referring to only when defining the object. Fortunately, C++ allows us to declare a class in this way, and the declaration for the example is

```
template<class genType>
class genClass {
    genType storage[50];
    ..................
};
```

Later, we make the decision about how to initialize `genType`:

```
genClass<int> intObject;
genClass<float> floatObject;
```

This generic class becomes a basis for generating two new classes, `genClass` of `int` and `genClass` of `float`, and then these two classes are used to create two objects, `intObject` and `floatObject`. In this way, the generic class manifests itself in different forms depending on the specific declaration. One generic declaration suffices for enabling such different forms.

We can go even further than that by not committing ourselves to 50 cells in storage and by delaying that decision until the object definition stage. But just in case, we may leave a default value so that the class declaration is now

```
template<class genType, int size = 50>
class genClass {
    genType storage[size];
    ................
};
```

The object definition is now

```
genClass<int> intObject1; // use the default size;
genClass<int,100> intObject2;
genClass<float,123> floatObject;
```

This method of using generic types is not limited to classes only; we can use them in function declarations as well. For example, the standard operation for swapping two values can be defined by the function

```
template<class genType>
void swap(genType& el1, genType& el2) {
    genType tmp = el1; el1 = el2; el2 = tmp;
}
```

This example also indicates the need for adapting built-in operators to specific situations. If `genType` is a number, a character, or a structure, then the assignment operator, =, performs its function properly. But if `genType` is an array, then we can expect a problem in `swap()`. The problem can be resolved by overloading the assignment operator by adding to it the functionality required by a specific data type.

After a generic function has been declared, a proper function can be generated at compilation time. For example, if the compiler sees two calls,

```
swap(n,m); // swap two integers;
swap(x,y); // swap two floats;
```

it generates two swap functions to be used during execution of the program.

1.3 INHERITANCE

OOLs allow for creating a hierarchy of classes so that objects do not have to be instantiations of a single class. Before discussing the problem of inheritance, consider the following class definitions:

```cpp
class BaseClass {
public:
    BaseClass() { }
    void f(char *s = "unknown") {
        cout << "Function f() in BaseClass called from " << s << endl;
        h();
    }
protected:
    void g(char *s = "unknown") {
        cout << "Function g() in BaseClass called from " << s << endl;
    }
private:
    void h() {
        cout << "Function h() in BaseClass\n";
    }
};
class Derived1Level1 : public virtual BaseClass {
public:
    void f(char *s = "unknown") {
        cout << "Function f() in Derived1Level1 called from " << s << endl;
        g("Derived1Level1");
        h("Derived1Level1");
    }
    void h(char *s = "unknown") {
        cout << "Function h() in Derived1Level1 called from " << s << endl;
    }
};
class Derived2Level1 : public virtual BaseClass {
public:
    void f(char *s = "unknown") {
        cout << "Function f() in Derived2Level1 called from " << s << endl;
        g("Derived2Level1");
//      h();   // error: BaseClass::h() is not accessible
    }
};
class DerivedLevel2 : public Derived1Level1, public Derived2Level1 {
public:
    void f(char *s = "unknown") {
        cout << "Function f() in DerivedLevel2 called from " << s << endl;
        g("DerivedLevel2");
```

```
        Derived1Level1::h("DerivedLevel2");
        BaseClass::f("DerivedLevel2");
    }
};
```

A sample program is

```
int main() {
    BaseClass bc;
    Derived1Level1 d1l1;
    Derived2Level1 d2l1;
    DerivedLevel2 dl2;
    bc.f("main(1)");
//  bc.g();  // error: BaseClass::g() is not accessible
//  bc.h();  // error: BaseClass::h() is not accessible
    d1l1.f("main(2)");
//  d1l1.g(); // error: BaseClass::g() is not accessible
    d1l1.h("main(3)");
    d2l1.f("main(4)");
//  d2l1.g(); // error: BaseClass::g() is not accessible
//  d2l1.h(); // error: BaseClass::h() is not accessible
    dl2.f("main(5)");
//  dl2.g();  // error: BaseClass::g() is not accessible
    dl2.h();
    return 0;
}
```

This sample produces the following output:

```
Function f() in BaseClass called from main(1)
Function h() in BaseClass
Function f() in Derived1Level1 called from main(2)
Function g() in BaseClass called from Derived1Level1
Function h() in Derived1Level1 called from Derived1Level1
Function h() in Derived1Level1 called from main(3)
Function f() in Derived2Level1 called from main(4)
Function g() in BaseClass called from Derived2Level1
Function f() in DerivedLevel2 called from main(5)
Function g() in BaseClass called from DerivedLevel2
Function h() in Derived1Level1 called from DerivedLevel2
Function f() in BaseClass called from DerivedLevel2
Function h() in BaseClass
Function h() in Derived1Level1 called from unknown
```

The class BaseClass is called a *base class* or a *superclass*, and other classes are called *subclasses* or *derived classes* because they are derived from the superclass in that they can use the data members and member functions specified in BaseClass as protected or public. They inherit all these members from their base class so that they do not have to repeat the same definitions. However, a derived class can override the definition of a member function by introducing its own definition. In this way, both the base class and the derived class have some measure of control over their member functions.

The base class can decide which member functions and data members can be revealed to derived classes so that the principle of information hiding holds not only with respect to the user of the base class, but also to the derived classes. Moreover, the derived class can decide which parts of the public and protected member functions and data members to retain and use and which to modify. For example, both Derived1Level1 and Derived2Level1 define their own versions of f(). However, the access to the member function with the same name in any of the classes higher up in the hierarchy is still possible by preceding the function with the name of the class and the scope operator, as shown in the call of BaseClass::f() from f() in DerivedLevel2.

A derived class can add some new members of its own. Such a class can become a base class for other classes that can be derived from it so that the inheritance hierarchy can be deliberately extended. For example, the class Derived1Level1 is derived from BaseClass, but at the same time, it is the base class for DerivedLevel2.

Inheritance in our examples is specified as public by using the word public after the semicolon in the heading of the definition of a derived class. Public inheritance means that public members of the base class are also public in the derived class, and protected members are also protected. In the case of protected inheritance (with the word protected in the heading of the definition), both public and protected members of the base class become protected in the derived class. Finally, for private inheritance, both public and protected members of the base class become private in the derived class. In all types of inheritance, private members of the base class are inaccessible to any of the derived classes. For example, an attempt to call h() from f() in Derived2Level1 causes a compilation error, "BaseClass::h() is not accessible." However, a call of h() from f() in Derived1Level1 causes no problem because it is a call to h() defined in Derived1Level1.

Protected members of the base class are accessible only to derived classes and not to nonderived classes. For this reason, both Derived1Level1 and Derived2Level1 can call BaseClass's protected member function g(), but a call to this function from main() is rendered illegal.

A derived class does not have to be limited to one base class only. It can be derived from more than one base class. For example, DerivedLevel2 is defined as a class derived from both Derived1Level1 and Derived2Level1, inheriting in this way all the member functions of Derived1Level1 and Derived2Level1. However, DerivedLevel2 also inherits the same member functions from BaseClass twice because both classes used in the definition of DerivedLevel2 are derived from BaseClass. This is redundant at best, and at worst can cause a compilation error, "member is ambiguous BaseClass::g() and BaseClass::g()." To prevent this from happening, the definitions of the two classes include the modifier virtual, which means that DerivedLevel2 contains only one copy of each member function from BaseClass. A similar problem arises if f() in DerivedLevel2 calls h() without the preceding scope operator and class name, Derived1Level1::h(). It does not matter that `h()` is private in `BaseClass` and inaccessible to `DerivedLevel2`. An error would be printed, "member is ambiguous `Derived1Level1::h()` and `BaseClass::h()`."

1.4 POINTERS

Variables used in a program can be considered as boxes that are never empty; they are filled with some content either by the programmer or, if uninitialized, by the operating system. Such a variable has at least two attributes: the content or value and the location of the box or variable in computer memory. This content can be a number, a character, or a compound item such as a structure or union. However, this content can also be the location of another variable; variables with such contents are called *pointers*. Pointers are usually auxiliary variables that allow us to access the values of other variables indirectly. A pointer is analogous to a road sign that leads us to a certain location or to a slip of paper on which an address has been jotted down. They are variables leading to variables, humble auxiliaries that point to some other variables as the focus of attention.

For example, in the declaration

```
int i = 15, j, *p, *q;
```

`i` and `j` are numerical variables and `p` and `q` are pointers to numbers; the star in front of `p` and `q` indicates their function. Assuming that the addresses of the variables `i`, `j`, `p`, and `q` are 1080, 1082, 1084, and 1086, then after assigning 15 to `i` in the declaration, the positions and values of the variables in computer memory are as in Figure 1.1a.

Now, we could make the assignment `p = i` (or `p = (int*) i` if the compiler does not accept it), but the variable `p` was created to store the address of an integer variable, not its value. Therefore, the proper assignment is `p = &i`, where the ampersand in front of `i` means that the address of `i` is meant and not its content. Figure 1.1b illustrates this situation. In Figure 1.1c, the arrow from `p` to `i` indicates that `p` is a pointer that holds the address of `i`.

We have to be able to distinguish the value of `p`, which is an address, from the value of the location whose address the pointer holds. For example, to assign 20 to the variable pointed to by `p`, the assignment statement is

```
*p = 20;
```

The star (*) here is an indirection operator that forces the system to first retrieve the contents of `p`, then access the location whose address has just been retrieved from `p`, and only afterward, assign 20 to this location (Figure 1.1d). Figures 1.1e through 1.1n give more examples of assignment statements and how the values are stored in computer memory.

In fact, pointers—like all variables—also have two attributes: a content and a location. This location can be stored in another variable, which then becomes a pointer to a pointer.

In Figure 1.1, addresses of variables were assigned to pointers. Pointers can, however, refer to anonymous locations that are accessible only through their addresses

FIGURE 1.1 Changes of values after assignments are made using pointer variables. Note that (b) and (c) show the same situation, and so do (d) and (e), (g) and (h), (i) and (j), (k) and (l), and (m) and (n).

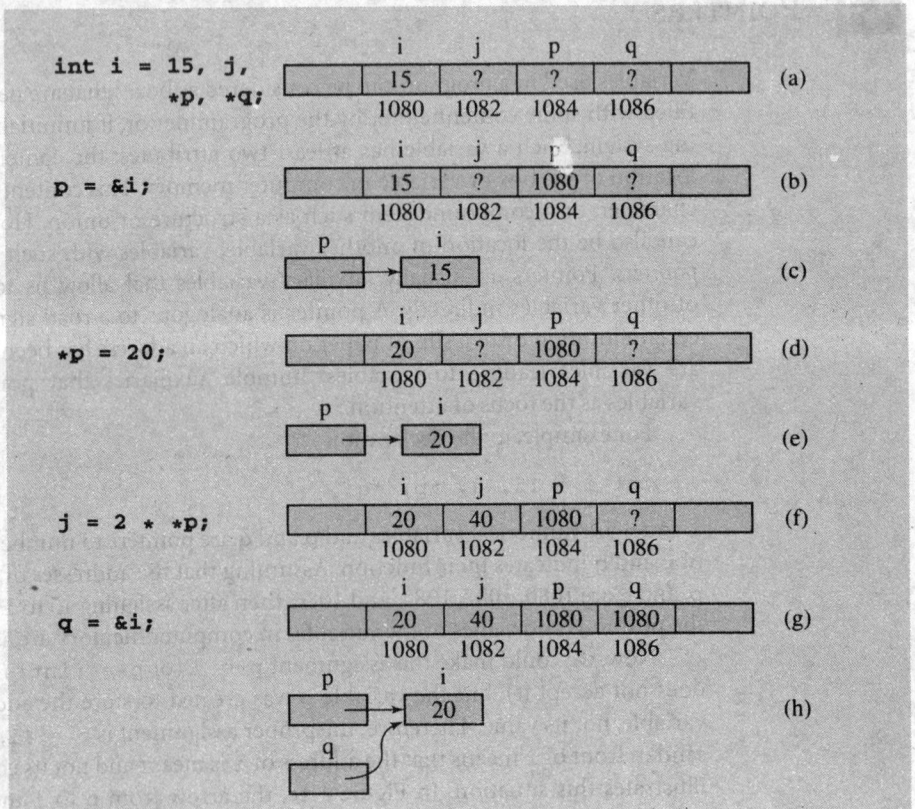

and not—like variables—by their names. These locations must be set aside by the memory manager, which is performed dynamically during the run of the program, unlike for variables, whose locations are allocated at compilation time.

To dynamically allocate and deallocate memory, two functions are used. One function, new, takes from memory as much space as needed to store an object whose type follows new. For example, with the instruction

```
p = new int;
```

the program requests enough space to store one integer, from the memory manager, and the address of the beginning of this portion of memory is stored in p. Now the values can be assigned to the memory block pointed to by p only indirectly through a pointer, either pointer p or any other pointer q that was assigned the address stored in p with the assignment q = p.

FIGURE 1.1 *(continued)*

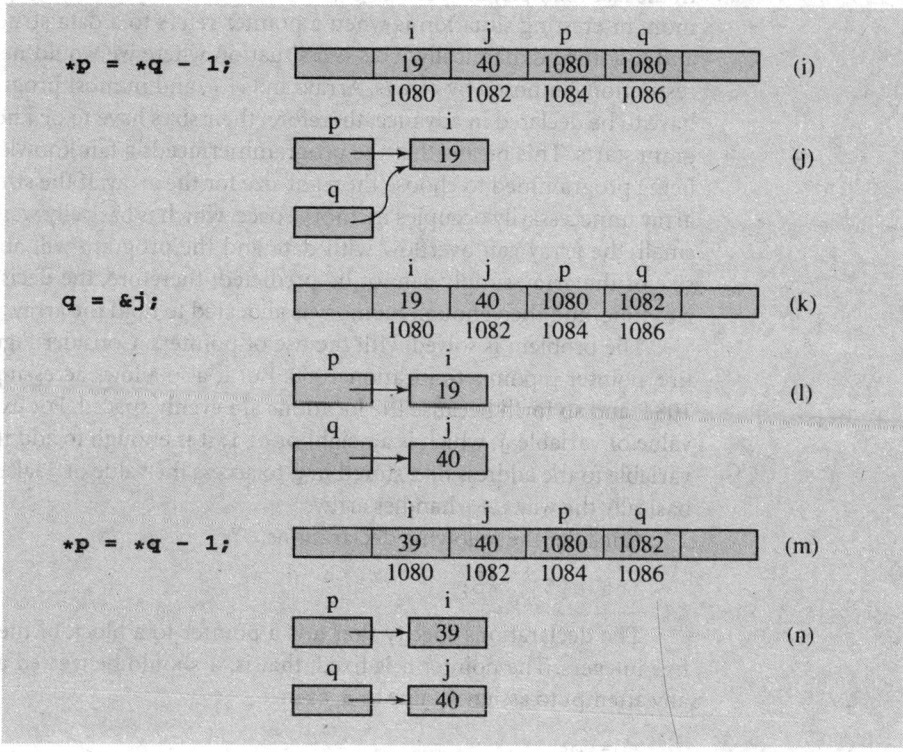

If the space occupied by the integer accessible from p is no longer needed, it can be returned to the pool of free memory locations managed by the operating system by issuing the instruction

```
delete p;
```

However, after executing this statement, the beginning addresses of the released memory block are still in p, although the block, as far as the program is concerned, does not exist anymore. It is like treating an address of a house that has been demolished as the address of the existing location. If we use this address to find someone, the result can be easily foreseen. Similarly, if after issuing the delete statement we do not erase the address from the pointer variable participating in deletion, the result is potentially dangerous, and we can crash the program when trying to access non-existing locations, particularly for more complex objects than numerical values. This is the *dangling reference problem*. To avoid this problem, an address has to be assigned to a pointer; if it cannot be the address of any location, it should be a null address, which is simply 0. After execution of the assignment

```
p = 0;
```

we may not say that p refers to null or points to null but that p becomes null or p is null.

1.4.1 Pointers and Arrays

In the last example, the pointer p refers to a block of memory that holds one integer. A more interesting situation is when a pointer refers to a data structure that is created and modified dynamically. This is a situation where we would need to overcome the restrictions imposed by arrays. Arrays in C++, and in most programming languages, have to be declared in advance; therefore, their sizes have to be known before the program starts. This means that the programmer needs a fair knowledge of the problem being programmed to choose the right size for the array. If the size is too big, then the array unnecessarily occupies memory space, which is basically wasted. If the size is too small, the array can overflow with data and the program will abort. Sometimes the size of the array simply cannot be predicted; therefore, the decision is delayed until run time, and then enough memory is allocated to hold the array.

The problem is solved with the use of pointers. Consider Figure 1.1b. In this figure, pointer p points to location 1080. But it also allows accessing of locations 1082, 1084, and so forth because the locations are evenly spaced. For example, to access the value of variable j, which is a neighbor of i, it is enough to add the size of an integer variable to the address of i stored in p to access the value of j, also from p. And this is basically the way C++ handles arrays.

Consider the following declarations:

```
int a[5], *p;
```

The declarations specify that a is a pointer to a block of memory that can hold five integers. The pointer a is fixed; that is, a should be treated as a constant so that any attempt to assign a value to a, as in

```
a = p;
```

or in

```
a++;
```

is considered a compilation error. Because a is a pointer, pointer notation can be used to access cells of the array a. For example, an array notation used in the loop that adds all the numbers in a,

```
for (sum = a[0], i = 1; i < 5; i++)
    sum += a[i];
```

can be replaced by a pointer notation

```
for (sum = *a, i = 1; i < 5; i++)
    sum += *(a + i);
```

or by

```
for (sum = *a, p = a+1; p < a+5; p++)
    sum += *p;
```

Note that a+1 is a location of the next cell of the array a so that a+1 is equivalent to &a[1]. Thus, if a equals 1020, then a+1 is not 1021 but 1022 because pointer arithmetic depends on the type of pointed entity. For example, after declarations

```
char b[5];
long c[5];
```

and assuming that b equals 1050 and c equals 1055, b+1 equals 1051 because one character occupies 1 byte, and c+1 equals 1059 because one long number occupies 4 bytes. The reason for these results of pointer arithmetic is that the expression c+i denotes the memory address c+i*sizeof(long).

In this discussion, the array a is declared statically by specifying in its declaration that it contains five cells. The size of the array is fixed for the duration of the program run. But arrays can also be declared dynamically. To that end, pointer variables are used. For example, the assignment

```
p = new int[n];
```

allocates enough room to store n integers. Pointer p can be treated as an array variable so that array notation can be used. For example, the sum of numbers in the array p can be found with the code that uses array notation,

```
for (sum = p[0], i = 1; i < n; i++)
    sum += p[i];
```

a pointer notation that is a direct rendition of the previous loop,

```
for (sum = *p, i = 1; i < n; i++)
    sum += *(p+i);
```

or a pointer notation that uses two pointers,

```
for (sum = *p, q = p+1; q < p+n; q++)
    sum += *q;
```

Because p is a variable, it can be assigned a new array. But if the array currently pointed to by p is no longer needed, it should be disposed of by the instruction

```
delete [] p;
```

Note the use of empty brackets in the instruction. The brackets indicate that p points to an array.

A very important type of array is a string, or an array of characters. Many predefined functions operate on strings. The names of these functions start with str, as in strlen(s) to find the length of the string s or strcpy(s1,s2) to copy string s2 to s1. It is important to remember that all these functions assume that strings end with the null character '\0'. For example, strcpy(s1,s2) continues copying until it finds this character in s2. If a programmer does not include this character in s2, copying stops when the first occurrence of this character is found somewhere in computer memory after location s2. This means that copying is performed to locations outside s1, which eventually may lead the program to crash.

1.4.2 Pointers and Copy Constructors

Some problems can arise when pointer data members are not handled properly when copying data from one object to another. Consider the following definition:

```
struct Node {
    char *name;
    int age;
    Node(char *n = "", int a = 0) {
        name = strdup(n);
        strcpy(name,n);
        age = a;
    }
};
```

The intention of the declarations

```
Node node1("Roger",20), node2(node1); //or node2 = node1;
```

is to create object node1, assign values to the two data members in node1, and then create object node2 and initialize its data members to the same values as in node1. These objects are to be independent entities so that assigning values to one of them should not affect values in the other. However, after the assignments

```
strcpy(node2.name,"Wendy");
node2.age = 30;
```

the printing statement

```
cout<<node1.name<<' '<<node1.age<<' '<<node2.name<<' '<<node2.age;
```

generates the output

```
Wendy 30 Wendy 20
```

The ages are different, but the names in the two objects are the same. What happened? The problem is that the definition of Node does not provide a copy constructor

```
Node(const Node&);
```

which is necessary to execute the declaration node2(node1) to initialize node1. If a user copy constructor is missing, the constructor is generated automatically by the compiler. But the compiler-generated copy constructor performs member-by-member copying. Because name is a pointer, the copy constructor copies the string address node1.name to node2.name, not the string content, so that right after execution of the declaration, the situation is as in Figure 1.2a. Now if the assignments

```
strcpy(node2.name,"Wendy");
node2.age = 30;
```

are executed, node2.age is properly updated, but the string "Roger" pointed to by the name member of both objects is overwritten by "Wendy", which is also pointed to by the two pointers (Figure 1.2b). To prevent this from happening, the user must define a proper copy constructor, as in

FIGURE 1.2 Illustrating the necessity of using a copy constructor for objects with pointer members.

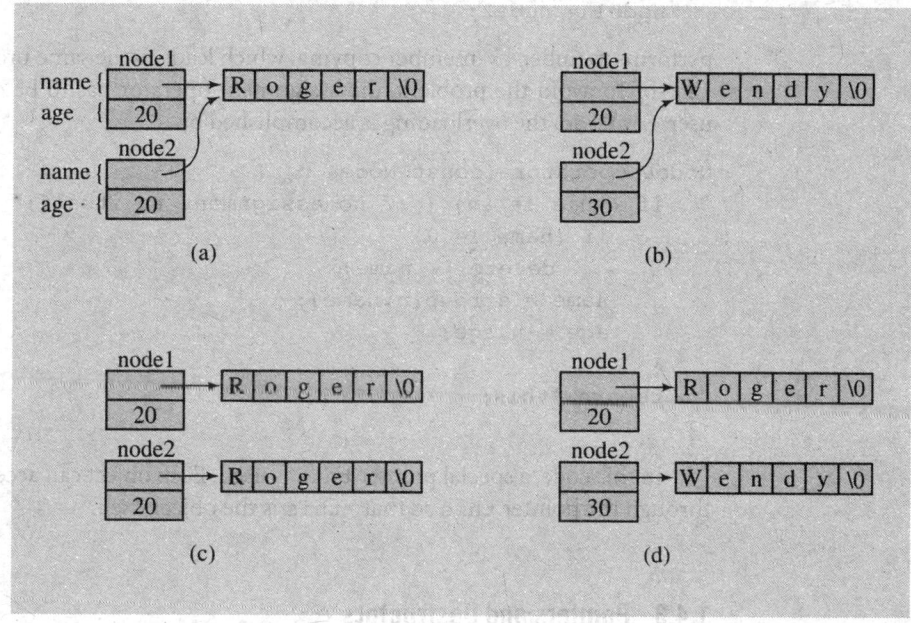

```
struct Node {
    char *name;
    int age;
    Node(char *n = 0, int a = 0) {
        name = strdup(n);
        age = a;
    }
    Node(const Node& n) { // copy constructor;
        name = strdup(n.name);
        age = n.age;
    }
};
```

With the new constructor, the declaration node2(node1) generates another copy of "Roger" pointed to by node2.name (Figure 1.2c), and the assignments to data members in one object have no effect on members in the other object, so that after execution of the assignments

```
strcpy(node2.name,"Wendy");
node2.age = 30;
```

the object node1 remains unchanged, as illustrated in Figure 1.2d.

Note that a similar problem is raised by the assignment operator. If a definition of the assignment operator is not provided by the user, an assignment

```
node1 = node2;
```

performs member-by-member copying, which leads to the same problem as in Figure 1.2a–b. To avoid the problem, the assignment operator has to be overloaded by the user. For Node, the overloading is accomplished by

```
Node& operator=(const Node& n) {
   if (this != &n) { // no assignment to itself;
      if (name != 0)
         delete [] name;
      name = strdup(n.name);
      age = n.age;
   }
   return *this;
}
```

In this code, a special pointer this is used. Each object can access its own address through the pointer this so that *this is the object itself.

1.4.3 Pointers and Destructors

What happens to locally defined objects of type Node? Like all local items, they are destroyed in the sense that they become unavailable outside the block in which they are defined, and memory occupied by them is also released. But although memory occupied by an object of type Node is released, not all the memory related to this object becomes available. One of the data members of this object is a pointer to a string; therefore, memory occupied by the pointer data member is released, but memory taken by the string is not. After the object is destroyed, the string previously available from its data member name becomes inaccessible (if not assigned to the name of some other object or to a string variable) and memory occupied by this string can no longer be released. This is a problem with objects that have data members pointing to dynamically allocated locations. To avoid the problem, the class definition should include a definition of a destructor. A *destructor* is a function that is automatically invoked when an object is destroyed, which takes place upon exit from the block in which the object is defined or upon the call of delete. Destructors take no arguments and return no values so that there can be only one destructor per class. For the class Node, a destructor can be defined as

```
~Node() {
   if (name != 0)
      delete [] name;
}
```

1.4.4 Pointers and Reference Variables

Consider the following declarations:

```
int n = 5, *p = &n, &r = n;
```

Variable p is declared as being of type int*, a pointer to an integer, and r is of type int&, an integer reference variable. A reference variable must be initialized in its declaration as a reference to a particular variable, and this reference cannot be changed. This means that a reference variable cannot be null. A reference variable r can be considered a different name for a variable n so that if n changes then r changes as well. This is because a reference variable is implemented as a constant pointer to the variable.

After the three declarations, the printing statement

```
cout << n << ' ' << *p << ' ' << r << endl;
```

outputs 5 5 5. After the assignment

```
n = 7;
```

the same printing statement outputs 7 7 7. Also, an assignment

```
*p = 9;
```

gives the output 9 9 9, and the assignment

```
r = 10;
```

leads to the output 10 10 10. These statements indicate that in terms of notation, what we can accomplish with dereferencing of pointer variables is accomplished without dereferencing of reference variables. This is no accident because, as mentioned, reference variables are implemented as constant pointers. Instead of the declaration

```
int& r = n;
```

we can use a declaration

```
int *const r = &n;
```

where r is a constant pointer to an integer, which means that the assignment

```
r = q;
```

where q is another pointer, is an error because the value of r cannot change. However, the assignment

```
*r = 1;
```

is acceptable if n is not a constant integer.

It is important to note the difference between the type int *const and the type const int *. The latter is a type of pointer to a constant integer:

```
const int *s = &m;
```

after which the assignment

```
s = &m;
```

where m in an integer (whether constant or not) is admissible, but the assignment

```
*s = 2;
```

is erroneous, even if m is not a constant.

Reference variables are used in passing arguments by reference to function calls. Passing by reference is required if an actual parameter should be changed permanently during execution of a function. This can be accomplished with pointers (and in C, this is the only mechanism available for passing by reference) or with reference variables. For example, after declaring a function

```
void f1(int i, int* j, int& k) {
    i = 1;
    *j = 2;
    k = 3;
}
```

the values of the variables

```
int n1 = 4, n2 = 5, n3 = 6;
```

after executing the call

```
f1(n1,&n2,n3);
```

are n1 = 4, n2 = 2, n3 = 3.

Reference type is also used in indicating the return type of functions. For example, having defined the function

```
int& f2(int a[], int i) {
    return a[i];
}
```

and declaring the array

```
int a[] = {1,2,3,4,5};
```

we can use f2() on any side of the assignment operator. For instance, on the right-hand side,

```
n = f2(a,3);
```

or on the left-hand side,

```
f2(a,3) = 6;
```

which assigns 6 to a[3] so that a = [1 2 3 6 5]. Note that we can accomplish the same with pointers, but dereferencing has to be used explicitly:

```
int* f3(int a[], int i) {
    return &a[i];
}
```

and then

```
*f3(a,3) = 6;
```

Reference variables and the reference return type have to be used with caution because there is a possibility of compromising the information-hiding principle when they are used improperly. Consider class C

```
class C {
public:
    int& getRefN() {
        return n;
    }
    int getN() {
        return n;
    }
private:
    int n;
} c;
```

and these assignments:

```
int& k = c.getRefN();
k = 7;
cout << c.getN();
```

Although n is declared private, after the first assignment it can be accessed at will from the outside through k and assigned any value. An assignment also can be made through getRefN():

```
c.getRefN() = 9;
```

1.4.5 Pointers to Functions

As indicated in Section 1.4.1, one of the attributes of a variable is its address indicating its position in computer memory. The same is true for functions: One of the attributes of a function is the address indicating the location of the body of the function in memory. Upon a function call, the system transfers control to this location to execute the function. For this reason, it is possible to use pointers to functions. These pointers are very useful in implementing functionals (that is, functions that take functions as arguments) such as an integral.

Consider a simple function

```
double f(double x) {
    return 2*x;
}
```

With this definition, f is the pointer to the function f(), *f is the function itself, and (*f)(7) is a call to the function.

Consider now writing a C++ function that computes the following sum:

$$\sum_{i=n}^{i=m} f(i)$$

To compute the sum, we have to supply not only limits *n* and *m*, but also a function *f*. Therefore, the desired implementation should allow for passing numbers not only as arguments, but also as functions. This is done in C++ in the following fashion:

```
double sum(double (*f)(double), int n, int m) {
    double result = 0;
    for (int i = n; i <= m; i++)
        result += f(i);
    return result;
}
```

In this definition of `sum()`, the declaration of the first formal argument

```
double (*f)(double)
```

means that `f` is a pointer to a function with one double argument and a double return value. Note the need for parentheses around `*f`. Because parentheses have precedence over the dereference operator `*`, the expression

```
double *f(double)
```

declares a function that returns a pointer to a double value.

The function `sum()` can be called now with any built-in or user-defined double function that takes one double argument, as in

```
cout << sum(f,1,5) << endl;
cout << sum(sin,3,7) << endl;
```

Another example is a function that finds a root of a continuous function in an interval. The root is found by repetitively bisecting an interval and finding a midpoint of the current interval. If the function value at the midpoint is zero or the interval is smaller than some small value, the midpoint is returned. If the values of the function on the left limit of the current interval and on the midpoint have opposite signs, the search continues in the left half of the current interval; otherwise, the current interval becomes its right half. Here is an implementation of this algorithm:

```
double root(double (*f)(double), double a, double b, double epsilon) {
    double middle = (a + b) / 2;
    while (f(middle) != 0 && fabs(b - a) > epsilon) {
        if (f(a) * f(middle) < 0)      // if f(a) and f(middle) have
            b = middle;                // opposite signs;
        else a = middle;
        middle = (a + b) / 2;
    }
    return middle;
}
```

1.5 POLYMORPHISM

Polymorphism refers to the ability of acquiring many forms. In the context of OOP this means that the same function name denotes many functions that are members of different objects. Consider the following example:

```cpp
class Class1 {
public:
    virtual void f() {
        cout << "Function f() in Class1\n";
    }
    void g() {
        cout << "Function g() in Class1\n";
    }
};
class Class2 {
public:
    virtual void f() {
        cout << "Function f() in Class2\n";
    }
    void g() {
        cout << "Function g() in Class2\n";
    }
};
class Class3 {
public:
    virtual void h() {
        cout << "Function h() in Class3\n";
    }
};
int main() {
    Class1 object1, *p;
    Class2 object2;
    Class3 object3;
    p = &object1;
    p->f();
    p->g();
    p = (Class1*) &object2;
    p->f();
    p->g();
    p = (Class1*) &object3;
//  p->f();   // Abnormal program termination;
    p->g();
//  p->h();   // h() is not a member of Class1;
    return 0;
}
```

The output of this program is as follows:

```
Function f() in Class1
Function g() in Class1
Function f() in Class2
Function g() in Class1
Function g() in Class1
```

We should not be surprised that when p is declared as a pointer to object1 of class type Class1, then the two function members are activated that are defined in Class1. But after p becomes a pointer to object2 of class type Class2, then p->f() activates a function defined in Class2, whereas p->g() activates a function defined in Class1. How is this possible? The difference lies in the moment at which a decision is made about the function to be called.

In the case of the so-called *static binding*, the decision concerning a function to be executed is determined at compilation time. In the case of *dynamic binding*, the decision is delayed until run time. In C++, dynamic binding is enforced by declaring a member function as virtual. In this way, if a virtual function member is called, then the function chosen for execution depends not on the type of pointer determined by its declaration, but on the type of value the pointer currently has. In our example, pointer p was declared to be of type Class1*. Therefore, if p points to function g() that is not virtual, then regardless of the place in the program in which the call instruction p->g() occurs, this is always considered a call to the function g() defined in Class1. This is due to the fact that the compiler makes this decision based on the type declaration of p and the fact that g() is not virtual. For virtual function members the situation drastically changes. This time, the decision is made during run time: If a function member is virtual, then the system looks at the type of the current pointer value and invokes the proper function member. After the initial declaration of p as being of type Class1*, virtual function f() belonging to Class1 is called, whereas after assigning to p the address of object2 of type Class2, f() belonging to Class2 is called.

Note that after p was assigned the address of object3, it still invokes g() defined in Class1. This is because g() is not redefined in Class3 and g() from Class1 is called. But an attempt to call p->f() results in a program crash, because f() is declared virtual in Class1 so the system tries to find, unsuccessfully, in Class3 a definition of f(). Also, notwithstanding the fact that p points to object3, instruction p->h() results in a compilation error, because the compiler does not find h() in Class1, where Class1* is still the type of pointer p. To the compiler, it does not matter that h() is defined in Class3 (be it virtual or not).

Polymorphism is a powerful tool in OOP. It is enough to send a standard message to many different objects without specifying how the message will be processed. There is no need to know of what type the objects are. The receiver is responsible for interpreting the message and following it. The sender does not have to modify the message depending on the type of receiver. There is no need for switch or if-else statements. Also, new units can be added to a complex program without needing to recompile the entire program.

1.6 C++ AND OBJECT-ORIENTED PROGRAMMING

The previous discussion presumed that C++ is an OOL, and all the features of OOLs that we discussed have been illustrated with C++ code. However, C++ is not a pure OOL. C++ is more object-oriented than C or Pascal, which have no object-oriented features, or Ada, which supports classes (packages) and instances; however, C++ is less object-oriented than pure OOLs such as Smalltalk or Eiffel.

C++ does not enforce the object-oriented approach. We can program in C++ without knowing that such features are a part of the language. The reason for this is the popularity of C. C++ is a superset of C, so a C programmer can easily switch to C++, adapting only to its friendlier features such as I/O, call-by-reference mechanism, default values for function parameters, operator overloading, inline functions, and the like. Using an OOL such as C++ does not guarantee that we are doing OOP. On the other hand, invoking the entire machinery of classes and member functions may not always be necessary, especially in small programs, so not enforcing OOP is not necessarily a disadvantage. Also, C++ is easier to integrate with existing C code than other OOLs.

C++ has an excellent encapsulation facility that allows for well-controlled information hiding. There is, however, a relaxation to this rule in the use of so-called friend functions. The problem is that private information of a certain class cannot be accessed by anyone, and the public information is accessible to every user. But sometimes we would like to allow only some users to have access to the private pool of information. This can be accomplished if the class lists the user functions as its friends. For example, if the definition is

```
class C {
    int n;
    friend int f();
} ob;
```

function f() has direct access to variable n belonging to the class C, as in

```
int f ()
{   return 10 * ob.n; }
```

This could be considered a violation of the information-hiding principle; however, the class C itself grants the right to make public to some users what is private and inaccessible to others. Thus, because the class has control over what to consider a friend function, the friend function mechanism can be considered an extension of the information-hiding principle. This mechanism, admittedly, is used to facilitate programming and speed up execution, because rewriting code without using friend functions can be a major problem. Such a relaxation of some rules is, by the way, not uncommon in computer science; other examples include the existence of loops in functional languages, such as LISP, or storing some information at the beginning of data files in violation of the relational database model, as in dBaseIII+.

1.7 The Standard Template Library

C++ is an object-oriented language, but recent extensions to the language bring C++ to a higher level. The most significant addition to the language is the Standard Template Library (STL), developed primarily by Alexander Stepanov and Meng Lee. The library includes three types of generic entities: containers, iterators, and algorithms. Algorithms are frequently used functions that can be applied to different data structures. The application is mediated through the iterators that determine which algorithms can be applied to which types of objects. The STL relieves programmers from writing their own implementations of various classes and functions. Instead, they can use prepackaged generic implementations accommodated to the problem at hand.

1.7.1 Containers

A container is a data structure that holds some objects that are usually of the same type. Different types of containers organize the objects within them differently. Although the number of different organizations is theoretically unlimited, only a small number of them have practical significance, and the most frequently used organizations are incorporated in the STL. The STL includes the following containers: `deque`, `list`, `map`, `multimap`, `set`, `multiset`, `stack`, `queue`, `priority_queue`, and `vector`.

The STL containers are implemented as template classes that include member functions specifying what operations can be performed on the elements stored in the data structure specified by the container or on the data structure itself. Some operations can be found in all containers, although they may be implemented differently. The member functions common to all containers include the default constructor, copy constructor, destructor, `empty()`, `max_size()`, `size()`, `swap()`, `operator=`, and, except in `priority_queue`, six overloaded relational operator functions (`operator<`, etc.). Moreover, the member functions common to all containers except `stack`, `queue`, and `priority_queue` include the functions `begin()`, `end()`, `rbegin()`, `rend()`, `erase()`, and `clear()`.

Elements stored in containers can be of any type, and they have to supply at least a default constructor, a destructor, and an assignment operator. This is particularly important for user-defined types. Some compilers may also require some relational operators to be overloaded (at least operators == and <, but maybe != and > as well) even though the program does not use them. Also, a copy constructor and the function operator = should be provided if data members are pointers because insertion operations use a copy of an element being inserted, not the element itself.

1.7.2 Iterators

An iterator is an object used to reference an element stored in a container. Thus, it is a generalization of the pointer. An iterator allows for accessing information included in a container so that desired operations can be performed on these elements.

As a generalization of pointers, iterators retain the same dereferencing notation. For example, `*i` is an element referenced by iterator `i`. Also, iterator arithmetic is similar to pointer arithmetic, although all operations on iterators are not allowed in all containers.

No iterators are supported for the `stack`, `queue`, and `priority_queue` containers. Iterator operations for classes `list`, `map`, `multimap`, `set`, and `multiset` are as follows (`i1` and `i2` are iterators, n is a number):

```
i1++, ++i1, i1--, --i1
i1 = i2
i1 == i2, i1 != i2
*i1
```

In addition to these operations, iterator operations for classes `deque` and `vector` are as follows:

```
i1 < i2, i1 <= i2, i1 > i2, i1 >= i2
i1 + n, i1 - n
i1 += n, i1 -= n
i1[n]
```

1.7.3 Algorithms

The STL provides some 70 generic functions, called algorithms, that can be applied to the STL containers and to arrays. A list of all the algorithms is in Appendix B. These algorithms are implementing operations that are very frequently used in most programs, such as locating an element in a container, inserting an element into a sequence of elements, removing an element from a sequence, modifying elements, comparing elements, finding a value based on a sequence of elements, sorting the sequence of elements, and so on. Almost all STL algorithms use iterators to indicate the range of elements on which they operate. The first iterator references the first element of the range, and the second iterator references an element *after* the last element of the range. Therefore, it is assumed that it is always possible to reach the position indicated by the second iterator by incrementing the first iterator. Here are some examples.

The call

```
random_shuffle(c.begin(), c.end());
```

randomly reorders all the elements of the container c. The call

```
i3 = find(i1, i2, el);
```

returns an iterator indicating the position of element `el` in the range `i1` up to, but not including, `i2`. The call

```
n = count_if(i1, i2, oddNum);
```

counts with the algorithm `count_if()` the elements in the range indicated by iterators `i1` and `i2` for which a one-argument user-defined Boolean function `oddNum()` returns `true`.

Algorithms are functions that are in addition to the member functions provided by containers. However, some algorithms are also defined as member functions to provide better performance.

1.7.4 Function Objects

In C++, the function call operator ( ) can be treated as any other operator; in particular, it can be overloaded. It can return any type and take any number of arguments, but like the assignment operator, it can be overloaded only as a member function. Any object that includes a definition of the function call operator is called a *function object*. A function object is an object, but it behaves as though it were a function. When the function object is called, its arguments become the arguments of the function call operator.

Consider the example of finding the sum of numbers resulting from applying a function f to integers in the interval $[n, m]$. An implementation sum( ) presented in Section 1.4.5 relied on using a function pointer as an argument of function sum( ). The same can be accomplished by first defining a class that overloads the function call operator:

```
class classf {
public:
    classf() {
    }
    double operator() (double x) {
        return 2*x;
    }
};
```

and defining

```
double sum2(classf f, int n, int m) {
    double result = 0;
    for (int i = n; i <= m; i++)
        result += f(i);
    return result;
}
```

which differs from sum( ) only in the first parameter, which is a function object, not a function; otherwise, it is the same. The new function can now be called, as in

```
classf cf;
cout << sum2(cf,2,5) << endl;
```

or simply

```
cout << sum2(classf(),2,5) << endl;
```

The latter way of calling requires a definition of constructor classf( ) (even if it has no body) to create an object of type classf( ) when sum2( ) is called.

The same can be accomplished without overloading the function call operator, as exemplified in these two definitions:

```
class classf2 {
public:
    classf2 () {
    }
```

```cpp
        double run (double x) {
            return 2*x;
        }
    };
    double sum3 (classf2 f, int n, int m) {
        double result = 0;
        for (int i = n; i <= m; i++)
            result += f.run(i);
        return result;
    }
```

and a call

```cpp
    cout << sum3(classf2(),2,5) << endl;
```

The STL relies very heavily on function objects. The mechanism of function pointers is insufficient for built-in operators. How could we pass a unary minus to `sum()`? The syntax `sum(-,2,5)` is illegal. To circumvent the problem, the STL defines function objects for the common C++ operators in `<functional>`. For example, unary minus is defined as

```cpp
    template<class T>
    struct negate : public unary_function<T,T> {
        T operator()(const T& x) const {
            return -x;
        }
    };
```

Now, after redefining function `sum()` so that it becomes a generic function:

```cpp
    template<class F>
    double sum(F f, int n, int m) {
        double result = 0;
        for (int i = n; i <= m; i++)
            result += f(i);
        return result;
    }
```

the function can also be called with the `negate` function object,

```cpp
    sum(negate<double>(),2,5);
```

1.8 VECTORS IN THE STANDARD TEMPLATE LIBRARY

The simplest STL container is the vector, which is a data structure with contiguous blocks of memory just like an array. Because memory locations are contiguous, they can be randomly accessed so that the access time of any element of the vector is constant. Storage is managed automatically so that on an attempt to insert an element into a full vector, a larger memory block is allocated for the vector, the vector elements are copied to the new block, and the old block is released. A vector is thus a flexible array; that is, an array whose size can be dynamically changed.

Figure 1.3 lists alphabetically all the vector member functions. An application of these functions is illustrated in Figure 1.4. The contents of affected vectors are shown as comments on the line in which member functions are called. The contents of a vector are output with the generic function `printVector()`, but in the program in Figure 1.4 only one call is shown.

FIGURE 1.3   An alphabetical list of member functions in the class `vector`.

Member Function	Operation
`void assign(iterator first, iterator last)`	Remove all the elements in the vector and insert in it the elements from the range indicated by iterators `first` and `last`.
`void assign(size_type n, const T& el = T())`	Remove all the elements in the vector and insert in it n copies of `el`.
`T& at(size_type n)`	Return the element in position n of the vector.
`const T& at(size_type n) const`	Return the element in position n of the vector.
`T& back()`	Return the last element of the vector.
`const T& back() const`	Return the last element of the vector.
`iterator begin()`	Return an iterator that references the first element of the vector.
`const_iterator begin() const`	Return an iterator that references the first element of the vector.
`size_type capacity() const`	Return the number of elements that can be stored in the vector.
`void clear()`	Remove all the elements in the vector.
`bool empty() const`	Return `true` if the vector includes no element and `false` otherwise.
`iterator end()`	Return an iterator that is past the last element of the vector.
`const_iterator end() const`	Return a `const` iterator that is past the last element of the vector.
`iterator erase(iterator i)`	Remove the element referenced by iterator `i` and return an iterator referencing the element after the one removed.
`iterator erase(iterator first, iterator last)`	Remove the elements in the range indicated by iterators `first` and `last` and return an iterator referencing the element after the last one removed.

FIGURE 1.3 *(continued)*

`T& front()`	Return the first element of the vector.
`const T& front() const`	Return the first element of the vector.
`iterator insert(iterator i, const T& el = T())`	Insert `el` before the element referenced by iterator `i` and return iterator referencing the newly inserted element.
`void insert(iterator i, size_type n, const T& el)`	Insert `n` copies of `el` before the element referenced by iterator `i`.
`void insert(iterator i, iterator first, iterator last)`	Insert elements from the range indicated by iterators `first` and `last` before the element referenced by iterator `i`.
`size_type max_size() const`	Return the maximum number of elements for the vector.
`T& operator[]`	Subscript operator.
`const T& operator[] const`	Subscript operator.
`void pop_back()`	Remove the last element of the vector.
`void push_back(const T& el)`	Insert `el` at the end of the vector.
`reverse_iterator rbegin()`	Return an iterator that references the last element of the vector.
`const_reverse_iterator rbegin() const`	Return an iterator that references the last element of the vector.
`reverse_iterator rend()`	Return an iterator that is before the first element of the vector.
`const_reverse_iterator rend() const`	Return a `const` iterator that is before the first element of the vector.
`void reserve(size_type n)`	Reserve enough room for the vector to hold `n` items if its capacity is less than `n`.
`void resize(size_type n, const T& el = T())`	Make the vector able to hold `n` elements by adding `n` - `size()` more positions with element `el` or by discarding overflowing `size()` - `n` positions from the end of the vector.
`size_type size() const`	Return the number of elements in the vector.
`void swap(vector<T>& v)`	Swap the content of the vector with the content of another vector `v`.
`vector()`	Construct an empty vector.
`vector(size_type n, const T& el = T())`	Construct a vector with `n` copies of `el` of type `T` (if `el` is not provided, a default constructor `T()` is used).
`vector(iterator first, iterator last)`	Construct a vector with the elements from the range indicated by iterators `first` and `last`.
`vector(const vector<T>& v)`	Copy constructor.

FIGURE 1.4 A program demonstrating the operation of vector member functions.

```cpp
#include <iostream>
#include <vector>
#include <algorithm>
#include <functional> // greater<T>

using namespace std;

template<class T>
    typename vector<T>::const_iterator i = v.begin();
    for ( ; i != v.end()-1; i++)
        cout << *i << ' ';
    cout << *i <<")\n";
    }
    typename vector<T>::const_iterator i = v.begin();
       for(;i ! = v.end()-1; i++)
         cout << *i << ' ';
    cout << *i << ")\n";
}

bool f1(int n) {
    return n < 4;
}

int main() {
    int a[] = {1,2,3,4,5};
    vector<int>  v1;         // v1 is empty, size = 0, capacity = 0
    printVector("v1",v1);
    for (int j = 1; j <= 5; j++)
        v1.push_back(j);   // v1 = (1 2 3 4 5), size = 5, capacity = 8
    vector<int>  v2(3,7); // v2 = (7 7 7)
    vector<int> ::iterator i1 = v1.begin()+1;
    vector<int>  v3(i1,i1+2);     // v3 = (2 3), size = 2, capacity = 2
    vector<int>  v4(v1); // v4 = (1 2 3 4 5), size = 5, capacity = 5
    vector<int>  v5(5);   // v5 = (0 0 0 0 0)
    v5[1] = v5.at(3) = 9;// v5 = (0 9 0 9 0)
    v3.reserve(6);        // v3 = (2 3), size = 2, capacity = 6
    v4.resize(7);         // v4 = (1 2 3 4 5 0 0), size = 7, capacity = 10
    v4.resize(3);         // v4 = (1 2 3), size = 3, capacity = 10
    v4.clear();           // v4 is empty, size = 0, capacity = 10 (!)
    v4.insert(v4.end(),v3[1]);                    // v4 = (3)
    v4.insert(v4.end(),v3.at(1));                 // v4 = (3 3)
    v4.insert(v4.end(),2,4);                      // v4 = (3 3 4 4)
```

FIGURE 1.4 (continued)

```
    v4.insert(v4.end(),v3.at(1));                     // v4 = (3 3)
    v4.insert(v4.end(),2,4);                          // v4 = (3 3 4 4)
    v4.insert(v4.end(),v1.begin()+1,v1.end()-1);      // v4 = (3 3 4 4 2 3 4)
    v4.erase(v4.end()-2);                             // v4 = (3 3 4 4 2 4)
    v4.erase(v4.begin(), v4.begin()+4);               // v4 = (2 4)
    v4.assign(3,8);                                   // v4 = (8 8 8)
    v4.assign(a,a+3);                                 // v4 = (1 2 3)
    vector<int> ::reverse_iterator i3 = v4.rbegin();
    for ( ; i3 != v4.rend(); i3++)
        cout << *i3 << ' ';                           // print: 3 2 1
    cout << endl;

//  algorithms

    v5[0] = 3;                                        // v5 = (3 9 0 9 0)
    replace_if(v5.begin(),v5.end(),f1,7);             // v5 = (7 9 7 9 7)
    v5[0] = 3; v5[2] = v5[4] = 0;                     // v5 = (3 9 0 9 0)
    replace(v5.begin(),v5.end(),0,7);                 // v5 = (3 9 7 9 7)
    sort(v5.begin(),v5.end());                        // v5 = (3 7 7 9 9)
    sort(v5.begin(),v5.end(),greater<int> ());        // v5 = (9 9 7 7 3)
    v5.front() = 2;                                   // v5 = (2 9 7 7 3)
    return 0;
```

To use the class vector, the program has to contain the `include` instruction

`#include <vector>`

The class vector has four constructors. The declaration

`vector<int> v5(5);`

uses the same constructor as declaration

`vector<int> v2(3,7);`

but for vector v5, the element with which it is filled is determined by the default integer constructor, which is zero.

Vector v1 is declared empty and then new elements are inserted with the function `push_back()`. Adding a new element to the vector is usually fast unless the vector is full and has to be copied to a new block. This situation occurs if the size of the vector equals its capacity. But if the vector has some unused cells, it can accommodate a new element immediately in constant time. The current values of the parameters can be tested with the function `size()`, which returns the number of elements currently in the vector, and the function `capacity()`, which returns the number of available cells in the vector. If necessary, the capacity can be changed with the function `reserve()`. For example, after executing

```
v3.reserve(6);
```

vector v3 = (2 3) retains the same elements and the same size = 2, but its capacity changes from 2 to 6. The function `reserve()` affects only the capacity of the vector, not its content. The function `resize()` affects contents and possibly the capacity. For example, vector v4 = (1 2 3 4 5) of size = capacity = 5 changes after execution of

```
v4.resize(7);
```

to v4 = (1 2 3 4 5 0 0), size = 7, capacity = 10, and after another call to `resize()`,

```
v4.resize(3);
```

to v4 = (1 2 3), size = 3, capacity = 10. These examples indicate that new space is allocated for vectors, but it is not returned.

Note that there is no member function `push_front()`. This reflects the fact that adding a new element in front of the vector is a complex operation because it requires that all of the elements are moved by one position to make room for the new element. It is thus a time-consuming operation and can be accomplished with the function `insert()`. This is also a function that automatically allocates more memory for the vector if necessary. Other functions that perform this task are the constructors, the function `reserve()`, and `operator=`.

Vector elements can be accessed with the subscript notation used for arrays, as in

```
v4[0] = n;
```

or with iterators with the dereferencing notation used for pointers, as in

```
vector<int>::iterator i4 = v4.begin();
*i4 = n;
```

Note that some member functions have `T&` return type (that is, a reference type). For example, for an integer vector, the signature of member function `front()` is

```
int& front();
```

This means that, for example, `front()` can be used on both the left side and the right side of the assignment operator:

```
v5.front() = 2;
v4[1] = v5.front();
```

Note that `front()` is an example of an overloaded member function because it can return a reference to a value or a constant reference to a value. To see the difference, consider these two assignments:

```
int& n1 = v5.front();         // use: T& front()
const int& n2 = v5.front();   // use: const T& front() const
```

Importantly, `front()` returns a value, not a reference to a value, notwithstanding the declaration of the return value as `T&`. To see the difference, consider one more assignment:

```
int n3 = v5.front();
```

At that point, the values of the three variables are:

```
n1 = 2, n2 = 2, n3 = 2.
```

But after the assignment

```
v5.front() = 3;
```

the values of the three variables are

```
n1 = 3, n2 = 3, n3 = 2.
```

All the STL algorithms can be applied to vectors. For example, the call

```
replace(v5.begin(),v5.end(),0,7);
```

replaces all 0s with 7s in vector v5 so that v5 = (3 9 0 9 0) turns into v5 = (3 9 7 9 7), and the call

```
sort(v5.begin(),v5.end());
```

sorts vector v5 in ascending order. Some algorithms allow the use of functional parameters. For example, if the program included the definition of function

```
bool f1(int n) {
    return n < 4;
}
```

then the call

```
replace_if(v5.begin(),v5.end(),f1,7);
```

applies `f1()` to all elements of v5 and replaces all elements less than 4 with 7. In this case, v5 = (3 9 0 9 0) turns into v5 = (7 9 7 9 7). Incidentally, a more cryptic way to accomplish the same outcome without the need of explicitly defining `f1` is given by

```
replace_if(v5.begin(),v5.end(),bind2nd(less<int>(),4),7);
```

In this expression `bind2nd(op,a)` is a generic function that behaves as though it converted a two-argument function object into a one-argument function object by supplying (binding) the second parameter. It does this by creating two-argument function objects in which a two-argument operation op takes a as the second argument.

The sorting algorithms allow for the same flexibility. In the example of sorting vector v5, v5 is sorted in ascending order. How can we sort it in descending order? One way is to sort it in ascending order and then reverse with the algorithm `reverse()`. Another way is to force `sort()` to apply the operator > in making its decisions. This is done directly by using a function object as a parameter, as in

```
sort(v5.begin(),v5.end(),greater<int>());
```

or indirectly, as in

```
sort(v5.begin(),v5.end(),f2);
```

where f2 is defined as

```
bool f2(int m, int n) {
    return m > n;
}
```

The first method is preferable, but this is possible only because the function object greater is already defined in the STL. This function object is defined as a template structure, which in essence generically overloads the operator >. Therefore, greater<int>() means that the operator should be applied to integers.

This version of the algorithm sort(), which takes a functional argument, is particularly useful when we need to sort objects more complex than integers and we need to use different criteria. Consider the following class definition:

```
class Person {
public:
    Person(char *n = "", int a = 0) {
       name = strdup(n);
       age = a;
    }
    bool operator==(const Person& p) const {
       return strcmp(name,p.name) == 0 && age == p.age;
    }
    bool operator<(const Person& p) const {
       return strcmp(name,p.name) < 0;
    }
    bool operator>(const Person& p) const {
       return !(*this == p) && !(*this < p);
    }
private:
    char *name;
    int age;
    friend bool lesserAge(const Person&, const Person&);
};
```

Now, with the declaration

```
vector<Person> v6(1,Person("Gregg",25));
```

adding to v6 two more objects

```
v6.push_back(Person("Ann",30));
v6.push_back(Person("Bill",20));
```

and executing

```
sort(v6.begin(),v6.end());
```

v6 changes from `v6` = (("Gregg", 25) ("Ann", 30) ("Bill", 20)) to `v6` = (("Ann", 30) ("Bill", 20) ("Gregg", 25)) sorted in ascending order because the version of `sort()` with only two iterator arguments uses the operator < overloaded in class `Person`. The call

```
sort(v6.begin(),v6.end(),greater<Person>());
```

changes `v6` = (("Ann", 30) ("Bill", 20) ("Gregg", 25)) to `v6` = (("Gregg", 25) ("Bill", 20) ("Ann", 30)) sorted in descending order because this version of `sort()` relies on the operator > overloaded for this class. What do we need to do to sort the objects by age? In this case, a function needs to be defined, as in

```
bool lesserAge(const Person& p1, const Person& p2) {
    return p1.age < p2.age;
}
```

and then used as an argument in the call to `sort()`,

```
sort(v6.begin(),v6.end(),lesserAge);
```

which causes `v6` = (("Gregg", 25) ("Bill", 20) ("Ann", 30)) to change to `v6` = (("Bill", 20) ("Gregg", 25) ("Ann", 30)).

1.9 DATA STRUCTURES AND OBJECT-ORIENTED PROGRAMMING

Although the computer operates on bits, we do not usually think in these terms; in fact, we would not like to. Although an integer is a sequence of, say, 16 bits, we prefer seeing an integer as an entity with its own individuality that is reflected in operations that can be performed on integers but not on variables of other types. And as an integer uses bits as its building blocks, other objects can use integers as their atomic elements. Some data types are already built into a particular language, but some data types can and need to be defined by the user. New data types have a distinctive structure, a new configuration of their elements, and this structure determines the behavior of objects of these new types. The task given to the data structures' domain is to explore such new structures and investigate their behavior in terms of time and space requirements. Unlike the object-oriented approach, where we start with behavior and then try to find the most suitable data type that allows for an efficient performance of desirable operations, we now start with a data type specification of some data structure and then look at what it can do, how it does it, and how efficiently. The data structures field is designed for building tools to be incorporated in and used by application programs and for finding data structures that can perform certain operations speedily and without imposing too much burden on computer memory. This field is interested in building classes by concentrating on the mechanics of these classes, on their gears and cogs, which in most cases are not visible to the user of the classes. The data structures field investigates the operability of these classes and its improvement by modifying the data structures found inside the classes, because it has direct access to them. It sharpens tools and advises the user to what purposes they can be applied. Because of inheritance, the user can add some more operations to these classes and try to squeeze from them more than the class designer did, but because the data structures are hidden from the user, these new operations can be tested by running and not by having access to the insides of the class, unless the user has access to the source code.

The data structures field performs best if done in the object-oriented fashion. In this way, it can build tools without the danger that these tools will be inadvertently misused in the application. By encapsulating the data structures into a class and making public only what is necessary for proper usage of the class, the data structures field can develop tools whose functions are not compromised by unnecessary tampering.

1.10 CASE STUDY: RANDOM ACCESS FILE

This case study is primarily designed to illustrate the use of generic classes and inheritance. The STL will be applied in the case studies in later chapters.

From the perspective of the operating systems, files are collections of bytes, regardless of their contents. From the user's perspective, files are collections of words, numbers, data sequences, records, and so on. If the user wants to access the fifth word in a text file, a searching procedure goes sequentially through the file starting at position 0, and checks all of the bytes along the way. It counts the number of sequences of blank characters, and after it skips four such sequences (or five if a sequence of blanks begins the file), it stops because it encounters the beginning of the fifth nonblank sequence or the fifth word. This word can begin at any position of the file. It is impossible to go to a particular position of any text file and be certain that this is a starting position of the fifth word of the file. Ideally, we want to go directly to a certain position of the file and be sure that the fifth word begins in it. The problem is caused by the lengths of the preceding words and sequences of blanks. If we know that each word occupies the same amount of space, then it is possible to go directly to the fifth word by going to the position 4·length(word). But because words are of various lengths, this can be accomplished by assigning the same number of bytes to each word; if a word is shorter, some padding characters are added to fill up the remaining space; if it is longer, then the word is trimmed. In this way, a new organization is imposed on the file. The file is now treated not merely as a collection of bytes, but as a collection of records; in our example, each record consists of one word. If a request comes to access the fifth word, the word can be directly accessed without looking at the preceding words. With the new organization, we created a random access file.

A random access file allows for direct access of each record. The records usually include more items than one word. The preceding example suggests one way of creating a random access file, namely, by using fixed-length records. Our task in this case study is to write a generic program that generates a random access file for any type of record. The workings of the program are illustrated for a file containing personal records, each record consisting of five data members (social security number, name, city, year of birth, and salary), and for a student file that stores student records. The latter records have the same data members as personal records, plus information about academic major. This allows us to illustrate inheritance.

In this case study, a generic random access file program inserts a new record into a file, finds a record in the file, and modifies a record. The name of the file has to be supplied by the user, and if the file is not found, it is created; otherwise, it is open for reading and writing. The program is shown in Figure 1.5.

FIGURE 1.5   Listing of a program to manage random access files.

```cpp
//********************  personal.h  ********************

#include <iostream>
#include <fstream>
#include <cstring>
using namespace std;

#ifndef PERSONAL
#define PERSONAL
class Personal {
public:
    Personal();
    Personal(char*,char*,char*,int,long);
    void writeToFile(fstream&) const;
    void readFromFile(fstream&);
    void readKey();
    int size() const {
        return 9 + nameLen + cityLen + sizeof(year) + sizeof(salary);
    }
    bool operator==(const Personal& pr) const {
        return strncmp(pr.SSN,SSN) == 0;
    }
protected:
    const int nameLen, cityLen;
    char SSN[10], *name, *city;
    int year;
    long salary;
    ostream& writeLegibly(ostream&);
    friend ostream& operator<<(ostream& out, Personal& pr) {
        return pr.writeLegibly(out);
    }
    istream& readFromConsole(istream&);
    friend istream& operator>>(istream& in, Personal& pr) {
        return pr.readFromConsole(in);
    }
};

#endif

//********************  personal.cpp  ********************

#include "personal.h"
Personal::Personal() : nameLen(10), cityLen(10) {
    name = new char[nameLen+1];
```

Continues

FIGURE 1.5   *(continued)*

```cpp
        city = new char[cityLen+1];
}

Personal::Personal(char *ssn, char *n, char *c, int y, long s) :
        nameLen(10), cityLen(10) {
    name = new char[nameLen+1];
    city = new char[cityLen+1];
    strcpy(SSN,ssn);
    strcpy(name,n);
    strcpy(city,c);
    year = y;
    salary = s;
}

void Personal::writeToFile(fstream& out) const {
    out.write(SSN,9);
    out.write(name,nameLen);
    out.write(city,cityLen);
    out.write(reinterpret_cast<const char*>(&year),sizeof(int));
    out.write(reinterpret_cast<const char*>(&salary),sizeof(int));
}

void Personal::readFromFile(fstream& in) {
    in.read(SSN,9);
    in.read(name,nameLen);
    in.read(city,cityLen);
    in.read(reinterpret_cast<char*>(&year),sizeof(int));
    in.read(reinterpret_cast<char*>(&salary),sizeof(int));
}

void Personal::readKey() {
    char s[80];
    cout << "Enter SSN: ";
    cin.getline(s,80);
    strncpy(SSN,s,9);
}

ostream& Personal::writeLegibly(ostream& out) {
    SSN[9] = name[nameLen] = city[cityLen] = '\0';
    out << "SSN = " << SSN << ", name = " << name
        << ", city = " << city << ", year = " << year
        << ", salary = " << salary;
    return out;
}
```

FIGURE 1.5 (continued)

```cpp
istream& Personal::readFromConsole(istream& in) {
    char s[80];
    cout << "SSN: ";
    in.getline(s,80);
    strncpy(SSN,s,9);
    cout << "Name: ";
    in.getline(s,80);
    strncpy(name,s,nameLen);
    cout << "City: ";
    in.getline(s,80);
    strncpy(city,s,cityLen);
    cout << "Birthyear: ";
    in >> year;
    cout << "Salary: ";
    in >> salary;
    in.ignore();
    return in;
}
//********************** student.h **********************
#ifndef STUDENT
#define STUDENT
#include "personal.h"

class Student : public Personal {
public:
    Student();
    Student(char*,char*,char*,int,long,char*);
    void writeToFile(fstream&) const;
    void readFromFile(fstream&);
    int size() const {
        return Personal::size() + majorLen;
    }
protected:
    char *major;
    const int majorLen;
    ostream& writeLegibly(ostream&);
    friend ostream& operator<<(ostream& out, Student& sr) {
        return sr.writeLegibly(out);
    }
    istream& readFromConsole(istream&);
    friend istream& operator>>(istream& in, Student& sr) {
        return sr.readFromConsole(in);
    }
};
#endif
```

Continues

FIGURE 1.5 (continued)

```cpp
//********************  student.cpp  ********************

#include "student.h"

Student::Student() : majorLen(10) {
    Personal();
    major = new char[majorLen+1];
}

Student::Student(char *ssn, char *n, char *c, int y, long s, char *m) :
        majorLen(11) {
    Personal(ssn,n,c,y,s);
    major = new char[majorLen+1];
    strcpy(major,m);
}

void Student::writeToFile(fstream& out) const {
    Personal::writeToFile(out);
    out.write(major,majorLen);
}

void Student::readFromFile(fstream& in) {
    Personal::readFromFile(in);
    in.read(major,majorLen);
}

ostream& Student::writeLegibly(ostream& out) {
    Personal::writeLegibly(out);
    major[majorLen] = '\0';
    out << ", major = " << major;
    return out;
}

istream& Student::readFromConsole(istream& in) {
    Personal::readFromConsole(in);
    char s[80];
    cout << "Major: ";
    in.getline(s,80);
    strncpy(major,s,9);
    return in;
}
```

FIGURE 1.5   (continued)

```cpp
//********************  database.h  ********************

#ifndef DATABASE
#define DATABASE

template<class T>
class Database {
public:
    Database();
    void run();
private:
    fstream database;
    char fName[20];
    ostream& print(ostream&);
    void add(T&);
    bool find(const T&);
    void modify(const T&);
    friend ostream& operator<<(ostream& out, Database& db) {
        return db.print(out);
    }
};
    #endif

//********************  database.cpp  ********************

#include <iostream>
#include "student.h"
#include "personal.h"
#include "database.h"

template<class T>
Database<T>::Database() {
}

template<class T>
void Database<T>::add(T& d) {
    database.open(fName,ios::in|ios::out|ios::binary);
    database.seekp(0,ios::end);
    d.writeToFile(database);
    database.close();
}
template<class T>
void Database<T>::modify(const T& d) {
```

Continues

FIGURE 1.5   (continued)

```
    T tmp;
    database.open(fName,ios::in|ios::out|ios::binary);
    while (!database.eof()) {
        tmp.readFromFile(database);
        if (tmp == d) {   // overloaded ==
            cin >> tmp; // overloaded >>
            database.seekp(-d.size(),ios::cur);
            tmp.writeToFile(database);
            database.close();
            return;
        }
    }
    database.close();
    cout << "The record to be modified is not in the database\n";
}

template<class T>
bool Database<T>::find(const T& d) {
    T tmp;
    database.open(fName,ios::in|ios::binary);
    while (!database.eof()) {
        tmp.readFromFile(database);
        if (tmp == d) { // overloaded ==
            database.close();
            return true;
        }
    }
    database.close();
    return false;
}

template<class T>
ostream& Database<T>::print(ostream& out) {
    T tmp;
    database.open(fName,ios::in|ios::binary);
    while (true) {
        tmp.readFromFile(database);
        if (database.eof())
            break;
        out << tmp << endl; // overloaded <<
    }
    database.close();
    return out;
}
```

FIGURE 1.5 (continued)

```
template<class T>
void Database<T>::run() {
    cout << "File name: ";
    cin >> fName;
    char option[5];
    T rec;
    cout << "1. Add 2. Find 3. Modify a record; 4. Exit\n";
    cout << "Enter an option: ";
    cin.getline(option,4); // get '\n';
    while (cin.getline(option,4)) {
        if (*option == '1') {
            cin >> rec;   // overloaded >>
            add(rec);
        }
        else if (*option == '2') {
            rec.readKey();
            cout << "The record is ";
            if (find(rec) == false)
                cout << "not ";
            cout << "in the database\n";
        }
        else if (*option == '3') {
            rec.readKey();
            modify(rec);
        }
        else if (*option != '4')
            cout << "Wrong option\n";
        else return;
        cout << *this;   // overloaded <<
        cout << "Enter an option: ";
    }
}

int main() {
    Database<Personal> ().run();
//  Database<Student> ().run();
    return0;
}
```

The function `find()` determines whether a record is in the file. It performs the search sequentially comparing each retrieved record `tmp` to the sought record `d` using an overloaded equality operator `==`. The function uses to some extent the fact that the file is random by scrutinizing it record by record, not byte by byte. To be sure, the records are built out of bytes and all the bytes belonging to a particular record have to be read, but only the bytes required by the equality operator are participating in the comparison.

The function `modify()` updates information stored in a particular record. The record is first retrieved from the file, also using sequential search, and the new information is read from the user using the overloaded input operator `>>`. To store the updated record `tmp` in the file, `modify()` forces the file pointer `database` to go back to the beginning of the record `tmp` that has just been read; otherwise, the record following `tmp` in the file would be overwritten. The starting position of `tmp` can be determined immediately because each record occupies the same number of bytes; therefore, it is enough to jump back the number of bytes occupied by one record. This is accomplished by calling `database.seekp(-d.size(),ios::cur)`, where `size()` must be defined in the class `T`, which is the class type for object `d`.

The generic `Database` class includes two more functions. Function `add()` places a record at the end of file. Function `print()` prints the contents of the file.

To see the class `Database` in action, we have to define a specific class that specifies the format of one record in a random access file. As an example, we define the class `Personal` with five data members, `SSN`, `name`, `city`, `year`, and `salary`. The first three data members are strings, but only `SSN` is always of the same size; therefore, the size is included in its declaration, `char SSN[10]`. To have slightly more flexibility with the other two strings, two constants, `nameLen` and `cityLen`, are used, whose values are in the constructors. For instance,

```
Personal::Personal() : nameLen(10), cityLen(10) {
    name = new char[nameLen+1];
    city = new char[cityLen+1];
}
```

Note that we cannot initialize constants with assignments, as in

```
Personal::Personal () {
    nameLen = cityLen = 10;
    char name[nameLen+1];
    char city[cityLen+1];
}
```

But this peculiar syntax used in C++ for initialization of constants in classes can be used to initialize variables.

Storing data from one object requires particular care, which is the task of function `writeToFile()`. The SSN data member is the simplest to handle. A social security number always includes nine digits; therefore, the output operator `<<` could be used. However, the lengths of names and cities and the sections of a record in the data file designated for these two data members should always have the same length. To guarantee this, the function `write()`, as in `out.write(name,nameLen)`, is used for outputting the two strings to the file because the function writes a specified number of characters from the string including the null characters '\0', which is not output with the operator <<.

Another problem is posed by the numerical data members, year and salary, particularly the latter data member. If salary is written to the file with the operator <<, then the salary 50,000 is written as a 5-byte-long string '50000', and the salary 100,000 as a 6-byte-long string '100000', which violates the condition that each record in the random access file should be of the same length. To avoid the problem, the numbers are stored in binary form. For example, 50,000 is represented in the data member salary as a string of 32 bits, 00000000000000001100001101010000 (assuming that long variables are stored in 4 bytes). We can now treat this sequence of bits as representing not a long number, but a string of four characters, 00000000, 00000000, 11000011, 01010000; that is, the characters whose ASCII codes are, in decimal, numbers 0, 0, 195, and 80. In this way, regardless of the value of salary, the value is always stored in 4 bytes. This is accomplished with the instruction

```
out.write(reinterpret_cast<const char*>(&salary),sizeof(long));
```

which forces function write() to treat salary as a 4-byte-long string by converting the address &salary to const char* and specifying the length of type long.

A similar approach is used to read data from a data file, which is the task of readFromFile(). In particular, strings that should be stored in numerical data members have to be converted from strings to numbers. For the salary member, this is done with the instruction

```
in.read(reinterpret_cast<char*>(&salary),sizeof(long));
```

which casts &salary to char* with the operator reinterpret_cast and specifies that 4 bytes (sizeof(long)) should be read into long data member salary.

This method of storing records in a data file poses a readability problem, particularly in the case of numbers. For example, 50,000 is stored as 4 bytes: two null characters, a special character, and a capital P. For a human reader, it is far from obvious that these characters represent 50,000. Therefore, a special routine is needed to output records in readable form. This is accomplished by overloading the operator <<, which uses an auxiliary function writeLegibly(). The database class also overloads the operator <<, which uses its own auxiliary function print(). The function repeatedly reads records from a data file into object tmp with readFromFile() and outputs tmp in readable form with operator <<. This explains why this program uses two functions for reading and two for writing: one is for maintaining data in a random access file, and the other is for reading and writing data in readable form.

To test the flexibility of the Database class, another user class is defined, class Student. This class is also used to show one more example of inheritance.

Class Student uses the same data members as class Personal by being defined as a class derived from Personal plus one more member, a string member major. Processing input and output on objects of class type Student is very similar to that for class Personal, but the additional member has to be accounted for. This is done by redefining functions from the base class and at the same time reusing them. Consider the function writeToFile() for writing student records in a data file in fixed-length format:

```
void Student::writeToFile(fstream& out) const {
    Personal::writeToFile(out);
    out.write(major,majorLen);
}
```

The function uses the base class's `writeToFile()` to initialize the five data members, SSN, name, city, year, and salary, and initializes the member major. Note that the scope resolution operator :: must be used to indicate clearly that `writeToFile()` being defined for class Student calls `writeToFile()` already defined in base class Personal. However, class Student inherits without any modification function `readKey()` and the overloaded operator == because the same key is used in both Personal and Student objects to uniquely identify any record, namely, SSN.

1.11 EXERCISES

1. If `i` is an integer and `p` and `q` are pointers to integers, which of the following assignments cause a compilation error?

   a. `p = &i;`
   b. `p = *&i;`
   c. `p = &*i;`
   d. `i = *&*p;`
   e. `i = *&p;`
   f. `i = &*p;`
   g. `p = &*&i;`
   h. `q = *&*p;`
   i. `q = **&p;`
   j. `q = *&p;`
   k. `q = &*p;`

2. Identify the errors; assume in (b) and (c) that `s2` has been declared and assigned a string:

   a.
   ```
   char* f(char *s) {
      char ch = 'A';
      return &ch;
   }
   ```

   b.
   ```
   char *s1;
   strcpy(s1,s2);
   ```

   c.
   ```
   char *s1;
   s1 = new char[strlen(s2)];
   strcpy(s1,s2);
   ```

3. Providing that the declarations

   ```
   int intArray[] = {1, 2, 3}, *p = intArray;
   ```

   have been made, what will be the content of `intArray` and `p` after executing individually (not in sequence)

   a. `*p++;`
   b. `(*p)++;`
   c. `*p++; (*p)++;`

4. Using only pointers (no array indexing), write

   a. A function to add all numbers in an integer array,

   b. A function to remove all odd numbers from an ordered array. The array should remain ordered. Would it be easier to write this function if the array were unordered?

5. Using pointers only, implement the following string functions:

   a. `strlen()`
   b. `strcmp()`
   c. `strcat()`
   d. `strchr()`

6. What is the difference between `if (p == q) { ... }` and `if (*p == *q) { ... }`?

7. Early versions of C++ did not support templates, but generic classes could be introduced using parameterized macros. In what respect is the use of templates better than the use of such macros?

8. What is the meaning of `private`, `protected`, and `public` parts of classes?

9. What should be the type of constructors and destructors defined in classes?

10. Assume the following class declaration:

    ```
    template<class T>
    class genClass {
        ...
        char aFunction(...);
        ... };
    ```

    What is wrong with this function definition?

    ```
    char genClass::aFunction(...) { ... };
    ```

11. Overloading is a powerful tool in C++, but there are some exceptions. What operators must not be overloaded?

12. If `classA` includes a `private` variable n, a `protected` variable m, and a `public` variable k, and `classB` is derived from `classA`, which of these variables can be used in `classB`? Can n become `private` in `classB`? `protected`? `public`? How about variables m and k? Does it make a difference whether the derivation of `classB` was `private`, `protected`, or `public`?

13. Transform the declaration

    ```
    template<class T, int size = 50>
    class genClass {
        T storage[size];
        .................
        void memberFun() {
            ............
            if (someVar < size) { ...... }
            ............
        }
    };
    ```

    which uses an integer variable `size` as a parameter to `template` to a declaration of `genClass`, which does not include `size` as a parameter to `template` and yet allows for flexibility of the value of `size`. Consider a declaration of `genClass`'s constructor. Is there any advantage of one version over another?

14. What is the difference between function members that are `virtual` and those that are not?

15. What happens if the declaration of `genClass`:

```
class genClass {
    .................
    virtual void process1(char);
    virtual void process2(char);
};
```

is followed by this declaration of `derivedClass`?

```
class derivedClass : public genClass {
    .................
    void process1(int);
    int process2(char);
};
```

Which member functions are invoked if the declaration of two pointers

```
genClass *objectPtr1 = &derivedClass,
         *objectPtr2 = &derivedClass;
```

is followed by these statements?

```
objectPtr1->process1(1000);
objectPtr2->process2('A');
```

1.12 PROGRAMMING ASSIGNMENTS

1. Write a class `Fraction` that defines adding, subtracting, multiplying, and dividing fractions by overloading standard operators for these operations. Write a function member for reducing factors and overload I/O operators to input and output fractions.

2. Write a class `Quaternion` that defines the four basic operations of quaternions and the two I/O operations. Quaternions, as defined in 1843 by William Hamilton and published in his *Lectures on Quaternions* in 1853, are an extension of complex numbers. Quaternions are quadruples of real numbers, $(a,b,c,d) = a + bi + cj + dk$, where $1 = (1,0,0,0)$, $i = (0,1,0,0)$, $j = (0,0,1,0)$, and $k = (0,0,0,1)$ and the following equations hold:

$$i^2 = j^2 = k^2 = -1$$
$$ij = k, jk = i, ki = j, ji = -k, kj = -i, ik = -j$$
$$(a + bi + cj + dk) + (p + qi + rj + sk)$$
$$= (a + p) + (b + q)i + (c + r)j + (d + s)k$$
$$(a + bi + cj + dk) \cdot (p + qi + rj + sk)$$
$$= (ap - bq - cr - ds) + (aq + bp + cs - dr)i$$
$$+ (ar + cp + dq - bs)j + (as + dp + br - cq)k$$

Use these equations in implementing a quaternion class.

3. Write a program to reconstruct a text from a concordance of words. This was a real problem of reconstructing some unpublished texts of the Dead Sea Scrolls using concordances. For example, here is William Wordsworth's poem, *Nature and the Poet*, and a concordance of words corresponding with the poem.

So pure the sky, so quiet was the air!
So like, so very like, was day to day!
Whene'er I look'd, thy image still was there;
It trembled, but it never pass'd away.

The 33-word concordance is as follows:

1:1 so quiet was the *air!
1:4 but it never pass'd *away.
1:4 It trembled, *but it never
1:2 was *day to day!
1:2 was day to *day!
1:3 thy *image still was there;
. . . .
1:2 so very like, *was day
1:3 thy image still *was there;
1:3 *Whene'er I look'd,

In this concordance, each word is shown in a context of up to five words, and the word referred to on each line is preceded with a star. For larger concordances, two numbers have to be included, a number corresponding with a poem and a number of the line where the words can be found. For example, assuming that 1 is the number of *Nature and the Poet*, line "1:4 but it never pass'd *away." means that the word "away" is found in this poem in line 4. Note that punctuation marks are included in the context.

Write a program that loads a concordance from a file and creates a vector where each cell is associated with one line of the concordance. Then, using a binary search, reconstruct the text.

4. Modify the program from the case study by maintaining an order during insertion of new records into the data file. This requires overloading the operator < in `Personal` and in `Student` to be used in a modified function `add()` in `Database`. The function finds a proper position for a record d, moves all the records in the file to make room for d, and writes d into the file. With the new organization of the data file, `find()` and `modify()` can also be modified. For example, `find()` stops the sequential search when it encounters a record greater than the record looked for (or reaches the end of file). A more efficient strategy can use binary search, discussed in Section 2.7.

5. Write a program that maintains an order in the data file indirectly. Use a vector of file position pointers (obtained through `tellg()` and `tellp()`) and keep the vector in sorted order without changing the order of records in the file.

6. Modify the program from the case study to remove records from the data file. Define function `isNull()` in classes `Personal` and `Student` to determine that a record is null. Define function `writeNullToFile()` in the two classes to overwrite a record to be deleted by a null record. A null record can be defined as having a nonnumeric character (a tombstone) in the first position of the `SSN` member. Then define function `remove()` in `Database` (very similar to `modify()`), which locates the position of a record to be deleted and overwrites it with the null record. After a session is finished, a `Database` destructor should be invoked, which copies nonnull records to a new data file, deletes the old data file, and renames the new data file with the name of the old data file.

BIBLIOGRAPHY

Breymann, Ulrich, *Designing Components with the C++ STL*, Harlow, England: Addison-Wesley, 2000.

Budd, Timothy, *Data Structures in C++ Using the Standard Template Library*, Reading, MA: Addison-Wesley, 1998.

Cardelli, Luca, and Wegner, Peter, "On Understanding Types, Data Abstraction, and Polymorphism," *Computing Surveys* 17 (1985), 471–522.

Deitel, Harvey M., and Deitel, P. J., *C++: How to Program*, Upper Saddle River, NJ: Prentice Hall, 2003.

Ege, Raimund K., *Programming in an Object-Oriented Environment*, San Diego: Academic Press, 1992.

Flaming, Bryan, *Practical Data Structures in C++*, New York: Wiley, 1993.

Johnsonbaugh, Richard, and Kalin, Martin, *Object-Oriented Programming in C++*, Upper Saddle River, NJ: Prentice Hall, 2000.

Khoshafian, Setrag, and Razmik, Abnous, *Object Orientation: Concepts, Languages, Databases, User Interfaces*, New York: Wiley, 1995.

Lippman, Stanley B., and Lajoie, Josée, *C++ Primer*, Reading, MA: Addison-Wesley, 1998.

Meyer, Bertrand, *Object-Oriented Software Construction*, Upper Saddle River, NJ: Prentice Hall, 1997.

Schildt, Herbert, *C++: The Complete Reference*, New York: McGraw-Hill, 2003.

Stroustrup, Bjarne, *The C++ Programming Language*, Boston, MA: Addison-Wesley, 2003.

Wang, Paul S., *Standard C++ with Object-Oriented Programming*, Pacific Grove, CA: Brooks/Cole, 2001.

2 Complexity Analysis

2.1 COMPUTATIONAL AND ASYMPTOTIC COMPLEXITY

The same problem can frequently be solved with algorithms that differ in efficiency. The differences between the algorithms may be immaterial for processing a small number of data items, but these differences grow with the amount of data. To compare the efficiency of algorithms, a measure of the degree of difficulty of an algorithm called *computational complexity* was developed by Juris Hartmanis and Richard E. Stearns.

Computational complexity indicates how much effort is needed to apply an algorithm or how costly it is. This cost can be measured in a variety of ways, and the particular context determines its meaning. This book concerns itself with the two efficiency criteria: time and space. The factor of time is usually more important than that of space, so efficiency considerations usually focus on the amount of time elapsed when processing data. However, the most inefficient algorithm run on a Cray computer can execute much faster than the most efficient algorithm run on a PC, so run time is always system-dependent. For example, to compare 100 algorithms, all of them would have to be run on the same machine. Furthermore, the results of run-time tests depend on the language in which a given algorithm is written, even if the tests are performed on the same machine. If programs are compiled, they execute much faster than when they are interpreted. A program written in C or Ada may be 20 times faster than the same program encoded in BASIC or LISP.

To evaluate an algorithm's efficiency, real-time units such as microseconds and nanoseconds should not be used. Rather, logical units that express a relationship between the size n of a file or an array and the amount of time t required to process the data should be used. If there is a linear relationship between the size n and time t—that is, $t_1 = cn_1$—then an increase of data by a factor of 5 results in the increase of the execution time by the same factor; if $n_2 = 5n_1$, then $t_2 = 5t_1$. Similarly, if $t_1 = \log_2 n$, then doubling n increases t by only one unit of time. Therefore, if $t_2 = \log_2(2n)$, then $t_2 = t_1 + 1$.

A function expressing the relationship between n and t is usually much more complex, and calculating such a function is important only in regard to large bodies of data; any terms that do not substantially change the function's magnitude should be eliminated from the function. The resulting function gives only an approximate measure of efficiency of the original function. However, this approximation is sufficiently close to the original, especially for a function that processes large quantities of data. This measure of efficiency is called *asymptotic complexity* and is used when disregarding certain terms of a function to express the efficiency of an algorithm or when calculating a function is difficult or impossible and only approximations can be found. To illustrate the first case, consider the following example:

$$f(n) = n^2 + 100n + \log_{10}n + 1{,}000 \tag{2.1}$$

For small values of n, the last term, 1,000, is the largest. When n equals 10, the second ($100n$) and last (1,000) terms are on equal footing with the other terms, making a small contribution to the function value. When n reaches the value of 100, the first and the second terms make the same contribution to the result. But when n becomes larger than 100, the contribution of the second term becomes less significant. Hence, for large values of n, due to the quadratic growth of the first term (n^2), the value of the function f depends mainly on the value of this first term, as Figure 2.1 demonstrates. Other terms can be disregarded for large n.

FIGURE 2.1 The growth rate of all terms of function $f(n) = n^2 + 100n + \log_{10}n + 1{,}000$.

n	f(n)	n^2		$100n$		$\log_{10}n$		1,000	
	Value	Value	%	Value	%	Value	%	Value	%
1	1,101	1	0.1	100	9.1	0	0.0	1,000	90.83
10	2,101	100	4.76	1,000	47.6	1	0.05	1,000	47.60
100	21,002	10,000	47.6	10,000	47.6	2	0.001	1,000	4.76
1,000	1,101,003	1,000,000	90.8	100,000	9.1	3	0.0003	1,000	0.09
10,000	101,001,004	100,000,000	99.0	1,000,000	0.99	4	0.0	1,000	0.001
100,000	10,010,001,005	10,000,000,000	99.9	10,000,000	0.099	5	0.0	1,000	0.00

2.2 BIG-O NOTATION

The most commonly used notation for specifying asymptotic complexity—that is, for estimating the rate of function growth—is the big-O notation introduced in 1894 by Paul Bachmann. Given two positive-valued functions f and g, consider the following definition:

Definition 1: $f(n)$ is $O(g(n))$ if there exist positive numbers c and N such that $f(n) \leq cg(n)$ for all $n \geq N$.

This definition reads: f is big-O of g if there is a positive number c such that f is not larger than cg for sufficiently large ns; that is, for all ns larger than some number N. The relationship between f and g can be expressed by stating either that $g(n)$ is an upper bound on the value of $f(n)$ or that, in the long run, f grows at most as fast as g.

The problem with this definition is that, first, it states only that there must exist certain c and N, but it does not give any hint of how to calculate these constants. Second, it does not put any restrictions on these values and gives little guidance in situations when there are many candidates. In fact, there are usually infinitely many pairs of cs and Ns that can be given for the same pair of functions f and g. For example, for

$$f(n) = 2n^2 + 3n + 1 = O(n^2) \tag{2.2}$$

where $g(n) = n^2$, candidate values for c and N are shown in Figure 2.2.

FIGURE 2.2 Different values of c and N for function $f(n) = 2n^2 + 3n + 1 = O(n^2)$ calculated according to the definition of big-O.

c	≥ 6	$\geq 3\frac{3}{4}$	$\geq 3\frac{1}{9}$	$\geq 2\frac{13}{16}$	$\geq 2\frac{16}{25}$	$\ldots$	$\rightarrow$	2
N	1	2	3	4	5	$\ldots$	$\rightarrow$	∞

We obtain these values by solving the inequality:

$$2n^2 + 3n + 1 \leq cn^2$$

or equivalently

$$2 + \frac{3}{n} + \frac{1}{n^2} \leq c$$

for different ns. The first inequality results in substituting the quadratic function from Equation 2.2 for $f(n)$ in the definition of the big-O notation and n^2 for $g(n)$. Because it is one inequality with two unknowns, different pairs of constants c and N for the same function $g(= n^2)$ can be determined. To choose the best c and N, it should be determined for which N a certain term in f becomes the largest and stays the largest. In Equation 2.2, the only candidates for the largest term are $2n^2$ and $3n$; these terms can be compared using the inequality $2n^2 > 3n$ that holds for $n > 1$. Thus, $N = 2$ and $c \geq 3\frac{3}{4}$, as Figure 2.2 indicates.

What is the practical significance of the pairs of constants just listed? All of them are related to the same function $g(n) = n^2$ and to the same $f(n)$. For a fixed g, an infinite number of pairs of cs and Ns can be identified. The point is that f and g grow at the same rate. The definition states, however, that g is almost always greater than or equal to f if it is multiplied by a constant c. "Almost always" means for all ns not less than a constant N. The crux of the matter is that the value of c depends on which N is chosen, and vice versa. For example, if 1 is chosen as the value of N—that is, if g is multiplied by c so that $cg(n)$ will not be less than f right away—then c has to be equal to 6 or greater. If $cg(n)$ is greater than or equal to $f(n)$ starting from $n = 2$, then it is enough that c is equal to 3.75.

The constant c has to be at least $3\frac{1}{9}$ if $cg(n)$ is not less than $f(n)$ starting from $n = 3$. Figure 2.3 shows the graphs of the functions f and g. The function g is plotted with different coefficients c. Also, N is always a point where the functions $cg(n)$ and f intersect each other.

FIGURE 2.3 Comparison of functions for different values of c and N from Figure 2.2.

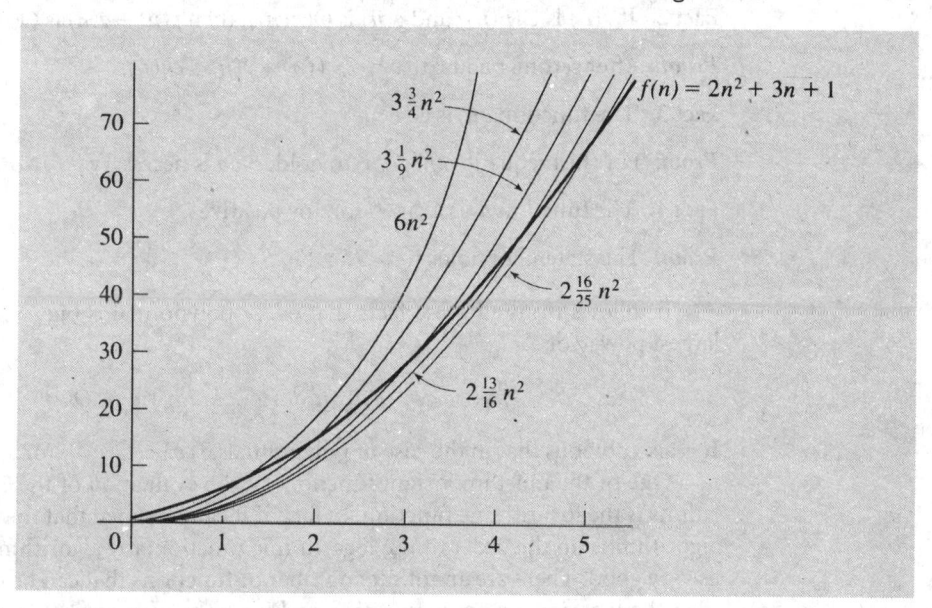

The inherent imprecision of the big-O notation goes even further, because there can be infinitely many functions g for a given function f. For example, the f from Equation 2.2 is big-O not only of n^2, but also of $n^3, n^4, \ldots, n^k, \ldots$ for any $k \geq 2$. To avoid this embarrassment of riches, the smallest function g is chosen, n^2 in this case.

The approximation of function f can be refined using big-O notation only for the part of the equation suppressing irrelevant information. For example, in Equation 2.1, the contribution of the third and last terms to the value of the function can be omitted (see Equation 2.3).

$$f(n) = n^2 + 100n + O(\log_{10} n) \qquad (2.3)$$

Similarly, the function f in Equation 2.2 can be approximated as

$$f(n) = 2n^2 + O(n) \qquad (2.4)$$

2.3 PROPERTIES OF BIG-O NOTATION

Big-O notation has some helpful properties that can be used when estimating the efficiency of algorithms.

Fact 1. (transitivity) If $f(n)$ is $O(g(n))$ and $g(n)$ is $O(h(n))$, then $f(n)$ if $O(h(n))$. (This can be rephrased as $O(O(g(n)))$ is $O(g(n))$.)

Proof: According to the definition, $f(n)$ is $O(g(n))$ if there exist positive numbers c_1 and N_1 such that $f(n) \leq c_1 g(n)$ for all $n \geq N_1$, and $g(n)$ is $O(h(n))$ if there exist positive numbers c_2 and N_2 such that $g(n) \leq c_2 h(n)$ for all $n \geq N_2$. Hence, $c_1 g(n) \leq c_1 c_2 h(n)$ for $n \geq N$ where N is the larger of N_1 and N_2. If we take $c = c_1 c_2$, then $f(n) \leq ch(n)$ for $n \geq N$, which means that f is $O(h(n))$.

Fact 2. If $f(n)$ is $O(h(n))$ and $g(n)$ is $O(h(n))$, then $f(n) + g(n)$ is $O(h(n))$.

Proof: After setting c equal to $c_1 + c_2$, $f(n) + g(n) \leq ch(n)$.

Fact 3. The function an^k is $O(n^k)$.

Proof: For the inequality $an^k \leq cn^k$ to hold, $c \geq a$ is necessary.

Fact 4. The function n^k is $O(n^{k+j})$ for any positive j.

Proof: The statement holds if $c = N = 1$.

It follows from all these facts that every polynomial is big-O of n raised to the largest power, or

$$f(n) = a_k n^k + a_{k-1} n^{k-1} + \cdots + a_1 n + a_0 \text{ is } O(n^k)$$

It is also obvious that in the case of polynomials, $f(n)$ is $O(n^{k+j})$ for any positive j.

One of the most important functions in the evaluation of the efficiency of algorithms is the logarithmic function. In fact, if it can be stated that the complexity of an algorithm is on the order of the logarithmic function, the algorithm can be regarded as very good. There are an infinite number of functions that can be considered better than the logarithmic function, among which only a few, such as $O(\lg \lg n)$ or $O(1)$, have practical bearing. Before we show an important fact about logarithmic functions, let us state without proof:

Fact 5. If $f(n) = cg(n)$, then $f(n)$ is $O(g(n))$.

Fact 6. The function $\log_a n$ is $O(\log_b n)$ for any positive numbers a and $b \neq 1$.

This correspondence holds between logarithmic functions. Fact 6 states that regardless of their bases, logarithmic functions are big-O of each other; that is, all these functions have the same rate of growth.

Proof: Letting $\log_a n = x$ and $\log_b n = y$, we have, by the definition of logarithm, $a^x = n$ and $b^y = n$.

Taking ln of both sides results in

$$x \ln a = \ln n \text{ and } y \ln b = \ln n$$

Thus

$$x \ln a = y \ln b,$$

$$\ln a \log_a n = \ln b \log_b n,$$

$$\log_a n = \frac{\ln b}{\ln a} \log_b n = c \log_b n$$

which proves that $\log_a n$ and $\log_b n$ are multiples of each other. By Fact 5, $\log_a n$ is $O(\log_b n)$.

Because the base of the logarithm is irrelevant in the context of big-O notation, we can always use just one base and Fact 6 can be written as

Fact 7. $\log_a n$ is $O(\lg n)$ for any positive $a \neq 1$, where $\lg n = \log_2 n$.

2.4 Ω AND Θ NOTATIONS

Big-O notation refers to the upper bounds of functions. There is a symmetrical definition for a lower bound in the definition of big-Ω:

Definition 2: The function $f(n)$ is $\Omega(g(n))$ if there exist positive numbers c and N such that $f(n) \geq cg(n)$ for all $n \geq N$.

This definition reads: f is Ω (big-omega) of g if there is a positive number c such that f is at least equal to cg for almost all ns. In other words, $cg(n)$ is a lower bound on the size of $f(n)$, or, in the long run, f grows at least at the rate of g.

The only difference between this definition and the definition of big-O notation is the direction of the inequality; one definition can be turned into the other by replacing "≥" with "≤." There is an interconnection between these two notations expressed by the equivalence

$$f(n) \text{ is } \Omega(g(n)) \text{ iff } g(n) \text{ is } O(f(n))$$

Ω notation suffers from the same profusion problem as does big-O notation: There is an unlimited number of choices for the constants c and N. For Equation 2.2, we are looking for such a c, for which $2n^2 + 3n + 1 \geq cn^2$, which is true for any $n \geq 0$, if $c \leq 2$, where 2 is the limit for c in Figure 2.2. Also, if f is an Ω of g and $h \leq g$, then f is an Ω of h; that is, if for f we can find one g such that f is an Ω of g, then we can find infinitely many. For example, the function 2.2 is an Ω of n^2 but also of n, $n^{1/2}$, $n^{1/3}$, $n^{1/4}, \ldots$, and also of $\lg n$, $\lg \lg n$, $\ldots$, and of many other functions. For practical purposes, only the closest Ωs are the most interesting (i.e., the largest lower bounds). This restriction is made implicitly each time we choose an Ω of a function f.

There are an infinite number of possible lower bounds for the function f; that is, there is an infinite set of gs such that $f(n)$ is $\Omega(g(n))$ as well as an unbounded number of possible upper bounds of f. This may be somewhat disquieting, so we restrict our attention to the smallest upper bounds and the largest lower bounds. Note that there is a common ground for big-O and Ω notations indicated by the equalities in the definitions of these notations: Big-O is defined in terms of "≤" and Ω in terms of "≥"; "=" is included in both inequalities. This suggests a way of restricting the sets of possible lower and upper bounds. This restriction can be accomplished by the following definition of Θ (theta) notation:

Definition 3: $f(n)$ is $\Theta(g(n))$ if there exist positive numbers c_1, c_2, and N such that $c_1 g(n) \leq f(n) \leq c_2 g(n)$ for all $n \geq N$.

This definition reads: f has an order of magnitude g, f is on the order of g, or both functions grow at the same rate in the long run. We see that $f(n)$ is $\Theta(g(n))$ if $f(n)$ is $O(g(n))$ and $f(n)$ is $\Omega(g(n))$.

The only function just listed that is both big-O and Ω of the function 2.2 is n^2. However, it is not the only choice, and there are still an infinite number of choices, because the functions $2n^2, 3n^2, 4n^2, \ldots$ are also Θ of function 2.2. But it is rather obvious that the simplest, n^2, will be chosen.

When applying any of these notations (big-O, Ω, and Θ), do not forget that they are approximations that hide some detail that in many cases may be considered important.

2.5 POSSIBLE PROBLEMS

All the notations serve the purpose of comparing the efficiency of various algorithms designed for solving the same problem. However, if only big-Os are used to represent the efficiency of algorithms, then some of them may be rejected prematurely. The problem is that in the definition of big-O notation, f is considered $O(g(n))$ if the inequality $f(n) \leq cg(n)$ holds in the long run for all natural numbers with a few exceptions. The number of ns violating this inequality is always finite. It is enough to meet the condition of the definition. As Figure 2.2 indicates, this number of exceptions can be reduced by choosing a sufficiently large c. However, this may be of little practical significance if the constant c in $f(n) \leq cg(n)$ is prohibitively large, say 10^8, although the function g taken by itself seems to be promising.

Consider that there are two algorithms to solve a certain problem and suppose that the number of operations required by these algorithms is $10^8 n$ and $10n^2$. The first function is $O(n)$ and the second is $O(n^2)$. Using just the big-O information, the second algorithm is rejected because the number of steps grows too fast. It is true but, again, in the long run, because for $n \leq 10^7$, which is 10 million, the second algorithm performs fewer operations than the first. Although 10 million is not an unheard-of number of elements to be processed by an algorithm, in many cases the number is much lower, and in these cases the second algorithm is preferable.

For these reasons, it may be desirable to use one more notation that includes constants which are very large for practical reasons. Udi Manber proposes a double-O (OO) notation to indicate such functions: f is $OO(g(n))$ if it is $O(g(n))$ and the constant c is too large to have practical significance. Thus, $10^8 n$ is $OO(n)$. However, the definition of "too large" depends on the particular application.

2.6 EXAMPLES OF COMPLEXITIES

Algorithms can be classified by their time or space complexities, and in this respect, several classes of such algorithms can be distinguished, as Figure 2.4 illustrates. Their growth is also displayed in Figure 2.5. For example, an algorithm is called *constant* if its execution time remains the same for any number of elements; it is called *quadratic* if its execution time is $O(n^2)$. For each of these classes, a number of operations is shown along with the real time needed for executing them on a machine able to perform 1 million operations per second, or one operation per microsecond (μsec). The table in

Section 2.6 Examples of Complexities   59

FIGURE 2.4 Classes of algorithms and their execution times on a computer executing 1 million operations per second (1 sec = 10^6 μsec = 10^3 msec).

Class	Complexity	Number of Operations and Execution Time (1 instr/μsec)					
n		10		10^2		10^3	
constant	$O(1)$	1	1 μsec	1	1 μsec	1	1 μsec
logarithimic	$O(\lg n)$	3.32	3 μsec	6.64	7 μsec	9.97	10 μsec
linear	$O(n)$	10	10 μsec	10^2	100 μsec	10^3	1 msec
$O(n \lg n)$	$O(n \lg n)$	33.2	33 μsec	664	664 μsec	9970	10 msec
quadratic	$O(n^2)$	10^2	100 μsec	10^4	10 msec	10^6	1 sec
cubic	$O(n^3)$	10^3	1 msec	10^6	1 sec	10^9	16.7 min
exponential	$O(2^n)$	1024	10 msec	10^{30}	$3.17 * 10^{17}$ yrs	10^{301}	
n		10^4		10^5		10^6	
constant	$O(1)$	1	1 μsec	1	1 μsec	1	1 μsec
logarithmic	$O(\lg n)$	13.3	13 μsec	16.6	7 μsec	19.93	20 μsec
linear	$O(n)$	10^4	10 msec	10^5	0.1 sec	10^6	1 sec
$O(n \lg n)$	$O(n \lg n)$	$133 * 10^3$	133 msec	$166 * 10^4$	1.6 sec	$199.3 * 10^5$	20 sec
quadratic	$O(n^2)$	10^8	1.7 min	10^{10}	16.7 min	10^{12}	11.6 days
cubic	$O(n^3)$	10^{12}	11.6 days	10^{15}	31.7 yr	10^{18}	31,709 yr
exponential	$O(2^n)$	10^{3010}		10^{30103}		10^{301030}	

FIGURE 2.5 Typical functions applied in big-O estimates.

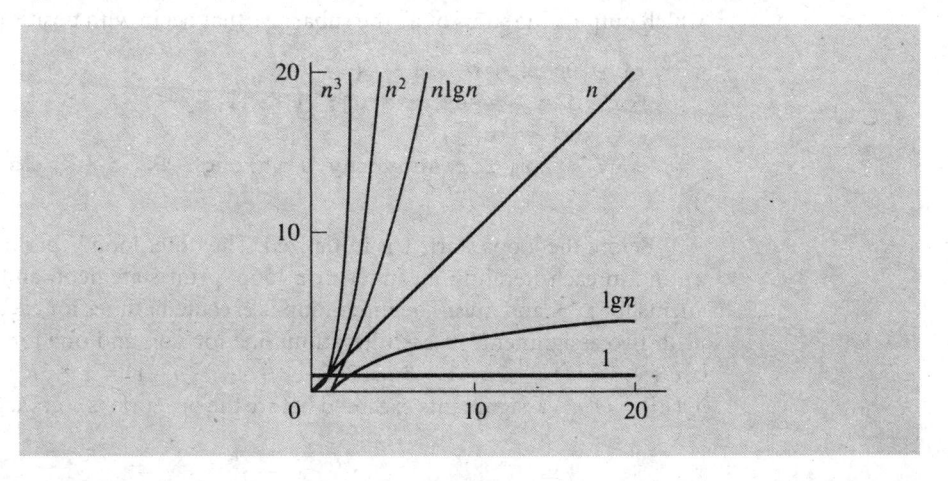

Figure 2.4 indicates that some ill-designed algorithms, or algorithms whose complexity cannot be improved, have no practical application on available computers. To process 1 million items with a quadratic algorithm, over 11 days are needed, and for a cubic algorithm, thousands of years. Even if a computer can perform one operation per nanosecond (1 billion operations per second), the quadratic algorithm finishes in only 16.7 seconds, but the cubic algorithm requires over 31 years. Even a 1,000-fold improvement in execution speed has very little practical bearing for this algorithm. Analyzing the complexity of algorithms is of extreme importance and cannot be abandoned on account of the argument that we have entered an era when, at relatively little cost, a computer on our desktop can execute millions of operations per second. The importance of analyzing the complexity of algorithms, in any context but in the context of data structures in particular, cannot be overstressed. The impressive speed of computers is of limited use if the programs that run on them use inefficient algorithms.

2.7 FINDING ASYMPTOTIC COMPLEXITY: EXAMPLES

Asymptotic bounds are used to estimate the efficiency of algorithms by assessing the amount of time and memory needed to accomplish the task for which the algorithms were designed. This section illustrates how this complexity can be determined.

In most cases, we are interested in time complexity, which usually measures the number of assignments and comparisons performed during the execution of a program. Chapter 9, which deals with sorting algorithms, considers both types of operations; this chapter considers only the number of assignment statements.

Begin with a simple loop to calculate the sum of numbers in an array:

```
for (i = sum = 0; i < n; i++)
    sum += a[i];
```

First, two variables are initialized, then the for loop iterates n times, and during each iteration, it executes two assignments, one of which updates sum and the other of which updates i. Thus, there are $2 + 2n$ assignments for the complete run of this for loop; its asymptotic complexity is $O(n)$.

Complexity usually grows if nested loops are used, as in the following code, which outputs the sums of all the subarrays that begin with position 0:

```
for (i = 0; i < n; i++) {
    for (j = 1, sum = a[0]; j <= i; j++)
        sum += a[j];
    cout<<"sum for subarray 0 through "<< i <<" is "<<sum<<endl;
}
```

Before the loops start, i is initialized. The outer loop is performed n times, executing in each iteration an inner for loop, print statement, and assignment statements for i, j, and sum. The inner loop is executed i times for each $i \in \{1, \ldots, n-1\}$ with two assignments in each iteration: one for sum and one for j. Therefore, there are $1 + 3n + \sum_{i=1}^{n-1} 2i = 1 + 3n + 2(1 + 2 + \cdots + n - 1) = 1 + 3n + n(n-1) = O(n) + O(n^2) = O(n^2)$ assignments executed before the program is completed.

Algorithms with nested loops usually have a larger complexity than algorithms with one loop, but it does not have to grow at all. For example, we may request printing sums of numbers in the last five cells of the subarrays starting in position 0. We adopt the foregoing code and transform it to

```
for (i = 4; i < n; i++) {
    for (j = i-3, sum = a[i-4]; j <= i; j++)
        sum += a[j];
    cout<<"sum for subarray "<<i-4<<" through "<< i <<" is ""<<sum<<endl;
}
```

The outer loop is executed $n - 4$ times. For each i, the inner loop is executed only four times; for each iteration of the outer loop there are eight assignments in the inner loop, and this number does not depend on the size of the array. With one initialization of i, $n - 4$ autoincrements of i, and $n - 4$ initializations of j and sum, the program makes $1 + 8 \cdot (n-4) + 3 \cdot (n-4) = O(n)$ assignments.

Analysis of these two examples is relatively uncomplicated because the number of times the loops executed did not depend on the ordering of the arrays. Computation of asymptotic complexity is more involved if the number of iterations is not always the same. This point can be illustrated with a loop used to determine the length of the longest subarray with the numbers in increasing order. For example, in [1 8 1 2 5 0 11 12], it is three, the length of subarray [1 2 5]. The code is

```
for (i = 0, length = 1; i < n-1; i++) {
    for (i1 = i2 = k = i; k < n-1 && a[k] < a[k+1]; k++, i2++);
    if (length < i2 - i1 + 1)
        length = i2 - i1 + 1;
}
```

Notice that if all numbers in the array are in decreasing order, the outer loop is executed $n - 1$ times, but in each iteration, the inner loop executes just one time. Thus, the algorithm is $O(n)$. The algorithm is least efficient if the numbers are in increasing order. In this case, the outer for loop is executed $n - 1$ times, and the inner loop is executed $n - 1 - i$ times for each $i \in \{0, \ldots, n - 2\}$. Thus, the algorithm is $O(n^2)$. In most cases, the arrangement of data is less orderly, and measuring the efficiency in these cases is of great importance. However, it is far from trivial to determine the efficiency in the average cases.

A fifth example used to determine the computational complexity is the *binary search algorithm*, which is used to locate an element in an ordered array. If it is an array of numbers and we try to locate number k, then the algorithm accesses the middle element of the array first. If that element is equal to k, then the algorithm returns its position; if not, the algorithm continues. In the second trial, only half of the original array is considered: the first half if k is smaller than the middle element, and the second otherwise. Now, the middle element of the chosen subarray is accessed and compared to k. If it is the same, the algorithm completes successfully. Otherwise, the subarray is divided into two halves, and if k is larger than this middle element, the first half is discarded; otherwise, the first half is retained. This process of halving and comparing continues until *k* is found or the array can no longer be divided into two subarrays. This relatively simple algorithm can be coded as follows:

```
template<class T>   // overloaded operator < is used;
int binarySearch(const T arr[], int arrSize, const T& key) {
    int lo = 0, mid, hi = arrSize-1;
    while (lo <= hi) {
        mid = (lo + hi)/2;
        if (key < arr[mid])
            hi = mid - 1;
        else if (arr[mid] < key)
            lo = mid + 1;
        else return mid;     // success: return the index of
    }                        //   the cell occupied by key;
    return -1;               // failure: key is not in the array;
}
```

If key is in the middle of the array, the loop executes only one time. How many times does the loop execute in the case where key is not in the array? First the algorithm looks at the entire array of size n, then at one of its halves of size $\frac{n}{2}$, then at one of the halves of this half, of size $\frac{n}{2^2}$, and so on, until the array is of size 1. Hence, we have the sequence $n, \frac{n}{2}, \frac{n}{2^2}, \ldots, \frac{n}{2^m}$, and we want to know the value of m. But the last term of this sequence $\frac{n}{2^m}$ equals 1, from which we have $m = \lg n$. So the fact that k is not in the array can be determined after $\lg n$ iterations of the loop.

2.8 THE BEST, AVERAGE, AND WORST CASES

The last two examples in the preceding section indicate the need for distinguishing at least three cases for which the efficiency of algorithms has to be determined. The *worst case* is when an algorithm requires a maximum number of steps, and the *best case* is when the number of steps is the smallest. The *average case* falls between these extremes. In simple cases, the average complexity is established by considering possible inputs to an algorithm, determining the number of steps performed by the algorithm for each input, adding the number of steps for all the inputs, and dividing by the number of inputs. This definition, however, assumes that the probability of occurrence of each input is the same, which is not always the case. To consider the probability explicitly, the average complexity is defined as the average over the number of steps executed when processing each input weighted by the probability of occurrence of this input, or,

$$C_{avg} = \sum_i p(input_i) steps(input_i)$$

This is the definition of expected value, which assumes that all the possibilities can be determined and that the probability distribution is known, which simply determines a probability of occurrence of each input, $p(input_i)$. The probability function p satisfies two conditions: It is never negative, $p(input_i) \geq 0$, and all probabilities add up to 1, $\sum_i p(input_i) = 1$.

As an example, consider searching sequentially an unordered array to find a number. The best case is when the number is found in the first cell. The worst case is when the number is in the last cell or is not in the array at all. In this case, all the cells are checked to determine this fact. And the average case? We may make the assumption that there is an equal chance for the number to be found in any cell of the array; that is, the probability distribution is uniform. In this case, there is a probability equal to $\frac{1}{n}$ that the number is in the first cell, a probability equal to $\frac{1}{n}$ that it is in the second cell, ..., and finally, a probability equal to $\frac{1}{n}$ that it is in the last, nth cell. This means that the probability of finding the number after one try equals $\frac{1}{n}$, the probability of having two tries equals $\frac{1}{n}$, ..., and the probability of having n tries also equals $\frac{1}{n}$. Therefore, we can average all these possible numbers of tries over the number of possibilities and conclude that it takes on the average

$$\frac{1 + 2 + \ldots + n}{n} = \frac{n+1}{2}$$

steps to find a number. But if the probabilities differ, then the average case gives a different outcome. For example, if the probability of finding a number in the first cell equals $\frac{1}{2}$, the probability of finding it in the second cell equals $\frac{1}{4}$, and the probability of locating it in any of the remaining cells is the same and equal to

$$\frac{1 - \frac{1}{2} - \frac{1}{4}}{n-2} = \frac{1}{4(n-2)}$$

then, on the average, it takes

$$\frac{1}{2} + \frac{2}{4} + \frac{3 + \ldots n}{4(n-2)} = 1 + \frac{n(n+1)-6}{8(n-2)} = 1 + \frac{n+3}{8}$$

steps to find a number, which is approximately four times better than $\frac{n+1}{2}$ found previously for the uniform distribution. Note that the probabilities of accessing a particular cell have no impact on the best and worst cases.

The complexity for the three cases was relatively easy to determine for sequential search, but usually it is not that straightforward. Particularly, the complexity of the average case can pose difficult computational problems. If the computation is very complex, approximations are used, and that is where we find the big-O, Ω, and Θ notations most useful.

As an example, consider the average case for binary search. Assume that the size of the array is a power of 2 and that a number to be searched has an equal chance to be in any of the cells of the array. Binary search can locate it either after one try in the middle of the array, or after two tries in the middle of the first half of the array, or after two tries in the middle of the second half, or after three tries in the middle of the first quarter of the array, or ... or after three tries in the middle of the fourth quarter, or after four tries in the middle of the first eighth of the array, or ... or after four tries in the middle of the eighth eighth of the array, or ... or after try $\lg n$ in the first cell, or after try $\lg n$ in the third cell, or ... or, finally, after try $\lg n$ in the last cell. That is, the number of all possible tries equals

$$1 \cdot 1 + 2 \cdot 2 + 4 \cdot 3 + 8 \cdot 4 + \ldots + \frac{n}{2} \lg n = \sum_{i=0}^{\lg n - 1} 2^i (i+1)$$

which has to be divided by $\frac{1}{n}$ to determine the average case complexity. What is this sum equal to? We know that it is between 1 (the best case result) and $\lg n$ (the worst case) determined in the preceding section. But is it closer to the best case—say, $\lg \lg n$—or to the worst case—for instance, $\frac{\lg n}{2}$, or $\lg \frac{n}{2}$? The sum does not lend itself to a simple conversion into a closed form; therefore, its estimation should be used. Our conjecture is that the sum is not less than the sum of powers of 2 in the specified range multiplied by a half of $\lg n$, that is,

$$s_1 = \sum_{i=0}^{\lg n - 1} 2^i (i+1) \geq \frac{\lg n}{2} \sum_{i=0}^{\lg n - 1} 2^i = s_2$$

The reason for this choice is that s_2 is a power series multiplied by a constant factor, and thus it can be presented in closed form very easily, namely,

$$s_2 = \frac{\lg n}{2} \sum_{i=0}^{\lg n - 1} 2^i = \frac{\lg n}{2} \left(1 + 2 \frac{2^{\lg n - 1} - 1}{2 - 1} \right) = \frac{\lg n}{2} (n-1)$$

which is $\Omega(n \lg n)$. Because s_2 is the lower bound for the sum s_1 under scrutiny—that is, s_1 is $\Omega(s_2)$—then so is $\frac{s_2}{n}$ the lower bound of the sought average case complexity $\frac{s_1}{n}$—that is, $\frac{s_1}{n} = \Omega(\frac{s_2}{n})$. Because $\frac{s_2}{n}$ is $\Omega(\lg n)$, so must be $\frac{s_1}{n}$. Because $\lg n$ is an assessment of the complexity of the worst case, the average case's complexity equals $\Theta(\lg n)$.

There is still one unresolved problem: Is $s_1 \geq s_2$? To determine this, we conjecture that the sum of each pair of terms positioned symmetrically with respect to the center of the sum s_1 is not less than the sum of the corresponding terms of s_2. That is,

$$2^0 \cdot 1 + 2^{\lg n - 1} \lg n \geq 2^0 \frac{\lg n}{2} + 2^{\lg n - 1} \frac{\lg n}{2}$$

$$2^1 \cdot 2 + 2^{\lg n - 2} (\lg n - 1) \geq 2^1 \frac{\lg n}{2} + 2^{\lg n - 2} \frac{\lg n}{2}$$

$$\ldots$$

$$2^j (j+1) + 2^{\lg n - 1 - j} (\lg n - j) \geq 2^j \frac{\lg n}{2} + 2^{\lg n - 1 - j} \frac{\lg n}{2}$$

$$\ldots$$

where $j \leq \frac{\lg n}{2} - 1$. The last inequality, which represents every other inequality, is transformed into

$$2^{\lg n - 1 - j} \left(\frac{\lg n}{2} - j \right) \geq 2^j \left(\frac{\lg n}{2} - j - 1 \right)$$

and then into

$$2^{\lg n - 1 - 2j} \geq \frac{\frac{\lg n}{2} - j - 1}{\frac{\lg n}{2} - j} = 1 - \frac{1}{\frac{\lg n}{2} - j} \qquad (2.5)$$

All of these transformations are allowed because all the terms that moved from one side of the conjectured inequality to another are nonnegative and thus do not change the direction of inequality. Is the inequality true? Because $j \leq \frac{\lg n}{2} - 1$, $2^{\lg n - 1 - 2j} \geq 2$, and the right-hand side of the inequality (2.5) is always less than 1, the conjectured inequality is true.

This concludes our investigation of the average case for binary search. The algorithm is relatively straightforward, but the process of finding the complexity for the average case is rather grueling, even for uniform probability distributions. For more complex algorithms, such calculations are significantly more challenging.

2.9 AMORTIZED COMPLEXITY

In many situations, data structures are subject to a sequence of operations rather than one operation. In this sequence, one operation possibly performs certain modifications that have an impact on the run time of the next operation in the sequence. One way of assessing the worst case run time of the entire sequence is to add worst case efficiencies for each operation. But this may result in an excessively large and unrealistic bound on the actual run time. To be more realistic, amortized analysis can be used to find the average complexity of a worst case sequence of operations. By analyzing sequences of operations rather than isolated operations, amortized analysis takes into account interdependence between operations and their results. For example, if an array is sorted and only a very few new elements are added, then re-sorting this array should be much faster than sorting it for the first time because, after the new additions, the array is nearly sorted. Thus, it should be quicker to put all elements in perfect order than in a completely disorganized array. Without taking this correlation into account, the run time of the two sorting operations can be considered twice the worst case efficiency. Amortized analysis, on the other hand, decides that the second sorting is hardly applied in the worst case situation so that the combined complexity of the two sorting operations is much less than double the worst case complexity. Consequently, the average for the worst case sequence of sorting, a few insertions, and sorting again is lower according to amortized analysis than according to worst case analysis, which disregards the fact that the second sorting is applied to an array operated on already by a previous sorting.

It is important to stress that amortized analysis is analyzing sequences of operations, or if single operations are analyzed, it is done in view of their being part of the sequence. The cost of operations in the sequence may vary considerably, but how frequently particular operations occur in the sequence is important. For example, for the sequence of operations $op_1, op_2, op_3, \ldots$, the worst case analysis renders the computational complexity for the entire sequence equal to

$$C(op_1, op_2, op_3, \ldots) = C_{worst}(op_1) + C_{worst}(op_2) + C_{worst}(op_3) + \ldots$$

whereas the average complexity determines it to be

$$C(op_1, op_2, op_3, \ldots) = C_{avg}(op_1) + C_{avg}(op_2) + C_{avg}(op_3) + \ldots$$

Although specifying complexities for a sequence of operations, neither worst case analysis nor average case analysis was looking at the position of a particular operation in the sequence. These two analyses considered the operations as executed in isolation and the sequence as a collection of isolated and independent operations. Amortized analysis changes the perspective by looking at what happened up until a particular point in the sequence of operations and then determines the complexity of a particular operation,

$$C(op_1, op_2, op_3, \ldots) = C(op_1) + C(op_2) + C(op_3) + \ldots$$

where C can be the worst, the average, the best case complexity, or very likely, a complexity other than the three depending on what happened before. To find amortized complexity in this way may be, however, too complicated. Therefore, another approach is used. The knowledge of the nature of particular processes and possible changes of a data structure is used to determine the function C, which can be applied to each operation of the sequence. The function is chosen in such a manner that it considers quick operations as slower than they really are and time-consuming operations as quicker than they actually are. It is as though the cheap (quick) operations are charged more time units to generate credit to be used for covering the cost of expensive operations that are charged below their real cost. It is like letting the government charge us more for income taxes than necessary so that at the end of the fiscal year the overpayment can be received back and used to cover the expenses of something else. The art of amortized analysis lies in finding an appropriate function C so that it overcharges cheap operations sufficiently to cover expenses of undercharged operations. The overall balance must be nonnegative. If a debt occurs, there must be a prospect of paying it.

Consider the operation of adding a new element to the vector-implemented as a flexible array. The best case is when the size of the vector is less than its capacity because adding a new element amounts to putting it in the first available cell. The cost of adding a new element is thus $O(1)$. The worst case is when size equals capacity, in which case there is no room for new elements. In this case, new space must be allocated, the existing elements are copied to the new space, and only then can the new element be added to the vector. The cost of adding a new element is $O(size(vector))$. It is clear that the latter situation is less frequent than the former, but this depends on another parameter, capacity increment, which refers to how much the vector is increased when overflow occurs. In the extreme case, it can be incremented by just one cell, so in the sequence of m consecutive insertions, each insertion causes overflow and requires $O(size(vector))$ time to finish. Clearly, this situation should be delayed. One solution is to allocate, say, 1 million cells for the vector, which in most cases does not cause an overflow, but the amount of space is excessively large and only a small percentage of space allocated for the vector may be expected to be in actual use. Another solution to the problem is to double the space allocated for the vector if overflow occurs. In this case, the pessimistic $O(size(vector))$ performance of the insertion operation may be expected to occur only infrequently. By using this estimate, it may be claimed that, in the best case, the cost of inserting m items is $O(m)$, but it is impossible to claim that, in the worst case, it is $O(m \cdot size(vector))$. Therefore, to see better what impact this performance has on the sequence of operations, the amortized analysis should be used.

In amortized analysis, the question is asked: What is the expected efficiency of a sequence of insertions? We know that the best case is $O(1)$ and the worst case is

Section 2.9 Amortized Complexity

$O(size(vector))$, but also we know that the latter case occurs only occasionally and leads to doubling the size of the vector. In this case, what is the expected efficiency of one insertion in the series of insertions? Note that we are interested only in sequences of insertions, excluding deletions and modifications, to have the worst case scenario. The outcome of amortized analysis depends on the assumed amortized cost of one insertion. It is clear that if

$$amCost(push(x)) = 1$$

where 1 represents the cost of one insertion, then we are not gaining anything from this analysis because easy insertions are paying for themselves right away, and the insertions causing overflow and thus copying have no credit to use to make up for their high cost. Is

$$amCost(push(x)) = 2$$

a reasonable choice? Consider the table in Figure 2.6a. It shows the change in vector capacity and the cost of insertion when size grows from 0 to 18; that is, the table indicates the changes in the vector during the sequence of 18 insertions into an initially

FIGURE 2.6   Estimating the amortized cost.

(a) Size	Capacity	Amortized Cost	Cost	Units Left	(b) Size	Capacity	Amortized Cost	Cost	Units Left
0	0				0	0			
1	1	2	0 + 1	1	1	1	3	0 + 1	2
2	2	2	1 + 1	1	2	2	3	1 + 1	3
3	4	2	2 + 1	0	3	4	3	2 + 1	3
4	4	2	1	1	4	4	3	1	5
5	8	2	4 + 1	−2	5	8	3	4 + 1	3
6	8	2	1	−1	6	8	3	1	5
7	8	2	1	0	7	8	3	1	7
8	8	2	1	1	8	8	3	1	9
9	16	2	8 + 1	−6	9	16	3	8 + 1	3
10	16	2	1	−5	10	16	3	1	5
⋮	⋮	⋮	⋮	⋮	⋮	⋮	⋮	⋮	⋮
16	16	2	1	1	16	16	3	1	17
17	32	2	16 + 1	−14	17	32	3	16 + 1	3
18	32	2	1	−13	18	32	3	1	5
⋮	⋮	⋮	⋮	⋮	⋮	⋮	⋮	⋮	⋮

empty vector. For example, if there are four elements in the vector (size = 4), then before inserting the fifth element, the four elements are copied at the cost of four units and then the new fifth element is inserted in the newly allocated space for the vector. Hence, the cost of the fifth insertion is 4 + 1. But to execute this insertion, two units allocated for the fifth insertion are available plus one unit left from the previous fourth insertion. This means that this operation is two units short to pay for itself. Thus, in the Units Left column, − 2 is entered to indicate the debt of two units. The table indicates that the debt decreases and becomes zero, one cheap insertion away from the next expensive insertion. This means that the operations are almost constantly executed in the red, and more important, if a sequence of operations finishes before the debt is paid off, then the balance indicated by amortized analysis is negative, which is inadmissible in the case of algorithm analysis. Therefore, the next best solution is to assume that

$$amCost(push(x)) = 3$$

The table in Figure 2.6b indicates that we are never in debt and that the choice of three units for amortized cost is not excessive because right after an expensive insertion, the accumulated units are almost depleted.

In this example, the choice of a constant function for amortized cost is adequate, but usually it is not. Define as *potential* a function that assigns a number to a particular state of a data structure ds that is a subject of a sequence of operations. The amortized cost is defined as a function

$$amCost(op_i) = cost(op_i) + potential(ds_i) - potential(ds_{i-1})$$

which is the real cost of executing the operation op_i plus the change in potential in the data structure ds as a result of execution of op_i. This definition holds for one single operation of a sequence of m operations. If amortized costs for all the operations are added, then the amortized cost for the sequence

$$amCost(op_1, \ldots, op_m) = \sum_{i=1}^{m}(cost(op_i) + potential(ds_i) - potential(ds_{i-1}))$$

$$= \sum_{i=1}^{m}(cost(op_i) + potential(ds_m) - potential(ds_0))$$

In most cases, the potential function is initially zero and is always nonnegative so that amortized time is an upper bound of real time. This form of amortized cost is used later in the book.

Amortized cost of including new elements in a vector can now be phrased in terms of the potential function defined as

$$potential(vector_i) = \begin{cases} 0 & \text{if } size_i = capacity_i \text{ (vector is full)} \\ 2size_i - capacity_i & \text{otherwise} \end{cases}$$

To see that the function works as intended, consider three cases. The first case is when a cheap pushing follows cheap pushing (vector is not extended right before the current push and is not extended as a consequence of the current push) and

$$amCost(push_i()) = 1 + 2size_{i-1} + 2 - capacity_{i-1} - 2size_{i-1} + capacity_i = 3$$

because the capacity does not change, $size_i = size_{i-1} + 1$, and the actual cost equals 1. For expensive pushing following cheap pushing,

$$amCost(push_i()) = size_{i-1} + 2 + 0 - 2size_{i-1} + capacity_{i-1} = 3$$

because $size_{i-1} + 1 = capacity_{i-1}$ and the actual cost equals $size_i + 1 = size_{i-1} + 2$, which is the cost of copying the vector elements plus adding the new element. For cheap pushing following expensive pushing,

$$amCost(push_i()) = 1 + 2size_i - capacity_i - 0 = 3$$

because $2(size_i - 1) = capacity_i$ and actual cost equals 1. Note that the fourth case, expensive pushing following expensive pushing, occurs only twice, when capacity changes from zero to one and from one to zero. In both cases, amortized cost equals 3.

2.10 NP-COMPLETENESS

A *deterministic* algorithm is a uniquely defined (determined) sequence of steps for a particular input; that is, given an input and a step during execution of the algorithm, there is only one way to determine the next step that the algorithm can make. A *nondeterministic* algorithm is an algorithm that can use a special operation that makes a guess when a decision is to be made. Consider the nondeterministic version of binary search.

If we try to locate number k in an unordered array of numbers, then the algorithm first accesses the middle element m of the array. If $m = k$, then the algorithm returns m's position; if not, the algorithm makes a guess concerning which way to go to continue: to the left of m or to its right. A similar decision is made at each stage: If number k is not located, continue in one of the two halves of the currently scrutinized subarray. It is easy to see that such a guessing very easily may lead us astray, so we need to endow the machine with the power of making correct guesses. However, an implementation of this nondeterministic algorithm would have to try, in the worst case, all the possibilities. One way to accomplish it is by requiring that the decision in each iteration is in reality this: if $m \neq k$, then go both to the right and to the left of m. In this way, a tree is created that represents the decisions made by the algorithm (Johnson & Papadimitriou p. 53). The algorithm solves the problem, if any of the branches allows us to locate k in the array that includes k and if no branch leads to such a solution when k is not in the array.

A *decision problem* has two answers, call them "yes" and "no." A decision problem is given by the set of all instances of the problem and the set of instances for which the answer is "yes." Many optimization problems do not belong to that category ("find the minimum x for which ...") but in most cases they can be converted to decision problems ("is x, for which ..., less than k?").

Generally, a nondeterministic algorithm solves a decision problem if it answers it in the affirmative and there is a path in the tree that leads to a yes answer, and it answers it in the negative if there is no such path. A nondeterministic algorithm is considered polynomial if a number of steps leading to an affirmative answer in a decision tree is $O(n^k)$, where n is the size of the problem instance.

Most of the algorithms analyzed in this book are polynomial-time algorithms; that is, their running time in the worst case is $O(n^k)$ for some k. Problems that can be solved with such algorithms are called *tractable* and the algorithms are considered *efficient*.

A problem belongs to the class of P problems if it can be solved in polynomial time with a deterministic algorithm. A problem belongs to the class of NP problems if it can be solved in polynomial time with a nondeterministic algorithm. P problems are obviously tractable. NP problems are also tractable, but only when nondeterministic algorithms are used.

Clearly, $P \subseteq NP$, because deterministic algorithms are those nondeterministic algorithms that do not use nondeterministic decisions. It is also believed that $P \neq NP$; that is, there exist problems with nondeterministic polynomial algorithms that cannot be solved with deterministic polynomial algorithms. This means that on deterministic Turing machines they are executed in nonpolynomial time and thus they are intractable. The strongest argument in favor of this conviction is the existence of NP-complete problems. But first we need to define the concept of reducibility of algorithms.

A problem P_1 is reducible to another problem P_2 if there is a way of encoding instances x of P_1 as instances $y = r(x)$ of P_2 using a *reduction function* r executed with a *reduction algorithm*; that is, for each x, x is an instance of P_1 iff $y = r(x)$ is an instance of P_2. Note that reducibility is not a symmetric relation: P_1 can be reducible to P_2 but not necessarily vice versa; that is, each instance x of P_1 should have a counterpart y of P_2, but there may be instances y of P_2 onto which no instances x of P_1 are mapped with the function r. Therefore, P_2 can be considered a harder problem than P_1.

The reason for the reduction is that if the value $r(x)$ for any x can be found efficiently (in polynomial time), then an efficient solution for y can be efficiently transformed into an efficient solution of x. Also, if there is no efficient algorithm for x, then there is no efficient solution for y.

A problem is called *NP-complete* if it is NP (it can be solved efficiently by a nondeterministic polynomial algorithm) and every NP problem can be polynomially reduced to this problem. Because reducibility is a transitive relation, we can also say that an NP problem P_1 is NP-complete if there is an NP-complete problem P_2 that is polynomially reducible to P_1. In this way, all NP-complete problems are computationally equivalent; that is, if an NP-complete problem can be solved with a deterministic polynomial algorithm, then so can be all NP-complete problems, and thus $P = NP$. Also, if any problem in NP is intractable, then all NP-complete problems are intractable.

The reduction process uses an NP-complete problem to show that another problem is also NP-complete. There must be, however, at least one problem directly proven to be NP-complete by other means than reduction to make the reduction process possible. A problem that was shown by Stephen Cook to be in that category is the satisfiability problem.

The *satisfiability problem* concerns Boolean expressions in conjunctive normal form (CNF). An expression is in CNF if it is a conjunction of alternatives where each alternative involves Boolean variables and their negations, and each variable is either true or false. For example,

$$(x \lor y \lor z) \land (w \lor x \lor \neg y \lor z) \land (\neg w \lor \neg y)$$

is in CNF. A Boolean expression is *satisfiable* if there exists an assignment of values true and false that renders the entire expression true. For example, our expression is satisfiable for x = false, y = false, and z = true. The satisfiability problem consists of determining whether a Boolean expression is satisfiable (the value assignments do not have to be given). The problem is NP, because assignments can be guessed and then the expression tested for satisfiability in polynomial time.

Cook proves that the satisfiability problem is NP-complete by using a theoretical concept of the Turing machine that can perform nondeterministic decisions (make good guesses). Operations of that machine are then described in terms of Boolean expressions, and it is shown that the expression is satisfiable iff the Turing machine terminates for a particular input (for the proof, see Appendix C).

To illustrate the reduction process, consider the *three-satisfiability problem*, which is the satisfiability problem in the case when each alternative in a Boolean expression in CNF includes only three different variables. We claim that the problem is NP-complete. The problem is NP, because a guessed assignment of truth values to variables in a Boolean expression can be verified in polynomial time. We show that the three-satisfiability problem is NP-complete by reducing it to the satisfiability problem. The reduction process involves showing that an alternative with any number of Boolean variables can be converted into a conjunction of alternatives, each alternative with three Boolean variables only. This is done by introducing new variables. Consider an alternative

$$A = (p_1 \vee p_2 \vee \ldots \vee p_k)$$

for $k \geq 4$ where $p_i \in \{x_i, \neg x_i\}$. With new variables $y_1, \ldots, y_{k-3}$, we transform A into

$$A' = (p_1 \vee p_2 \vee y_1) \wedge (p_3 \vee \neg y_1 \vee y_2) \wedge (p_4 \vee \neg y_2 \vee y_3) \wedge \ldots \ldots$$
$$(p_{k-2} \vee \neg y_{k-4} \vee y_{k-3}) \wedge (p_{k-1} \vee p_k \vee \neg y_{k-3})$$

If the alternative A is satisfiable, then at least one term p_i is true, so the values of y_j's can be so chosen that A' is true: if p_i is true, then we set $y_1, \ldots, y_{i-2}$ to true and the remaining $y_{i-1}, \ldots, y_{k-3}$ to false. Conversely, if A' is satisfiable, then at least one p_i must be true, because if all p_i's are false, then the expression

$$A' = (\text{false} \vee \text{false} \vee y_1) \wedge (\text{false} \vee \neg y_1 \vee y_2) \wedge (\text{false} \vee \neg y_2 \vee y_3) \wedge \ldots$$
$$\wedge (\text{false} \vee \text{false} \vee \neg y_{k-3})$$

has the same truth value as the expression

$$(y_1) \wedge (\neg y_1 \vee y_2) \wedge (\neg y_2 \vee y_3) \wedge \ldots \wedge (\neg y_{k-3})$$

which cannot be true for any choice of values for y_j's, thus is not satisfiable.

2.11 EXERCISES

1. Explain the meaning of the following expressions:
   a. $f(n)$ is $O(1)$.
   b. $f(n)$ is $\Theta(1)$.
   c. $f(n)$ is $n^{O(1)}$.

2. Assuming that $f_1(n)$ is $O(g_1(n))$ and $f_2(n)$ is $O(g_2(n))$, prove the following statements:
   a. $f_1(n) + f_2(n)$ is $O(\max(g_1(n),g_2(n)))$.
   b. If a number k can be determined such that for all $n > k$, $g_1(n) \leq g_2(n)$, then $O(g_1(n)) + O(g_2(n))$ is $O(g_2(n))$.
   c. $f_1(n) * f_2(n)$ is $O(g_1(n) * g_2(n))$ (rule of product).
   d. $O(cg(n))$ is $O(g(n))$.
   e. c is $O(1)$.

3. Prove the following statements:
   a. $\sum_{i=1}^{n} i^2$ is $O(n^3)$ and more generally, $\sum_{i=1}^{n} i^k$ is $O(n^{k+1})$.
   b. $an^k/\lg n$ is $O(n^k)$ but $an^k/\lg n$ is not $\Theta(n^k)$.
   c. $n^{1.1} + n \lg n$ is $\Theta(n^{1.1})$.
   d. 2^n is $O(n!)$ and $n!$ is not $O(2^n)$.
   e. 2^{n+a} is $O(2^n)$.
   f. 2^{2n+a} is not $O(2^n)$.
   g. $2^{\sqrt{\lg n}}$ is $O(n^a)$.

4. Make the same assumptions as in Exercise 2 and, by finding counterexamples, refute the following statements:
   a. $f_1(n) - f_2(n)$ is $O(g_1(n) - g_2(n))$.
   b. $f_1(n)/f_2(n)$ is $O(g_1(n)/g_2(n))$.

5. Find functions f_1 and f_2 such that both $f_1(n)$ and $f_2(n)$ are $O(g(n))$, but $f_1(n)$ is not $O(f_2)$.

6. Is it true that
   a. if $f(n)$ is $\Theta(g(n))$, then $2^{f(n)}$ is $\Theta(2^{g(n)})$?
   b. $f(n) + g(n)$ is $\Theta(\min(f(n),g(n)))$?
   c. 2na is $O(2^n)$?

7. The algorithm presented in this chapter for finding the length of the longest subarray with the numbers in increasing order is inefficient, because there is no need to continue to search for another array if the length already found is greater than the length

of the subarray to be analyzed. Thus, if the entire array is already in order, we can discontinue the search right away, converting the worst case into the best. The change needed is in the outer loop, which now has one more test:

```
for (i = 0, length = 1; i < n-1 && length < n==i; i++)
```

What is the worst case now? Is the efficiency of the worst case still $O(n^2)$?

8. Find the complexity of the function used to find the kth smallest integer in an unordered array of integers

```
int selectkth(int a[], int k, int n) {
    int i, j, mini, tmp;
    for (i = 0; i < k; i++) {
        mini = i;
        for (j = i+1; j < n; j++)
            if (a[j]<a[mini])
                mini = j;
        tmp = a[i];
        a[i] = a[mini];
        a[mini] = tmp;
    }
    return a[k-1];
}
```

9. Determine the complexity of the following implementations of the algorithms for adding, multiplying, and transposing $n \times n$ matrices:

```
for (i = 0; i < n; i++)
    for (j = 0; j < n; j++)
        a[i][j] = b[i][j] + c[i][j];

for (i = 0; i < n; i++)
    for (j = 0; j < n; j++)
        for (k = a[i][j] = 0; k < n; k++)
            a[i][j] += b[i][k] * c[k][j];

for (i = 0; i < n - 1; i++)
    for (j = i+1; j < n; j++) {
        tmp = a[i][j];
        a[i][j] = a[j][i];
        a[j][i] = tmp;
    }
```

10. Find the computational complexity for the following four loops:

   a. ```
      for (cnt1 = 0, i = 1; i <= n; i++)
          for (j = 1; j <= n; j++)
              cnt1++;
      ```
   b. ```
      for (cnt2 = 0, i = 1; i <= n; i++)
          for (j = 1; j <= i; j++)
              cnt2++;
      ```
   c. ```
      for (cnt3 = 0, i = 1; i <= n; i *= 2)
          for (j = 1; j <= n; j++)
              cnt3++;
      ```
   d. ```
      for (cnt4 = 0, i = 1; i <= n; i *= 2)
          for (j = 1; j <= i; j++)
              cnt4++;
      ```

11. Find the average case complexity of sequential search in an array if the probability of accessing the last cell equals $\frac{1}{2}$, the probability of the next to last cell equals $\frac{1}{4}$, and the probability of locating a number in any of the remaining cells is the same and equal to $\frac{1}{4(n-2)}$.

12. Consider a process of incrementing a binary n-bit counter. An increment causes some bits to be flipped: Some 0s are changed to 1s, and some 1s to 0s. In the best case, counting involves only one bit switch; for example, when 000 is changed to 001, sometimes all the bits are changed, as when incrementing 011 to 100.

Number	Flipped Bits
000	
001	1
010	2
011	1
100	3
101	1
110	2
111	1

    Using worst case assessment, we may conclude that the cost of executing $m = 2^n - 1$ increments is $O(mn)$. Use amortized analysis to show that the cost of executing m increments is $O(m)$.

13. How can you convert a satisfiability problem into a three-satisfiability problem for an instance when an alternative in a Boolean expression has two variables? One variable?

BIBLIOGRAPHY

Computational Complexity

Hartmanis, Juris, and Hopcroft, John E., "An Overview of the Theory of Computational Complexity," *Journal of the ACM* 18 (1971), 444–475.

Hartmanis, Juris, and Stearns, Richard E., "On the Computational Complexity of Algorithms," *Transactions of the American Mathematical Society* 117 (1965), 284–306.

Preparata, Franco P., "Computational Complexity," in Pollack, S. V. (ed.), *Studies in Computer Science*, Washington, DC: The Mathematical Association of America, 1982, 196–228.

Big-O, Ω, and Θ Notations

Brassard, G., "Crusade for a Better Notation," *SIGACT News* 17 (1985), 60–64.

Knuth, Donald, *The Art of Computer Programming, Vol. 2: Seminumerical Algorithms*, Reading, MA: Addison-Wesley, 1998.

Knuth, Donald, "Big Omicron and Big Omega and Big Theta," *SIGACT News,* April–June, 8 (1976), 18–24.

Vitanyi, P. M. B., and Meertens, L., "Big Omega versus the Wild Functions," *SIGACT News* 16 (1985), 56–59.

OO Notation

Manber, Udi, *Introduction to Algorithms: A Creative Approach*, Reading, MA: Addison-Wesley, 1989.

Amortized Analysis

Heileman, Gregory L., *Discrete Structures, Algorithms, and Object-Oriented Programming*, New York: McGraw-Hill, 1996, chs. 10–11.

Tarjan, Robert E., "Amortized Computational Complexity," *SIAM Journal on Algebraic and Discrete Methods* 6 (1985), 306–318.

NP-Completeness

Cook, Stephen A., "The Complexity of Theorem-Proving Procedures," *Proceedings of the Third Annual ACM Symposium on Theory of Computing*, 1971, 151–158.

Garey, Michael R., and Johnson, David S., *Computers and Intractability: A Guide to the Theory of NP-Completeness*, San Francisco: Freeman, 1979.

Johnson, D. S., and Papadimitriou, C. H., Computational Complexity, in Lawler, E. L., Lenstra, J. K., Rinnoy, Kan A. H. G., and Shmoys, D. B. (eds.), *The Traveling Salesman Problem*, New York: Wiley 1985, 37–85.

Karp, Richard M., "Reducibility Among Combinatorial Problems," in R. E. Miller and J. W. Thatcher (eds.), *Complexity of Computer Computations*, New York: Plenum Press, 1972, 85–103.

3 Linked Lists

An array is a very useful data structure provided in programming languages. However, it has at least two limitations: (1) its size has to be known at compilation time and (2) the data in the array are separated in computer memory by the same distance, which means that inserting an item inside the array requires shifting other data in this array. This limitation can be overcome by using *linked structures*. A linked structure is a collection of nodes storing data and links to other nodes. In this way, nodes can be located anywhere in memory, and passing from one node of the linked structure to another is accomplished by storing the addresses of other nodes in the linked structure. Although linked structures can be implemented in a variety of ways, the most flexible implementation is by using pointers.

3.1 SINGLY LINKED LISTS

If a node contains a data member that is a pointer to another node, then many nodes can be strung together using only one variable to access the entire sequence of nodes. Such a sequence of nodes is the most frequently used implementation of a *linked list*, which is a data structure composed of nodes, each node holding some information and a pointer to another node in the list. If a node has a link only to its successor in this sequence, the list is called a *singly linked list*. An example of such a list is shown in Figure 3.1. Note that only one variable p is used to access any node in the list. The last node on the list can be recognized by the null pointer.

Each node in the list in Figure 3.1 is an instance of the following class definition:

```
class IntSLLNode {
public:
    IntSLLNode() {
        next = 0;
    }
    IntSLLNode(int i, IntSLLNode *in = 0) {
        info = i; next = in;
    }
```

Section 3.1 Singly Linked Lists ■ 77

FIGURE 3.1 A singly linked list.

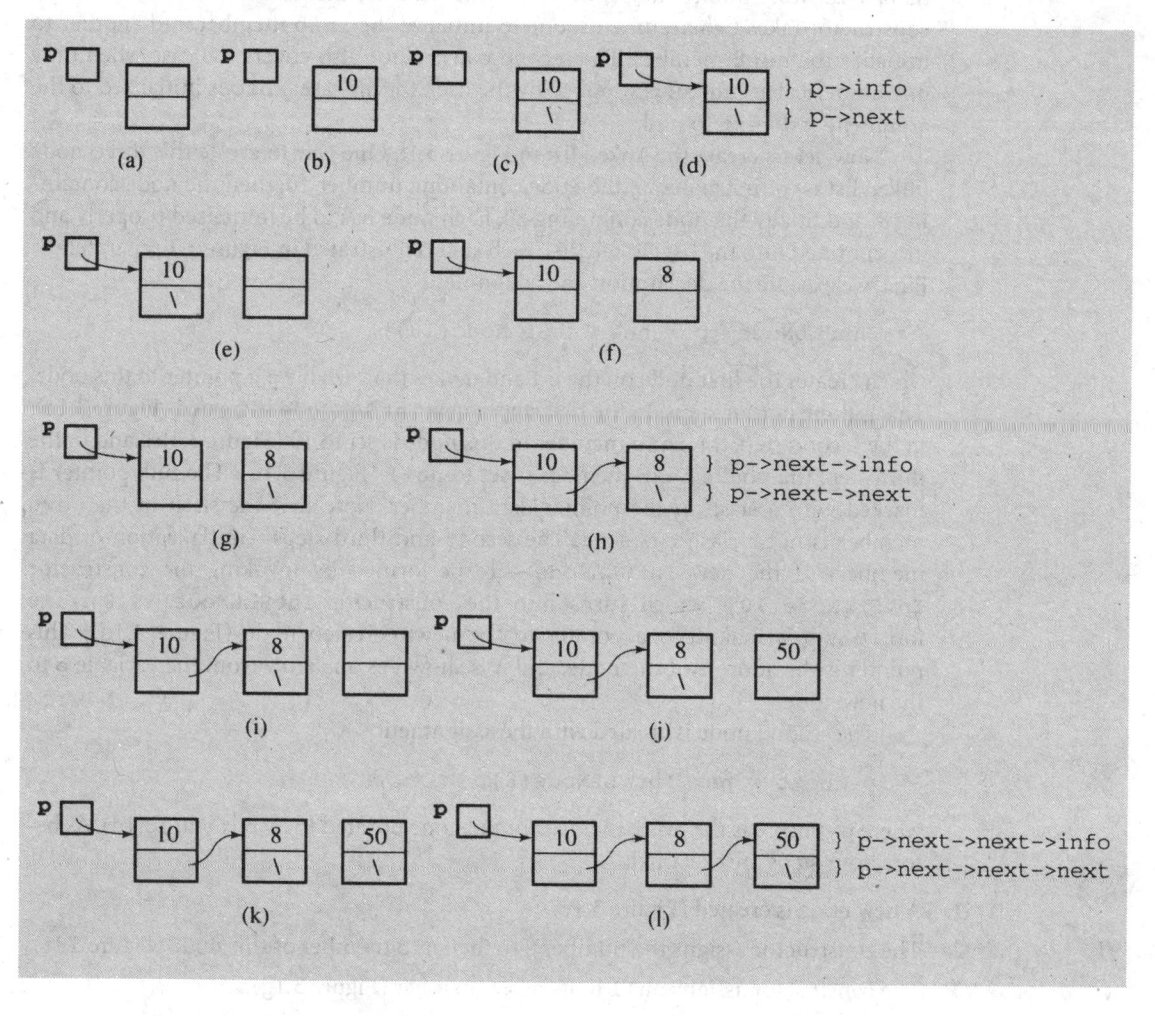

```
                                              } p->info
                                              } p->next
   (a)        (b)         (c)        (d)

   (e)                    (f)

   (g)                    (h)  } p->next->info
                               } p->next->next

   (i)                    (j)

   (k)                    (l)  } p->next->next->info
                               } p->next->next->next
```

```
    int info;
    IntSLLNode *next;
};
```

A node includes two data members: `info` and `next`. The `info` member is used to store information, and this member is important to the user. The `next` member is used to link nodes to form a linked list. It is an auxiliary data member used to maintain the linked list. It is indispensable for implementation of the linked list, but less important (if at all) from the user's perspective. Note that `IntSLLNode` is defined in terms of itself because one data member, `next`, is a pointer to a node of the same type that is just being defined. Objects that include such a data member are called self-referential objects.

The definition of a node also includes two constructors. The first constructor initializes the next pointer to null and leaves the value of info unspecified. The second constructor takes two arguments, one to initialize the info member and another to initialize the next member. The second constructor also covers the case when only one numerical argument is supplied by the user. In this case, info is initialized to the argument and next to null.

Now, let us create the linked list in Figure 3.1l. One way to create this three-node linked list is to first generate the node containing number 10, then the node containing 8, and finally the node containing 50. Each node has to be initialized properly and incorporated into the list. To see this, each step is illustrated in Figure 3.1 separately. First, we execute the declaration and assignment

```
IntSLLNode *p = new IntSLLNode(10);
```

which creates the first node on the list and makes the variable p a pointer to this node. This is done in four steps. In the first step, a new IntSLLNode is created (Figure 3.1a), in the second step, the info member of this node is set to 10 (Figure 3.1b), and in the third step, the node's next member is set to null (Figure 3.1c). The null pointer is marked with a slash in the pointer data member. Note that the slash in the next member is not a slash character. The second and third steps—initialization of data members of the new IntSLLNode—are performed by invoking the constructor IntSLLNode(10), which turns into the constructor IntSLLNode(10,0). The fourth step is making p a pointer to the newly created node (Figure 3.1d). This pointer is the address of the node, and it is shown as an arrow from the variable p to the new node.

The second node is created with the assignment

```
p->next = new IntSLLNode(8);
```

where p->next is the next member of the node pointed to by p (Figure 3.1d). As before, four steps are executed:

1. A new node is created (Figure 3.1e).
2. The constructor assigns the number 8 to the info member of this node (Figure 3.1f).
3. The constructor assigns null to its next member (Figure 3.1g).
4. The new node is included in the list by making the next member of the first node a pointer to the new node (Figure 3.1h).

Note that the data members of nodes pointed to by p are accessed using the arrow notation, which is clearer than using a dot notation, as in (*p).next.

The linked list is now extended by adding a third node with the assignment

```
p->next->next = new IntSLLNode(50);
```

where p->next->next is the next member of the second node. This cumbersome notation has to be used because the list is accessible only through the variable p.

In processing the third node, four steps are also executed: creating the node (Figure 3.1i), initializing its two data members (Figure 3.1j–k), and then incorporating the node in the list (Figure 3.1l).

Our linked list example illustrates a certain inconvenience in using pointers: The longer the linked list, the longer the chain of nexts to access the nodes at the end of the list. In this example, p->next->next->next allows us to access the next member of the 3rd node on the list. But what if it were the 103rd or, worse, the 1,003rd node on the list? Typing 1,003 nexts, as in p->next->...->next, would be daunting. If we missed one next in this chain, then a wrong assignment is made. Also, the flexibility of using linked lists is diminished. Therefore, other ways of accessing nodes in linked lists are needed. One way is always to keep two pointers to the linked list: one to the first node and one to the last, as shown in Figure 3.2.

FIGURE 3.2 An implementation of a singly linked list of integers.

```
//*********************   intSLLst.h   *************************
//          singly-linked list class to store integers

#ifndef INT_LINKED_LIST
#define INT_LINKED_LIST

class IntSLLNode {
public:
    int info;
    IntSLLNode *next;
    IntSLLNode(int el, IntSLLNode *ptr = 0) {
        info = el; next = ptr;
    }
};

class IntSLList {
public:
    IntSLList() {
        head = tail = 0;
    }
    ~IntSLList();
    int isEmpty() {
        return head == 0;
    }
    void addToHead(int);
    void addToTail(int);
    int  deleteFromHead(); // delete the head and return its info;
    int  deleteFromTail(); // delete the tail and return its info;
    void deleteNode(int);
    bool isInList(int) const;
```

Continues

FIGURE 3.2 *(continued)*

```cpp
private:
    IntSLLNode *head, *tail;
};

#endif

//*********************** intSLLst.cpp ***************************

#include <iostream.h>
#include "intSLLst.h"

IntSLLList::~IntSLLList() {
    for (IntSLLNode *p; !isEmpty(); ) {
        p = head->next;
        delete head;
        head = p;
    }
}
void IntSLLList::addToHead(int el) {
    head = new IntSLLNode(el,head);
    if (tail == 0)
       tail = head;
}
void IntSLLList::addToTail(int el) {
    if (tail != 0) {    // if list not empty;
        tail->next = new IntSLLNode(el);
        tail = tail->next;
    }
    else head = tail = new IntSLLNode(el);
}
int IntSLLList::deleteFromHead() {
    int el = head->info;
    IntSLLNode *tmp = head;
    if (head == tail)   // if only one node in the list;
        head = tail = 0;
    else head = head->next;
    delete tmp;
    return el;
}
int IntSLLList::deleteFromTail() {
    int el = tail->info;
    if (head == tail) { // if only one node in the list;
```

FIGURE 3.2 (continued)

```
            delete head;
            head = tail = 0;
        }
        else {                    // if more than one node in the list,
            IntSLLNode *tmp;      // find the predecessor of tail;
            for (tmp = head; tmp->next != tail; tmp = tmp->next);
            delete tail;
            tail = tmp;           // the predecessor of tail becomes tail;
            tail->next = 0;
        }
    return el;
}
void IntSLList::deleteNode(int el) {
    if (head != 0)                              // if nonempty list;
        if (head == tail && el == head->info) { // if only one
            delete head;                        // node in the list;
            head = tail = 0;
        }
        else if (el == head->info) {// if more than one node in the list
            IntSLLNode *tmp = head;
            head = head->next;
            delete tmp;             // and old head is deleted;
        }
        else {                      // if more than one node in the list
            IntSLLNode *pred, *tmp;
            for (pred = head, tmp = head->next; // and a nonhead node
                 tmp != 0 && !(tmp->info == el);// is deleted;
                 pred = pred->next, tmp = tmp->next);
            if (tmp != 0) {
                pred->next = tmp->next;
                if (tmp == tail)
                    tail = pred;
                delete tmp;
            }
        }
}
bool IntSLList::isInList(int el) const {
    IntSLLNode *tmp;
    for (tmp = head; tmp != 0 && !(tmp->info == el); tmp = tmp->next);
    return tmp != 0;
}
```

The singly linked list implementation in Figure 3.2 uses two classes: one class, `IntSLLNode`, for nodes of the list, and another, `IntSLList`, for access to the list. The class `IntSLList` defines two data members, head and tail, which are pointers to the first and the last nodes of a list. This explains why all members of `IntSLLNode` are declared public. Because particular nodes of the list are accessible through pointers, nodes are made inaccessible to outside objects by declaring head and tail private so that the information-hiding principle is not really compromised. If some of the members of `IntSLLNode` were declared nonpublic, then classes derived from `IntSLList` could not access them.

An example of a list is shown in Figure 3.3. The list is declared with the statement

`IntSLList list;`

The first object in Figure 3.3a is not part of the list; it allows for having access to the list. For simplicity, in subsequent figures, only nodes belonging to the list are shown, the access node is omitted, and the head and tail members are marked as in Figure 3.3b.

FIGURE 3.3 A singly linked list of integers.

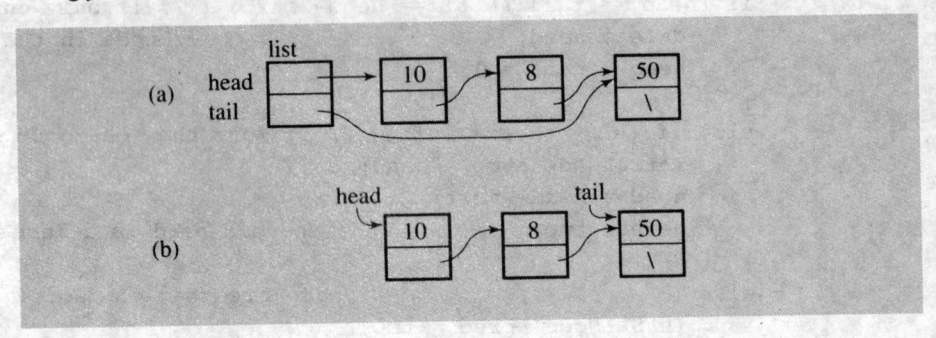

Besides the head and tail members, the class `IntSLList` also defines member functions that allow us to manipulate the lists. We now look more closely at some basic operations on linked lists presented in Figure 3.2.

3.1.1 Insertion

Adding a node at the beginning of a linked list is performed in four steps.

1. An empty node is created. It is empty in the sense that the program performing insertion does not assign any values to the data members of the node (Figure 3.4a).
2. The node's `info` member is initialized to a particular integer (Figure 3.4b).
3. Because the node is being included at the front of the list, the `next` member becomes a pointer to the first node on the list; that is, the current value of head (Figure 3.4c).
4. The new node precedes all the nodes on the list, but this fact has to be reflected in the value of head; otherwise, the new node is not accessible. Therefore, head is updated to become the pointer to the new node (Figure 3.4d).

FIGURE 3.4   Inserting a new node at the beginning of a singly linked list.

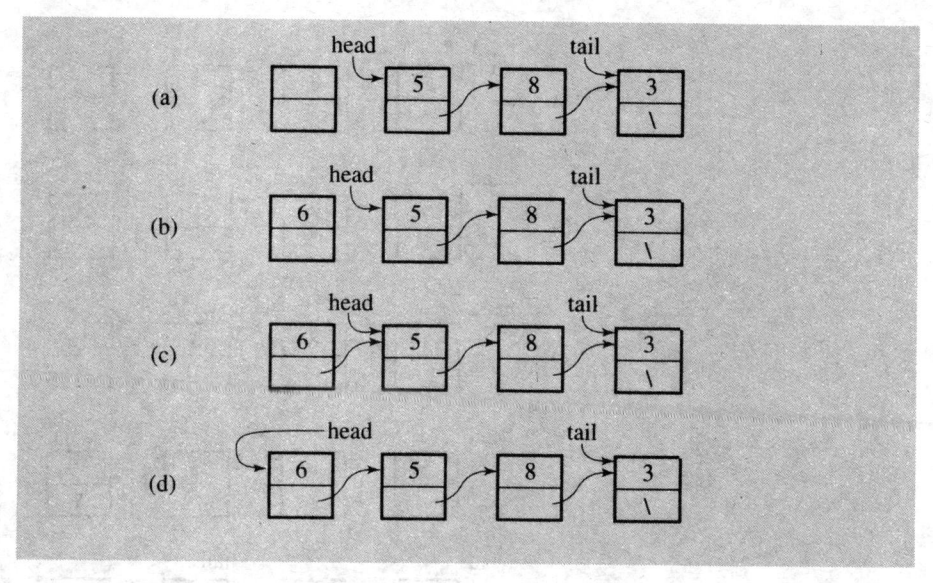

(a) head → 5 → 8 → 3 → \ (tail)

(b) head → 6, 5 → 8 → 3 → \ (tail)

(c) head → 6 → 5 → 8 → 3 → \ (tail)

(d) head → 6 → 5 → 8 → 3 → \ (tail)

The four steps are executed by the member function `addToHead()` (Figure 3.2). The function executes the first three steps indirectly by calling the constructor `IntSLLNode(el,head)`. The last step is executed directly in the function by assigning the address of the newly created node to `head`.

The member function `addToHead()` singles out one special case, namely, inserting a new node in an empty linked list. In an empty linked list, both `head` and `tail` are null; therefore, both become pointers to the only node of the new list. When inserting in a nonempty list, only `head` needs to be updated.

The process of adding a new node to the end of the list has five steps.

1. An empty node is created (Figure 3.5a).
2. The node's `info` member is initialized to an integer `el` (Figure 3.5b).
3. Because the node is being included at the end of the list, the `next` member is set to null (Figure 3.5c).
4. The node is now included in the list by making the `next` member of the last node of the list a pointer to the newly created node (Figure 3.5d).
5. The new node follows all the nodes of the list, but this fact has to be reflected in the value of `tail`, which now becomes the pointer to the new node (Figure 3.5e).

All these steps are executed in the `if` clause of `addToTail()` (Figure 3.2). The `else` clause of this function is executed only if the linked list is empty. If this case were not included, the program may crash because in the `if` clause we make an assignment to the `next` member of the node referred by `tail`. In the case of an empty linked list, it is a pointer to a nonexisting data member of a nonexisting node.

FIGURE 3.5 Inserting a new node at the end of a singly linked list.

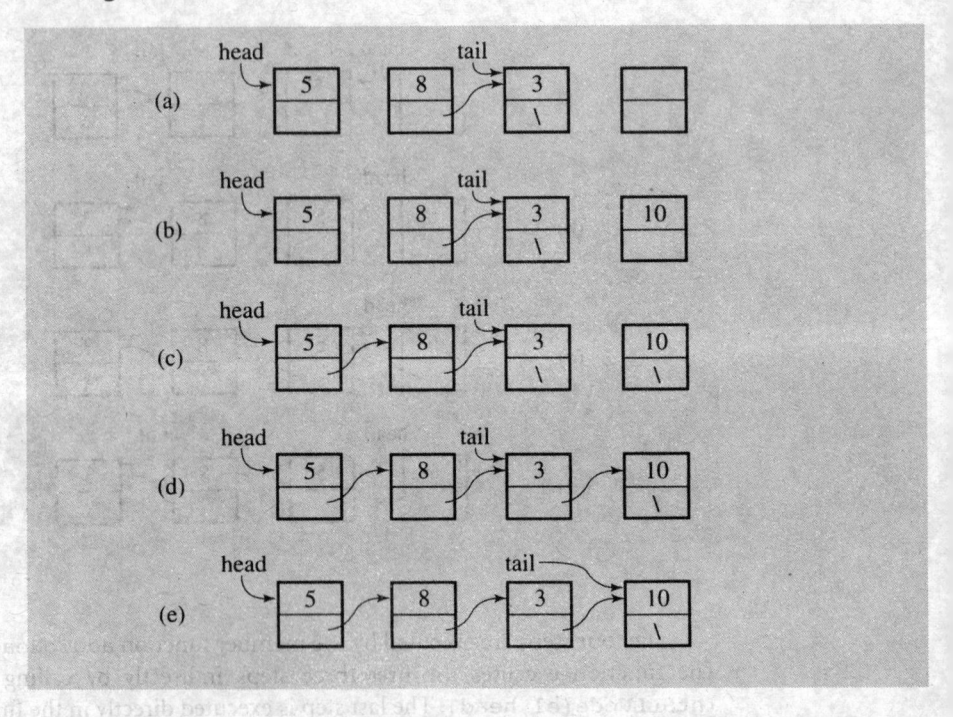

(a) head → 5 → 8 → tail → 3 \

(b) head → 5 → 8 → tail → 3 \ , 10

(c) head → 5 → 8 → tail → 3 → 10 \

(d) head → 5 → 8 → tail → 3 → 10 \

(e) head → 5 → 8 → 3 → tail → 10 \

The process of inserting a new node at the beginning of the list is very similar to the process of inserting a node at the end of the list. This is because the implementation of `IntSLList` uses two pointer members: `head` and `tail`. For this reason, both `addToHead()` and `addToTail()` can be executed in constant time $O(1)$; that is, regardless of the number of nodes in the list, the number of operations performed by these two member functions does not exceed some constant number c. Note that because the `head` pointer allows us to have access to a linked list, the `tail` pointer is not indispensable; its only role is to have immediate access to the last node of the list. With this access, a new node can be added easily at the end of the list. If the `tail` pointer were not used, then adding a node at the end of the list would be more complicated because we would first have to reach the last node in order to attach a new node to it. This requires scanning the list and requires $O(n)$ steps to finish; that is, it is linearly proportional to the length of the list. The process of scanning lists is illustrated when discussing deletion of the last node.

3.1.2 Deletion

One deletion operation consists of deleting a node at the beginning of the list and returning the value stored in it. This operation is implemented by the member function `deleteFromHead()`. In this operation, the information from the first node is temporarily stored in a local variable `el`, and then `head` is reset so that what was the second node becomes the first node. In this way, the former first node can be deleted in constant time $O(1)$ (Figure 3.6).

FIGURE 3.6 Deleting a node at the beginning of a singly linked list.

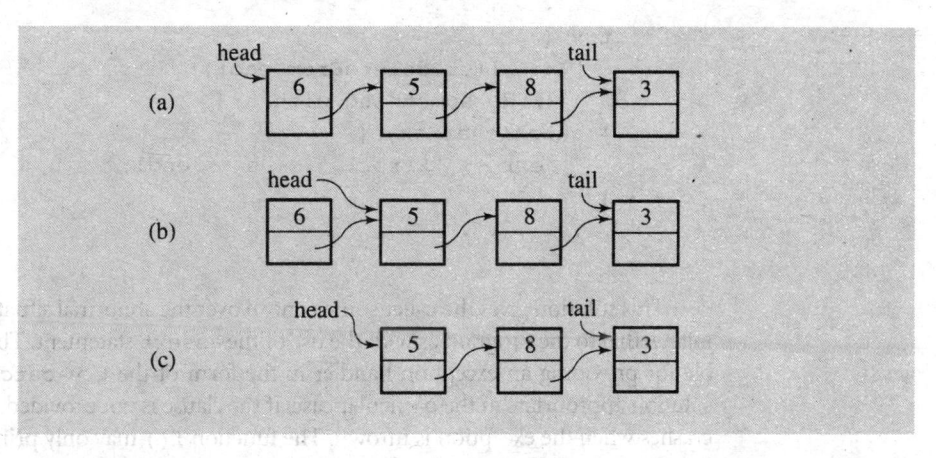

Unlike before, there are now two special cases to consider. One case is when we attempt to remove a node from an empty linked list. If such an attempt is made, the program is very likely to crash, which we don't want to happen. The caller should also know that such an attempt is made to perform a certain action. After all, if the caller expects a number to be returned from the call to deleteFromHead() and no number can be returned, then the caller may be unable to accomplish some other operations.

There are several ways to approach this problem. One way is to use an assert statement:

```
int IntSLList::deleteFromHead() {
    assert(!isEmpty());  // terminate the program if false;
    int el = head->info;
    . . . . . . . . .
    return el;
}
```

The assert statement checks the condition !isEmpty(), and if the condition is false, the program is aborted. This is a crude solution because the caller may wish to continue even if no number is returned from deleteFromHead().

Another solution is to throw an exception and catch it by the user, as in:

```
int IntSLList::deleteFromHead() {
    if (isEmpty())
        throw("Empty");
    int el = head->info;
    . . . . . . . . .
    return el;
}
```

The throw clause with the string argument is expected to have a matching try-catch clause in the caller (or caller's caller, etc.) also with the string argument, which catches the exception, as in

```
void f() {
    . . . . . . . . . .
    try {
        n = list.deleteFromHead();
        // do something with n;
    } catch(char *s) {
        cerr << "Error: " << s << endl;
    }
    . . . . . . . . . .
}
```

This solution gives the caller some control over the abnormal situation without making it lethal to the program as with the use of the assert statement. The user is responsible for providing an exception handler in the form of the try-catch clause, with the solution appropriate to the particular case. If the clause is not provided, then the program crashes when the exception is thrown. The function f() may only print a message that a list is empty when an attempt is made to delete a number from an empty list, another function g() may assign a certain value to n in such a case, and yet another function h() may find such a situation detrimental to the program and abort the program altogether.

The idea that the user is responsible for providing an action in the case of an exception is also presumed in the implementation given in Figure 3.2. The member function assumes that the list is not empty. To prevent the program from crashing, the member function isEmpty() is added to the IntSLList class, and the user should use it as in:

```
if (!list.isEmpty())
    n = list.deleteFromHead();
else do not remove;
```

Note that including a similar if statement in deleteFromHead() does not solve the problem. Consider this code:

```
int IntSLList::deleteFromHead() {
    if (!isEmpty()) {            // if nonempty list;
        int el = head->info;
        . . . . . . . . . .
        return el;
    }
    else return 0;
}
```

If an if statement is added, then the else clause must also be added; otherwise, the program does not compile because "not all control paths return a value." But now, if 0 is returned, the caller does not know whether the returned 0 is a sign of failure or if it is a literal 0 retrieved from the list. To avoid any confusion, the caller must use an if statement to test whether the list is empty before calling `deleteFromHead()`. In this way, one `if` statement would be redundant.

To maintain uniformity in the interpretation of the return value, the last solution can be modified so that instead of returning an integer, the function returns the pointer to an integer:

```
int* IntSLList::deleteFromHead() {
    if (!isEmpty()) {          // if nonempty list;
        int *el = new int(head->info);
        . . . . . . . . . .
        return el;
    }
    else return 0;
}
```

where 0 in the `else` clause is the null pointer, not the number 0. In this case, the function call

```
n = *list.deleteFromHead();
```

results in a program crash if `deleteFromHead()` returns the null pointer.

Therefore, a test must be performed by the caller before calling `deleteFrom-Head()` to check whether `list` is empty or a pointer variable has to be used,

```
int *p = list.deleteFromHead();
```

and then a test is performed after the call to check whether `p` is null. In either case, this means that the `if` statement in `deleteFromHead()` is redundant.

The second special case is when the list has only one node to be removed. In this case, the list becomes empty, which requires setting `tail` and `head` to null.

The second deletion operation consists of deleting a node from the end of the list, and it is implemented as the member function `deleteFromTail()`. The problem is that after removing a node, `tail` should refer to the new tail of the list; that is, `tail` has to be moved backward by one node. But moving backward is impossible because there is no direct link from the last node to its predecessor. Hence, this predecessor has to be found by searching from the beginning of the list and stopping right before `tail`. This is accomplished with a temporary variable `tmp` that scans the list within the `for` loop. The variable `tmp` is initialized to the head of the list, and then in each iteration of the loop it is advanced to the next node. If the list is as in Figure 3.7a, then `tmp` first refers to the head node holding number 6; after executing the assignment `tmp = tmp->next`, `tmp` refers to the second node (Figure 3.7b). After the second iteration and executing the same assignment, `tmp` refers to the third node (Figure 3.7c). Because this node is also the next to last node, the loop is exited, after which the last node is deleted (Figure 3.7d). Because `tail` is now pointing to a nonexisting node, it is immediately set to point to the next to last node currently pointed to by `tmp` (Figure 3.7e). To mark the fact that it is the last node of the list, the `next` member of this node is set to null (Figure 3.7f).

Note that in the `for` loop, a temporary variable is used to scan the list. If the loop were simplified to

```
for ( ; head->next != tail; head = head->next);
```

FIGURE 3.7 Deleting a node from the end of a singly linked list.

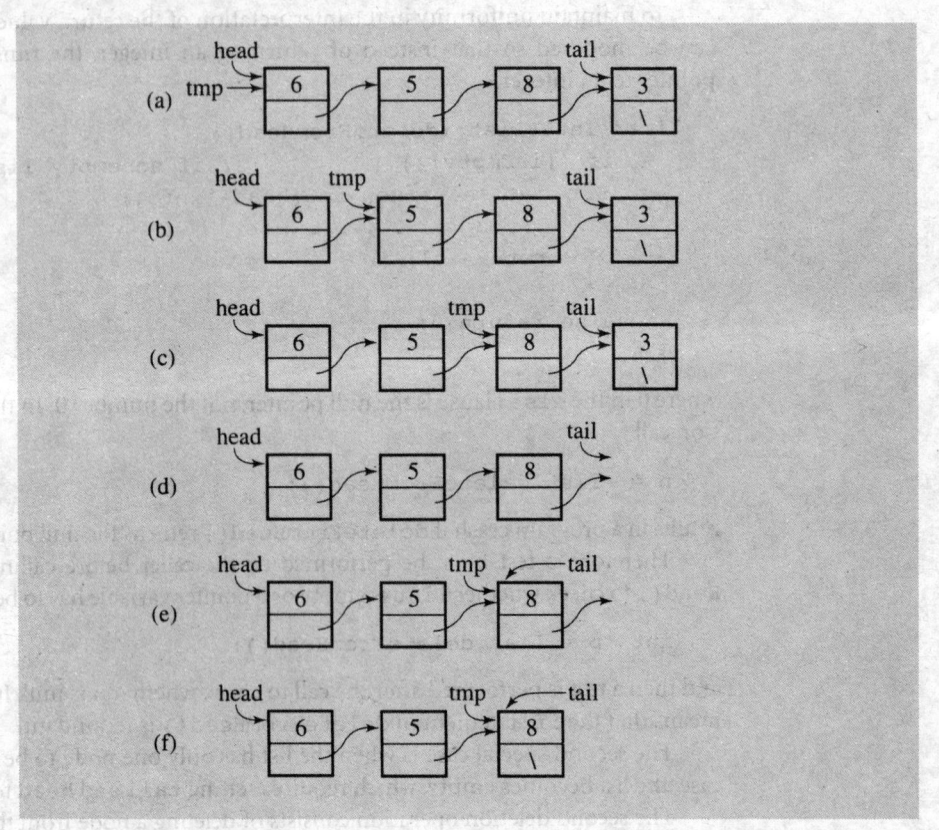

then the list is scanned only once, and the access to the beginning of the list is lost because `head` was permanently updated to the next to last node, which is about to become the last node. It is absolutely critical that, in cases such as this, a temporary variable is used so that the access to the beginning of the list is kept intact.

In removing the last node, the two special cases are the same as in `deleteFromHead()`. If the list is empty, then nothing can be removed, but what should be done in this case is decided in the user program just as in the case of `deleteFromHead()`. The second case is when a single-node list becomes empty after removing its only node, which also requires setting `head` and `tail` to null.

The most time-consuming part of `deleteFromTail()` is finding the next to last node performed by the `for` loop. It is clear that the loop performs $n-1$ iterations in a list of n nodes, which is the main reason this member function takes $O(n)$ time to delete the last node.

The two discussed deletion operations remove a node from the head or from the tail (that is, always from the same position) and return the integer that happens to be in the node being removed. A different approach is when we want to delete a node that holds a particular integer regardless of the position of this node in the list. It may be right at the beginning, at the end, or anywhere inside the list. Briefly, a node has to be located first and then detached from the list by linking the predecessor of this node directly to its successor. Because we do not know where the node may be, the process of finding and deleting a node with a certain integer is much more complex than the deletion operations discussed so far. The member function `deleteNode()` (Figure 3.2) is an implementation of this process.

A node is removed from inside a list by linking its predecessor to its successor. But because the list has only forward links, the predecessor of a node is not reachable from the node. One way to accomplish the task is to find the node to be removed by first scanning the list and then scanning it again to find its predecessor. Another way is presented in `deleteNode()`, as shown in Figure 3.8. Assume that we want to delete a node that holds number 8. The function uses two pointer variables, `pred` and `tmp`, which are initialized in the `for` loop so that they point to the first and second nodes of the list, respectively (Figure 3.8a). Because the node `tmp` has the number 5, the first iteration is executed in which both `pred` and `tmp` are advanced to the next nodes (Figure 3.8b). Because the condition of the `for` loop is now true (`tmp` points to the node with 8), the loop is exited and an assignment `pred->next = tmp->next` is executed (Figure 3.8c). This assignment effectively excludes the node with 8 from the list. The node is still accessible from variable `tmp`, and this access is used to return space occupied by this node to the pool of free memory cells by executing `delete` (Figure 3.8d).

FIGURE 3.8 Deleting a node from a singly linked list.

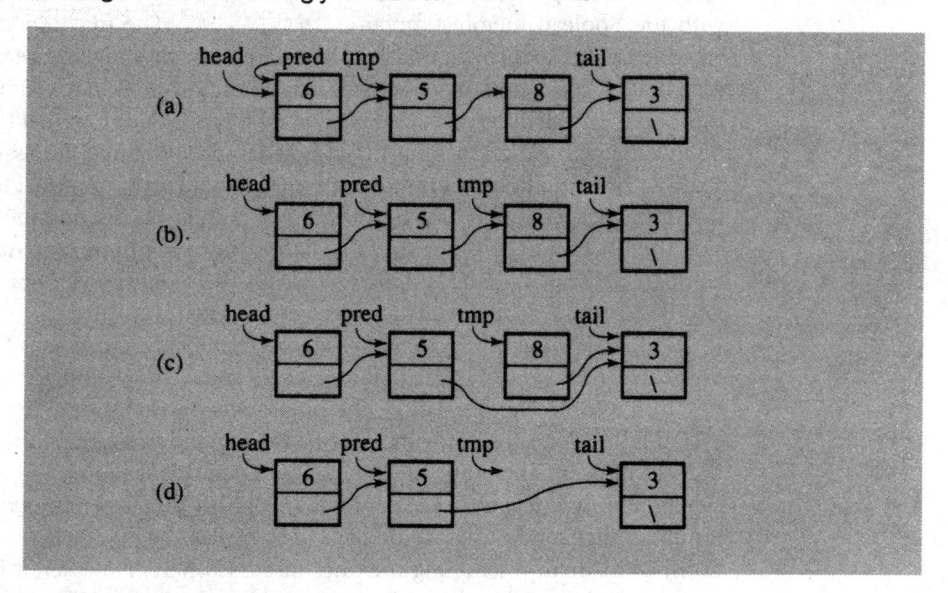

The preceding paragraph discusses only one case. Here are the remaining cases:

1. An attempt to remove a node from an empty list, in which case the function is immediately exited.
2. Deleting the only node from a one-node linked list: Both `head` and `tail` are set to null.

3. Removing the first node of the list with at least two nodes, which requires updating head.
4. Removing the last node of the list with at least two nodes, leading to the update of tail.
5. An attempt to delete a node with a number that is not in the list: Do nothing.

It is clear that the best case for deleteNode() is when the head node is to be deleted, which takes $O(1)$ time to accomplish. The worst case is when the last node needs to be deleted, which reduces deleteNode() to deleteFromTail() and to its $O(n)$ performance. What is the average case? It depends on how many iterations the for loop executes. Assuming that any node on the list has an equal chance to be deleted, the loop performs no iteration if it is the first node, one iteration if it is the second node, ..., and finally $n-1$ iterations if it is the last node. For a long sequence of deletions, one deletion requires on the average

$$\frac{0+1+\ldots+(n-1)}{n} = \frac{\frac{(n-1)n}{2}}{n} = \frac{n-1}{2}$$

That is, on the average, deleteNode() executes $O(n)$ steps to finish, just like in the worst case.

3.1.3 Search

The insertion and deletion operations modify linked lists. The searching operation scans an existing list to learn whether a number is in it. We implement this operation with the Boolean member function isInList(). The function uses a temporary variable tmp to go through the list starting from the head node. The number stored in each node is compared to the number being sought, and if the two numbers are equal, the loop is exited; otherwise, tmp is updated to tmp->next so that the next node can be investigated. After reaching the last node and executing the assignment tmp = tmp->next, tmp becomes null, which is used as an indication that the number el is not in the list. That is, if tmp is not null, the search was discontinued somewhere inside the list because el was found. That is why isInList() returns the result of comparison tmp != null: If tmp is not null, el was found and true is returned. If tmp is null, the search was unsuccessful and false is returned.

With reasoning similar to that used to determine the efficiency of deleteNode(), isInList() takes $O(1)$ time in the best case and $O(n)$ in the worst and average cases.

In the foregoing discussion, the operations on nodes have been stressed. However, a linked list is built for the sake of storing and processing information, not for the sake of itself. Therefore, the approach used in this section is limited in that the list can store only integers. If we wanted a linked list for float numbers or for arrays of numbers, then a new class would have to be declared with a new set of member functions, all of them resembling the ones discussed here. However, it is more advantageous to declare such a class only once without deciding in advance what type of data will be stored in it. This can be done very conveniently in C++ with templates. To illustrate the use of templates for list processing, the next section uses them to define lists, although examples of list operations are still limited to lists that store integers.

3.2 DOUBLY LINKED LISTS

The member function `deleteFromTail()` indicates a problem inherent to singly linked lists. The nodes in such lists contain only pointers to the successors; therefore, there is no immediate access to the predecessors. For this reason, `deleteFromTail()` was implemented with a loop that allowed us to find the predecessor of `tail`. Although this predecessor is, so to speak, within sight, it is out of reach. We have to scan the entire list to stop right in front of `tail` to delete it. For long lists and for frequent executions of `deleteFromTail()`, this may be an impediment to swift list processing. To avoid this problem, the linked list is redefined so that each node in the list has two pointers, one to the successor and one to the predecessor. A list of this type is called a *doubly linked list*, and is illustrated in Figure 3.9. Figure 3.10 contains a fragment of implementation for a generic `DoublyLinkedList` class.

FIGURE 3.9 A doubly linked list.

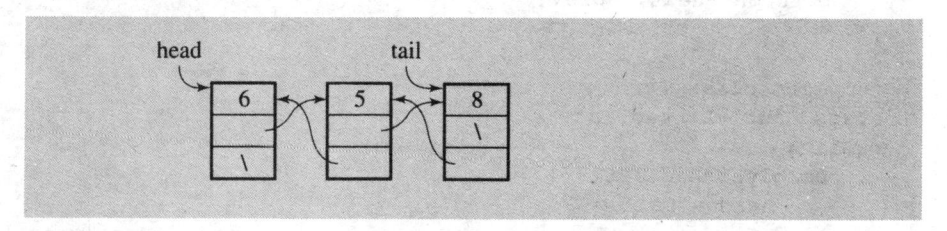

Member functions for processing doubly linked lists are slightly more complicated than their singly linked counterparts because there is one more pointer member to be maintained. Only two functions are discussed: a function to insert a node at the end of the doubly linked list and a function to remove a node from the end (Figure 3.10).

To add a node to a list, the node has to be created, its data members properly initialized, and then the node needs to be incorporated into the list. Inserting a node at the end of a doubly linked list performed by `addToDLLTail()` is illustrated in Figure 3.11. The process is performed in six steps:

1. A new node is created (Figure 3.11a), and then its three data members are initialized:
2. the `info` member to the number `el` being inserted (Figure 3.11b),
3. the `next` member to null (Figure 3.11c),
4. and the `prev` member to the value of `tail` so that this member points to the last node in the list (Figure 3.11d). But now, the new node should become the last node; therefore,
5. `tail` is set to point to the new node (Figure 3.11e). But the new node is not yet accessible from its predecessor; to rectify this,
6. the `next` member of the predecessor is set to point to the new node (Figure 3.11f).

FIGURE 3.10 An implementation of a doubly linked list.

```cpp
//***************************     genDLLst.h     ***************************
#ifndef DOUBLY_LINKED_LIST
#define DOUBLY_LINKED_LIST

template<class T>
class DLLNode {
public:
    DLLNode() {
        next = prev = 0;
    }
    DLLNode(const T& el, DLLNode *n = 0, DLLNode *p = 0) {
        info = el; next = n; prev = p;
    }
    T info;
    DLLNode *next, *prev;
};

template<class T>
class DoublyLinkedList {
public:
    DoublyLinkedList() {
        head = tail = 0;
    }
    void addToDLLTail(const T&);
    T deleteFromDLLTail();
    . . . . . . . . . . . . . .
protected:
    DLLNode<T> *head, *tail;
};
template<class T>
void DoublyLinkedList<T>::addToDLLTail(const T& el) {
    if (tail != 0) {
         tail = new DLLNode<T>(el,0,tail);
         tail->prev->next = tail;
    }
    else head = tail = new DLLNode<T>(el);
}
template<class T>
T DoublyLinkedList<T>::deleteFromDLLTail() {
    T el = tail->info;
    if (head == tail) { // if only one node in the list;
         delete head;
         head = tail = 0;
    }
    else {                      // if more than one node in the list;
         tail = tail->prev;
```

FIGURE 3.10 (continued)

```
        delete tail->next;
        tail->next = 0;
    }
    return el;
}
. . . . . . . . . . . . .
#endif
```

FIGURE 3.11 Adding a new node at the end of a doubly linked list.

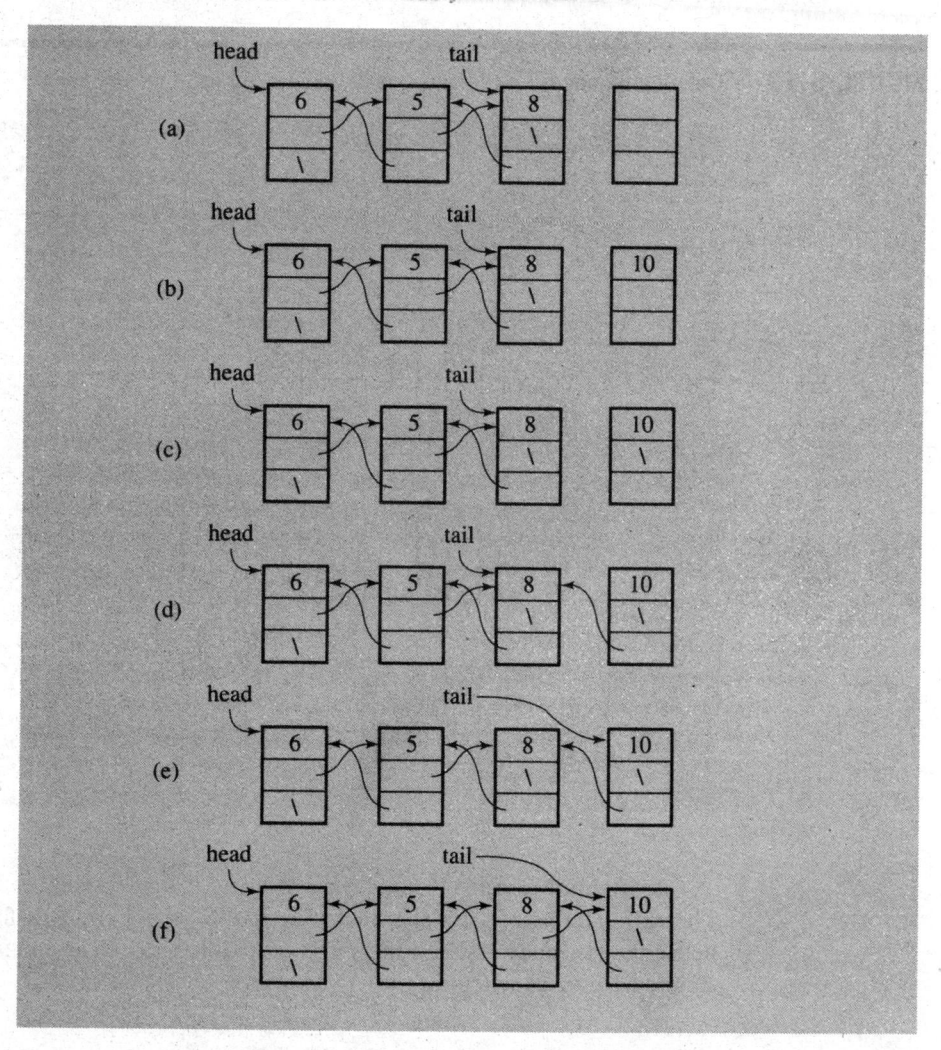

A special case concerns the last step. It is assumed in this step that the newly created node has a predecessor, so it accesses its `prev` member. It should be obvious that for an empty linked list, the new node is the only node in the list and it has no predecessor. In this case, both `head` and `tail` refer to this node, and the sixth step is now setting `head` to point to this node. Note that step four—setting the `prev` member to the value of `tail`—is executed properly because for an initially empty list, `tail` is null. Thus, null becomes the value of the `prev` member of the new node.

Deleting the last node from the doubly linked list is straightforward because there is direct access from the last node to its predecessor, and no loop is needed to remove the last node. When deleting the last node from the list in Figure 3.12a, the temporary variable `el` is set to the value in the node, then `tail` is set to its predecessor (Figure 3.12b), and the last and now redundant node is deleted (Figure 3.12c). In this way, the next to last node becomes the last node. The `next` member of the tail node is a dangling reference; therefore, `next` is set to null (Figure 3.12d). The last step is returning the copy of the object stored in the removed node.

FIGURE 3.12 Deleting a node from the end of a doubly linked list.

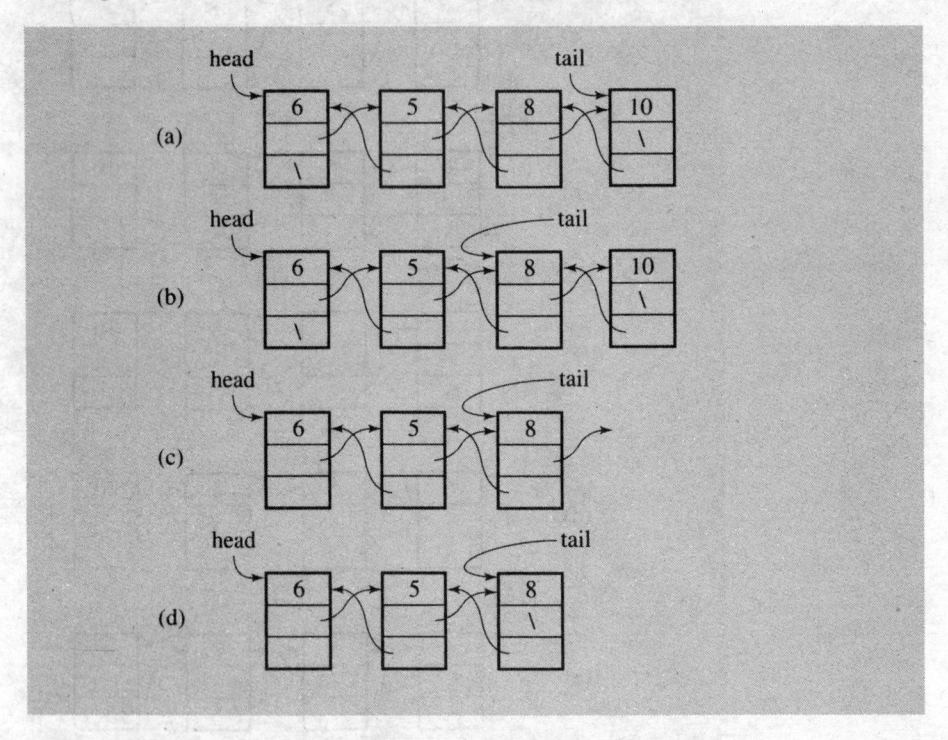

An attempt to delete a node from an empty list may result in a program crash. Therefore, the user has to check whether the list is not empty before attempting to delete the last node. As with the singly linked list's `deleteFromHead()`, the caller should have an `if` statement

```
if (!list.isEmpty())
    n = list.deleteFromDLLTail();
else do not remove;
```

Another special case is the deletion of the node from a single-node linked list. In this case, both `head` and `tail` are set to null.

Because of the immediate accessibility of the last node, both `addToDLLTail()` and `deleteFromDLLTail()` execute in constant time $O(1)$.

Functions for operating at the beginning of the doubly linked list are easily obtained from the two functions just discussed by changing `head` to `tail` and vice versa, changing `next` to `prev` and vice versa, and exchanging the order of parameters when executing `new`.

3.3 CIRCULAR LISTS

In some situations, a *circular list* is needed in which nodes form a ring: The list is finite and each node has a successor. An example of such a situation is when several processes are using the same resource for the same amount of time, and we have to assure that each process has a fair share of the resource. Therefore, all processes—let their numbers be 6, 5, 8, and 10, as in Figure 3.13—are put on a circular list accessible through the pointer `current`. After one node in the list is accessed and the process number is retrieved from the node to activate this process, `current` moves to the next node so that the next process can be activated the next time.

FIGURE 3.13   A circular singly linked list.

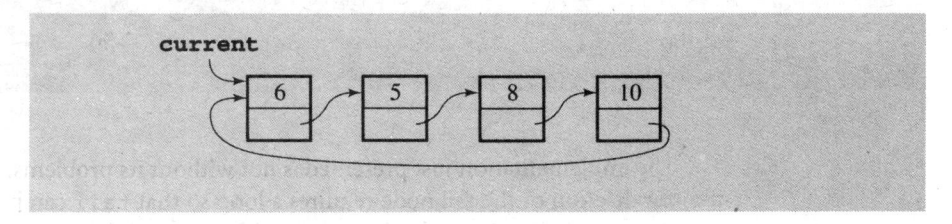

In an implementation of a circular singly linked list, we can use only one permanent pointer, `tail`, to the list even though operations on the list require access to the tail and its successor, the head. To that end, a linear singly linked list as discussed in Section 3.1 uses two permanent pointers, `head` and `tail`.

Figure 3.14a shows a sequence of insertions at the front of the circular list, and Figure 3.14b illustrates insertions at the end of the list. As an example of a member function operating on such a list, we present a function to insert a node at the tail of a circular singly linked list in $O(1)$:

```
void addToTail(int el) {
    if (isEmpty()) {
        tail = new IntSLLNode(el);
        tail->next = tail;
```

```
            }
            else {
                 tail->next = new IntSLLNode(el,tail->next);
                 tail = tail->next;
            }
    }
```

FIGURE 3.14 Inserting nodes at the front of a circular singly linked list (a) and at its end (b).

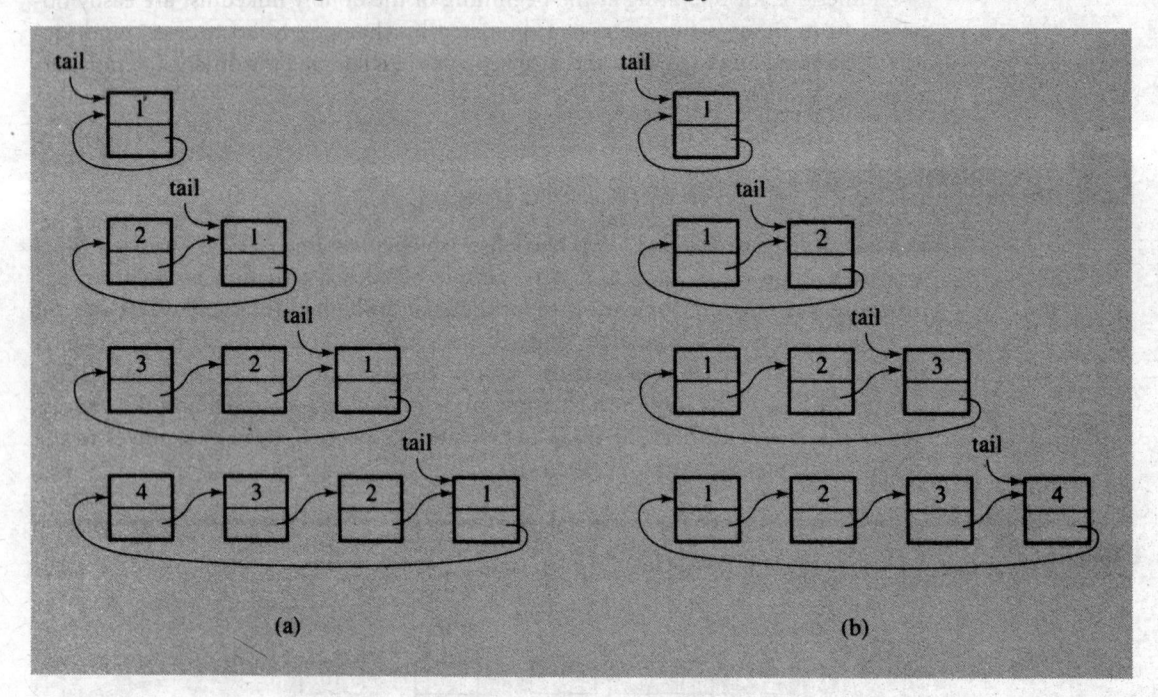

The implementation just presented is not without its problems. A member function for deletion of the tail node requires a loop so that `tail` can be set to its predecessor after deleting the node. This makes this function delete the tail node in $O(n)$ time. Moreover, processing data in the reverse order (printing, searching, etc.) is not very efficient. To avoid the problem and still be able to insert and delete nodes at the front and at the end of the list without using a loop, a doubly linked circular list can be used. The list forms two rings: one going forward through `next` members and one going backward through `prev` members. Figure 3.15 illustrates such a list accessible through the last node. Deleting the node from the end of the list can be done easily because there is direct access to the next to last node that needs to be updated in the case of such a deletion. In this list, both insertion and deletion of the tail node can be done in $O(1)$ time.

FIGURE 3.15 A circular doubly linked list.

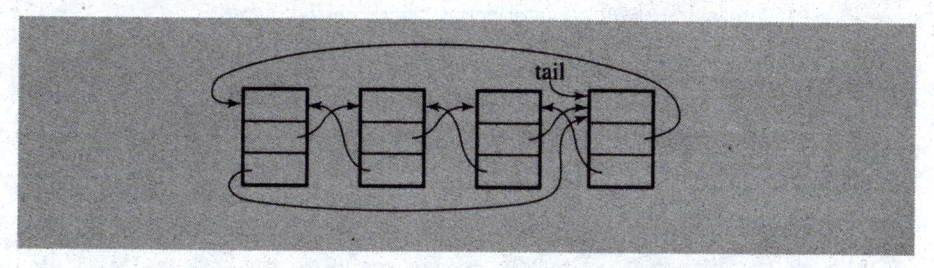

3.4 SKIP LISTS

Linked lists have one serious drawback: They require sequential scanning to locate a searched-for element. The search starts from the beginning of the list and stops when either a searched-for element is found or the end of the list is reached without finding this element. Ordering elements on the list can speed up searching, but a sequential search is still required. Therefore, we may think about lists that allow for skipping certain nodes to avoid sequential processing. A *skip list* is an interesting variant of the ordered linked list that makes such a nonsequential search possible (Pugh 1990).

In a skip list of n nodes, for each k and i such that $1 \leq k \leq \lfloor \lg n \rfloor$ and $1 \leq i \leq \lfloor n/2^{k-1} \rfloor - 1$, the node in position $2^{k-1} \cdot i$ points to the node in position $2^{k-1} \cdot (i+1)$. This means that every second node points to the node two positions ahead, every fourth node points to the node four positions ahead, and so on, as shown in Figure 3.16a. This is accomplished by having different numbers of pointers in nodes on the list: Half of the nodes have just one pointer, one-fourth of the nodes have two pointers, one-eighth of the nodes have three pointers, and so on. The number of pointers indicates the *level* of each node, and the number of levels is $maxLevel = \lfloor \lg n \rfloor + 1$.

Searching for an element `el` consists of following the pointers on the highest level until an element is found that finishes the search successfully. In the case of reaching the end of the list or encountering an element `key` that is greater than `el`, the search is restarted from the node preceding the one containing `key`, but this time starting from a pointer on a lower level than before. The search continues until `el` is found, or the first-level pointers are followed to reach the end of the list or to find an element greater than `el`. Here is a pseudocode for this algorithm:

```
find(element el)
    p = the nonnull list on the highest level i;
    while el not found and i ≥ 0
        if p->key < el
            p = a sublist that begins in the predecessor of p on level --i;
        else if p->key > el
            if p is the last node on level i
                p = a nonnull sublist that begins in p on the highest level < i;
                i = the number of the new level;
        else p = p->next;
```

FIGURE 3.16 A skip list with (a) evenly and (b) unevenly spaced nodes of different levels; (c) the skip list with pointer nodes clearly shown.

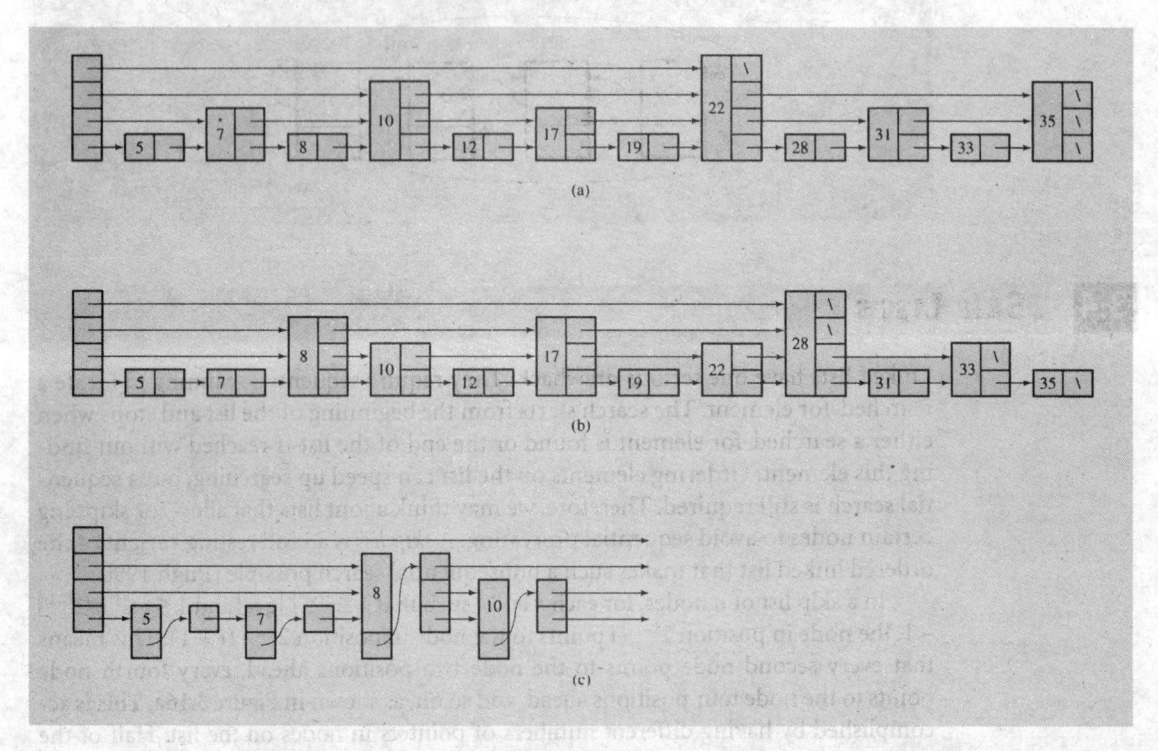

For example, if we look for number 16 in the list in Figure 3.16b, then level four is tried first, which is unsuccessful because the first node on this level has 28. Next, we try the third-level sublist starting from the root: It first leads to 8, and then to 17. Hence, we try the second-level sublist that originates in the node holding 8: It leads to 10 and then again to 17. The last try is by starting the first-level sublist, which begins in node 10; this sublist's first node has 12, the next number is 17, and because there is no lower level, the search is pronounced unsuccessful. Code for the searching procedure is given in Figure 3.17.

Searching appears to be efficient; however, the design of skip lists can lead to very inefficient insertion and deletion procedures. To insert a new element, all nodes following the node just inserted have to be restructured; the number of pointers and the values of pointers have to be changed. In order to retain some of the advantages that skip lists offer with respect to searching and avoid problems with restructuring the lists when inserting and deleting nodes, the requirement on the positions of nodes of different levels is now abandoned and only the requirement on the number of nodes of different levels is kept. For example, the list in Figure 3.16a becomes the list in Figure 3.16b: Both lists have six nodes with only one pointer, three nodes with two pointers, two nodes with three pointers, and one node with four pointers. The

FIGURE 3.17 An implementation of a skip list.

```
//*********************** genSkipL.h ***********************
//                  generic skip list class

const int maxLevel = 4;

template<class T>
class SkipListNode {
public:
    SkipListNode() {
    }
    T key;
    SkipListNode **next;
};

template<class T>
class SkipList {
public:
    SkipList();
    bool isEmpty() const;
    void choosePowers();
    int  chooseLevel();
    T* skipListSearch(const T&);
    void skipListInsert(const T&);
private:
    typedef SkipListNode<T> *nodePtr;
    nodePtr root[maxLevel];
    int powers[maxLevel];
};

template<class T>
SkipList<T>::SkipList() {
    for (int i = 0; i < maxLevel; i++)
        root[i] = 0;
}
template<class T>
bool SkipList<T>::isEmpty() const {
        return root[0] == 0;
}
template<class T>
void SkipList<T>::choosePowers() {
    powers[maxLevel-1] = (2 << (maxLevel-1)) - 1;  // 2^maxLevel - 1
    for (int i = maxLevel - 2, j = 0; i >= 0; i--, j++)
```

Continues

FIGURE 3.17 *(continued)*

```cpp
            powers[i] = powers[i+1] - (2 << j);        // 2^(j+1)
}
template<class T>
int SkipList<T>::chooseLevel() {
    int i, r = rand() % powers[maxLevel-1] + 1;
    for (i = 1; i < maxLevel; i++)
        if (r < powers[i])
            return i-1; // return a level < the highest level;
    return i-1;         // return the highest level;
}
template<class T>
T* SkipList<T>::skipListSearch(const T& key) {
    if (isEmpty()) return 0;
    nodePtr prev, curr;
    int lvl;                            // find the highest non-null
    for (lvl = maxLevel-1; lvl >= 0 && !root[lvl]; lvl--);  // level;
    prev = curr = root[lvl];
    while (true) {
        if (key == curr->key)                       // success if equal;
            return &curr->key;
        else if (key < curr->key) {                 // if smaller, go down
            if (lvl == 0)                           // if possible,
                return 0;
            else if (curr == root[lvl])             // by one level
                curr = root[--lvl];                 // starting from the
            else curr = *(prev->next + --lvl); // predecessor which
        }                                           // can be the root;
        else {                                      // if greater,
            prev = curr;                            // go to the next
            if (*(curr->next + lvl) != 0)           // non-null node
                curr = *(curr->next + lvl);         // on the same level
            else {                                  // or to a list on a
                                                    // lower level;
                for (lvl--; lvl >= 0 && *(curr->next + lvl)==0; lvl--);
                if (lvl >= 0)
                    curr = *(curr->next + lvl);
                else return 0;
            }
        }
    }
}
```

FIGURE 3.17 (continued)

```
template<class T>
void SkipList<T>::skipListInsert(const T& key) {
    nodePtr curr[maxLevel], prev[maxLevel], newNode;
    int lvl, i;
    curr[maxLevel-1] = root[maxLevel-1];
    prev[maxLevel-1] = 0;
    for (lvl = maxLevel - 1; lvl >= 0; lvl--) {
        while (curr[lvl] && curr[lvl]->key < key) { // go to the next
            prev[lvl] = curr[lvl];                   // if smaller;
            curr[lvl] = *(curr[lvl]->next + lvl);
        }
        if (curr[lvl] && curr[lvl]->key == key)      // don't include
            return;                                   // duplicates;
        if (lvl > 0)                                  // go one level down
            if (prev[lvl] == 0) {                     // if not the lowest
                curr[lvl-1] = root[lvl-1];            // level, using a link
                prev[lvl-1] = 0;                      // either from the root
            }
            else {                                    // or from the predecessor;
                curr[lvl-1] = *(prev[lvl]->next + lvl-1);
                prev[lvl-1] = prev[lvl];
            }
    }
    lvl = chooseLevel();         // generate randomly level for newNode;
    newNode = new SkipListNode<T>;
    newNode->next = new nodePtr[sizeof(nodePtr) * (lvl+1)];
    newNode->key  = key;
    for (i = 0; i <= lvl; i++) {              // initialize next fields of
        *(newNode->next + i) = curr[i];       // newNode and reset to newNode
        if (prev[i] == 0)                     // either fields of the root
            root[i] = newNode;                // or next fields of newNode's
        else *(prev[i]->next + i) = newNode;  // predecessors;
    }
}
```

new list is searched exactly the same way as the original list. Inserting does not require list restructuring, and nodes are generated so that the distribution of the nodes on different levels is kept adequate. How can this be accomplished?

Assume that *maxLevel* = 4. For 15 elements, the required number of one-pointer nodes is eight, two-pointer nodes is four, three-pointer nodes is two, and four-pointer nodes is one. Each time a node is inserted, a random number r between 1 and 15 is generated, and if $r < 9$, then a node of level one is inserted. If $r < 13$, a second-level node is inserted, if $r < 15$, it is a third-level node, and if $r = 15$, the node of level four is generated and inserted. If $maxLevel = 5$, then for 31 elements the correspondence between the value of r and the level of node is as follows:

r	Level of Node to Be Inserted
31	5
29–30	4
25–28	3
17–24	2
1–16	1

To determine such a correspondence between r and the level of node for any *maxLevel*, the function `choosePowers()` initializes the array `powers[]` by putting lower bounds on each range. For example, for $maxLevel = 4$, the array is [1 9 13 15]; for $maxLevel = 5$, it is [1 17 25 29 31]. `chooseLevel()` uses `powers[]` to determine the level of the node about to be inserted. Figure 3.17 contains the code for `choosePowers()` and `chooseLevel()`. Note that the levels range between 0 and $maxLevel–1$ (and not between 1 and *maxLevel*) so that the array indexes can be used as levels. For example, the first level is level zero.

But we also have to address the question of implementing a node. The easiest way is to make each node have *maxLevel* pointers, but this is wasteful. We need only as many pointers per node as the level of the node requires. To accomplish this, the `next` member of each node is not a pointer to the next node, but to an array of pointer(s) to the next node(s). The size of this array is determined by the level of the node. The `SkipListNode` and `SkipList` classes are declared, as in Figure 3.17. In this way, the list in Figure 3.16b is really a list whose first four nodes are shown in Figure 3.16c. Only now can an inserting procedure be implemented, as in Figure 3.17.

How efficient are skip lists? In the ideal situation, which is exemplified by the list in Figure 3.16a, the search time is $O(\lg n)$. In the worst situation, when all lists are on the same level, the skip list turns into a regular singly linked list, and the search time is $O(n)$. However, the latter situation is unlikely to occur; in the random skip list, the search time is of the same order as the best case; that is, $O(\lg n)$. This is an improvement over the efficiency of search in regular linked lists. It also turns out that skip lists fare extremely well in comparison with more sophisticated data structures, such as self-adjusting trees or AVL trees (see Sections 6.7.2 and 6.8), and therefore they are a viable alternative to these data structures (see also the table in Figure 3.20).

3.5 SELF-ORGANIZING LISTS

The introduction of skip lists was motivated by the need to speed up the searching process. Although singly and doubly linked lists require sequential search to locate an element or to see that it is not in the list, we can improve the efficiency of the search by dynamically organizing the list in a certain manner. This organization depends on the configuration of data;

thus, the stream of data requires reorganizing the nodes already on the list. There are many different ways to organize the lists, and this section describes four of them:

1. *Move-to-front method.* After the desired element is located, put it at the beginning of the list (Figure 3.18a).
2. *Transpose method.* After the desired element is located, swap it with its predecessor unless it is at the head of the list (Figure 3.18b).

FIGURE 3.18 Accessing an element on a linked list and changes on the list depending on the self-organization technique applied: (a) move-to-front method, (b) transpose method, (c) count method, and (d) ordering method, in particular, alphabetical ordering, which leads to no change. In the case when the desired element is not in the list, (e) the first three methods add a new node with this element at the end of the list and (f) the ordering method maintains an order on the list.

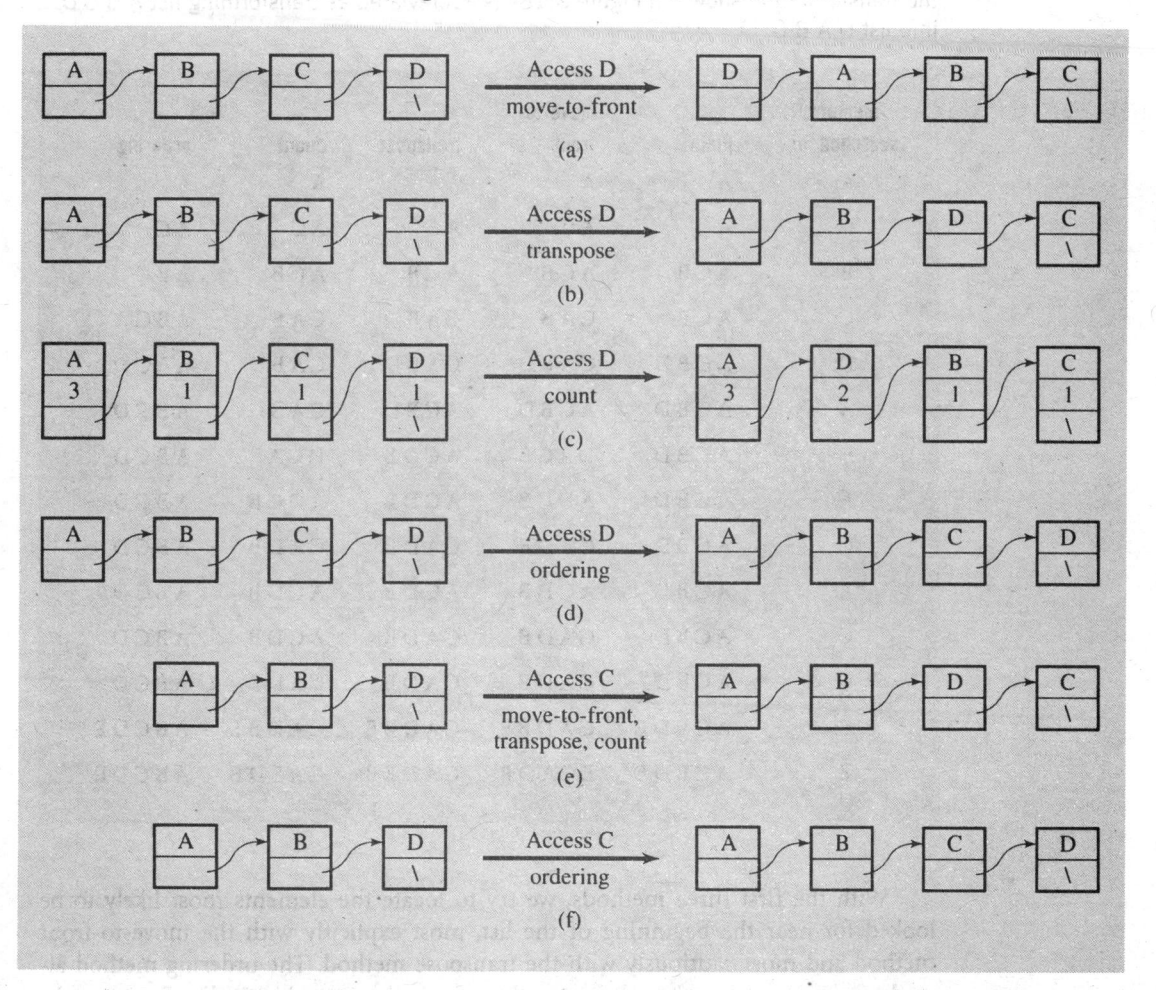

3. *Count method.* Order the list by the number of times elements are being accessed (Figure 3.18c).
4. *Ordering method.* Order the list using certain criteria natural for the information under scrutiny (Figure 3.18d).

In the first three methods, new information is stored in a node added to the end of the list (Figure 3.18e); in the fourth method, new information is stored in a node inserted somewhere in the list to maintain the order of the list (Figure 3.18f). An example of searching for elements in a list organized by these different methods is shown in Figure 3.19.

FIGURE 3.19 Processing the stream of data, A C B C D A D A C A C C E E, by different methods of organizing linked lists. Linked lists are presented in an abbreviated form; for example, the transformation shown in Figure 3.18a is abbreviated as transforming list A B C D into list D A B C.

element searched for	plain	move-to-front	transpose	count	ordering
A:	A	A	A	A	A
C:	A C	A C	A C	A C	A C
B:	A C B	A C B	A C B	A C B	A B C
C:	A C B	C A B	C A B	C A B	A B C
D:	A C B D	C A B D	C A B D	C A B D	A B C D
A:	A C B D	A C B D	A C B D	C A B D	A B C D
D:	A C B D	D A C B	A C D B	D C A B	A B C D
A:	A C B D	A D C B	A C D B	A D C B	A B C D
C:	A C B D	C A D B	C A D B	C A D B	A B C D
A:	A C B D	A C D B	A C D B	A C D B	A B C D
C:	A C B D	C A D B	C A D B	A C D B	A B C D
C:	A C B D	C A D B	C A D B	C A D B	A B C D
E:	A C B D E	C A D B E	C A D B E	C A D B E	A B C D E
E:	A C B D E	E C A D B	C A D E B	C A E D B	A B C D E

With the first three methods, we try to locate the elements most likely to be looked for near the beginning of the list, most explicitly with the move-to-front method and most cautiously with the transpose method. The ordering method already uses some properties inherent to the information stored in the list. For example, if we are storing nodes pertaining to people, then the list can be organized alphabetically by the name of the person or the city or in ascending or descending order using, say, birthday or salary. This is particularly advantageous when searching for information that is not in the list, because the search can terminate without scanning the entire list. Searching all the nodes of the list, however, is necessary in such cases using the other three methods. The count method can be subsumed in the category of the ordering methods if frequency is part of the information. In many cases, however, the count itself is an additional piece of information required solely to maintain the list; hence, it may not be considered "natural" to the information at hand.

Analyses of the efficiency of these methods customarily compare their efficiency to that of *optimal static ordering*. With this ordering, all the data are already ordered by the frequency of their occurrence in the body of data so that the list is used only for searching, not for inserting new items. Therefore, this approach requires two passes through the body of data, one to build the list and another to use the list for search alone.

To experimentally measure the efficiency of these methods, the number of all actual comparisons was compared to the maximum number of possible comparisons. The latter number is calculated by adding the lengths of the list at the moment of processing each element. For example, in the table in Figure 3.19, the body of data contains 14 letters, 5 of them being different, which means that 14 letters were processed. The length of the list before processing each letter is recorded, and the result, $0 + 1 + 2 + 3 + 3 + 4 + 4 + 4 + 4 + 4 + 4 + 4 + 4 + 5 = 46$, is used to compare the number of all comparisons made to this combined length. In this way, we know what percentage of the list was scanned during the entire process. For all the list organizing methods except optimal ordering, this combined length is the same; only the number of comparisons can change. For example, when using the move-to-front technique for the data in the table in Figure 3.19, 33 comparisons were made, which is 71.7% when compared to 46. The latter number gives the worst possible case, the combined length of intermediate lists every time all the nodes in the list are looked at. Plain search, with no reorganization, required only 30 comparisons, which is 65.2%.

These samples are in agreement with theoretical analyses that indicate that count and move-to-front methods are, in the long run, at most twice as costly as the optimal static ordering; the transpose method approaches, in the long run, the cost of the move-to-front method. In particular, with amortized analysis, it can be established that the cost of accessing a list element with the move-to-front method is at most twice the cost of accessing this element on the list that uses optimal static ordering.

In a proof of this statement, the concept of inversion is used. For two lists containing the same elements, an inversion is defined to be a pair of elements (x, y) such that on one list x precedes y and on the other list y precedes x. For example, the list (C, B, D, A) has four inversions with respect to list (A, B, C, D): (C, A), (B, A), (D, A), and (C, B). Define the amortized cost to be the sum of actual cost and the difference between the number of inversions before accessing an element and after accessing it,

$$amCost(x) = cost(x) + (inversionsBeforeAccess(x) - inversionsAfterAccess(x))$$

To assess this number, consider an optimal list OL = (A, B, C, D) and a move-to-front list MTF = (C, B, D, A). The access of elements usually changes the balance of inversions. Let *displaced*(x) be the number of elements preceding x in MTF but following x in OL. For example, *displaced*$(A) = 3$, *displaced*$(B) = 1$, *displaced*$(C) = 0$, and *displaced*$(D) = 0$. If $pos_{MTF}(x)$ is the current position of x in MTF, then $pos_{MTF}(x) - 1 - displaced(x)$ is the number of elements preceding x in both lists. It is easy to see that for D

this number equals 2, and for the remaining elements it is 0. Now, accessing an element x and moving it to the front of MTF creates $pos_{MTF}(x) - 1 - displaced(x)$ new inversions and removes $displaced(x)$ other inversions so that the amortized time to access x is

$$amCost(x) = pos_{MTF}(x) + pos_{MTF}(x) - 1 - displaced(x) - displaced(x) =$$
$$2(pos_{MTF}(x) - displaced(x)) - 1$$

where $cost(x) = pos_{MTF}(x)$. Accessing A transforms MTF = (C, B, D, A) into (A, C, B, D) and $amCost(A) = 2(4-3) - 1 = 1$. For B, the new list is (B, C, D, A) and $amCost(B) = 2(2-1) - 1 = 1$. For C, the list does not change and $amCost(C) = 2(1-0) - 1 = 1$. Finally, for D, the new list is (D, C, B, A) and $amCost(D) = 2(3-0) - 1 = 5$. However, the number of common elements preceding x on the two lists cannot exceed the number of all the elements preceding x on OL; therefore, $pos_{MTF}(x) - 1 - displaced(x) \leq pos_{OL}(x) - 1$, so that

$$amCost(x) \leq 2pos_{OL}(x) - 1$$

The amortized cost of accessing an element x in MTF is in excess of $pos_{OL}(x) - 1$ units to its actual cost of access on OL. This excess is used to cover an additional cost of accessing elements in MTF for which $pos_{MTF}(x) > pos_{OL}(x)$; that is, elements that require more accesses on MTF than on OL.

It is important to stress that the amortized costs of single operations are meaningful in the context of sequences of operations. A cost of an isolated operation may seldom equal its amortized cost; however, in a sufficiently long sequence of accesses, each access on the average takes at most $2pos_{OL}(x) - 1$ time.

Figure 3.20 contains sample runs of the self-organizing lists. The second and the fourth columns of numbers refer to files containing programs, and the remaining columns refer to files containing English text. There is a general tendency for all methods to improve their efficiency with the size of the file. The move-to-front and count

FIGURE 3.20 Measuring the efficiency of different methods using formula (number of data comparison)/(combined length) expressed in percentages.

Different Words/ All Words	156/347	149/423	609/1,510	550/2,847	1,163/5,866	2,013/23,065
Optimal	28.5	26.4	24.5	17.6	16.2	10.0
Plain	70.3	71.2	67.1	56.3	51.7	35.4
Move-to-Front	61.3	49.5	54.5	31.3	30.5	18.4
Transpose	68.8	69.5	66.1	53.3	49.4	32.9
Count	61.2	51.6	54.7	34.0	32.0	19.8
Alphabetical Order	50.9	45.6	48.0	55.7	50.4	50.0
Skip List	15.1	12.3	6.6	5.5	4.8	3.8

methods are almost the same in their efficiency, and both outperform the transpose, plain, and ordering methods. The poor performance for smaller files is due to the fact that all of the methods are busy including new words in the lists, which requires an exhaustive search of the lists. Later, the methods concentrate on organizing the lists to reduce the number of searches. The table in Figure 3.20 also includes data for a skip list. There is an overwhelming difference between the skip list's efficiency and that of the other methods. However, keep in mind that in the table in Figure 3.20, only comparisons of data are included, with no indication of the other operations needed for execution of the analyzed methods. In particular, there is no indication of how many pointers are used and relinked, which, when included, may make the difference between various methods less dramatic.

These sample runs show that for lists of modest size, the linked list suffices. With the increase in the amount of data and in the frequency with which they have to be accessed, more sophisticated methods and data structures need to be used.

3.6 SPARSE TABLES

In many applications, the choice of a table seems to be the most natural one, but space considerations may preclude this choice. This is particularly true if only a small fraction of the table is actually used. A table of this type is called a *sparse table* because the table is populated sparsely by data and most of its cells are empty. In this case, the table can be replaced by a system of linked lists.

As an example, consider the problem of storing grades for all students in a university for a certain semester. Assume that there are 8,000 students and 300 classes. A natural implementation is a two-dimensional array `grades` where student numbers are indexes of the columns and class numbers the indexes of the rows (see Figure 3.21). An association of student names and numbers is represented by the one-dimensional array `students` and an association of class names and numbers by the array `classes`. The names do not have to be ordered. If order is required, then another array can be used where each array element is occupied by a record with two fields, name and number,[1] or the original array can be sorted each time an order is required. This, however, leads to the constant reorganization of `grades`, and is not recommended.

Each cell of `grades` stores a grade obtained by each student after finishing a class. If signed grades such as A–, B+, or C+ are used, then 2 bytes are require to store each grade. To reduce the table size by one-half, the array `gradeCodes` in Figure 3.21c associates each grade with a letter that requires only one byte of storage.

The entire table (Figure 3.21d) occupies 8,000 students · 300 classes · 1 byte = 2.4 million bytes. This table is very large but is sparsely populated by grades. Assuming that, on the average, students take four classes a semester, each column of the table has only four cells occupied by grades, and the rest of the cells, 296 cells or 98.7%, are unoccupied and wasted.

---

[1]This is called an *index-inverted table*.

FIGURE 3.21 Arrays and sparse table used for storing student grades.

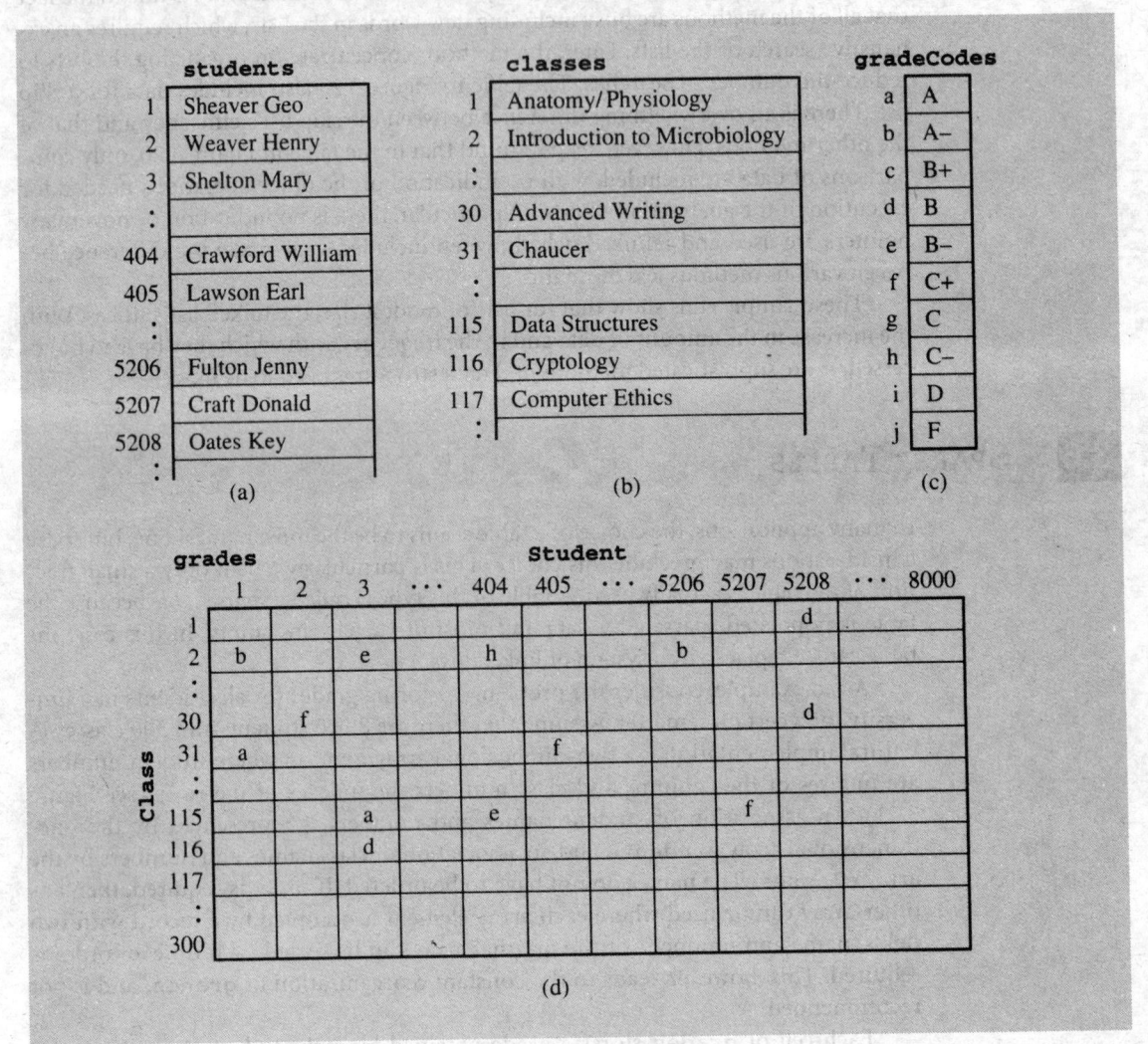

students

1	Sheaver Geo
2	Weaver Henry
3	Shelton Mary
:	
404	Crawford William
405	Lawson Earl
:	
5206	Fulton Jenny
5207	Craft Donald
5208	Oates Key
:	

(a)

classes

1	Anatomy/Physiology
2	Introduction to Microbiology
:	
30	Advanced Writing
31	Chaucer
:	
115	Data Structures
116	Cryptology
117	Computer Ethics
:	

(b)

gradeCodes

a	A
b	A−
c	B+
d	B
e	B−
f	C+
g	C
h	C−
i	D
j	F

(c)

grades — Student

Class	1	2	3	...	404	405	...	5206	5207	5208	...	8000
1										d		
2	b		e		h			b				
:												
30		f								d		
31	a					f						
:												
115		a			e				f			
116		d										
117												
:												
300												

(d)

A better solution is to use two two-dimensional arrays. `classesTaken` represents all the classes taken by every student, and `studentsInClasses` represents all students participating in each class (see Figure 3.22). A cell of each table is an object with two data members: a student or class number and a grade. We assume that a student can take at most 8 classes and that there can be at most 250 students signed up for a class. We need two arrays, because with only one array it is very time-consuming to produce lists. For example, if only `classesTaken` is used, then printing a list of all students taking a particular class requires an exhaustive search of `classesTaken`.

Assume that the computer on which this program is being implemented requires 2 bytes to store an integer. With this new structure, 3 bytes are needed for each cell.

FIGURE 3.22 Two-dimensional arrays for storing student grades.

classesTaken

	1	2	3	...	404	405	...	5206	5207	5208	...	8000
1	2 b	30 f	2 e		2 h	31 f		2 b	115 f	1 d		
2	31 a	115 a			115 e	64 f		33 b	121 a	30 d		
3	124 g	116 d			218 b	120 a		86 c	146 b	208 a		
4	136 g				221 b			121 d	156 b	211 b		
5					285 h			203 a		234 d		
6					292 b							
7												
8												

(a)

studentsInClasses

	1	2	...	30	31	...	115	116	...	300
1	5208 d	1 b		2 f	1 a		3 a	3 d		
2		3 e		5208 d	405 f		404 e			
3		404 h					5207 f			
4		5206 b								
⋮										
250										

(b)

Therefore, classesTaken occupies 8,000 students · 8 classes · 3 bytes = 192,000 bytes, studentsInClasses occupies 300 classes · 250 students · 3 bytes = 225,000 bytes; both tables require a total of 417,000 bytes, less than one-fifth the number of bytes required for the sparse table in Figure 3.21.

Although this is a much better implementation than before, it still suffers from a wasteful use of space; seldom, if ever, will both arrays be full because most classes have fewer than 250 students, and most students take fewer than 8 classes. This structure is also inflexible: If a class can be taken by more than 250 students, a problem occurs that has to be circumvented in an artificial way. One way is to create a nonexistent class that holds students for the overflowing class. Another way is to recompile the program with a new table size, which may not be practical at a future time. Another more flexible solution is needed that uses space frugally.

Two one-dimensional arrays of linked lists can be used as in Figure 3.23. Each cell of the array class is a pointer to a linked list of students taking a class, and each cell of the array student indicates a linked list of classes taken by a student. The linked lists contain nodes of five data members: student number, class number, grade, a pointer to the next student, and a pointer to the next class. Assuming that each pointer requires only 2 bytes and one node occupies 9 bytes, the entire structure can be stored in 8,000 students · 4

FIGURE 3.23 Student grades implemented using linked lists.

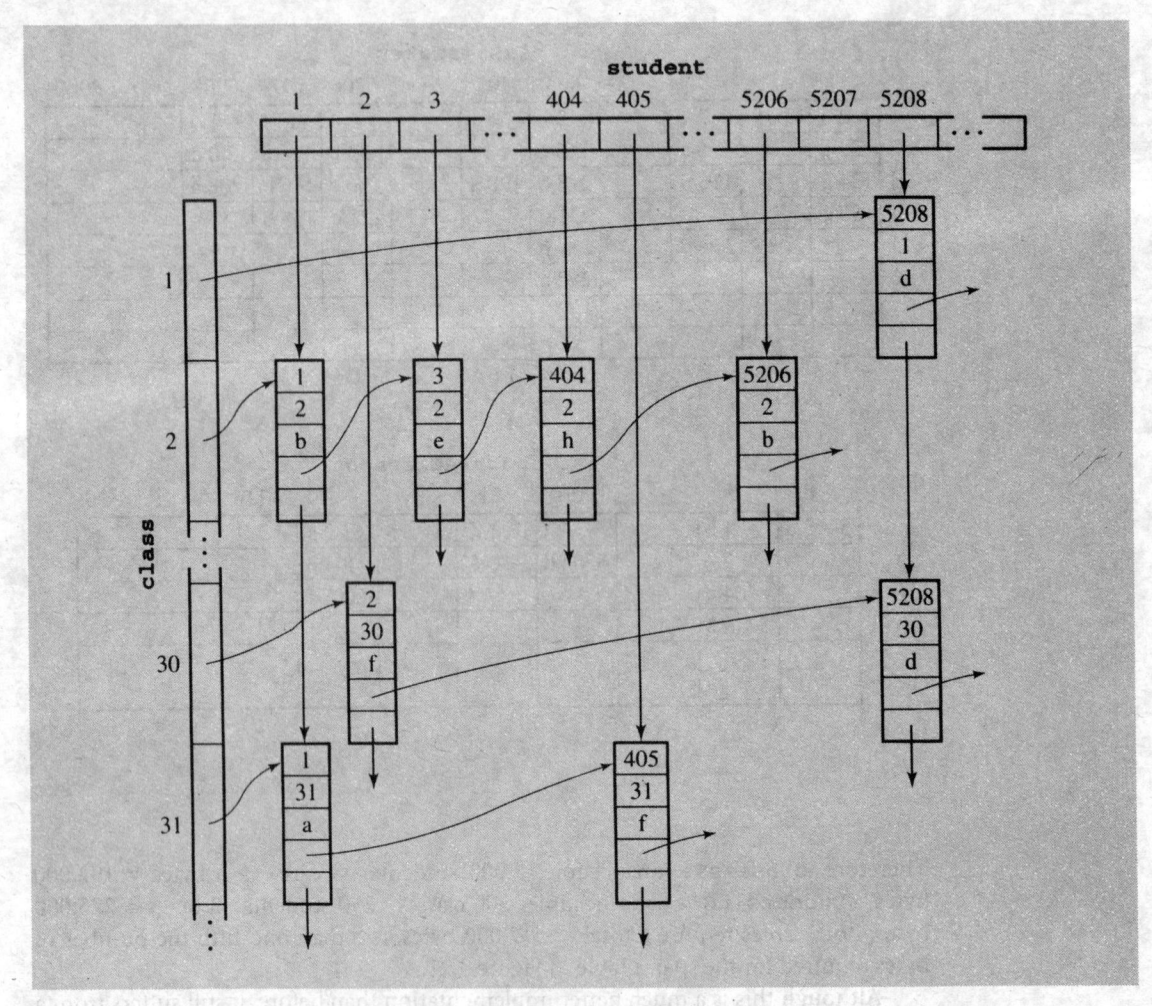

classes (on the average) · 9 bytes = 288,000 bytes, which is approximately 10% of the space required for the first implementation and about 70% of the space for the second. No space is used unnecessarily, there is no restriction imposed on the number of students per class, and the lists of students taking a class can be printed immediately.

3.7 LISTS IN THE STANDARD TEMPLATE LIBRARY

The list sequence container is an implementation of various operations on the nodes of a linked list. The STL implements a list as a generic doubly linked list with pointers to the head and to the tail. An instance of such a list that stores integers is presented in Figure 3.9.

The class `list` can be used in a program only if it is included with the instruction

```
#include <list>
```

The member functions included in the list container are presented in Figure 3.24. A new list is generated with the instruction

```
list<T> lst;
```

FIGURE 3.24 An alphabetical list of member functions in the class `list`.

Member Function	Action and Return Value
`void assign(iterator first, iterator last)`	Remove all the nodes in the list and insert into it the elements from the range indicated by iterators `first` and `last`.
`void assign(size_type n, const T& el = T())`	Remove all the nodes in the list and insert into it n copies of `el` (if `el` is not provided, a default constructor `T()` is used).
`T& back()`	Return the element in the last node of the list.
`const T& back() const`	Return the element in the last node of the list.
`iterator begin()`	Return an iterator that references the first node of the list.
`const_iterator begin() const`	Return an iterator that references the first node of the list.
`void clear()`	Remove all the nodes in the list.
`bool empty() const`	Return `true` if the list includes no nodes and false otherwise.
`iterator end()`	Return an iterator that is past the last node of the list.
`const_iterator end() const`	Return an iterator that is past the last node of the list.
`iterator erase(iterator i)`	Remove the node referenced by iterator `i` and return an iterator referencing the element after the one removed.
`iterator erase(iterator first, iterator last)`	Remove the nodes in the range indicated by iterators `first` and `last` and return an iterator referencing the element after the last one removed.
`T& front()`	Return the element in the first node of the list.
`const T& front() const`	Return the element in the first node of the list.
`iterator insert(iterator i, const T& el = T())`	Insert `el` before the node referenced by iterator `i` and return an iterator referencing the new node.
`void insert(iterator i, size_type n, const T& el)`	Insert n copies of `el` before the node referenced by iterator `i`.
`void insert(iterator i, iterator first, iterator last)`	Insert elements from location referenced by `first` to location referenced by `last` before the node referenced by iterator `i`.
`list()`	Construct an empty list.
`list(size_type n, const T& el = T())`	Construct a list with n copies of `el` of type `T`.
`list(iterator first, iterator last)`	Construct a list with the elements from the range indicated by iterators `first` and `last`.

Continues

FIGURE 3.24 *(continued)*

`list(const list<T>& lst)`	Copy constructor.
`size_type max_size() const`	Return the maximum number of nodes for the list.
`void merge(list<T>& lst)`	For the sorted current list and `lst`, remove all nodes from `lst` and insert them in sorted order in the current list.
`void merge(list<T>& lst, Comp f)`	For the sorted current list and `lst`, remove all nodes from `lst` and insert them in the current list in the sorted order specified by a two-argument Boolean function `f()`.
`void pop_back()`	Remove the last node of the list.
`void pop_front()`	Remove the first node of the list.
`void push_back(const T& el)`	Insert `el` at the end of the list.
`void push_front(const T& el)`	Insert `el` at the head of the list.
`void remove(const T& el)`	Remove from the list all the nodes that include `el`.
`void remove_if(Pred f)`	Remove the nodes for which a one-argument Boolean function `f()` returns `true`.
`void resize(size_type n, const T& el = T())`	Make the list have `n` nodes by adding `n - size()` more nodes with element `el` or by discarding overflowing `size() - n` nodes from the end of the list.
`void reverse()`	Reverse the list.
`reverse_iterator rbegin()`	Return an iterator that references the last node of the list.
`const_reverse_iterator rbegin() const`	Return an iterator that references the last node of the list.
`reverse_iterator rend()`	Return an iterator that is before the first node of the list.
`const_reverse_iterator rend() const`	Return an iterator that is before the first node of the list.
`size_type size() const`	Return the number of nodes in the list.
`void sort()`	Sort elements of the list in ascending order.
`void sort(Comp f)`	Sort elements of the list in the order specified by a two-argument Boolean function `f()`.
`void splice(iterator i, list<T>& lst)`	Remove the nodes of list `lst` and insert them into the list before the position referenced by iterator `i`.
`void splice(iterator i, list<T>& lst, iterator j)`	Remove from list `lst` the node referenced by iterator `j` and insert it into the list before the position referenced by iterator `i`.
`void splice(iterator i, list<T>& lst, iterator first, iterator last)`	Remove from list `lst` the nodes in the range indicated by iterators `first` and `last` and insert them into the list before the position referenced by iterator `i`.

FIGURE 3.24 (continued)

```
void swap(list<T>& lst)      Swap the content of the list with the content of another list lst.
void unique()                Remove duplicate elements from the sorted list.
void unique(Comp f)          Remove duplicate elements from the sorted list where being a
                             duplicate is specified by a two-argument Boolean function f().
```

where T can be any data type. If it is a user-defined type, the type must also include a default constructor, which is required for initialization of new nodes. Otherwise, the compiler is unable to compile the member functions with arguments initialized by the default constructor. These include one constructor and functions `resize()`, `assign()`, and one version of `insert()`. Note that this problem does not arise when creating a list of pointers to user-defined types, as in

```
list<T*> ptrLst;
```

The working of most of the member functions has already been illustrated in the case of the vector container (see Figure 1.4 and the discussion of these functions in Section 1.8). The vector container has only three member functions not found in the list container (`at()`, `capacity()`, and `reserve()`), but there are a number of list member functions that are not found in the vector container. Examples of their operation are presented in Figure 3.25.

FIGURE 3.25 A program demonstrating the operation of list member functions.

```
#include <iostream>
#include <list>
#include <algorithm>
#include <functional>

using namespace std;

int main() {
    list<int> lst1;               // lst1 is empty
    list<int> lst2(3,7);          // lst2 = (7 7 7)
    for (int j = 1; j <= 5; j++)  // lst1 = (1 2 3 4 5)
        lst1.push_back(j);
    list<int>::iterator i1 = lst1.begin(), i2 = i1, i3;
    i2++; i2++; i2++;
    list<int> lst3(++i1,i2);      // lst3 = (2 3)
    list<int> lst4(lst1);         // lst4 = (1 2 3 4 5)
```

Continues

FIGURE 3.25 (continued)

```
    i1 = lst4.begin();
    lst4.splice(++i1,lst2);    // lst2 is empty,
                               // lst4 = (1 7 7 7 2 3 4 5)
    lst2 = lst1;               // lst2 = (1 2 3 4 5)
    i2 = lst2.begin();
    lst4.splice(i1,lst2,++i2); // lst2 = (1 3 4 5),
                               // lst4 = (1 7 7 7 2 2 3 4 5)
    i2 = lst2.begin();
    i3 = i2;
    lst4.splice(i1,lst2,i2,++i3); // lst2 = (3 4 5),
                               // lst4 = (1 7 7 7 2 1 2 3 4 5)
    lst4.remove(1);            // lst4 = (7 7 7 2 2 3 4 5)
    lst4.sort();               // lst4 = (2 2 3 4 5 7 7 7)
    lst4.unique();             // lst4 = (2 3 4 5 7)
    lst1.merge(lst2);          // lst1 = (1 2 3 3 4 4 5 5),
                               // lst2 is empty
    lst3.reverse();            // lst3 = (3 2)
    lst4.reverse();            // lst4 = (7 5 4 3 2)
    lst3.merge(lst4,greater<int>());    // lst3 = (7 5 4 3 3 2 2),
                               // lst4 is empty
    lst3.remove_if(bind2nd(not_equal_to<int>(),3));// lst3 = (3 3)
    lst3.unique(not_equal_to<int>());   // lst3 = (3 3)
    return 0;
}
```

3.8 DEQUES IN THE STANDARD TEMPLATE LIBRARY

A *deque* (double-ended queue) is a list that allows for direct access to both ends of the list, particularly to insert and delete elements. Hence, a deque can be implemented as a doubly linked list with pointer data members head and tail, as discussed in Section 3.2. Moreover, as pointed out in the previous section, the container list uses a doubly linked list already. The STL, however, adds another functionality to the deque, namely, random access to any position of the deque, just as in arrays and vectors. Vectors, as discussed in Section 1.8, have poor performance for insertion and deletion of elements at the front, but these operations are quickly performed for doubly linked lists. This means that the STL deque should combine the behavior of a vector and a list.

The member functions of the STL container deque are listed in Figure 3.26. The functions are basically the same as those available for lists, with few exceptions. deque does not include function splice(), which is specific to list, and functions merge(), remove(), sort(), and unique(), which are also available as algorithms,

FIGURE 3.26 A list of member functions in the class `deque`.

Member Function	Operation
`void assign(iterator first, iterator last)`	Remove all the elements in the deque and insert into it the elements from the range indicated by iterators `first` and `last`.
`void assign(size_type n, const T& el = T())`	Remove all the elements in the deque and insert into it n copies of `el`.
`T& at(size_type n)`	Return the element in position n of the deque.
`const T& at(size_type n) const`	Return the element in position n of the deque.
`T& back()`	Return the last element in the deque.
`const T& back() const`	Return the last element in the deque.
`iterator begin()`	Return an iterator that references the first element of the deque.
`const_iterator begin() const`	Return an iterator that references the first element of the deque.
`void clear()`	Remove all the elements in the deque.
`deque()`	Construct an empty deque.
`deque(size_type n, const T& el = T())`	Construct a deque with n copies of `el` of type `T` (if `el` is not provided, a default constructor `T()` is used).
`deque(const deque <T>& dq)`	Copy constructor.
`deque(iterator first, iterator last)`	Construct a deque and initialize it with values from the range indicated by iterators `first` and `last`.
`bool empty() const`	Return `true` if the deque includes no elements and `false` otherwise.
`iterator end()`	Return an iterator that is past the last element of the deque.
`const_iterator end() const`	Return an iterator that is past the last element of the deque.
`iterator erase(iterator i)`	Remove the element referenced by iterator `i` and return an iterator referencing the element after the one removed.
`iterator erase(iterator first, iterator last)`	Remove the elements in the range indicated by iterators `first` and `last` and return an iterator referencing the element after the last one removed.
`T& front()`	Return the first element in the deque.
`const T& front() const`	Return the first element in the deque.
`iterator insert(iterator i, const T& el = T())`	Insert `el` before the element indicated by iterator `i` and return an iterator referencing the newly inserted element.

Continues

FIGURE 3.26 (continued)

`void insert(iterator i, size_type n, const T& el)`	Insert n copies of `el` before the element referenced by iterator `i`.
`void insert(iterator i, iterator first, iterator last)`	Insert elements from the location referenced by `first` to the location referenced by `last` before the element referenced by iterator `i`.
`size_type max_size() const`	Return the maximum number of elements for the deque.
`T& operator[]`	Subscript operator.
`void pop_back()`	Remove the last element of the deque.
`void pop_front()`	Remove the first element of the deque.
`void push_back(const T& el)`	Insert `el` at the end of the deque.
`void push_front(const T& el)`	Insert `el` at the beginning of the deque.
`reverse_iterator rbegin()`	Return an iterator that references the last element of the deque.
`const_reverse_iterator rbegin() const`	Return an iterator that references the last element of the deque.
`reverse_iterator rend()`	Return an iterator that is before the first element of the deque.
`const_reverse_iterator rend() const`	Return an iterator that is before the first element of the deque.
`void resize(size_type n, const T& el = T())`	Make the deque have n positions by adding `n - size()` more positions with element `el` or by discarding overflowing `size() - n` positions from the end of the deque.
`size_type size() const`	Return the number of elements in the deque.
`void swap(deque<T>& dq)`	Swap the content of the deque with the content of another deque `dq`.

and `list` only reimplements them as member functions. The most significant difference is the function `at()` (and its equivalent, `operator[]`), which is unavailable in `list`. The latter function is available in `vector`, and if we compare the set of member functions in `vector` (Figure 1.3) and in `deque`, we see only a few differences. `vector` does not have `pop_front()` and `push_front()`, as does `deque`, but `deque` does not include functions `capacity()` and `reserve()`, which are available in `vector`. A few operations are illustrated in Figure 3.27. Note that for lists only autoincrement and autodecrement were possible for iterators, but for deques we can add any number to iterators. For example, `dq1.begin()+1` is legal for deques, but not for lists.

A very interesting aspect of the STL deque is its implementation. Random access can be simulated in doubly linked lists that have the definition of `operator[]`

FIGURE 3.27 A program demonstrating the operation of deque member functions.

```cpp
#include <iostream>
#include <algorithm>
#include <deque>

using namespace std;

int main() {
    deque<int> dq1;
    dq1.push_front(1);                          // dq1 = (1)
    dq1.push_front(2);                          // dq1 = (2 1)
    dq1.push_back(3);                           // dq1 = (2 1 3)
    dq1.push_back(4);                           // dq1 = (2 1 3 4)
    deque<int> dq2(dq1.begin()+1,dq1.end()-1);  // dq2 = (1 3)
    dq1[1] = 5;                                 // dq1 = (2 5 3 4)
    dq1.erase(dq1.begin());                     // dq1 = (5 3 4)
    dq1.insert(dq1.end()-1,2,6);                // dq1 = (5 3 6 6 4)
    sort(dq1.begin(),dq1.end());                // dq1 = (3 4 5 6 6)
    deque<int> dq3;
    dq3.resize(dq1.size()+dq2.size());          // dq3 = (0 0 0 0 0 0 0)
    merge(dq1.begin(),dq1.end(),dq2.begin(),dq2.end(),dq3.begin());
    // dq1 = (3 4 5 6 6) and dq2 = (1 3) ==> dq3 = (1 3 3 4 5 6 6)
    return 0;
}
```

(int n) which includes a loop that sequentially scans the list and stops at the *n*th node. The STL implementation solves this problem differently. An STL deque is not implemented as a linked list, but as an array of pointers to blocks or arrays of data. The number of blocks changes dynamically depending on storage needs, and the size of the array of pointers changes accordingly. (We encounter a similar approach applied in extendible hashing in Section 10.5.1.)

To discuss one possible implementation, assume that the array of pointers has four cells and an array of data has three cells; that is, blockSize = 3. An object deque includes the fields head, tail, headBlock, tailBlock, and blocks. After execution of push_front(e1) and push_front(e2) with an initially empty deque, the situation is as in Figure 3.28a. First, the array blocks is created, and then one data block is accessible from a middle cell of blocks. Next, e1 is inserted in the middle of the data block. The subsequent calls place elements consecutively in the first half of the data array. The third call to push_front() cannot successfully place e3 in the current data array; therefore, a new data array is created and e3 is located in the last cell (Figure 3.28b). Now we execute push_back() four times. Element e4 is placed in an existing data array accessible from deque through tailBlock. Elements e5, e6, and e7 are placed in a new data block, which also becomes accessible through tailBlock (Figure 3.28c).

FIGURE 3.28 Changed on the deque in the process of pushing new elements.

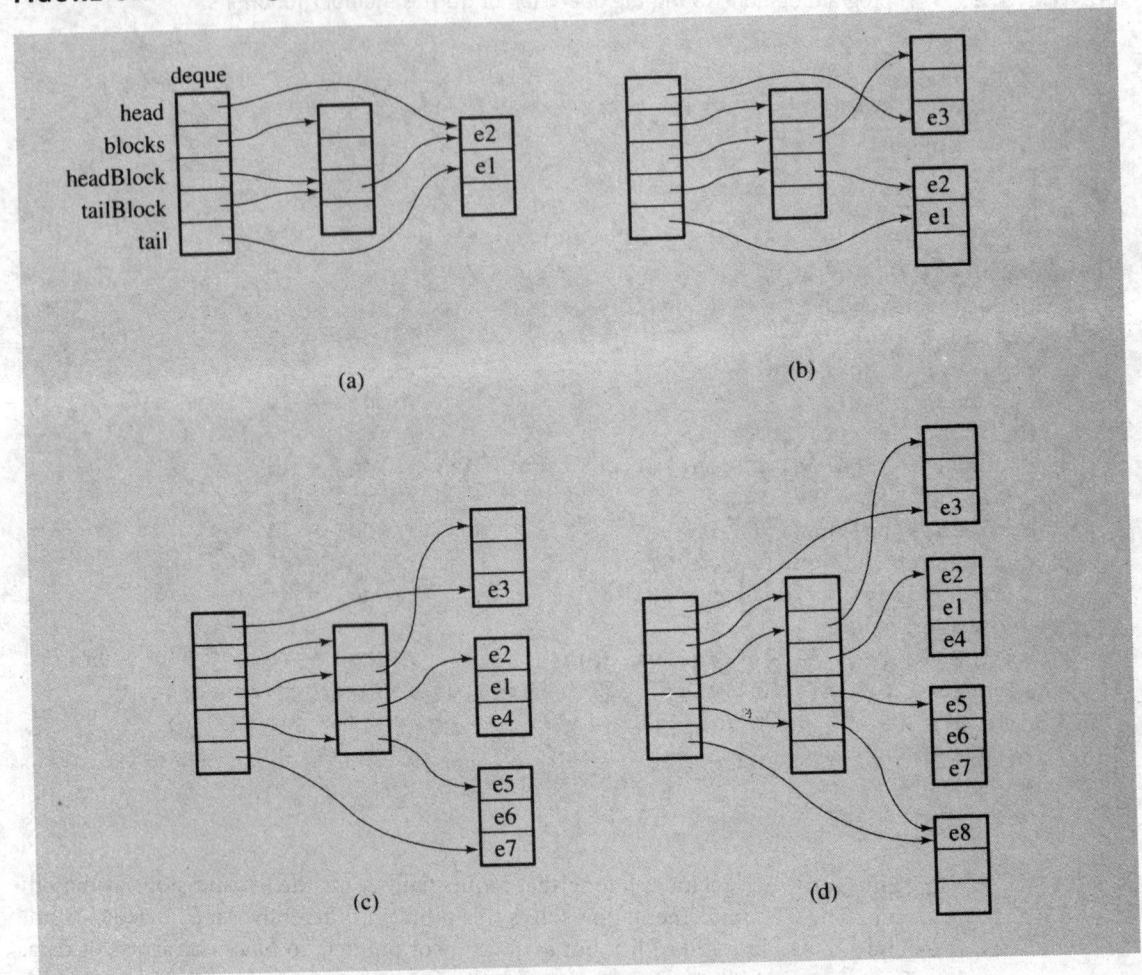

The next call to `push_back()` affects the pointer array `blocks` because the last data block is full and the block is accessible for the last cell of `blocks`. In this case, a new pointer array is created that contains (in this implementation) twice as many cells as the number of data blocks. Next, the pointers from the old array `blocks` are copied to the new array, and then a new data block can be created to accommodate element `e8` being inserted (Figure 3.28d). This is an example of the worst case for which between $n/\text{blockSize}$ and $n/\text{blockSize} + 2$ cells have to be copied from the old array to the new one; therefore, in the worst case, the pushing operation takes $O(n)$ time to perform. But assuming that `blockSize` is a large number, the worst case can be expected to occur very infrequently. Most of the time, the pushing operation requires constant time.

Inserting an element into a deque is very simple conceptually. To insert an element in the first half of the deque, the front element is pushed onto the deque, and all elements that should precede the new element are copied to the preceding cell. Then the new element can be placed in the desired position. To insert an element into the second half of the deque, the last element is pushed onto the deque, and elements that should follow the new element in the deque are copied to the next cell.

With the discussed implementation, a random access can be performed in constant time. For the situation illustrated in Figure 3.28—that is, with declarations

```
T **blocks;
T **headBlock;
T *head;
```

the subscript operator can be overloaded as follows:

```
T& operator[] (int n) {
    if (n < blockSize - (head - *headBlock))    // if n is
         return *(head + n);                    // in the first
    else {                                       // block;
         n = n - (blockSize - (head - *headBlock));
         int q = n / blockSize + 1;
         int r = n % blockSize;
         return *(*(headBlock + q) + r);
    }
}
```

Although access to a particular position requires several arithmetic, dereferencing, and assignment operations, the number of operations is constant for any size of the deque.

3.9 CONCLUDING REMARKS

Linked lists have been introduced to overcome limitations of arrays by allowing dynamic allocation of necessary amounts of memory. Also, linked lists allow easy insertion and deletion of information, because such operations have a local impact on the list. To insert a new element at the beginning of an array, all elements in the array have to be shifted to make room for the new item; hence, insertion has a global impact on the array. Deletion is the same. So should we always use linked lists instead of arrays?

Arrays have some advantages over linked lists, namely that they allow random accessing. To access the tenth node in a linked list, all nine preceding nodes have to be passed. In the array, we can go to the tenth cell immediately. Therefore, if an immediate access of any element is necessary, then an array is a better choice. This was the case with binary search, and it will be the case with most sorting algorithms (see Chapter 9). But if we are constantly accessing only some elements—the first, the second, the last, and the like—and if changing the structure is the core of an algorithm, then using a linked list is a better option. A good example is a queue, which is discussed in the next chapter.

Another advantage in the use of arrays is space. To hold items in arrays, the cells have to be of the size of the items. In linked lists, we store one item per node, and the node also includes at least one pointer; in doubly linked lists, the node contains two pointers. For large linked lists, a significant amount of memory is needed to store the pointers. Therefore, if a problem does not require many shifts of data, then having an oversized array may not be wasteful at all if its size is compared to the amount of space needed for the linked structure storing the same data as the array.

3.10 CASE STUDY: A LIBRARY

This case study is a program that can be used in a small library to include new books in the library, to check out books to people, and to return them.

As this program is a practice in the use of linked lists, almost everything is implemented in terms of such lists. But to make the program more interesting, it uses linked lists of linked lists that also contain cross-references (see Figure 3.29).

First, there could be a list including all authors of all books in the library. However, searching through such a list can be time-consuming, so the search can be sped up by choosing at least one of the two following strategies:

- The list can be ordered alphabetically, and the search can be interrupted if we find the name, if we encounter an author's name greater than the one we are searching for, or if we reach the end of list.
- We can use an array of pointers to the author structures indexed with letters; each slot of the array points to the linked list of authors whose names start with the same letter.

The best strategy is to combine both approaches. However, in this case study, only the second approach is used, and the reader is urged to amplify the program by adding the first approach. Note the articles *a*, *an*, and *the* at the beginning of the titles should be disregarded during the sorting operation.

The program uses an array `catalog` of all the authors of the books included in the library and an array `people` of all the people who have used the library at least once. Both arrays are indexed with letters so that, for instance, position `catalog['F']` refers to a linked list of all the authors whose names start with F.

Because we can have several books by the same author, one of the data members of the author node refers to the list of books by this author that can be found in the library. Similarly, because each patron can check out several books, the node corresponding to this patron contains a reference to the list of books currently checked out by this patron. This fact is also indicated by setting the `patron` member of the checked-out book to the node pertaining to the patron who is taking the book out.

Books can be returned, and that fact should be reflected by removing the appropriate nodes from the list of the checked-out books of the patron who returns them. The `patron` member in the node related to the book that is being returned has to be reset to null.

Section 3.10 Case Study: A Library 121

FIGURE 3.29 Linked lists indicating library status.

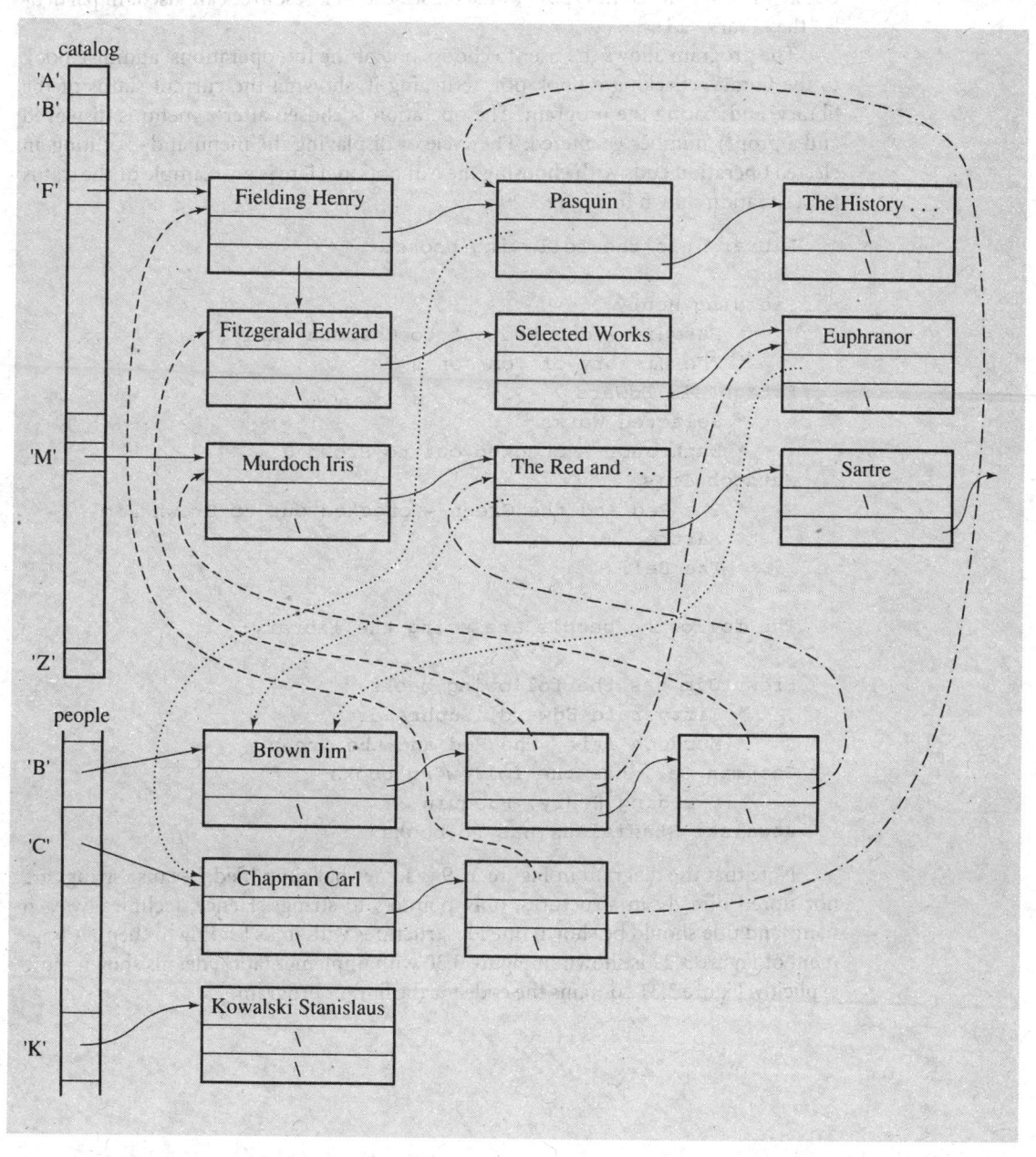

The program defines four classes: Author, Book, Patron, and CheckedOut-Book. To define different types of linked lists, the STL resources are used, in particular, the library <list>.

The program allows the user to choose one of the five operations: adding a book to the library, checking a book out, returning it, showing the current status of the library, and exiting the program. The operation is chosen after a menu is displayed and a proper number is entered. The cycle of displaying the menu and executing an elected operation ends with choosing the exit option. Here is an example of the status for a situation shown in Figure 3.29.

```
Library has the following books:

Fielding Henry
     * Pasquin - checked out to Chapman Carl
     * The History of Tom Jones
Fitzgerald Edward
     * Selected Works
     * Euphranor - checked out to Brown Jim
Murdoch Iris
     * The Red and the Green - checked out to Brown Jim
     * Sartre
     * The Bell

The following people are using the library:

Brown Jim has the following books
     * Fitzgerald Edward, Euphranor
     * Murdoch Iris, The Red and the Green
Chapman Carl has the following books
     * Fielding Henry, Pasquin
Kowalski Stanislaus has no books
```

Note that the diagram in Figure 3.29 is somewhat simplified, because strings are not stored directly in structures, only pointers to strings. Hence, technically, each name and title should be shown outside structures with links leading to them. A fragment of Figure 3.29 is shown in Figure 3.30 with implementation details shown more explicitly. Figure 3.31 contains the code for the library program.

FIGURE 3.30 Fragment of structure from Figure 3.29 with all the objects used in the implementation.

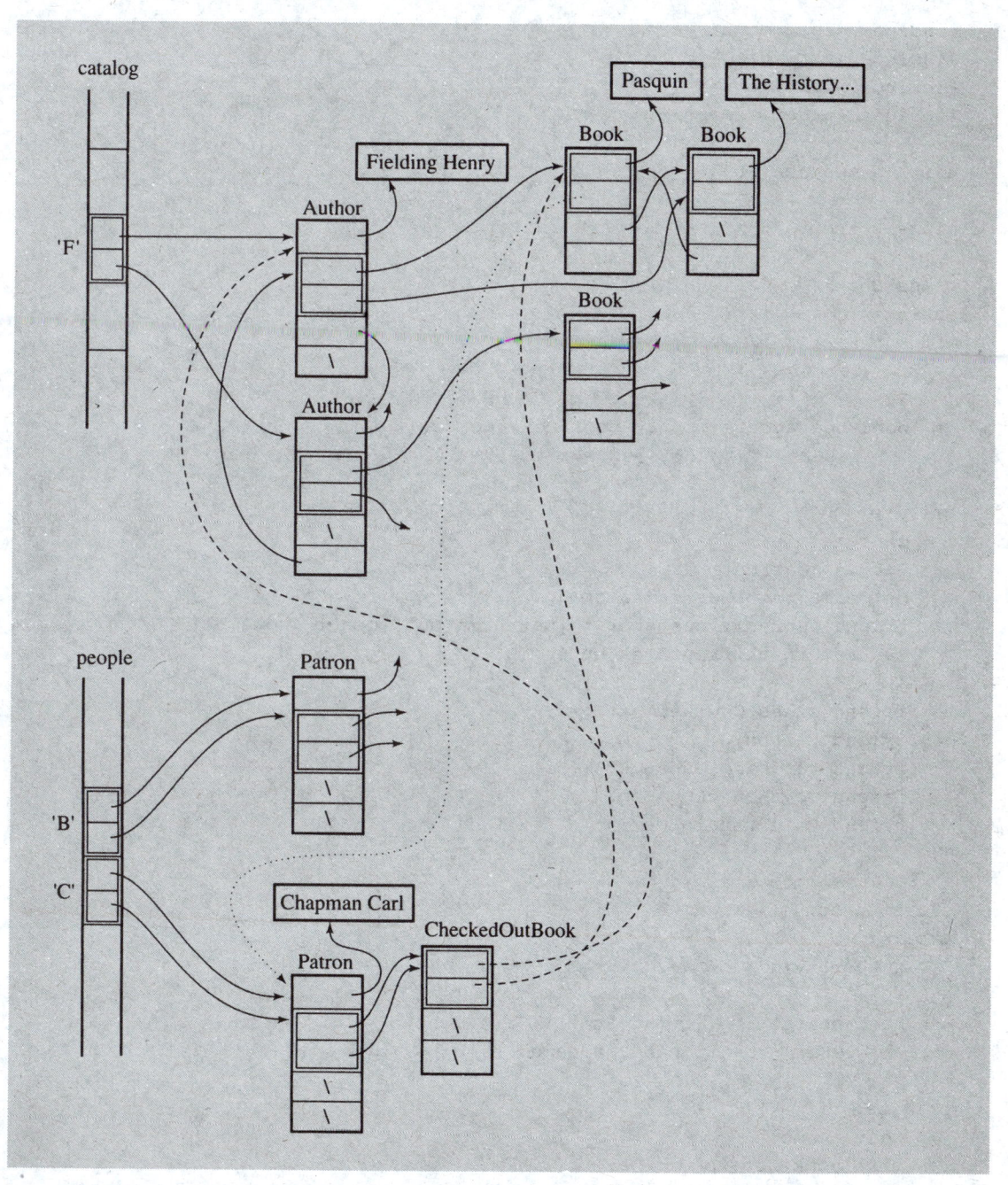

FIGURE 3.31 The library program.

```cpp
#include <iostream>
#include <string>
#include <list>
#include <algorithm>

using namespace std;

class Patron;        // forward declaration;

class Book {
public:
    Book() {
        patron = 0;
    }
    bool operator== (const Book& bk) const {
        return strcmp(title,bk.title) == 0;
    }
private:
    char *title;
    Patron *patron;
    ostream& printBook(ostream&) const;
    friend ostream& operator<< (ostream& out, const Book& bk) {
        return bk.printBook(out);
    }
    friend class CheckedOutBook;
    friend Patron;
    friend void includeBook();
    friend void checkOutBook();
    friend void returnBook();
};

class Author {
public:
    Author() {
    }
    bool operator== (const Author& ar) const {
        return strcmp(name,ar.name) == 0;
    }
private:
    char *name;
    list<Book> books;
    ostream& printAuthor(ostream&) const;
```

FIGURE 3.31 *(continued)*

```cpp
        friend ostream& operator<< (ostream& out,const Author& ar) {
            return ar.printAuthor(out);
        }
        friend void includeBook();
        friend void checkOutBook();
        friend void returnBook();
        friend class CheckedOutBook;
        friend Patron;
};

class CheckedOutBook {
public:
    CheckedOutBook(list<Author>::iterator ar = 0,
                   list<Book>::iterator bk = 0) {
        author = ar;
        book = bk;
    }
    bool operator== (const CheckedOutBook& bk) const {
        return strcmp(author->name,bk.author->name) == 0 &&
               strcmp(book->title,bk.book->title) == 0;
    }
private:
    list<Author>::iterator author;
    list<Book>::iterator book;
    friend void checkOutBook();
    friend void returnBook();
    friend Patron;
};

class Patron {
public:
    Patron() {
    }
    bool operator== (const Patron& pn) const {
        return strcmp(name,pn.name) == 0;
    }
private:
    char *name;
    list<CheckedOutBook> books;
    ostream& printPatron(ostream&) const;
    friend ostream& operator<< (ostream& out, const Patron& pn) {
```

Continues

FIGURE 3.31 *(continued)*

```
            return pn.printPatron(out);
    }
    friend void checkOutBook();
    friend void returnBook();
            friend Book;
};

list<Author> catalog['Z'+1];
list<Patron> people['Z'+1];

ostream& Author::printAuthor(ostream& out) const {
    out << name << endl;
    list<Book>::const_iterator ref = books.begin();
    for ( ; ref != books.end(); ref++)
        out << *ref; // overloaded <<
    return out;
}

ostream& Book::printBook(ostream& out) const {
    out << "    * " << title;
    if (patron != 0)
        out << " - checked out to " << patron->name; // overloaded <<
    out << endl;
    return out;
}

ostream& Patron::printPatron(ostream& out) const {
    out << name;
    if (!books.empty()) {
        out << " has the following books:\n";
        list<CheckedOutBook>::const_iterator bk = books.begin();
        for ( ; bk != books.end(); bk++)
            out << "    * " << bk->author->name << ", "
                << bk->book->title << endl;
    }
    else out << " has no books\n";
    return out;
}

template<class T>
ostream& operator<< (ostream& out, const list<T>& lst) {
```

FIGURE 3.31   *(continued)*

```cpp
    for (list<T>::const_iterator ref = lst.begin(); ref != lst.end();
         ref++)
        out << *ref; // overloaded <<
    return out;
}

char* getString(char *msg) {
    char s[82], i, *destin;
    cout << msg;
    cin.get(s,80);
    while (cin.get(s[81]) && s[81] != '\n');    // discard overflowing
    destin = new char[strlen(s)+1];             // characters;
    for (i = 0; destin[i] = toupper(s[i]); i++);
    return destin;
}

void status() {
    register int i;
    cout << "Library has the following books:\n\n";
    for (i = 'A'; i <= 'Z'; i++)
        if (!catalog[i].empty())
            cout << catalog[i];
    cout << "\nThe following people are using the library:\n\n";
    for (i = 'A'; i <= 'Z'; i++)
        if (!people[i].empty())
            cout << people[i];
}

void includeBook() {
    Author newAuthor;
    Book newBook;
    newAuthor.name = getString("Enter author's name: ");
    newBook.title  = getString("Enter the title of the book: ");
    list<Author>::iterator oldAuthor =
                find(catalog[newAuthor.name[0]].begin(),
                     catalog[newAuthor.name[0]].end(),newAuthor);
    if (oldAuthor == catalog[newAuthor.name[0]].end()) {
        newAuthor.books.push_front(newBook);
        catalog[newAuthor.name[0]].push_front(newAuthor);
    }
    else (*oldAuthor).books.push_front(newBook);
}
```

Continues

FIGURE 3.31 *(continued)*

```
void checkOutBook() {
    Patron patron;
    Author author;
    Book book;
    list<Author>::iterator authorRef;
    list<Book>::iterator bookRef;
    patron.name = getString("Enter patron's name: ");
    while (true) {
        author.name = getString("Enter author's name: ");
        authorRef = find(catalog[author.name[0]].begin(),
                         catalog[author.name[0]].end(),author);
        if (authorRef == catalog[author.name[0]].end())
             cout << "Misspelled author's name\n";
        else break;
    }
    while (true) {
        book.title = getString("Enter the title of the book: ");
        bookRef = find((*authorRef).books.begin(),
                       (*authorRef).books.end(),book);
        if (bookRef == (*authorRef).books.end())
             cout << "Misspelled title\n";
        else break;
    }
    list<Patron>::iterator patronRef;
    patronRef = find(people[patron.name[0]].begin(),
                     people[patron.name[0]].end(),patron);
    CheckedOutBook checkedOutBook(authorRef,bookRef);
    if (patronRef == people[patron.name[0]].end()) { // a new patron
         patron.books.push_front(checkedOutBook);    // in the library;
         people[patron.name[0]].push_front(patron);
         (*bookRef).patron = &*people[patron.name[0]].begin();
    }
    else {
         (*patronRef).books.push_front(checkedOutBook);
         (*bookRef).patron = &*patronRef;
    }
}

void returnBook() {
    Patron patron;
```

FIGURE 3.31   (continued)

```cpp
    Book book;
    Author author;
    list<Patron>::iterator patronRef;
    list<Book>::iterator bookRef;
    list<Author>::iterator authorRef;
    while (true) {
        patron.name = getString("Enter patron's name: ");
        patronRef = find(people[patron.name[0]].begin(),
                    people[patron.name[0]].end(),patron);
        if (patronRef == people[patron.name[0]].end())
            cout << "Patron's name misspelled\n";
        else break;
    }
    while (true) {
        author.name = getString("Enter author's name: ");
        authorRef = find(catalog[author.name[0]].begin(),
                    catalog[author.name[0]].end(),author);
        if (authorRef == catalog[author.name[0]].end())
            cout << "Misspelled author's name\n";
        else break;
    }
    while (true) {
        book.title = getString("Enter the title of the book: ");
        bookRef = find((*authorRef).books.begin(),
                    (*authorRef).books.end(),book);
        if (bookRef == (*authorRef).books.end())
            cout << "Misspelled title\n";
        else break;
    }
    CheckedOutBook checkedOutBook(authorRef,bookRef);
    (*bookRef).patron = 0;
    (*patronRef).books.remove(checkedOutBook);
}

int menu() {
    int option;
    cout << "\nEnter one of the following options:\n"
         << "1. Include a book in the catalog\n2. Check out a book\n"
         << "3. Return a book\n4. Status\n5. Exit\n"
         << "Your option? ";
    cin >> option;
```

Continues

FIGURE 3.31   *(continued)*

```
    cin.get();              // discard '\n';
    return option;
}

void main() {
    while (true)
        switch (menu()) {
            case 1: includeBook();   break;
            case 2: checkOutBook();  break;
            case 3: returnBook();    break;
            case 4: status();        break;
            case 5: return 0;
            default: cout << "Wrong option, try again: ";
        }
}
```

3.11 EXERCISES

1. Assume that a circular doubly linked list has been created, as in Figure 3.32. After each of the following assignments, indicate changes made in the list by showing which links have been modified. The second assignment should make changes in the list modified by the first assignment, and so on.

FIGURE 3.32 A circular doubly linked list.

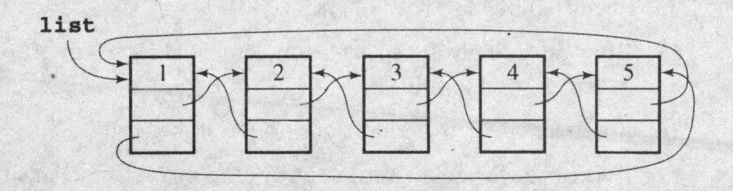

```
list->next->next->next = list->prev;

list->prev->prev->prev = list->next->next->next->prev;

list->next->next->next->prev = list->prev->prev->prev;

list->next = list->next->next;

list->next->prev->next = list->next->next->next;
```

2. How many nodes does the shortest linked list have? The longest linked list?

3. The linked list in Figure 3.11 was created in Section 3.2 with three assignments. Create this list with only one assignment.

4. Merge two ordered singly linked lists of integers into one ordered list.

5. Delete an *i*th node on a linked list. Be sure that such a node exists.

6. Delete from list L_1 nodes whose positions are to be found in an ordered list L_2. For instance, if L_1 = (A B C D E) and L_2 = (2 4 8), then the second and the fourth nodes are to be deleted from list L_1 (the eighth node does not exist), and after deletion, L_1 = (A C E).

7. Delete from list L_1 nodes occupying positions indicated in ordered lists L_2 and L_3. For instance, if L_1 = (A B C D E), L_2 = (2 4 8), and L_3 = (2 5), then after deletion, L_1 = (A C).

8. Delete from an ordered list L nodes occupying positions indicated in list L itself. For instance, if L = (1 3 5 7 8), then after deletion, L = (1 7).

9. A linked list does not have to be implemented with pointers. Suggest other implementations of linked lists.

10. Write a member function to check whether two singly linked lists have the same contents.

11. Write a member function to reverse a singly linked list using only one pass through the list.

12. Insert a new node into a singly linked list (a) before and (b) after a node pointed to by p in this list (possibly the first or the last). Do not use a loop in either operation.

13. Attach a singly linked list to the end of another singly linked list.

14. Put numbers in a singly linked list in ascending order. Use this operation to find the median in the list of numbers.

15. How can a singly linked list be implemented so that insertion requires no test for whether head is null?

16. Insert a node in the middle of a doubly linked list.

17. Write code for class `IntCircularSLList` for a circular singly linked list that includes equivalents of the member functions listed in Figure 3.2.

18. Write code for class `IntCircularDLList` for a circular doubly linked list that includes equivalents of the member functions listed in Figure 3.2.

19. How likely is the worst case for searching a skip list to occur?

20. Consider the move-to-front, transpose, count, and ordering methods.
    a. In what case is a list maintained by these methods not changed?
    b. In what case do these methods require an exhaustive search of lists for each search, assuming that only elements in the list are searched for?

21. In the discussion of self-organizing lists, only the number of comparisons was considered as the measure of different methods' efficiency. This measure can, however, be greatly affected by a particular implementation of the list. Discuss how the efficiency of the move-to-front, transpose, count, and ordering methods are affected in the case when the list is implemented as
    a. an array
    b. a singly linked list
    c. a doubly linked list

22. For doubly linked lists, there are two variants of the move-to-front and transpose methods (Matthews, Rotem, & Bretholz 1980). A *move-to-end* method moves an element being accessed to the end from which the search started. For instance, if the doubly linked list is a list of items *A B C D* and the search starts from the right end to access node *C,* then the reorganized list is *A B D C*. If the search for *C* started from the left end, the resulting list is *C A B D*.

    The *swapping* technique transposes a node with this predecessor also with respect to the end from which the search started (Ng & Oommen 1989). Assuming that only elements of the list are in the data, what is the worst case for a move-to-end doubly linked list when the search is made alternately from the left and from the right? For a swapping list?

23. What is the maximum number of comparisons for optimal search for the 14 letters shown in Figure 3.19?

24. Adapt the binary search to linked lists. How efficient can this search be?

3.12 PROGRAMMING ASSIGNMENTS

1. Farey fractions of level one are defined as sequence $\left(\frac{0}{1}, \frac{1}{1}\right)$. This sequence is extended in level two to form a sequence $\left(\frac{0}{1}, \frac{1}{2}, \frac{1}{1}\right)$, sequence $\left(\frac{0}{1}, \frac{1}{3}, \frac{1}{2}, \frac{2}{3}, \frac{1}{1}\right)$ at level three, sequence $\left(\frac{0}{1}, \frac{1}{4}, \frac{1}{3}, \frac{1}{2}, \frac{2}{3}, \frac{3}{4}, \frac{1}{1}\right)$ at level four, so that at each level n, a new fraction $\frac{a+b}{c+d}$ is inserted between two neighbor fractions $\frac{a}{c}$ and $\frac{b}{d}$ only if $c + d \leq n$. Write a program which for a number n entered by the user creates—by constantly extending it—a linked list of fractions at level n and then displays them.

2. Write a simple airline ticket reservation program. The program should display a menu with the following options: reserve a ticket, cancel a reservation, check whether a ticket is reserved for a particular person, and display the passengers. The information is maintained on an alphabetized linked list of names. In a simpler version of the program, assume that tickets are reserved for only one flight. In a fuller version, place no limit on the number of flights. Create a linked list of flights with each node including a pointer to a linked list of passengers.

3. Read Section 12.1 about sequential-fit methods. Implement the discussed methods with linked lists and compare their efficiency.

4. Write a program to simulate managing files on disk. Define the disk as a one-dimensional array `disk` of size `numOfSectors*sizeOfSector`, where `sizeOfSector` indicates the number of characters stored in one sector. (For the sake of debugging, make it a very small number.) A pool of available sectors is kept in a linked list `sectors` of three field structures: two fields to indicate ranges of available sectors and one `next` field. Files are kept in a linked list `files` of four field structures: file name, the number of characters in the file, a pointer to a linked list of sectors where the contents of the file can be found, and the `next` field.

   a. In the first part, implement functions to save and delete files. Saving files requires claiming a sufficient number of sectors from `pool`, if available. The sectors may not be contiguous, so the linked list assigned to the file may contain several nodes. Then the contents of the file have to be written to the sectors assigned to the file. Deletion of a file only requires removing the nodes corresponding with this file (one from `files` and the rest from its own linked list of sectors) and transferring the sectors assigned to this file back to `pool`. No changes are made in `disk`.

   b. File fragmentation slows down file retrieval. In the ideal situation, one cluster of sectors is assigned to one file. However, after many operations with files, it may not be possible. Extend the program to include a function `together()` to transfer files to contiguous sectors; that is, to create a situation illustrated in Figure 3.33. Fragmented files `file1` and `file2` occupy only one cluster of sectors after `together()` is finished.

FIGURE 3.33 Linked lists used to allocate disk sectors for files: (a) a pool of available sectors; two files (b) before and (c) after putting them in contiguous sectors; the situation in sectors of the disk (d) before and (e) after this operation.

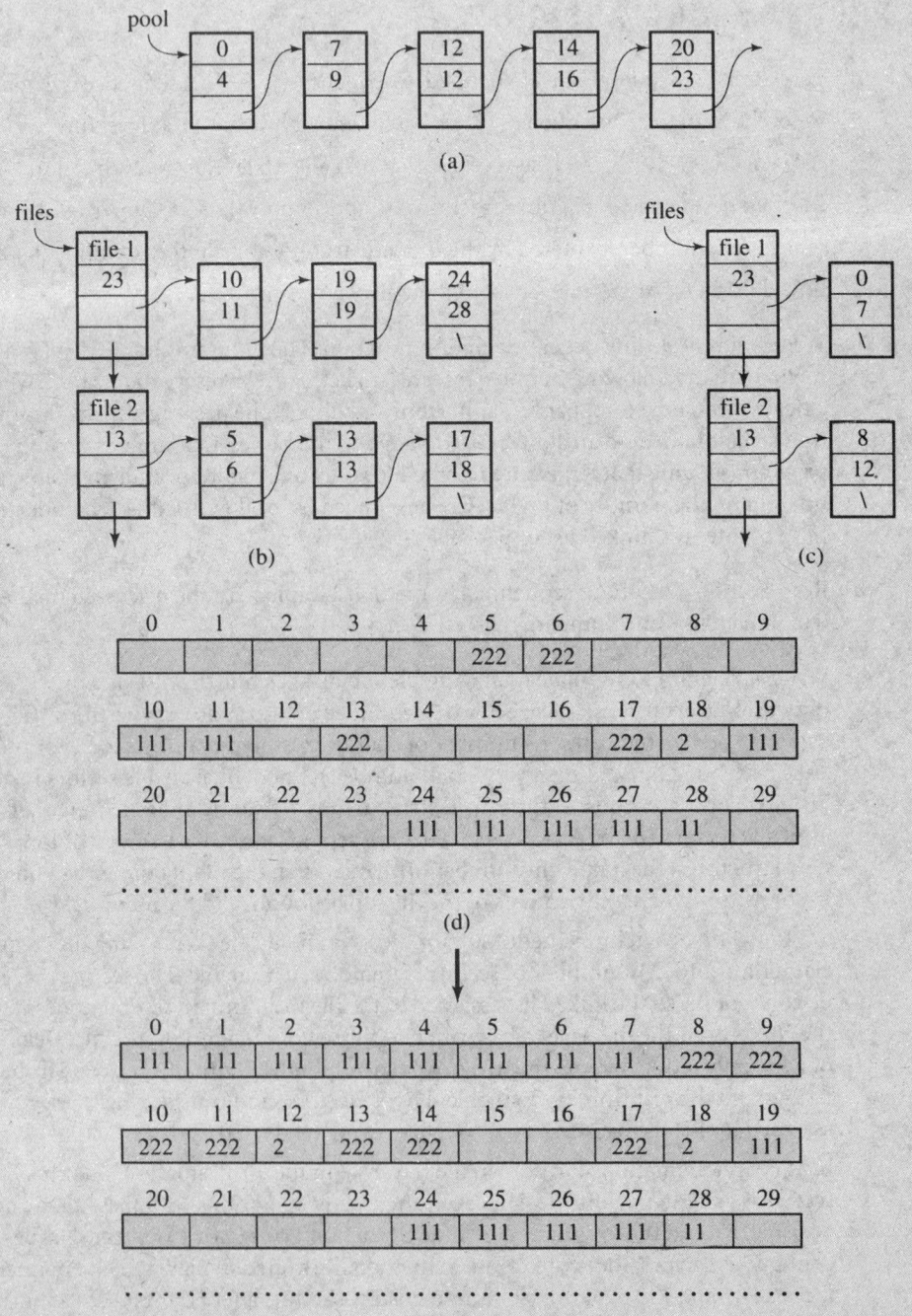

However, particular care should be taken not to overwrite sectors occupied by other files. For example, `file1` requires eight sectors; five sectors are free at the beginning of `pool`, but sectors 5 and 6 are occupied by `file2`. Therefore, a file f occupying such sectors has to be located first by scanning `files`. The contents of these sectors must be transferred to unoccupied positions, which requires updating the sectors belonging to f in the linked list; only then can the released sectors be utilized. One way of accomplishing this is by copying from the area into which one file is copied chunks of sectors of another file into an area of the disk large enough to accommodate these chunks. In the example in Figure 3.33, contents of `file1` are first copied to sectors 0 through 4, and then copying is temporarily suspended because sector 5 is occupied. Thus, contents of sectors 5 and 6 are moved to sector 12 and 14, and the copying of `file1` is resumed.

5. Write a simple line editor. Keep the entire text on a linked list, one line in a separate node. Start the program with entering `EDIT file`, after which a prompt appears along with the line number. If the letter I is entered with a number n following it, then insert the text to be followed before line n. If I is not followed by a number, then insert the text before the current line. If D is entered with two numbers n and m, one n, or no number following it, then delete lines n through m, line n, or the current line. Do the same with the command L, which stands for listing lines. If A is entered, then append the text to the existing lines. Entry E signifies exit and saving the text in a file. Here is an example:

```
 EDIT testfile
1> The first line
2>
3> And another line
4> I 3
3> The second line
4> One more line
5> L
1> The first line
2>
3> The second line
4> One more line
5> And another line     // This is now line 5, not 3;
5> D 2                  // line 5, since L was issued from line 5;
4> L                    // line 4, since one line was deleted;
1> The first line
2> The second line      // this and the following lines
3> One more line        // now have new numbers;
4> And another line
4> E
```

6. Extend the case study program in this chapter to have it store all the information in the file `Library` at exit and initialize all the linked lists using this information at the invocation of the program. Also, extend it by adding more error checking, such as not

allowing the same book to be checked out at the same time to more than one patron or not including the same patron more than once in the library.

7. Test the efficiency of skip lists. In addition to the functions given in this chapter, implement `skipListDelete()` and then compare the number of node accesses in searching, deleting, and inserting for large numbers of elements. Compare this efficiency with the efficiency of linked lists and ordered linked lists. Test your program on a randomly generated order of operations to be executed on the elements. These elements should be processed in random order. Then try your program on nonrandom samples.

8. Write a rudimentary link program to check whether all variables have been initialized and whether local variables have the same names as global variables. Create a linked list of global variables and, for each function, create a linked list of local variables. In both lists, store information on the first initialization of each variable and check if any initialization has been made before a variable is used for the first time. Also, compare both lists to detect possible matches and issue a warning if a match is found. The list of local variables is removed after the processing of one function is finished and created anew when a new function is encountered. Consider the possibility of maintaining alphabetical order on both lists.

BIBLIOGRAPHY

Bentley, Jon L., and McGeoch, Catharine C., "Amortized Analyses of Self-Organizing Sequential Search Heuristics," *Communications of the ACM* 28 (1985), 404–411.

Foster, John M., *List Processing,* London: McDonald, 1967.

Hansen, Wilfred J., "A Predecessor Algorithm for Ordered Lists," *Information Processing Letters* 7 (1978), 137–138.

Hester, James H. and Hirschberg, Daniel S., "Self-Organizing Linear Search," *Computing Surveys* 17 (1985), 295–311.

Matthews, D., Rotem, D., and Bretholz, E., "Self-Organizing Doubly Linked Lists," *International Journal of Computer Mathematics* 8 (1980), 99–106.

Ng, D. T. H., and Oommen, B. J., "Generalizing Singly-Linked List Reorganizing Heuristics for Doubly-Linked Lists," in Kreczmar, A. and Mirkowska, G. (eds.), *Mathematical Foundations of Computer Science 1989,* Berlin: Springer, 1989, 380–389.

Pugh, William, "Skip Lists: a Probabilistic Alternative to Balanced Trees," *Communications of the ACM* 33 (1990), 668–676.

Rivest, Ronald, "On Self-Organizing Sequential Search Heuristics," *Communications of the ACM* 19 (1976), No. 2, 63–67.

Sleator, Daniel D., and Tarjan, Robert E., "Amortized Efficiency of List Update and Paging Rules," *Communications of the ACM* 28 (1985), 202–208.

Wilkes, Maurice V., "Lists and Why They Are Useful," *Computer Journal* 7 (1965), 278–281.

Stacks and Queues

4

As the first chapter explained, abstract data types allow us to delay the specific implementation of a data type until it is well understood what operations are required to operate on the data. In fact, these operations determine which implementation of the data type is most efficient in a particular situation. This situation is illustrated by two data types, stacks and queues, which are described by a list of operations. Only after the list of the required operations is determined do we present some possible implementations and compare them.

4.1 STACKS

A *stack* is a linear data structure that can be accessed only at one of its ends for storing and retrieving data. Such a stack resembles a stack of trays in a cafeteria: New trays are put on the top of the stack and taken off the top. The last tray put on the stack is the first tray removed from the stack. For this reason, a stack is called an *LIFO* structure: last in/first out.

A tray can be taken only if there are trays on the stack, and a tray can be added to the stack only if there is enough room; that is, if the stack is not too high. Therefore, a stack is defined in terms of operations that change its status and operations that check this status. The operations are as follows:

- *clear()*—Clear the stack.
- *isEmpty()*—Check to see if the stack is empty.
- *push(el)*—Put the element *el* on the top of the stack.
- *pop()*—Take the topmost element from the stack.
- *topEl()*—Return the topmost element in the stack without removing it.

A series of push and pop operations is shown in Figure 4.1. After pushing number 10 onto an empty stack, the stack contains only this number. After pushing 5 on the stack, the number is placed on top of 10 so that, when the popping operation is executed, 5 is removed from the stack, because it arrived after 10, and 10 is left on the

137

FIGURE 4.1 A series of operations executed on a stack.

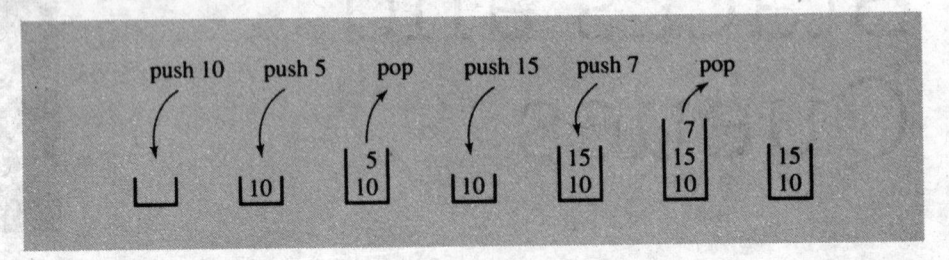

stack. After pushing 15 and then 7, the topmost element is 7, and this number is removed when executing the popping operation, after which the stack contains 10 at the bottom and 15 above it.

Generally, the stack is very useful in situations when data have to be stored and then retrieved in reverse order. One application of the stack is in matching delimiters in a program. This is an important example because delimiter matching is part of any compiler: No program is considered correct if the delimiters are mismatched.

In C++ programs, we have the following delimiters: parentheses "(" and ")", square brackets "[" and "]", curly brackets "{" and "}", and comment delimiters "/*" and "*/". Here are examples of C++ statements that use delimiters properly:

```
a = b + (c - d) * (e - f);
g[10] = h[i[9]] + (j + k) * l;
while (m < (n[8] + o)) { p = 7; /* initialize p */ r = 6; }
```

These examples are statements in which mismatching occurs:

```
a = b + (c - d) * (e - f));
g[10] = h[i[9]] + j + k) * l;
while (m < (n[8] + o]) { p = 7; /* initialize p */ r = 6; }
```

A particular delimiter can be separated from its match by other delimiters; that is, delimiters can be nested. Therefore, a particular delimiter is matched up only after all the delimiters following it and preceding its match have been matched. For example, in the condition of the loop

```
while (m < (n[8] + o))
```

the first opening parenthesis must be matched with the last closing parenthesis, but this is done only after the second opening parenthesis is matched with the next to last closing parenthesis; this, in turn, is done after the opening square bracket is matched with the closing bracket.

The delimiter matching algorithm reads a character from a C++ program and stores it on a stack if it is an opening delimiter. If a closing delimiter is found, the delimiter is compared to a delimiter popped off the stack. If they match, processing continues; if not, processing discontinues by signaling an error. The processing of the

C++ program ends successfully after the end of the program is reached and the stack is empty. Here is the algorithm:

```
delimiterMatching(file)
   read character ch from file;
   while not end of file
      if ch is '(', '[', or '{'
         push(ch);
      else if ch is '/'
         read the next character;
         if this character is '*'
            skip all characters until "*/" is found and report an error
               if the end of file is reached before "*/" is encountered;
         else ch = the character read in;
            continue;  // go to the beginning of the loop;
      else if ch is ')', ']', or '}'
         if ch and popped off delimiter do not match
            failure;
   // else ignore other characters;
      read next character ch from file;
   if stack is empty
      success;
   else failure;
```

Figure 4.2 shows the processing that occurs when applying this algorithm to the statement

```
s=t[5]+u/(v*(w+y));
```

The first column in Figure 4.2 shows the contents of the stack at the end of the loop before the next character is input from the program file. The first line shows the initial situation in the file and on the stack. Variable ch is initialized to the first character of the file, letter s, and in the first iteration of the loop, the character is simply ignored. This situation is shown in the second row in Figure 4.2. Then the next character, equal sign, is read. It is also ignored and so is the letter t. After reading the left bracket, the bracket is pushed onto the stack so that the stack now has one element, the left bracket. Reading digit 5 does not change the stack, but after the right bracket becomes the value of ch, the topmost element is popped off the stack and compared with ch. Because the popped off element (left bracket) matches ch (right bracket), the processing of input continues. After reading and discarding the letter u, a slash is read and the algorithm checks whether it is part of the comment delimiter by reading the next character, a left parenthesis. Because the character read is not an asterisk, the slash is not a beginning of a comment, so ch is set to left parenthesis. In the next iteration, this parenthesis is pushed onto the stack and processing continues, as shown in Figure 4.2. After reading the last character, a semicolon, the loop is exited and the stack is checked. Because it is empty (no unmatched delimiters are left), success is pronounced.

FIGURE 4.2 Processing the statement `s=t[5]+u/(v*(w+y));` with the algorithm `delimiterMatching()`.

Stack	Nonblank Character Read	Input Left
empty		s = t[5] + u / (v * (w + y));
empty	s	= t[5] + u / (v * (w + y));
empty	=	t[5] + u / (v * (w + y));
empty	t	[5] + u / (v * (w + y));
[	[	5] + u / (v * (w + y));
[	5	] + u / (v * (w + y));
empty	]	+ u / (v * (w + y));
empty	+	u / (v * (w + y));
empty	u	/ (v * (w + y));
empty	/	(v * (w + y));
(	(	v * (w + y));
(	v	* (w + y));
(	*	(w + y));
((	(	w + y));
((	w	+ y));
((	+	y));
((	y	));
((	)	);
empty	)	;
empty		

As another example of stack application, consider adding very large numbers. The largest magnitude of integers is limited, so we are not able to add 18,274,364,583,929,273,748,459,595,684,373 and 8,129,498,165,026,350,236, because integer variables cannot hold such large values, let alone their sum. The problem can be solved if we treat these numbers as strings of numerals, store the numbers corresponding to these numerals on two stacks, and then perform addition by popping numbers from the stacks. The pseudocode for this algorithm is as follows:

Section 4.1 Stacks ■ 141

```
addingLargeNumbers()
    read the numerals of the first number and store the numbers corresponding to
        them on one stack;
    read the numerals of the second number and store the numbers corresponding
        to them on another stack;
    result = 0;
    while at least one stack is not empty
        pop a number from each nonempty stack and add them to result;
        push the unit part on the result stack;
        store carry in result;
    push carry on the result stack if it is not zero;
    pop numbers from the result stack and display them;
```

Figure 4.3 shows an example of the application of this algorithm. In this example, numbers 592 and 3,784 are added.

1. Numbers corresponding to digits composing the first number are pushed onto `operandStack1`, and numbers corresponding to the digits of 3,784 are pushed onto `operandStack2`. Note the order of digits on the stacks.

FIGURE 4.3 An example of adding numbers 592 and 3,784 using stacks.

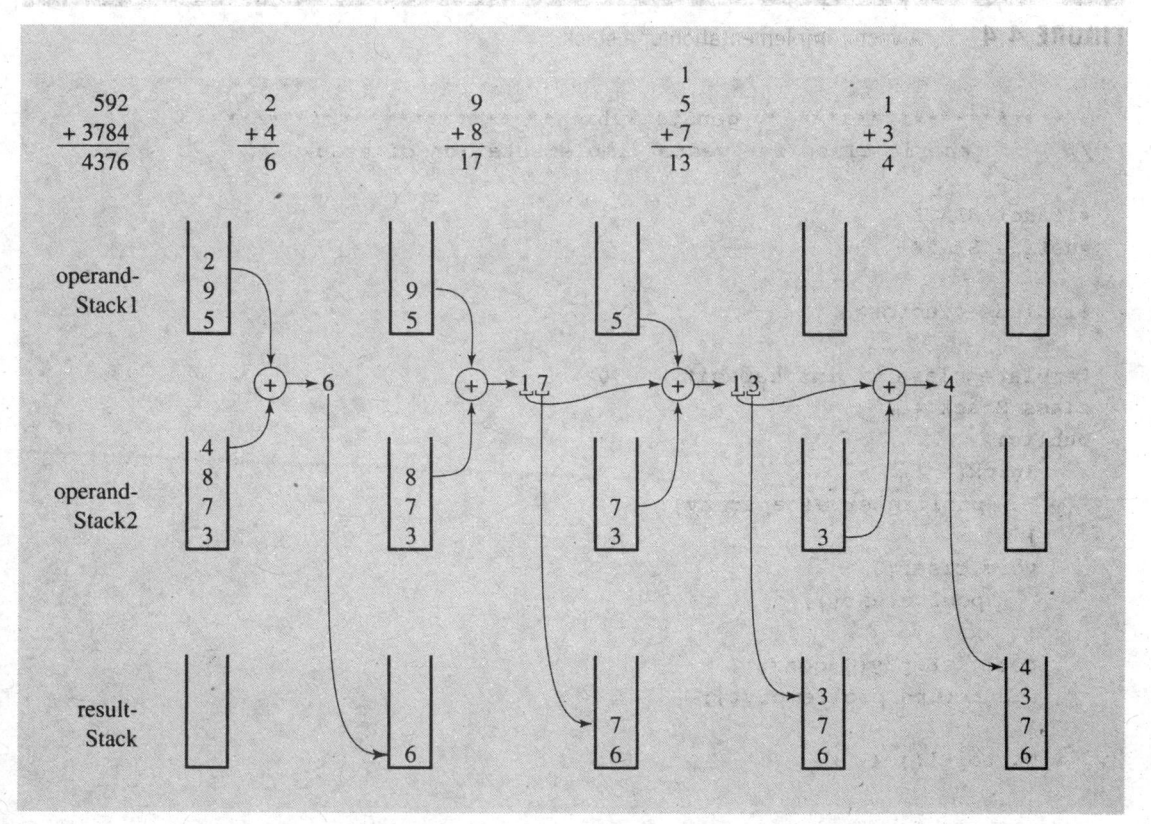

2. Numbers 2 and 4 are popped from the stacks, and the result, 6, is pushed onto `resultStack`.
3. Numbers 9 and 8 are popped from the stacks, and the unit part of their sum, 7, is pushed onto `resultStack`; the tens part of the result, number 1, is retained as a carry in the variable `result` for subsequent addition.
4. Numbers 5 and 7 are popped from the stacks, added to the carry, and the unit part of the result, 3, is pushed onto `resultStack`, and the carry, 1, becomes a value of the variable `result`.
5. One stack is empty, so a number is popped from the nonempty stack, added to carry, and the result is stored on `resultStack`.
6. Both operand stacks are empty, so the numbers from `resultStack` are popped and printed as the final result.

Consider now implementation of our abstract stack data structure. We used push and pop operations as though they were readily available, but they also have to be implemented as functions operating on the stack.

A natural implementation for a stack is a flexible array, that is, a vector. Figure 4.4 contains a generic stack class definition that can be used to store any type of objects. Also, a linked list can be used for implementation of a stack (Figure 4.5).

FIGURE 4.4 A vector implementation of a stack.

```
//******************** genStack.h ************************
//      generic class for vector implementation of stack

#ifndef STACK
#define STACK

#include <vector>

template<class T, int capacity = 30>
class Stack {
public:
    Stack() {
        pool.reserve(capacity);
    }
    void clear() {
        pool.clear();
    }
    bool isEmpty() const {
        return pool.empty();
    }
    T& topEl() {
```

FIGURE 4.4 *(continued)*

```
            return pool.back();
    }
    T pop() {
        T el = pool.back();
        pool.pop_back();
        return el;
    }
    void push(const T& el) {
        pool.push_back(el);
    }
private:
    vector<T> pool;
};

#endif
```

FIGURE 4.5 Implementing a stack as a linked list.

```
//********************  genListStack.h  *************************
//    generic stack defined as a doubly linked list

#ifndef LL_STACK
#define LL_STACK

#include <list>

template<class T>
class LLStack {
public:
    LLStack() {
    }
    void clear() {
        lst.clear();
    }
    bool isEmpty() const {
        return lst.empty();
```

Continues

FIGURE 4.5 (continued)

```
    }
    T& topEl() {
        return lst.back();
    }
    T pop() {
        T el = lst.back();
        lst.pop_back();
        return el;
    }
    void push(const T& el) {
        lst.push_back(el);
    }
private:
    list<T> lst;
};

#endif
```

Figure 4.6 shows the same sequence of push and pop operations as Figure 4.1 with the changes that take place in the stack implemented as a vector (Figure 4.6b) and as a linked list (Figure 4.6c).

The linked list implementation matches the abstract stack more closely in that it includes only the elements that are on the stack because the number of nodes in the list is the same as the number of stack elements. In the vector implementation, the capacity of the stack can often surpass its size.

The vector implementation, like the linked list implementation, does not force the programmer to make a commitment at the beginning of the program concerning the size of the stack. If the size can be reasonably assessed in advance, then the predicted size can be used as a parameter for the stack constructor to create in advance a vector of the specified capacity. In this way, an overhead is avoided to copy the vector elements to a new larger location when pushing a new element to the stack for which size equals capacity.

It is easy to see that in the vector and linked list implementations, popping and pushing are executed in constant time $O(1)$. However, in the vector implementation, pushing an element onto a full stack requires allocating more memory and copies the elements from the existing vector to a new vector. Therefore, in the worst case, pushing takes $O(n)$ time to finish.

FIGURE 4.6 A series of operations executed on an abstract stack (a) and the stack implemented with a vector (b) and with a linked list (c).

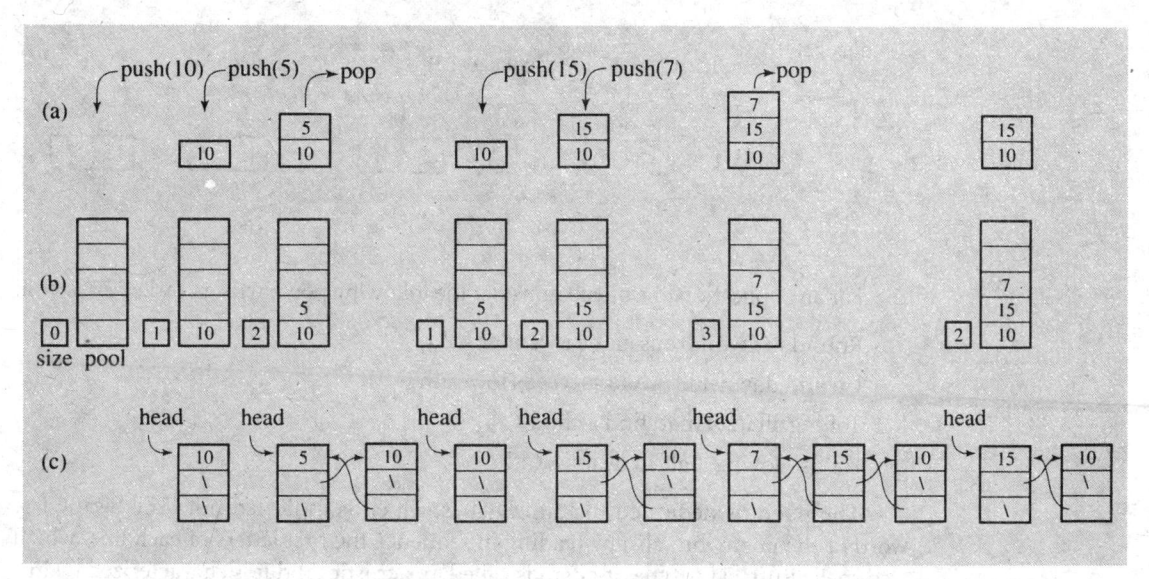

4.2 QUEUES

A *queue* is simply a waiting line that grows by adding elements to its end and shrinks by taking elements from its front. Unlike a stack, a queue is a structure in which both ends are used: one for adding new elements and one for removing them. Therefore, the last element has to wait until all elements preceding it on the queue are removed. A queue is an *FIFO* structure: first in/first out.

Queue operations are similar to stack operations. The following operations are needed to properly manage a queue:

- *clear()*—Clear the queue.
- *isEmpty()*—Check to see if the queue is empty.
- *enqueue(el)*—Put the element *el* at the end of the queue.
- *dequeue()*—Take the first element from the queue.
- *firstEl()*—Return the first element in the queue without removing it.

A series of enqueue and dequeue operations is shown in Figure 4.7. This time—unlike for stacks—the changes have to be monitored both at the beginning of the queue and at the end. The elements are enqueued on one end and dequeued from the other. For example, after enqueuing 10 and then 5, the dequeue operation removes 10 from the queue (Figure 4.7).

FIGURE 4.7 A series of operations executed on a queue.

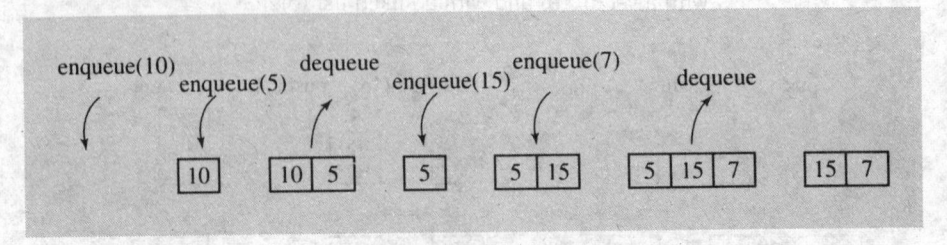

For an application of a queue, consider the following poem written by Lewis Carroll:

Round the wondrous globe I wander wild,

Up and down-hill—Age succeeds to youth—

Toiling all in vain to find a child

Half so loving, half so dear as Ruth.

The poem is dedicated to Ruth Dymes, which is indicated not only by the last word of the poem, but also by reading in sequence the first letters of each line, which also spells Ruth. This type of poem is called an acrostic, and it is characterized by initial letters that form a word or phrase when taken in order. To see whether a poem is an acrostic, we devise a simple algorithm that reads a poem, echoprints it, retrieves and stores the first letter from each line on a queue, and after the poem is processed, all the stored first letters are printed in order. Here is an algorithm:

```
acrosticIndicator()
    while not finished
        read a line of poem;
        enqueue the first letter of the line;
        output the line;
    while queue is not empty
        dequeue and print a letter;
```

There is a more significant example to follow, but first consider the problem of implementation.

One possible queue implementation is an array, although this may not be the best choice. Elements are added to the end of the queue, but they may be removed from its beginning, thereby releasing array cells. These cells should not be wasted. Therefore, they are utilized to enqueue new elements, whereby the end of the queue may occur at the beginning of the array. This situation is better pictured as a circular array, as Figure 4.8c illustrates. The queue is full if the first element immediately precedes in the counterclockwise direction the last element. However, because a circular array is implemented with a "normal" array, the queue is full if either the first element is in the first cell and the last element is in the last cell (Figure 4.8a) or if the first element is right after the last (Figure 4.8b). Similarly, *enqueue()* and *dequeue()* have to consider the possibility of wrapping around the array when adding or removing elements. For

FIGURE 4.8 (a–b) Two possible configurations in an array implementation of a queue when the queue is full. (c) The same queue viewed as a circular array. (f) Enqueuing number 6 to a queue storing 2, 4, and 8. (d–e) The same queue seen as a one-dimensional array with the last element (d) at the end of the array and (e) in the middle.

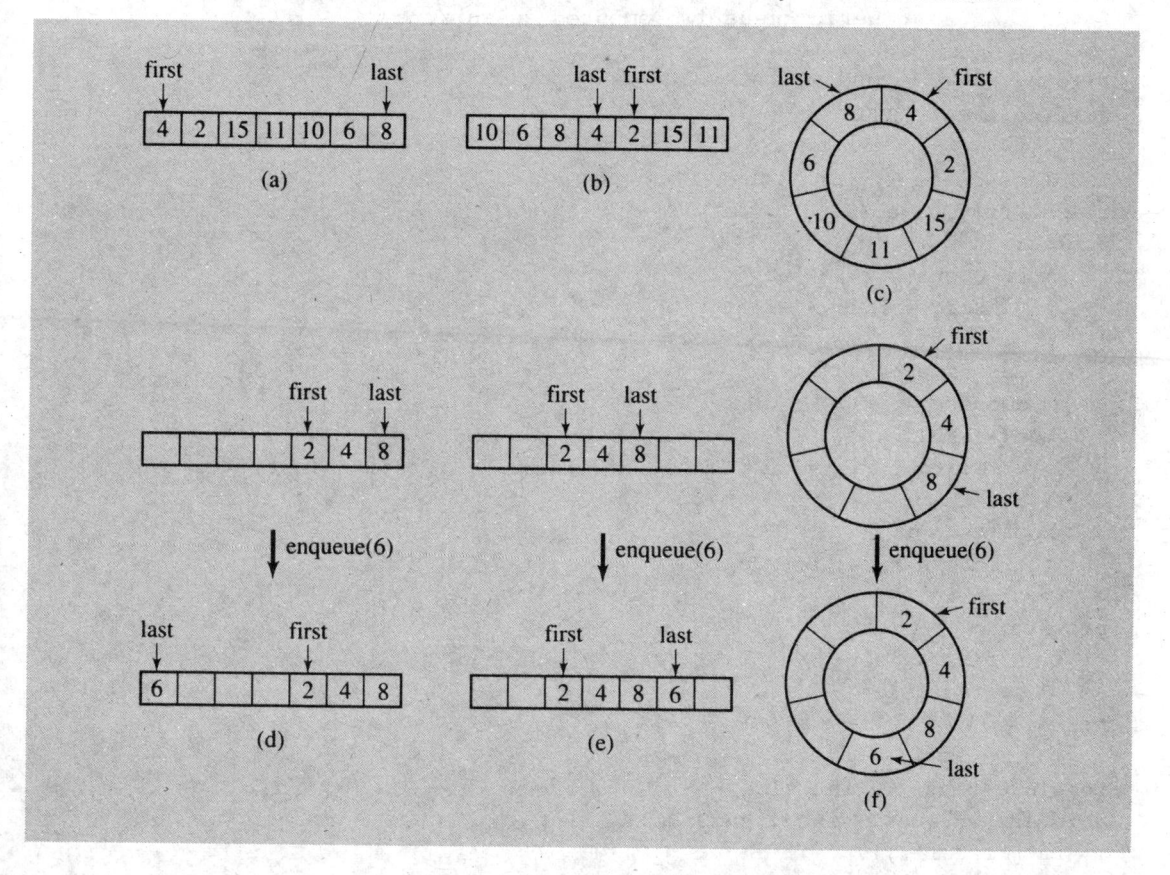

example, *enqueue()* can be viewed as operating on a circular array (Figure 4.8c), but in reality, it is operating on a one-dimensional array. Therefore, if the last element is in the last cell and if any cells are available at the beginning of the array, a new element is placed there (Figure 4.8d). If the last element is in any other position, then the new element is put after the last, space permitting (Figure 4.8e). These two situations must be distinguished when implementing a queue viewed as a circular array (Figure 4.8f).

Figure 4.9 contains possible implementations of member functions that operate on queues.

A more natural queue implementation is a doubly linked list, as offered in the previous chapter and also in STL's `list` (Figure 4.10).

In both suggested implementations enqueuing and dequeuing can be executed in constant time $O(1)$, provided a doubly linked list is used in the list implementation. In the singly linked list implementation, dequeuing requires $O(n)$ operations

FIGURE 4.9 Array implementation of a queue.

```
//******************* genArrayQueue.h *********************
//            generic queue implemented as an array

#ifndef ARRAY_QUEUE
#define ARRAY_QUEUE

template<class T, int size = 100>
class ArrayQueue {
public:
    ArrayQueue() {
        first = last = -1;
    }
    void enqueue(T);
    T dequeue();
    bool isFull()  {
        return first == 0 && last == size-1 || first == last + 1;
    }
    bool isEmpty() {
        return first == -1;
    }
private:
    int first, last;
    T storage[size];
};

template<class T, int size>
void ArrayQueue<T,size>::enqueue(T el) {
   if (!isFull())
        if (last == size-1 || last == -1) {
            storage[0] = el;
            last = 0;
            if (first == -1)
                first = 0;
        }
        else storage[++last] = el;
   else cout << "Full queue.\n";
}

template<class T, int size>
T ArrayQueue<T,size>::dequeue() {
```

FIGURE 4.9 (continued)

```
    T tmp;
    tmp = storage[first];
    if (first == last)
        last = first = -1;
    else if (first == size-1)
        first = 0;
    else first++;
    return tmp;
}

#endif
```

FIGURE 4.10   Linked list implementation of a queue.

```
//********************   genQueue.h   *************************
//     generic queue implemented with doubly linked list

#ifndef DLL_QUEUE
#define DLL_QUEUE

#include <list>

template<class T>
class Queue {
public:
    Queue() {
    }
    void clear() {
        lst.clear();
    }
    bool isEmpty() const {
        return lst.empty();
    }
    T& front() {
        return lst.front();
    }
```

Continues

FIGURE 4.10 (continued)

```
    T dequeue() {
        T el = lst.front();
        lst.pop_front();
        return el;
    }
    void enqueue(const T& el) {
        lst.push_back(el);
    }
private:
    list<T> lst;
};

#endif
```

primarily to scan the list and stop at the next to last node (see the discussion of `deleteFromTail()` in Section 3.1.2).

Figure 4.11 shows the same sequence of enqueue and dequeue operations as Figure 4.7, and indicates the changes in the queue implemented as an array (Figure 4.11b) and as a linked list (Figure 4.11c). The linked list keeps only the numbers that the logic of the queue operations indicated by Figure 4.11a requires. The array includes all the numbers until it fills up, after which new numbers are included starting from the beginning of the array.

Queues are frequently used in simulations to the extent that a well-developed and mathematically sophisticated theory of queues exists, called *queuing theory*, in which various scenarios are analyzed and models are built that use queues. In queuing processes there are a number of customers coming to servers to receive service. The throughput of the server may be limited. Therefore, customers have to wait in queues before they are served, and they spend some amount of time while they are being served. By customers, we mean not only people, but also objects. For example, parts on an assembly line in the process of being assembled into a machine, trucks waiting for service at a weighing station on an interstate, or barges waiting for a sluice to be opened so they can pass through a channel also wait in queues. The most familiar examples are lines in stores, post offices, or banks. The types of problems posed in simulations are: How many servers are needed to avoid long queues? How large must the waiting space be to put the entire queue in it? Is it cheaper to increase this space or to open one more server?

As an example, consider Bank One which, over a period of three months, recorded the number of customers coming to the bank and the amount of time needed to serve them. The table in Figure 4.12a shows the number of customers who arrived during one-minute intervals throughout the day. For 15% of such intervals, no customer arrived, for 20%, only one arrived, and so on. Six clerks were employed, no lines were ever observed, and the bank management wanted to know whether six clerks were too

FIGURE 4.11 A series of operations executed on an abstract queue (a) and the stack implemented with an array (b) and with a linked list (c).

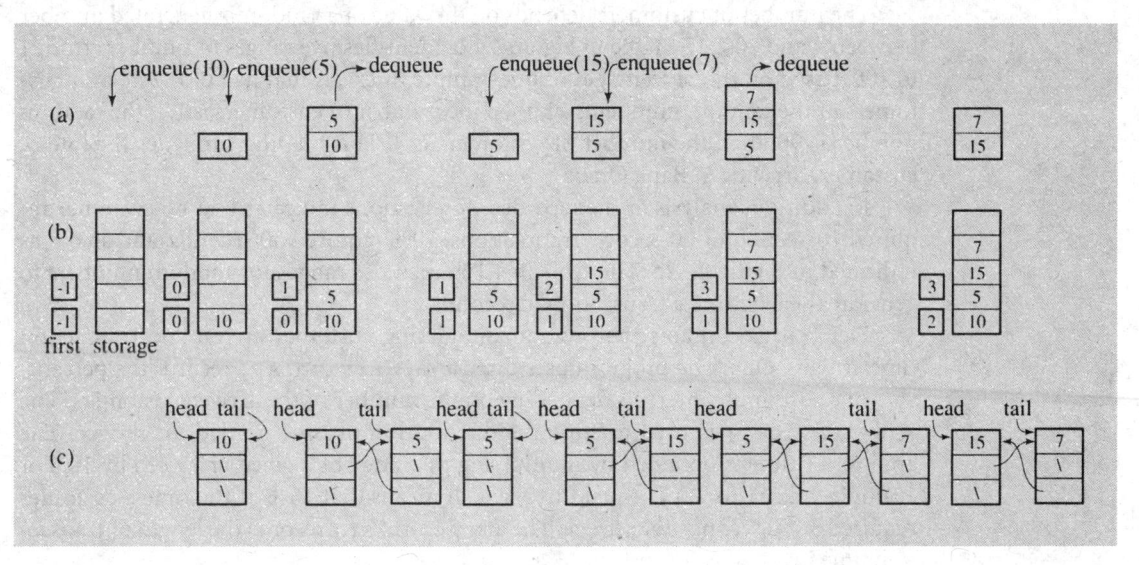

FIGURE 4.12 Bank One example: (a) data for number of arrived customers per one-minute interval and (b) transaction time in seconds per customer.

Number of Customers Per Minute	Percentage of One-Minute Intervals	Range
0	15	1–15
1	20	16–35
2	25	36–60
3	10	61–70
4	30	71–100

(a)

Amount of Time Needed for Service in Seconds	Percentage of Customers	Range
0	0	—
10	0	—
20	0	—
30	10	1–10
40	5	11–15
50	10	16–25
60	10	26–35
70	0	—
80	15	36–50
90	25	51–75
100	10	76–85
110	15	86–100

(b)

many. Would five suffice? Four? Maybe even three? Can lines be expected at any time? To answer these questions, a simulation program was written that applied the recorded data and checked different scenarios.

The number of customers depends on the value of a randomly generated number between 1 and 100. The table in Figure 4.12a identifies five ranges of numbers from 1 to 100, based on the percentages of one-minute intervals that had 0, 1, 2, 3, or 4 customers. If the random number is 21, then the number of customers is 1; if the random number is 90, then the number of customers is 4. This method simulates the rate of customers arriving at Bank One.

In addition, analysis of the recorded observations indicates that no customer required 10-second or 20-second transactions, 10% required 30 seconds, and so on, as indicated in Figure 4.12b. The table in 4.12b includes ranges for random numbers to generate the length of a transaction in seconds.

Figure 4.13 contains the program simulating customer arrival and transaction time at Bank One. The program uses three arrays. `arrivals[]` records the percentages of one-minute intervals depending on the number of the arrived customers. The array `service[]` is used to store the distribution of time needed for service. The amount of time is obtained by multiplying the index of a given array cell by 10. For example, `service[3]` is equal to 10, which means that 10% of the time a customer required 3 · 10 seconds for service. The array `clerks[]` records the length of transaction time in seconds.

FIGURE 4.13   Bank One example: implementation code.

```
#include <iostream>
#include <cstdlib>

using namespace std;

#include "genQueue.h"
int option(int percents[]) {
    register int i = 0, choice = rand()%100+1, perc;
    for (perc = percents[0]; perc < choice; perc += percents[i+1], i++);
    return i;
}

int main() {
    int arrivals[] = {15,20,25,10,30};
    int service[] = {0,0,0,10,5,10,10,0,15,25,10,15};
    int clerks[] = {0,0,0,0}, numOfClerks = sizeof(clerks)/sizeof(int);
    int customers, t, i, numOfMinutes = 100, x;
    double maxWait = 0.0, currWait = 0.0, thereIsLine = 0.0;
    Queue<int> simulQ;
    cout.precision(2);
    for (t = 1; t <= numOfMinutes; t++) {
```

FIGURE 4.13   (continued)

```
    cout << " t = " << t;
    for (i = 0; i < numOfClerks; i++)// after each minute subtract
        if (clerks[i] < 60)            // at most 60 seconds from time
            clerks[i] = 0;             // left to service the current
        else clerks[i] -= 60;          // customer by clerk i;
    customers = option(arrivals);
    for (i = 0; i < customers; i++) {// enqueue all new customers
        x = option(service)*10;        // (or rather service time
        simulQ.enqueue(x);             // they require);
        currWait += x;
    }
    // dequeue customers when clerks are available:
    for (i = 0; i < numOfClerks && !simulQ.isEmpty(); )
        if (clerks[i] < 60)        {
            x = simulQ.dequeue();      // assign more than one customer
            clerks[i] += x;            // to a clerk if service time
            currWait  -= x;            // is still below 60 sec;
        }
        else i++;
    if (!simulQ.isEmpty()) {
        thereIsLine++;
        cout << " wait = " << currWait/60.0;
        if (maxWait < currWait)
            maxWait = currWait;
    }
    else cout << " wait = 0;";
    }
    cout << "\nFor " << numOfClerks << " clerks, there was a line "
         << thereIsLine/numOfMinutes*100.0 << "% of the time;\n"
         << "maximum wait time was " << maxWait/60.0 << " min.";
    return 0;
}
```

For each minute (represented by the variable `t`), the number of arriving customers is randomly chosen, and for each customer, the transaction time is also randomly determined. The function `option()` generates a random number, finds the range into which it falls, and then outputs the position, which is either the number of customers or a tenth the number of seconds.

Executions of this program indicate that six and five clerks are too many. With four clerks, service is performed smoothly; 25% of the time there is a short line of waiting customers. However, three clerks are always busy and there is always a long line of customers waiting. Bank management would certainly decide to employ four clerks.

4.3 PRIORITY QUEUES

In many situations, simple queues are inadequate, because first in/first out scheduling has to be overruled using some priority criteria. In a post office example, a handicapped person may have priority over others. Therefore, when a clerk is available, a handicapped person is served instead of someone from the front of the queue. On roads with tollbooths, some vehicles may be put through immediately, even without paying (police cars, ambulances, fire engines, and the like). In a sequence of processes, process P_2 may need to be executed before process P_1 for the proper functioning of a system, even though P_1 was put on the queue of waiting processes before P_2. In situations like these, a modified queue, or *priority queue,* is needed. In priority queues, elements are dequeued according to their priority and their current queue position.

The problem with a priority queue is in finding an efficient implementation that allows relatively fast enqueuing and dequeuing. Because elements may arrive randomly to the queue, there is no guarantee that the front elements will be the most likely to be dequeued and that the elements put at the end will be the last candidates for dequeuing. The situation is complicated because a wide spectrum of possible priority criteria can be used in different cases such as frequency of use, birthday, salary, position, status, and others. It can also be the time of scheduled execution on the queue of processes, which explains the convention used in priority queue discussions in which higher priorities are associated with lower numbers indicating priority.

Priority queues can be represented by two variations of linked lists. In one type of linked list, all elements are entry ordered; and in another, order is maintained by putting a new element in its proper position according to its priority. In both cases, the total operational times are $O(n)$ because, for an unordered list, adding an element is immediate but searching is $O(n)$, and in a sorted list, taking an element is immediate but adding an element is $O(n)$.

Another queue representation uses a short ordered list and an unordered list, and a threshold priority is determined (Blackstone et al. 1981). The number of elements in the sorted list depends on a threshold priority. This means that in some cases this list can be empty and the threshold may change dynamically to have some elements in this list. Another way is always having the same number of elements in the sorted list; the number $\sqrt{n}$ is a good candidate. Enqueuing takes on the average $O(\sqrt{n})$ time and dequeuing is immediate.

Another implementation of queues was proposed by J. O. Hendriksen (1977, 1983). It uses a simple linked list with an additional array of pointers to this list to find a range of elements in the list in which a newly arrived element should be included.

Experiments by Douglas W. Jones (1986) indicate that a linked list implementation, in spite of its $O(n)$ efficiency, is best for 10 elements or less. The efficiency of the two-list version depends greatly on the distribution of priorities, and it may be excellent or as poor as that of the simple list implementation for large numbers of elements. Hendriksen's implementation, with its $O(\sqrt{n})$ complexity, operates consistently well with queues of any size.

4.4 STACKS IN THE STANDARD TEMPLATE LIBRARY

A generic stack class is implemented in the STL as a container adaptor: It uses a container to make it behave in a specified way. The stack container is not created anew; it is an adaptation of an already existing container. By default, `deque` is the underlying container, but the user can also choose either `list` or `vector` with the following declarations:

```
stack<int> stack1;                    // deque by default
stack<int,vector<int> > stack2;       // vector
stack<int,list<int> > stack3;         // list
```

Member functions in the container `stack` are listed in Figure 4.14. Note that the return type of `pop()` is void; that is, `pop()` does not return a popped off element. To have access to the top element, the member function `top()` has to be used. Therefore, the popping operation discussed in this chapter has to be implemented with a call to `top()` followed by the call to `pop()`. Because popping operations in user programs are intended for capturing the popped off element most of the time and not only for removing it, the desired popping operation is really a sequence of the two member functions from the container `stack`. To contract them to one operation, a new class can be created that inherits all operations from `stack` and redefines `pop()`. This is a solution used in the case study at the end of the chapter.

FIGURE 4.14 A list of `stack` member functions.

Member Function	Operation
`bool empty() const`	Return `true` if the stack includes no element and `false` otherwise.
`void pop()`	Remove the top element of the stack.
`void push(const T& el)`	Insert `el` at the top of the stack.
`size_type size() const`	Return the number of elements on the stack.
`stack()`	Create an empty stack.
`T& top()`	Return the top element on the stack.
`const T& top() const`	Return the top element on the stack.

4.5 QUEUES IN THE STANDARD TEMPLATE LIBRARY

The queue container is implemented by default as the container `deque`, and the user may opt for using the container `list` instead. An attempt to use the container `vector` results in a compilation error because `pop()` is implemented as a call to `pop_front()`, which is assumed to be a member function of the underlying container, and `vector` does not include such a member function. For the list of queue's member functions, see Figure 4.15. A short program in Figure 4.16 illustrates the operations of the member functions. Note that the dequeueing operation discussed in this chapter is implemented by `front()` followed by `pop()`, and the enqueueing operation is implemented with the function `push()`.

FIGURE 4.15   A list of `queue` member functions.

Member Function	Operation
`T& back()`	Return the last element in the queue.
`const T& back() const`	Return the last element in the queue.
`bool empty() const`	Return `true` if the queue includes no element and `false` otherwise.
`T& front()`	Return the first element in the queue.
`const T& front() const`	Return the first element in the queue.
`void pop()`	Remove the first element in the queue.
`void push(const T& el)`	Insert `el` at the end of the queue.
`queue()`	Create an empty queue.
`size_type size() const`	Return the number of elements in the queue.

FIGURE 4.16   An example application of `queue`'s member functions.

```
#include <iostream>
#include <queue>
#include <list>

using namespace std;

int main() {
    queue<int> q1;
    queue<int,list<int> > q2; //leave space between angle brackets > >
    q1.push(1); q1.push(2); q1.push(3);
    q2.push(4); q2.push(5); q2.push(6);
    q1.push(q2.back());
    while (!q1.empty()) {
        cout << q1.front() << ' ';      // 1 2 3 6
        q1.pop();
    }
    while (!q2.empty()) {
        cout << q2.front() << ' ';      // 4 5 6
        q2.pop();
    }
    return 0;
}
```

4.6 Priority Queues in the Standard Template Library

The `priority_queue` container (Figure 4.17) is implemented with the container `vector` by default, and the user may choose the container `deque`. The `priority_queue` container maintains an order in the queue by keeping an element with the highest priority in front of the queue. To accomplish this, a two-argument Boolean function is used by the insertion operation `push()`, which reorders the elements in the queue to satisfy this requirement. The function can be supplied by the user; otherwise, the operation < is used and the element with the highest value is considered to have the highest priority. If the highest priority is determined by the smallest value, then the function object `greater` needs to be used to indicate that `push()` should apply the operator > rather than < in making its decisions when inserting new elements to the priority queue. An example is shown in Figure 4.18. The priority queue `pq1` is defined as a vector-based queue that uses the operation < to determine the priority of integers in the queue. The second queue, `pq2`, uses the operation > during insertion. Finally, the queue `pq3` is of the same type as `pq1`, but it is also initialized with the numbers from the array `a`. The three `while` loops show in which order the elements from the three queues are dequeued.

FIGURE 4.17 A list of `priority_queue` member functions.

Member Function	Operation
`bool empty() const`	Return `true` if the queue includes no element and `false` otherwise.
`void pop()`	Remove an element in the queue with the highest priority.
`void push(const T& el)`	Insert `el` in a proper location on the priority queue.
`priority_queue(comp f())`	Create an empty priority queue that uses a two-argument Boolean function `f` to order elements on the queue.
`priority_queue(iterator first, iterator last, comp f())`	Create a priority queue that uses a two-argument Boolean Function `f` to order elements on the queue; initialize the queue with elements from the range indicated by iterators `first` and `last`.
`size_type size() const`	Return the number of elements in the priority queue.
`T& top()`	Return the element in the priority queue with the highest priority.
`const T& top() const`	Return the element in the priority queue with the highest priority.

FIGURE 4.18 A program that uses member functions of the container `priority_queue`.

```cpp
#include <iostream>
#include <queue>
#include <functional>

using namespace std;

int main() {
    priority_queue<int> pq1; // plus vector<int> and less<int>
    priority_queue<int,vector<int>,greater<int> > pq2;
    pq1.push(3); pq1.push(1); pq1.push(2);
    pq2.push(3); pq2.push(1); pq2.push(2);
    int a[] = {4,6,5};
    priority_queue<int> pq3(a,a+3);
    while (!pq1.empty()) {
        cout << pq1.top() << ' ';    // 3 2 1
        pq1.pop();
    }
    while (!pq2.empty()) {
        cout << pq2.top() << ' ';    // 1 2 3
        pq2.pop();
    }
    while (!pq3.empty()) {
        cout << pq3.top() << ' ';    // 6 5 4
        pq3.pop();
    }
    return 0;
}
```

It is more interesting to see an application of the user-defined objects. Consider the class Person defined in Section 1.8:

```cpp
class Person {
public:
    . . . . .
    bool operator<(const Person& p) const {
        return strcmp(name,p.name) < 0;
    }
    bool operator>(const Person& p) const {
        return !(*this == p) && !(*this < p);
    }
```

```
    private:
        char *name;
        int age;
};
```

Our intention now is to create three priority queues. In the first two queues, the priority is determined by lexicographical order, but in pqName1 it is the descending order and in pqName2 the ascending order. To that end, pqName1 uses the overloaded operator <. The queue pqName2 uses the overloaded operator >, as made known by defining with the function object greater<Person>:

```
Person p[] = {Person("Gregg",25),Person("Ann",30),Person("Bill",20)};
priority_queue<Person> pqName1(p,p+3);
priority_queue<Person,vector<Person>,greater<Person> > pqName2(p,p+3);
```

In these two declarations, the two priority queues are also initialized with objects from the array p.

In Section 1.8, there is also a Boolean function lesserAge used to determine the order of Person objects by age, not by name. How can we create a priority queue in which the highest priority is determined by age? One way to accomplish this is to define a function object,

```
class lesserAge {
public:
bool operator()(const Person& p1, const Person& p2) const {
    return p1.age < p2.age;
}
};
```

and then declare a new priority queue

```
priority_queue<Person,vector<Person>,lesserAge> pqAge(p,p+3);
```

initialized with the same objects as pqName1 and pqName2. Printing elements from the three queues indicates the different priorities of the objects in different queues:

```
pqName1:    (Gregg,25) (Bill,20) (Ann,30)
pqName2:    (Ann,30) (Bill,20) (Gregg,25)
pqAge:      (Ann,30) (Gregg,25) (Bill,20)
```

4.7 CASE STUDY: EXITING A MAZE

Consider the problem of a trapped mouse that tries to find its way to an exit in a maze (Figure 4.19a). The mouse hopes to escape from the maze by systematically trying all the routes. If it reaches a dead end, it retraces its steps to the last position and begins at least one more untried path. For each position, the mouse can go in one of four directions: right, left, down, up. Regardless of how close it is to the exit, it always tries the open paths in this order, which may lead to some unnecessary detours. By retaining information that allows for resuming the search after a dead end is reached, the mouse uses a method called *backtracking*. This method is discussed further in the next chapter.

FIGURE 4.19 (a) A mouse in a maze; (b) two-dimensional character array representing the situation.

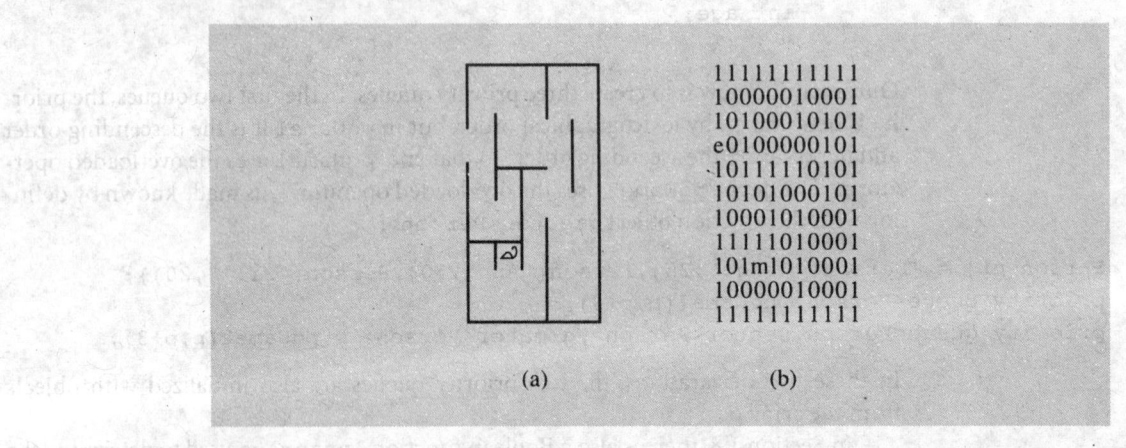

```
11111111111
10000010001
10100010101
e0100000101
10111110101
10101000101
10001010001
11111010001
101m1010001
10000010001
11111111111
```

(a)   (b)

The maze is implemented as a two-dimensional character array in which passages are marked with 0s, walls by 1s, exit position by the letter e, and the initial position of the mouse by the letter m (Figure 4.19b). In this program, the maze problem is slightly generalized by allowing the exit to be in any position of the maze (picture the exit position as having an elevator that takes the mouse out of the trap) and allowing passages to be on the borderline. To protect itself from falling off the array by trying to continue its path when an open cell is reached on one of the borderlines, the mouse also has to constantly check whether it is in such a borderline position or not. To avoid it, the program automatically puts a frame of 1s around the maze entered by the user.

The program uses two stacks: one to initialize the maze and another to implement backtracking.

The user enters a maze one line at a time. The maze entered by the user can have any number of rows and any number of columns. The only assumption the program makes is that all rows are of the same length and that it uses only these characters: any number of 1s, any number of 0s, one e, and one m. The rows are pushed on the stack `mazeRows` in the order they are entered after attaching one 1 at the beginning and one 1 at the end. After all rows are entered, the size of the array store can be determined, and then the rows from the stack are transferred to the array.

A second stack, `mazeStack`, is used in the process of escaping the maze. To remember untried paths for subsequent tries, the positions of the untried neighbors of the current position (if any) are stored on a stack and always in the same order, first upper neighbor, then lower, then left, and finally right. After stacking the open avenues on the stack, the mouse takes the topmost position and tries to follow it by first storing untried neighbors and then trying the topmost position, and so forth, until it reaches the exit or exhausts all possibilities and finds itself trapped. To avoid falling into an infinite loop of trying paths that have already been investigated, each visited position of the maze is marked with a period.

Here is a pseudocode of an algorithm for escaping a maze:

Section 4.7 Case Study: Exiting a Maze

```
exitMaze()
    initialize stack, exitCell, entryCell, currentCell = entryCell;
    while currentCell is not exitCell
        mark currentCell as visited;
        push onto the stack the unvisited neighbors of currentCell;
        if stack is empty
            failure;
        else pop off a cell from the stack and make it currentCell;
    success;
```

The stack stores coordinates of positions of cells. This could be done, for instance, by using two integer stacks for *x* and *y* coordinates. Another possibility is to use one integer stack with both coordinates stored in one integer variable with the help of the shifting operation. In the program in Figure 4.21, a class `MazeCell` is used with two data fields, x and y, so that one `mazeStack` is used for storing `MazeCell` objects.

Consider an example shown in Figure 4.20. The program actually prints out the maze after each step made by the mouse.

0. After the user enters the maze

```
1100
000e
00m1
```

the maze is immediately surrounded with a frame of 1s

```
111111
111001
1000e1
100m11
111111
```

entryCell and currentCell are initialized to (3 3) and exitCell to (2 4) (Figure 4.20a).

FIGURE 4.20   An example of processing a maze.

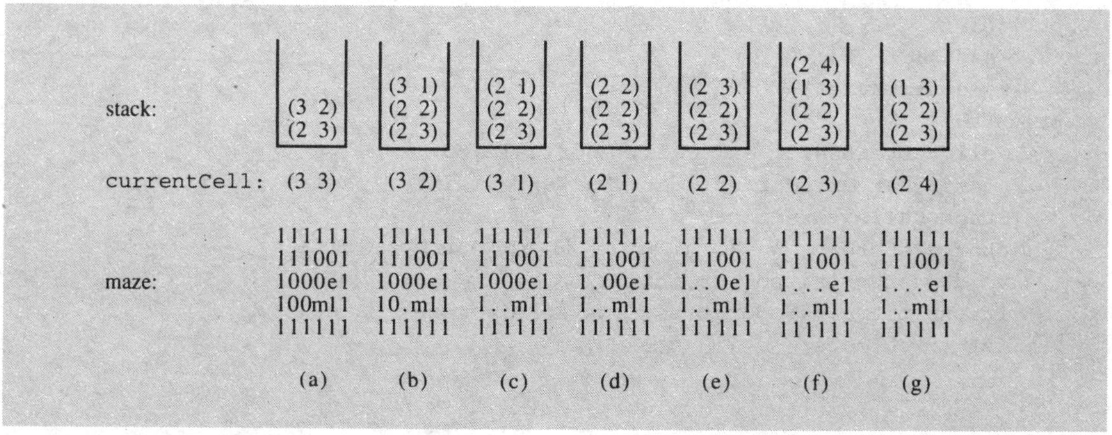

							(2 4)	
		(3 1)	(2 1)	(2 2)	(2 3)	(1 3)	(1 3)
stack:	(3 2)	(2 2)	(2 2)	(2 2)	(2 2)	(2 2)	(2 2)
	(2 3)	(2 3)	(2 3)	(2 3)	(2 3)	(2 3)	(2 3)
currentCell:	(3 3)	(3 2)	(3 1)	(2 1)	(2 2)	(2 3)	(2 4)
maze:	111111 111001 1000e1 100m11 111111	111111 111001 1000e1 10.m11 111111	111111 111001 1000e1 1..m11 111111	111111 111001 1.00e1 1..m11 111111	111111 111001 1..0e1 1..m11 111111	111111 111001 1...e1 1..m11 111111	111111 111001 1...e1 1..m11 111111
	(a)	(b)	(c)	(d)	(e)	(f)	(g)

FIGURE 4.21 Listing the program for maze processing.

```cpp
#include <iostream>
#include <string>
#include <stack>

using namespace std;

template<class T>
class Stack : public stack<T> {
public:
    T pop() {
        T tmp = top();
        stack<T>::pop();
        return tmp;
    }
};

class Cell {
public:
    Cell(int i = 0, int j = 0) {
        x = i; y = j;
    }
    bool operator== (const Cell& c) const {
        return x == c.x && y == c.y;
    }
private:
    int x, y;
    friend class Maze;
};

class Maze {
public:
    Maze();
    void exitMaze();
private:
    Cell currentCell, exitCell, entryCell;
    const char exitMarker, entryMarker, visited, passage, wall;
    Stack<Cell> mazeStack;
    char **store;           // array of strings;
    void pushUnvisited(int,int);
    friend ostream& operator<< (ostream&, const Maze&);
    int rows, cols;
};
```

FIGURE 4.21 (continued)

```cpp
Maze::Maze() : exitMarker('e'), entryMarker('m'), visited('.'),
               passage('0'), wall('1') {
    Stack<char*> mazeRows;
    char str[80], *s;
    int col, row = 0;
    cout << "Enter a rectangular maze using the following "
         << "characters:\nm - entry\ne - exit\n1 - wall\n0 - passage\n"
         << "Enter one line at at time; end with Ctrl-z:\n";
    while (cin >> str) {
        row++;
        cols = strlen(str);
        s = new char[cols+3];     // two more cells for borderline
                                  // columns;
        mazeRows.push(s);
        strcpy(s+1,str);
        s[0] = s[cols+1] = wall; // fill the borderline cells with 1s;
        s[cols+2] = '\0';
        if (strchr(s,exitMarker) != 0) {
            exitCell.x = row;
            exitCell.y = strchr(s,exitMarker) - s;
        }
        if (strchr(s,entryMarker) != 0) {
            entryCell.x = row;
            entryCell.y = strchr(s,entryMarker) - s;
        }
    }
    rows = row;
    store = new char*[rows+2];         // create a 1D array of pointers;
    store[0] = new char[cols+3];       // a borderline row;
    for ( ; !mazeRows.empty(); row--) {
        store[row] = mazeRows.pop();
    }
    store[rows+1] = new char[cols+3]; // another borderline row;
    store[0][cols+2] = store[rows+1][cols+2] = '\0';
    for (col = 0; col <= cols+1; col++) {
        store[0][col] = wall;          // fill the borderline rows with 1s;
        store[rows+1][col] = wall;
    }
}
```

Continues

FIGURE 4.21   (continued)

```cpp
void Maze::pushUnvisited(int row, int col) {
    if (store[row][col] == passage || store[row][col] == exitMarker) {
        mazeStack.push(Cell(row,col));
    }
}
void Maze::exitMaze() {
    int row, col;
    currentCell = entryCell;
    while (!(currentCell == exitCell)) {
        row = currentCell.x;
        col = currentCell.y;
        cout << *this;         // print a snapshot;
        if (!(currentCell == entryCell))
            store[row][col] = visited;
        pushUnvisited(row-1,col);
        pushUnvisited(row+1,col);
        pushUnvisited(row,col-1);
        pushUnvisited(row,col+1);
        if (mazeStack.empty()) {
            cout << *this;
            cout << "Failure\n";
            return;
        }
        else currentCell = mazeStack.pop();
    }
    cout << *this;
    cout << "Success\n";
}
ostream& operator<< (ostream& out, const Maze& maze) {
    for (int row = 0; row <= maze.rows+1; row++)
        out << maze.store[row] << endl;
    out << endl;
    return out;
}
int main() {
    Maze().exitMaze();
    return 0;
}
```

1. Because `currentCell` is not equal to `exitCell`, all four neighbors of the current cell (3 3) are tested, and only two of them are candidates for processing, namely, (3 2) and (2 3); therefore, they are pushed onto the stack. The stack is checked to see whether it contains any position, and because it is not empty, the topmost position (3 2) becomes current (Figure 4.20b).

2. `currentCell` is still not equal to `exitCell`; therefore, the two viable options accessible from (3 2) are pushed onto the stack, namely, positions (2 2) and (3 1). Note that the position holding the mouse is not included in the stack. After the current position is marked as visited, the situation in the maze is as in Figure 4.20c. Now, the topmost position, (3 1), is popped off the stack, and it becomes the value of `currentCell`. The process continues until the exit is reached, as shown step by step in Figure 4.20d-f.

Note that in step four (Figure 4.20d), the position (2 2) is pushed onto the stack, although it is already there. However, this poses no danger, because when the second instance of this position is popped from the stack, all the paths leading from this position have already been investigated using the first instance of this position on the stack. Note also that the mouse makes a detour, although there is a shorter path from its initial position to the exit.

Figure 4.21 contains code implementing the maze exiting algorithm. Note that the program defines a class `Stack` derived from `stack`. `Stack` inherits from `stack` all the member functions, but it redefines `pop()` so that a call to a new `pop()` results in both removing the top element from the stack and returning it to the caller.

4.8 EXERCISES

1. Reverse the order of elements on stack S
   a. using two additional stacks
   b. using one additional queue
   c. using one additional stack and some additional non-array variables

2. Put the elements on the stack S in ascending order using one additional stack and some additional non-array variables.

3. Transfer elements from stack S_1 to stack S_2 so that the elements from S_2 are in the same order as on S_1
   a. using one additional stack
   b. using no additional stack but only some additional non-array variables

4. Suggest an implementation of a stack to hold elements of two different types, such as structures and float numbers.

5. Using additional non-array variables, order all elements on a queue using also
   a. two additional queues
   b. one additional queue

6. In this chapter, two different implementations were developed for a stack: class Stack and class LLStack. The names of member functions in both classes suggest that the same data structure is meant; however, a tighter connection between these two classes can be established. Define an abstract base class for a stack and derive from it both class Stack and class LLStack.

7. Define a stack in terms of a queue; that is, create a class

    ```
    template <class T>
    class StackQ {
        Queue<T> pool;
        . . . . . . . . . . .
        void push(const T& el) {
            pool.enqueue(el);
        . . . . . . . . . .
    ```

8. Define a queue in terms of a stack.

9. A generic queue class could be defined in terms of a vector:

    ```
    template<class T, int capacity = 30>
    class QueueV {
        . . . . . . . . . .
    private:
        vector<T> pool;
    }
    ```

    Is this a viable solution?

10. Modify the program from the case study to print out the path without dead ends and, possibly, with detours. For example, for an input maze

    ```
    1111111
    1e00001
    1110111
    1000001
    100m001
    1111111
    ```

    the program from the case study outputs the processed maze

    ```
    1111111
    1e....1
    111.111
    1.....1
    1..m..1
    1111111
    Success
    ```

    The modified program should, in addition, generate the path from the exit to the mouse:

    `[1 1] [1 2] [1 3] [2 3] [3 3] [3 4] [3 5] [4 5] [4 4] [4 3]`

    which leaves out two dead ends, `[1 4] [1 5]` and `[3 2] [3 1] [4 1] [4 2]`, but retains a detour, `[3 4] [3 5] [4 5] [4 4]`.

11. Modify the program from the previous exercise so that it prints the maze with the path without dead ends; the path is indicated by dashes and vertical bars to indicate the changes of direction of the path; for the input maze from the previous exercise, the modified program should output

    ```
    1111111
    1e--..1
    111|111
    1..|--1
    1..m-|1
    1111111
    ```

4.9 PROGRAMMING ASSIGNMENTS

1. Write a program that determines whether an input string is a palindrome; that is, whether it can be read the same way forward and backward. At each point, you can read only one character of the input string; do not use an array to first store this string and then analyze it (except, possibly, in a stack implementation). Consider using multiple stacks.

2. Write a program to convert a number from decimal notation to a number expressed in a number system whose base (or radix) is a number between 2 and 9. The conversion is performed by repetitious division by the base to which a number is being converted and then taking the remainders of division in the reverse order. For example, in converting to binary, number 6 requires three such divisions: 6/2 = 3 remainder 0, 3/2 = 1 remainder 1, and finally, 1/2 = 0 remainder 1. The remainders 0, 1, and 1 are put in reverse order so that the binary equivalent of 6 is equal to 110.

   Modify your program so that it can perform a conversion in the case when the base is a number between 11 and 27. Number systems with bases greater than 10 require more symbols. Therefore, use capital letters. For example, a hexadecimal system requires 16 digits: 0, 1, ..., 9, A, B, C, D, E, F. In this system, decimal number 26 is equal to 1A in hexadecimal notation because 26/16 = 1 remainder 10 (that is, A), and 1/16 = 0 remainder 1.

3. Write a program that implements the algorithm `delimiterMatching()` from Section 4.1.

4. Write a program that implements the algorithm `addingLargeNumbers()` from Section 4.1.

5. Write a program to add any number of large integers. The problem can be approached in at least two ways.

   a. First, add two numbers and then repeatedly add the next number with the result of the previous addition.

   b. Create a vector of stacks and then use a generalized version of `addingLargeNumbers()` to all stacks at the same time.

6. Write a program to perform the four basic arithmetic operations, +, −, ·, and /, on very large integers; the result of division should also be an integer. Apply these operations to compute 123^{45}, or the hundredth number in the sequence $1 * 2 + 3, 2 * 3^2 + 4, 3 * 4^3 + 5, \ldots$. Also apply them to compute the Gödel numbers of arithmetical expressions.

   The Gödel numbering function GN first establishes a correspondence between basic elements of language and numbers:

Symbol	Gödel Number GN
=	1
+	2
*	3
−	4
/	5
(	6
)	7
^	8
0	9
S	10
x_i	$11 + 2 * i$
X_i	$12 + 2 * i$

where S is the successor function. Then, for any formula $F = s_1 s_2 \ldots s_n$:

$$GN('s_1 s_2 \ldots s_n') = 2^{GN(s_1)} * 3^{GN(s_2)} * \ldots * p_n^{GN(s_n)}$$

where p_n is the nth prime. For example,

$$GN(1) = GN(S0) = 2^{10} * 3^9$$

and

$$GN('x_1 + x_3 = x_4') = 2^{11+2} * 3^2 * 5^{11+6} * 7^1 * 11^{11+8}$$

In this way, every arithmetic expression can be assigned a unique number. This method has been used by Gödel to prove theorems, known as Gödel's theorems, which are of extreme importance for the foundations of mathematics.

7. Write a program for adding very large floating-point numbers. Extend this program to other arithmetic operations.

BIBLIOGRAPHY

Queues

Sloyer, Clifford, Copes, Wayne, Sacco, William, and Starck, Robert, *Queues: Will This Wait Never End!* Providence, RI: Janson, 1987.

Priority Queues

Blackstone, John H., Hogg, Gary L., and Phillips, Don T., "A Two-List Synchronization Procedure for Discrete Event Simulation," *Communications of the ACM* 24 (1981), 825–829.

Hendriksen, James O., "An Improved Events List Algorithm," *Proceedings of the 1977 Winter Simulation Conference,* Piscataway, NJ: IEEE, 1977, 547–557.

Hendriksen, James O., "Event List Management—A Tutorial," *Proceedings of the 1983 Winter Simulation Conference,* Piscataway, NJ: IEEE, 1983, 543–551.

Jones, Douglas W., "An Empirical Comparison of Priority-Queue and Event-Set Implementations," *Communications of the ACM* 29 (1986), 300–311.

5 Recursion

5.1 RECURSIVE DEFINITIONS

One of the basic rules for defining new objects or concepts is that the definition should contain only such terms that have already been defined or that are obvious. Therefore, an object that is defined in terms of itself is a serious violation of this rule—a vicious circle. On the other hand, there are many programming concepts that define themselves. As it turns out, formal restrictions imposed on definitions such as existence and uniqueness are satisfied and no violation of the rules takes place. Such definitions are called *recursive definitions,* and are used primarily to define infinite sets. When defining such a set, giving a complete list of elements is impossible, and for large finite sets, it is inefficient. Thus, a more efficient way has to be devised to determine if an object belongs to a set.

A recursive definition consists of two parts. In the first part, called the *anchor* or the *ground case*, the basic elements that are the building blocks of all other elements of the set are listed. In the second part, rules are given that allow for the construction of new objects out of basic elements or objects that have already been constructed. These rules are applied again and again to generate new objects. For example, to construct the set of natural numbers, one basic element, 0, is singled out, and the operation of incrementing by 1 is given as:

1. $0 \in \mathbf{N}$;
2. if $n \in \mathbf{N}$, then $(n + 1) \in \mathbf{N}$;
3. there are no other objects in the set $\mathbf{N}$.

(More axioms are needed to ensure that only the set that we know as the natural numbers can be constructed by these rules.)

According to these rules, the set of natural numbers $\mathbf{N}$ consists of the following items: $0, 0 + 1, 0 + 1 + 1, 0 + 1 + 1 + 1$, and so on. Although the set $\mathbf{N}$ contains objects (and only such objects) that we call natural numbers, the definition results in a somewhat unwieldy list of elements. Can you imagine doing arithmetic on large numbers using such a specification? Therefore, it is more convenient to use the following definition, which encompasses the whole range of Arabic numeric heritage:

1. $0, 1, 2, 3, 4, 5, 6, 7, 8, 9 \in \mathbf{N}$;
2. if $n \in \mathbf{N}$, then $n0, n1, n2, n3, n4, n5, n6, n7, n8, n9 \in \mathbf{N}$;
3. these are the only natural numbers.

Then the set **N** includes all possible combinations of the basic building blocks 0 through 9.

Recursive definitions serve two purposes: *generating* new elements, as already indicated, and *testing* whether an element belongs to a set. In the case of testing, the problem is solved by reducing it to a simpler problem, and if the simpler problem is still too complex it is reduced to an even simpler problem, and so on, until it is reduced to a problem indicated in the anchor. For instance, is 123 a natural number? According to the second condition of the definition introducing the set **N**, $123 \in \mathbf{N}$ if $12 \in \mathbf{N}$ and the first condition already says that $3 \in \mathbf{N}$; but $12 \in \mathbf{N}$ if $1 \in \mathbf{N}$ and $2 \in \mathbf{N}$, and they both belong to **N**.

The ability to decompose a problem into simpler subproblems of the same kind is sometimes a real blessing, as we shall see in the discussion of quicksort in Section 9.3.3, or a curse, as we shall see shortly in this chapter.

Recursive definitions are frequently used to define functions and sequences of numbers. For instance, the factorial function, !, can be defined in the following manner:

$$n! = \begin{cases} 1 & \text{if } n = 0 \text{ (anchor)} \\ n \cdot (n-1)! & \text{if } n > 0 \text{ (inductive step)} \end{cases}$$

Using this definition, we can generate the sequence of numbers

$$1, 1, 2, 6, 24, 120, 720, 5040, 40320, 362880, 3628800, \ldots$$

which includes the factorials of the numbers $0, 1, 2, \ldots, 10, \ldots$

Another example is the definition

$$f(n) = \begin{cases} 1 & \text{if } n = 0 \\ f(n-1) + \dfrac{1}{f(n-1)} & \text{if } n > 0 \end{cases}$$

which generates the sequence of rational numbers

$$1, 2, \frac{5}{2}, \frac{29}{10}, \frac{941}{290}, \frac{969{,}581}{272{,}890}, \ldots$$

Recursive definitions of sequences have one undesirable feature: To determine the value of an element s_n of a sequence, we first have to compute the values of some or all of the previous elements, $s_1, \ldots, s_{n-1}$. For example, calculating the value of 3! requires us to first compute the values of 0!, 1!, and 2!. Computationally, this is undesirable because it forces us to make calculations in a roundabout way. Therefore, we want to find an equivalent definition or formula that makes no references to other elements of the sequence. Generally, finding such a formula is a difficult problem that cannot always be solved. But the formula is preferable to a recursive definition because it simplifies the computational process and allows us to find the answer for an integer n without computing the values for integers $0, 1, \ldots, n-1$. For example, a definition of the sequence g,

$$g(n) = \begin{cases} 1 & \text{if } n = 0 \\ 2 \cdot g(n-1) & \text{if } n > 0 \end{cases}$$

can be converted into the simple formula

$$g(n) = 2^n$$

In the foregoing discussion, recursive definitions have been dealt with only theoretically, as a definition used in mathematics. Naturally, our interest is in computer science. One area where recursive definitions are used extensively is in the specification of the grammars of programming languages. Every programming language manual contains—either as an appendix or throughout the text—a specification of all valid language elements. Grammar is specified either in terms of block diagrams or in terms of the Backus-Naur form (BNF). For example, the syntactic definition of a statement in the C++ language can be presented in the block diagram form:

statement ─┬─ while ─(─ expression ─)─ statement ─────────────────
           │
           └─ if ─────(─ expression ─)─ statement ─┬─ else ─ statement ─

or in BNF:

                                                                                               <statement> ::= while (<expression>) <statement> |
                                                                                               if (<expression>) <statement> |
                                                                                               if (<expression>) <statement>else<statement> |
                                                                                               ...

The language element <statement> is defined recursively, in terms of itself. Such definitions naturally express the possibility of creating such syntactic constructs as nested statements or expressions.

Recursive definitions are also used in programming. The good news is that virtually no effort is needed to make the transition from a recursive definition of a function to its implementation in C++. We simply make a translation from the formal definition into C++ syntax. Hence, for example, a C++ equivalent of factorial is the function

```
unsigned int factorial (unsigned int n) {
   if (n == 0)
        return 1;
   else return n * factorial (n - 1);
}
```

The problem now seems to be more critical because it is far from clear how a function calling itself can possibly work, let alone return the correct result. This chapter shows that it is possible for such a function to work properly. Recursive definitions on most computers are eventually implemented using a run-time stack, although the whole work of implementing recursion is done by the operating system, and the source code includes no indication of how it is performed. E. W. Dijkstra introduced the idea of using a stack to implement recursion. To better understand recursion and to see how it works, it is necessary to discuss the processing of function calls and to look at operations carried out by the system at function invocation and function exit.

5.2 FUNCTION CALLS AND RECURSION IMPLEMENTATION

What happens when a function is called? If the function has formal parameters, they have to be initialized to the values passed as actual parameters. In addition, the system has to know where to resume execution of the program after the function has finished. The function can be called by other functions or by the main program (the function `main()`). The information indicating where it has been called from has to be remembered by the system. This could be done by storing the return address in main memory in a place set aside for return addresses, but we do not know in advance how much space might be needed, and allocating too much space for that purpose alone is not efficient.

For a function call, more information has to be stored than just a return address. Therefore, dynamic allocation using the run-time stack is a much better solution. But what information should be preserved when a function is called? First, automatic (local) variables must be stored. If function `f1()`, which contains a declaration of an automatic variable x, calls function `f2()`, which locally declares the variable x, the system has to make a distinction between these two variables x. If `f2()` uses a variable x, then its own x is meant; if `f2()` assigns a value to x, then x belonging to `f1()` should be left unchanged. When `f2()` is finished, `f1()` can use the value assigned to its private x before `f2()` was called. This is especially important in the context of the present chapter, when `f1()` is the same as `f2()`, when a function calls itself recursively. How does the system make a distinction between these two variables x?

The state of each function, including `main()`, is characterized by the contents of all automatic variables, by the values of the function's parameters, and by the return address indicating where to restart in the calling function. The data area containing all this information is called an *activation record* or a *stack frame* and is allocated on the run-time stack. An activation record exists for as long as a function owning it is executing. This record is a private pool of information for the function, a repository that stores all information necessary for its proper execution and how to return to where it was called from. Activation records usually have a short lifespan because they are dynamically allocated at function entry and deallocated upon exiting. Only the activation record of `main()` outlives every other activation record.

An activation record usually contains the following information:

- Values for all parameters to the function, location of the first cell if an array is passed or a variable is passed by reference, and copies of all other data items.
- Local (automatic) variables that can be stored elsewhere, in which case, the activation record contains only their descriptors and pointers to the locations where they are stored.

- The return address to resume control by the caller, the address of the caller's instruction immediately following the call.
- A dynamic link, which is a pointer to the caller's activation record.
- The returned value for a function not declared as void. Because the size of the activation record may vary from one call to another, the returned value is placed right above the activation record of the caller.

As mentioned, if a function is called either by main() or by another function, then its activation record is created on the run-time stack. The run-time stack always reflects the current state of the function. For example, suppose that main() calls function f1(), f1() calls f2(), and f2() in turn calls f3(). If f3() is being executed, then the state of the run-time stack is as shown in Figure 5.1. By the nature of the stack, if the activation record for f3() is popped by moving the stack pointer right below the return value of f3(), then f2() resumes execution and now has free access to the private pool of information necessary for reactivation of its execution. On the other hand, if f3() happens to call another function f4(), then the run-time stack increases its height because the activation record for f4() is created on the stack and the activity of f3() is suspended.

FIGURE 5.1 Contents of the run-time stack when main() calls function f1(), f1() calls f2(), and f2() calls f3().

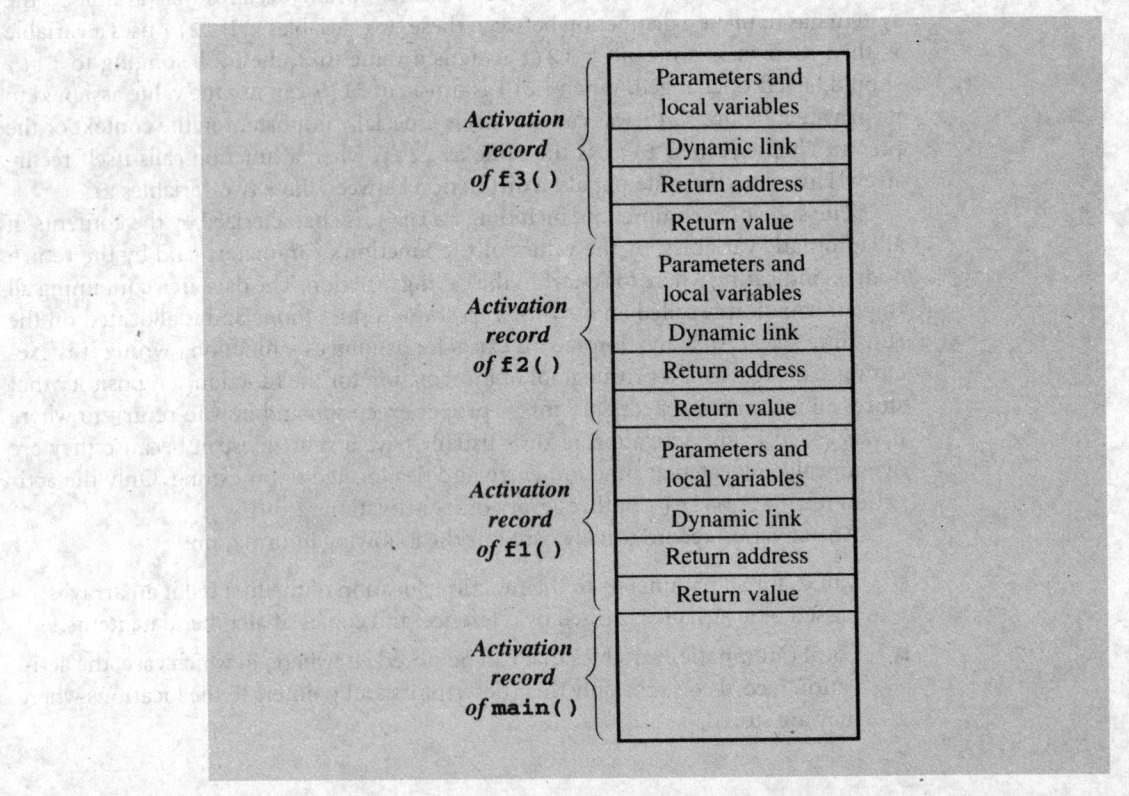

Activation record of f3()	Parameters and local variables
	Dynamic link
	Return address
	Return value
Activation record of f2()	Parameters and local variables
	Dynamic link
	Return address
	Return value
Activation record of f1()	Parameters and local variables
	Dynamic link
	Return address
	Return value
Activation record of main()	

Creating an activation record whenever a function is called allows the system to handle recursion properly. Recursion is calling a function that happens to have the same name as the caller. Therefore, a recursive call is not literally a function calling itself, but rather an instantiation of a function calling another instantiation of the same original. These invocations are represented internally by different activation records and are thus differentiated by the system.

5.3 ANATOMY OF A RECURSIVE CALL

The function that defines raising any number x to a nonnegative integer power n is a good example of a recursive function. The most natural definition of this function is given by:

$$x^n = \begin{cases} 1 & \text{if } n = 0 \\ x \cdot x^{n-1} & \text{if } n > 0 \end{cases}$$

A C++ function for computing x^n can be written directly from the definition of a power:

```
/* 102 */ double power (double x, unsigned int n) {
/* 103 */     if (n == 0)
/* 104 */         return 1.0;
          // else
/* 105 */     return x * power(x,n-1);
          }
```

Using this definition, the value of x^4 can be computed in the following way:

$$x^4 = x \cdot x^3 = x \cdot (x \cdot x^2) = x \cdot (x \cdot (x \cdot x^1)) = x \cdot (x \cdot (x \cdot (x \cdot x^0)))$$
$$= x \cdot (x \cdot (x \cdot (x \cdot 1))) = x \cdot (x \cdot (x \cdot (x))) = x \cdot (x \cdot (x \cdot x))$$
$$= x \cdot (x \cdot x \cdot x) = x \cdot x \cdot x \cdot x$$

The repetitive application of the inductive step eventually leads to the anchor, which is the last step in the chain of recursive calls. The anchor produces 1 as a result of raising x to the power of zero; the result is passed back to the previous recursive call. Now, that call, whose execution has been pending, returns its result, $x \cdot 1 = x$. The third call, which has been waiting for this result, computes its own result, namely, $x \cdot x$, and returns it. Next, this number $x \cdot x$ is received by the second call, which multiplies it by x and returns the result, $x \cdot x \cdot x$, to the first invocation of power(·). This call receives $x \cdot x \cdot x$, multiplies it by x, and returns the final result. In this way, each new call increases the level of recursion, as follows:

call 1 $x^4 = x \cdot x^3$ $= x \cdot x \cdot x \cdot x$
call 2 $x \cdot x^2$ $= x \cdot x \cdot x$
call 3 $x \cdot x^1$ $= x \cdot x$
call 4 $x \cdot x^0$ $= x \cdot 1 = x$
call 5   1

or alternatively, as

call 1	power(x,4)
call 2	power(x,3)
call 3	power(x,2)
call 4	power(x,1)
call 5	power(x,0)
call 5	1
call 4	x
call 3	$x \cdot x$
call 2	$x \cdot x \cdot x$
call 1	$x \cdot x \cdot x \cdot x$

What does the system do as the function is being executed? As we already know, the system keeps track of all calls on its run-time stack. Each line of code is assigned a number by the system,[1] and if a line is a function call, then its number is a return address. The address is used by the system to remember where to resume execution after the function has completed. For this example, assume that the lines in the function power() are assigned the numbers 102 through 105 and that it is called main() from the statement

```
          int main()
          { ...
/* 136 */    y = power(5.6,2);
          ...
          }
```

A trace of the recursive calls is relatively simple, as indicated by this diagram

call 1	power(5.6,2)
call 2	power(5.6,1)
call 3	power(5.6,0)
call 3	1
call 2	5.6
call 1	31.36

because most of the operations are performed on the run-time stack.

When the function is invoked for the first time, four items are pushed onto the run-time stack: the return address 136, the actual parameters 5.6 and 2, and one location reserved for the value returned by power(). Figure 5.2a represents this situation. (In this and subsequent diagrams, SP is a stack pointer, AR is an activation record, and question marks stand for locations reserved for the returned values. To distinguish values from addresses, the latter are parenthesized, although addresses are numbers exactly like function arguments.)

Now the function power() is executed. First, the value of the second argument, 2, is checked, and power() tries to return the value of $5.6 \cdot$ power(5.6,1) because

---

[1]This is not quite precise because the system uses machine code rather than source code to execute programs. This means that one line of source program is usually implemented by several machine instructions.

FIGURE 5.2 Changes to the run-time stack during execution of `power(5.6,2)`.

Third call to `power()`			0 ← SP 5.6 (105) ?	0 ← SP 5.6 (105) 1.0	0 5.6 (105) 1.0			
Second call to `power()`		1 ← SP 5.6 (105) ?	1 5.6 (105) ?	1 5.6 (105) ?	1 ← SP 5.6 (105) ?	1 ← SP 5.6 (105) 5.6	1 5.6 (105) 5.6	
First call to `power()`	2 ← SP 5.6 (136) ?	2 5.6 (136) ?	2 5.6 (136) ?	2 5.6 (136) ?	2 5.6 (136) ?	2 5.6 (136) ?	2 ← SP 5.6 (136) ?	2 ← SP 5.6 (136) 31.36
AR for `main()`	⋮ y ⋮	⋮ y ⋮	⋮ y ⋮	⋮ y ⋮	⋮ y ⋮	⋮ y ⋮	⋮ y ⋮	⋮ y ⋮
	(a)	(b)	(c)	(d)	(e)	(f)	(g)	(h)

Key: SP Stack pointer
AR Activation record
? Location reserved for returned value

that argument is not 0. This cannot be done immediately because the system does not know the value of `power(5.6,1)`; it must be computed first. Therefore, `power()` is called again with the arguments 5.6 and 1. But before this call is executed, the run-time stack receives new items, and its contents are shown in Figure 5.2b.

Again, the second argument is checked to see if it is 0. Because it is equal to 1, `power()` is called for the third time, this time with the arguments 5.6 and 0. Before the function is executed, the system remembers the arguments and the return address by putting them on the stack, not forgetting to allocate one cell for the result. Figure 5.2c contains the new contents of the stack.

Again, the question arises: Is the second argument equal to zero? Because it finally is, a concrete value—namely, 1.0—can be returned and placed on the stack, and the function is finished without making any additional calls. At this point, there are two pending calls on the run-time stack—the calls to `power()`—that have to be completed. How is this done? The system first eliminates the activation record of `power()` that has just finished. This is performed logically by popping all its fields (the result, two arguments, and the return address) off the stack. We say "logically" because physically all these fields remain on the stack and only the SP is decremented appropriately. This is important because we do not want the result to be destroyed since it has not been used yet. Before and after completion of the last call of `power()`, the stack looks the same, but the SP's value is changed (see Figures 5.2d and 5.2e).

Now the second call to `power()` can complete because it waited for the result of the call `power(5.6,0)`. This result, 1.0, is multiplied by 5.6 and stored in the field allocated for the result. After that, the system can pop the current activation record off the stack by decrementing the SP, and it can finish the execution of the first call to `power()` that needed the result for the second call. Figure 5.2f shows the contents of the stack before changing the SP's value, and Figure 5.2g shows the contents of the stack after this change. At this moment, `power()` can finish its first call by multiplying the result of its second call, 5.6, by its first argument, also 5.6. The system now returns to the function that invoked `power()`, and the final value, 31.36, is assigned to `y`. Right before the assignment is executed, the content of the stack looks like Figure 5.2h.

The function `power()` can be implemented differently, without using any recursion, as in the following loop:

```
double nonRecPower(double x, unsigned int n) {
    double result = 1;
    for (result = x; n > 1; --n)
        result *= x;
    return result;
}
```

Do we gain anything by using recursion instead of a loop? The recursive version seems to be more intuitive because it is similar to the original definition of the power function. The definition is simply expressed in C++ without losing the original structure of the definition. The recursive version increases program readability, improves self-documentation, and simplifies coding. In our example, the code of the nonrecursive version is not substantially larger than in the recursive version, but for most recursive implementations, the code is shorter than it is in the nonrecursive implementations.

5.4 TAIL RECURSION

All recursive definitions contain a reference to a set or function being defined. There are, however, a variety of ways such a reference can be implemented. This reference can be done in a straightforward manner or in an intricate fashion, just once or many times. There may be many possible levels of recursion or different levels of complexity. In the following sections, some of these types are discussed, starting with the simplest case, *tail recursion.*

Tail recursion is characterized by the use of only one recursive call at the very end of a function implementation. In other words, when the call is made, there are no statements left to be executed by the function; the recursive call is not only the last statement but there are no earlier recursive calls, direct or indirect. For example, the function `tail()` defined as

```
void tail (int i) {
    if (i > 0) {
        cout << i << ' ';
        tail(i-1);
    }
}
```

is an example of a function with tail recursion, whereas the function `nonTail()` defined as

```cpp
void nonTail (int i) {
  if (i > 0) {
    nonTail(i-1);
    cout << i << ' ';
    nonTail(i-1);
  }
}
```

is not. Tail recursion is simply a glorified loop and can be easily replaced by one. In this example, it is replaced by substituting a loop for the `if` statement and decrementing the variable `i` in accordance with the level of recursive call. In this way, `tail()` can be expressed by an iterative function:

```cpp
void iterativeEquivalentOfTail (int i) {
  for ( ; i > 0; i--)
    cout << i << ' ';
}
```

Is there any advantage in using tail recursion over iteration? For languages such as C++, there may be no compelling advantage, but in a language such as Prolog, which has no explicit loop construct (loops are simulated by recursion), tail recursion acquires a much greater weight. In languages endowed with a loop or its equivalents, such as an `if` statement combined with a `goto` statement, tail recursion should not be used.

5.5 NONTAIL RECURSION

Another problem that can be implemented in recursion is printing an input line in reverse order. Here is a simple recursive implementation:

```cpp
/* 200 */ void reverse() {
              char ch;
/* 201 */     cin.get(ch);
/* 202 */     if (ch != '\n') {
/* 203 */         reverse();
/* 204 */         cout.put(ch);
          }
}
```

Where is the trick? It does not seem possible that the function does anything. But it turns out that, by the power of recursion, it does exactly what it was designed for. `main()` calls `reverse()` and the input is the string: "ABC." First, an activation record is created with cells for the variable `ch` and the return address. There is no need to reserve a cell for a result, because no value is returned, which is indicated by using `void` in front of the function's name. The function `get()` reads in the first character, "A." Figure 5.3a shows the contents of the run-time stack right before `reverse()` calls itself recursively for the first time.

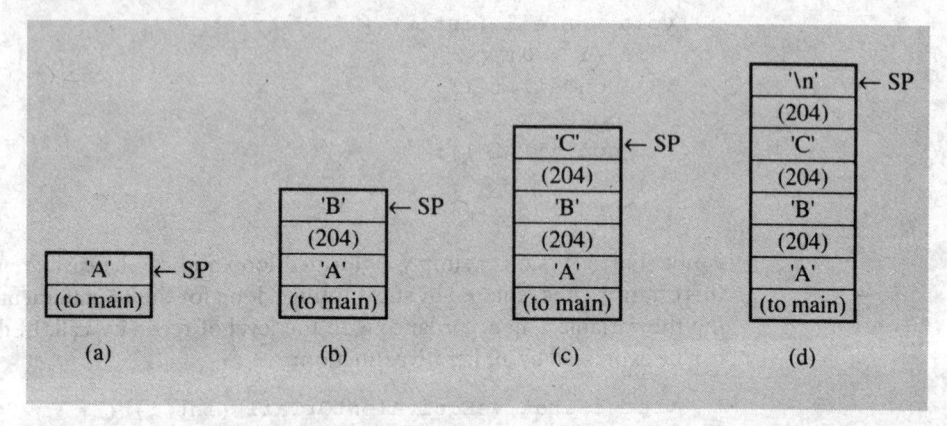

FIGURE 5.3 Changes on the run-time stack during the execution of reverse().

			'\n' ← SP
			(204)
		'C' ← SP	'C'
		(204)	(204)
	'B' ← SP	'B'	'B'
	(204)	(204)	(204)
'A' ← SP	'A'	'A'	'A'
(to main)	(to main)	(to main)	(to main)
(a)	(b)	(c)	(d)

The second character is read in and checked to see if it is the end-of-line character, and if not, reverse() is called again. But in either case, the value of ch is pushed onto the run-time stack along with the return address. Before reverse() is called for a third time (the second time recursively), there are two more items on the stack (see Figure 5.3b).

Note that the function is called as many times as the number of characters contained in the input string, including the end-of-line character. In our example, reverse() is called four times, and the run-time stack during the last call is shown in Figure 5.3d.

On the fourth call, get() finds the end-of-line character and reverse() executes no other statement. The system retrieves the return address from the activation record and discards this record by decrementing SP by the proper number of bytes. Execution resumes from line 204, which is a print statement. Because the activation record of the third call is now active, the value of ch, the letter "C," is output as the first character. Next, the activation record of the third call to reverse() is discarded and now SP points to where "B" is stored. The second call is about to be finished, but first, "B" is assigned to ch and then the statement on line 204 is executed, which results in printing "B" on the screen right after "C." Finally, the activation record of the first call to reverse() is reached. Then "A" is printed, and what can be seen on the screen is the string "CBA." The first call is finally finished and the program continues execution in main().

Compare the recursive implementation with a nonrecursive version of the same function:

```
void simpleIterativeReverse() {
  char stack[80];
  register int top = 0;
  cin.getline(stack,80);
  for (top = strlen(stack) - 1; top >= 0; cout.put(stack[top--]));
}
```

The function is quite short and, perhaps, a bit more cryptic than its recursive counterpart. What is the difference then? Keep in mind that the brevity and relative simplicity of the second version are due mainly to the fact that we want to reverse a string or array of characters. This means that functions like `strlen()` and `getline()` from the standard C++ library can be used. If we are not supplied with such functions, then our iterative function has to be implemented differently:

```cpp
void iterativeReverse() {
    char stack[80];
    register int top = 0;
    cin.get(stack[top]);
    while(stack[top]!='\n')
        cin.get(stack[++top]);
    for (top -= 2; top >= 0; cout.put(stack[top--]));
}
```

The `while` loop replaces `getline()` and the autoincrement of variable `top` replaces `strlen()`. The `for` loop is about the same as before. This discussion is not purely theoretical because reversing an input line consisting of integers uses the same implementation as `iterativeReverse()` after changing the data type of `stack` from `char` to `int` and modifying the `while` loop.

Note that the variable name `stack` used for the array is not accidental. We are just making explicit what is done implicitly by the system. Our stack takes over the run-time stack's duty. Its use is necessary here because one simple loop does not suffice, as in the case of tail recursion. In addition, the statement `put()` from the recursive version has to be accounted for. Note also that the variable `stack` is local to the function `iterativeReverse()`. However, if it were a requirement to have a global stack object `st`, then this implementation can be written as

```cpp
void nonRecursiveReverse() {
    int ch;
    cin.get(ch);
    while (ch != '\n') {
        st.push(ch);
        cin.get(ch);
    }
    while (!st.empty())
        cout.put(st.pop());
}
```

with the declaration `Stack<char> st` outside the function.

After comparing `iterativeReverse()` to `nonRecursiveReverse()`, we can conclude that the first version is better because it is faster, no function calls are made, and the function is self-sufficient, whereas `nonRecursiveReverse()` calls at least one function during each loop iteration, slowing down execution.

One way or the other, the transformation of nontail recursion into iteration usually involves the explicit handling of a stack. Furthermore, when converting a function from a recursive into an iterative version, program clarity can be diminished and the brevity of program formulation lost. Iterative versions of recursive C++ functions are not as verbose as in other programming languages, so program brevity may not be an issue.

To conclude this section, consider a construction of the von Koch snowflake. The curve was constructed in 1904 by Swedish mathematician Helge von Koch as an example of a continuous and nondifferentiable curve with an infinite length and yet encompassing a finite area. Such a curve is a limit of an infinite sequence of snowflakes, of which the first three are presented in Figure 5.4. As in real snowflakes, two of these curves have six petals, but to facilitate the algorithm, it is treated as a combination of three identical curves drawn in different angles and joined together. One such curve is drawn in the following fashion:

1. Divide an interval *side* into three even parts.
2. Move one-third of *side* in the direction specified by *angle*.
3. Turn to the right 60° (i.e., turn –60°) and go forward one-third of *side*.
4. Turn to the left 120° and proceed forward one-third of *side*.
5. Turn right 60° and again draw a line one-third of *side* long.

The result of these five steps is summarized in Figure 5.5. This line, however, becomes more jagged if every one of the four intervals became a miniature of the whole curve; that is, if the process of drawing four lines were made for each of these *side*/3 long intervals. As a result, 16 intervals *side*/9 long would be drawn. The process may be continued indefinitely—at least in theory. Computer graphics resolution prevents us from going too far because if lines are smaller than the diameter of a pixel, we just see one dot on the screen.

FIGURE 5.4 Examples of von Koch snowflakes.

FIGURE 5.5 The process of drawing four sides of one segment of the von Koch snowflake.

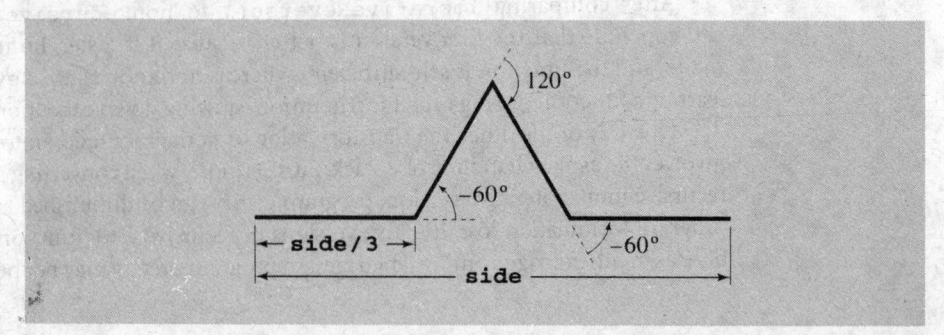

The five steps that, instead of drawing one line of length *side*, draw four lines each of length one-third of *side* form one cycle only. Each of these four lines can also be compound lines drawn by the use of the described cycle. This is a situation in which recursion is well suited, which is reflected by the following pseudocode:

```
drawFourLines (side, level)
    if (level == 0)
        draw a line;
    else
        drawFourLines(side/3, level-1);
        turn left 60°;
        drawFourLines(side/3, level-1);
        turn right 120°;
        drawFourLines(side/3, level-1);
        turn left 60°;
        drawFourLines(side/3, level-1);
```

This pseudocode can be rendered almost without change into C++ code (Figure 5.6).

FIGURE 5.6 Recursive implementation of the von Koch snowflake.

```
//*********************** snowflake.h ***************************
//                   Visual C++ program

#include <cmath>

double PI = 3.14159;

class vonKoch {
public:
    vonKoch(int,int,CDC*);
    void snowflake();
private:
    double side, angle;
    int level;
    CPoint currPt, pt;
    CDC *pen;
    void right(double x) {
        angle += x;
    }
    void left (double x) {
        angle -= x;
    }
    void drawFourLines(double side, int level);
};
```

Continues

FIGURE 5.6 *(continued)*

```
vonKoch::vonKoch(int s, int lvl, CDC *pDC) {
    pen = pDC;
    currPt.x = 200;
    currPt.y = 100;
    pen->MoveTo(currPt);
    angle = 0.0;
    side = s;
    level = lvl;
}

void vonKoch::drawFourLines(double side, int level) {
    // arguments to sin() and cos() are angles
    // specified in radians, i.e., the coefficient
    // PI/180 is necessary;
    if (level == 0) {
        pt.x = int(cos(angle*PI/180)*side) + currPt.x;
        pt.y = int(sin(angle*PI/180)*side) + currPt.y;
        pen->LineTo(pt);
        currPt.x = pt.x;
        currPt.y = pt.y;
    }
    else {
        drawFourLines(side/3,level-1);
        left (60);
        drawFourLines(side/3,level-1);
        right(120);
        drawFourLines(side/3,level-1);
        left (60);
        drawFourLines(side/3,level-1);
    }
}

void vonKoch::snowflake() {
    for (int i = 1; i <= 3; i++) {
        drawFourLines(side,level);
            right(120);
    }
}

// The function OnDraw() is generated by Visual C++ when creating
// a snowflake project of type MFC AppWizard (exe), Single Document;

#include "a:snowflake.h"
```

FIGURE 5.6   *(continued)*

```
void CSnowflakeView::OnDraw(CDC* pDC)
{
    CSnowflakeDoc* pDoc = GetDocument();
    ASSERT_VALID(pDoc);

    // TODO: add draw code for native data here

    vonKoch(200,4,pDC).snowflake();
}
```

5.6 INDIRECT RECURSION

The preceding sections discussed only direct recursion, where a function `f()` called itself. However, `f()` can call itself indirectly via a chain of other calls. For example, `f()` can call `g()`, and `g()` can call `f()`. This is the simplest case of indirect recursion. The chain of intermediate calls can be of an arbitrary length, as in:

$$f() \rightarrow f1() \rightarrow f2() \rightarrow \cdots \rightarrow fn() \rightarrow f()$$

There is also the situation when `f()` can call itself indirectly through different chains. Thus, in addition to the chain just given, another chain might also be possible. For instance

$$f() \rightarrow g1() \rightarrow g2() \rightarrow \cdots \rightarrow gm() \rightarrow f()$$

This situation can be exemplified by three functions used for decoding information. `receive()` stores the incoming information in a buffer, `decode()` converts it into legible form, and `store()` stores it in a file. `receive()` fills the buffer and calls `decode()`, which in turn, after finishing its job, submits the buffer with decoded information to `store()`. After `store()` accomplishes its tasks, it calls `receive()` to intercept more encoded information using the same buffer. Therefore, we have the chain of calls

`receive() -> decode() -> store() -> receive() -> decode() -> ...`

which is finished when no new information arrives. These three functions work in the following manner:

```
receive(buffer)
    while buffer is not filled up
        if information is still incoming
            get a character and store it in buffer;
        else exit();
    decode(buffer);
```

```
decode(buffer)
    decode information in buffer;
    store(buffer);

store(buffer)
    transfer information from buffer to file;
    receive(buffer);
```

A more mathematically oriented example concerns formulas calculating the trigonometric functions sine, cosine, and tangent:

$$\sin(x) = \sin\left(\frac{x}{3}\right) \cdot \frac{\left(3 - \tan^2\left(\frac{x}{3}\right)\right)}{\left(1 + \tan^2\left(\frac{x}{3}\right)\right)}$$

$$\tan(x) = \frac{\sin(x)}{\cos(x)}$$

$$\cos(x) = 1 - \sin\left(\frac{x}{2}\right)$$

As usual in the case of recursion, there has to be an anchor in order to avoid falling into an infinite loop of recursive calls. In the case of sine, we can use the following approximation:

$$\sin(x) \approx x - \frac{x^3}{6}$$

where small values of x give a better approximation. To compute the sine of a number x such that its absolute value is greater than an assumed tolerance, we have to compute $\sin\left(\frac{x}{3}\right)$ directly, $\sin\left(\frac{x}{3}\right)$ indirectly through tangent, and also indirectly, $\sin\left(\frac{x}{6}\right)$ through tangent and cosine. If the absolute value of $\frac{x}{3}$ is sufficiently small, which does not require other recursive calls, we can represent all the calls as a tree, as in Figure 5.7.

FIGURE 5.7 A tree of recursive calls for sin (x).

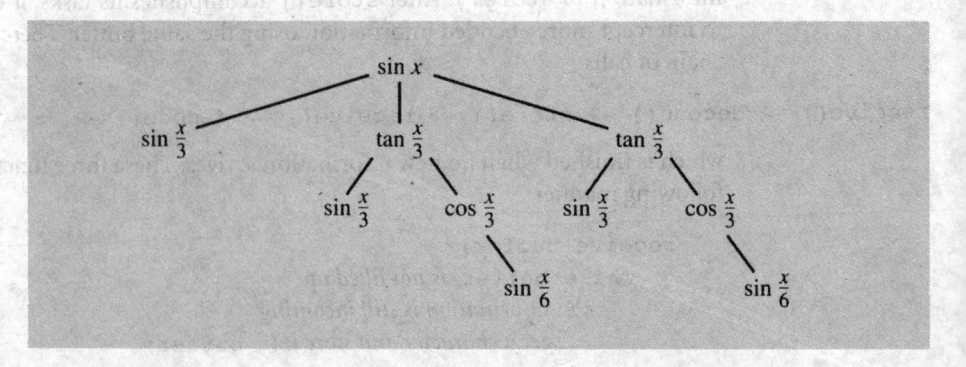

5.7 NESTED RECURSION

A more complicated case of recursion is found in definitions in which a function is not only defined in terms of itself, but also is used as one of the parameters. The following definition is an example of such a nesting:

$$h(n) = \begin{cases} 0 & \text{if } n = 0 \\ n & \text{if } n > 4 \\ h(2 + h(2n)) & \text{if } n \leq 4 \end{cases}$$

Function h has a solution for all $n \geq 0$. This fact is obvious for all $n > 4$ and $n = 0$, but it has to be proven for $n = 1, 2, 3,$ and 4. Thus, $h(2) = h(2 + h(4)) = h(2 + h(2 + h(8))) = 12$. (What are the values of $h(n)$ for $n = 1, 3,$ and 4?)

Another example of nested recursion is a very important function originally suggested by Wilhelm Ackermann in 1928 and later modified by Rozsa Peter:

$$A(n,m) = \begin{cases} m+1 & \text{if } n = 0 \\ A(n-1,1) & \text{if } n > 0, m = 0 \\ A(n-1, A(n, m-1)) & \text{otherwise} \end{cases}$$

This function is interesting because of its remarkably rapid growth. It grows so fast that it is guaranteed not to have a representation by a formula that uses arithmetic operations such as addition, multiplication, and exponentiation. To illustrate the rate of growth of the Ackermann function, we need only show that

$$A(3,m) = 2^{m+3} - 3$$
$$A(4,m) = 2^{2^{2^{\cdot^{\cdot^{2^{16}}}}}} - 3$$

with a stack of m 2s in the exponent; $A(4,1) = 2^{2^{16}} - 3 = 2^{65536} - 3$, which exceeds even the number of atoms in the universe (which is 10^{80} according to current theories).

The definition translates very nicely into C++, but the task of expressing it in a nonrecursive form is truly troublesome.

5.8 EXCESSIVE RECURSION

Logical simplicity and readability are used as an argument supporting the use of recursion. The price for using recursion is slowing down execution time and storing on the run-time stack more things than required in a nonrecursive approach. If recursion is too deep (for example, computing $5.6^{100,000}$), then we can run out of space on the stack and our program crashes. But usually, the number of recursive calls is much smaller than 100,000, so the danger of overflowing the stack may not be imminent.[2]

---

[2] Even if we try to compute the value of $5.6^{100,000}$ using an iterative algorithm, we are not completely free from a troublesome situation because the number is much too large to fit even a variable of double length. Thus, although the program would not crash, the computed value would be incorrect, which may be even more dangerous than a program crash.

However, if some recursive function repeats the computations for some parameters, the run time can be prohibitively long even for very simple cases.

Consider Fibonacci numbers. A sequence of Fibonacci numbers is defined as follows:

$$\text{Fib}(n) = \begin{cases} n & \text{if } n < 2 \\ \text{Fib}(n-2) + \text{Fib}(n-1) & \text{otherwise} \end{cases}$$

The definition states that if the first two numbers are 0 and 1, then any number in the sequence is the sum of its two predecessors. But these predecessors are in turn sums of their predecessors, and so on, to the beginning of the sequence. The sequence produced by the definition is

$$0, 1, 1, 2, 3, 5, 8, 13, 21, 34, 55, 89, \ldots$$

How can this definition be implemented in C++? It takes almost term-by-term translation to have a recursive version, which is

```
unsigned int Fib (unsigned int n) {
  if (n < 2)
    return n;
  // else
    return Fib(n-2) + Fib(n-1);
}
```

The function is simple and easy to understand but extremely inefficient. To see it, compute `Fib(6)`, the seventh number of the sequence, which is 8. Based on the definition, the computation runs as follows:

```
Fib(6) =              Fib(4)              + Fib(5)
       =     Fib(2)      +     Fib(3)     + Fib(5)
       = Fib(0)+Fib(1)   +     Fib(3)     + Fib(5)
       =     0 + 1       +     Fib(3)     + Fib(5)
       =        1        + Fib(1)+ Fib(2) + Fib(5)
       =        1        + Fib(1)+Fib(0)+Fib(1) + Fib(5)
```

etc.

This is just the beginning of our calculation process, and even here there are certain shortcuts. All these calculations can be expressed more concisely in the form of the tree shown in Figure 5.8. Tremendous inefficiency results because `Fib()` is called 25 times to determine the seventh element of the Fibonacci sequence. The source of this inefficiency is the repetition of the same calculations because the system forgets what has already been calculated. For example, `Fib()` is called eight times with parameter n = 1 to decide that 1 can be returned. For each number of the sequence, the function computes all its predecessors without taking into account that it suffices to do this only once. To find `Fib(6)` = 8, it computes `Fib(5)`, `Fib(4)`, `Fib(3)`, `Fib(2)`, `Fib(1)`, and `Fib(0)` first. To determine these values, `Fib(4)`,..., `Fib(0)` have to be computed to know the value of `Fib(5)`. Independently of this, the chain of computations `Fib(3)`, ..., `Fib(0)` is executed to find `Fib(4)`.

FIGURE 5.8 The tree of calls for `Fib(6)`.

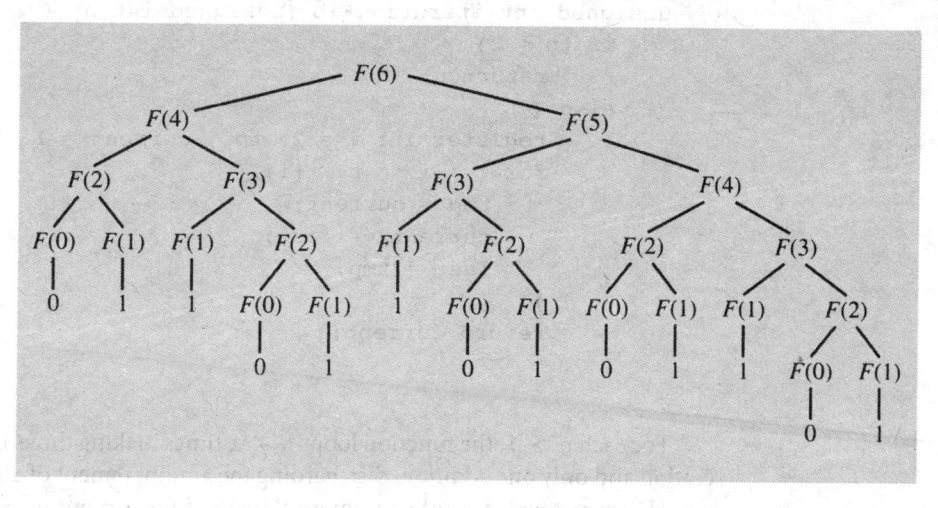

We can prove that the number of additions required to find `Fib(n)` using a recursive definition is equal to Fib($n + 1$) − 1. Counting two calls per one addition plus the very first call means that `Fib()` is called 2 · Fib($n + 1$) − 1 times to compute `Fib(n)`. This number can be exceedingly large for fairly small ns, as the table in Figure 5.9 indicates.

It takes almost a quarter of a million calls to find the twenty-sixth Fibonacci number, and nearly 3 million calls to determine the thirty-first! This is too heavy a price for the simplicity of the recursive algorithm. As the number of calls and the run time grow exponentially with n, the algorithm has to be abandoned except for very small numbers.

FIGURE 5.9 Number of addition operations and number of recursive calls to calculate Fibonacci numbers.

n	Fib(n+1)	Number of Additions	Number of Calls
6	13	12	25
10	89	88	177
15	987	986	1,973
20	10,946	10,945	21,891
25	121,393	121,392	242,785
30	1,346,269	1,346,268	2,692,537

An iterative algorithm may be produced rather easily as follows:

```
unsigned int iterativeFib (unsigned int n) {
    if (n < 2)
        return n;
    else {
        register int i = 2, tmp, current = 1, last = 0;
        for ( ; i <= n; ++i) {
            tmp = current;
            current += last;
            last = tmp;
        }
        return current;
    }
}
```

For each n > 1, the function loops n − 1 times making three assignments per iteration and only one addition, disregarding the autoincrement of i (see Figure 5.10).

However, there is another, numerical method for computing Fib(n), using a formula discovered by A. de Moivre:

$$\text{Fib}(n) = \frac{\phi^n - \hat{\phi}^n}{\sqrt{5}}$$

where $\phi = \frac{1}{2}(1 + \sqrt{5})$ and $\hat{\phi} = 1 - \phi = \frac{1}{2}(1 - \sqrt{5}) \approx -0.618034$. Because $-1 < \hat{\phi} < 0$, $\hat{\phi}^n$ becomes very small when n grows. Therefore, it can be omitted from the formula and

$$\text{Fib}(n) = \frac{\phi^n}{\sqrt{5}}$$

approximated to the nearest integer. This leads us to the third implementation for computing a Fibonacci number. To round the result to the nearest integer, we use the function `ceil` (for ceiling):

FIGURE 5.10 Comparison of iterative and recursive algorithms for calculating Fibonacci numbers.

		Assignments	
n	Number of Additions	Iterative Algorithm	Recursive Algorithm
6	5	15	25
10	9	27	177
15	14	42	1,973
20	19	57	21,891
25	24	72	242,785
30	29	87	2,692,537

```
unsigned int deMoivreFib (unsigned int n) {
    return ceil(exp(n*log(1.6180339897) - log(2.2360679775)) - .5);
}
```

Try to justify this implementation using the definition of logarithm.

5.9 BACKTRACKING

In solving some problems, a situation arises where there are different ways leading from a given position, none of them known to lead to a solution. After trying one path unsuccessfully, we return to this crossroads and try to find a solution using another path. However, we must ensure that such a return is possible and that all paths can be tried. This technique is called *backtracking,* and it allows us to systematically try all available avenues from a certain point after some of them lead to nowhere. Using backtracking, we can always return to a position that offers other possibilities for successfully solving the problem. This technique is used in artificial intelligence, and one of the problems in which backtracking is very useful is the eight queens problem.

The eight queens problem attempts to place eight queens on a chessboard in such a way that no queen is attacking any other. The rules of chess say that a queen can take another piece if it lies on the same row, on the same column, or on the same diagonal as the queen (see Figure 5.11). To solve this problem, we try to put the first queen on the board, then the second so that it cannot take the first, then the third so that it is not in conflict with the two already placed, and so on, until all of the queens are placed. What happens if, for instance, the sixth queen cannot be placed in a nonconflicting position? We choose another position for the fifth queen and try again with the sixth. If this does not work, the fifth queen is moved again. If all the possible positions for the fifth queen have been tried, the fourth queen is moved and then the

FIGURE 5.11 The eight queens problem.

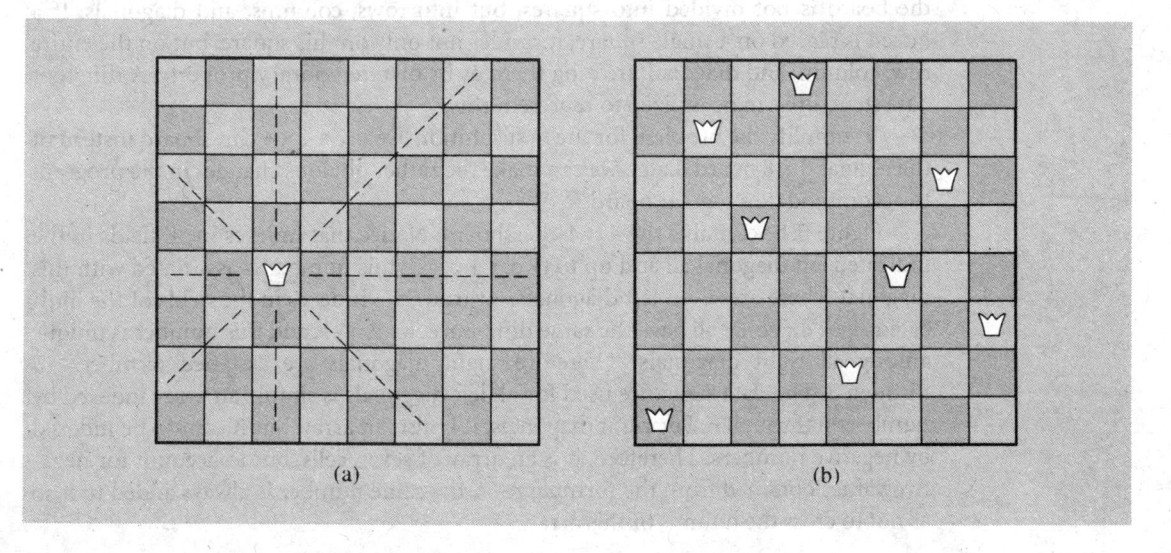

process restarts. This process requires a great deal of effort, most of which is spent backtracking to the first crossroads offering some untried avenues. In terms of code, however, the process is rather simple due to the power of recursion, which is a natural implementation of backtracking. Pseudocode for this backtracking algorithm is as follows (the last line pertains to backtracking):

```
putQueen(row)
    for every position col on the same row
        if position col is available
            place the next queen in position col;
            if (row < 8)
                putQueen(row+1);
            else success;
            remove the queen from position col;
```

This algorithm finds all possible solutions without regard to the fact that some of them are symmetrical.

The most natural approach for implementing this algorithm is to declare an 8 × 8 array board of 1s and 0s representing a chessboard. The array is initialized to 1s, and each time a queen is put in a position (r, c), board[r][c] is set to 0. Also, a function must set to 0, as not available, all positions on row r, in column c, and on both diagonals that cross each other in position (r, c). When backtracking, the same positions (that is, positions on corresponding row, column, and diagonals) have to be set back to 1, as again available. Because we can expect hundreds of attempts to find available positions for queens, the setting and resetting process is the most time-consuming part of the implementation; for each queen, between 22 and 28 positions have to be set and then reset, 15 for row and column, and between 7 and 13 for diagonals.

In this approach, the board is viewed from the perspective of the player who sees the entire board along with all the pieces at the same time. However, if we focus solely on the queens, we can consider the chessboard from their perspective. For the queens, the board is not divided into squares, but into rows, columns, and diagonals. If a queen is placed on a single square, it resides not only on this square, but on the entire row, column, and diagonal, treating them as its own temporary property. A different data structure can be utilized to represent this.

To simplify the problem for the first solution, we use a 4 × 4 chessboard instead of the regular 8 × 8 board. Later, we can make the rather obvious changes in the program to accommodate a regular board.

Figure 5.12 contains the 4 × 4 chessboard. Notice that indexes in all fields in the indicated left diagonal all add up to two, $r + c = 2$; this number is associated with this diagonal. There are seven left diagonals, 0 through 6. Indexes in the fields of the indicated right diagonal all have the same difference, $r - c = -1$, and this number is unique among all right diagonals. Therefore, right diagonals are assigned numbers −3 through 3. The data structure used for all left diagonals is simply an array indexed by numbers 0 through 6. For right diagonals, it is also an array, but it cannot be indexed by negative numbers. Therefore, it is an array of seven cells, but to account for negative values obtained from the formula $r - c$, the same number is always added to it so as not to cross the bounds of this array.

FIGURE 5.12 A 4 × 4 chessboard.

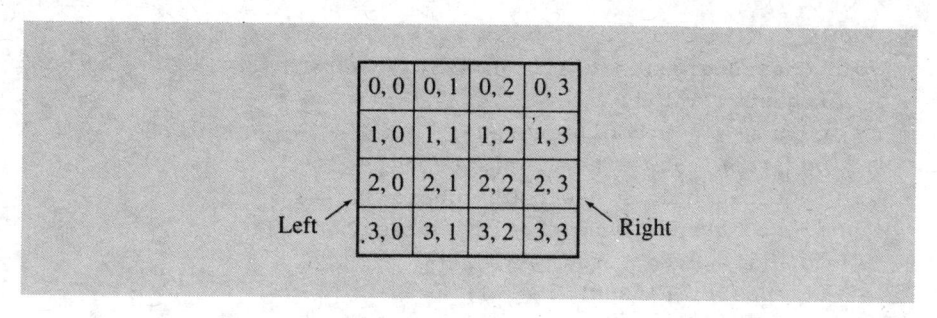

0,0	0,1	0,2	0,3
1,0	1,1	1,2	1,3
2,0	2,1	2,2	2,3
3,0	3,1	3,2	3,3

Left ↗   ↖ Right

An analogous array is also needed for columns, but not for rows, because a queen i is moved along row i and all queens $< i$ have already been placed in rows $< i$. Figure 5.13 contains the code to implement these arrays. The program is short due to recursion, which hides some of the goings-on from the user's sight.

FIGURE 5.13 Eight queens problem implementation.

```
class ChessBoard {
public:
    ChessBoard();      // 8 x 8 chessboard;
    ChessBoard(int);   // n x n chessboard;
    void findSolutions();
private:
    const bool available;
    const int squares, norm;
    bool *column, *leftDiagonal, *rightDiagonal;
    int  *positionInRow, howMany;
    void putQueen(int);
    void printBoard(ostream&);
    void initializeBoard();
};

ChessBoard::ChessBoard() : available(true), squares(8), norm(squares-1)
{
    initializeBoard();
}
ChessBoard::ChessBoard(int n) : available(true), squares(n),
norm(squares-1) {
    initializeBoard();
}
```

Continues

FIGURE 5.13   (continued)

```
void ChessBoard::initializeBoard() {
    register int i;
    column = new bool[squares];
    positionInRow = new int[squares];
    leftDiagonal  = new bool[squares*2 - 1];
    rightDiagonal = new bool[squares*2 - 1];
    for (i = 0; i < squares; i++)
        positionInRow[i] = -1;
    for (i = 0; i < squares; i++)
        column[i] = available;
    for (i = 0; i < squares*2 - 1; i++)
        leftDiagonal[i] = rightDiagonal[i] = available;
    howMany = 0;
}
void ChessBoard::putQueen(int row) {
    for (int col = 0; col < squares; col++)
        if (column[col] == available &&
            leftDiagonal [row+col] == available &&
            rightDiagonal[row-col+norm] == available) {
            positionInRow[row] = col;
            column[col] = !available;
            leftDiagonal[row+col] = !available;
            rightDiagonal[row-col+norm] = !available;
            if (row < squares-1)
                putQueen(row+1);
            else printBoard(cout);
            column[col] = available;
            leftDiagonal[row+col] = available;
            rightDiagonal[row-col+norm] = available;
        }
}
void ChessBoard::findSolutions() {
    putQueen(0);
    cout << howMany << " solutions found.\n";
}
```

Figures 5.14 through 5.17 document the steps taken by putQueen() to place four queens on the chessboard. Figure 5.14 contains the move number, queen number, and row and column number for each attempt to place a queen. Figure 5.15 contains the changes to the arrays positionInRow, column, leftDiagonal, and

Section 5.9 Backtracking ■ 195

FIGURE 5.14 Steps leading to the first successful configuration of four queens as found by the function putQueen().

Move	Queen	row	col	
{1}	1	0	0	
{2}	2	1	2	failure
{3}	2	1	3	
{4}	3	2	1	failure
{5}	1	0	1	
{6}	2	1	3	
{7}	3	2	0	
{8}	4	3	2	

FIGURE 5.15 Changes in the four arrays used by function putQueen().

positionInRow	column	leftDiagonal	rightDiagonal	row
(0, 2, ,)	(!a, a, !a, a)	(!a, a, a, !a, a, a, a)	(a, a, !a, !a, a, a, a)	0, 1
{1}{2}	{1} {2}	{1} {2}	{2}{1}	{1}{2}
(0, 3, 1,)	(!a, !a, a, !a)	(!a, a, a, !a, !a, a, a)	(a, !a, a, !a, !a, a, a)	1, 2
{1}{3}{4}	{1} {4} {3}	{1} {4}{3}	{3} {1} {4}	{3}{4}
(1, 3, 0, 2)	(!a, !a, !a, !a)	(a, !a, !a, a, !a, !a, a)	(a, !a, !a, a, !a, !a, a)	0, 1, 2, 3
{5} {6} {7} {8}	{7} {5} {8} {6}	{5} {7} {6} {8}	{6} {5} {8} {7}	{5}{6}{7}{8}

rightDiagonal. Figure 5.16 shows the changes to the run-time stack during the eight steps. All changes to the run-time stack are depicted by an activation record for each iteration of the for loop, which mostly lead to a new invocation of putQueen(). Each activation record stores a return address and the values of row and col. Figure 5.17 illustrates the changes to the chessboard. A detailed description of each step follows.

{1} We start by trying to put the first queen in the upper left corner (0, 0). Because it is the very first move, the condition in the if statement is met, and the queen is placed in this square. After the queen is placed, the column 0, the main right diagonal, and the leftmost diagonal are marked as unavailable. In Figure 5.15, {1} is put underneath cells reset to !available in this step.

FIGURE 5.16 Changes on the run-time stack for the first successful completion of `putQueen()`.

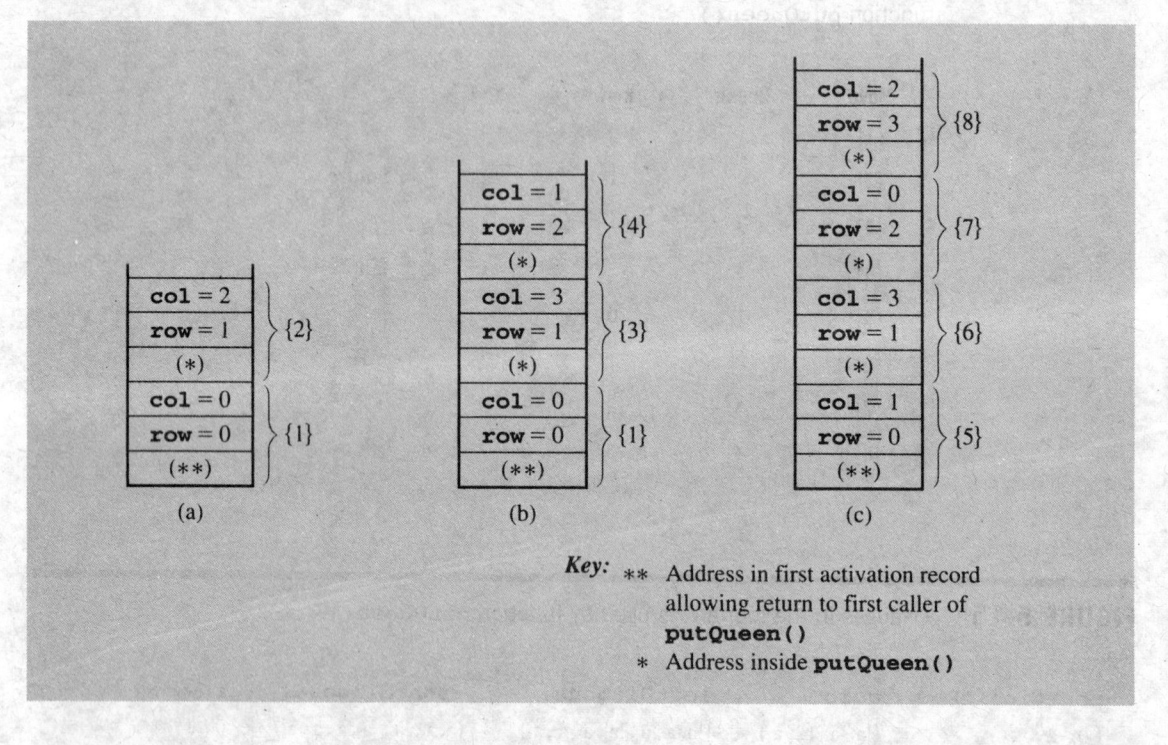

		col = 2	
		row = 3	{8}
		(*)	
	col = 1	col = 0	
	row = 2 {4}	row = 2	{7}
	(*)	(*)	
col = 2	col = 3	col = 3	
row = 1 {2}	row = 1 {3}	row = 1	{6}
(*)	(*)	(*)	
col = 0	col = 0	col = 1	
row = 0 {1}	row = 0 {1}	row = 0	{5}
(**)	(**)	(**)	
(a)	(b)	(c)	

Key: ** Address in first activation record allowing return to first caller of `putQueen()`
* Address inside `putQueen()`

FIGURE 5.17 Changes to the chessboard leading to the first successful configuration.

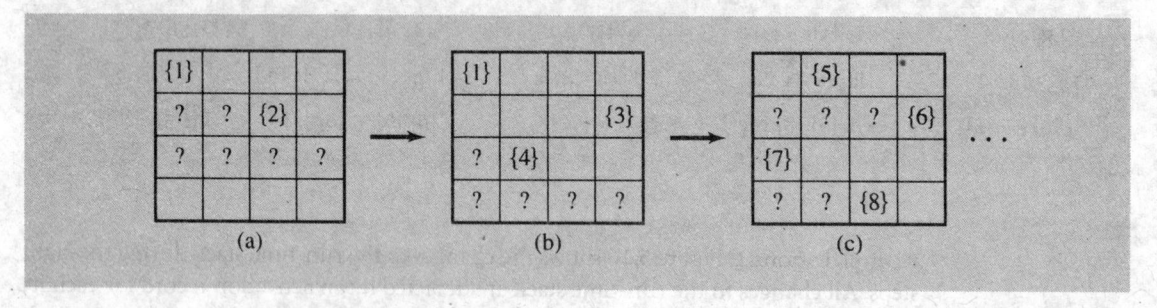

| {1} | | | | | {1} | | | | | | {5} | | * |
| ? | ? | {2} | | | | | | {3} | | ? | ? | ? | {6} | ...
?	?	?	?		?	{4}				{7}			
					?	?	?	?		?	?	{8}	
(a)					(b)					(c)			

{2} Since `row<3`, `putQueen()` calls itself with `row+1`, but before its execution, an activation record is created on the run-time stack (see Figure 5.16a). Now we check the availability of a field on the second row (i.e., `row==1`). For `col==0`, column 0 is guarded, for `col==1`, the main right diagonal is checked, and for `col==2`, all three parts of the `if` statement condition are true. Therefore, the second queen is placed in position (1, 2), and this fact is immediately reflected in the proper cells of all four arrays. Again, `row<3`. `putQueen()` is called trying to locate the third queen in row 2. After all the positions in this row, 0 through 3, are tested, no available position is found, the for loop is exited without executing the body of the if statement, and this call to putQueen() is complete. But this call was executed by putQueen() dealing with the second row, to which control is now returned.

{3} Values of col and row are restored and the execution of the second call of putQueen() continues by resetting some fields in three arrays back to available, and since col==2, the for loop can continue iteration. The test in the if statement allows the second queen to be placed on the board, this time in position (1, 3).

{4} Afterward, putQueen() is called again with row==2, the third queen is put in (2, 1), and after the next call to putQueen(), an attempt to place the fourth queen is unsuccessful (see Figure 5.17b). No calls are made, the call from step {3} is resumed, and the third queen is once again moved, but no position can be found for it. At the same time, col becomes 3, and the for loop is finished.

{5} As a result, the first call of putQueen() resumes execution by placing the first queen in position (0, 1).

{6–8} This time execution continues smoothly and we obtain a complete solution.

Figure 5.18 contains a trace of all calls leading to the first successful placement of four queens on a 4 × 4 chessboard.

FIGURE 5.18 Trace of calls to putQueen() to place four queens.

```
putQueen(0)
  col = 0;
  putQueen(1)
    col = 0;
    col = 1;
    col = 2;
    putQueen(2)
      col = 0;
      col = 1;
      col = 2;
      col = 3;
    col = 3;
    putQueen(2)
      col = 0;
      col = 1;
      putQueen(3)
        col = 0;
        col = 1;
        col = 2;
        col = 3;
```

Continues

FIGURE 5.18 (continued)

```
        col = 2;
        col = 3;
    col = 1;
    putQueen(1)
      col = 0;
      col = 1;
      col = 2;
      col = 3;
      putQueen(2)
        col = 0;
        putQueen(3)
          col = 0;
          col = 1;
          col = 2;
          success
```

5.10 CONCLUDING REMARKS

After looking at all these examples (and one more to follow), what can be said about recursion as a programming tool? Like any other topic in data structures, it should be used with good judgment. There are no general rules for when to use it and when not to use it. Each particular problem decides. Recursion is usually less efficient than its iterative equivalent. But if a recursive program takes 100 milliseconds (ms) for execution, for example, and the iterative version only 10 ms, then although the latter is 10 times faster, the difference is hardly perceivable. If there is an advantage in the clarity, readability, and simplicity of the code, the difference in the execution time between these two versions can be disregarded. Recursion is often simpler than the iterative solution and more consistent with the logic of the original algorithm. The factorial and power functions are such examples, and we will see more interesting cases in chapters to follow.

Although every recursive procedure can be converted into an iterative version, the conversion is not always a trivial task. In particular, it may involve explicitly manipulating a stack. That is where the time–space trade-off comes into play: Using iteration often necessitates the introduction of a new data structure to implement a stack, whereas recursion relieves the programmer of this task by handing it over to the system. One way or the other, if nontail recursion is involved, very often a stack has to be maintained by the programmer or by the system. But the programmer decides who carries the load.

Two situations can be presented in which a nonrecursive implementation is preferable even if recursion is a more natural solution. First, iteration should be used in the so-called real-time systems where an immediate response is vital for proper functioning of the program. For example, in military environments, in the space shuttle, or in certain types of scientific experiments, it may matter whether the response time is 10 ms or 100 ms. Second, the programmer is encouraged to avoid recursion in programs that are executed hundreds of times. The best example of this kind of program is a compiler.

But these remarks should not be treated too stringently, because sometimes a recursive version is faster than a nonrecursive implementation. Hardware may have built-in stack operations that considerably speed up functions operating on the runtime stack, such as recursive functions. Running a simple routine implemented recursively and iteratively and comparing the two run times can help to decide if recursion is advisable—in fact, recursion can execute faster than iteration. Such a test is especially important if tail recursion comes into play. However, when a stack cannot be eliminated from the iterative version, the use of recursion is usually recommended, because the execution time for both versions does not differ substantially—certainly not by a factor of 10.

Recursion should be eliminated if some part of the work is unnecessarily repeated to compute the answer. The Fibonacci series computation is a good example of such a situation. It shows that the ease of using recursion can sometimes be deceptive, and this is where iteration can grapple effectively with run-time limitations and inefficiencies. Whether a recursive implementation leads to unnecessary repetitions may not be immediately apparent; therefore, drawing a tree of calls similar to Figure 5.8 can be very helpful. This tree shows that `Fib(n)` is called many times with the same argument n. A tree drawn for power or factorial functions is reduced to a linked list with no repetitions in it. If such a tree is very deep (that is, it has many levels), then the program can endanger the run-time stack with an overflow. If the tree is shallow and bushy, with many nodes on the same level, then recursion seems to be a good approach—but only if the number of repetitions is very moderate.

5.11 CASE STUDY: A RECURSIVE DESCENT INTERPRETER

All programs written in any programming language have to be translated into a representation that the computer system can execute. However, this is not a simple process. Depending on the system and programming language, the process may consist of translating one executable statement at a time and immediately executing it, which is called *interpretation,* or translating the entire program first and then executing it, which is called *compilation.* Whichever strategy is used, the program should not contain sentences or formulas that violate the formal specification of the programming language in which the program is written. For example, if we want to assign a value to a variable, we must put the variable first, then the equal sign, and then a value after it.

Writing an interpreter is by no means a trivial task. As an example, this case study is a sample interpreter for a limited programming language. Our language consists only of assignment statements; it contains no declarations, `if-else` statements, loops, functions, or the like. For this limited language, we would like to write a program that accepts any input and

- determines if it contains valid assignment statements (this process is known as parsing); and simultaneously,
- evaluates all expressions.

Our program is an interpreter in that it not only checks whether the assignment statements are syntactically correct, but also executes the assignments.

The program is to work in the following way. If we enter the assignment statements

```
var1 = 5;
var2 = var1;
var3 = 44/2.0 * (var2 + var1);
```

then the system can be prompted for the value of each variable separately. For instance, after entering

```
print var3
```

the system should respond by printing

```
var3 = 220
```

Evaluation of all variables stored so far may be requested by entering

```
status
```

and the following values should be printed in our example:

```
var3 = 220
var2 = 5
var1 = 5
```

All current values are to be stored on `idList` and updated if necessary. Thus, if

```
var2 = var2 * 5;
```

is entered, then

```
print var2
```

should return

```
var2 = 25
```

The interpreter prints a message if any undefined identifier is used and if statements and expressions do not conform to common grammatical rules such as unmatched parentheses, two identifiers in a row, and so on.

The program can be written in a variety of ways, but to illustrate recursion, we chose a method known as *recursive descent*. This consists of several mutually recursive functions according to the diagrams in Figure 5.19.

These diagrams serve to define a statement and its parts. For example, a term is a factor or a factor followed by either the multiplication symbol "*" or the division symbol "/" and then another factor. A factor, in turn, is either an identifier, a number, an expression enclosed in a pair of matching parentheses, or a negated factor. In this

FIGURE 5.19 Diagrams of functions used by the recursive descent interpreter.

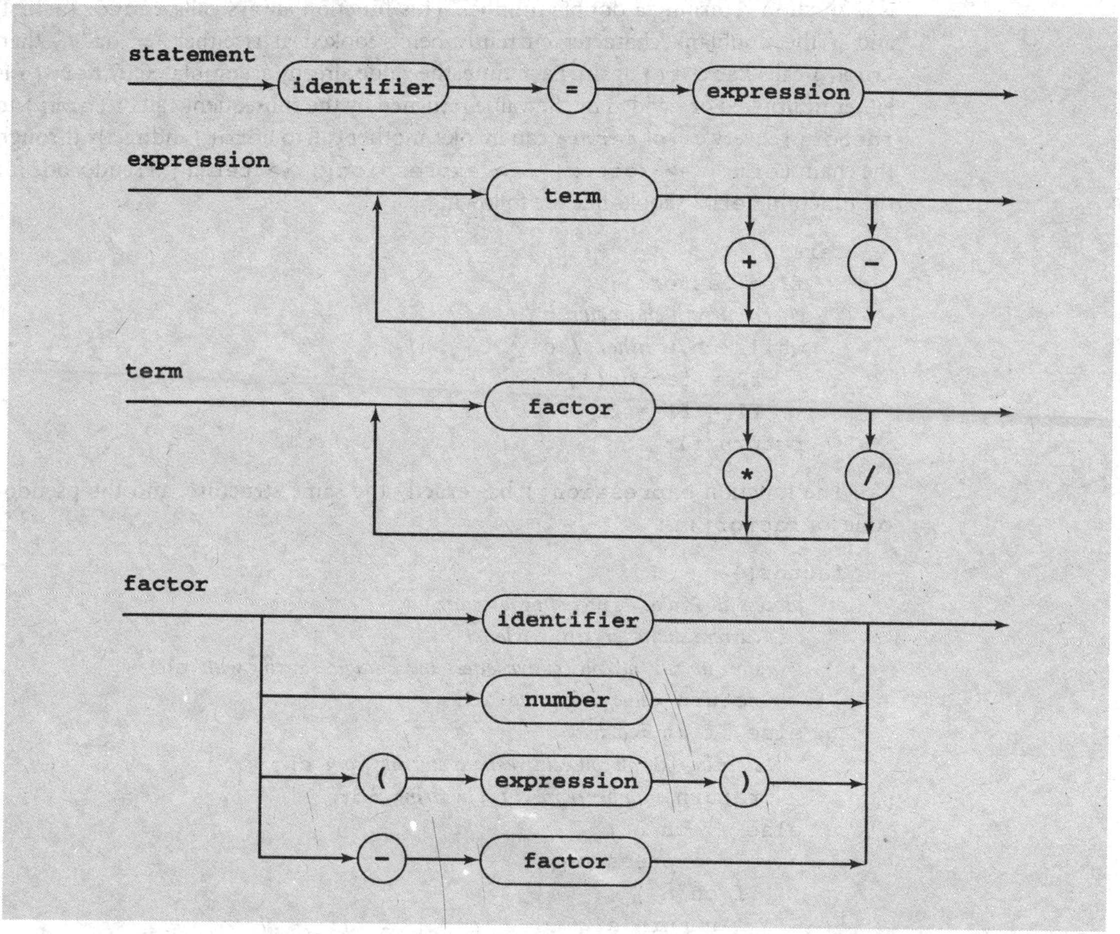

```
statement
    ──────▶( identifier )──▶( = )──▶( expression )──────▶

expression
    ──────────────┬──▶( term )──┬─────────────────▶
                  │             │
                  │           ( + )  ( − )
                  └─────────────┘

term
    ──────────────┬──▶( factor )──┬─────────────────▶
                  │               │
                  │            ( * )  ( / )
                  └───────────────┘

factor
    ──────┬──────────────▶( identifier )──────────────┬──▶
          │                                           │
          ├──────────────▶(  number  )────────────────┤
          │                                           │
          ├──▶( ( )──▶( expression )──▶( ) )──────────┤
          │                                           │
          └──▶( − )──▶(  factor  )───────────────────┘
```

method, a statement is looked at in more and more detail. It is broken down into its components, and if the components are compound, they are separated into their constituent parts until the simplest language elements are found: numbers, variable names, operators, and parentheses. Thus, the program recursively descends from a global overview of the statement to more detailed elements.

The diagrams in Figure 5.19 indicate that recursive descent is a combination of direct and indirect recursion. For example, a factor can be a factor preceded by a minus, an expression can be a term, a term can be a factor, a factor can be an expression that, in turn, can be a term, until the level of identifiers or numbers is found. Thus, an expression can be composed of expressions, a term of terms, and a factor of factors.

How can the recursive descent interpreter be implemented? The simplest approach is to treat every word in the diagrams as a function name. For instance, `term()` is a function returning a double number. This function always calls `factor()` first, and if the nonblank character currently being looked at is either "*" or "/", then `term()` calls `factor()` again. Each time, the value already accumulated by `term()` is either multiplied or divided by the value returned by the subsequent call of `term()` to `factor()`. Every call of `term()` can invoke another call to `term()` indirectly through the chain `term() -> factor() -> expression() -> term()`. Pseudocode for the function `term()` looks like the following:

```
term()
    f1 = factor();
    check current character ch;
    while ch is either / or *
        f2 = factor();
        f1 = f1 * f2 or f1 / f2;
    return f1;
```

The function `expression()` has exactly the same structure, and the pseudocode for `factor()` is:

```
factor()
    process all +s and –s preceding a factor;
    if current character ch is a letter
        store in id all consecutive letters and/or digits starting with ch;
        return value assigned to id;
    else if ch is a digit
        store in id all consecutive digits starting from ch;
        return number represented by string id;
    else if ch is (
        e = expression();
        if ch is )
            return e;
```

We have tacitly assumed that `ch` is a global variable that is used for scanning an input character by character.

However, in the pseudocode, we assumed that only valid statements are entered for evaluation. What happens if a mistake is made, such as entering two equal signs, mistyping a variable name, or forgetting an operator? In the interpreter, parsing is simply discontinued after printing an error message. Figure 5.20 contains the complete code for our interpreter.

FIGURE 5.20 Implementation of a simple language interpreter.

```cpp
//************************** interpreter.h **********************

#ifndef INTERPRETER
#define INTERPRETER

#include <iostream>
#include <list>
#include <algorithm> // find()

using namespace std;

class IdNode {
public:
    IdNode(char *s = "", double e = 0) {
        id = strdup(s);
        value = e;
    }
    bool operator== (const IdNode& node) const {
        return strcmp(id,node.id) == 0;
    }
private:
    char *id;
    double value;
    friend class Statement;
    friend ostream& operator<< (ostream&, const IdNode&);
};

class Statement {
public:
    Statement() {
    }
    void getStatement();
private:
    list<IdNode> idList;
    char ch;
    double factor();
    double term();
    double expression();
    void readId(char*);
    void issueError(char *s) {
```

Continues

FIGURE 5.20   (continued)

```
            cerr << s << endl; exit(1);
        }
        double findValue(char*);
        void    processNode(char*, double);
        friend ostream& operator<< (ostream&, const Statement&);
};

#endif

//************************* interpreter.cpp **********************

#include <cctype>
#include "interpreter.h"

double Statement::findValue(char *id) {
    IdNode tmp(id);
    list<IdNode>::iterator i = find(idList.begin(),idList.end(),tmp);
    if (i != idList.end())
        return i->value;
    else issueError("Unknown variable");
    return 0;  // this statement will never be reached;
}

void Statement::processNode(char* id,double e) {
    IdNode tmp(id,e);
    list<IdNode>::iterator i = find(idList.begin(),idList.end(),tmp);
    if (i != idList.end())
        i->value = e;
    else idList.push_front(tmp);
}

// readId() reads strings of letters and digits that start with
// a letter, and stores them in array passed to it as an actual
// parameter.
// Examples of identifiers are: var1, x, pqr123xyz, aName, etc.

void Statement::readId(char *id) {
    int i = 0;
    if (isspace(ch))
        cin >> ch;        // skip blanks;
    if (isalpha(ch)) {
        while (isalnum(ch)) {
```

FIGURE 5.20   (continued)

```
                id[i++] = ch;
                cin.get(ch); // don't skip blanks;
            }
            id[i] = '\0';
        }
        else issueError("Identifier expected");
}

double Statement::factor() {
    double var, minus = 1.0;
    static char id[200];
    cin >> ch;
    while (ch == '+' || ch == '-') {           // take all '+'s and '-'s.
        if (ch == '-')
            minus *= -1.0;
        cin >> ch;
    }
    if (isdigit(ch) || ch == '.') {            // Factor can be a number
        cin.putback(ch);
        cin >> var >> ch;
    }
    else if (ch == '(') {                      // or a parenthesized
                                               // expression,
        var = expression();
        if (ch == ')')
            cin >> ch;
        else issueError("Right paren left out");
    }
    else {
        readId(id);                            // or an identifier.
        if (isspace(ch))
            cin >> ch;
        var = findValue(id);
    }
    return minus * var;
}

double Statement::term() {
    double f = factor();
    while (true) {
        switch (ch) {
```

Continues

FIGURE 5.20 (continued)

```
                case '*' : f *= factor(); break;
                case '/' : f /= factor(); break;
                default  : return f;
            }
        }
}

double Statement::expression() {
    double t = term();
    while (true) {
        switch (ch) {
            case '+' : t += term(); break;
            case '-' : t -= term(); break;
            default  : return t;
        }
    }
}

void Statement::getStatement() {
    char id[20], command[20];
    double e;
    cout << "Enter a statement: ";
    cin  >> ch;
    readId(id);
    strupr(strcpy(command,id));
    if (strcmp(command,"STATUS") == 0)
         cout << *this;
    else if (strcmp(command,"PRINT") == 0) {
         readId(id);
         cout << id << " = " << findValue(id) << endl;
    }
    else if (strcmp(command,"END") == 0)
         exit(0);
    else {
        if (isspace(ch))
           cin >> ch;
        if (ch == '=') {
            e = expression();
            if (ch != ';')
                 issueError("There are some extras in the statement");
            else processNode(id,e);
        }
```

FIGURE 5.20 *(continued)*

```
            else issueError("'=' is missing");
    }
}

ostream& operator<< (ostream& out, const Statement& s) {
    list<IdNode>::iterator i = s.idList.begin();
    for ( ; i != s.idList.end(); i++)
        out << *i;
    out << endl;
    return out;
}

ostream& operator<< (ostream& out, const IdNode& r) {
    out << r.id << " = " << r.value << endl;
    return out;
}

//*********************** useInterpreter.cpp ***********************

#include "interpreter.h"

using namespace std;

int main() {
    Statement statement;
    cout << "The program processes statements of the following format:\n"
         << "\t<id> = <expr>;\n\tprint <id>\n\tstatus\n\tend\n\n";
    while (true)                    // This infinite loop is broken by
                                    //                       exit(1)
        statement.getStatement();   // in getStatement() or upon finding an
                                    //                       error.
    return 0;
}
```

5.12 EXERCISES

1. The set of natural numbers **N** defined at the beginning of this chapter includes the numbers 10, 11, ..., 20, 21, ..., and also the numbers 00, 000, 01, 001, ... Modify this definition to allow only numbers with no leading zeros.

2. Write a recursive function that calculates and returns the length of a linked list.

3. What is the output for the following version of reverse():

    ```
    void reverse() {
      int ch;
      cin.get(ch);
      if (ch != '\n')
          reverse();
      cout.put(ch);
    }
    ```

4. What is the output of the same function if ch is declared as

    ```
    static char ch;
    ```

5. Write a recursive method that for a positive integer n prints odd numbers
    a. between 1 and n
    b. between n and 1

6. Write a recursive method that for a positive integer returns a string with commas in the appropriate places, for example, putCommas(1234567) returns the string "1,234,567."

7. Write a recursive method to print a *Syracuse sequence* that begins with a number n_0 and each element n_i of the sequence is $n_{i-1}/2$ if n_{i-1} is even and $3n_{i-1} + 1$ otherwise. The sequence ends with 1.

8. Write a recursive method that uses only addition, subtraction, and comparison to multiply two numbers.

9. Write a recursive function to compute the binomial coefficient according to the definition

$$\binom{n}{k} = \begin{cases} 1 & \text{if } k = 0 \text{ or } k = n \\ \binom{n-1}{k-1} + \binom{n-1}{k} & \text{otherwise} \end{cases}$$

10. Write a recursive function to add the first n terms of the series

$$1 + \frac{1}{2} - \frac{1}{3} + \frac{1}{4} - \frac{1}{5} \cdots$$

11. Write a recursive function GCD(n,m) that returns the greatest common divisor of two integers n and m according to the following definition:

$$GCD(n,m) = \begin{cases} m & \text{if } m \leq n \text{ and } n \bmod m = 0 \\ GCD(m,n) & \text{if } n < m \\ GCD(m, n \bmod m) & \text{otherwise} \end{cases}$$

12. Give a recursive version of the following function:

    ```
    void cubes(int n) {
        for (int i = 1; i <=n; i++)
            cout << i * i * i<<' ';
    }
    ```

13. An early application of recursion can be found in the seventeenth century in John Napier's method of finding logarithms. The method was as follows:

    start with two numbers n, m *and their logarithms* logn, logm *if they are known;*
    while *not done*
        for a geometric mean of two earlier numbers find a logarithm which is an arithmetic mean of two earlier logarithms, that is, logk = (logn+logm)/2 *for* k = √nm;
        proceed recursively for pairs (n, √nm) *and* (√nm, m);

    For example, the 10-based logarithms of 100 and 1,000 are numbers 2 and 3, the geometric mean of 100 and 1,000 is 316.23, and the arithmetic mean of their logarithms, 2 and 3, is 2.5. Thus, the logarithm of 316.23 equals 2.5. The process can be continued: The geometric mean of 100 and 316.23 is 177.83, whose logarithm is equal to (2 + 2.5)/2 = 2.25.

    a. Write a recursive function logarithm() that outputs logarithms until the difference between adjacent logarithms is smaller than a certain small number.
    b. Modify this function so that a new function logarithmOf() finds a logarithm of a specific number x between 100 and 1,000. Stop processing if you reach a number y such that $y - x < \epsilon$ for some ϵ.
    c. Add a function that calls logarithmOf() after determining between what powers of 10 a number x falls so that it does not have to be a number between 100 and 1,000.

14. The algorithms for both versions of the power function given in this chapter are rather simpleminded. Is it really necessary to make eight multiplications to compute x^8? It can be observed that $x^8 = (x^4)^2$, $x^4 = (x^2)^2$, and $x^2 = x \cdot x$; that is, only three multiplications are needed to find the value of x^8. Using this observation, improve both algorithms for computing x^n. Hint: A special case is needed for odd exponents.

15. Execute by hand the functions tail() and nonTail() for the parameter values of 0, 2, and 4. Definitions of these functions are given in Section 5.4.

16. Check recursively if the following objects are palindromes:
    a. a word
    b. a sentence (ignoring blanks, lower- and uppercase differences, and punctuation marks so that "Madam, I'm Adam" is accepted as a palindrome)

17. For a given character recursively, without using strchr() or strrchr(),
    a. Check if it is in a string.
    b. Count all of its occurrences in a string.
    c. Remove all of its occurrences from a string.

18. Write equivalents of the last three functions for substrings (do not use strstr()).

19. What changes have to be made in the program in Figure 5.6 to draw a line as in Figure 5.21? Try it and experiment with other possibilities to generate other curves.

FIGURE 5.21 Lines to be drawn with modified program in Figure 5.6.

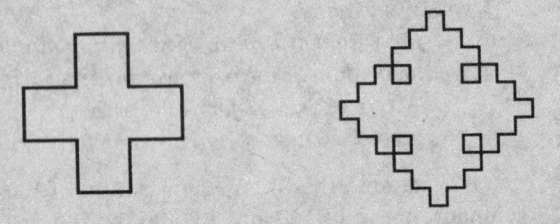

20. Create a tree of calls for sin(x) assuming that only $\frac{x}{18}$ (and smaller values) do not trigger other calls.

21. Write recursive and nonrecursive functions to print out a nonnegative integer in binary. The functions should not use bitwise operations.

22. The nonrecursive version of the function for computing Fibonacci numbers uses information accumulated during computation, whereas the recursive version does not. However, it does not mean that no recursive implementation can be given that can collect the same information as the nonrecursive counterpart. In fact, such an implementation can be obtained directly from the nonrecursive version. What would it be? Consider using two functions instead of one; one would do all the work, and the other would only invoke it with the proper parameters.

23. The function putQueen() does not recognize that certain configurations are symmetric. Adapt function putQueen() for a full 8 × 8 chessboard, write the function printBoard(), and run a program for solving the eight queens problem so that it does not print symmetric solutions.

24. Finish the trace of execution of putQueen() shown in Figure 5.18.

25. Execute the following program by hand from the case study, using these two entries:
    a. `v = x + y*w - z`
    b. `v = x * (y - w) --z`

    Indicate clearly which functions are called at which stage of parsing these sentences.

26. Extend our interpreter so that it can also process exponentiation, ^. Remember that exponentiation has precedence over all other operations so that `2 - 3^4 * 5` is the same as `2 - ((3^4) * 5)`. Notice also that exponentiation is a right-associative operator (unlike addition or multiplication); that is, `2^3^4` is the same as `2^(3^4)` and not `(2^3)^4`.

27. In C++, the division operator, /, returns an integer result when it is applied to two integers; for instance, `11/5` equals 2. However, in our interpreter, the result is `2.2`. Modify this interpreter so that division works the same way as in C++.

28. Our interpreter is unforgiving when a mistake is made by the user, because it finishes execution if a problem is detected. For example, when the name of a variable is mistyped when requesting its value, the program notifies the user and exits and destroys the list of identifiers. Modify the program so that it continues execution after finding an error.

29. Write the shortest program you can that uses recursion.

5.13 PROGRAMMING ASSIGNMENTS

1. Compute the standard deviation σ for n values x_k stored in an array `data` and for the equal probabilities $\frac{1}{n}$ associated with them. The standard deviation is defined as

$$\sigma = \sqrt{V}$$

   where the variance, V, is defined by

$$V = \frac{1}{n-1} \Sigma_k (x_k - \bar{x})^2$$

   and the mean, $\bar{x}$, by

$$\bar{x} = \frac{1}{n} \Sigma_k x_k$$

   Write recursive and iterative versions of both V and $\bar{x}$ and compute the standard deviation using both versions of the mean and variance. Run your program for $n = 500$, 1,000, 1,500, and 2,000 and compare the run times.

2. Write a program to do symbolic differentiation. Use the following formulas:

   Rule 1: $(fg)' = fg' + f'g$

   Rule 2: $(f + g)' = f' + g'$

   Rule 3: $\left(\frac{f}{g}\right)' = \frac{f'g - fg'}{g^2}$

   Rule 4: $(ax^n)' = nax^{n-1}$

An example of application of these rules is given below for differentiation with respect to x:

$$\left(5x^3 + \frac{6x}{y} - 10x^2y + 100\right)'$$

$$= (5x^3)' + \left(\frac{6x}{y}\right)' + (-10x^2y)' + (100)' \qquad \text{by Rule 2}$$

$$= 15x^2 + \left(\frac{6x}{y}\right)' + (-10x^2y)' \qquad \text{by Rule 4}$$

$$= 15x^2 + \frac{(6x)'y - (6x)y'}{y^2} + (-10x^2y)' \qquad \text{by Rule 3}$$

$$= 15x^2 + \frac{6y}{y^2} + (-10x^2y)' \qquad \text{by Rule 4}$$

$$= 15x^2 + \frac{6y}{y^2} + (-10x^2)y' + (-10x^2)'y \qquad \text{by Rule 1}$$

$$= 15x^2 + \frac{6}{y} - 20xy \qquad \text{by Rule 4}$$

First, run your program for polynomials only, and then add formulas for derivatives for trigonometric functions, logarithms, and so on, that extend the range of functions handled by your program.

3. An $n \times n$ square consists of black and white cells arranged in a certain way. The problem is to determine the number of white areas and the number of white cells in each area. For example, a regular 8×8 chessboard has 32 one-cell white areas; the square in Figure 5.22a consists of 10 areas, 2 of them of 10 cells, and 8 of 2 cells; the square in Figure 5.22b has 5 white areas of 1, 3, 21, 10, and 2 cells.

   Write a program that, for a given $n \times n$ square, outputs the number of white areas and their sizes. Use an $(n + 2) \times (n + 2)$ array with properly marked cells. Two additional rows and columns constitute a frame of black cells surrounding the entered square to simplify your implementation. For instance, the square in Figure 5.22b is stored as the square in Figure 5.22c.

FIGURE 5.22 (a–b) Two $n \times n$ squares of black and white cells and (c) an $(n + 2) \times (n + 2)$ array implementing square (b).

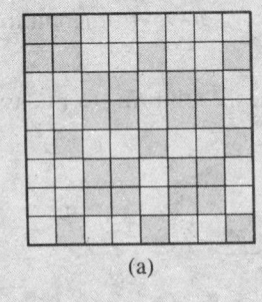

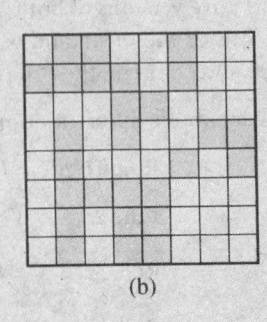

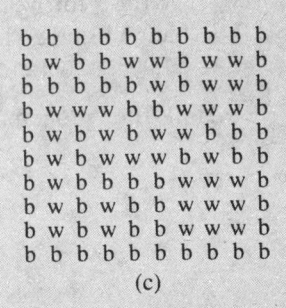

Traverse the square row by row and, for the first unvisited cell encountered, invoke a function that processes one area. The secret is in using four recursive calls in this function for each unvisited white cell and marking it with a special symbol as visited (counted).

4. Write a program for *pretty printing* C++ programs; that is, for printing programs with consistent use of indentation, the number of spaces between tokens such as key words, parentheses, brackets, operators, the number of blank lines between blocks of code (classes, functions), aligning braces with key words, aligning `else` statements with the corresponding `if` statements, and so on. The program takes as input a C++ file and prints code in this file according to the rules incorporated in the pretty printing program. For example, the code

```
if (n == 1) { n = 2 * m;
if (m < 10)
f(n,m-1); else f(n,m-2); } else n = 3 * m;
```

should be transformed into

```
if (n == 1) {
    n = 2 * m;
    if (m < 10)
        f(n,m-1);
    else f(n,m-2);
}
else n = 3 * m;
```

5. An excellent example of a program that can be greatly simplified by the use of recursion is the Chapter 4 case study, escaping a maze. As already explained, in each maze cell the mouse stores on the maze stack up to four cells neighboring the cell in which it is currently located. The cells put on the stack are the ones that should be investigated after reaching a dead end. It does the same for each visited cell. Write a program that uses recursion to solve the maze problem. Use the following pseudocode:

```
exitCell(currentCell)
    if currentCell is the exit
        success;
    else exitCell(the passage above currentCell);
        exitCell(the passage below currentCell);
        exitCell(the passage left to currentCell);
        exitCell(the passage right to currentCell);
```

BIBLIOGRAPHY

Recursion and Applications of Recursion

Barron, David W., *Recursive Techniques in Programming*, New York: Elsevier, 1975.

Berlioux, Pierre, and Bizard, Philippe, *Algorithms: The Construction, Proof, and Analysis of Programs*, New York: Wiley, 1986, Chs. 4–6.

Bird, Richard S., *Programs and Machines*, New York: Wiley, 1976.

Burge, William H., *Recursive Programming Techniques*, Reading, MA: Addison-Wesley, 1975.

Lorentz, Richard, *Recursive Algorithms*, Norwood, NJ: Ablex, 1994.

Roberts, Eric, *Thinking Recursively*, New York: Wiley, 1986.

Rohl, Jeffrey S., *Recursion via Pascal*, Cambridge: Cambridge University Press, 1984.

Transformations between Recursion and Iteration

Auslander, M. A., and Strong, H. R., "Systematic Recursion Removal," *Communications of the ACM* 21 (1978), 127–134.

Bird, R. S., "Notes on Recursion Elimination," *Communications of the ACM* 20 (1977), 434–439.

Dijkstra, Edsger W., "Recursive Programming," *Numerische Mathematik* 2 (1960), 312–318.

Algorithm to Solve the Eight Queens Problem

Wirth, Niklaus, *Algorithms and Data Structures*, Englewood Cliffs, NJ: Prentice Hall, 1986.

Binary Trees

6.1 TREES, BINARY TREES, AND BINARY SEARCH TREES

Linked lists usually provide greater flexibility than arrays, but they are linear structures and it is difficult to use them to organize a hierarchical representation of objects. Although stacks and queues reflect some hierarchy, they are limited to only one dimension. To overcome this limitation, we create a new data type called a *tree* that consists of *nodes* and *arcs*. Unlike natural trees, these trees are depicted upside down with the *root* at the top and the *leaves* (*terminal nodes*) at the bottom. The root is a node that has no parent; it can have only child nodes. Leaves, on the other hand, have no children, or rather, their children are empty structures. A tree can be defined recursively as the following:

1. An empty structure is an empty tree.
2. If $t_1, \ldots, t_k$ are disjointed trees, then the structure whose root has as its children the roots of $t_1, \ldots, t_k$ is also a tree.
3. Only structures generated by rules 1 and 2 are trees.

Figure 6.1 contains examples of trees. Each node has to be reachable from the root through a unique sequence of arcs, called a *path*. The number of arcs in a path is called the *length* of the path. The *level* of a node is the length of the path from the root to the node plus 1, which is the number of nodes in the path. The *height* of a nonempty tree is the maximum level of a node in the tree. The empty tree is a legitimate tree of height 0 (by definition), and a single node is a tree of height 1. This is the only case in which a node is both the root and a leaf. The level of a node must be between 1 (the level of the root) and the height of the tree, which in the extreme case is the level of the only leaf in a degenerate tree resembling a linked list.

Figure 6.2 contains an example of a tree that reflects the hierarchy of a university. Other examples are genealogical trees, trees reflecting the grammatical structure of sentences, and trees showing the taxonomic structure of organisms, plants, or characters. Virtually all areas of science make use of trees to represent hierarchical structures.

FIGURE 6.1 Examples of trees.

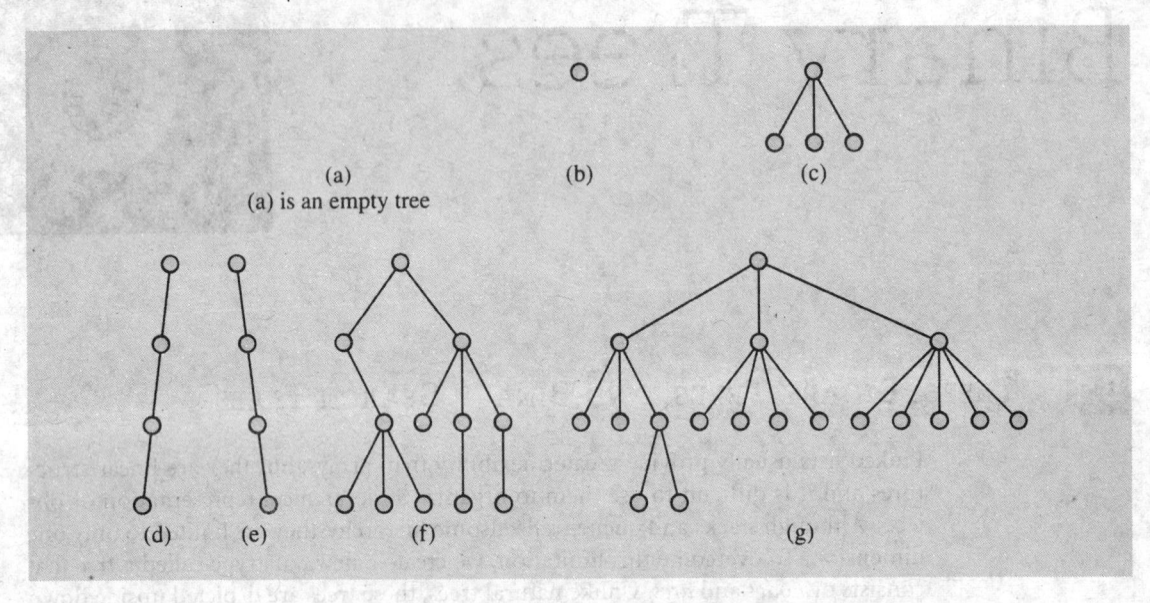

(a)
(a) is an empty tree
(b)
(c)
(d) (e) (f) (g)

FIGURE 6.2 Hierarchical structure of a university shown as a tree.

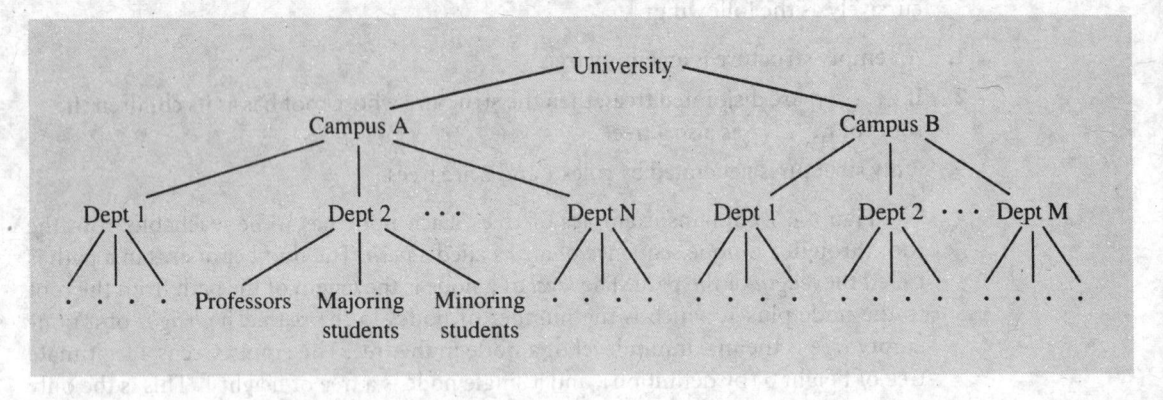

    The definition of a tree does not impose any condition on the number of children of a given node. This number can vary from 0 to any integer. In hierarchical trees, this is a welcome property. For example, the university has only two branches, but each campus can have a different number of departments. Such trees are used in database management systems, especially in the hierarchical model. But representing hierarchies is not the only reason for using trees. In fact, in the discussion to follow, that aspect of trees is treated rather lightly, mainly in the discussion of expression trees. This chapter focuses on tree operations that allow us to accelerate the search process.

Consider a linked list of n elements. To locate an element, the search has to start from the beginning of the list, and the list must be scanned until the element is found or the end of the list is reached. Even if the list is ordered, the search of the list always has to start from the first node. Thus, if the list has 10,000 nodes and the information in the last node is to be accessed, then all 9,999 of its predecessors have to be traversed, an obvious inconvenience. If all the elements are stored in an *orderly tree*, a tree where all elements are stored according to some predetermined criterion of ordering, the number of tests can be reduced substantially even when the element to be located is the one furthest away. For example, the linked list in Figure 6.3a can be transformed into the tree in Figure 6.3b.

FIGURE 6.3   Transforming (a) a linked list into (b) a tree.

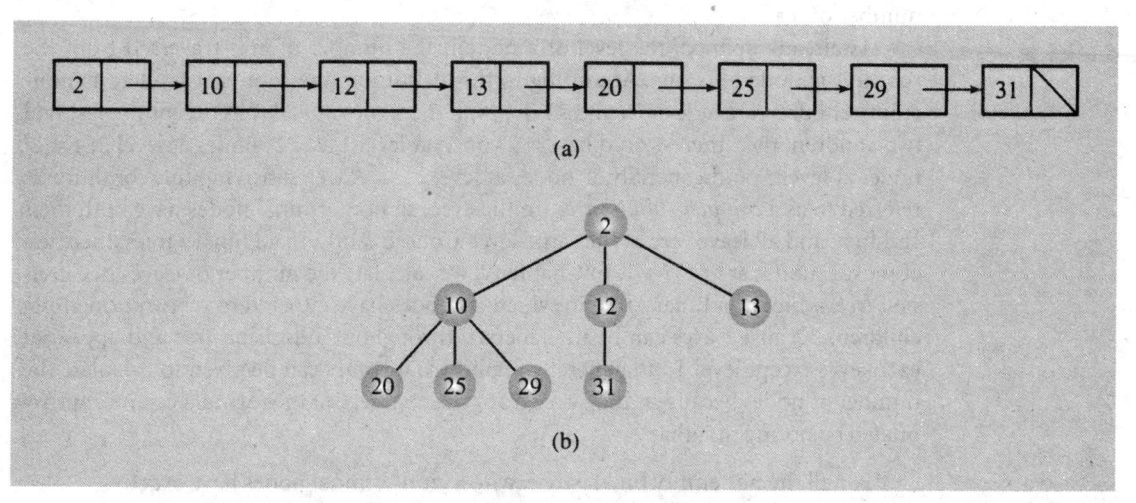

Was a reasonable criterion of ordering applied to construct this tree? To test whether 31 is in the linked list, eight tests have to be performed. Can this number be reduced further if the same elements are ordered from top to bottom and from left to right in the tree? What would an algorithm be like that forces us to make three tests only: one for the root, 2; one for its middle child, 12; and one for the only child of this child, 31? The number 31 could be located on the same level as 12, or it could be a child of 10. With this ordering of the tree, nothing really interesting is achieved in the context of searching. (The heap discussed later in this chapter uses this approach.) Consequently, a better criterion must be chosen.

Again, note that each node can have any number of children. In fact, there are algorithms developed for trees with a deliberate number of children (see the next chapter), but this chapter discusses only binary trees. A *binary tree* is a tree whose nodes have two children (possibly empty), and each child is designated as either a left child or a right child. For example, the trees in Figure 6.4 are binary trees, whereas the university tree in Figure 6.2 is not. An important characteristic of binary trees,

FIGURE 6.4 Examples of binary trees.

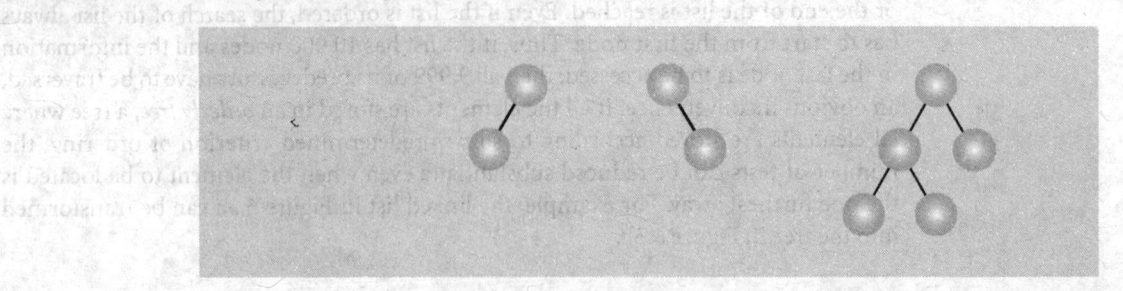

which is used later in assessing an expected efficiency of sorting algorithms, is the number of leaves.

As already defined, the level of a node is the number of arcs traversed from the root to the node plus one. According to this definition, the root is at level 1, its nonempty children are at level 2, and so on. If all the nodes at all levels except the last had two children, then there would be $1 = 2^0$ node at level 1, $2 = 2^1$ nodes at level 2, $4 = 2^2$ nodes at level 3, and generally, 2^i nodes at level $i + 1$. A tree satisfying this condition is referred to as a *complete binary tree*. In this tree, all nonterminal nodes have both their children, and all leaves are at the same level. Consequently, in all binary trees, there are at most 2^i nodes at level $i + 1$. In Chapter 9, we calculate the number of leaves in a *decision tree*, which is a binary tree in which all nodes have either zero or two nonempty children. Because leaves can be interspersed throughout a decision tree and appear at each level except level 1, no generally applicable formula can be given to calculate the number of nodes because it may vary from tree to tree. But the formula can be approximated by noting first that

> For all the nonempty binary trees whose nonterminal nodes have exactly two nonempty children, the number of leaves m is greater than the number of nonterminal nodes k and $m = k + 1$.

If a tree has only a root, this observation holds trivially. If it holds for a certain tree, then after attaching two leaves to one of the already existing leaves, this leaf turns into a nonterminal node, whereby m is decremented by 1 and k is incremented by 1. However, because two new leaves have been grafted onto the tree, m is incremented by 2. After these two increments and one decrement, the equation $(m - 1) + 2 = (k + 1) + 1$ is obtained and $m = k + 1$, which is exactly the result aimed at (see Figure 6.5). It implies that an $i + 1$–level complete decision tree has 2^i leaves, and due to the preceding observation, it also has $2^i - 1$ nonterminal nodes, which makes $2^i + 2^i - 1 = 2^{i+1} - 1$ nodes in total (see also Figure 6.35).

In this chapter, the *binary search trees,* also called *ordered binary trees,* are of particular interest. A binary search tree has the following property: For each node n of the tree, all values stored in its left subtree (the tree whose root is the left child) are less than value v stored in n, and all values stored in the right subtree are greater than v. For reasons to be discussed later, storing multiple copies of the same value in the same

FIGURE 6.5
Adding a leaf to tree (a), preserving the relation of the number of leaves to the number of nonterminal nodes (b).

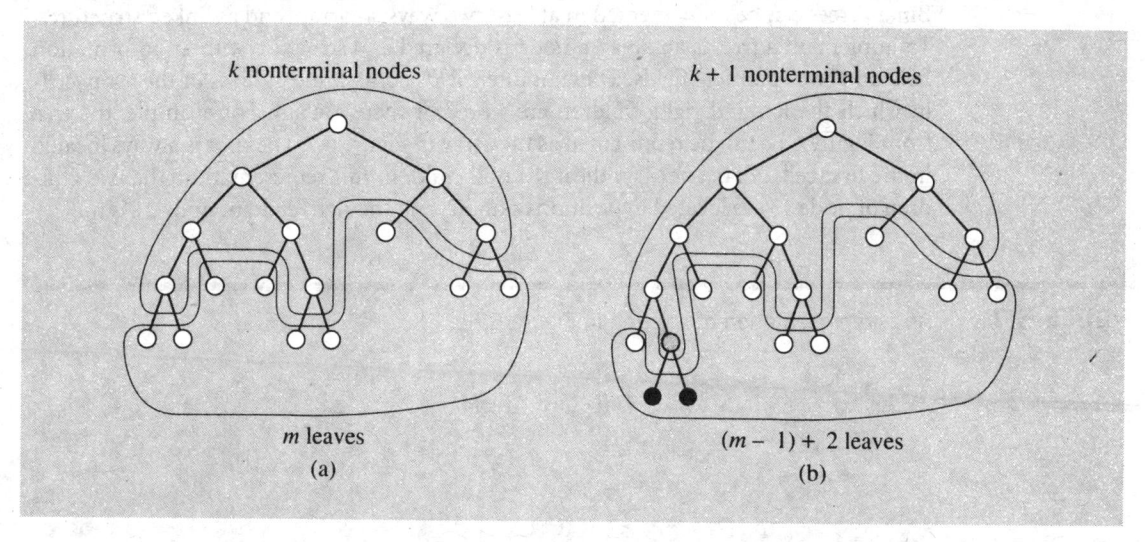

tree is avoided. An attempt to do so can be treated as an error. The meanings of "less than" or "greater than" depend on the type of values stored in the tree. We use operators "<" and ">", which can be overloaded depending on the content. Alphabetical order is also used in the case of strings. The trees in Figure 6.6 are binary search trees. Note that Figure 6.6c contains a tree with the same data as the linked list in Figure 6.3a whose searching was to be optimized.

FIGURE 6.6
Examples of binary search trees.

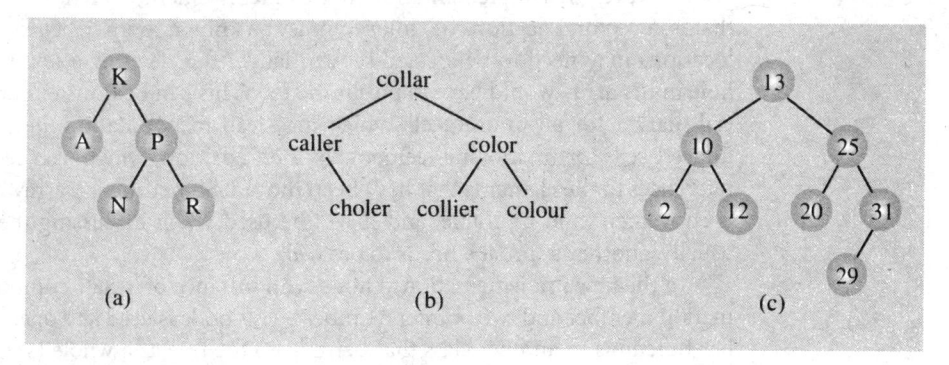

6.2 IMPLEMENTING BINARY TREES

Binary trees can be implemented in at least two ways: as arrays and as linked structures. To implement a tree as an array, a node is declared as a structure with an information field and two "pointer" fields. These pointer fields contain the indexes of the array cells in which the left and right children are stored, if there are any. For example, the tree from Figure 6.6c can be represented as the array in Figure 6.7. The root is always located in the first cell, cell 0, and –1 indicates a null child. In this representation, the two children of node 13 are located in positions 4 and 2, and the right child of node 31 is null.

FIGURE 6.7 Array representation of the tree in Figure 6.6c.

Index	Info	Left	Right
0	13	4	2
1	31	6	−1
2	25	7	1
3	12	−1	−1
4	10	5	3
5	2	−1	−1
6	29	−1	−1
7	20	−1	−1

However, this implementation may be inconvenient, even if the array is flexible—that is, a vector. Locations of children must be known to insert a new node, and these locations may need to be located sequentially. After deleting a node from the tree, a hole in the array would have to be eliminated. This can be done either by using a special marker for an unused cell, which may lead to populating the array with many unused cells, or by moving elements by one position, which also requires updating references to the elements that have been moved. Sometimes an array implementation is convenient and desirable, and it will be used when discussing the heap sort. But usually, another approach needs to be used.

In the new implementation, a node is an instance of a class composed of an information member and two pointer members. This node is used and operated on by member functions in another class that pertains to the tree as a whole (see Figure 6.8). For this reason, members of `BSTNode` are declared public because they can be accessible only from nonpublic members of objects of type `BST` so that the information-hiding principle still stands. It is important to have members of `BSTNode` be public because otherwise they are not accessible to classes derived from `BST`.

FIGURE 6.8 Implementation of a generic binary search tree.

```
//********************* genBST.h *************************
//               generic binary search tree

#include <queue>
#include <stack>

using namespace std;

#ifndef BINARY_SEARCH_TREE
#define BINARY_SEARCH_TREE

template<class T>
class Stack : public stack<T> { ... } // as in Figure 4.21

template<class T>
class Queue : public queue<T> {
public:
    T dequeue() {
        T tmp = front();
        queue<T>::pop();
        return tmp;
    }
    void enqueue(const T& el) {
        push(el);
    }
};

template<class T>
class BSTNode {
public:
    BSTNode() {
        left = right = 0;
    }
    BSTNode(const T& el, BSTNode *l = 0, BSTNode *r = 0) {
        key = el; left = l; right = r;
    }
    T key;
    BSTNode *left, *right;
};
```

Continues

FIGURE 6.8   (continued)

```
template<class T>
class BST {
public:
    BST() {
        root = 0;
    }
    ~BST() {
        clear();
    }
    void clear() {
        clear(root); root = 0;
    }
    bool isEmpty() const {
        return root == 0;
    }
    void preorder() {
        preorder(root);                          // Figure 6.11
    }
    void inorder() {
        inorder(root);                           // Figure 6.11
    }
    void postorder() {
        postorder(root);                         // Figure 6.11
    }
    T* search(const T& el) const {
        return search(root,el);                  // Figure 6.9
    }
    void breadthFirst();                         // Figure 6.10
    void iterativePreorder();                    // Figure 6.15
    void iterativeInorder();                     // Figure 6.17
    void iterativePostorder();                   // Figure 6.16
    void MorrisInorder();                        // Figure 6.20
    void insert(const T&);                       // Figure 6.23
    void deleteByMerging(BSTNode<T>*&);          // Figure 6.29
    void findAndDeleteByMerging(const T&);       // Figure 6.29
    void deleteByCopying(BSTNode<T>*&);          // Figure 6.32
    void balance(T*,int,int);                    // Section 6.7
    . . . . . . . . . . . . . . .
protected:
    BSTNode<T>* root;
```

FIGURE 6.8 (continued)

```
    void clear(BSTNode<T>*);
    T* search(BSTNode<T>*, const T&) const;   // Figure 6.9
    void preorder(BSTNode<T>*);                // Figure 6.11
    void inorder(BSTNode<T>*);                 // Figure 6.11
    void postorder(BSTNode<T>*);               // Figure 6.11
    virtual void visit(BSTNode<T>* p) {
        cout << p->key << ' ';
    }
    . . . . . . . . . . . . . . .
};

#endif
```

6.3 SEARCHING A BINARY SEARCH TREE

An algorithm for locating an element in this tree is quite straightforward, as indicated by its implementation in Figure 6.9. For every node, compare the key to be located with the value stored in the node currently pointed at. If the key is less than the value, go to the left subtree and try again. If it is greater than that value, try the right subtree. If it is the same, obviously the search can be discontinued. The search is also aborted if there is no way to go, indicating that the key is not in the tree. For example, to locate the number 31 in the tree in Figure 6.6c, only three tests are performed. First, the tree is checked to see if the number is in the root node. Next, because 31 is greater than 13, the root's right child containing the value 25 is tried. Finally, because 31 is again

FIGURE 6.9 A function for searching a binary search tree.

```
template<class T>
T* BST<T>::search(BSTNode<T>* p, const T& el) const {
    while (p != 0)
        if (el == p->key)
            return &p->key;
        else if (el < p->key)
            p = p->left;
        else p = p->right;
    return 0;
}
```

greater than the value of the currently tested node, the right child is tried again, and the value 31 is found.

The worst case for this binary tree is when it is searched for the numbers 26, 27, 28, 29, or 30 because those searches each require four tests (why?). In the case of all other integers, the number of tests is fewer than four. It can now be seen why an element should only occur in a tree once. If it occurs more than once, then two approaches are possible. One approach locates the first occurrence of an element and disregards the others. In this case, the tree contains redundant nodes that are never used for their own sake; they are accessed only for testing. In the second approach, all occurrences of an element may have to be located. Such a search always has to finish with a leaf. For example, to locate all instances of 13 in the tree, the root node 13 has to be tested, then its right child 25, and finally the node 20. The search proceeds along the worst-case scenario: when the leaf level has to be reached in expectation that some more occurrences of the desired element can be encountered.

The complexity of searching is measured by the number of comparisons performed during the searching process. This number depends on the number of nodes encountered on the unique path leading from the root to the node being searched for. Therefore, the complexity is the length of the path leading to this node plus 1. Complexity depends on the shape of the tree and the position of the node in the tree.

The *internal path length (IPL)* is the sum of all path lengths of all nodes, which is calculated by summing $\sum(i-1)l_i$ over all levels i, where l_i is the number of nodes on level i. A depth of a node in the tree is determined by the path length. An average depth, called an *average path length*, is given by the formula IPL/n, which depends on the shape of the tree. In the worst case, when the tree turns into a linked list, $path_{worst} = \frac{1}{n}\sum_{i=1}^{n}(i-1) = \frac{n-1}{2} = O(n)$, and a search can take n time units.

The best case occurs when all leaves in the tree of height h are in at most two levels, and only nodes in the next to last level can have one child. To simplify the computation, we approximate the average path length for such a tree, $path_{best}$, by the average path of a complete binary tree of the same height.

By looking at simple examples, we can determine that for the complete binary tree of height h, IPL $= \sum_{i=1}^{h-1} i 2^i$. From this and from the fact that $\sum_{i=1}^{h-1} 2^i = 2^h - 2$, we have

$$\text{IPL} = 2\text{IPL} - \text{IPL} = (h-1)2^h - \sum_{i=1}^{h-1} 2^i = (h-2)2^h + 2$$

As has already been established, the number of nodes in the complete binary tree $n = 2^h - 1$, so

$$path_{best} = \text{IPL}/n = \left((h-2)2^h + 2\right)/(2^h - 1) \approx h - 2$$

which is in accordance with the fact that, in this tree, one-half of the nodes are in the leaf level with path length $h-1$. Also, in this tree, the height $h = \lg(n+1)$, so $path_{best} = \lg(n+1) - 2$; the average path length in a perfectly balanced tree is $\lceil \lg(n+1) \rceil - 2 = O(\lg n)$ where $\lceil x \rceil$ is the closest integer greater than x.

The average case in an average tree is somewhere between $\frac{n-1}{2}$ and $\lg(n+1) - 2$. Is a search for a node in an average position in a tree of average shape closer to $O(n)$ or $O(\lg n)$? First, the average shape of the tree has to be represented computationally.

The root of a binary tree can have an empty left subtree and a right subtree with all $n-1$ nodes. It also can have one node in the left subtree and $n-2$ nodes in the right, and so on. Finally, it can have an empty right subtree with all remaining nodes in the left. The same reasoning can be applied to both subtrees of the root, to the subtrees of these subtrees, down to the leaves. The average internal path length is the average over all these differently shaped trees.

Assume that the tree contains nodes 1 through n. If i is the root, then its left subtree has $i - 1$ nodes, and its right subtree has $n - i$ nodes. If $path_{i-1}$ and $path_{n-i}$ are average paths in these subtrees, then the average path of this tree is

$$path_n(i) = ((i-1)(path_{i-1} + 1) + (n-i)(path_{n-i} + 1))/n$$

Assuming that elements are coming randomly to the tree, the root of the tree can be any number i, $1 \le i \le n$. Therefore, the average path of an average tree is obtained by averaging all values of $path_n(i)$ over all values of i. This gives the formula

$$path_n = \frac{1}{n}\sum_{i=1}^{n} path_n(i) = \frac{1}{n^2}\sum_{i=1}^{n}((i-1)(path_{i-1} + 1) + (n-i)(path_{n-i} + 1))$$

$$= \frac{2}{n^2}\sum_{i=1}^{n-1} i(path_i + 1)$$

from which, and from $path_1 = 0$, we obtain $2 \ln n = 2 \ln 2 \lg n = 1.386 \lg n$ as an approximation for $path_n$ (see section A.4 in Appendix A). This is an approximation for the average number of comparisons in an average tree. This number is $O(\lg n)$, which is closer to the best case than to the worst case. This number also indicates that there is little room for improvement, because $path_{best}/path_n \approx .7215$, and the average path length in the best case is different by only 27.85% from the expected path length in the average case. Searching in a binary tree is, therefore, very efficient in most cases, even without balancing the tree. However, this is true only for randomly created trees because, in highly unbalanced and elongated trees whose shapes resemble linked lists, search time is $O(n)$, which is unacceptable considering that $O(\lg n)$ efficiency can be achieved.

6.4 TREE TRAVERSAL

Tree traversal is the process of visiting each node in the tree exactly one time. Traversal may be interpreted as putting all nodes on one line or linearizing a tree.

The definition of traversal specifies only one condition—visiting each node only one time—but it does not specify the order in which the nodes are visited. Hence, there are as many tree traversals as there are permutations of nodes; for a tree with n nodes, there are $n!$ different traversals. Most of them, however, are rather chaotic and do not indicate much regularity so that implementing such traversals lacks generality: For each n, a separate set of traversal procedures must be implemented, and only a few of them can be used for a different number of data. For example, two possible traversals of the tree in Figure 6.6c that may be of some use are the sequence 2, 10, 12, 20, 13, 25, 29, 31 and the sequence 29, 31, 20, 12, 2, 25, 10, 13. The first sequence lists even numbers and then odd numbers in ascending order. The second sequence lists all nodes from level to level right to left, starting from the lowest level up to the root. The sequence 13, 31, 12, 2, 10, 29, 20, 25 does not indicate any regularity in the order of numbers or in the order of the traversed nodes. It is just a random jumping from node to node that in all likelihood is of no use. Nevertheless, all these sequences are the results of three legitimate traversals out of 8! = 40,320. In the face of such an abundance of traversals and the apparent uselessness of most of them, we would like to restrict our attention to two classes only, namely, breadth-first and depth-first traversals.

6.4.1 Breadth-First Traversal

Breadth-first traversal is visiting each node starting from the highest (or lowest) level and moving down (or up) level by level, visiting nodes on each level from left to right (or from right to left). There are thus four possibilities, and one such possibility—a top-down, left-to-right breadth-first traversal of the tree in Figure 6.6c—results in the sequence 13, 10, 25, 2, 12, 20, 31, 29.

Implementation of this kind of traversal is straightforward when a queue is used. Consider a top-down, left-to-right, breadth-first traversal. After a node is visited, its children, if any, are placed at the end of the queue, and the node at the beginning of the queue is visited. Considering that for a node on level n, its children are on level n + 1, by placing these children at the end of the queue, they are visited after all nodes from level n are visited. Thus, the restriction that all nodes on level n must be visited before visiting any nodes on level n + 1 is accomplished.

An implementation of the corresponding member function is shown in Figure 6.10.

FIGURE 6.10 Top-down, left-to-right, breadth-first traversal implementation.

```
template<class T>
void BST<T>::breadthFirst() {
    Queue<BSTNode<T>*> queue;
    BSTNode<T> *p = root;
    if (p != 0) {
        queue.enqueue(p);
        while (!queue.empty()) {
            p = queue.dequeue();
            visit(p);
            if (p->left != 0)
                queue.enqueue(p->left);
            if (p->right != 0)
                queue.enqueue(p->right);
        }
    }
}
```

6.4.2 Depth-First Traversal

Depth-first traversal proceeds as far as possible to the left (or right), then backs up until the first crossroad, goes one step to the right (or left), and again as far as possible to the left (or right). We repeat this process until all nodes are visited. This definition, however, does not clearly specify exactly when nodes are visited: before proceeding down the tree or after backing up? There are some variations of the depth-first traversal.

There are three tasks of interest in this type of traversal:

V—visiting a node

L—traversing the left subtree

R—traversing the right subtree

An orderly traversal takes place if these tasks are performed in the same order for each node. The three tasks can themselves be ordered in 3! = 6 ways, so there are six possible ordered depth-first traversals:

VLR VRL

LVR RVL

LRV RLV

If the number of different orders still seems like a lot, it can be reduced to three traversals where the move is always from left to right and attention is focused on the first column. The three traversals are given these standard names:

VLR—preorder tree traversal

LVR—inorder tree traversal

LRV—postorder tree traversal

Short and elegant functions can be implemented directly from the symbolic descriptions of these three traversals, as shown in Figure 6.11.

These functions may seem too simplistic, but their real power lies in recursion, in fact, double recursion. The real job is done by the system on the run-time stack. This simplifies coding but lays a heavy burden upon the system. To better understand this process, inorder tree traversal is discussed in some detail.

In inorder traversal, the left subtree of the current node is visited first, then the node itself, and finally, the right subtree. All of this, obviously, holds if the tree is not empty. Before analyzing the run-time stack, the output given by the inorder traversal is determined by referring to Figure 6.12. The following steps correspond to the letters in that figure:

(a) Node 15 is the root on which `inorder()` is called for the first time. The function calls itself for node 15's left child, node 4.

(b) Node 4 is not null, so `inorder()` is called on node 1. Because node 1 is a leaf (that is, both its subtrees are empty), invocations of `inorder()` on the subtrees do not result in other recursive calls of `inorder()`, as the condition in the `if` statement is not met. Thus, after `inorder()` called for the null left subtree is finished, node 1 is visited and then a quick call to `inorder()` is executed for the null right subtree of node 1. After

FIGURE 6.11 Depth-first traversal implementation.

```cpp
template<class T>
void BST<T>::inorder(BSTNode<T> *p) {
    if (p != 0) {
        inorder(p->left);
        visit(p);
        inorder(p->right);
    }
}

template<class T>
void BST<T>::preorder(BSTNode<T> *p) {
    if (p != 0) {
        visit(p);
        preorder(p->left);
        preorder(p->right);
    }
}

template<class T>
void BST<T>::postorder(BSTNode<T>* p) {
    if (p != 0) {
        postorder(p->left);
        postorder(p->right);
        visit(p);
    }
}
```

resuming the call for node 4, node 4 is visited. Node 4 has a null right subtree; hence, `inorder()` is called only to check that, and right after resuming the call for node 15, node 15 is visited.

(c) Node 15 has a right subtree, so `inorder()` is called for node 20.

(d) `inorder()` is called for node 16, the node is visited, and then on its null left subtree, which is followed by visiting node 16. After a quick call to `inorder()` on the null right subtree of node 16 and return to the call on node 20, node 20 is also visited.

(e) `inorder()` is called on node 25, then on its empty left subtree, then node 25 is visited, and finally `inorder()` is called on node 25's empty right subtree.

If the visit includes printing the value stored in a node, then the output is:

```
1 4 15 16 20 25
```

Section 6.4 Tree Traversal 229

FIGURE 6.12 Inorder tree traversal.

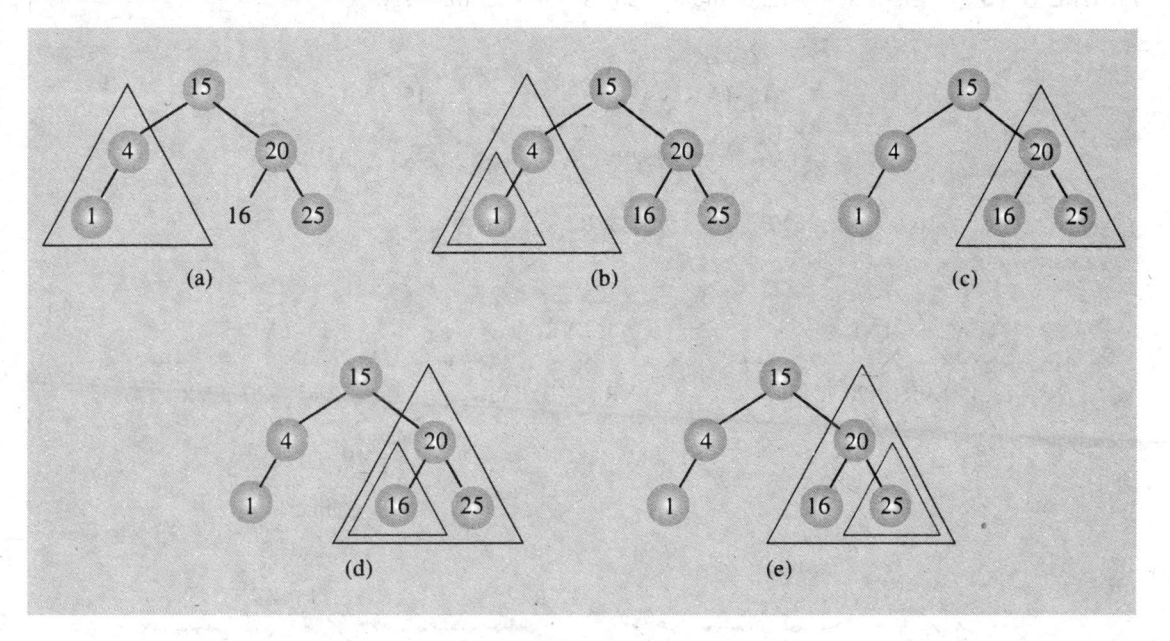

The key to the traversal is that the three tasks, L, V, and R, are performed for each node separately. This means that the traversal of the right subtree of a node is held pending until the first two tasks, L and V, are accomplished. If the latter two are finished, they can be crossed out as in Figure 6.13.

To present the way `inorder()` works, the behavior of the run-time stack is observed. The numbers in parentheses in Figure 6.14 indicate return addresses shown on the left-hand side of the code for `inorder()`.

```
           template<class T>
           void BST<T>::inorder(BSTnode<T> *node) {
               if (node != 0) {
/* 1 */            inorder(node->left);
/* 2 */            visit(node);
/* 3 */            inorder(node->right);
/* 4 */        }
           }
```

A rectangle with an up arrow and a number indicates the current value of `node` pushed onto the stack. For example, ↑4 means that `node` points to the node of the tree whose value is the number 4. Figure 6.14 shows the changes of the run-time stack when `inorder()` is executed for the tree in Figure 6.12.

(a) Initially, the run-time stack is empty (or rather it is assumed that the stack is empty by disregarding what has been stored on it before the first call to `inorder()`).

FIGURE 6.13 Details of several of the first steps of inorder traversal.

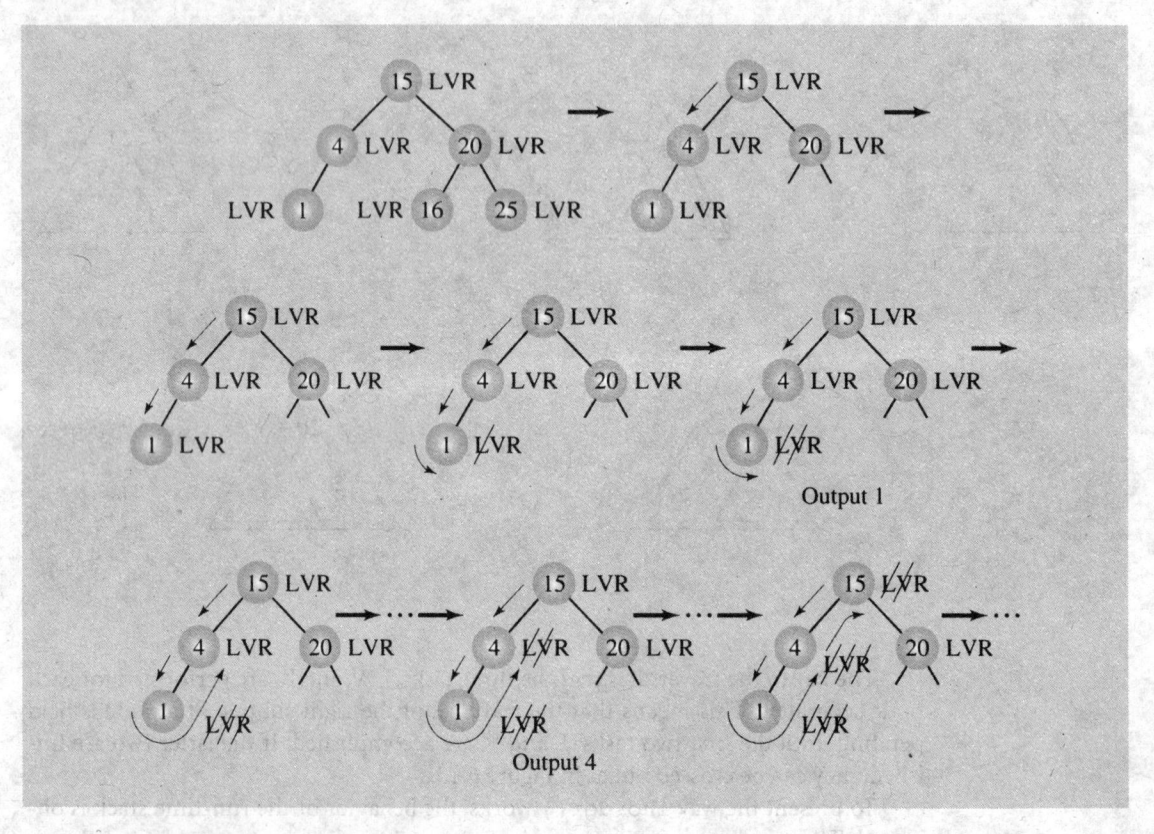

(b) Upon the first call, the return address of inorder() and the value of node, ↑15, are pushed onto the run-time stack. The tree, pointed to by node, is not empty, the condition in the if statement is satisfied, and inorder() is called again with node 4.

(c) Before it is executed, the return address, (2), and current value of node, ↑4, are pushed onto the stack. Because node is not null, inorder() is about to be invoked for node's left child, ↑1.

(d) First, the return address, (2), and the node's value are stored on the stack.

(e) inorder() is called with node 1's left child. The address (2) and the current value of parameter node, null, are stored on the stack. Because node is null, inorder() is exited immediately; upon exit, the activation record is removed from the stack.

(f) The system goes now to its run-time stack, restores the value of the node, ↑1, executes the statement under (2), and prints the number 1. Because node is not completely processed, the value of node and address (2) are still on the stack.

(g) With the right child of node ↑1, the statement under (3) is executed, which is the next call to inorder(). First, however, the address (4) and node's current value, null, are pushed onto the stack. Because node is null, inorder() is exited; upon exit, the stack is updated.

Section 6.4 Tree Traversal 231

FIGURE 6.14 Changes in the run-time stack during inorder traversal.

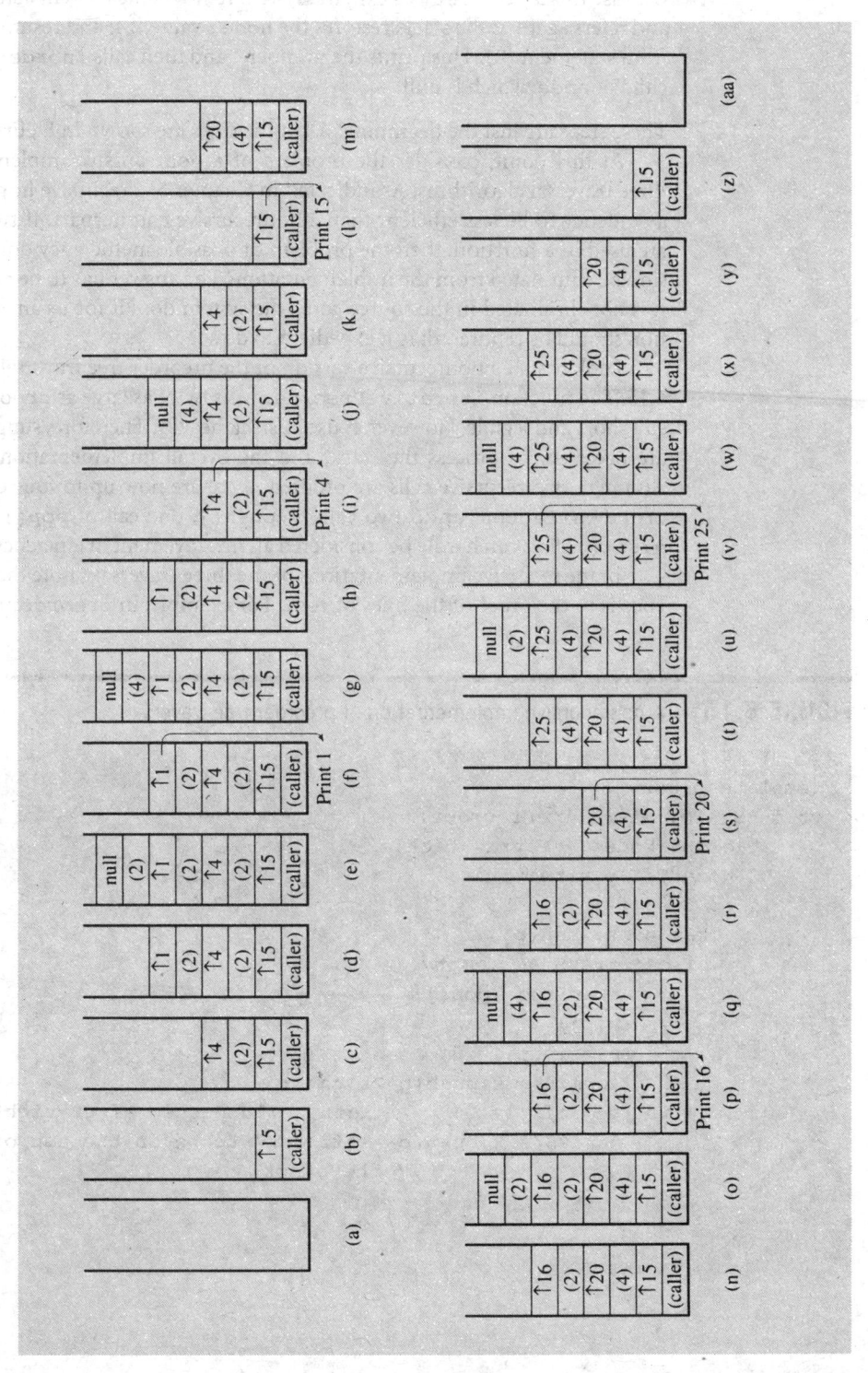

(h) The system now restores the old value of the node, ↑1, and executes statement (4).

(i) Because this is inorder()'s exit, the system removes the current activation record and refers again to the stack; restores the node's value, ↑4; and resumes execution from statement (2). This prints the number 4 and then calls inorder() for the right child of node, which is null.

These steps are just the beginning. All of the steps are shown in Figure 6.14.

At this point, consider the problem of a nonrecursive implementation of the three traversal algorithms. As indicated in Chapter 5, a recursive implementation has a tendency to be less efficient than a nonrecursive counterpart. If two recursive calls are used in a function, then the problem of possible inefficiency doubles. Can recursion be eliminated from the implementation? The answer has to be positive because if it is not eliminated in the source code, the system does it for us anyway. So the question should be rephrased: Is it expedient to do so?

Look first at a nonrecursive version of the preorder tree traversal shown in Figure 6.15. The function iterativePreorder() is twice as large as preorder(), but it is still short and legible. However, it uses a stack heavily. Therefore, supporting functions are necessary to process the stack, and the overall implementation is not so short. Although two recursive calls are omitted, there are now up to four calls per iteration of the while loop: up to two calls of push(), one call of pop(), and one call of visit(). This can hardly be considered an improvement in efficiency.

In the recursive implementations of the three traversals, note that the only difference is in the order of the lines of code. For example, in preorder(), first a node is

FIGURE 6.15 A nonrecursive implementation of preorder tree traversal.

```
template<class T>
void BST<T>::iterativePreorder() {
    Stack<BSTNode<T>*> travStack;
    BSTNode<T> *p = root;
    if (p != 0) {
        travStack.push(p);
        while (!travStack.empty()) {
            p = travStack.pop();
            visit(p);
            if (p->right != 0)
                travStack.push(p->right);
            if (p->left != 0)    // left child pushed after right
                travStack.push(p->left); // to be on the top of
        }                                // the stack;
    }
}
```

visited, and then there are calls for the left and right subtrees. On the other hand, in `postorder()`, visiting a node succeeds both calls. Can we so easily transform the nonrecursive version of a left-to-right preorder traversal into a nonrecursive left-to-right postorder traversal? Unfortunately, no. In `iterativePreorder()`, visiting occurs before both children are pushed onto the stack. But this order does not really matter. If the children are pushed first and then the node is visited—that is, if `visit(p)` is placed after both calls to `push()`—the resulting implementation is still a preorder traversal. What matters here is that `visit()` has to follow `pop()`, and the latter has to precede both calls of `push()`. Therefore, nonrecursive implementations of inorder and postorder traversals have to be developed independently.

A nonrecursive version of postorder traversal can be obtained rather easily if we observe that the sequence generated by a left-to-right postorder traversal (an LRV order) is the same as the reversed sequence generated by a right-to-left preorder traversal (a VRL order). In this case, the implementation of `iterativePreorder()` can be adopted to create `iterativePostorder()`. This means that two stacks have to be used, one to visit each node in the reverse order after a right-to-left preorder traversal is finished. It is, however, possible to develop a function for postorder traversal that pushes onto the stack a node that has two descendants, once before traversing its left subtree and once before traversing its right subtree. An auxiliary pointer q is used to distinguish between these two cases. Nodes with one descendant are pushed only once, and leaves do not need to be pushed at all (Figure 6.16).

FIGURE 6.16 A nonrecursive implementation of postorder tree traversal.

```
template<class T>
void BST<T>::iterativePostorder() {
    Stack<BSTNode<T>*> travStack;
    BSTNode<T>* p = root, *q = root;
    while (p != 0) {
        for ( ; p->left != 0; p = p->left)
            travStack.push(p);
        while (p != 0 && (p->right == 0 || p->right == q)) {
            visit(p);
            q = p;
            if (travStack.empty())
                return;
            p = travStack.pop();
        }
        travStack.push(p);
        p = p->right;
    }
}
```

A nonrecursive inorder tree traversal is also a complicated matter. One possible implementation is given in Figure 6.17. In cases like this, we can clearly see the power of recursion: `iterativeInorder()` is almost unreadable, and without thorough explanation, it is not easy to determine the purpose of this function. On the other hand, recursive `inorder()` immediately demonstrates a purpose and logic. Therefore, `iterativeInorder()` can be defended in one case only: if it is shown that there is a substantial gain in execution time and that the function is called often in a program. Otherwise, `inorder()` is preferable to its iterative counterpart.

FIGURE 6.17 A nonrecursive implementation of inorder tree traversal.

```
template<class T>
void BST<T>::iterativeInorder() {
    Stack<BSTNode<T>*> travStack;
    BSTNode<T> *p = root;
    while (p != 0) {
        while (p != 0) {              // stack the right child (if any)
            if (p->right)             // and the node itself when going
                travStack.push(p->right); // to the left;
            travStack.push(p);
            p = p->left;
        }
        p = travStack.pop();          // pop a node with no left child
        while (!travStack.empty() && p->right == 0) { // visit it
            visit(p);                 // and all nodes with no right
            p = travStack.pop();      // child;
        }
        visit(p);                     // visit also the first node with
        if (!travStack.empty())       // a right child (if any);
            p = travStack.pop();
        else p = 0;
    }
}
```

6.4.3 Stackless Depth-First Traversal

Threaded Trees

The traversal functions analyzed in the preceding section were either recursive or nonrecursive, but both kinds used a stack either implicitly or explicitly to store information about nodes whose processing has not been finished. In the case of recursive functions, the run-time stack was utilized. In the case of nonrecursive variants, an explicitly defined and user-maintained stack was used. The concern is that some additional time has to be spent to maintain the stack, and some more space has to be set aside for the stack itself. In the worst case, when the tree is unfavorably skewed, the stack may hold information about almost every node of the tree, a serious concern for very large trees.

It is more efficient to incorporate the stack as part of the tree. This is done by incorporating *threads* in a given node. Threads are pointers to the predecessor and successor of the node according to an inorder traversal, and trees whose nodes use threads are called *threaded trees*. Four pointers are needed for each node in the tree, which again takes up valuable space.

The problem can be solved by overloading existing pointers. In trees, left or right pointers are pointers to children, but they can also be used as pointers to predecessors and successors, thereby being overloaded with meaning. How do we distinguish these meanings? For an overloaded operator, context is always a disambiguating factor. In trees, however, a new data member has to be used to indicate the current meaning of the pointers.

Because a pointer can point to one node at a time, the left pointer is either a pointer to the left child or to the predecessor. Analogously, the right pointer points either to the right subtree or to the successor (Figure 6.18a).

FIGURE 6.18 (a) A threaded tree and (b) an inorder traversal's path in a threaded tree with right successors only.

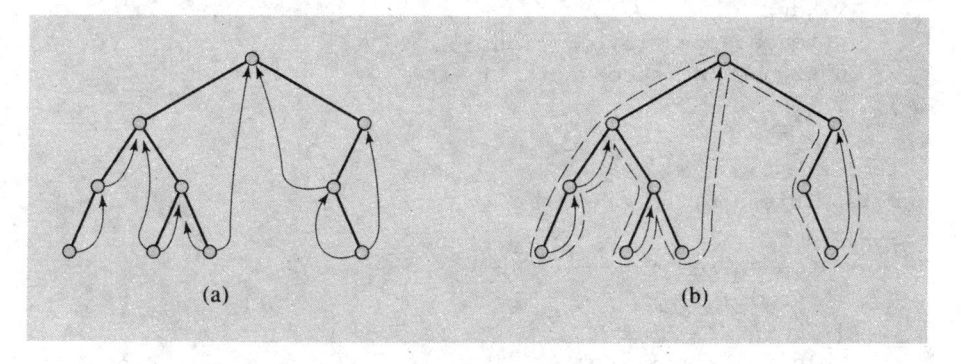

Figure 6.18a suggests that pointers to both predecessors and successors have to be maintained, which is not always the case. It may be sufficient to use only one thread, as shown in the implementation of the inorder traversal of a threaded tree, which requires only pointers to successors (Figure 6.18b).

The function is relatively simple. The dashed line in Figure 6.18b indicates the order in which p accesses nodes in the tree. Note that only one variable, p, is needed to traverse the tree. No stack is needed; therefore, space is saved. But is it really? As indicated, nodes require a data member indicating how the right pointer is being used. In the implementation of `threadedInorder()`, the Boolean data member `successor` plays this role, as shown in Figure 6.19. Hence, `successor` requires only one bit of computer memory, insignificant in comparison to other fields. However, the exact details are highly dependent on the implementation. The operating system almost certainly pads a bit structure with additional bits for proper alignment of machine words. If so, `successor` needs at least one byte, if not an entire word, defeating the argument about saving space by using threaded trees.

FIGURE 6.19 Implementation of the generic threaded tree and the inorder traversal of a threaded tree.

```cpp
//*********************    genThreaded.h    *********************
//               Generic binary search threaded tree

#ifndef THREADED_TREE
#define THREADED_TREE

template<class T>
class ThreadedNode {
public:
    ThreadedNode() {
        left = right = 0;
    }
    ThreadedNode(const T& el, ThreadedNode *l = 0, ThreadedNode *r = 0) {
        key = el; left = l; right = r; successor = 0;
    }
    T key;
    ThreadedNode *left, *right;
    unsigned int successor : 1;
};

template<class T>
class ThreadedTree {
public:
    ThreadedTree() {
        root = 0;
    }
    void insert(const T&);                       // Figure 6.24
    void inorder();
    . . . . . . . . . . . . . . . .
protected:
    ThreadedNode<T>* root;
    . . . . . . . . . . . . . . . .
};

#endif

template<class T>
void ThreadedTree<T>::inorder() {
```

FIGURE 6.19 *(continued)*

```
ThreadedNode<T> *prev, *p = root;
if (p != 0) {                    // process only nonempty trees;
    while (p->left != 0)         // go to the leftmost node;
        p = p->left;
    while (p != 0) {
        visit(p);
        prev = p;
        p = p->right;            // go to the right node and only
        if (p != 0 && prev->successor == 0) // if it is a
            while (p->left != 0)// descendant go to the
                p = p->left;    // leftmost node, otherwise
    }                            // visit the successor;
}
```

Threaded trees can also be used for preorder and postorder traversals. In preorder traversal, the current node is visited first and then traversal continues with its left descendant, if any, or right descendant, if any. If the current node is a leaf, threads are used to go through the chain of its already visited inorder successors to restart traversal with the right descendant of the last successor.

Postorder traversal is only slightly more complicated. First, a dummy node is created that has the root as its left descendant. In the traversal process, a variable can be used to check the type of the current action. If the action is left traversal and the current node has a left descendant, then the descendant is traversed; otherwise, the action is changed to right traversal. If the action is right traversal and the current node has a right nonthread descendant, then the descendant is traversed and the action is changed to left traversal; otherwise, the action changes to visiting a node. If the action is visiting a node, then the current node is visited, and afterward, its postorder successor has to be found. If the current node's parent is accessible through a thread (that is, current node is parent's left child), then traversal is set to continue with the right descendant of the parent. If the current node has no right descendant, then it is the end of the right-extended chain of nodes. First, the beginning of the chain is reached through the thread of the current node, then the right references of nodes in the chain are reversed, and finally, the chain is scanned backward, each node is visited, and then right references are restored to their previous setting.

Traversal through Tree Transformation

The first set of traversal algorithms analyzed earlier in this chapter needed a stack to retain some information necessary for successful processing. Threaded trees incorporated a stack as part of the tree at the cost of extending the nodes by one field to make a distinction between the interpretation of the right pointer as a pointer to the child or to the successor. Two such tag fields are needed if both successor and predecessor are considered. However, it is possible to traverse a tree without using any stack or threads. There are many such algorithms, all of them made possible by making temporary changes in the tree during traversal. These changes consist of reassigning new values to some pointers. However, the tree may temporarily lose its tree structure, which needs to be restored before traversal is finished. The technique is illustrated by an elegant algorithm devised by Joseph M. Morris applied to inorder traversal.

First, it is easy to notice that inorder traversal is very simple for degenerate trees, in which no node has a left child (see Figure 6.1e). No left subtree has to be considered for any node. Therefore, the usual three steps, LVR (visit left subtree, visit node, visit right subtree), for each node in inorder traversal turn into two steps, VR. No information needs to be retained about the current status of the node being processed before traversing its left child, simply because there is no left child. Morris's algorithm takes into account this observation by temporarily transforming the tree so that the node being processed has no left child; hence, this node can be visited and its right subtree processed. The algorithm can be summarized as follows:

```
MorrisInorder()
    while not finished
        if node has no left descendant
            visit it;
            go to the right;
        else make this node the right child of the rightmost node in its left descendant;
            go to this left descendant;
```

This algorithm successfully traverses the tree, but only once, because it destroys its original structure. Therefore, some information has to be retained to allow the tree to restore its original form. This is achieved by retaining the left pointer of the node moved down its right subtree, as in the case of nodes 10 and 5 in Figure 6.21.

An implementation of the algorithm is shown in Figure 6.20, and the details of the execution are illustrated in Figure 6.21. The following description is divided into actions performed in consecutive iterations of the outer while loop:

1. Initially, p points to the root, which has a left child. As a result, the inner while loop takes tmp to node 7, which is the rightmost node of the left child of node 10, pointed to by p (Figure 6.21a). Because no transformation has been done, tmp has no right child, and in the inner if statement, the root, node 10, is made the right child of tmp. Node 10 retains its left pointer to node 5, its original left child. Now, the tree is not a tree anymore, because it contains a cycle (Figure 6.21b). This completes the first iteration.

2. Pointer p points to node 5, which also has a left child. First, tmp reaches the largest node in this subtree, which is 3 (Figure 6.21c), and then the current root, node 5, becomes the right child of node 3 while retaining contact with node 3 through its left pointer (Figure 6.21d).

3. Because node 3, pointed to by p, has no left child, in the third iteration, this node is visited, and p is reassigned to its right child, node 5 (Figure 6.21e).

4. Node 5 has a nonnull left pointer, so tmp finds a temporary parent of node 5, which is the same node currently pointed to by tmp (Figure 6.21f). Next, node 5 is visited, and

FIGURE 6.20 Implementation of the Morris algorithm for inorder traversal.

```
template<class T>
void BST<T>::MorrisInorder() {
    BSTNode<T> *p = root, *tmp;
    while (p != 0)
        if (p->left == 0) {
            visit(p);
            p = p->right;
        }
        else {
            tmp = p->left;
            while (tmp->right != 0 && // go to the rightmost node
                   tmp->right != p)   // of the left subtree or
                tmp = tmp->right;     // to the temporary parent
            if (tmp->right == 0) {    // of p; if 'true'
                tmp->right = p;       // rightmost node was
                p = p->left;          // reached, make it a
            }                         // temporary parent of the
            else {                    // current root, else
                                      // a temporary parent has
                visit(p);             // been found; visit node p
                tmp->right = 0;       // and then cut the right
                                      // pointer of the current
                p = p->right;         // parent, whereby it
            }                         // ceases to be a parent;
        }
}
```

configuration of the tree in Figure 6.21b is reestablished by setting the right pointer of node 3 to null (Figure 6.21g).

5. Node 7, pointed to now by p, is visited, and p moves down to its right child (6.21h).
6. tmp is updated to point to the temporary parent of node 10 (Figure 6.21i). Next, node 10 is visited and then reestablished to its status of root by nullifying the right pointer of node 7 (Figure 6.21j).
7. Finally, node 20 is visited without further ado, because it has no left child, nor has its position been altered.

This completes the execution of Morris's algorithm. Notice that there are seven iterations of the outer while loop for only five nodes in the tree in Figure 6.21. This is due to the fact that there are two left children in the tree, so the number of extra iterations depends on the number of left children in the entire tree. The algorithm performs worse for trees with a large number of such children.

FIGURE 6.21 Tree traversal with the Morris method.

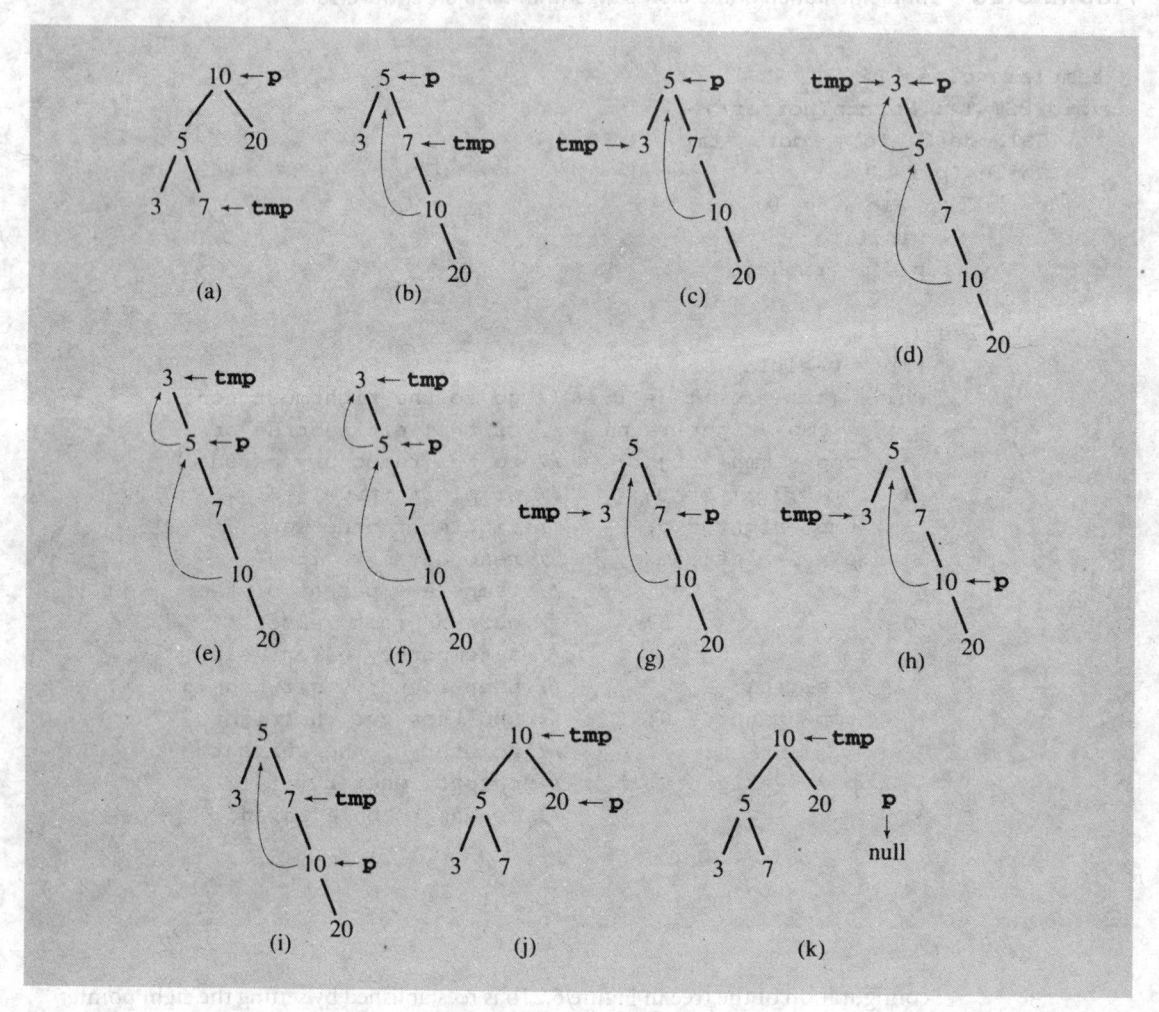

Preorder traversal is easily obtainable from inorder traversal by moving `visit()` from the inner `else` clause to the inner `if` clause. In this way, a node is visited before a tree transformation.

Postorder traversal can also be obtained from inorder traversal by first creating a dummy node whose left descendant is the tree being processed and whose right descendant is null. Then this temporarily extended tree is the subject of traversal as in inorder traversal except that in the inner `else` clause, after finding a temporary parent, nodes between `p->left` (included) and `p` (excluded) extended to the right in a modified tree are processed in the reverse order. To process them in constant time, the chain of nodes is scanned down and right pointers are reversed to point to parents of nodes. Then the same chain is scanned upward, each node is visited, and the right pointers are restored to their original setting.

How efficient are the traversal procedures discussed in this section? All of them run in $\Theta(n)$ time, threaded implementation requires $\Theta(n)$ more space for threads than nonthreaded binary search trees, and both recursive and iterative traversals require $O(n)$ additional space (on the run-time stack or user-defined stack). Several dozens of runs on randomly generated trees of 5,000 nodes indicate that for preorder and inorder traversal routines (recursive, iterative, Morris, and threaded), the difference in the execution time is only on the order of 5–10%. Morris traversals have one undeniable advantage over other types of traversals: They do not require additional space. Recursive traversals rely on the run-time stack, which can be overflowed when traversing trees of large height. Iterative traversals also use a stack, and although the stack can be overflowed as well, the problem is not as imminent as in the case of the run-time stack. Threaded trees use nodes that are larger than the nodes used by nonthreaded trees, which usually should not pose a problem. But both iterative and threaded implementations are much less intuitive than their recursive counterparts; therefore, the clarity of implementation and comparable run time clearly favors, in most situations, recursive implementations over other implementations.

6.5 INSERTION

Searching a binary tree does not modify the tree. It scans the tree in a predetermined way to access some or all of the keys in the tree, but the tree itself remains undisturbed after such an operation. Tree traversals can change the tree but they may also leave it in the same condition. Whether or not the tree is modified depends on the actions prescribed by `visit()`. There are certain operations that always make some systematic changes in the tree, such as adding nodes, deleting them, modifying elements, merging trees, and balancing trees to reduce their height. This section deals only with inserting a node into a binary search tree.

To insert a new node with key `el`, a tree node with a dead end has to be reached, and the new node has to be attached to it. Such a tree node is found using the same technique that tree searching used: The key `el` is compared to the key of a node currently being examined during a tree scan. If `el` is less than that key, the left child (if any) of `p` is tried; otherwise, the right child (if any) is tested. If the child of `p` to be tested is empty, the scanning is discontinued and the new node becomes this child. The procedure is illustrated in Figure 6.22. Figure 6.23 contains an implementation of the algorithm to insert a node.

In analyzing the problem of traversing binary trees, three approaches have been presented: traversing with the help of a stack, traversing with the aid of threads, and traversing through tree transformation. The first approach does not change the tree during the process. The third approach changes it, but restores it to the same condition as before it started. Only the second approach needs some preparatory operations on the tree to become feasible: It requires threads. These threads may be created each time before the traversal procedure starts its task and removed each time it is finished. If the traversal is performed infrequently, this becomes a viable option. Another approach is to maintain the threads in all operations on the tree when inserting a new element in the binary search tree.

FIGURE 6.22 Inserting nodes into binary search trees.

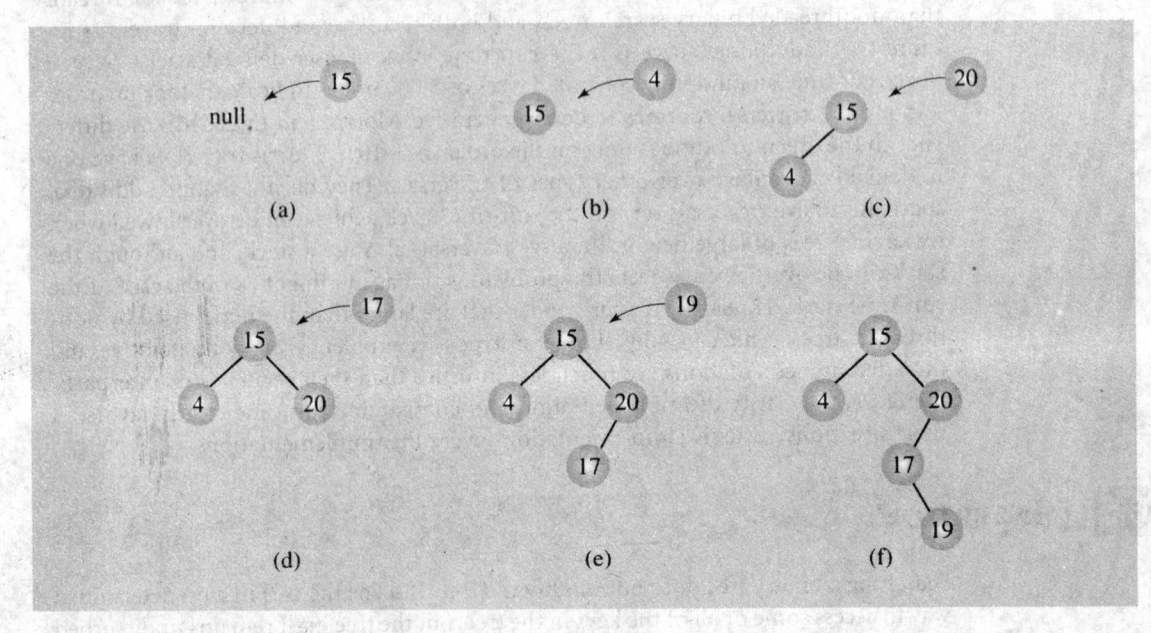

(a) 15 → null

(b) 4 → 15

(c) 20 → 15, 4

(d) 17 → 15 (4, 20)

(e) 19 → 15 (4, 20 (17))

(f) 15 (4, 20 (17 (19)))

FIGURE 6.23 Implementation of the insertion algorithm.

```
template<class T>
void BST<T>::insert(const T& el) {
    BSTNode<T> *p = root, *prev = 0;
    while (p != 0) {           // find a place for inserting new node;
        prev = p;
        if (p->key < el)
            p = p->right;
        else p = p->left;
    }
    if (root == 0)    // tree is empty;
        root = new BSTNode<T>(el);
    else if (prev->key < el)
        prev->right = new BSTNode<T>(el);
    else prev->left  = new BSTNode<T>(el);
}
```

The function for inserting a node in a threaded tree is a simple extension of insert() for regular binary search trees to adjust threads whenever applicable. This function is for inorder tree traversal and it only takes care of successors, not predecessors.

A node with a right child has a successor some place in its right subtree. Therefore, it does not need a successor thread. Such threads are needed to allow climbing the tree, not going down it. A node with no right child has its successor somewhere above it. Except for one node, all nodes with no right children will have threads to their successors. If a node becomes the right child of another node, it inherits the successor from its new parent. If a node becomes a left child of another node, this parent becomes its successor. Figure 6.24 contains the implementation of this algorithm. The first few insertions are shown in Figure 6.25.

FIGURE 6.24 Implementation of the algorithm to insert a node into a threaded tree.

```
template<class T>
void ThreadedTree<T>::insert(const T& el) {
    ThreadedNode<T> *p, *prev = 0, *newNode;
    newNode = new ThreadedNode<T>(el);
    if (root == 0) {              // tree is empty;
        root = newNode;
        return;
    }
    p = root;                     // find a place to insert newNode;
    while (p != 0) {
        prev = p;
        if (p->key > el)
            p = p->left;
        else if (p->successor == 0)// go to the right node only if it
            p = p->right;    // is a descendant, not a successor;
        else break;          // don't follow successor link;
    }
    if (prev->key > el) {    // if newNode is left child of
        prev->left  = newNode;     // its parent, the parent
        newNode->successor = 1;    // also becomes its successor;
        newNode->right = prev;
    }
    else if (prev->successor == 1) {// if the parent of newNode
        newNode->successor = 1;    // is not the rightmost node,
        prev->successor = 0;       // make parent's successor
        newNode->right = prev->right; // newNode's successor,
        prev->right = newNode;
    }
    else prev->right = newNode;    // otherwise it has no successor;
}
```

FIGURE 6.25 Inserting nodes into a threaded tree.

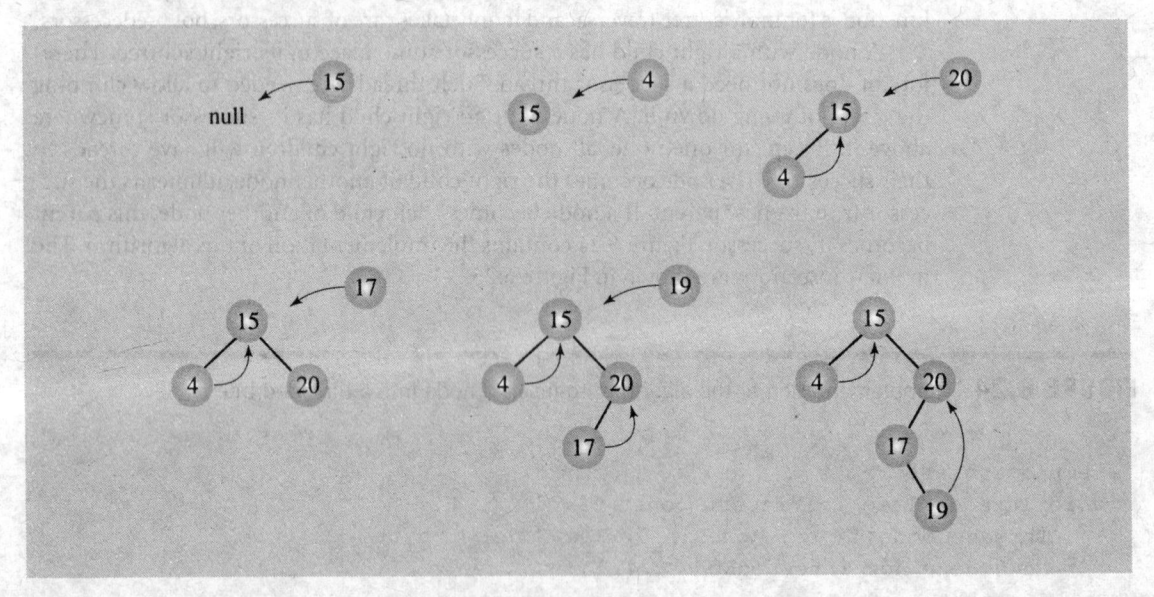

6.6 DELETION

Deleting a node is another operation necessary to maintain a binary search tree. The level of complexity in performing the operation depends on the position of the node to be deleted in the tree. It is by far more difficult to delete a node having two subtrees than to delete a leaf; the complexity of the deletion algorithm is proportional to the number of children the node has. There are three cases of deleting a node from the binary search tree:

1. The node is a leaf; it has no children. This is the easiest case to deal with. The appropriate pointer of its parent is set to null and the node is disposed of by `delete` as in Figure 6.26.

FIGURE 6.26 Deleting a leaf.

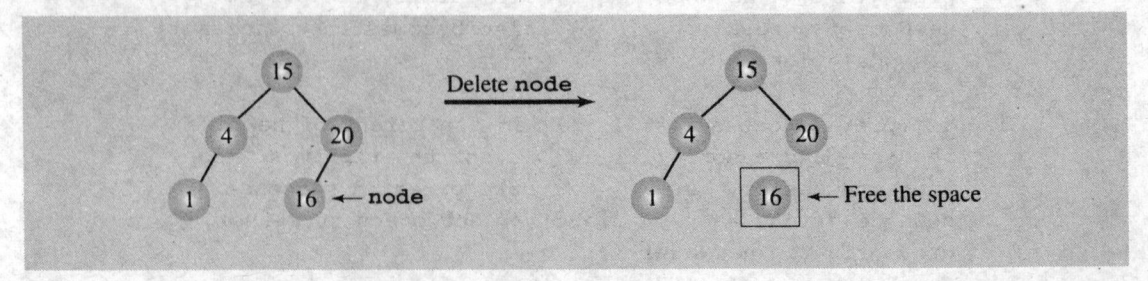

2. The node has one child. This case is not complicated. The parent's pointer to the node is reset to point to the node's child. In this way, the node's children are lifted up by one

level and all great-great-. . . grandchildren lose one "great" from their kinship designations. For example, the node containing 20 (see Figure 6.27) is deleted by setting the right pointer of its parent containing 15 to point to 20's only child, which is 16.

FIGURE 6.27 Deleting a node with one child.

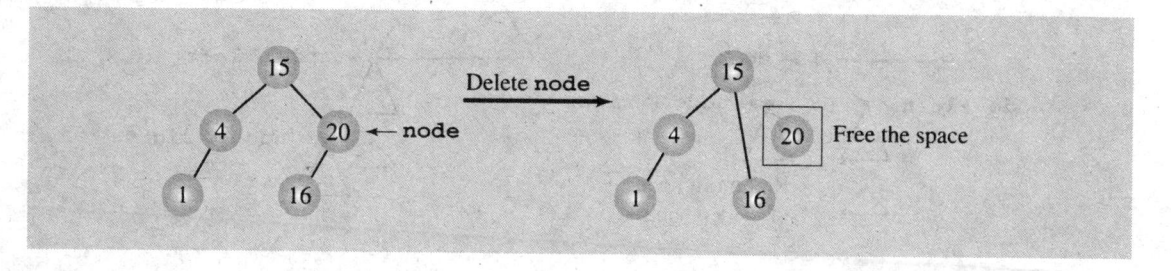

3. The node has two children. In this case, no one-step operation can be performed because the parent's right or left pointer cannot point to both the node's children at the same time. This section discusses two different solutions to this problem.

6.6.1 Deletion by Merging

This solution makes one tree out of the two subtrees of the node and then attaches it to the node's parent. This technique is called *deleting by merging*. But how can we merge these subtrees? By the nature of binary search trees, every key of the right subtree is greater than every key of the left subtree, so the best thing to do is to find in the left subtree the node with the greatest key and make it a parent of the right subtree. Symmetrically, the node with the lowest key can be found in the right subtree and made a parent of the left subtree.

The desired node is the rightmost node of the left subtree. It can be located by moving along this subtree and taking right pointers until null is encountered. This means that this node will not have a right child, and there is no danger of violating the property of binary search trees in the original tree by setting that rightmost node's right pointer to the right subtree. (The same could be done by setting the left pointer of the leftmost node of the right subtree to the left subtree.) Figure 6.28 depicts this operation. Figure 6.29 contains the implementation of the algorithm.

It may appear that `findAndDeleteByMerging()` contains redundant code. Instead of calling `search()` before invoking `deleteByMerging()`, `findAndDeleteByMerging()` seems to forget about `search()` and searches for the node to be deleted using its private code. But using `search()` in function `findAndDeleteByMerging()` is a treacherous simplification. `search()` returns a pointer to the node containing key. In `findAndDeleteByMerging()`, it is important to have this pointer stored specifically in one of the pointers of the node's parent. In other words, a caller to `search()` is satisfied if it can access the node from any direction, whereas `findAndDeleteByMerging()` wants to access it from either its parent's left or right pointer data member. Otherwise, access to the entire subtree having this node as its root would be lost. One reason for this is the fact that `search()` focuses on the

246 ■ Chapter 6 Binary Trees

FIGURE 6.28 Summary of deleting by merging.

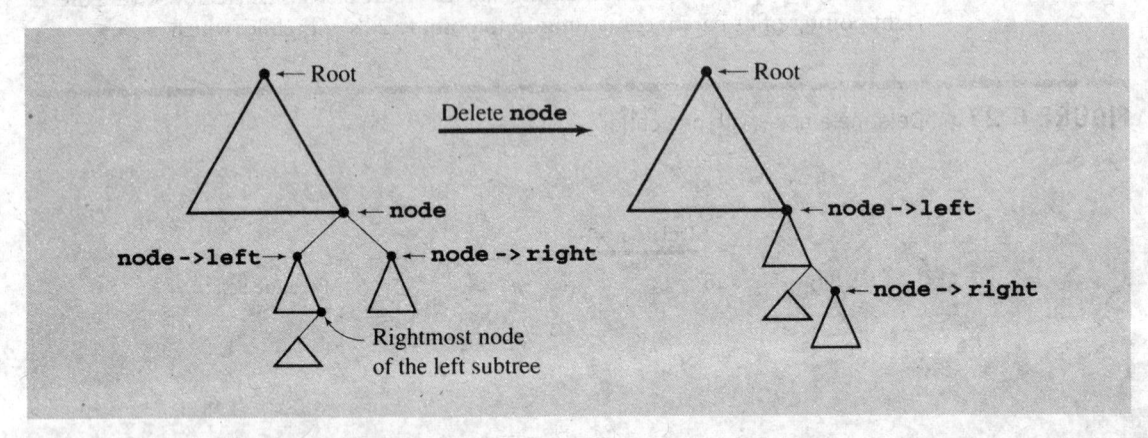

FIGURE 6.29 Implementation of an algorithm for deleting by merging.

```
template<class T>
void BST<T>::deleteByMerging(BSTNode<T>*& node) {
    BSTNode<T> *tmp = node;
    if (node != 0) {
        if (!node->right)            // node has no right child: its left
            node = node->left;       // child (if any) is attached to its
                                     // parent;
        else if (node->left == 0)    // node has no left child: its right
            node = node->right;      // child is attached to its parent;
        else {                       // be ready for merging subtrees;
            tmp = node->left;        // 1. move left
            while (tmp->right != 0)  // 2. and then right as far as
                                     //    possible;
                tmp = tmp->right;
            tmp->right =             // 3. establish the link between
                node->right;         //    the rightmost node of the left
                                     //    subtree and the right subtree;
            tmp = node;              // 4.
            node = node->left;       // 5.
        }
        delete tmp;                  // 6.
    }
}
```

FIGURE 6.29 (continued)

```
template<class T>
void BST<T>::findAndDeleteByMerging(const T& el) {
    BSTNode<T> *node = root, *prev = 0;
    while (node != 0) {
        if (node->key == el)
            break;
        prev = node;
        if (node->key < el)
            node = node->right;
        else node = node->left;
    }
    if (node != 0 && node->key == el)
        if (node == root)
            deleteByMerging(root);
        else if (prev->left == node)
            deleteByMerging(prev->left);
        else deleteByMerging(prev->right);
    else if (root != 0)
        cout << "key " << el << " is not in the tree\n";
    else cout << "the tree is empty\n";
}
```

node's key, and `findAndDeleteByMerging()` focuses on the node itself as an element of a larger structure, namely, a tree.

Figure 6.30 shows each step of this operation. It shows what changes are made when `findAndDeleteByMerging()` is executed. The numbers in this figure correspond to numbers put in comments in the code in Figure 6.29.

The algorithm for deletion by merging may result in increasing the height of the tree. In some cases, the new tree may be highly unbalanced, as Figure 6.31a illustrates. Sometimes the height may be reduced (see Figure 6.31b). This algorithm is not necessarily inefficient, but it is certainly far from perfect. There is a need for an algorithm that does not give the tree the chance to increase its height when deleting one of its nodes.

6.6.2 Deletion by Copying

Another solution, called *deletion by copying*, was proposed by Thomas Hibbard and Donald Knuth. If the node has two children, the problem can be reduced to one of two simple cases: The node is a leaf or the node has only one nonempty child. This can be done by replacing the key being deleted with its immediate predecessor (or successor). As already indicated in the algorithm deletion by merging, a key's predecessor is the key in the rightmost node in the left subtree (and analogically, its immediate successor is the key in the leftmost node in the right subtree). First, the predecessor has to be

248 ■ Chapter 6 Binary Trees

FIGURE 6.30 Details of deleting by merging.

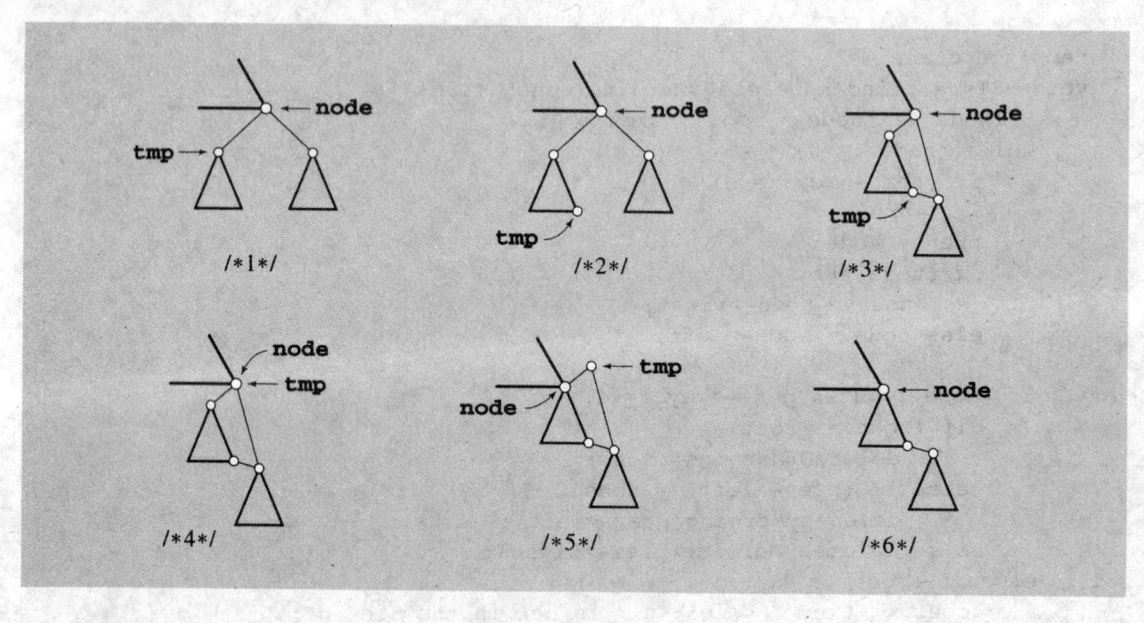

FIGURE 6.31 The height of a tree can be (a) extended or (b) reduced after deleting by merging.

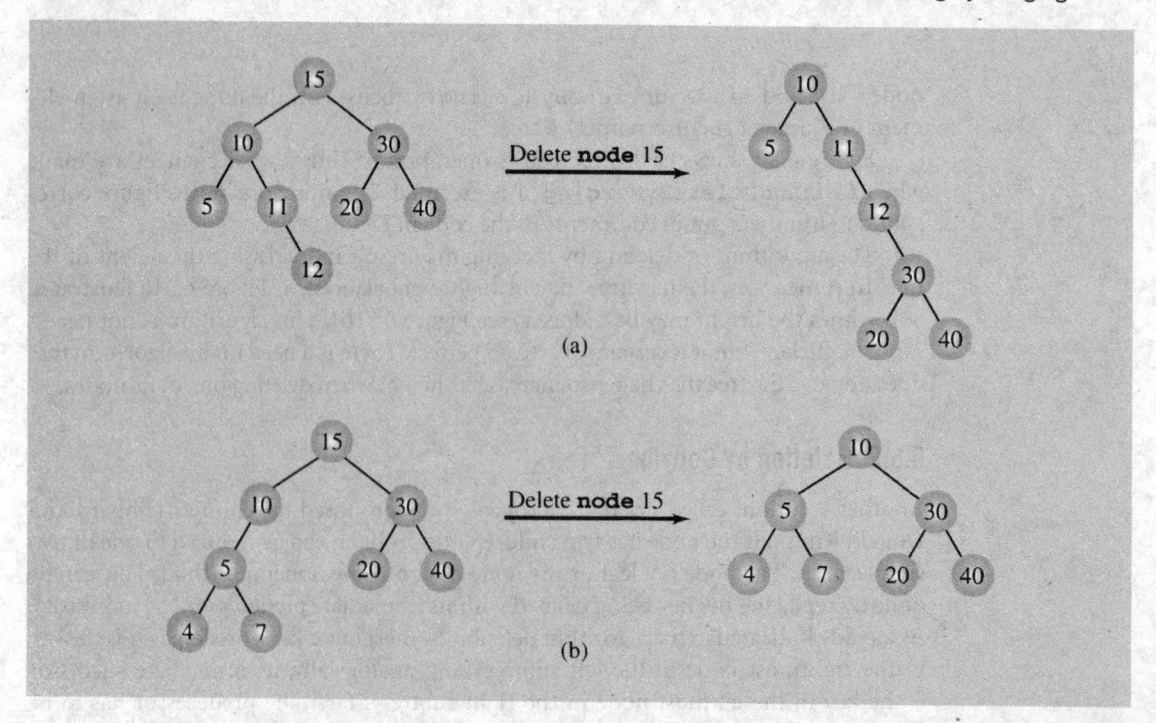

located. This is done, again, by moving one step to the left by first reaching the root of the node's left subtree and then moving as far to the right as possible. Next, the key of the located node replaces the key to be deleted. And that is where one of two simple cases comes into play. If the rightmost node is a leaf, the first case applies; however, if it has one child, the second case is relevant. In this way, deletion by copying removes a key k_1 by overwriting it by another key k_2 and then removing the node that holds k_2, whereas deletion by merging consisted of removing a key k_1 along with the node that holds it.

To implement the algorithm, two functions can be used. One function, `deleteByCopying()`, is illustrated in Figure 6.32. The second function, `findAndDeleteByCopying()`, is just like `findAndDeleteByMerging()`, but it calls `deleteByCopying()` instead of `deleteByMerging()`. A step-by-step trace is shown in Figure 6.33, and the numbers under the diagrams refer to the numbers indicated in comments included in the implementation of `deleteByCopying()`.

FIGURE 6.32 Implementation of an algorithm for deleting by copying.

```
template<class T>
void BST<T>::deleteByCopying(BSTNode<T>*& node) {
    BSTNode<T> *previous, *tmp = node;
    if (node->right == 0)                           // node has no right child;
         node = node->left;
    else if (node->left == 0)                       // node has no left child;
         node = node->right;
    else {
         tmp = node->left;                          // node has both children;
         previous = node;                           // 1.
         while (tmp->right != 0) {                  // 2.
              previous = tmp;
              tmp = tmp->right;
         }
         node->key = tmp->key;                      // 3.
         if (previous == node)
              previous ->left  = tmp->left;
         else previous ->right = tmp->left;         // 4.
    }
    delete tmp;                                     // 5.
}
```

This algorithm does not increase the height of the tree, but it still causes a problem if it is applied many times along with insertion. The algorithm is asymmetric; it always deletes the node of the immediate predecessor of the key in node, possibly reducing the height of the left subtree and leaving the right subtree unaffected. Therefore, the right subtree of node can grow after later insertions, and if the key in node is again deleted, the height of the right tree remains the same. After many insertions and

FIGURE 6.33 Deleting by copying.

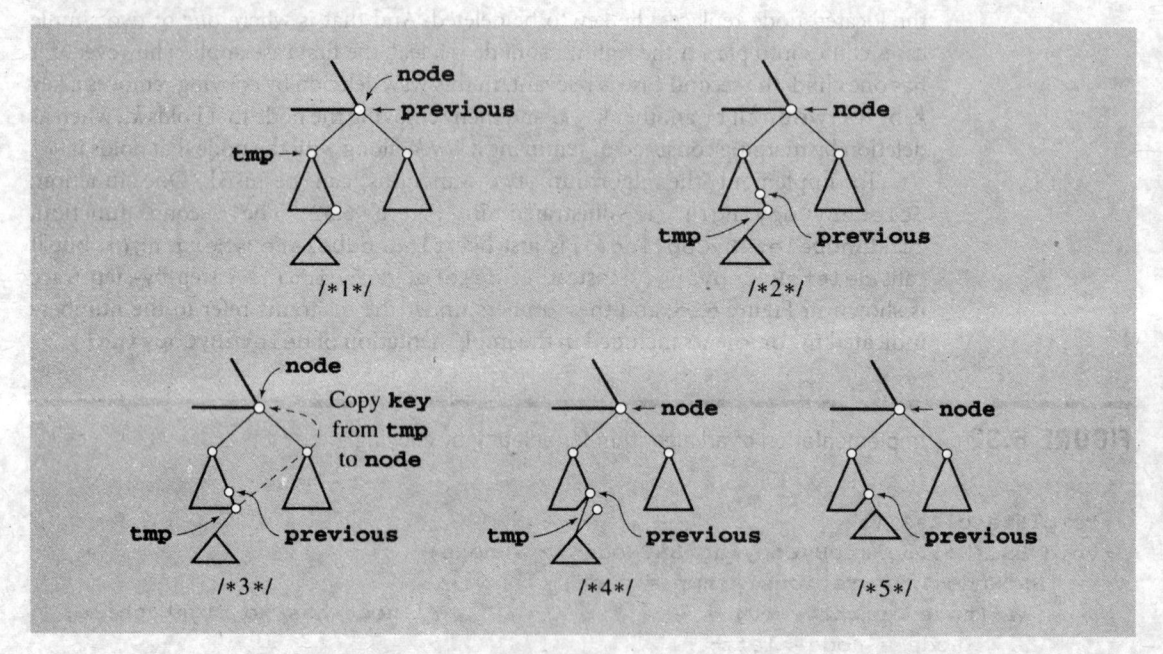

deletions, the entire tree becomes right unbalanced, with the right tree bushier and larger than the left subtree.

To circumvent this problem, a simple improvement can make the algorithm symmetrical. The algorithm can alternately delete the predecessor of the key in node from the left subtree and delete its successor from the right subtree. The improvement is significant. Simulations performed by Jeffrey Eppinger show that an expected internal path length for many insertions and asymmetric deletions is $\Theta(n \lg^3 n)$ for n nodes, and when symmetric deletions are used, the expected IPL becomes $\Theta(n \lg n)$. Theoretical results obtained by J. Culberson confirm these conclusions. According to Culberson, insertions and asymmetric deletions give $\Theta(n\sqrt{n})$ for the expected IPL and $\Theta(\sqrt{n})$ for the average search time (average path length), whereas symmetric deletions lead to $\Theta(\lg n)$ for the average search time, and as before, $\Theta(n \lg n)$ for the average IPL.

These results may be of moderate importance for practical applications. Experiments show that for a 2,048-node binary tree, only after 1.5 million insertions and asymmetric deletions does the IPL become worse than in a randomly generated tree.

Theoretical results are only fragmentary because of the extraordinary complexity of the problem. Arne Jonassen and Donald Knuth analyzed the problem of random insertions and deletions for a tree of only three nodes, which required using Bessel functions and bivariate integral equations, and the analysis turned out to rank among "the more difficult of all exact analyses of algorithms that have been carried out to date." Therefore, the reliance on experimental results is not surprising.

6.7 BALANCING A TREE

At the beginning of this chapter, two arguments were presented in favor of trees: They are well suited to represent the hierarchical structure of a certain domain, and the search process is much faster using trees instead of linked lists. The second argument, however, does not always hold. It all depends on what the tree looks like. Figure 6.34 shows three binary search trees. All of them store the same data, but obviously, the tree in Figure 6.34a is the best and Figure 6.34c is the worst. In the worst case, three tests are needed in the former and six tests are needed in the latter to locate an object. The problem with the trees in Figures 6.34b and 6.34c is that they are somewhat unsymmetrical, or lopsided; that is, objects in the tree are not distributed evenly to the extent that the tree in Figure 6.34c practically turned into a linked list, although, formally, it is still a tree. Such a situation does not arise in balanced trees.

FIGURE 6.34 Different binary search trees with the same information.

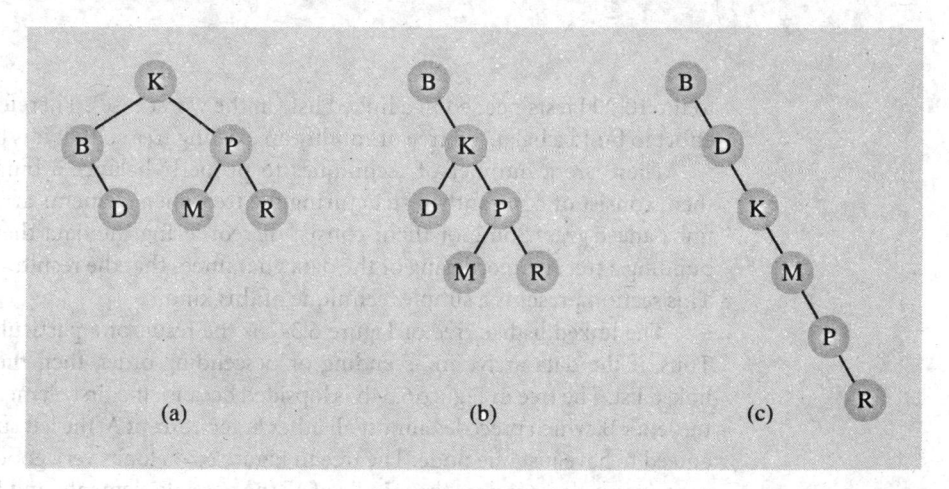

A binary tree is *height-balanced* or simply *balanced* if the difference in height of both subtrees of any node in the tree is either zero or one. For example, for node K in Figure 6.34b, the difference between the heights of its subtrees being equal to one is acceptable. But for node B this difference is three, which means that the entire tree is unbalanced. For the same node B in 6.34c, the difference is the worst possible, namely, five. Also, a tree is considered *perfectly balanced* if it is balanced and all leaves are to be found on one level or two levels.

Figure 6.35 shows how many nodes can be stored in binary trees of different heights. Because each node can have two children, the number of nodes on a certain level is double the number of parents residing on the previous level (except, of course, the root). For example, if 10,000 elements are stored in a perfectly balanced tree, then the tree is of height $\lceil \lg(10,001) \rceil = \lceil 13.289 \rceil = 14$. In practical terms, this means that if 10,000 elements are stored in a perfectly balanced tree, then at most 14 nodes have to be checked to locate a particular element. This is a substantial difference compared

FIGURE 6.35 Maximum number of nodes in binary trees of different heights.

Height	Nodes at One Level	Nodes at All Levels
1	$2^0 = 1$	$1 = 2^1 - 1$
2	$2^1 = 2$	$3 = 2^2 - 1$
3	$2^2 = 4$	$7 = 2^3 - 1$
4	$2^3 = 8$	$15 = 2^4 - 1$
⋮		
11	$2^{10} = 1{,}024$	$2{,}047 = 2^{11} - 1$
⋮		
14	$2^{13} = 8{,}192$	$16{,}383 = 2^{14} - 1$
⋮		
h	2^{h-1}	$n = 2^h - 1$
⋮		

to the 10,000 tests needed in a linked list (in the worst case). Therefore, it is worth the effort to build a balanced tree or modify an existing tree so that it is balanced.

There are a number of techniques to properly balance a binary tree. Some of them consist of constantly restructuring the tree when elements arrive and lead to an unbalanced tree. Some of them consist of reordering the data themselves and then building a tree, if an ordering of the data guarantees that the resulting tree is balanced. This section presents a simple technique of this kind.

The linked listlike tree of Figure 6.34c is the result of a particular stream of data. Thus, if the data arrive in ascending or descending order, then the tree resembles a linked list. The tree in Figure 6.34b is lopsided because the first element that arrived was the letter B, which precedes almost all other letters, except A; the left subtree of B is guaranteed to have just one node. The tree in Figure 6.34a looks very good, because the root contains an element near the middle of all the possible elements, and P is more or less in the middle of K and Z. This leads us to an algorithm based on binary search technique.

When data arrive, store all of them in an array. After all the data arrive, sort the array using one of the efficient algorithms discussed in Chapter 9. Now, designate for the root the middle element in the array. The array now consists of two subarrays: one between the beginning of the array and the element just chosen for the root and one between the root and the end of the array. The left child of the root is taken from the middle of the first subarray, its right child an element in the middle of the second subarray. Now, building the level containing the children of the root is finished. The next level, with children of children of the root, is constructed in the same fashion using four subarrays and the middle elements from each of them.

In this description, first the root is inserted into an initially empty tree, then its left child, then its right child, and so on level by level. An implementation of this algorithm is greatly simplified if the order of insertion is changed: First insert the root, then its left child, then the left child of this left child, and so on. This allows for using the following simple recursive implementation:

```
template<class T>
void BST<T>::balance(T data[], int first, int last) {
    if (first <= last) {
        int middle = (first + last)/2;
        insert(data[middle]);
        balance (data,first,middle-1);
        balance (data,middle+1,last);
    }
}
```

An example of the application of `balance()` is shown in Figure 6.36. First, number 4 is inserted (Figure 6.36a), then 1 (Figure 6.36b), then 0 and 2 (Figure 6.36c), and finally, 3, 6, 5, 7, 8, and 9 (Figure 6.36d).

This algorithm has one serious drawback: All data must be put in an array before the tree can be created. They can be stored in the array directly from the input. In this case, the algorithm may be unsuitable when the tree has to be used while the data to be included in the tree are still coming. But the data can be transferred from an unbalanced tree to the array using inorder traversal. The tree can now be deleted and re-created using `balance()`. This, at least, does not require using any sorting algorithm to put data in order.

FIGURE 6.36 Creating a binary search tree from an ordered array.

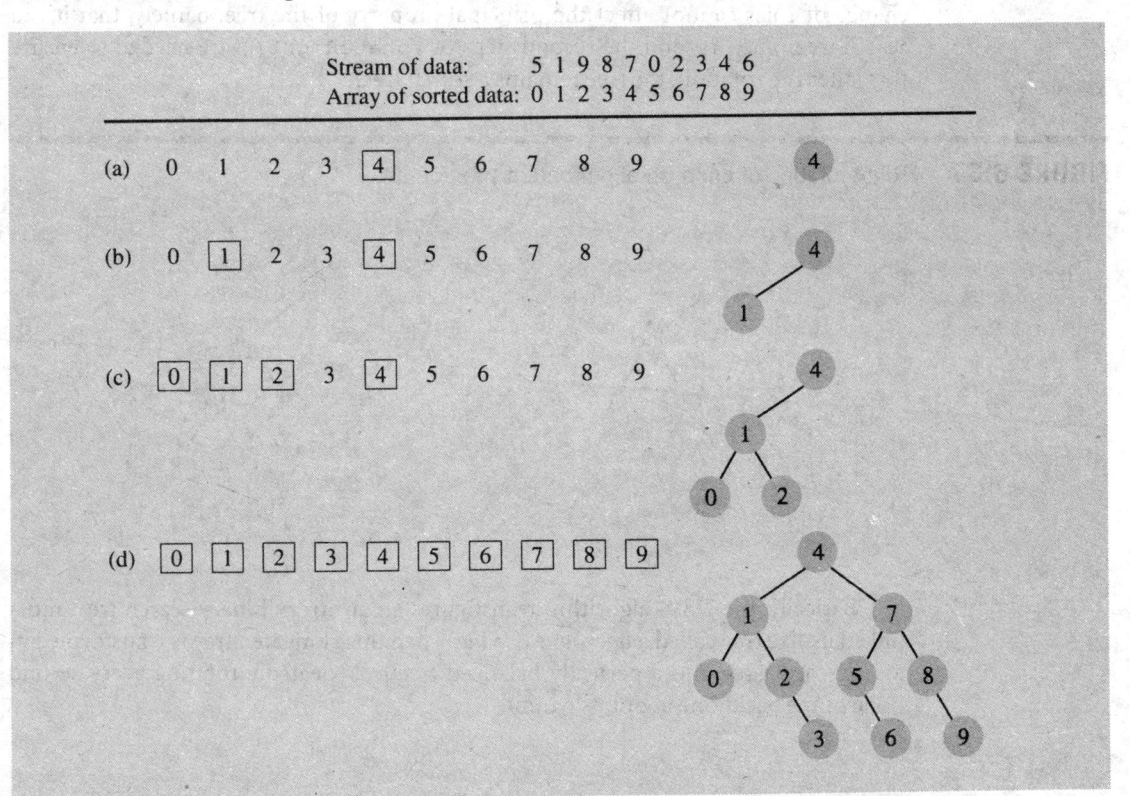

6.7.1 The DSW Algorithm

The algorithm discussed in the previous section was somewhat inefficient in that it required an additional array that needed to be sorted before the construction of a perfectly balanced tree began. To avoid sorting, it required deconstructing the tree after placing elements in the array using the inorder traversal and then reconstructing the tree, which is inefficient except for relatively small trees. There are, however, algorithms that require little additional storage for intermediate variables and use no sorting procedure. The very elegant DSW algorithm was devised by Colin Day and later improved by Quentin F. Stout and Bette L. Warren.

The building block for tree transformations in this algorithm is the *rotation*. There are two types of rotation, left and right, which are symmetrical to one another. The right rotation of the node Ch about its parent Par is performed according to the following algorithm:

```
rotateRight (Gr, Par, Ch)
    if Par is not the root of the tree    // i.e., if Gr is not null
        grandparent Gr of child Ch becomes Ch's parent;
    right subtree of Ch becomes left subtree of Ch's parent Par;
    node Ch acquires Par as its right child;
```

The steps involved in this compound operation are shown in Figure 6.37. The third step is the core of the rotation, when Par, the parent node of child Ch, becomes the child of Ch, when the roles of a parent and its child change. However, this exchange of roles cannot affect the principal property of the tree, namely, that it is a search tree. The first and the second steps of rotateRight() are needed to ensure that, after the rotation, the tree remains a search tree.

FIGURE 6.37 Right rotation of child Ch about parent Par.

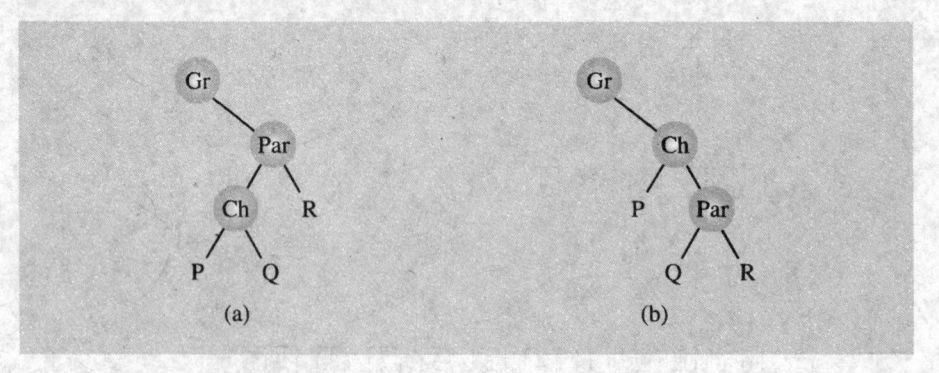

Basically, the DSW algorithm transfigures an arbitrary binary search tree into a linked listlike tree called a *backbone* or *vine*. Then this elongated tree is transformed in a series of passes into a perfectly balanced tree by repeatedly rotating every second node of the backbone about its parent.

In the first phase, a backbone is created using the following routine:

```
createBackbone(root, n)
    tmp = root;
    while (tmp != 0)
        if tmp has a left child
            rotate this child about tmp;  // hence the left child
                                          // becomes parent of tmp;
            set tmp to the child that just became parent;
        else set tmp to its right child;
```

This algorithm is illustrated in Figure 6.38. Note that a rotation requires knowledge about the parent of tmp, so another pointer has to be maintained when implementing the algorithm.

FIGURE 6.38 Transforming a binary search tree into a backbone.

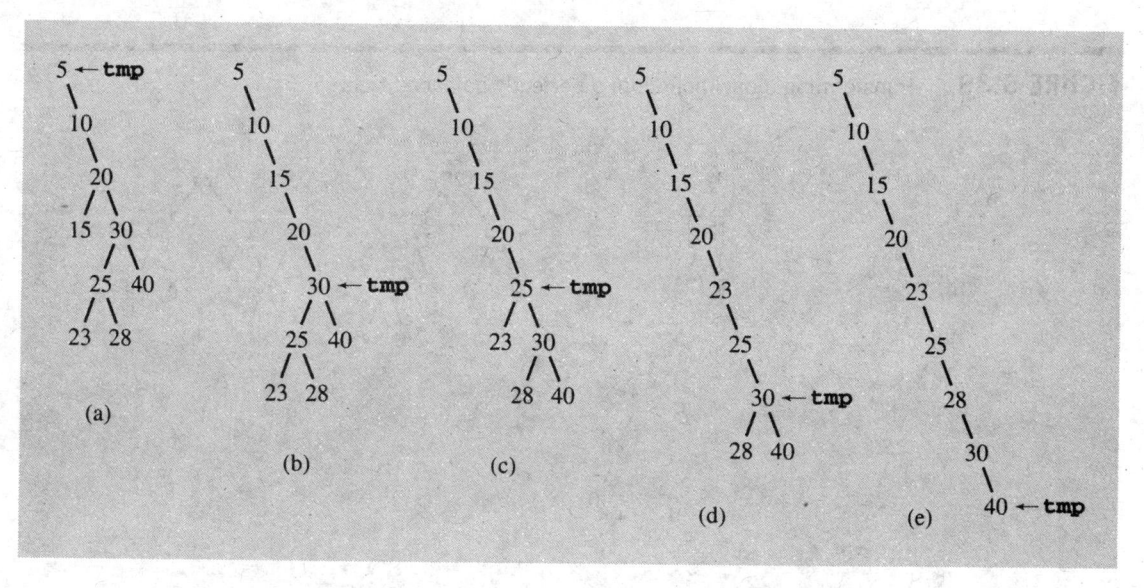

In the best case, when the tree is already a backbone, the while loop is executed n times and no rotation is performed. In the worst case, when the root does not have a right child, the while loop executes $2n - 1$ times with $n - 1$ rotations performed, where n is the number of nodes in the tree; that is, the run time of the first phase is $O(n)$. In this case, for each node except the one with the smallest value, the left child of tmp is rotated about tmp. After all rotations are finished, tmp points to the root, and after n iterations, it descends down the backbone to become null.

In the second phase, the backbone is transformed into a tree, but this time, the tree is perfectly balanced by having leaves only on two adjacent levels. In each pass down the backbone, every second node down to a certain point is rotated about its parent. The first pass is used to account for the difference between the number n of nodes in the current tree and the number $2^{\lfloor \lg(n+1) \rfloor} - 1$ of nodes in the closest complete binary tree where $\lfloor x \rfloor$ is the closest integer less than x. That is, the overflowing nodes are treated separately.

```
createPerfectTree(n)
    m = 2^⌊lg(n+1)⌋-1;
    make n-m rotations starting from the top of backbone;
    while (m > 1)
        m = m/2;
        make m rotations starting from the top of backbone;
```

Figure 6.39 contains an example. The backbone in Figure 6.38e has nine nodes and is preprocessed by one pass outside the loop to be transformed into the backbone shown in Figure 6.39b. Now, two passes are executed. In each backbone, the nodes to be promoted by one level by left rotations are shown as squares; their parents, about which they are rotated, are circles.

FIGURE 6.39 Transforming a backbone into a perfectly balanced tree.

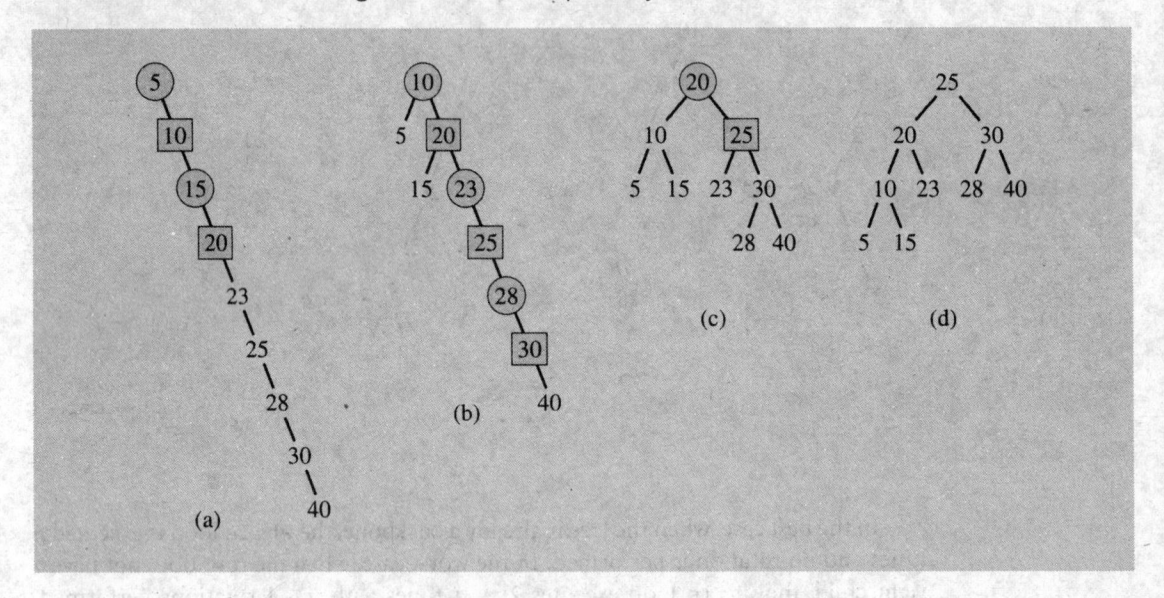

To compute the complexity of the tree building phase, observe that the number of iterations performed by the `while` loop equals

$$(2^{\lg(m+1)-1} - 1) + \cdots + 15 + 7 + 3 + 1 = \sum_{i=1}^{\lg(m+1)-1} (2^i - 1) = m - \lg(m+1)$$

The number of rotations can now be given by the formula

$$n - m + (m - \lg(m+1)) = n - \lg(m+1) = n - \lfloor \lg(n+1) \rfloor$$

that is, the number of rotations is $O(n)$. Because creating a backbone also required at most $O(n)$ rotations, the cost of global rebalancing with the DSW algorithm is optimal in terms of time because it grows linearly with n and requires a very small and fixed amount of additional storage.

6.7.2 AVL Trees

The previous two sections discussed algorithms that rebalanced the tree globally; each and every node could have been involved in rebalancing either by moving data from nodes or by reassigning new values to pointers. Tree rebalancing, however, can be performed locally if only a portion of the tree is affected when changes are required after an element is inserted into or deleted from the tree. One classical method has been proposed by Adel'son-Vel'skii and Landis, which is commemorated in the name of the tree modified with this method: the AVL tree.

An *AVL tree* (originally called an *admissible tree*) is one in which the height of the left and right subtrees of every node differ by at most one. For example, all the trees in Figure 6.40 are AVL trees. Numbers in the nodes indicate the *balance factors* that are the differences between the heights of the left and right subtrees. A balance factor is the height of the right subtree minus the height of the left subtree. For an AVL tree, all balance factors should be +1, 0, or –1. Notice that the definition of the AVL tree is the same as the definition of the balanced tree. However, the concept of the AVL tree always implicitly includes the techniques for balancing the tree. Moreover, unlike the two methods previously discussed, the technique for balancing AVL trees does not guarantee that the resulting tree is perfectly balanced.

FIGURE 6.40 Examples of AVL trees.

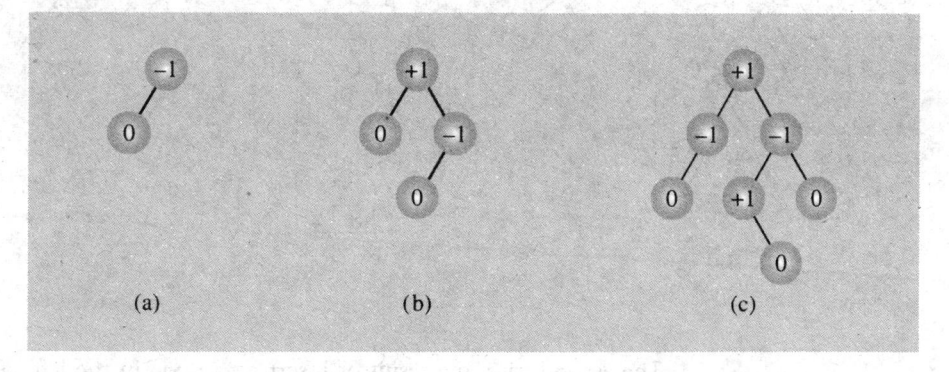

The definition of an AVL tree indicates that the minimum number of nodes in a tree is determined by the recurrence equation

$$AVL_h = AVL_{h-1} + AVL_{h-2} + 1$$

where $AVL_0 = 0$ and $AVL_1 = 1$ are the initial conditions.[1] This formula leads to the following bounds on the height h of an AVL tree depending on the number of nodes n (see Appendix A.5):

$$\lg(n + 1) \le h < 1.44\lg(n + 2) - 0.328$$

Therefore, h is bounded by $O(\lg n)$; the worst case search requires $O(\lg n)$ comparisons. For a perfectly balanced binary tree of the same height, $h = \lceil\lg(n + 1)\rceil$. Therefore, the search time in the worst case in an AVL tree is 44% worse (it requires 44% more comparisons) than in the best case tree configuration. Empirical studies indicate that the average number of searches is much closer to the best case than to the worst and is equal to $\lg n + 0.25$ for large n (Knuth 1998). Therefore, AVL trees are definitely worth studying.

If the balance factor of any node in an AVL tree becomes less than −1 or greater than 1, the tree has to be balanced. An AVL tree can become out of balance in four situations, but only two of them need to be analyzed; the remaining two are symmetrical. The first case, the result of inserting a node in the right subtree of the right child, is illustrated in Figure 6.41. The heights of the participating subtrees are indicated within these subtrees. In the AVL tree in Figure 6.41a, a node is inserted somewhere in the right subtree of Q (Figure 6.41b), which disturbs the balance of the tree P. In this case, the problem can be easily rectified by rotating node Q about its parent P (Figure 6.41c) so that the balance factor of both P and Q becomes zero, which is even better than at the outset.

FIGURE 6.41 Balancing a tree after insertion of a node in the right subtree of node Q.

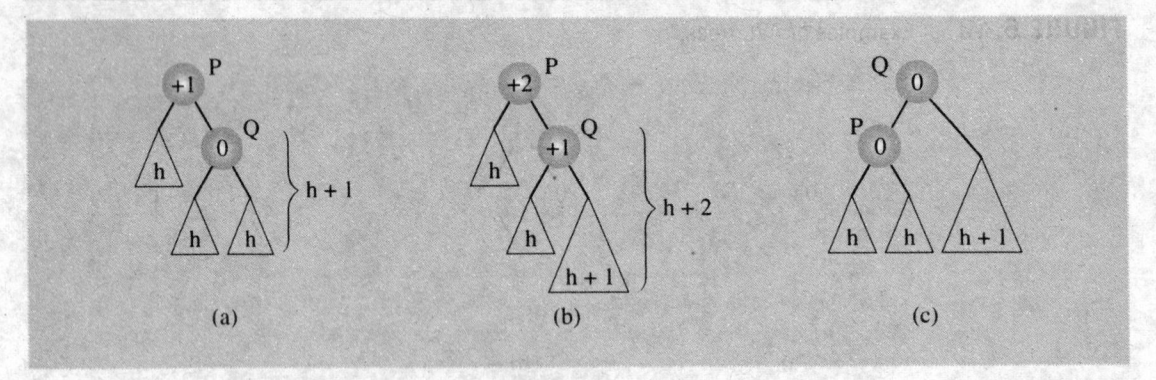

The second case, the result of inserting a node in the left subtree of the right child, is more complex. A node is inserted into the tree in Figure 6.42a; the resulting tree is shown in Figure 6.42b and in more detail in Figure 6.42c. Note that R's balance factor can also be −1. To bring the tree back into balance, a double rotation is performed. The balance of the tree P is restored by rotating R about node Q (Figure 6.42d) and then by rotating R again, this time about node P (Figure 6.42e).

---

[1] Numbers generated by this recurrence formula are called *Leonardo numbers*.

FIGURE 6.42 Balancing a tree after insertion of a node in the left subtree of node Q.

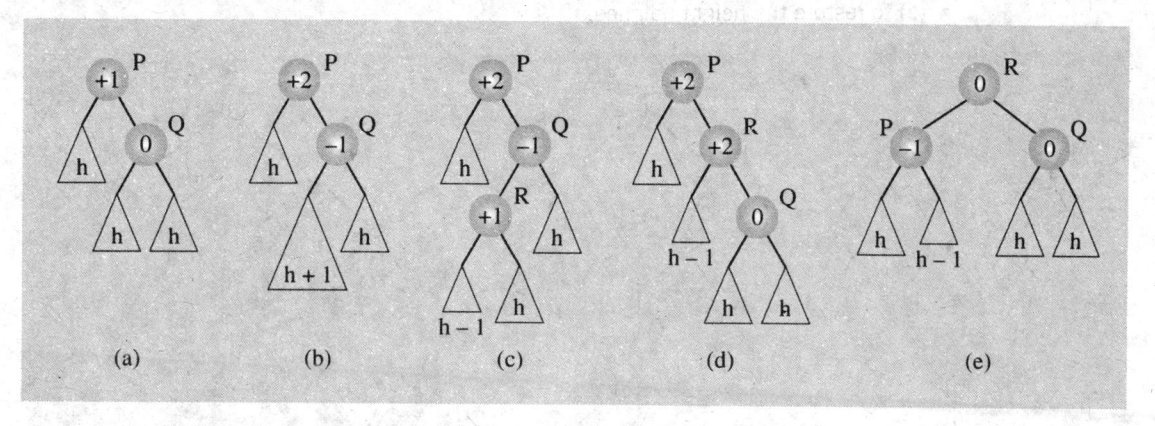

In these two cases, the tree P is considered a stand-alone tree. However, P can be part of a larger AVL tree; it can be a child of some other node in the tree. If a node is entered into the tree and the balance of P is disturbed and then restored, does extra work need to be done to the predecessor(s) of P? Fortunately not. Note that the heights of the trees in Figures 6.41c and 6.42e resulting from the rotations are the same as the heights of the trees before insertion (Figures 6.41a and 6.42a) and are equal to $h + 2$. This means that the balance factor of the parent of the new root (Q in Figure 6.41c and R in Figure 6.42e) remains the same as it was before the insertion, and the changes made to the subtree P are sufficient to restore the balance of the entire AVL tree. The problem is in finding a node P for which the balance factor becomes unacceptable after a node has been inserted into the tree.

This node can be detected by moving up toward the root of the tree from the position in which the new node has been inserted and by updating the balance factors of the nodes encountered. Then, if a node with a ±1 balance factor is encountered, the balance factor may be changed to ±2, and the first node whose balance factor is changed in this way becomes the root P of a subtree for which the balance has to be restored. Note that the balance factors do not have to be updated above this node because they remain the same.

In Figure 6.43a, a path is marked with one balance factor equal to +1. Insertion of a new node at the end of this path results in an unbalanced tree (Figure 6.43b), and the balance is restored by one left rotation (Figure 6.43c).

However, if the balance factors on the path from the newly inserted node to the root of the tree are all zero, all of them have to be updated, but no rotation is needed for any of the encountered nodes. In Figure 6.44a, the AVL tree has a path of all zero balance factors. After a node has been appended to the end of this path (Figure 6.44b), no changes are made in the tree except for updating the balance factors of all nodes along this path.

FIGURE 6.43 An example of inserting a new node (b) in an AVL tree (a), which requires one rotation (c) to restore the height balance.

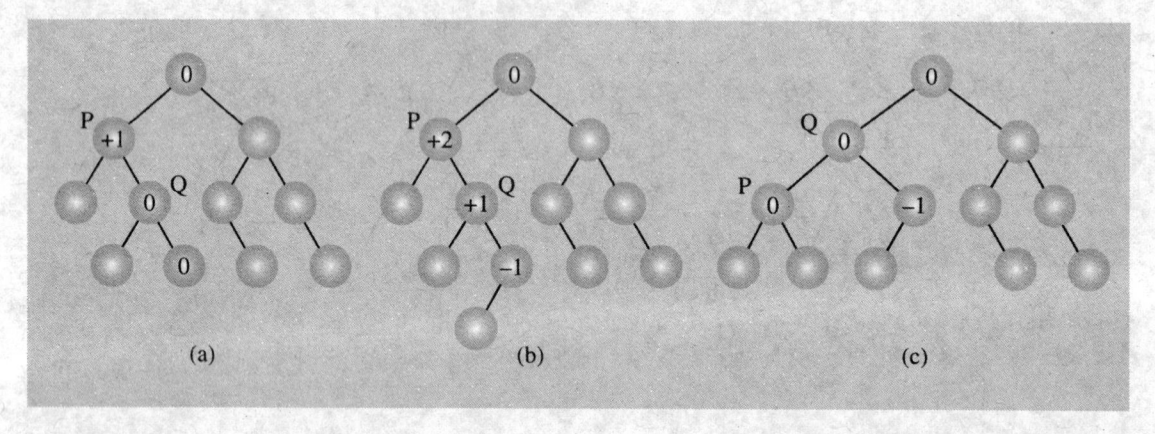

FIGURE 6.44 In an AVL tree (a) a new node is inserted (b) requiring no height adjustments.

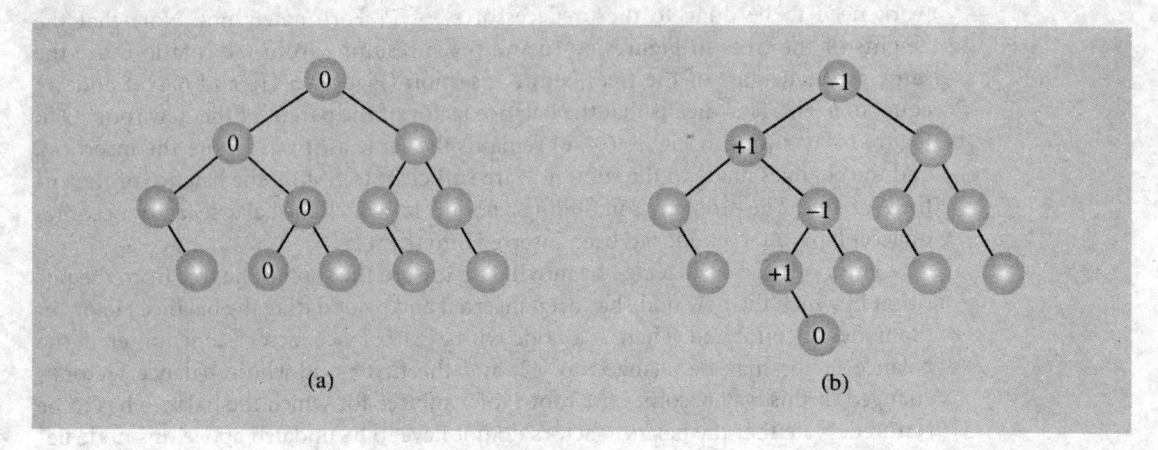

Deletion may be more time-consuming than insertion. First, we apply `deleteByCopying()` to delete a node. This technique allows us to reduce the problem of deleting a node with two descendants to deleting a node with at most one descendant.

After a node has been deleted from the tree, balance factors are updated from the parent of the deleted node up to the root. For each node in this path whose balance factor becomes ±2, a single or double rotation has to be performed to restore the balance of the tree. Importantly, the rebalancing does not stop after the first node P is found for which the balance factor would become ±2, as is the case with insertion. This also means that deletion leads to at most $O(\lg n)$ rotations, because in the worst case, every node on the path from the deleted node to the root may require rebalancing.

Deletion of a node does not have to necessitate an immediate rotation because it may improve the balance factor of its parent (by changing it from ±1 to 0), but it may also worsen the balance factor for the grandparent (by changing it from ±1 to ±2). We illustrate only those cases that require immediate rotation. There are four such cases (plus four symmetric cases). In each of these cases, we assume that the left child of node *P* is deleted.

In the first case, the tree in Figure 6.45a turns, after deleting a node, into the tree in Figure 6.45b. The tree is rebalanced by rotating *Q* about *P* (Figure 6.45c). In the second case, *P* has a balance factor equal to +1, and its right subtree *Q* has a balance

FIGURE 6.45 Rebalancing an AVL tree after deleting a node.

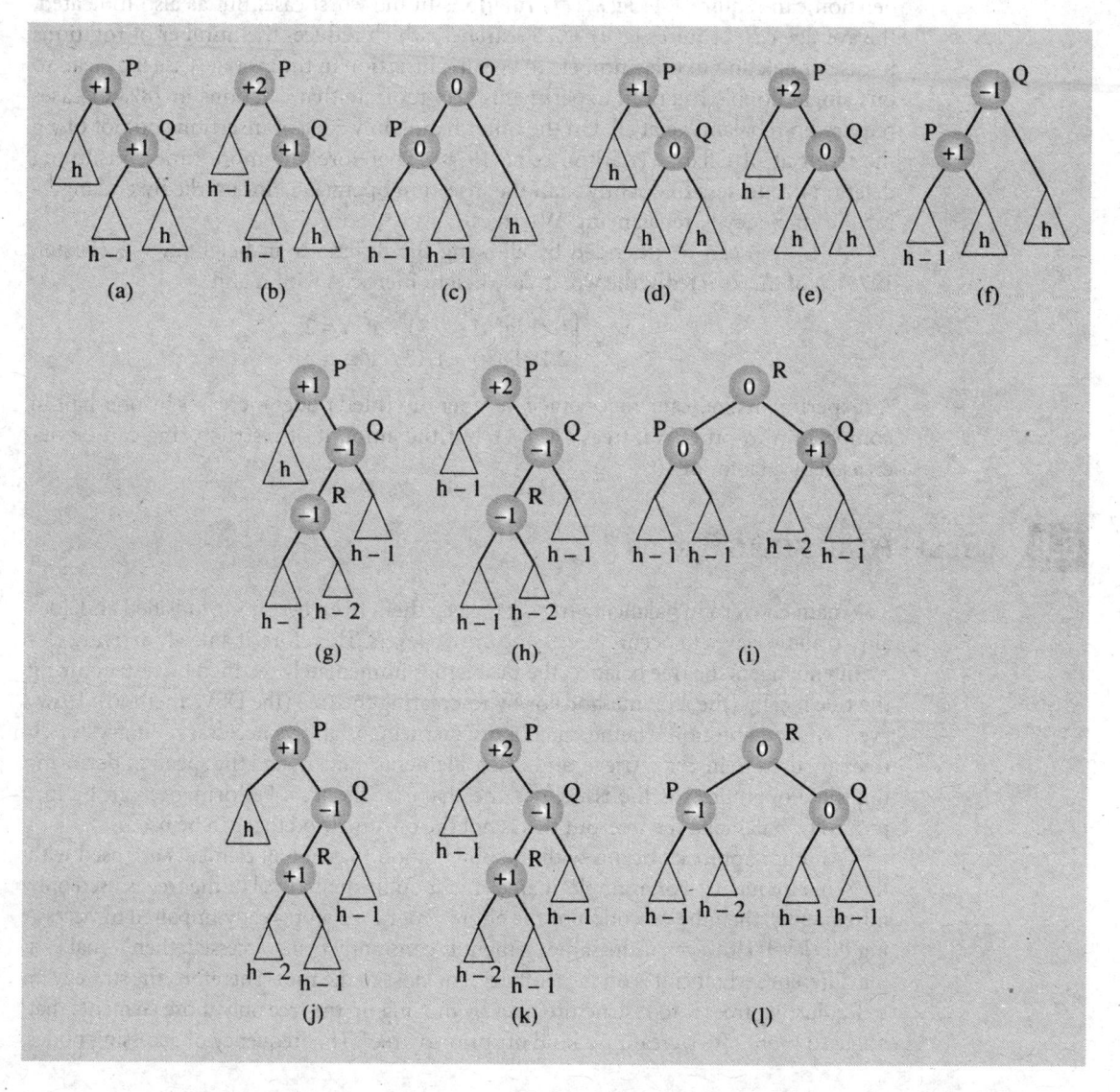

factor equal to 0 (Figure 6.45d). After deleting a node in the left subtree of P (Figure 6.45e), the tree is rebalanced by the same rotation as in the first case (Figure 6.45f). In this way, cases one and two can be processed together in an implementation after checking that the balance factor of Q is +1 or 0. If Q is −1, we have two other cases, which are more complex. In the third case, the left subtree R of Q has a balance factor equal to −1 (Figure 6.45g). To rebalance the tree, first R is rotated about Q and then about P (Figures 6.45h–i). The fourth case differs from the third in that R's balance factor equals +1 (Figure 6.45j), in which case the same two rotations are needed to restore the balance factor of P (Figures 6.45k–l). Cases three and four can be processed together in a program processing AVL trees.

The previous analyses indicate that insertions and deletions require at most $1.44 \lg(n+2)$ searches. Also, insertion can require one single or one double rotation, and deletion can require $1.44 \lg(n+2)$ rotations in the worst case. But as also indicated, the average case requires $\lg(n) + .25$ searches, which reduces the number of rotations in case of deletion to this number. To be sure, insertion in the average case may lead to one single/double rotation. Experiments also indicate that deletions in 78% of cases require no rebalancing at all. On the other hand, only 53% of insertions do not bring the tree out of balance (Karlton et al. 1976). Therefore, the more time-consuming deletion occurs less frequently than the insertion operation, not markedly endangering the efficiency of rebalancing AVL trees.

AVL trees can be extended by allowing the difference in height $\Delta > 1$ (Foster, 1973). Not unexpectedly, the worst-case height increases with Δ and

$$h = \begin{cases} 1.81 \lg(n) - 0.71 & \text{if } \Delta = 2 \\ 2.15 \lg(n) - 1.13 & \text{if } \Delta = 3 \end{cases}$$

As experiments indicate, the average number of visited nodes increases by one-half in comparison to pure AVL trees ($\Delta = 1$), but the amount of restructuring can be decreased by a factor of 10.

6.8 SELF-ADJUSTING TREES

The main concern in balancing trees is to keep them from becoming lopsided and, ideally, to allow leaves to occur only at one or two levels. Therefore, if a newly arriving element endangers the tree balance, the problem is immediately rectified by restructuring the tree locally (the AVL method) or by re-creating the tree (the DSW method). However, we may question whether such a restructuring is always necessary. Binary search trees are used to insert, retrieve, and delete elements quickly, and the speed of performing these operations is the issue, not the shape of the tree. Performance can be improved by balancing the tree, but this is not the only method that can be used.

Another approach begins with the observation that not all elements are used with the same frequency. For example, if an element on the tenth level of the tree is used only infrequently, then the execution of the entire program is not greatly impaired by accessing this level. However, if the same element is constantly being accessed, then it makes a big difference whether it is on the tenth level or close to the root. Therefore, the strategy in self-adjusting trees is to restructure trees by moving up the tree only those elements that are used more often, creating a kind of "priority tree." The frequency of accessing nodes can be determined in a variety of ways. Each node can have a counter field that records the number of times the element has been used for any operation. Then the tree can be scanned to move the most frequently accessed elements toward the root. In a less sophisticated approach, it is assumed that an element being accessed has a good chance of being accessed again soon. Therefore, it is moved up the tree. No restructuring is performed for new elements. This assumption may lead to promoting elements that are occasionally accessed, but the overall tendency is to move up elements with a high frequency of access, and for the most part, these elements will populate the first few levels of the tree.

6.8.1 Self-Restructuring Trees

A strategy proposed by Brian Allen and Ian Munro and by James Bitner consists of two possibilities:

1. *Single rotation.* Rotate a child about its parent if an element in a child is accessed, unless it is the root (Figure 6.46a).
2. *Moving to the root.* Repeat the child–parent rotation until the element being accessed is in the root (Figure 6.46b).

FIGURE 6.46 Restructuring a tree by using (a) a single rotation or (b) moving to the root when accessing node R.

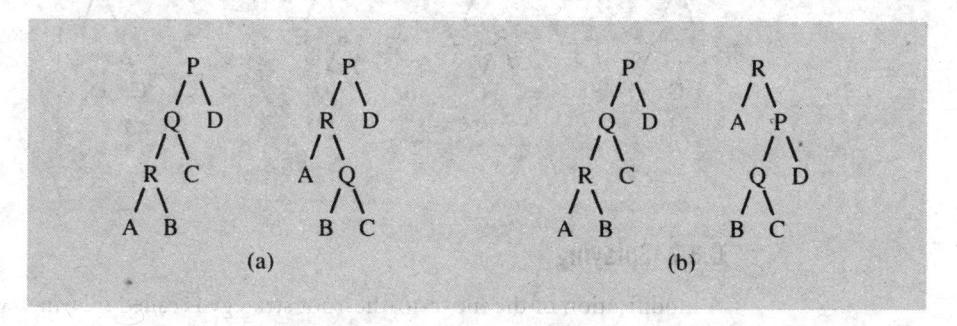

```
          P                  P                    P                R
         / \                / \                  / \              / \
        Q   D              R   D                Q   D            A   P
       / \                / \                  / \                  / \
      R   C              A   Q                R   C                Q   D
     / \                    / \              / \                  / \
    A   B                  B   C            A   B                B   C
         (a)                                      (b)
```

Using the single-rotation strategy, frequently accessed elements are eventually moved up close to the root so that later accesses are faster than previous ones. In the move-to-the-root strategy, it is assumed that the element being accessed has a high probability to be accessed again, so it percolates right away up to the root. Even if it is not used in the next access, the element remains close to the root. These strategies, however, do not work very well in unfavorable situations, when the binary tree is elongated as in Figure 6.47. In this case, the shape of the tree improves slowly. Nevertheless, it has been determined that the cost of moving a node to the root converges to the cost of accessing the node in an optimal tree times $2 \ln 2$; that is, it converges to $(2 \ln 2) \lg n$. The result holds for any probability distribution (that is, independently of the probability that a particular request is issued). However, the average search time when all requests are equally likely is, for the single rotation technique, equal to $\sqrt{\pi n}$.

FIGURE 6.47 (a–e) Moving element T to the root and then (e–i) moving element S to the root.

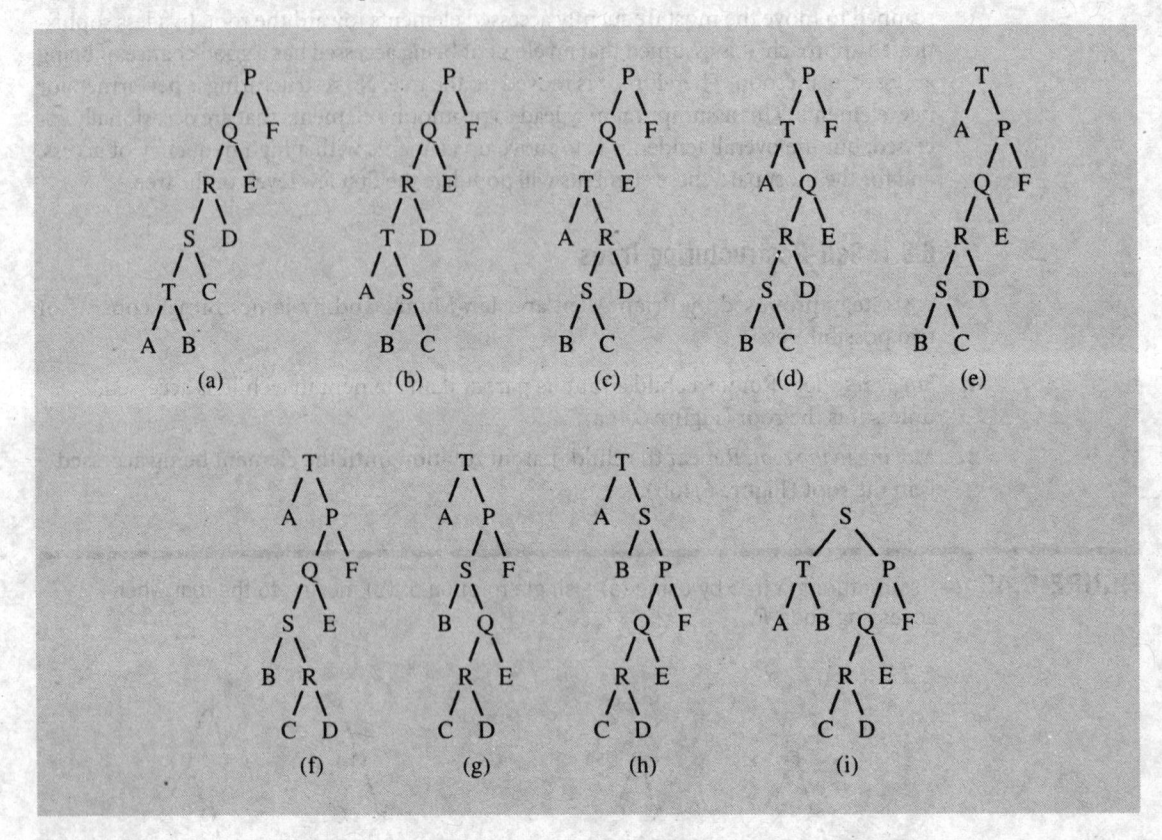

6.8.2 Splaying

A modification of the move-to-the-root strategy is called *splaying*, which applies single rotations in pairs in an order depending on the links between the child, parent, and grandparent (Sleator and Tarjan 1985). First, three cases are distinguished depending on the relationship between a node R being accessed and its parent Q and grandparent P (if any) nodes:

Case 1: Node R's parent is the root.

Case 2: *Homogeneous configuration.* Node R is the left child of its parent Q, and Q is the left child of its parent P, or R and Q are both right children.

Case 3: *Heterogeneous configuration.* Node R is the right child of its parent Q, and Q is the left child of its parent P, or R is the left child of Q, and Q is the right child of P.

The algorithm to move a node R being accessed to the root of the tree is as follows:

Section 6.8 Self-Adjusting Trees

```
splaying(P,Q,R)
    while R is not the root
        if R's parent is the root
            perform a singular splay, rotate R about its parent (Figure 6.48a);
        else if R is in a homogeneous configuration with its predecessors
            perform a homogeneous splay, first rotate Q about P
            and then R about Q (Figure 6.48b);
        else  // if R is in a heterogeneous configuration
              // with its predecessors
            perform a heterogeneous splay, first rotate R about Q
            and then about P (Figure 6.48c);
```

FIGURE 6.48 Examples of splaying.

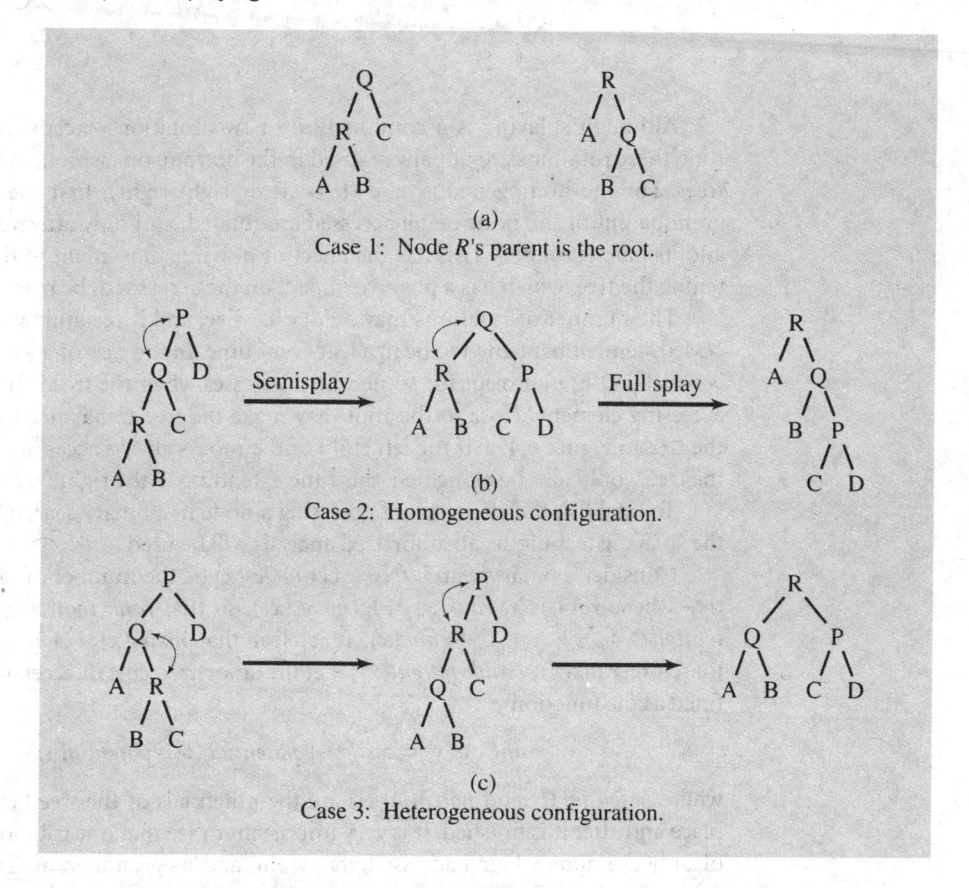

```
        Q                    R
       / \                  / \
      R   C                A   Q
     / \                      / \
    A   B                    B   C
              (a)
    Case 1: Node R's parent is the root.
```

```
      P                    Q                         R
     / \                  / \                       / \
    Q   D   Semisplay    R   P      Full splay     A   Q
   / \        ──►       / \ / \        ──►            / \
  R   C                A  B C  D                     B   P
 / \                                                    / \
A   B                                                  C   D
                    (b)
        Case 2: Homogeneous configuration.
```

```
      P                    P                         R
     / \                  / \                       / \
    Q   D      ──►       R   D         ──►        Q    P
   / \                  / \                      / \  / \
  A   R                Q   C                    A  B C   D
     / \              / \
    B   C            A   B
                    (c)
        Case 3: Heterogeneous configuration.
```

The difference in restructuring a tree is illustrated in Figure 6.49, where the tree from Figure 6.47a is used to access node T located at the fifth level. The shape of the tree is immediately improved. Then, node R is accessed (Figure 6.49c) and the shape of the tree becomes even better (Figure 6.49d).

FIGURE 6.49 Restructuring a tree with splaying (a–c) after accessing T and (c–d) then R.

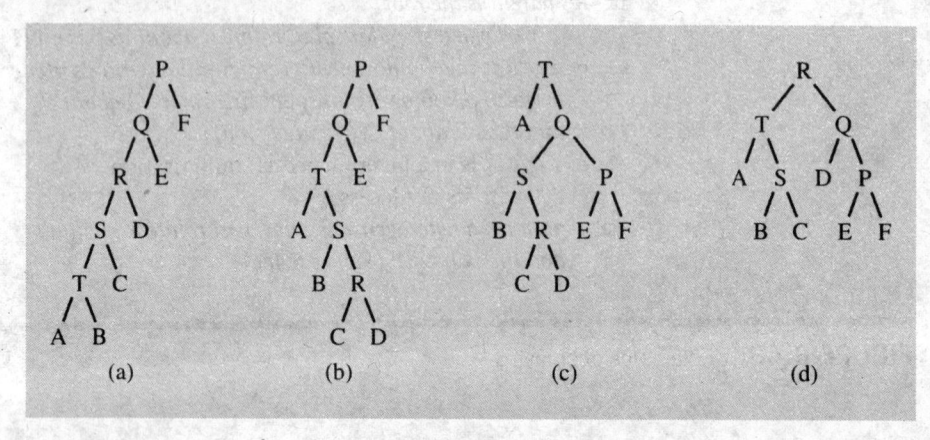

Although splaying is a combination of two rotations except when next to the root, these rotations are not always used in the bottom-up fashion, as in self-adjusting trees. For the homogeneous case (left-left or right-right), first the parent and the grandparent of the node being accessed are rotated, and only afterward are the node and its parent rotated. This has the effect of moving an element to the root and flattening the tree, which has a positive impact on the accesses to be made.

The number of rotations may seem excessive, and it certainly would be if an accessed element happened to be in a leaf every time. In the case of a leaf, the access time is usually $O(\lg n)$, except for some initial accesses when the tree is not balanced. But accessing elements close to the root may make the tree unbalanced. For example, in the tree in Figure 6.49a, if the left child of the root is always accessed, then eventually, the tree would also be elongated, this time extending to the right.

To establish the efficiency of accessing a node in a binary search tree that utilizes the splaying technique, an amortized analysis will be used.

Consider a binary search tree t. Let $nodes(x)$ be the number of nodes in the subtree whose root is x, $rank(x) = \lg(nodes(x))$, so that $rank(root(t)) = \lg(n)$, and $potential(t) = \sum_{x \text{ is a node of } t} rank(x)$. It is clear that $nodes(x) + 1 \leq nodes(parent(x))$; therefore, $rank(x) < rank(parent(x))$. Let the amortized cost of accessing node x be defined as the function

$$amCost(x) = cost(x) + potential_s(t) - potential_0(t)$$

where $potential_s(t)$ and $potential_0(t)$ are the potentials of the tree before access takes place and after it is finished. It is very important to see that one rotation changes ranks of only the node x being accessed, its parent, and its grandparent. This is the reason for basing the definition of the amortized cost of accessing node x on the change in the potential of the tree, which amounts to the change of ranks of the nodes involved in splaying operations that promote x to the root. We can state now a lemma specifying the amortized cost of one access.

Access lemma (Sleator and Tarjan 1985). For the amortized time to splay the tree t at a node x,

$$amCost(x) < 3(\lg(n) - rank(x)) + 1$$

The proof of this conjecture is divided into three parts, each dealing with the different case indicated in Figure 6.48. Let $par(x)$ be a parent of x and $gpar(x)$ a grandparent of x (in Figure 6.48, $x = R$, $par(x) = Q$, and $gpar(x) = P$).

Case 1: One rotation is performed. This can be only the last splaying step in the sequence of such steps that move node x to the root of the tree t, and if there are a total of s splaying steps in the sequence, then the amortized cost of the last splaying step s is

$$amCost_s(x) = cost_s(x) + potential_s(t) - potential_{s-1}(t)$$
$$= 1 + (rank_s(x) - rank_{s-1}(x)) + (rank_s(par(x)) - rank_{s-1}(par(x)))$$

where $cost_s(x) = 1$ represents the actual cost, the cost of the one splaying step (which in this step is limited to one rotation); $potential_{s-1}(t) = rank_{s-1}(x) + rank_{s-1}(par(x)) + C$ and $potential_s(t) = rank_s(x) + rank_s(par(x)) + C$, because x and $par(x)$ are the only nodes whose ranks are modified. Now because $rank_s(x) = rank_{s-1}(par(x))$

$$amCost_s(x) = 1 - rank_{s-1}(x) + rank_s(par(x))$$

and because $rank_s(par(x)) < rank_s(x)$

$$amCost_s(x) < 1 - rank_{s-1}(x) + rank_s(x).$$

Case 2: Two rotations are performed during a homogeneous splay. As before, number 1 represents the actual cost of one splaying step.

$$amCost_i(x) = 1 + (rank_i(x) - rank_{i-1}(x)) + (rank_i(par(x)) - rank_{i-1}(par(x))) + (rank_i(gpar(x)) - rank_{i-1}(gpar(x)))$$

Because $rank_i(x) = rank_{i-1}(gpar(x))$

$$amCost_i(x) = 1 - rank_{i-1}(x) + rank_i(par(x)) - rank_{i-1}(par(x)) + rank_i(gpar(x))$$

Because $rank_i(gpar(x)) < rank_i(par(x)) < rank_i(x)$

$$amCost_i(x) < 1 - rank_{i-1}(x) - rank_{i-1}(par(x)) + 2rank_i(x)$$

and because $rank_{i-1}(x) < rank_{i-1}(par(x))$, that is, $-rank_{i-1}(par(x)) < -rank_{i-1}(x)$

$$amCost_i(x) < 1 - 2rank_{i-1}(x) + 2rank_i(x).$$

To eliminate number 1, consider the inequality $rank_{i-1}(x) < rank_{i-1}(gpar(x))$; that is, $1 \leq rank_{i-1}(gpar(x)) - rank_{i-1}(x)$. From this, we obtain

$$amCost_i(x) < rank_{i-1}(gpar(x)) - rank_{i-1}(x) - 2rank_{i-1}(x) + 2rank_i(x)$$
$$amCost_i(x) < rank_{i-1}(gpar(x)) - 3rank_{i-1}(x) + 2rank_i(x)$$

and because $rank_i(x) = rank_{i-1}(gpar(x))$

$$amCost_i(x) < -3rank_{i-1}(x) + 3rank_i(x)$$

Case 3: Two rotations are performed during a heterogeneous splay. The only difference in this proof is making the assumption that $rank_i(gpar(x)) < rank_i(x)$ and $rank_i(par(x)) < rank_i(x)$ instead of $rank_i(gpar(x)) < rank_i(par(x)) < rank_i(x)$, which renders the same result.

The total amortized cost of accessing a node x equals the sum of amortized costs of all the spaying steps executed during this access. If the number of steps equals s, then at most one (the last) step requires only one rotation (case 1) and thus

$$amCost(x) = \sum_{i=1}^{s} amCost_i(x) = \sum_{i=1}^{s-1} amCost_i(x) + amCost_s(x)$$

$$< \sum_{i=1}^{s-1} 3(rank_i(x) - rank_{i-1}(x)) + rank_s(x) - rank_{s-1}(x) + 1$$

Because $rank_s(x) > rank_{s-1}(x)$,

$$amCost(x) < \sum_{i=1}^{s-1} 3(rank_i(x) - rank_{i-1}(x)) + 3(rank_s(x) - rank_{s-1}(x)) + 1$$

$$= 3(rank_s(x) - rank_0(x)) + 1 = 3(\lg n - rank_0(x)) + 1 = O(\lg n)$$

This indicates that the amortized cost of an access to a node in a tree that is restructured with the splaying technique equals $O(\lg n)$, which is the same as the worst case in balanced trees. However, to make the comparison more adequate, we should compare a sequence of m accesses to nodes rather than one access because, with the amortize cost, one isolated access can still be on the order of $O(n)$. The efficiency of a tree that applies splaying is thus comparable to that of a balanced tree for a sequence of accesses and equals $O(m \lg n)$. ❑

Splaying is a strategy focusing upon the elements rather than the shape of the tree. It may perform well in situations in which some elements are used much more frequently than others. If elements near the root are accessed with about the same frequency as elements on the lowest levels, then splaying may not be the best choice. In this case, a strategy that stresses balancing the tree rather than frequency is better; a modification of the splaying method is a more viable option.

Semisplaying is a modification that requires only one rotation for a homogeneous splay and continues splaying with the parent of the accessed node, not with the node itself. It is illustrated in Figure 6.48b. After R is accessed, its parent Q is rotated about P and splaying continues with Q, not with R. A rotation of R about Q is not performed, as would be the case for splaying.

Figure 6.50 illustrates the advantages of semisplaying. The elongated tree from Figure 6.49a becomes more balanced with semisplaying after accessing T (Figures 6.50a–c), and after T is accessed again, the tree in Figure 6.50d has basically the same number of levels as the tree in Figure 6.46a. (It may have one more level if E or F was a subtree higher than any of subtrees A, B, C, or D.) For implementation of this tree strategy, see the case study at the end of this chapter.

It is interesting that although the theoretical bounds obtained from self-organizing trees compare favorably with the bounds for AVL trees and random binary search

FIGURE 6.50 (a–c) Accessing T and restructuring the tree with semisplaying; (c–d) accessing T again.

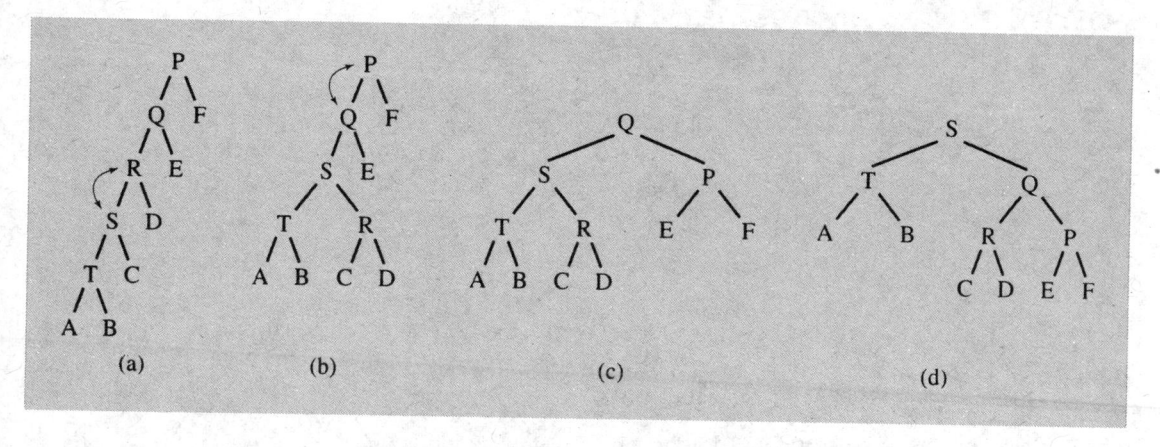

trees—that is, with no balancing technique applied to it—experimental runs for trees of various sizes and different ratios of accessing keys indicate that almost always the AVL tree outperforms self-adjusting trees, and many times even a regular binary search tree performs better than a self-organizing tree (Bell, Gupta 1993). At best, this result indicates that computational complexity and amortized performance should not always be considered as the only measures of algorithm performance.

6.9 HEAPS

A particular kind of binary tree, called a *heap*, has the following two properties:

1. The value of each node is greater than or equal to the values stored in each of its children.
2. The tree is perfectly balanced, and the leaves in the last level are all in the leftmost positions.

To be exact, these two properties define a *max heap*. If "greater" in the first property is replaced with "less," then the definition specifies a *min heap*. This means that the root of a max heap contains the largest element, whereas the root of a min heap contains the smallest. A tree has the *heap property* if each nonleaf has the first property. Due to the second condition, the number of levels in the tree is $O(\lg n)$.

The trees in Figure 6.51a are all heaps; the trees in Figure 6.51b violate the first property, and the trees in Figure 6.51c violate the second.

Interestingly, heaps can be implemented by arrays. For example, the array data = [2 8 6 1 10 15 3 12 11] can represent a nonheap tree in Figure 6.52. The elements are placed at sequential locations representing the nodes from top to bottom and in each level from left to right. The second property reflects the fact that the array is packed, with no gaps. Now, a heap can be defined as an array heap of length n in which

$$\text{heap[i]} \geq \text{heap[2} \cdot \text{i + 1]}, \text{ for } 0 \leq i < \frac{n-1}{2}$$

FIGURE 6.51 Examples of (a) heaps and (b–c) nonheaps.

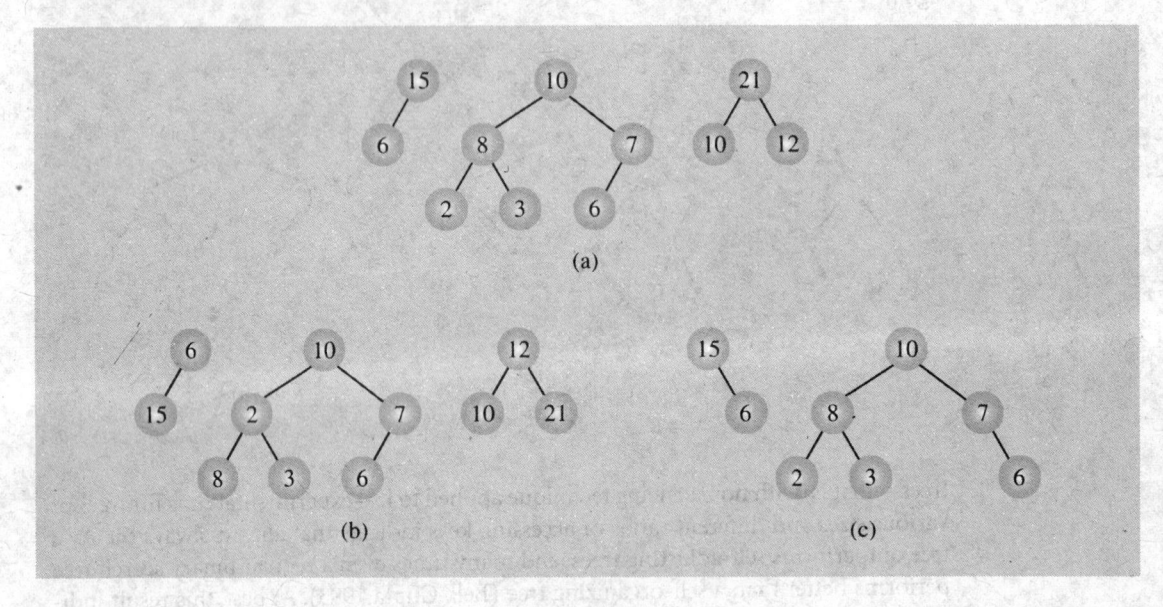

FIGURE 6.52 The array [2 8 6 1 10 15 3 12 11] seen as a tree.

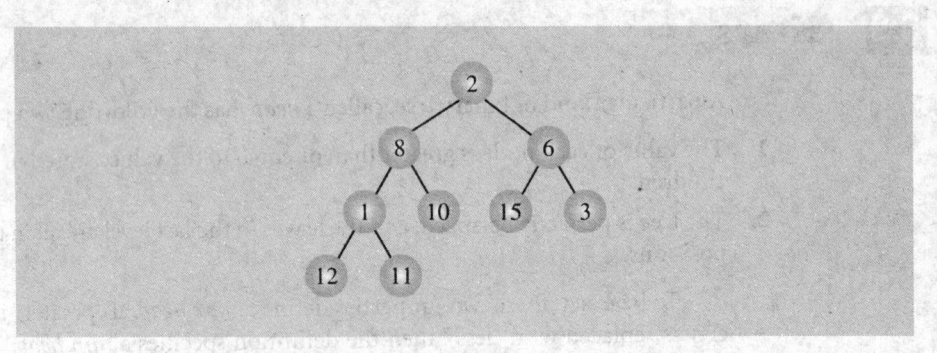

and

$$\text{heap}[i] \geq \text{heap}[2 \cdot i + 2], \text{ for } 0 \leq i < \frac{n-2}{2}$$

Elements in a heap are not perfectly ordered. We know only that the largest element is in the root node and that, for each node, all its descendants are less than or equal to that node. But the relation between sibling nodes or, to continue the kinship terminology, between uncle and nephew nodes is not determined. The order of the elements obeys a linear line of descent, disregarding lateral lines. For this reason, all the trees in Figure 6.53 are legitimate heaps, although the heap in Figure 6.53b is ordered best.

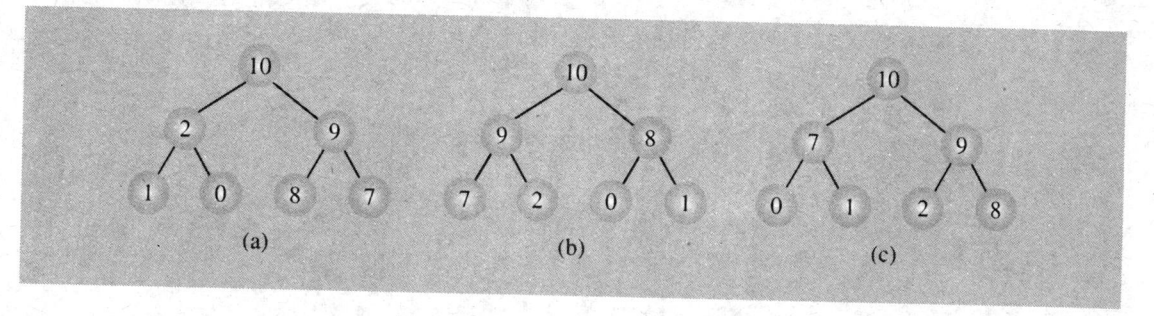

FIGURE 6.53 Different heaps constructed with the same elements.

[Figure showing three different heaps (a), (b), (c) constructed with the same elements, all with root 10]

6.9.1 Heaps as Priority Queues

A heap is an excellent way to implement a priority queue. Section 4.3 used linked lists to implement priority queues, structures for which the complexity was expressed in terms of $O(n)$ or $O(\sqrt{n})$. For large n, this may be too ineffective. On the other hand, a heap is a perfectly balanced tree; hence, reaching a leaf requires $O(\lg n)$ searches. This efficiency is very promising. Therefore, heaps can be used to implement priority queues. To this end, however, two procedures have to be implemented to enqueue and dequeue elements on a priority queue.

To enqueue an element, the element is added at the end of the heap as the last leaf. Restoring the heap property in the case of enqueuing is achieved by moving from the last leaf toward the root.

The algorithm for enqueuing is as follows:

```
heapEnqueue(el)
    put el at the end of heap;
    while el is not in the root and el > parent(el)
        swap el with its parent;
```

For example, the number 15 is added to the heap in Figure 6.54a as the next leaf (Figure 6.54b), which destroys the heap property of the tree. To restore this property, 15 has to be moved up the tree until it either ends up in the root or finds a parent that is not less than 15. In this example, the latter case occurs, and 15 has to be moved only twice without reaching the root.

Dequeuing an element from the heap consists of removing the root element from the heap, because by the heap property it is the element with the greatest priority. Then the last leaf is put in its place, and the heap property almost certainly has to be restored, this time by moving from the root down the tree.

The algorithm for dequeuing is as follows:

```
heapDequeue()
    extract the element from the root;
    put the element from the last leaf in its place;
    remove the last leaf;
    // both subtrees of the root are heaps;
```

FIGURE 6.54 Enqueuing an element to a heap.

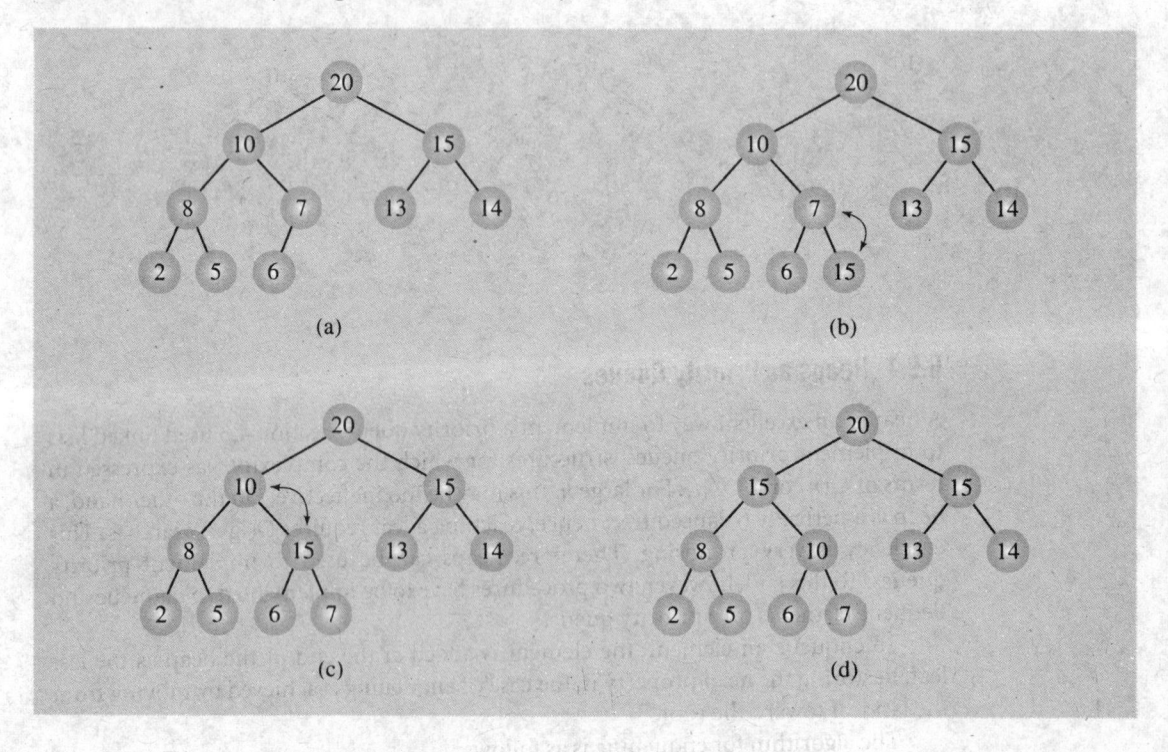

```
p = the root;
while p is not a leaf and p < any of its children
    swap p with the larger child;
```

For example, 20 is dequeued from the heap in Figure 6.55a and 6 is put in its place (Figure 6.55b). To restore the heap property, 6 is swapped first with its larger child, number 15 (Figure 6.55c), and once again with the larger child, 14 (Figure 6.55d).

The last three lines of the dequeuing algorithm can be treated as a separate algorithm that restores the heap property only if it has been violated by the root of the tree. In this case, the root element is moved down the tree until it finds a proper position. This algorithm, which is the key to the heap sort, is presented in one possible implementation in Figure 6.56.

6.9.2 Organizing Arrays as Heaps

Heaps can be implemented as arrays, and in that sense, each heap is an array, but all arrays are not heaps. In some situations, however, most notably in heap sort (see Section 9.3.2), we need to convert an array into a heap (that is, reorganize the data in the array so that the resulting organization represents a heap). There are several ways to do this, but in light of the preceding section the simplest way is to start with an empty heap and sequentially include elements into a growing heap. This is a top-down

FIGURE 6.55 Dequeuing an element from a heap.

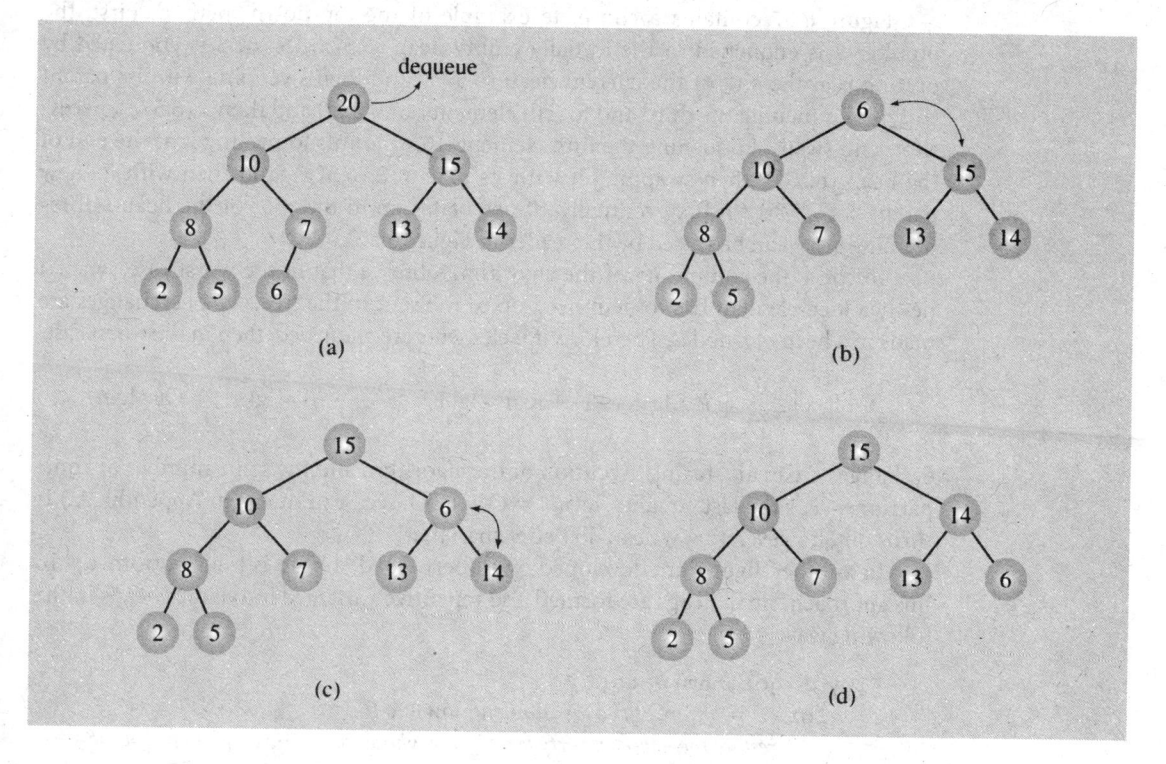

FIGURE 6.56 Implementation of an algorithm to move the root element down a tree.

```
template<class T>
void moveDown (T data[], int first, int last) {
    int largest = 2*first + 1;
    while (largest <= last) {
        if (largest < last && // first has two children (at 2*first+1 and
            data[largest] < data[largest+1]) // 2*first+2) and the second
            largest++;                        // is larger than the first;

        if (data[first] < data[largest]) {   // if necessary,
            swap(data[first],data[largest]); // swap child and parent,
            first = largest;                 // and move down;
            largest = 2*first+1;
        }
        else largest = last+1;   // to exit the loop: the heap property
    }                            // isn't violated by data[first];
}
```

method and it was proposed by John Williams; it extends the heap by enqueuing new elements in the heap.

Figure 6.57 contains a complete example of the top-down method. First, the number 2 is enqueued in the initially empty heap (6.57a). Next, 8 is enqueued by putting it at the end of the current heap (6.57b) and then swapping with its parent (6.57c). Enqueuing the third and fourth elements, 6 (6.57d) and then 1 (6.57e), necessitates no swaps. Enqueuing the fifth element, 10, amounts to putting it at the end of the heap (6.57f), then swapping it with its parent, 2 (6.57g), and then with its new parent, 8 (6.57h) so that eventually 10 percolates up to the root of the heap. All remaining steps can be traced by the reader in Figure 6.57.

To check the complexity of the algorithm, observe that in the worst case, when a newly added element has to be moved up to the root of the tree, $\lfloor \lg k \rfloor$ exchanges are made in a heap of k nodes. Therefore, if n elements are enqueued, then in the worst case

$$\sum_{k=1}^{n} \lfloor \lg k \rfloor \leq \sum_{k=1}^{n} \lg k = \lg 1 + \cdots + \lg n = \lg(1 \cdot 2 \cdots \cdots n) = \lg(n!) = O(n \lg n)$$

exchanges are made during execution of the algorithm and the same number of comparisons. (For the last equality, $\lg(n!) = O(n \lg n)$, see Section A.2 in Appendix A.) It turns out, however, that we can do better than that.

In another algorithm, developed by Robert Floyd, a heap is built bottom-up. In this approach, small heaps are formed and repetitively merged into larger heaps in the following way:

```
FloydAlgorithm(data[])
    for i = index of the last nonleaf down to 0
        restore the heap property for the tree whose root is data[i] by calling
        moveDown(data,i,n-1);
```

Figure 6.58 contains an example of transforming the array data[ ] = [2 8 6 1 10 15 3 12 11] into a heap.

We start from the last nonleaf node, which is data[n/2-1], n being the array size. If data[n/2-1] is less than one of its children, it is swapped with the larger child. In the tree in Figure 6.58a, this is the case for data[3] = 1 and data[7] = 12. After exchanging the elements, a new tree is created, shown in Figure 6.58b. Next the element data[n/2-2] = data[2] = 6 is considered. Because it is smaller than its child data[5] = 15, it is swapped with that child and the tree is transformed to that in Figure 6.58c. Now data[n/2-3] = data[1] = 8 is considered. Because it is smaller than one of its children, which is data[3] = 12, an interchange occurs, leading to the tree in Figure 6.58d. But now it can be noticed that the order established in the subtree whose root was 12 (Figure 6.58c) has been somewhat disturbed because 8 is smaller than its new child 11. This simply means that it does not suffice to compare a node's value with its children's, but a similar comparison needs to be done with grandchildren's, great-grandchildren's, and so on, until the node finds its proper position. Taking this into consideration, the next swap is made, after which the tree in Figure 6.58e is created. Only now is the element data[n/2-4] = data[0] = 2 compared with its children, which leads to two swaps (Figures 6.58f–g).

Section 6.9 Heaps   275

FIGURE 6.57 Organizing an array as a heap with a top-down method.

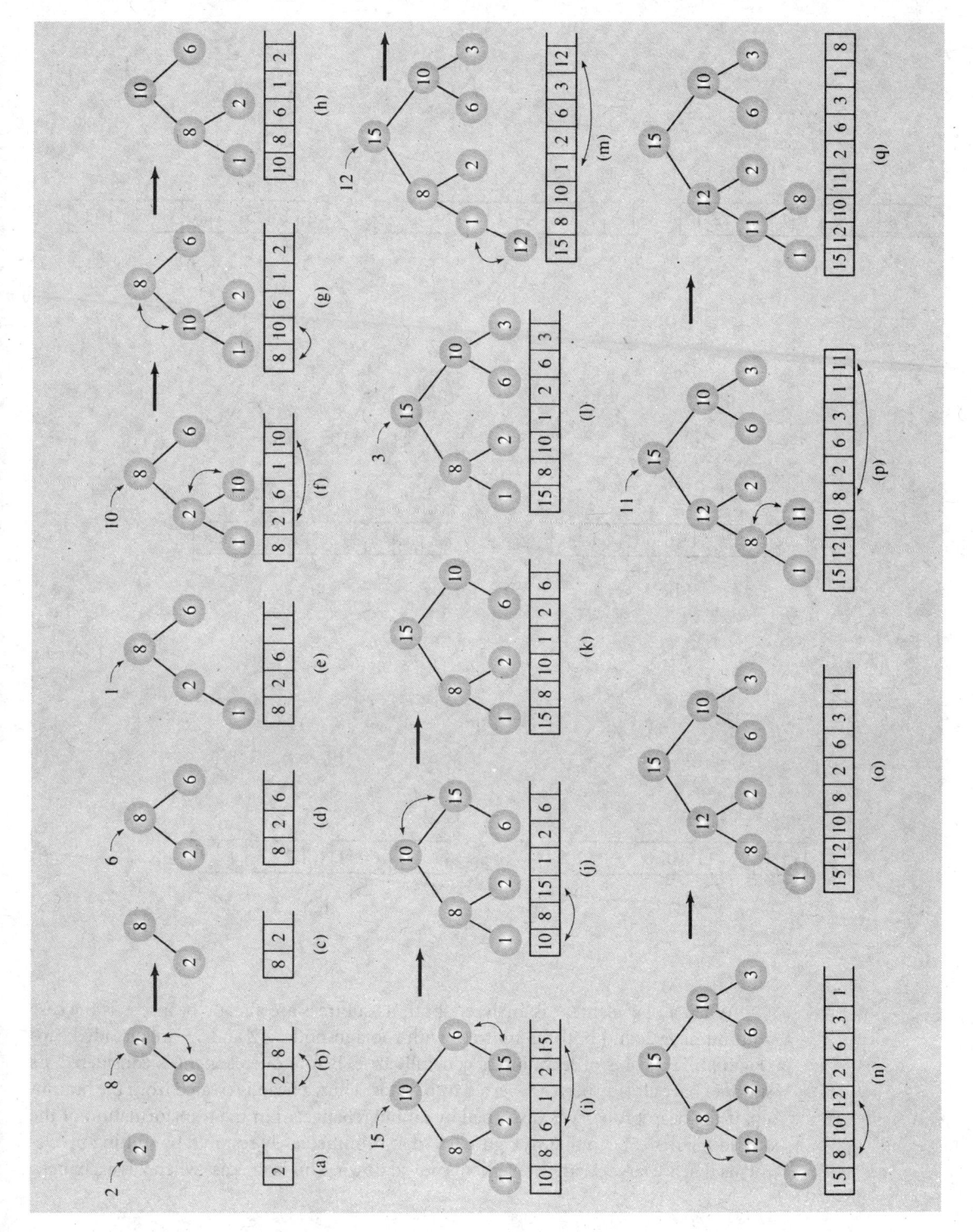

FIGURE 6.58 Transforming the array [2 8 6 1 10 15 3 12 11] into a heap with a bottom-up method.

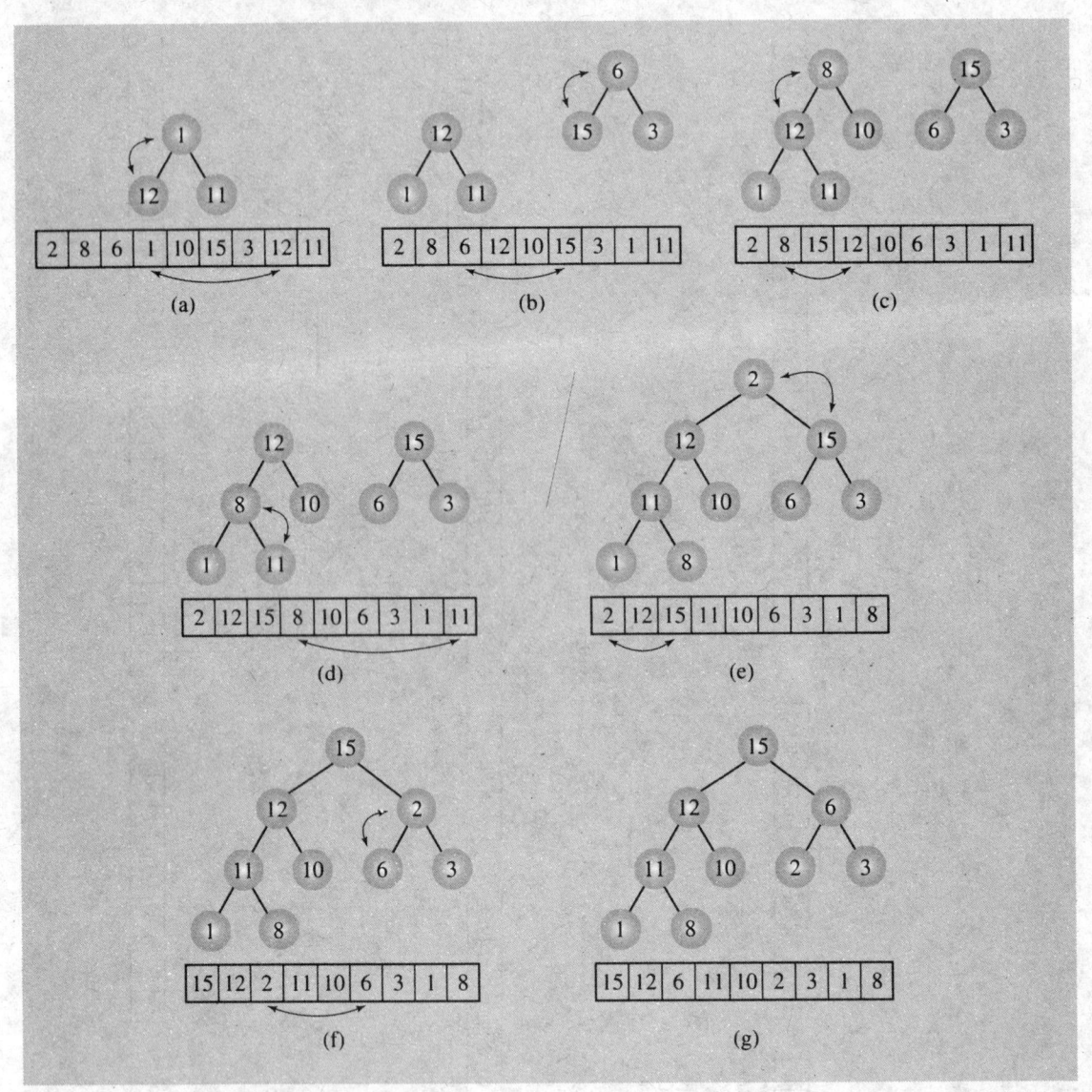

When a new element is analyzed, both its subtrees are already heaps, as is the case with number 2, and both its subtrees with roots in nodes 12 and 15 are already heaps (Figure 6.58e). This observation is generally true: Before one element is considered, its subtrees have already been converted into heaps. Thus, a heap is created from the bottom up. If the heap property is disturbed by an interchange, as in the transformation of the tree in Figure 6.58c to that in Figure 6.58d, it is immediately restored by shifting up elements that are larger than the element moved down. This is the case when 2 is exchanged with 15. The new tree is not a heap because node 2 has still larger children (Figure 6.58f). To remedy this problem, 6 is shifted up and 2 is moved down. Figure 6.58g is a heap.

We assume that a complete binary tree is created; that is, $n = 2^k - 1$ for some k. To create the heap, `moveDown()` is called $\frac{n+1}{2}$ times, once for each nonleaf. In the worst case, `moveDown()` moves data from the next to last level, consisting of $\frac{n+1}{4}$ nodes, down by one level to the level of leaves performing $\frac{n+1}{4}$ swaps. Therefore, all nodes from this level make $1 \cdot \frac{n+1}{4}$ moves. Data from the second to last level, which has $\frac{n+1}{8}$ nodes, are moved two levels down to reach the level of the leaves. Thus, nodes from this level perform $2 \cdot \frac{n+1}{8}$ moves and so on up to the root. The root of the tree as the tree becomes a heap is moved, again in the worst case, $\lg(n+1) - 1 = \lg \frac{n+1}{2}$ levels down the tree to end up in one of the leaves. Because there is only one root, this contributes $\lg \frac{n+1}{2} \cdot 1$ moves. The total number of movements can be given by this sum

$$\sum_{i=2}^{\lg(n+1)} \frac{n+1}{2^i}(i-1) = (n+1) \sum_{i=2}^{\lg(n+1)} \frac{i-1}{2^i}$$

which is $O(n)$ because the series $\sum_{i=2}^{\infty} \frac{i}{2^i}$ converges to 1.5 and $\sum_{i=2}^{\infty} \frac{1}{2^i}$ converges to 0.5. For an array that is not a complete binary tree, the complexity is all the more bounded by $O(n)$. The worst case for comparisons is twice this value, which is also $O(n)$, because for each node in `moveDown()`, both children of the node are compared to each other to choose the larger. That, in turn, is compared to the node. Therefore, for the worst case, Williams's method performs better than Floyd's.

The performance for the average case is much more difficult to establish. It has been found that Floyd's heap construction algorithm requires, on average, $1.88n$ comparisons (Doberkat 1984; Knuth 1998), and the number of comparisons required by Williams's algorithm in this case is between $1.75n$ and $2.76n$, and the number of swaps is $1.3n$ (Hayward and McDiarmid 1991; McDiarmid and Reed 1989). Thus, in the average case, the two algorithms perform at the same level.

6.10 POLISH NOTATION AND EXPRESSION TREES

One of the applications of binary trees is an unambiguous representation of arithmetical, relational, or logical expressions. In the early 1920s, a Polish logician, Jan Łukasiewicz (pronounced: wook-a-sie-vich), invented a special notation for propositional logic that allows us to eliminate all parentheses from formulas. However, Łukasiewicz's notation, called *Polish notation*, results in less readable formulas than the parenthesized originals and it was not widely used. It proved useful after the emergence of computers, however, especially for writing compilers and interpreters.

To maintain readability and prevent the ambiguity of formulas, extra symbols such as parentheses have to be used. However, if avoiding ambiguity is the only goal, then these symbols can be omitted at the cost of changing the order of symbols used in the formulas. This is exactly what the compiler does. It rejects everything that is not essential to retrieve the proper meaning of formulas, rejecting it as "syntactic sugar."

How does this notation work? Look first at the following example. What is the value of this algebraic expression?

$$2 - 3 \cdot 4 + 5$$

The result depends on the order in which the operations are performed. If we multiply first and then subtract and add, the result is –5 as expected. If subtraction is done first, then addition and multiplication, as in

$$(2 - 3) \cdot (4 + 5)$$

the result is –9. But if we subtract after we multiply and add, as in

$$2 - (3 \cdot 4 + 5)$$

then the result of evaluation is –15. If we see the first expression, then we know in what order to evaluate it. But the computer does not know that, in such a case, multiplication has precedence over addition and subtraction. If we want to override the precedence, then parentheses are needed.

Compilers need to generate assembly code in which one operation is executed at a time and the result is retained for other operations. Therefore, all expressions have to be broken down unambiguously into separate operations and put into their proper order. That is where Polish notation is useful. It allows us to create an *expression tree*, which imposes an order on the execution of operations. For example, the first expression, $2 - 3 \cdot 4 + 5$, which is the same as $2 - (3 \cdot 4) + 5$, is represented by the tree in Figure 6.59a. The second and the third expressions correspond to the trees in Figures 6.59b and 6.59c. It is obvious now that in both Figure 6.59a and Figure 6.59c we have to first multiply 3 by 4 to obtain 12. But 12 is subtracted from 2, according to the tree in Figure 6.59a, and added to 5, according to Figure 6.59c. There is no ambiguity involved in this tree representation. The final result can be computed only if intermediate results are calculated first.

FIGURE 6.59 Examples of three expression trees and results of their traversals.

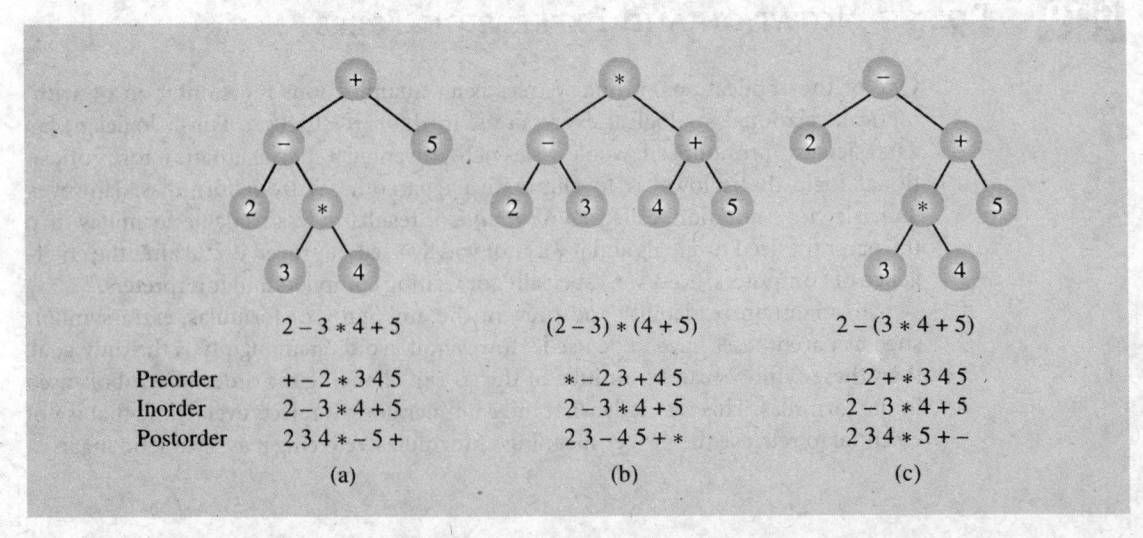

	(a)	(b)	(c)
	$2 - 3 * 4 + 5$	$(2 - 3) * (4 + 5)$	$2 - (3 * 4 + 5)$
Preorder	$+ - 2 * 3 4 5$	$* - 2 3 + 4 5$	$- 2 + * 3 4 5$
Inorder	$2 - 3 * 4 + 5$	$2 - 3 * 4 + 5$	$2 - 3 * 4 + 5$
Postorder	$2 3 4 * - 5 +$	$2 3 - 4 5 + *$	$2 3 4 * 5 + -$

Notice also that trees do not use parentheses and yet no ambiguity arises. We can maintain this parentheses-free situation if the expression tree is linearized (that is, if the tree is transformed into an expression using a tree traversal method). The three traversal methods relevant in this context are preorder, inorder, and postorder tree traversals. Using these traversals, nine outputs are generated, as shown in Figure 6.59. Interestingly, inorder traversal of all three trees results in the same output, which is the initial expression that caused all the trouble. What it means is that inorder tree traversal is not suitable for generating unambiguous output. But the other two traversals are. They are different for different trees and are therefore useful for the purpose of creating unambiguous expressions and sentences.

Because of the importance of these different conventions, special terminology is used. Preorder traversal generates *prefix notation,* inorder traversal generates *infix notation,* and postorder traversal generates *postfix notation.* Note that we are accustomed to infix notation. In infix notation, an operator is surrounded by its two operands. In prefix notation, the operator precedes the operands, and in postfix notation, the operator follows the operands. Some programming languages are using Polish notation. For example, Forth and PostScript use postfix notation. LISP and, to a large degree, LOGO use prefix notation.

6.10.1 Operations on Expression Trees

Binary trees can be created in two different ways: top-down or bottom-up. In the implementation of insertion, the first approach was used. This section applies the second approach by creating expression trees bottom-up while scanning infix expressions from left to right.

The most important part of this construction process is retaining the same precedence of operations as in the expression being scanned, as exemplified in Figure 6.59. If parentheses are not allowed, the task is simple, as parentheses allow for many levels of nesting. Therefore, an algorithm should be powerful enough to process any number of nesting levels in an expression. A natural approach is a recursive implementation. We modify the recursive descent interpreter discussed in Chapter 5's case study and outline a recursive descent expression tree constructor.

As Figure 6.59 indicates, a node contains either an operator or an operand, the latter being either an identifier or a number. To simplify the task, all of them can be represented as strings in an instance of the class defined as

```
class ExprTreeNode {
public:
    ExprTreeNode(char *k, ExprTreeNode *l, ExprTreeNode *r){
        key = new char[strlen(k)+1];
        strcpy(key,k);
        left = l; right = r;
    }
    . . . . . . .
private:
    char *key;
    ExprTreeNode *left, *right;
}
```

Expressions that are converted to trees use the same syntax as expressions in the case study in Chapter 5. Therefore, the same syntax diagrams can be used. Using these diagrams, a class `ExprTree` can be created in which member functions for processing a factor and term have the following pseudocode (a function for processing an expression has the same structure as the function processing a term):

```
factor()
    if (token is a number, id or operator)
        return new ExprTreeNode(token);
    else if (token is '(')
        ExprTreeNode *p = expr();
        if (token is ')')
            return p;
        else error;

term()
    ExprTreeNode *p1, *p2;
    p1 = factor();
    while (token is '*' or '/')
        oper = token;
        p2 = factor();
        p1 = new ExprTreeNode(oper,p1,p2);
    return p1;
```

The tree structure of expressions is very suitable for generating assembly code or intermediate code in compilers, as shown in this pseudocode of a function from `ExprTree` class:

```
void generateCode() {
    generateCode(root);
}
generateCode(ExprTreeNode *p) {
    if (p->key is a number or id)
        return p->key;
    else if (p->key is an addition operator)
        result = newTemporaryVar();
        output << "add\t" << generateCode(p->left) << "\t"
               << generateCode(p->right) << "\t"
               <<result<<endl;
        return result;
    . . . . . . . . .
}
```

With these member functions, an expression

$$(var2 + n) * (var2 + var1)/5$$

is transformed into an expression tree shown in Figure 6.60, and from this tree, `generateCode()` generates the following intermediate code:

FIGURE 6.60 An expression tree.

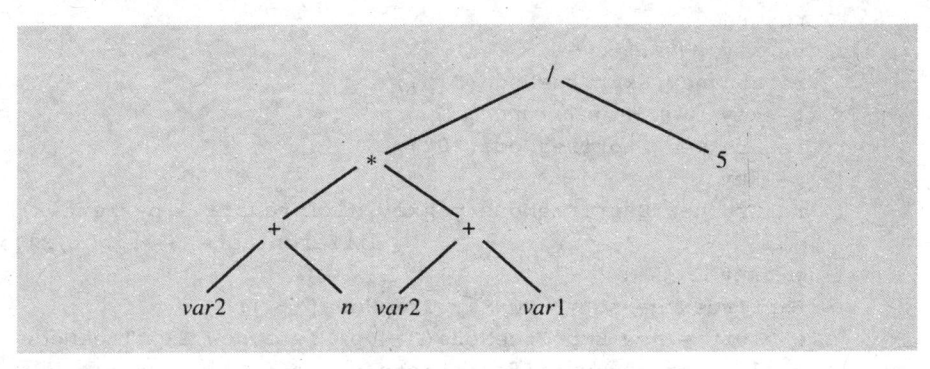

```
add    var2    n       _tmp_3
add    var2    var1    _tmp_4
mul    _tmp_3  _tmp_4  _tmp_2
div    _tmp_2  5       _tmp_1
```

Expression trees are also very convenient for performing other symbolic operations, such as differentiation. Rules for differentiation (given in the programming assignments in Chapter 5) are shown in the form of tree transformations in Figure 6.61 and in the following pseudocode:

FIGURE 6.61 Tree transformations for differentiation of multiplication and division.

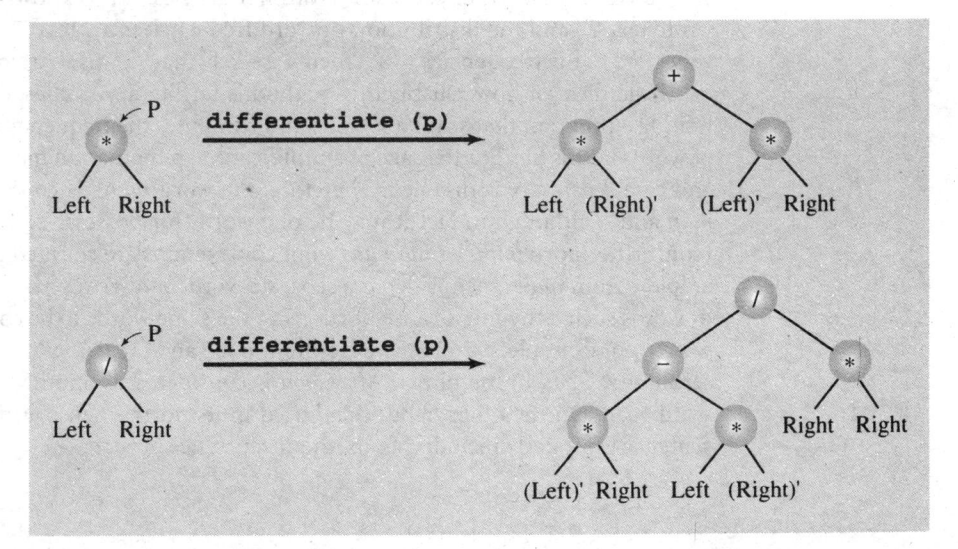

```
differentiate(p,x) {
    if (p == 0)
        return 0;
    if (p->key is the id x)
        return new ExprTreeNode("1");
    if (p->key is another id or a number)
        return new ExprTreeNode("0");
    if (p->key is '+' or '-')
        return new ExprTreeNode(p->key,differentiate(p->left,x),
                                       differentiate(p->right,x));
    if (p->key is '*')
        ExprTreeNode *q = new ExprTreeNode("+");
        q->left = new ExprTreeNode("*",p->left,new ExprTreeNode(*p->right));
        q->left->right = differentiate(q->left->right,x);
        q->right = new ExprTreeNode("*",new ExprTreeNode(*p->left),p->right);
        q->right->left = differentiate(q->right->left,x);
        return q;
    . . . . . . . . .
}
```

Here p is a pointer to the expression to be differentiated with respect to x.

The rule for division is left as an exercise.

6.11 CASE STUDY: COMPUTING WORD FREQUENCIES

One tool in establishing authorship of text in cases when the text is not signed, or it is attributed to someone else, is using word frequencies. If it is known that an author A wrote text T_1 and the distribution of word frequencies in a text T_2 under scrutiny is very close to the frequencies in T_1, then it is likely that T_2 was written by author A.

Regardless of how reliable this method is for literary studies, our interest lies in writing a program that scans a text file and computes the frequency of the occurrence of words in this file. For the sake of simplification, punctuation marks are disregarded and case sensitivity is disabled. Therefore, the word *man's* is counted as two words, *man* and *s*, although in fact it may be one word (for possessive) and not two words (contraction for *man is* or *man has*). But contractions are counted separately; for example, *s* from *man's* is considered a separate word. Similarly, separators in the middle of words such as hyphens cause portions of the same words to be considered separate words. For example, *pre-existence* is split into *pre* and *existence*. Also, by disabling case sensitivity, *Good* in the phrase *Mr. Good* is considered as another occurrence of the word *good*. On the other hand, *Good* used in its normal sense at the beginning of a sentence is properly included as another occurrence of *good*.

This program focuses not so much on linguistics as on building a self-adjusting binary search tree using the semisplaying technique. If a word is encountered in the file for the first time, it is inserted in the tree; otherwise, the semisplaying is started from the node corresponding to this word.

Another concern is storing all predecessors when scanning the tree. It is achieved by using a pointer to the parent. In this way, from each node we can access any predecessor of this node up to the root of the tree.

Figure 6.62 shows the structure of the tree using the content of a short file, and Figure 6.63 contains the complete code. The program reads a word, which is any sequence of alphanumeric characters that starts with a letter (spaces, punctuation marks, and the like are discarded) and checks whether the word is in the tree. If so, the semisplaying technique is used to reorganize the tree and then the word's frequency count is incremented. Note that this movement toward the root is accomplished by changing links of the nodes involved, not by physically transferring information from one node to its parent and then to its grandparent and so on. If a word is not found in the tree, it is inserted in the tree by creating a new leaf for it. After all words are processed, an inorder tree traversal goes through the tree to count all the nodes and add all frequency counts to print as the final result the number of words in the tree and the number of words in the file.

FIGURE 6.62 Semisplay tree used for computing word frequencies.

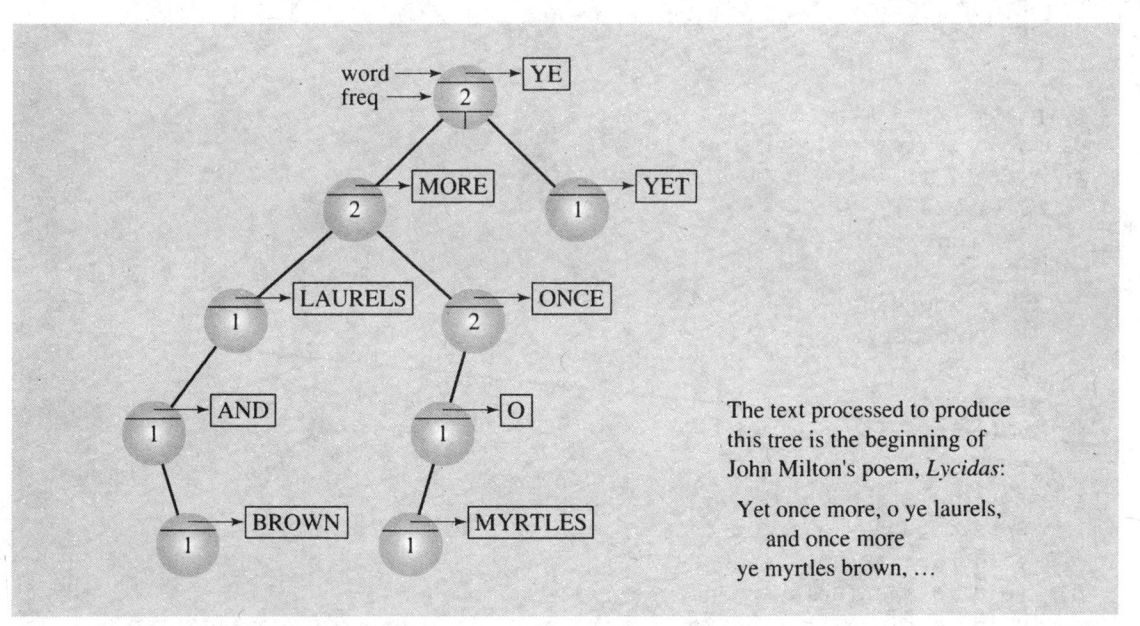

The text processed to produce this tree is the beginning of John Milton's poem, *Lycidas*:

Yet once more, o ye laurels,
and once more
ye myrtles brown, ...

FIGURE 6.63 Implementation of word frequency computation.

```cpp
//************************ genSplay.h ************************
//                    generic splaying tree class

#ifndef SPLAYING
#define SPLAYING

template<class T> class SplayTree;

template<class T>
class SplayingNode {
public:
    SplayingNode() {
        left = right = parent = 0;
    }
    SplayingNode(const T& el, SplayingNode *l = 0, SplayingNode *r = 0,
                 SplayingNode *p = 0) {
        info = el; left = l; right = r; parent = p;
    }
    T info;
    SplayingNode *left, *right, *parent;
};

template<class T>
class SplayTree {
public:
    SplayTree() {
        root = 0;
    }
    void inorder() {
        inorder(root);
    }
    T* search(const T&);
    void insert(const T&);
}
protected:
    SplayingNode<T> *root;
    void rotateR(SplayingNode<T>*);
    void rotateL(SplayingNode<T>*);
    void continueRotation(SplayingNode<T>* gr, SplayingNode<T>* par,
                          SplayingNode<T>* ch, SplayingNode<T>* desc);
    void semisplay(SplayingNode<T>*);
    void inorder(SplayingNode<T>*);
```

FIGURE 6.63 *(continued)*

```
    void virtual visit(SplayingNode<T>*) {
    }
};

template<class T>
void SplayTree<T>::continueRotation(SplayingNode<T>* gr,
SplayingNode<T>* par, SplayingNode<T>* ch, SplayingNode<T>* desc) {
    if (gr != 0) { // if par has a grandparent;
        if (gr->right == ch->parent)
             gr->right = ch;
        else gr->left  = ch;
    }
    else root = ch;
    if (desc != 0)
        desc->parent = par;
    par->parent = ch;
    ch->parent = gr;
}

template<class T>
void SplayTree<T>::rotateR(SplayingNode<T>* p) {
    p->parent->left = p->right;
    p->right = p->parent;
    continueRotation(p->parent->parent,p->right,p,p->right->left);
}

template<class T>
void SplayTree<T>::rotateL(SplayingNode<T>* p) {
    p->parent->right = p->left;
    p->left = p->parent;
    continueRotation(p->parent->parent,p->left,p,p->left->right);
}

template<class T>
void SplayTree<T>::semisplay(SplayingNode<T>* p) {
    while (p != root) {
        if (p->parent->parent == 0)    // if p's parent is the root;
            if (p->parent->left == p)
                rotateR(p);
            else rotateL(p);
        else if (p->parent->left == p) // if p is a left child;
```

Continues

FIGURE 6.63 *(continued)*

```
                if (p->parent->parent->left == p->parent) {
                    rotateR(p->parent);
                    p = p->parent;
                }
                else {
                    rotateR(p); // rotate p and its parent;
                    rotateL(p); // rotate p and its new parent;
                }
        else                            // if p is a right child;
            if (p->parent->parent->right == p->parent) {
                rotateL(p->parent);
                p = p->parent;
            }
            else {
                rotateL(p); // rotate p and its parent;
                rotateR(p); // rotate p and its new parent;
            }
        if (root == 0)                  // update the root;
            root = p;
    }
}

template<class T>
T* SplayTree<T>::search(const T& el) {
    SplayingNode<T> *p = root;
    while (p != 0)
        if (p->info == el) {            // if el is in the tree,
            semisplay(p);               // move it upward;
            return &p->info;
        }
        else if (el < p->info)
            p = p->left;
        else p = p->right;
    return 0;
}

template<class T>
void SplayTree<T>::insert(const T& el) {
    SplayingNode<T> *p = root, *prev = 0, *newNode;
    while (p != 0) {  // find a place for inserting a new node;
        prev = p;
        if (el < p->info)
```

FIGURE 6.63 (continued)

```
              p = p->left;
         else p = p->right;
    }
    if ((newNode = new SplayingNode<T>(el,0,0,prev)) == 0) {
         cerr << "No room for new nodes\n";
         exit(1);
    }
    if (root == 0)     // the tree is empty;
         root = newNode;
    else if (el < prev->info)
         prev->left  = newNode;
    else prev->right = newNode;
}

template<class T>
void SplayTree<T>::inorder(SplayingNode<T> *p) {
    if (p != 0) {
         inorder(p->left);
         visit(p);
         inorder(p->right);
    }
}

#endif

//******************** splay.cpp ***********************

#include <iostream>
#include <fstream>
#include <cctype>
#include <cstring>
#include <cstdlib> // exit()
#include "genSplay.h"
using namespace std;

class Word {
public:
    Word() {
         freq = 1;
    }
    int operator== (const Word& ir) const {
```

Continues

FIGURE 6.63   *(continued)*

```
            return strcmp(word,ir.word) == 0;
    }
    int operator< (const Word& ir) const {
            return strcmp(word,ir.word) < 0;
    }
private:
    char *word;
    int freq;
    friend class WordSplay;
    friend ostream& operator<< (ostream&,const Word&);
};

class WordSplay : public SplayTree<Word> {
public:
    WordSplay() {
        differentWords = wordCnt = 0;
    }
    void run(ifstream&,char*);
private:
    int differentWords, // counter of different words in a text file;
        wordCnt;        // counter of all words in the same file;
    void visit(SplayingNode<Word>*);
};

void WordSplay::visit(SplayingNode<Word> *p) {
    differentWords++;
    wordCnt += p->info.freq;
}

void WordSplay::run(ifstream& fIn, char *fileName) {
    char ch = ' ', i;
    char s[100];
    Word rec;
    while (!fIn.eof()) {
        while (1)
            if (!fIn.eof() && !isalpha(ch)) // skip nonletters
                fIn.get(ch);
            else break;
        if (fIn.eof())        // spaces at the end of fIn;
            break;
        for (i = 0; !fIn.eof() && isalpha(ch); i++) {
            s[i] = toupper(ch);
```

FIGURE 6.63   (continued)

```
            fIn.get(ch);
        }
        s[i] = '\0';
        if (!(rec.word = new char[strlen(s)+1])) {
            cerr << "No room for new words.\n";
            exit(1);
        }
        strcpy(rec.word,s);
        Word *p = search(rec);
        if (p == 0)
            insert(rec);
        else p->freq++;
    }
    inorder();
    cout << "\n\nFile " << fileName
         << " contains " << wordCnt << " words among which "
         << differentWords << " are different\n";
}

int main(int argc, char* argv[]) {
    char fileName[80];
    WordSplay splayTree;
    if (argc != 2) {
        cout << "Enter a file name: ";
        cin >> fileName;
    }
    else strcpy(fileName,argv[1]);
    ifstream fIn(fileName);
    if (fIn.fail()) {
        cerr << "Cannot open " << fileName << endl;
        return 0;
    }
    splayTree.run(fIn,fileName);
    fIn.close();
    return 0;
}
```

6.12 EXERCISES

1. The function `search()` given in Section 6.3 is well suited for searching binary *search* trees. Try to adopt all four traversal algorithms so that they become search procedures for any binary tree.

2. Write functions
   a. to count the number of nodes in a binary tree
   b. to count the number of leaves
   c. to count the number of right children
   d. to find the height of the tree
   e. to delete all leaves from a binary tree

3. Write a function that checks whether a binary tree is perfectly balanced.

4. Design an algorithm to test whether a binary tree is a binary search tree.

5. Apply `preorder()`, `inorder()`, and `postorder()` to the tree in Figure 6.64, if `visit(p)` is defined as:

   a. `if (p->left != 0 && p->key - p->left->key < 2)`
        `p->left->key += 2;`
   b. `if (p->left == 0)`
        `p->right = 0;`
   c. `if (p->left == 0)`
        `p->left = new IntBSTNode(p->key - 1)`
   d. `{ tmp = p->right;`
        `p->right = p->left;`
        `p->left = tmp;`
      `}`

FIGURE 6.64 An example of a binary search tree.

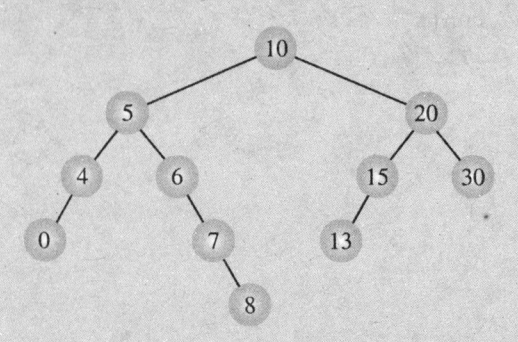

6. For which trees do the preorder and inorder traversals generate the same sequence?

7. Figure 6.59 indicates that the inorder traversal for different trees can result in the same sequence. Is this possible for the preorder or postorder traversals? If it is, show an example.

8. Draw all possible binary search trees for the three elements A, B, and C.

9. What are the minimum and maximum numbers of leaves in a balanced tree of height h?

10. Write a function to create a mirror image of a binary tree.

11. Consider an operation R that for a given traversal method t processes nodes in the opposite order than t and an operation C that processes nodes of the mirror image of a given tree using traversal method t. For the tree traversal methods—preorder, inorder, and postorder—determine which of the following nine equalities are true:

$$R(\text{preorder}) = C(\text{preorder})$$
$$R(\text{preorder}) = C(\text{inorder})$$
$$R(\text{preorder}) = C(\text{postorder})$$
$$R(\text{inorder}) = C(\text{preorder})$$
$$R(\text{inorder}) = C(\text{inorder})$$
$$R(\text{inorder}) = C(\text{postorder})$$
$$R(\text{postorder}) = C(\text{preorder})$$
$$R(\text{postorder}) = C(\text{inorder})$$
$$R(\text{postorder}) = C(\text{postorder})$$

12. Using inorder, preorder, and postorder tree traversal, visit only leaves of a tree. What can you observe? How can you explain this phenomenon?

13. (a) Write a function that prints each binary tree rotated to the left with proper indentation, as in Figure 6.65a. (b) Adopt this function to print a threaded tree sideways; if appropriate, print the key in the successor node, as in Figure 6.65b.

FIGURE 6.65 Printing a binary search tree (a) and a threaded tree (b) growing from left to right.

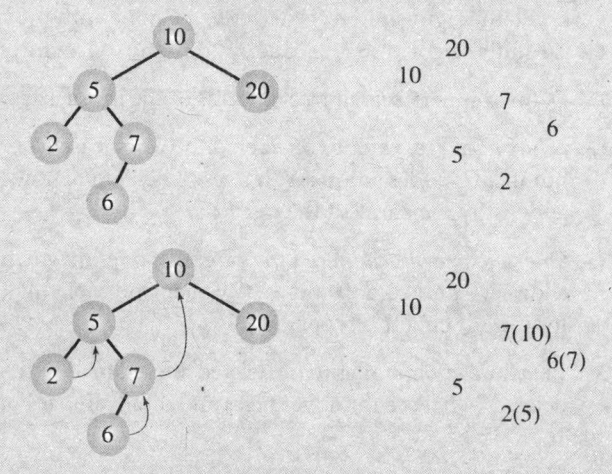

14. Outline functions for inserting and deleting a node in a threaded tree in which threads are put only in the leaves in the way illustrated by Figure 6.66.

FIGURE 6.66 Examples of threaded trees.

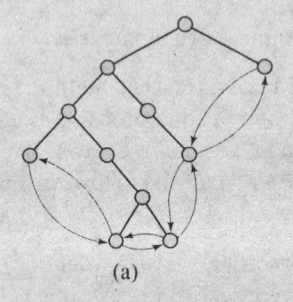

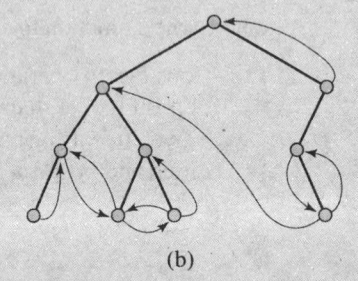

(a)

(b)

15. The tree in Figure 6.66b includes threads linking predecessors and successors according to the postorder traversal. Are these threads adequate to perform threaded preorder, inorder, and postorder traversals?

16. Apply the function `balance()` to the English alphabet to create a balanced tree.

17. A sentence *Dpq* that uses a Sheffer's alternative is false only if both *p* and *q* are true. In 1925, J. Łukasiewicz simplified Nicod's axiom from which all theses of propositional logic can be derived. Transform the Nicod-Łukasiewicz axiom into an infix parenthesized sentence and build a binary tree for it. The axiom is *DDpDqrDDsDssDDsqDDpsDps*.

18. Write an algorithm for printing a parenthesized infix expression from an expression tree. Do not include redundant parentheses.

19. Hibbard's (1962) algorithm to delete a key from a binary search tree requires that if the node containing the key has a right child, then the key is replaced by the smallest key in the right subtree; otherwise, the node with the key is removed. In what respect is Knuth's algorithm (`deleteByCopying()`) an improvement?

20. Define a binary search tree in terms of the inorder traversal.

21. A *Fibonacci tree* can be considered the worst case AVL tree in that it has the smallest number of nodes among AVL trees of height *h*. Draw Fibonacci trees for $h = 1, 2, 3, 4$ and justify the name of the tree.

22. *One-sided height-balanced trees* are AVL trees in which only two balance factors are allowed: -1 and 0 or 0 and $+1$ (Zweben and McDonald 1978). What is the rationale for introducing this type of tree?

23. In *lazy deletion*, nodes to be deleted are retained in the tree and only marked as deleted. What are the advantages and disadvantages of this approach?

24. What is the number of comparisons and swaps in the best case for creating a heap using
    a. Williams's method?
    b. Floyd's method?

25. A crossover between Floyd's and Williams's methods for constructing a heap is a method in which an empty position occupied by an element is moved down to the bottom of the tree and then the element is moved up the tree, as in Williams's method, from the position that was just moved down. A pseudocode of this function is as follows:

    ```
    i = n/2-1;  // position of the last parent in the array of n elements;
    while (i >= 0)
        // Floyd's phase:
        tmp = data[i];
        consider element data[i] empty and move it down to the bottom
              swapping it every time with larger child;
        put tmp in the leaf at which this process ended;
        // Williams's phase:
        while tmp is not the root data[i] of the current tree
              and it is larger than its parent
            swap tmp with its parent;
        i--;  // go to the preceding parent;
    ```

    It has been shown that this algorithm requires $1.65n$ comparisons in the average case (McDiarmid and Reed 1989). Show changes in the array [2 8 6 1 10 15 3 12 11] during execution of the algorithm. What is the worst case?

6.13 PROGRAMMING ASSIGNMENTS

1. Write a program that accepts an arithmetic expression written in prefix (Polish) notation, builds an expression tree, and then traverses the tree to evaluate the expression. The evaluation should start after a complete expression has been entered.

2. A binary tree can be used to sort n elements of an array data. First, create a complete binary tree, a tree with all leaves at one level, whose height $h = \lceil \lg n \rceil + 1$, and store all elements of the array in the first n leaves. In each empty leaf, store an element E greater than any element in the array. Figure 6.67a shows an example for data = {8, 20, 41, 7, 2}, $h = \lceil \lg(5) \rceil + 1 = 4$, and $E = 42$. Then, starting from the bottom of the tree, assign to each node the minimum of its two children values, as in Figure 6.67b, so that the smallest element e_{min} in the tree is assigned to the root. Next, until the element E is assigned to the root, execute a loop that in each iteration stores E in the leaf, with the value of e_{min}, and that, also starting from the bottom, assigns to each node the minimum of its two children. Figure 6.67c displays this tree after one iteration of the loop.

FIGURE 6.67 Binary tree used for sorting.

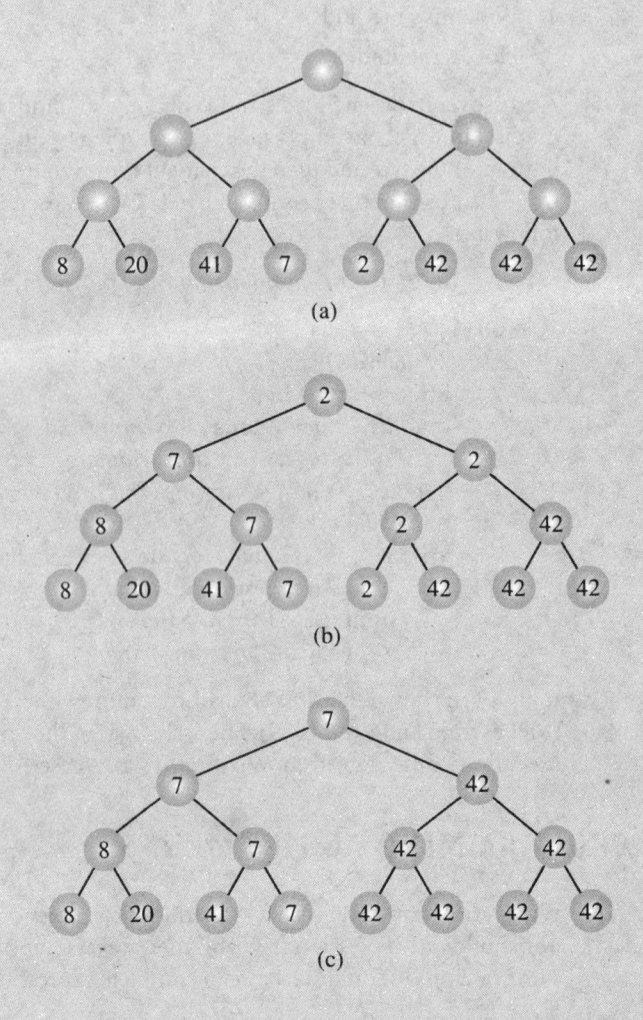

3. Implement a menu-driven program for managing a software store. All information about the available software is stored in a file `software`. This information includes the name, version, quantity, and price of each package. When it is invoked, the program automatically creates a binary search tree with one node corresponding to one software package and includes as its key the name of the package and its version. Another field in this node should include the position of the record in the file `software`. The only access to the information stored in `software` should be through this tree.

The program should allow the file and tree to be updated when new software packages arrive at the store and when some packages are sold. The tree is updated in the usual way. All packages are entry ordered in the file `software`; if a new package arrives, then it is put at the end of the file. If the package already has an entry in the tree (and the file), then only the quantity field is updated. If a package is sold out, the corresponding node is deleted from the tree, and the quantity field in the file is changed to 0. For example, if the file has these entries:

```
Adobe Photoshop              7.0      21       580
Norton Utilities                      10        30
Norton SystemWorks 2003                6        50
Visual J++ Professional      6.0      19       100
Visual J++ Standard          6.0      27        40
```

then after selling all six copies of Norton SystemWorks 2003, the file is

```
Adobe Photoshop              7.0      21       580
Norton Utilities                      10        30
Norton SystemWorks 2003                0        50
Visual J++ Professional      6.0      19       100
Visual J++ Standard          6.0      27        40
```

If an exit option is chosen from the menu, the program cleans up the file by moving entries from the end of the file to the positions marked with 0 quantities. For example, the previous file becomes

```
Adobe Photoshop              7.0      21       580
Norton Utilities                      10        30
Visual J++ Standard          6.0      27        40
Visual J++ Professional      6.0      19       100
```

4. Implement algorithms for constructing expression trees and for differentiating the expressions they represent. Extend the program to simplify expression trees. For example, two nodes can be eliminated from the subtrees representing $a \pm 0, a \cdot 1$, or $\frac{a}{1}$.

5. Write a cross-reference program that constructs a binary search tree with all words included from a text file and records the line numbers on which these words were used. These line numbers should be stored on linked lists associated with the nodes of the tree. After the input file has been processed, print in alphabetical order all words of the text file along with the corresponding list of numbers of the lines in which the words occur.

6. Perform an experiment with alternately applying insertion and deletion of random elements in a randomly created binary search tree. Apply asymmetric and symmetric deletions (discussed in this chapter); for both these variants of the deletion algorithm, alternate deletions strictly with insertions and alternate these operations randomly. This gives four different combinations. Also, use two different random number generators to ensure randomness. This leads to eight combinations. Run all of these combinations for trees of heights 500, 1,000, 1,500, and 2,000. Plot the results and compare them with the expected IPLs indicated in this chapter.

7. Each unit in a Latin textbook contains a Latin-English vocabulary of words that have been used for the first time in a particular unit. Write a program that converts a set of such vocabularies stored in file Latin into a set of English-Latin vocabularies. Make the following assumptions:

   a. Unit names are preceded by a percentage symbol.
   b. There is only one entry per line.
   c. A Latin word is separated by a colon from its English equivalent(s); if there is more than one equivalent, they are separated by a comma.

   To output English words in alphabetical order, create a binary search tree for each unit containing English words and linked lists of Latin equivalents. Make sure that there is only one node for each English word in the tree. For example, there is only one node for *and*, although *and* is used twice in unit 6: with words *ac* and *atque*. After the task has been completed for a given unit (that is, the content of the tree has been stored in an output file), delete the tree along with all linked lists from computer memory before creating a tree for the next unit.

   Here is an example of a file containing Latin-English vocabularies:

   ```
   %Unit 5
   ante : before, in front of, previously
   antiquus : ancient
   ardeo : burn, be on fire, desire
   arma : arms, weapons
   aurum : gold
   aureus : golden, of gold

   %Unit 6
   animal : animal
   Athenae : Athens
   atque : and
   ac : and
   aurora : dawn

   %Unit 7
   amo : love
   amor : love
   annus : year
   Asia : Asia
   ```

   From these units, the program should generate the following output:

   ```
   %Unit 5
   ancient : antiquus
   arms : arma
   be on fire : ardeo
   before : ante
   burn : ardeo
   desire : ardeo
   ```

```
            gold: aurum
            golden : aureus
            in front of : ante
            of gold : aureus
            previously : ante
            weapons : arma

            %Unit 6
            Athens : Athenae
            and : ac, atque
            animal : animal
            dawn : aurora

            %Unit 7
            Asia : Asia
            love : amor, amo
            year : annus
```

BIBLIOGRAPHY

Insertions and Deletions

Culberson, Joseph, "The Effect of Updates in Binary Search Trees," *Proceedings of the 17th Annual Symposium on Theory of Computing* (1985), 205–212.

Eppinger, Jeffrey L., "An Empirical Study of Insertion and Deletion in Binary Search Trees," *Communications of the ACM* 26 (1983), 663–669.

Hibbard, Thomas N., "Some Combinatorial Properties of Certain Trees with Applications to Searching and Sorting," *Journal of the ACM* 9 (1962), 13–28.

Jonassen, Arne T., and Knuth, Donald E., "A Trivial Algorithm Whose Analysis Isn't," *Journal of Computer and System Sciences* 16 (1978), 301–322.

Knuth, Donald E., "Deletions That Preserve Randomness," *IEEE Transactions of Software Engineering*, SE-3 (1977), 351–359.

Tree Traversals

Berztiss, Alfs, "A Taxonomy of Binary Tree Traversals," *BIT* 26 (1986), 266–276.

Burkhard, W. A., "Nonrecursive Tree Traversal Algorithms," *Computer Journal* 18 (1975), 227–230.

Morris, Joseph M., "Traversing Binary Trees Simply and Cheaply," *Information Processing Letters* 9 (1979), 197–200.

Balancing Trees

Baer, J. L., and Schwab, B., "A Comparison of Tree-Balancing Algorithms," *Communications of the ACM* 20 (1977), 322–330.

Chang, Hsi, and Iyengar, S. Sitharama, "Efficient Algorithms to Globally Balance a Binary Search Tree," *Communications of the ACM* 27 (1984), 695–702.

Day, A. Colin, "Balancing a Binary Tree," *Computer Journal* 19 (1976), 360–361.

Martin, W. A., and Ness, D. N., "Optimizing Binary Trees Grown with a Sorting Algorithm," *Communications of the ACM* 1 (1972), 88–93.

Stout, Quentin F., and Warren, Bette L., "Tree Rebalancing in Optimal Time and Space," *Communications of the ACM* 29 (1986), 902–908.

AVL Trees

Adel'son-Vel'skii, G. M., and Landis, E. M., "An Algorithm for the Organization of Information," *Soviet Mathematics* 3 (1962), 1259–1263.

Foster, Caxton C., "A Generalization of AVL Trees," *Communications of the ACM* 16 (1973), 512–517.

Karlton, P. L., Fuller, S. H., Scroggs, R. E., and Kaehler, E. B., "Performance of Height-Balanced Trees," *Communications of the ACM* 19 (1976), 23–28.

Knuth, Donald, *The Art of Computer Programming, Vol. 3: Sorting and Searching*, Reading, MA: Addison-Wesley, 1998.

Zweben, S. H., and McDonald, M. A., "An Optimal Method for Deletion in One-Sided Height Balanced Trees," *Communications of the ACM* 21 (1978), 441–445.

Self-Adjusting Trees

Allen, Brian, and Munro, Ian, "Self-Organizing Binary Search Trees," *Journal of the ACM* 25 (1978), 526–535.

Bell, Jim, and Gupta, Gopal, "An Evaluation of Self-Adjusting Binary Search Tree Techniques," *Software—Practice and Experience* 23 (1993), 369–382.

Bitner, James R., "Heuristics That Dynamically Organize Data Structures," *SIAM Journal on Computing* 8 (1979), 82–110.

Sleator, Daniel D., and Tarjan, Robert E., "Self-Adjusting Binary Search Trees," *Journal of the ACM* 32 (1985), 652–686.

Heaps

Bollobés, B., and Simon, I., "Repeated Random Insertion into a Priority Queue Structure," *Journal of Algorithms* 6 (1985), 466–477.

Doberkat, E. E., "An Average Case of Floyd's Algorithm to Construct Heaps," *Information and Control* 61 (1984), 114–131.

Floyd, Robert W., "Algorithm 245: Treesort 3," *Communications of the ACM* 7 (1964), 701.

Frieze, A., "On the Random Construction of Heaps," *Information Processing Letters* 27 (1988), 103.

Gonnett, Gaston H., and Munro, Ian, "Heaps on Heaps," *SIAM Journal on Computing* 15 (1986), 964–971.

Hayward, Ryan, and McDiarmid, Colin, "Average Case Analysis of Heap Building by Repeated Insertion," *Journal of Algorithms* 12 (1991), 126–153.

McDiarmid, C. J. H., and Reed, B. A., "Building Heaps Fast," *Journal of Algorithms* 10 (1989), 351–365.

Weiss, Mark A., *Data Structures and Algorithm Analysis,* Redwood City, CA: Benjamin Cummings, 1992, Ch. 6.

Williams, J. W. J., "Algorithm 232: Heapsort," *Communications of the ACM* 7 (1964), 347–348.

7 Multiway Trees

At the beginning of the preceding chapter, a general definition of a tree was given, but the thrust of that chapter was binary trees, in particular, binary search trees. A tree was defined as either an empty structure or a structure whose children are disjoint trees $t_1, \ldots, t_m$. According to this definition, each node of this kind of tree can have more than two children. This tree is called a *multiway tree of order m*, or an *m-way tree*.

In a more useful version of a multiway tree, an order is imposed on the keys residing in each node. A *multiway search tree of order m*, or an *m-way search tree*, is a multiway tree in which

1. Each node has m children and $m - 1$ keys.
2. The keys in each node are in ascending order.
3. The keys in the first i children are smaller than the ith key.
4. The keys in the last $m - i$ children are larger than the ith key.

The m-way search trees play the same role among m-way trees that binary search trees play among binary trees, and they are used for the same purpose: fast information retrieval and update. The problems they cause are similar. The tree in Figure 7.1 is a 4-way tree in which accessing the keys can require a different number of tests for different keys: The number 35 can be found in the second node tested, and 55 is in the fifth node checked. The tree, therefore, suffers from a known malaise: It is unbalanced. This problem is of particular importance if we want to use trees to process data on secondary storage such as disks or tapes where each access is costly. Constructing such trees requires a more careful approach.

FIGURE 7.1 A 4-way tree.

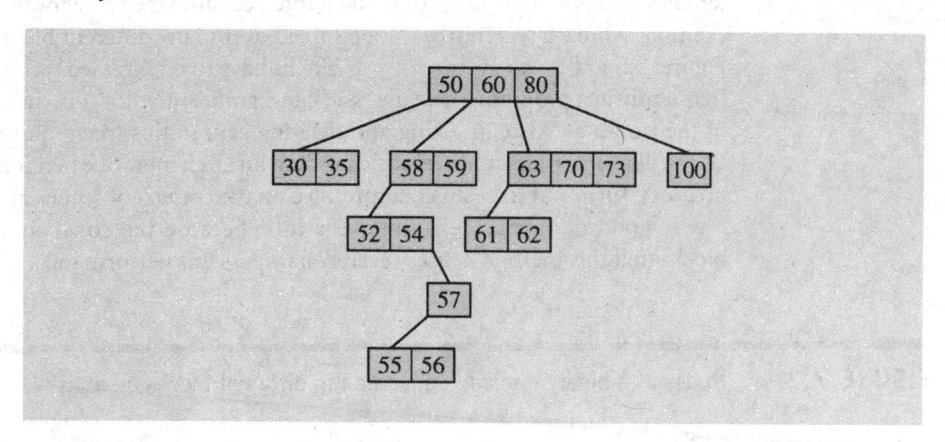

7.1 THE FAMILY OF B-TREES

The basic unit of I/O operations associated with a disk is a block. When information is read from a disk, the entire block containing this information is read into memory, and when information is stored on a disk, an entire block is written to the disk. Each time information is requested from a disk, this information has to be located on the disk, the head has to be positioned above the part of the disk where the information resides, and the disk has to be spun so that the entire block passes underneath the head to be transferred to memory. This means that there are several time components for data access:

access time = seek time + rotational delay (latency) + transfer time

This process is extremely slow compared to transferring information within memory. The first component, *seek time,* is particularly slow because it depends on the mechanical movement of the disk head to position the head at the correct track of the disk. *Latency* is the time required to position the head above the correct block, and on the average, it is equal to the time needed to make one-half of a revolution. For example, the time needed to transfer 5KB (kilobytes) from a disk requiring 40 ms (milliseconds) to locate a track, making 3,000 revolutions per minute and with a data transfer rate of 1,000KB per second, is

access time = 40 ms + 10 ms + 5 ms = 55 ms

This example indicates that transferring information to and from the disk is on the order of milliseconds. On the other hand, the CPU processes data on the order of microseconds, 1,000 times faster, or on the order of nanoseconds, 1 million times faster, or even faster. We can see that processing information on secondary storage can significantly decrease the speed of a program.

If a program constantly uses information stored in secondary storage, the characteristics of this storage have to be taken into account when designing the program. For example, a binary search tree can be spread over many different blocks on a disk, as in Figure 7.2, so that an average of two blocks have to be accessed. When the tree is used frequently in a program, these accesses can significantly slow down the execution time of the program. Also, inserting and deleting keys in this tree require many block accesses. The binary search tree, which is such an efficient tool when it resides entirely in memory, turns out to be an encumbrance. In the context of secondary storage, its otherwise good performance counts very little because the constant accessing of disk blocks that this method causes severely hampers this performance.

FIGURE 7.2 Nodes of a binary tree can be located in different blocks on a disk.

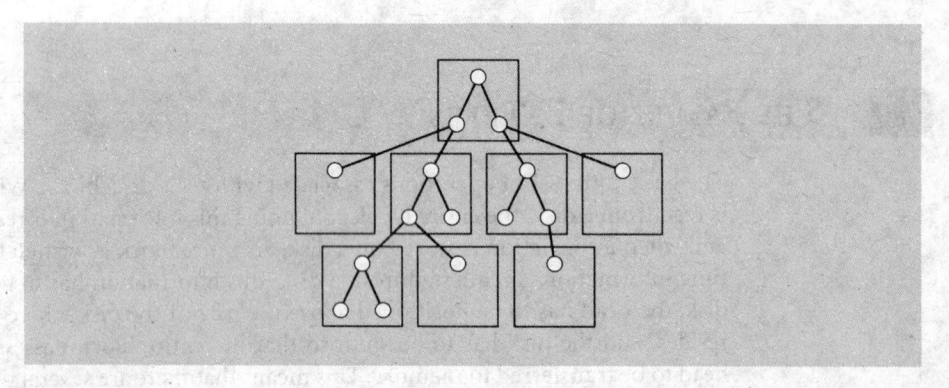

It is also better to access a large amount of data at one time than to jump from one position on the disk to another to transfer small portions of data. For example, if 10KB have to be transferred, then using the characteristics of the disk given earlier, we see that

$$access\ time = 40\ ms + 10\ ms + 10\ ms = 60\ ms$$

However, if this information is stored in two 5KB pieces, then

$$access\ time = 2 \cdot (40\ ms + 10\ ms + 5\ ms) = 110\ ms$$

which is nearly twice as long as in the previous case. The reason is that each disk access is very costly; if possible, the data should be organized to minimize the number of accesses.

7.1.1 B-Trees

In database programs where most information is stored on disks or tapes, the time penalty for accessing secondary storage can be significantly reduced by the proper choice of data structures. *B-trees* (Bayer and McCreight, 1972) are one such approach.

A B-tree operates closely with secondary storage and can be tuned to reduce the impediments imposed by this storage. One important property of B-trees is the size of each node, which can be made as large as the size of a block. The number of keys in one node can vary depending on the sizes of the keys, organization of the data (are only keys kept in the nodes or entire records?), and of course, on the size of a block. Block size varies for each system. It can be 512 bytes, 4KB, or more; block size is the size of each node of a B-tree. The amount of information stored in one node of the B-tree can be rather large.

A *B-tree of order m* is a multiway search tree with the following properties:

1. The root has at least two subtrees unless it is a leaf.
2. Each nonroot and each nonleaf node holds $k - 1$ keys and k pointers to subtrees where $\lceil m/2 \rceil \leq k \leq m$.
3. Each leaf node holds $k - 1$ keys where $\lceil m/2 \rceil \leq k \leq m$.
4. All leaves are on the same level.[1]

According to these conditions, a B-tree is always at least half full, has few levels, and is perfectly balanced.

A node of a B-tree is usually implemented as a `class` containing an array of $m - 1$ cells for keys, an m-cell array of pointers to other nodes, and possibly other information facilitating tree maintenance, such as the number of keys in a node and a leaf/nonleaf flag, as in

```
template <class T, int M>
class BTreeNode {
public:
    BTreeNode();
    BTreeNode(const T&);
private:
    bool leaf;
    int keyTally;
    T keys[M-1];
    BTreeNode *pointers[M];
    friend BTree<T,M>;
};
```

Usually, m is large (50–500) so that information stored in one page or block of secondary storage can fit into one node. Figure 7.3a contains an example of a B-tree of order 7 that stores codes for some items. In this B-tree, the keys appear to be the only objects of interest. In most cases, however, such codes would only be fields of larger structures, possibly variant records (unions). In these cases, the array `keys` is an array of objects, each having a unique identifier field (such as the identifying code in Figure 7.3a) and an address of the entire record on secondary storage, as in Figure

---

[1] In this definition, the order of a B-tree specifies the *maximum* number of children. Sometimes nodes of a B-tree of order m are defined as having k keys and $k + 1$ pointers where $m \leq k \leq 2m$, which specifies the *minimum* number of children.

FIGURE 7.3 One node of a B-tree of order 7 (a) without and (b) with an additional indirection.

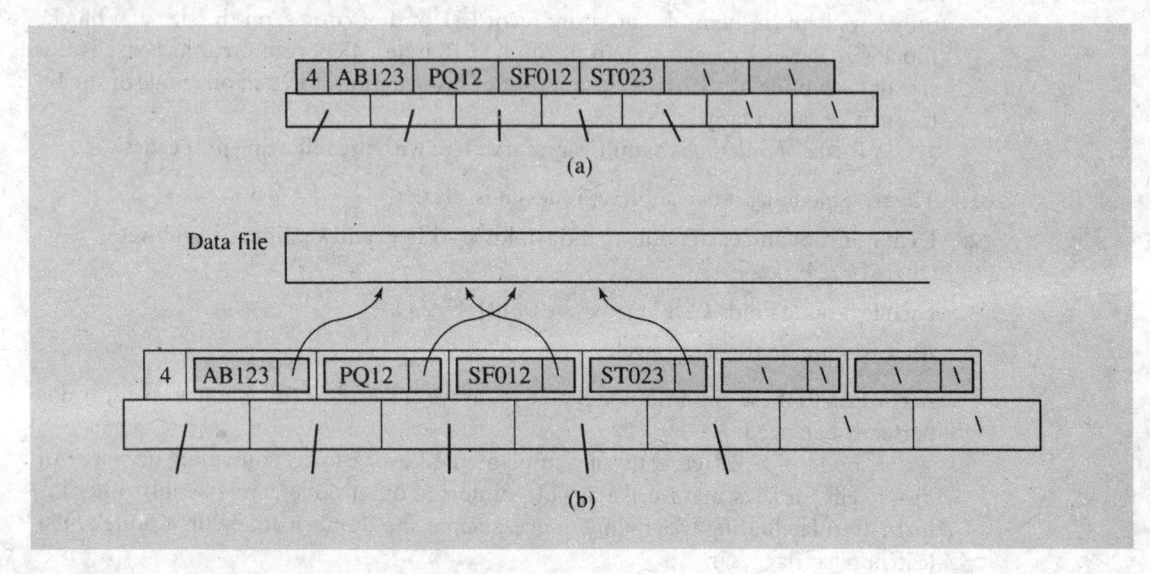

7.3b. If the contents of one such node also reside in secondary storage, each key access would require two secondary storage accesses. In the long run, this is better than keeping the entire records in the nodes, because in this case, the nodes can hold a very small number of such records. The resulting B-tree is much deeper, and search paths through it are much longer than in a B-tree with the addresses of records.

From now on, B-trees will be shown in an abbreviated form without explicitly indicating `keyTally` or the pointer fields, as in Figure 7.4.

FIGURE 7.4 A B-tree of order 5 shown in an abbreviated form.

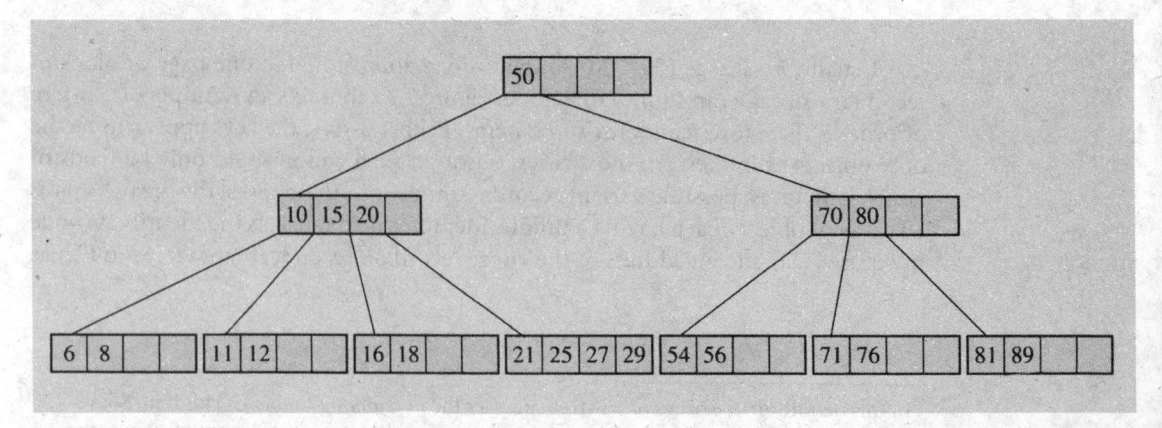

Searching in a B-Tree

An algorithm for finding a key in a B-tree is simple, and is coded as follows:

```
BTreeNode *BTreeSearch(keyType K, BTreeNode *node){
    if (node != 0) {
        for (i=1; i <= node->keyTally && node->keys[i-1] < K; i++);
        if (i > node->keyTally || node->keys[i-1] > K)
            return BTreeSearch(K,node->pointers[i-1]);
        else return node;
    }
    else return 0;
}
```

The worst case of searching is when a B-tree has the smallest allowable number of pointers per nonroot node, $q = \lceil m/2 \rceil$, and the search has to reach a leaf (for either a successful or an unsuccessful search). In this case, in a B-tree of height h, there are

$$1 \text{ key in the root } +$$
$$2(q-1) \text{ keys on the second level } +$$
$$2q(q-1) \text{ keys on the third level } +$$
$$2q^2(q-1) \text{ keys on the fourth level } +$$
$$\vdots$$
$$2q^{h-2}(q-1) \text{ keys in the leaves (level } h) =$$
$$1 + \left(\sum_{i=0}^{h-2} 2q^i \right)(q-1) \text{ keys in the B-tree}$$

With the formula for the sum of the first n elements in a geometric progression,

$$\sum_{i=0}^{n} q^i = \frac{q^{n+1} - 1}{q - 1}$$

the number of keys in the worst-case B-tree can be expressed as

$$1 + 2(q-1)\left(\sum_{i=0}^{h-2} q^i\right) = 1 + 2(q-1)\left(\frac{q^{h-1} - 1}{q - 1}\right) = -1 + 2q^{h-1}$$

The relation between the number n of keys in any B-tree and the height of the B-tree is then expressed as

$$n \geq -1 + 2q^{h-1}$$

Solving this inequality for the height h results in

$$h \leq \log_q \frac{n+1}{2} + 1$$

This means that for a sufficiently large order m, the height is small even for a large number of keys stored in the B-tree. For example, if $m = 200$ and $n = 2,000,000$, then $h \leq 4$; in the worst case, finding a key in this B-tree requires four seeks. If the root can be kept in memory at all times, this number can be reduced to only three seeks into secondary storage.

Inserting a Key into a B-Tree

Both the insertion and deletion operations appear to be somewhat challenging if we remember that all leaves have to be at the last level. Not even balanced binary trees require that. Implementing insertion becomes easier when the strategy of building a tree is changed. When inserting a node into a binary search tree, the tree is always built from top to bottom, resulting in unbalanced trees. If the first incoming key is the smallest, then this key is put in the root, and the root does not have a left subtree unless special provisions are made to balance the tree.

But a tree can be built from the bottom up so that the root is an entity always in flux, and only at the end of all insertions can we know for sure the contents of the root. This strategy is applied to inserting keys into B-trees. In this process, given an incoming key, we go directly to a leaf and place it there, if there is room. When the leaf is full, another leaf is created, the keys are divided between these leaves, and one key is promoted to the parent. If the parent is full, the process is repeated until the root is reached and a new root created.

To approach the problem more systematically, there are three common situations encountered when inserting a key into a B-tree.

1. A key is placed in a leaf that still has some room, as in Figure 7.5. In a B-tree of order 5, a new key, 7, is placed in a leaf, preserving the order of the keys in the leaf so that key 8 must be shifted to the right by one position.

FIGURE 7.5 A B-tree (a) before and (b) after insertion of the number 7 into a leaf that has available cells.

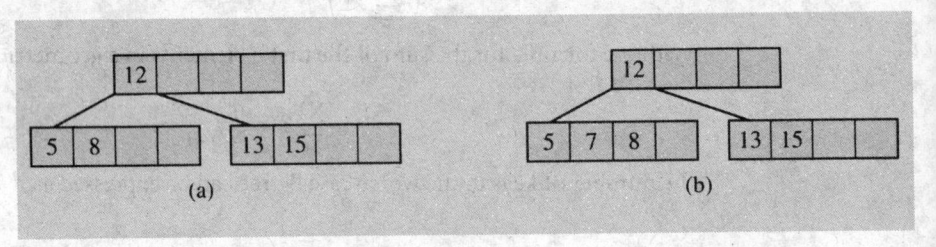

2. The leaf in which a key should be placed is full, as in Figure 7.6. In this case, the leaf is *split*, creating a new leaf, and half of the keys are moved from the full leaf to the new leaf. But the new leaf has to be incorporated into the B-tree. The middle key is moved to the parent, and a pointer to the new leaf is placed in the parent as well. The same procedure can be repeated for each internal node of the B-tree so that each such split adds one more node to the B-tree. Moreover, such a split guarantees that each leaf never has less than $\lceil m/2 \rceil - 1$ keys.

3. A special case arises if the root of the B-tree is full. In this case, a new root and a new sibling of the existing root have to be created. This split results in two new nodes in the B-tree. For example, after inserting the key 13 in the third leaf in Figure 7.7a, the leaf is split (as in case 2), a new leaf is created, and the key 15 is about to be moved to the parent, but the parent has no room for it (7.7b). So the parent is split (7.7c), but

FIGURE 7.6 Inserting the number 6 into a full leaf.

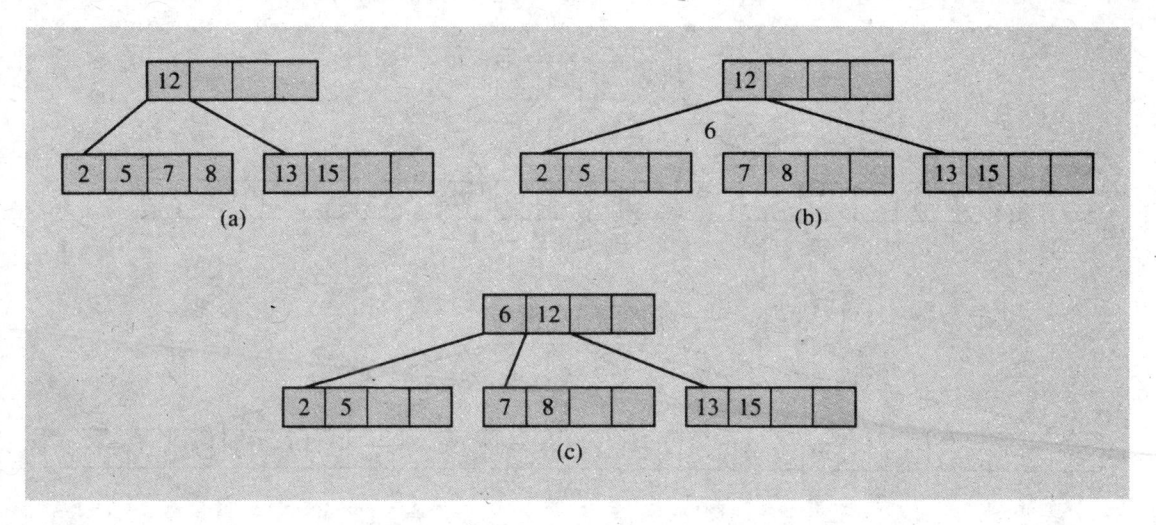

now two B-trees have to be combined into one. This is achieved by creating a new root and moving the middle key to it (7.7d). It should be obvious that it is the only case in which the B-tree increases in height.

An algorithm for inserting keys in B-trees follows:

```
BTreeInsert(K)
  find a leaf node to insert K;
  while (true)
      find a proper position in array keys for K;
      if node is not full
         insert K and increment keyTally;
         return;
      else split node into node1 and node2; // node1 = node, node2 is new;
         distribute keys and pointers evenly between node1 and node2 and
         initialize properly their keyTally's;
         K = middle key;
         if node was the root
            create a new root as parent of node1 and node2;
            put K and pointers to node1 and node2 in the root, and set its keyTally to 1;
            return;
         else node = its parent; // and now process the node's parent;
```

Figure 7.8 shows the growth of a B-tree of order 5 in the course of inserting new keys. Note that at all times the tree is perfectly balanced.

A variation of this insertion strategy uses *presplitting:* When a search is made from the top down for a particular key, each visited node that is already full is split. In this way, no split has to be propagated upward.

FIGURE 7.7 Inserting the number 13 into a full leaf.

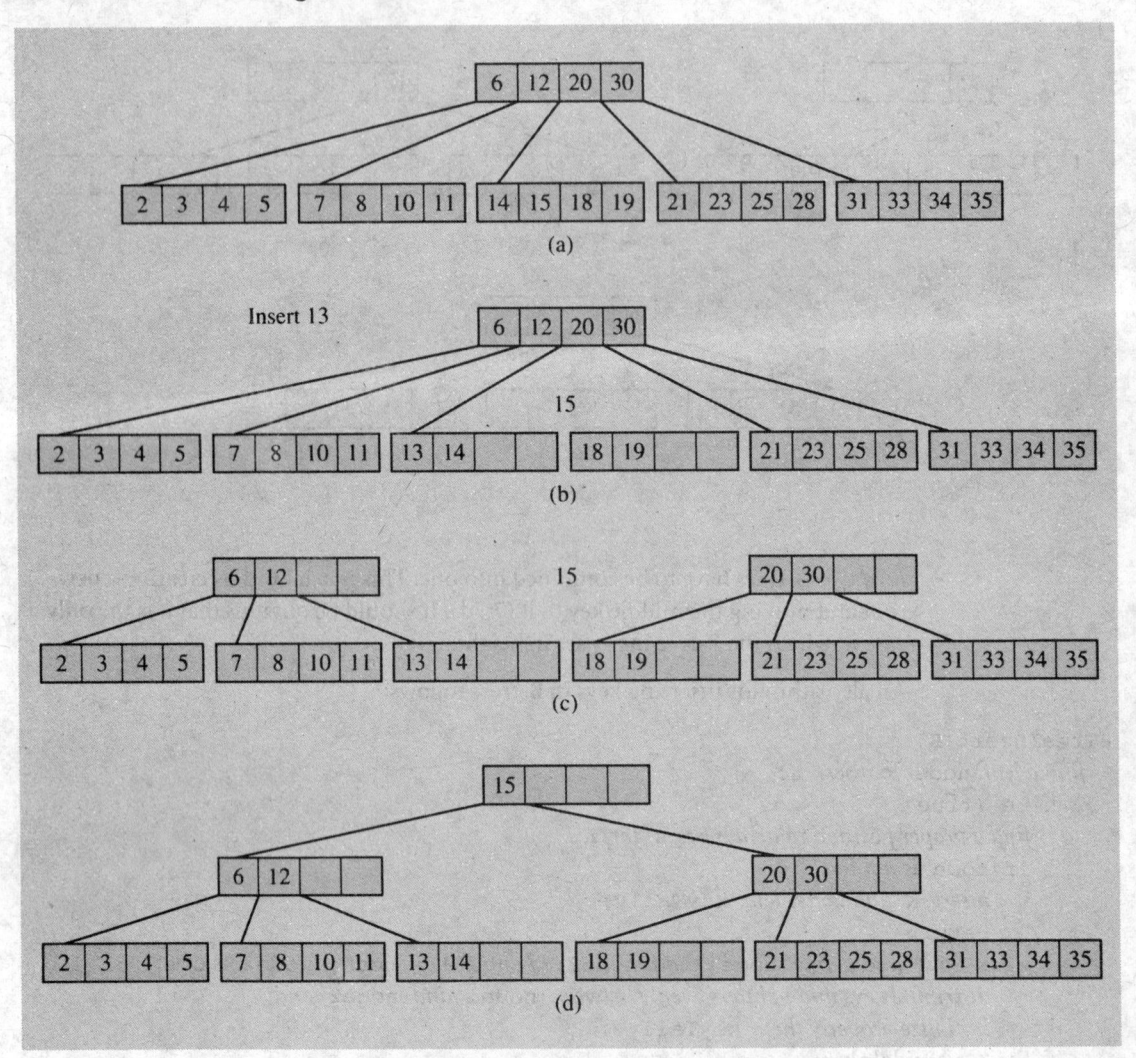

How often are node splits expected to occur? A split of the root node of a B-tree creates two new nodes. All other splits add only one more node to the B-tree. During the construction of a B-tree of p nodes, $p - h$ splits have to be performed, where h is the height of the B-tree. Also, in a B-tree of p nodes, there are at least

$$1 + (\lceil m/2 \rceil - 1)(p - 1)$$

keys. The rate of splits with respect to the number of keys in the B-tree can be given by

$$\frac{p - h}{1 + (\lceil m/2 \rceil - 1)(p - 1)}$$

FIGURE 7.8 Building a B-tree of order 5 with the `BTreeInsert()` algorithm.

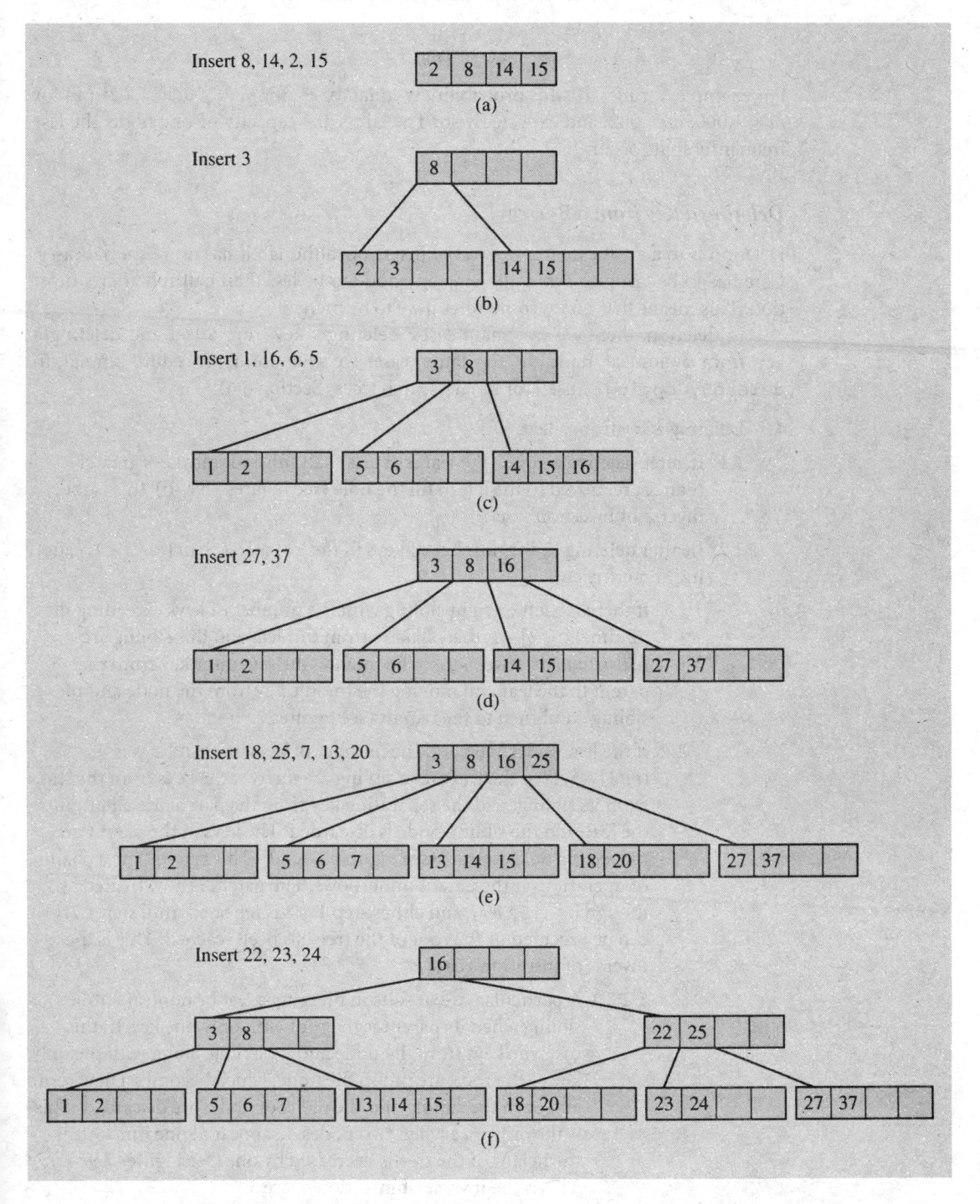

After dividing the numerator and denominator by $p - h$ and observing that $\frac{1}{p-h} \to 0$ and $\frac{p-1}{p-h} \to 1$ with the increase of p, the average probability of a split is

$$\frac{1}{\lceil m/2 \rceil - 1}$$

For example, for $m = 10$, this probability is equal to .25; for $m = 100$, it is .02; and for $m = 1,000$, it is .002, and expectedly so: The larger the capacity of one node, the less frequently splits occur.

Deleting a Key from a B-Tree

Deletion is to a great extent a reversal of insertion, although it has more special cases. Care has to be taken to avoid allowing any node to be less than half full after a deletion. This means that nodes sometimes have to be merged.

In deletion, there are two main cases: deleting a key from a leaf and deleting a key from a nonleaf node. In the latter case, we will use a procedure similar to `deleteByCopying()` used for binary search trees (Section 6.6).

1. Deleting a key from a leaf.
    1.1 If, after deleting a key K, the leaf is at least half full and only keys greater than K are moved to the left to fill the hole (see Figures 7.9a–b), this is the inverse of insertion's case 1.
    1.2 If, after deleting K, the number of keys in the leaf is less than $\lceil m/2 \rceil - 1$, causing an *underflow*:
        1.2.1 If there is a left or right sibling with the number of keys exceeding the minimal $\lceil m/2 \rceil - 1$, then all keys from this leaf and this sibling are *redistributed* between them by moving the separator key from the parent to the leaf and moving the middle key from the node and the sibling combined to the parent (see Figures 7.9b–c).
        1.2.2 If the leaf underflows and the number of keys in its siblings is $\lceil m/2 \rceil - 1$, then the leaf and a sibling are *merged;* the keys from the leaf, from its sibling, and the separating key from the parent are all put in the leaf, and the sibling node is discarded. The keys in the parent are moved if a hole appears (see Figures 7.9c–d). This can initiate a chain of operations if the parent underflows. The parent is now treated as though it were a leaf, and either step 1.2.2 is repeated until step 1.2.1 can be executed or the root of the tree has been reached. This is the inverse of insertion's case 2.
            1.2.2.1 A particular case results in merging a leaf or nonleaf with its sibling when its parent is the root with only one key. In this case, the keys from the node and its sibling, along with the only key of the root, are put in the node, which becomes a new root, and both the sibling and the old root nodes are discarded. This is the only case when two nodes disappear at one time. Also, the height of the tree is decreased by one (see Figures 7.9c–e). This is the inverse of insertion's case 3.

Section 7.1 The Family of B-Trees ■ 311

FIGURE 7.9 Deleting keys from a B-tree.

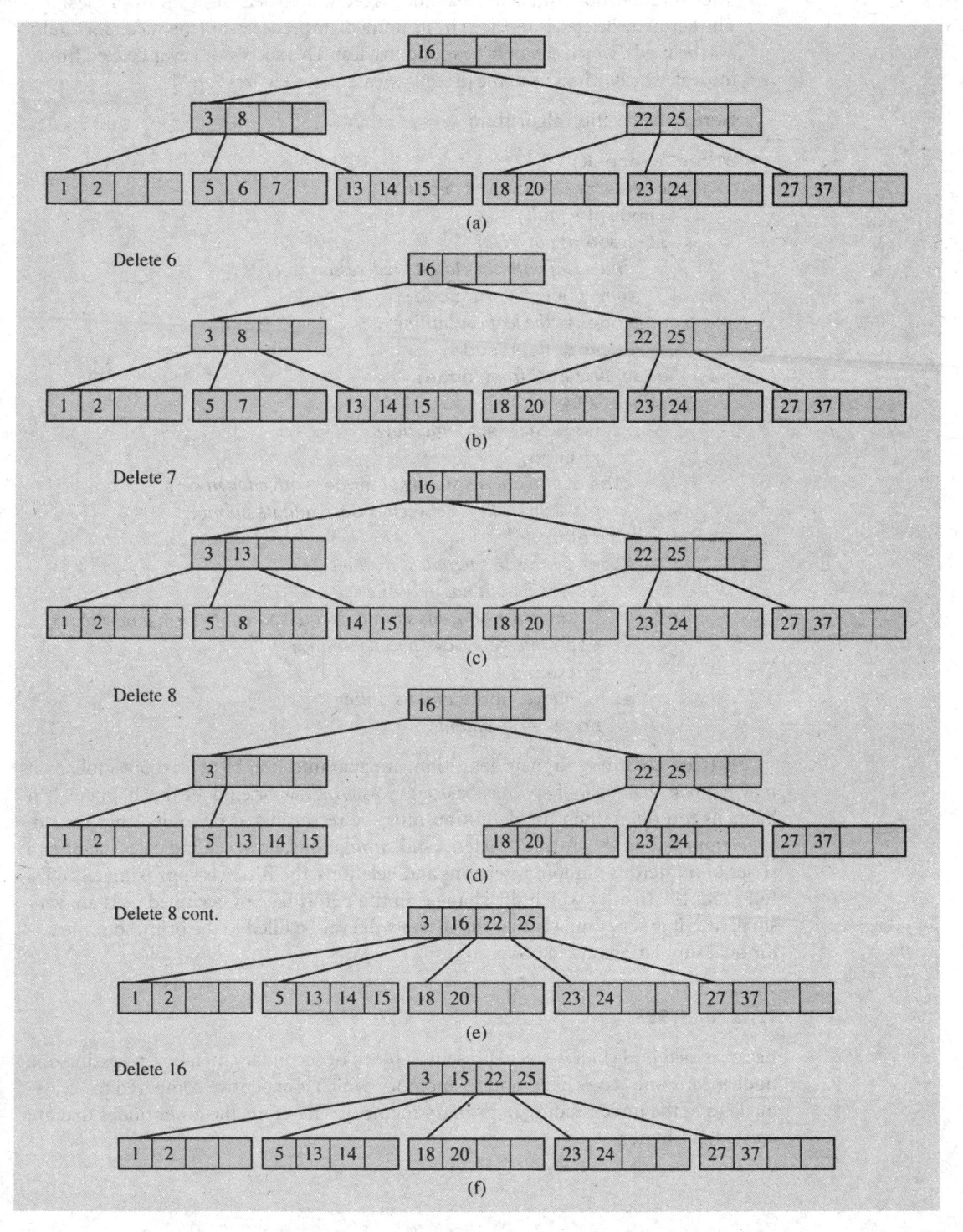

2. Deleting a key from a nonleaf. This may lead to problems with tree reorganization. Therefore, deletion from a nonleaf node is reduced to deleting a key from a leaf. The key to be deleted is replaced by its immediate predecessor (the successor could also be used), which can only be found in a leaf. This successor key is deleted from the leaf, which brings us to the preceding case 1 (see Figures 7.9e–f).

Here is the deletion algorithm:

```
BTreeDelete(K)
    node = BTreeSearch(K,root);
    if (node != null)
        if node is not a leaf
            find a leaf with the closest predecessor S of K;
            copy S over K in node;
            node = the leaf containing S;
            delete S from node;
        else delete K from node;
        while (1)
            if node does not underflow
                return;
            else if there is a sibling of node with enough keys
                redistribute keys between node and its sibling;
                return;
            else if node's parent is the root
                if the parent has only one key
                    merge node, its sibling, and the parent to form a new root;
                else merge node and its sibling;
                return;
            else merge node and its sibling;
            node = its parent;
```

B-trees, according to their definition, are guaranteed to be at least 50% full, so it may happen that 50% of space is basically wasted. How often does this happen? If it happens too often, then the definition must be reconsidered or some other restrictions imposed on this B-tree. Analyses and simulations, however, indicate that after a series of numerous random insertions and deletions, the B-tree is approximately 69% full (Yao, 1978), after which the changes in the percentage of occupied cells are very small. But it is very unlikely that the B-tree will ever be filled to the brim, so some additional stipulations are in order.

7.1.2 B*-Trees

Because each node of a B-tree represents a block of secondary memory, accessing one node means one access of secondary memory, which is expensive compared to accessing keys in the node residing in primary memory. Therefore, the fewer nodes that are created, the better.

A *B*-tree* is a variant of the B-tree, and was introduced by Donald Knuth and named by Douglas Comer. In a B*-tree, all nodes except the root are required to be at least two-thirds full, not just half full as in a B-tree. More precisely, the number of keys in all nonroot nodes in a B-tree of order m is now k for $\lfloor \frac{2m-1}{3} \rfloor \leq k \leq m - 1$. The frequency of node splitting is decreased by delaying a split, and when the time comes, by splitting two nodes into three, not one into two. The average utilization of B*-tree is 81% (Leung, 1984).

A split in a B*-tree is delayed by attempting to redistribute the keys between a node and its sibling when the node overflows. Figure 7.10 contains an example of a B*-tree of order 9. The key 6 is to be inserted into the left node, which is already full. Instead of splitting the node, all keys from this node and its sibling are evenly divided and the median key, key 10, is put into the parent. Notice that this evenly divides not only the keys, but also the free spaces so that the node which was full is now able to accommodate one more key.

FIGURE 7.10 Overflow in a B*-tree is circumvented by redistributing keys between an overflowing node and its sibling.

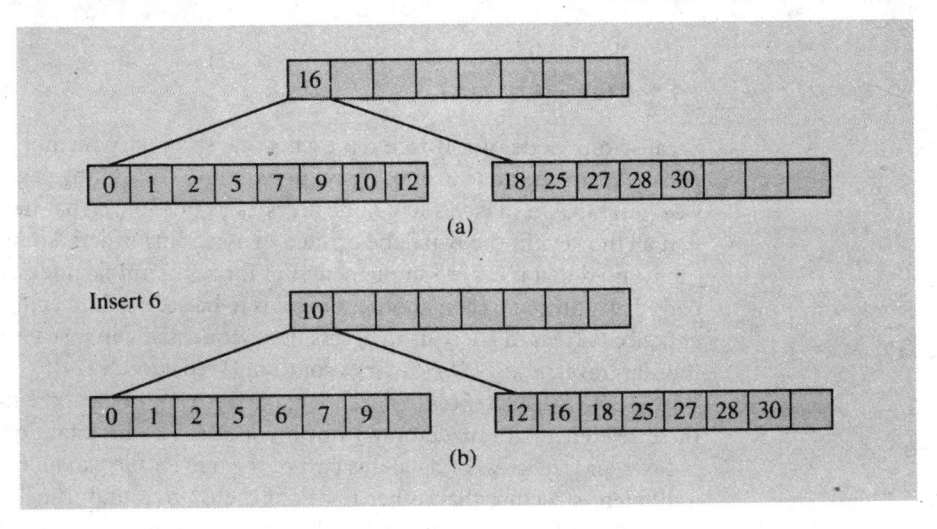

If the sibling is also full, a split occurs: One new node is created, the keys from the node and its sibling (along with the separating key from the parent) are evenly divided among three nodes, and two separating keys are put into the parent (see Figure 7.11). All three nodes participating in the split are guaranteed to be two-thirds full.

Note that, as may be expected, this increase of a *fill factor* can be done in a variety of ways, and some database systems allow the user to choose a fill factor between .5 and 1. In particular, a B-tree whose nodes are required to be at least 75% full is called a B**-tree (McCreight, 1977). The latter suggests a generalization: A B^n-*tree* is a B-tree whose nodes are required to be $\frac{n+1}{n+2}$ full.

FIGURE 7.11 If a node and its sibling are both full in a B*-tree, a split occurs: A new node is created and keys are distributed between three nodes.

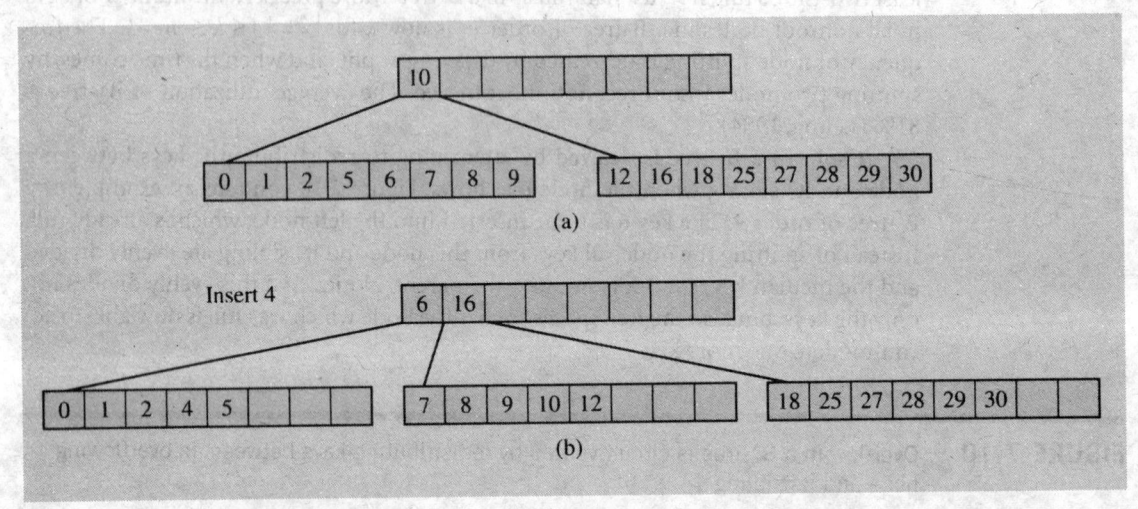

7.1.3 B+-Trees

Because one node of a B-tree represents one secondary memory page or block, the passage from one node to another requires a time-consuming page change. Therefore, we would like to make as few node accesses as possible. What happens if we request that all the keys in the B-tree be printed in ascending order? An inorder tree traversal can be used that is easy to implement, but for nonterminal nodes, only one key is displayed at a time and then another page has to be accessed. Therefore, we would like to enhance B-trees to allow us to access data sequentially in a faster manner than using inorder traversal. A *B+-tree* offers a solution (Wedekind, 1974).[2]

In a B-tree, references to data are made from any node of the tree, but in a B+-tree, these references are made only from the leaves. The internal nodes of a B+-tree are indexes for fast access of data; this part of the tree is called an *index set*. The leaves have a different structure than other nodes of the B+-tree, and usually they are linked sequentially to form a *sequence set* so that scanning this list of leaves results in data given in ascending order. Hence, a B+-tree is truly a B plus tree: It is an index implemented as a regular B-tree plus a linked list of data. Figure 7.12 contains an example of a B+-tree. Note that internal nodes store keys, pointers, and a key count. Leaves store keys, references to records in a data file associated with the keys, and pointers to the next leaf.

Operations on B+-trees are not very different from operations on B-trees. Inserting a key into a leaf that still has some room requires putting the keys of this leaf in

---

[2]Wedekind, who considered these trees to be only "a slight variation" of B-trees, called them B*-trees.

FIGURE 7.12   An example of a B⁺-tree of order 4.

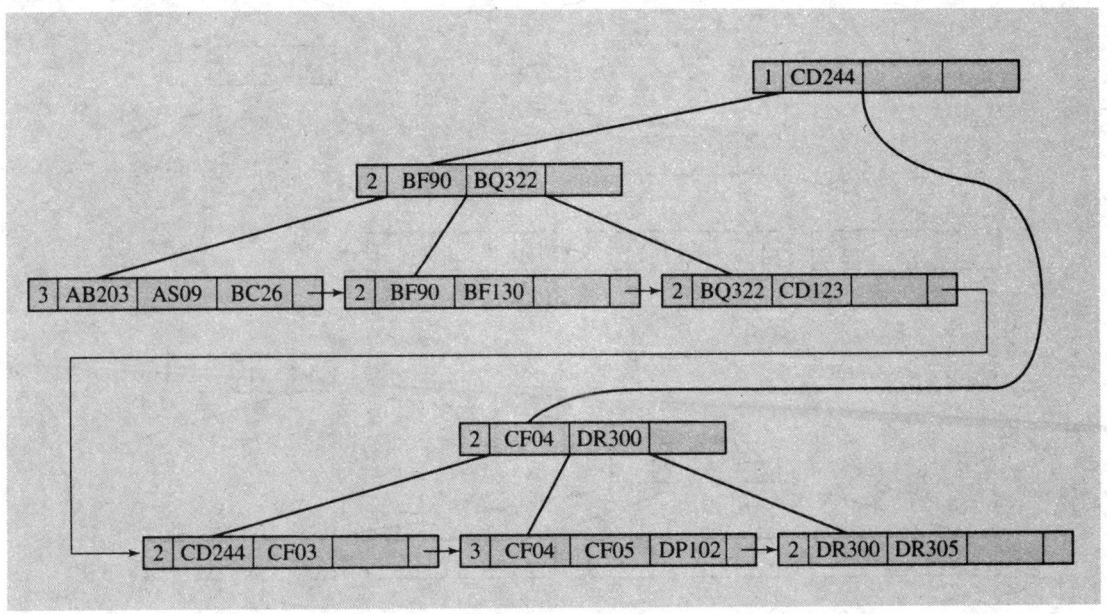

order. No changes are made in the index set. If a key is inserted into a full leaf, the leaf is split, the new leaf node is included in the sequence set, all keys are distributed evenly between the old and the new leaves, and the first key from the new node is copied (not moved, as in a B-tree) to the parent. If the parent is not full, this may require local reorganization of the keys of the parent (see Figure 7.13). If the parent is full, the splitting process is performed the same way as in B-trees. After all, the index set is a B-tree. In particular, keys are moved, not copied, in the index set.

Deleting a key from a leaf leading to no underflow requires putting the remaining keys in order. No changes are made to the index set. In particular, if a key that occurs only in a leaf is deleted, then it is simply deleted from the leaf but can remain in the internal node. The reason is that it still serves as a proper guide when navigating down the B⁺-tree because it still properly separates keys between two adjacent children, even if the separator itself does not occur in either of the children. The deletion of key 6 from the tree in Figure 7.13b results in the tree in Figure 7.14a. Note that the number 6 is not deleted from an internal node.

When the deletion of a key from a leaf causes an underflow, then either the keys from this leaf and the keys of a sibling are redistributed between this leaf and its sibling or the leaf is deleted and the remaining keys are included in its sibling. Figure 7.14b illustrates the latter case. After deleting the number 2, an underflow occurs and two leaves are combined to form one leaf. The separating key is removed from the parent, and keys in the parent are put in order. Both these operations require updating the separator in the parent. Also, removing a leaf may trigger merges in the index set.

FIGURE 7.13 An attempt to insert the number 6 into the first leaf of a B+-tree.

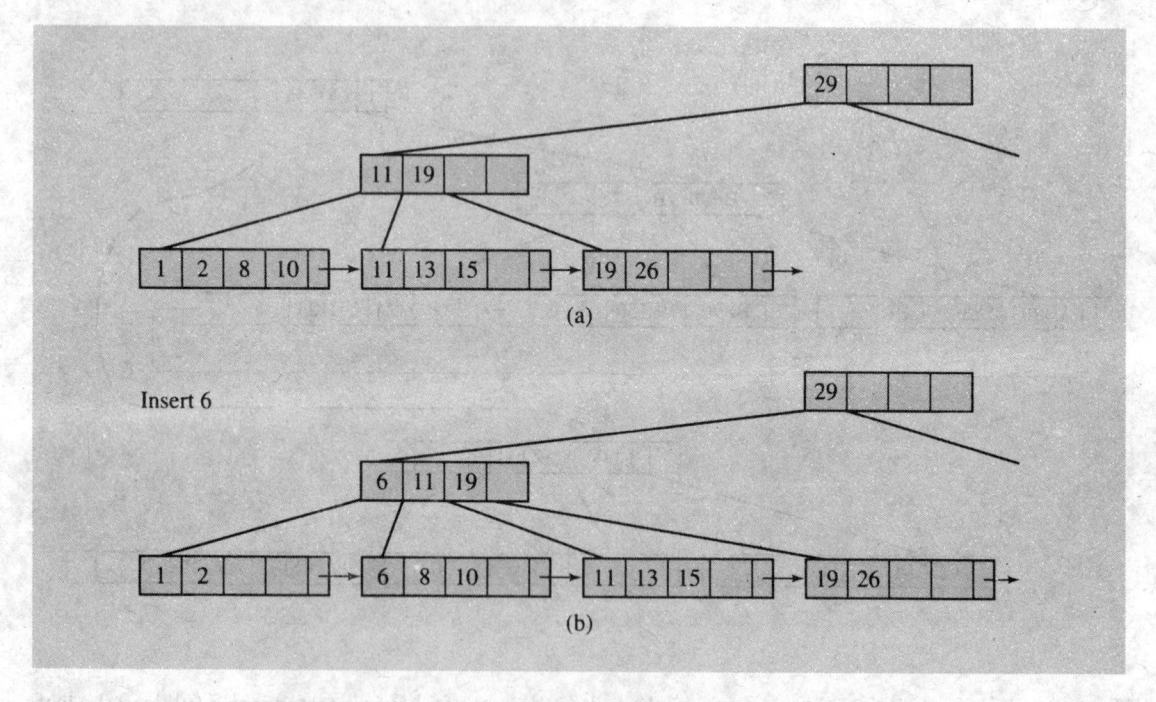

FIGURE 7.14 Actions after deleting the number 6 from the B+-tree in Figure 7.13b.

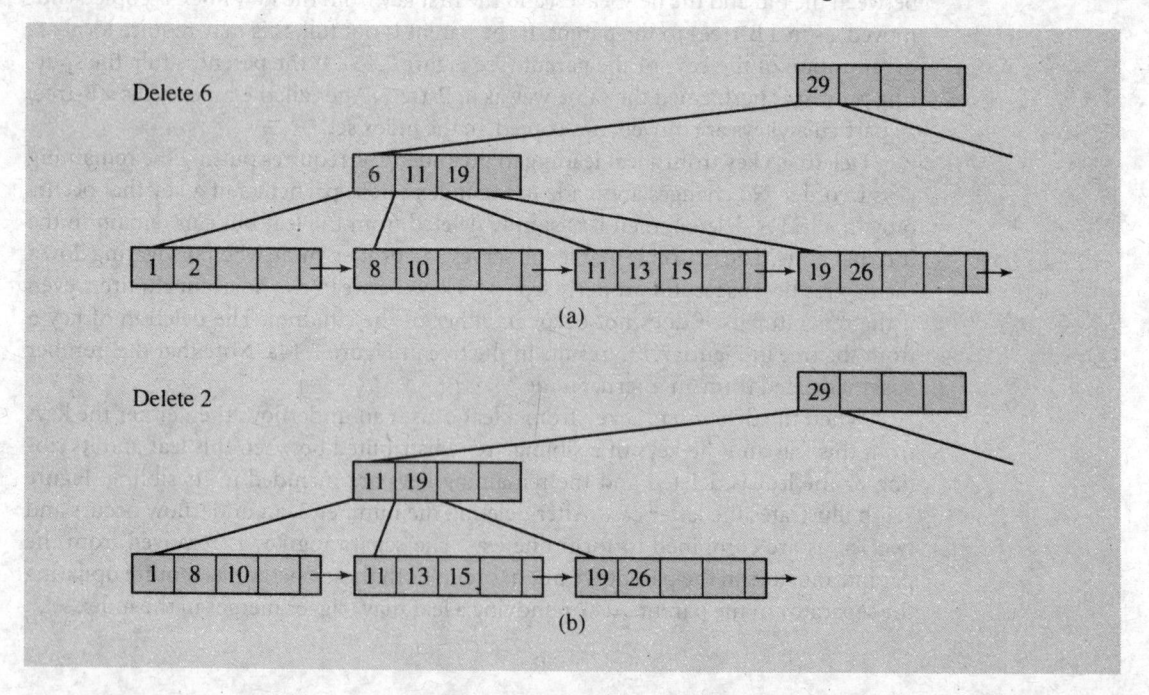

7.1.4 Prefix B⁺-Trees

If a key occurred in a leaf and in an internal node of a B⁺-tree, then it is enough to delete it only from the leaf because the key retained in the node is still a good guide in subsequent searches. So it really does not matter whether a key in an internal node is in any leaf or not. What counts is that it is an acceptable separator for keys in adjacent children; for example, for two keys K_1 and K_2, the separator s must meet the condition $K_1 < s \leq K_2$. This property of the separator keys is also retained if we make keys in internal nodes as small as possible by removing all redundant contents from them and still have a properly working B⁺-tree.

A *simple prefix B⁺-tree* (Bayer and Unterauer 1977) is a B⁺-tree in which the chosen separators are the shortest prefixes that allow us to distinguish two neighboring index keys. For example, in Figure 7.12, the left child of the root has two keys, BF90 and BQ322. If a key is less than BF90, the first leaf is chosen; if it is less than BQ322, the second leaf is the right pick. But observe that we also have the same results, if instead of BF90, keys BF9 or just BF are used and instead of BQ322, one of three prefixes of this key is used: BQ32, BQ3, or just BQ. After choosing the shortest prefixes of both keys, if any key is less than BF, the search ends up in the first leaf, and if the key is less than BQ, the second leaf is chosen; the result is the same as before. Reducing the size of the separators to the bare minimum does not change the result of the search. It only makes separators smaller. As a result, more separators can be placed in the same node, whereby such a node can have more children. The entire B⁺-tree can have fewer levels, which reduces the branching factor and makes processing the tree faster.

This reasoning does not stop at the level of parents of the leaves. It is carried over to any other level so that the entire index set of a B⁺-tree is filled with prefixes (see Figure 7.15).

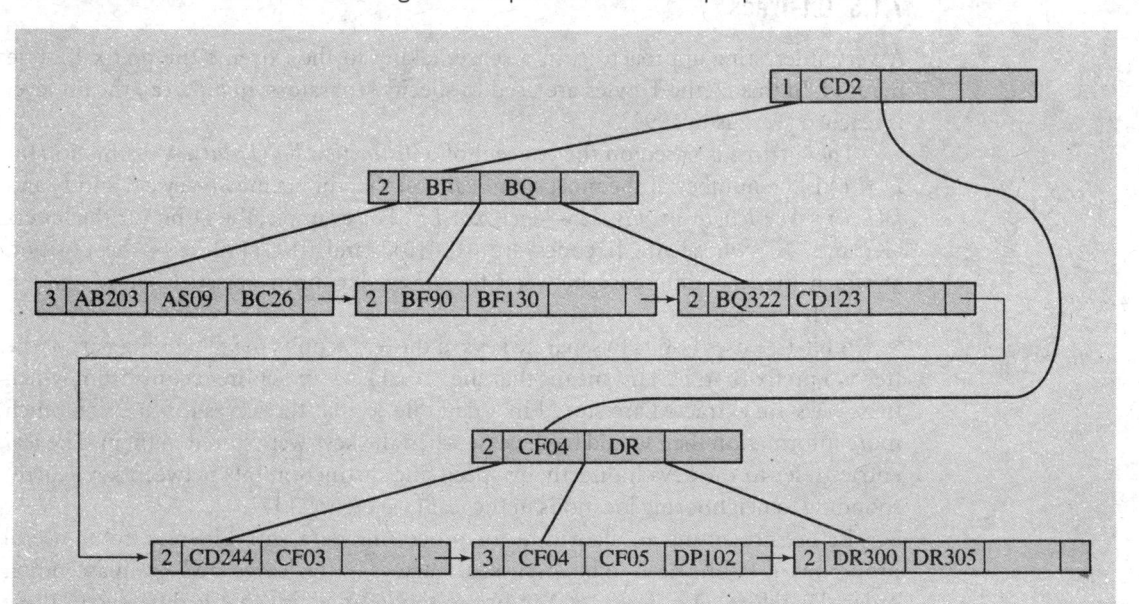

FIGURE 7.15   A B⁺-tree from Figure 7.12 presented as a simple prefix B⁺-tree.

The operations on simple prefix B⁺-trees are much the same as the operations on B⁺-trees with certain modifications to account for prefixes used as separators. In particular, after a split, the first key from the new node is neither moved nor copied to the parent, but the shortest prefix is found that differentiates it from the prefix of the last key in the old node, and the shortest prefix is then placed in the parent. For deletion, however, some separators retained in the index set may turn out to be too long, but to make deletion faster, they do not have to be immediately shortened.

The idea of using prefixes as separators can be carried even further if we observe that prefixes of prefixes can be omitted in lower levels of the tree, which is the idea behind a *prefix B⁺-tree*. This method works particularly well if prefixes are long and repetitious. Figure 7.16 contains an example. Each key in the tree has a prefix AB12XY, and this prefix appears in all internal nodes. This is redundant; Figure 7.16b shows the same tree with "AB12XY" stripped from prefixes in children of the root. To restore the original prefix, the key from the parent node, except for its last character, becomes the prefix of the key found in the current node. For example, the first cell of the child of the root in Figure 7.16b has the key "08." The last character of the key in the root is discarded and the obtained prefix, "AB12XY," is put in front of "08." The new prefix, "AB12XY08," is used to determine the direction of the search.

How efficient are prefix B⁺-trees? Experimental runs indicate that there is almost no difference in the time needed to execute algorithms in B⁺-trees and simple prefix B⁺-trees, but prefix B⁺-trees need 50–100% more time. In terms of disk accesses, there is no difference between these trees in the number of times the disk is accessed for trees of 400 nodes or less. For trees of 400–800 nodes, both simple prefix B⁺-trees and prefix B⁺-trees require 20–25% fewer accesses (Bayer and Unterauer, 1977). This indicates that simple prefix B⁺-trees are a viable option, but prefix B⁺-trees remain largely of theoretical interest.

7.1.5 Bit-Trees

A very interesting approach is, in a sense, taking to the extreme the prefix B⁺-tree method. In this method, bytes are used to specify separators. In *bit-trees*, the bit level is reached (Ferguson 1992).

The bit-tree is based on the concept of a *distinction bit* (*D-bit*). A distinction bit $D(K,L)$ is the number of the most significant bit that differs in two keys, K and L, and $D(K,L) = $ *key-length-in-bits* $- 1 - \lfloor \lg(K \text{ xor } L) \rfloor$. For example, the D-bit for the letters "K" and "N", whose ASCII codes are 01001011 and 01001110, is 5, the position at which the first difference between these keys has been detected; $D(\text{"K"},\text{"N"}) = 8 - 1 - \lfloor \lg 5 \rfloor = 5$.

A bit-tree uses D-bits to separate keys in the leaves only; the remaining part of the tree is a prefix B⁺-tree. This means that the actual keys and entire records from which these keys are extracted are stored in a data file so that the leaves can include much more information than would be the case when the keys were stored in them. The leaf entries refer to the keys indirectly by specifying distinction bits between keys corresponding to neighboring locations in the leaf (see Figure 7.17).

Before presenting an algorithm for processing data with bit-trees, some useful properties of D-bits need to be discussed. All keys in the leaves are kept in ascending order. Therefore, $D_i = D(K_{i-1}, K_i)$ indicates the leftmost bit that is different in these

FIGURE 7.16 (a) A simple prefix B+-tree and (b) its abbreviated version presented as a prefix B+-tree.

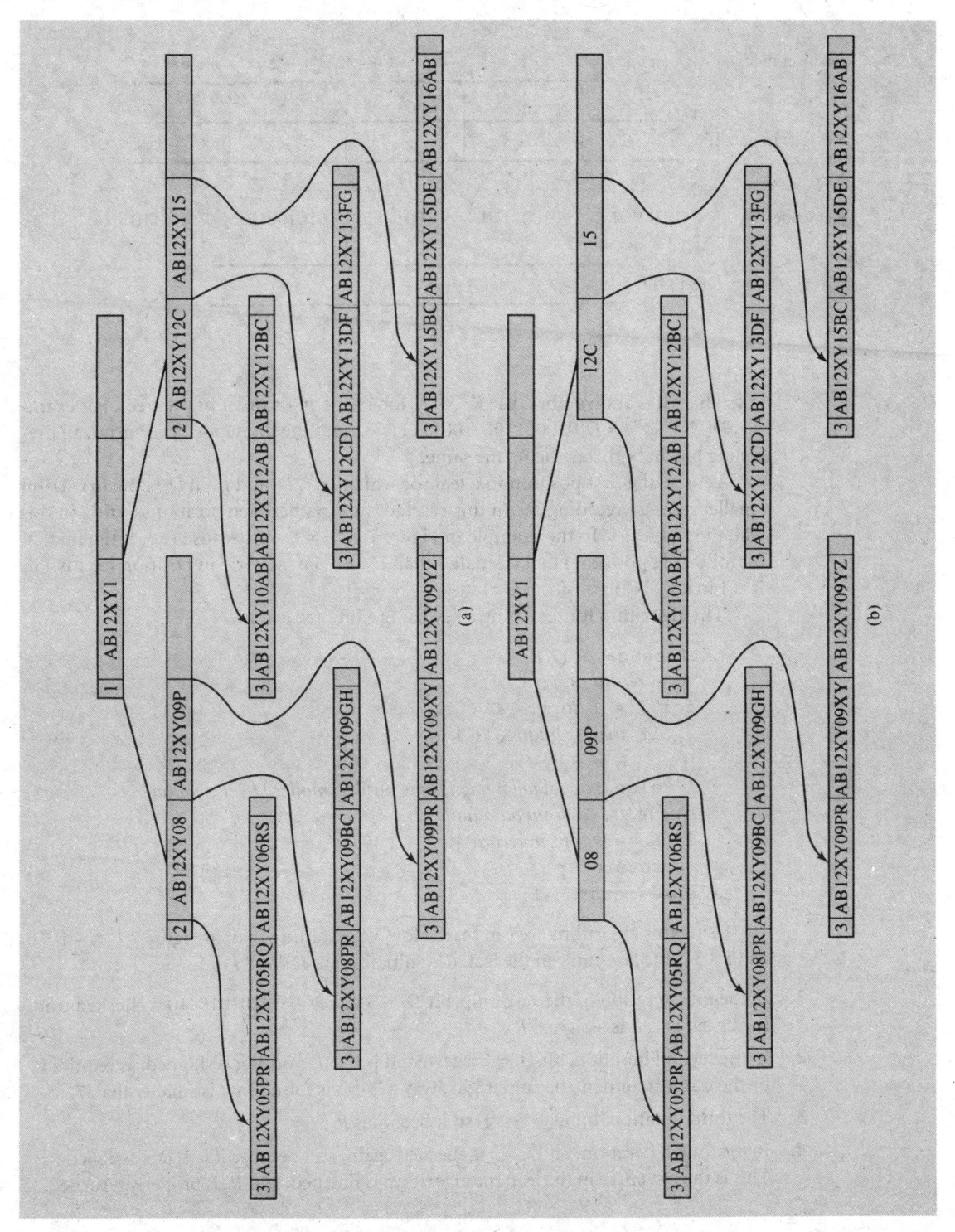

FIGURE 7.17 A leaf of a bit-tree.

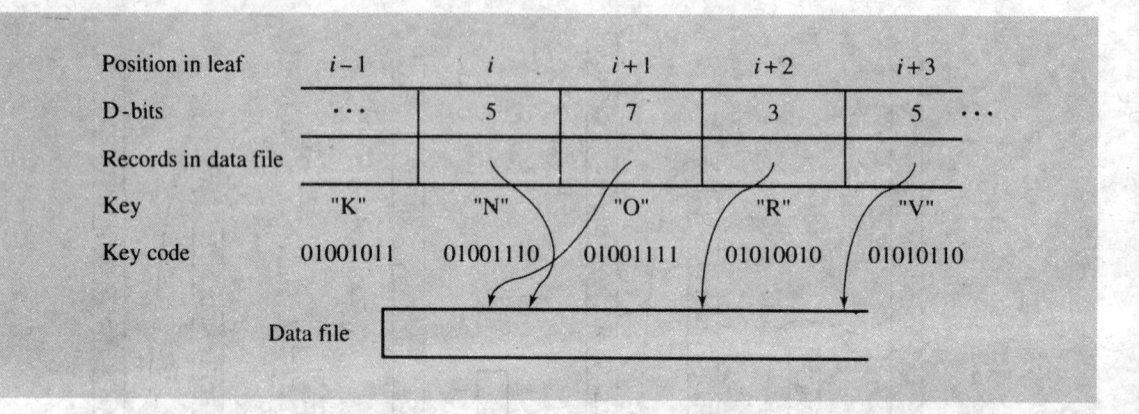

Position in leaf	$i-1$	i	$i+1$	$i+2$	$i+3$
D-bits	...	5	7	3	5 ...
Records in data file					
Key	"K"	"N"	"O"	"R"	"V"
Key code	01001011	01001110	01001111	01010010	01010110

Data file

keys; this bit is always 1 because $K_{i-1} < K_i$ for $1 \le i < m$ (= order of the tree). For example, $D(\text{"N"},\text{"O"}) = D(01001110, 01001111) = 7$, and the bit in position 7 is on, all preceding bits in both keys being the same.

Let j be the first position in a leaf for which $D_j < D_i$ and $j > i$; D_j is the first D-bit smaller than a preceding D_i. In this case, for all keys between positions i and j in this leaf, the D_i bit is 1. In the example in Figure 7.17, $j = i + 2$, because D_{i+2} is the first D-bit following position i that is smaller than D_i. Bit 5 in key "O" in position $i + 1$ is 1 as it is 1 in key "N" in position i.

The algorithm for searching a key using a bit-tree leaf is

```
bitTreeSearch(K)
    R = record R_0;
    for i = 1 to m - 1
        if the D_i bit in K is 1
            R = R_i;
        else skip all following D-bits until a smaller D-bit is found;
    read record R from data file;
    if K == key from record R
        return R;
    else return -1;
```

Using this algorithm, we can search for "V" assuming that, in Figure 7.17, $i - 1 = 0$ and $i + 3$ is the last entry in the leaf. R is initialized to R_0, and i to 1.

1. In the first iteration of the `for` loop, bit $D_1 = 5$ in key "V" = 01010110 is checked, and because it is 1, R is assigned R_1.
2. In the second iteration, bit $D_2 = 7$ is tested. It is 0, but nothing is skipped, as required by the `else` statement, because right away a D-bit is found that is smaller than 7.
3. The third iteration: bit $D_3 = 3$ is 1, so R becomes R_3.
4. In the fourth iteration, bit $D_4 = 5$ is checked again, and because it is 1, R is assigned R_5. This is the last entry in the leaf; the algorithm is finished, and R_5 is properly returned.

What happens if the desired key is not in the data file? We can try to locate "S" = 01010011 using the same assumptions on $i-1$ and $i+3$. Bit $D_1 = 5$ is 0, so the position with D-bit 7 is skipped, and because bit $D_3 = 3$ in "S" is 1, the algorithm would return record R_3. To prevent this, `bitTreeSearch()` checks whether the record it found really corresponds with the desired key. If not, a negative number is returned to indicate failure.

7.1.6 R-Trees

Spatial data are the kind of objects that are utilized frequently in many areas. Computer-assisted design, geographical data, and VLSI design are examples of domains in which spatial data are created, searched, and deleted. This type of data requires special data structures to be processed efficiently. For example, we may request that all counties in an area specified by geographical coordinates be printed or that all buildings in walking distance from city hall be identified. Many different data structures have been developed to accommodate this type of data. One example is an *R-tree* (Guttman, 1984).

An R-tree of order m is a B-treelike structure containing at least m entries in one node for some $m \le$ maximum number allowable per one node (except the root). Hence, an R-tree is not required to be at least half full.

A leaf in an R-tree contains entries of the form (*rect,id*) where *rect* = $([c_1^1, c_1^2], \ldots, [c_n^1, c_n^2])$ is an *n*-dimensional rectangle, c_i^1 and c_i^2 are coordinates along the same axis, and *id* is a pointer to a record in a data file. *rect* is the smallest rectangle containing object *id*, for example, the entry in a leaf corresponding to an object X on a Cartesian plane as in Figure 7.18 is the pair $(([10,100], [5,52]), X)$.

FIGURE 7.18 An area X on the Cartesian plane enclosed tightly by the rectangle ([10,100], [5,52]). The rectangle parameters and the area identifier are stored in a leaf of an R-tree.

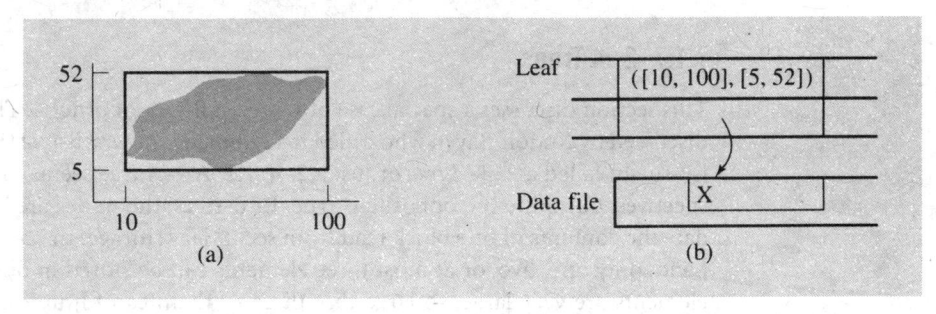

A nonleaf node cell entry has the form (*rect,child*) where *rect* is the smallest rectangle encompassing all the rectangles found in *child*. The structure of an R-tree is not identical to the structure of a B-tree: The former can be viewed as a series of *n* keys and *n* pointers corresponding to these keys.

Inserting new rectangles in an R-tree is made in B-tree fashion, with splits and redistribution. A crucial operation is finding a proper leaf in which to insert a rectangle *rect*. When moving down the R-tree, the subtree chosen in the current node is the one that corresponds to the rectangle requiring the least enlargement to include *rect*. If a split occurs, new encompassing rectangles have to be created. The detailed algorithm is more involved because, among other things, it is not obvious how to divide rectangles of a node being split. The algorithm should generate rectangles that enclose rectangles of the two resulting nodes and are minimal in size.

Figure 7.19 contains an example of inserting four rectangles into an R-tree. After inserting the first three rectangles, R_1, R_2, and R_3, only the root is full (Figure 7.19a). Inserting R_4 causes a split, resulting in the creation of two encompassing rectangles (Figure 7.19b). Inserting R_7 changes nothing, and inserting R_8 causes rectangle R_6 to be extended to accommodate R_8 (Figure 7.19c). Figure 7.19d shows another split after entering R_9 in the R-tree. R_6 is discarded, and R_{10} and R_{11} are created.

A rectangle R can be contained in many other encompassing rectangles, but it can be stored only once in a leaf. Therefore, a search procedure may take a wrong path at some level h when it sees that R is enclosed by another rectangle found in a node on this level. For example, rectangle R_3 in Figure 7.19d is enclosed by both R_{10} and R_{11}. Because R_{10} is before R_{11} in the root, the search accesses the middle leaf when looking for R_3. However, if R_{11} preceded R_{10} in the root, following the path corresponding with R_{11} would be unsuccessful. For large and high R-trees, this overlapping becomes excessive.

A modification of R-trees, called an R^+-tree, removes this overlap (Sellis, Roussopoulos, and Faloutsos, 1987; Stonebraker, Sellis, and Hanson, 1986). The encompassing rectangles are no longer overlapping, and each encompassing rectangle is associated with all the rectangles it intersects. But now the data rectangle can be found in more than one leaf. For example, Figure 7.20 shows an R^+-tree constructed after the data rectangle R_9 was inserted into the R-tree in Figure 7.19c. Figure 7.20 replaces Figure 7.19d. Note that R_8 can be found in two leaves, because it is intersected by two encompassing rectangles, R_{10} and R_{11}. Operations on an R^+-tree make it difficult to assure without further manipulation that nodes are at least half full.

7.1.7 2–4 Trees

This section discusses a special case of B-tree, a B-tree of order 4. This B-tree was first discussed by Rudolf Bayer, who called it a *symmetric binary B-tree* (Bayer, 1972), but it is usually called a *2–3–4 tree* or just a *2–4 tree*. A 2–4 tree seems to offer no new perspectives, but quite the opposite is true. In B-trees, the nodes are large to accommodate the contents of one block read from secondary storage. In 2–4 trees, on the other hand, only one, two, or at most three elements can be stored in one node. Unless the elements are very large, so large that three of them can fill up one block on a disk, there seems to be no reason for even mentioning B-trees of such a small order. Although B-trees have been introduced in the context of handling data on secondary storage, it does not mean that they have to be used only for that purpose.

We spent an entire chapter discussing binary trees, in particular, binary search trees, and developing algorithms that allow quick access to the information stored in these trees. Can B-trees offer a better solution to the problem of balancing or traversing binary trees? We now return to the topics of binary trees and processing data in memory.

B-trees are well suited to challenge the algorithms used for binary search trees, because a B-tree by its nature has to be balanced. No special treatment is needed in addition to building a tree: Building a B-tree balances it at the same time. Instead of using

Section 7.1 The Family of B-Trees   323

FIGURE 7.19   Building an R-tree.

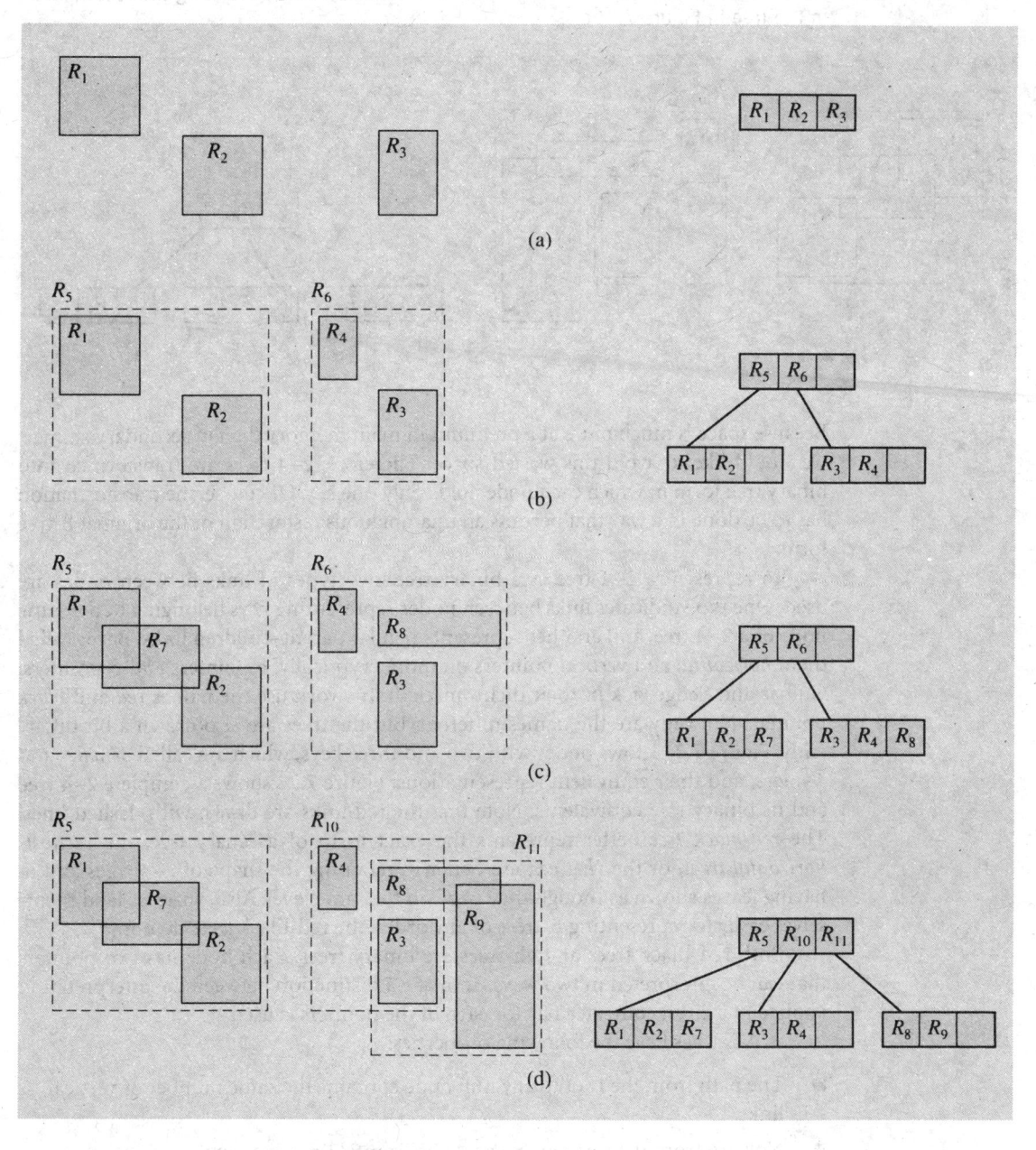

(a)

(b)

(c)

(d)

binary search trees, we may use B-trees of small order such as 2–4 trees. However, if these trees are implemented as structures similarly to B-trees, there are three locations per node to store up to three keys and four locations per node to store up to four pointers. In the worst case, half of these cells are unused, and on the average, 69% are used.

FIGURE 7.20 An R+-tree representation of the R-tree in Figure 7.19d after inserting the rectangle R_9 in the tree in Figure 7.19c.

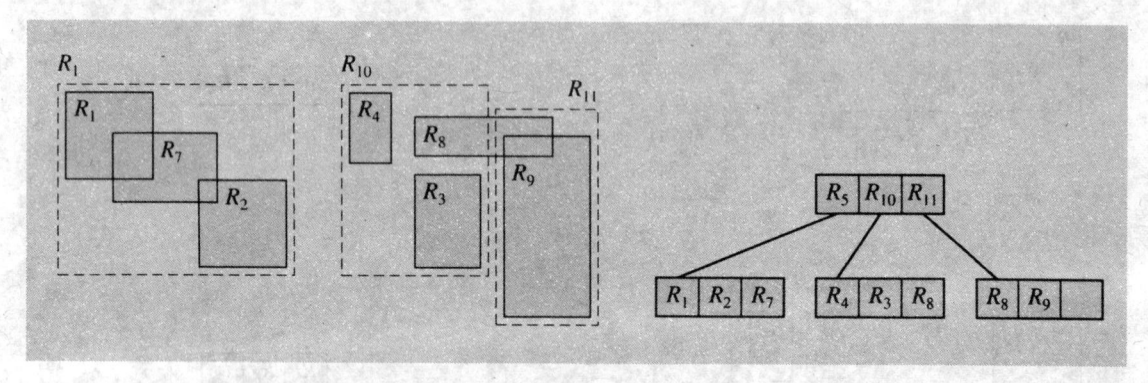

Because space is much more at a premium in main memory than in secondary storage, we would like to avoid this wasted space. Therefore, 2–4 trees are transformed into binary tree form in which each node holds only one key. Of course, the transformation has to be done in a way that permits an unambiguous restoration of the original B-tree form.

To represent a 2–4 tree as a binary tree, two types of links between nodes are used: One type indicates links between nodes representing keys belonging to the same node of a 2–4 tree, and another represents regular parent–children links. Bayer called them *horizontal* and *vertical* pointers or, more cryptically, ρ-pointers and δ-pointers; Guibas and Sedgewick in their dichromatic framework use the names *red* and *black* pointers. Not only are the names different, but the trees are also drawn a bit differently. Figure 7.21 shows nodes with two and three keys, which are called *3-nodes* and *4-nodes*, and their equivalent representations. Figure 7.22 shows a complete 2–4 tree and its binary tree equivalents. Note that the red links are drawn with dashed lines. The *red-black tree* better represents the exact form of a binary tree; the *vertical-horizontal trees*, or the *vh-trees*, are better in retaining the shape of 2–4 trees and in having leaves shown as though they were on the same level. Also, vh-trees lend themselves easily to representing B-trees of any order; the red-black trees do not.

Both red-black trees and vh-trees are binary trees. Each node has two pointers that can be interpreted in two ways. To make a distinction between the interpretation applied in a given context, a flag for each of the pointers is used.

Vh-trees have the following properties:

- The path from the root to any null node contains the same number of vertical links.

- No path from the root can have two horizontal links in a row.

The operations performed on vh-trees should be the same as on binary trees, although their implementation is much more involved. Only searching is the same: To find a key in a vh-tree, no distinction is made between the different types of pointers. We can use the same searching procedure as for binary search trees: If the key

FIGURE 7.21 (a) A 3-node represented (b–c) in two possible ways by red-black trees and (d–e) in two possible ways by vh-trees. (f) A 4-node represented (g) by a red-black tree and (h) by a vh-tree.

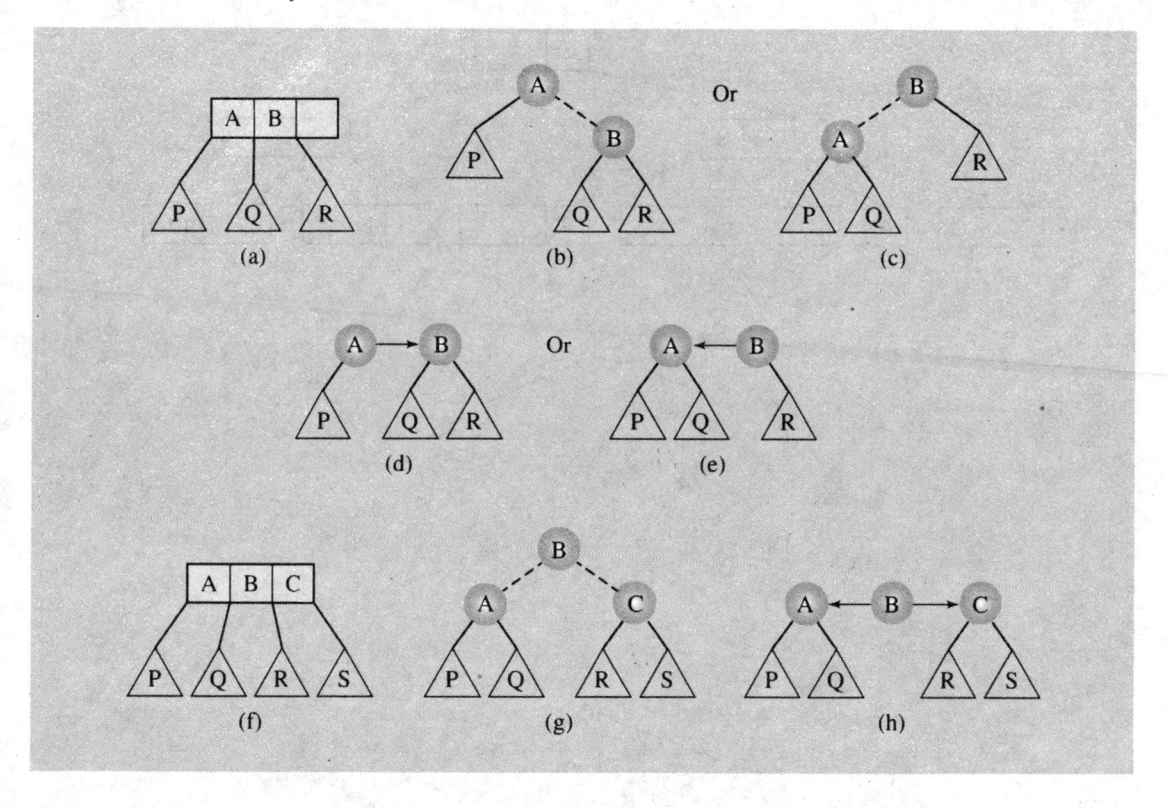

is found, stop. If the key in the current node is larger than the one we are looking for, we go to the left subtree; otherwise, we go to the right subtree.

To find the cost of searching a key in the worst case in a vh-tree, observe that in each such tree we would like to find a correspondence between the number of nodes in the tree and its height. First, observe that if the shortest path to a leaf consists of vertical links only, then the longest path to another leaf can begin and end with horizontal links and have vertical and horizontal links used interchangeably. Therefore,

$$path_{longest} \leq 2 \cdot path_{shortest} + 1$$

with equality if the shortest and the longest paths are as just described. Now we would like to find the minimum number of nodes n_{min} in a vh-tree of a particular height h. Consider first vh-trees of odd height. Figure 7.23a shows a vh-tree of height 7 and, implicitly, of heights 1, 3, and 5. Beginning with $h = 3$, we can observe a geometric progression in the number of nodes added to the tree of previous odd height

$$h = 3\ 5\ 7\ 9 \ldots$$
$$\text{number of new nodes} = 3\ 6\ 12\ 24 \ldots$$

FIGURE 7.22 (a) A 2–4 tree represented (b) by a red-black tree and (c) by a binary tree with horizontal and vertical pointers.

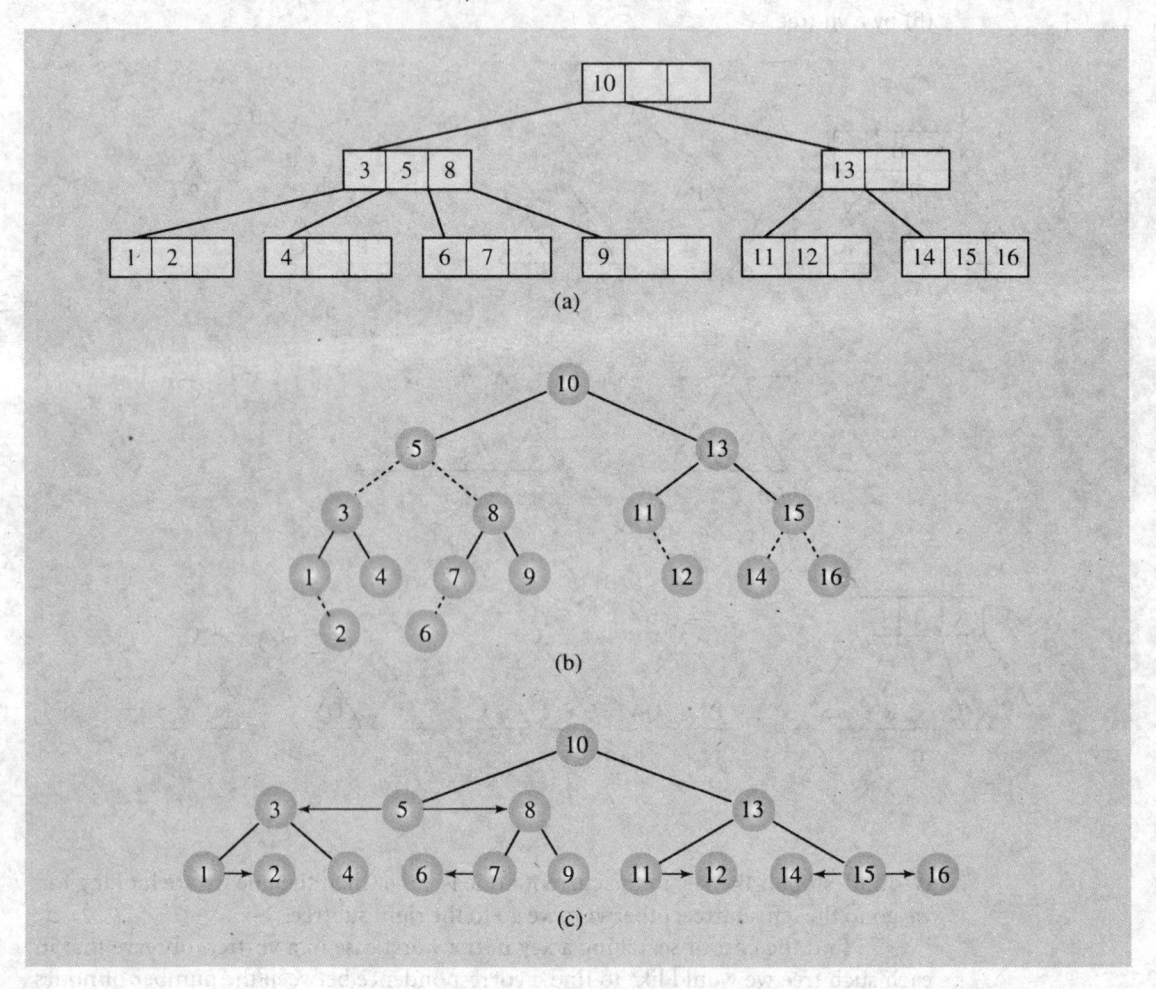

The sum of the first m terms of a geometric sequence is expressed with the formula $a_1 \frac{q^m - 1}{q - 1}$ and thus after adding 1 representing the root,

$$n_{min} = 3 \frac{2^{\frac{h-1}{2}} - 1}{2 - 1} + 1 = 3 \cdot 2^{\frac{h-1}{2}} - 2$$

From this we have

$$n \geq 3 \cdot 2^{\frac{h-1}{2}} - 2$$

and so

$$2 \lg \frac{n + 2}{3} + 1 \geq h$$

FIGURE 7.23 (a) A vh-tree of height 7; (b) a vh-tree of height 8.

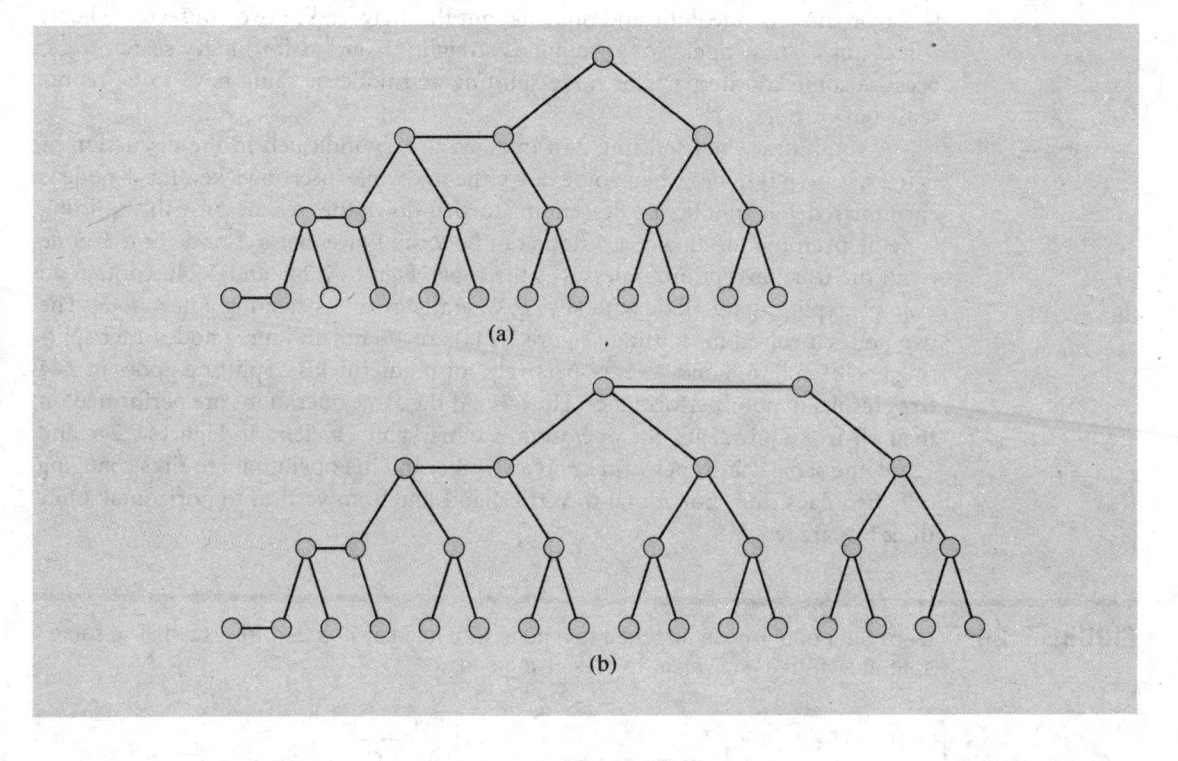

For even heights, as exemplified in Figure 7.23b for a vh-tree of height 8, we obtain

$$h = 2\ 4\ 6\ 8\ldots$$
$$\text{number of new nodes} = 2\ 4\ 8\ 16\ \ldots$$

$$n_{min} = 2(2^{\frac{h}{2}} - 1)$$

and consequently

$$n \geq n_{min} = 2(2^{\frac{h}{2}} - 1)$$

from which

$$2\lg(n + 2) - 2 \geq h$$

It is simple to check that for any n, the bound for even heights is larger, so it can be used as an upper bound for all heights. The lower bound is given by the height of a complete binary tree. The number of nodes in such a tree of height h was found to be $n = 2^h - 1$ (see Figure 6.35), from which

$$\lg(n + 1) \leq h \leq 2\lg(n + 2) - 2$$

This is the worst case of search, when searching has to reach the leaf level.

Insertions restructure the tree by adding one more node and one more link to the tree. Should it be a horizontal or vertical link? Deletions restructure the tree as well by removing one node and one link, but this may lead to two consecutive horizontal links. These operations are not as straightforward as for binary search trees, because some counterparts of node splitting and node merging have to be represented in vh-trees.

A good idea when splitting 2–4 trees, as already indicated in the discussion of B-trees, is to split nodes when going down the tree while inserting a key. If a 4-node is encountered, it is split before descending further down the tree. Because this splitting is made from the top down, a 4-node can be a child of either a 2-node or a 3-node (with the usual exception: unless it is the root). Figures 7.24a and 7.24b contain an example. Splitting the node with keys B, C, and D requires creating a new node. The two nodes involved in splitting (Figure 7.24a) are 4/6 full and three nodes after splitting are 4/9 full (6/8 and 7/12, respectively, for pointer fields). Splitting nodes in 2–4 trees results in poor performance. However, if the same operations are performed on their vh-tree equivalents, the operation is remarkably efficient. In Figures 7.24c and 7.24d, the same split is performed on a vh-tree, and the operation requires changing only two flags from horizontal to vertical and one from vertical to horizontal: Only three bits are reset!

FIGURE 7.24 (a–b) Split of a 4-node attached to a node with one key in a 2–4 tree. (c–d) The same split in a vh-tree equivalent to these two nodes.

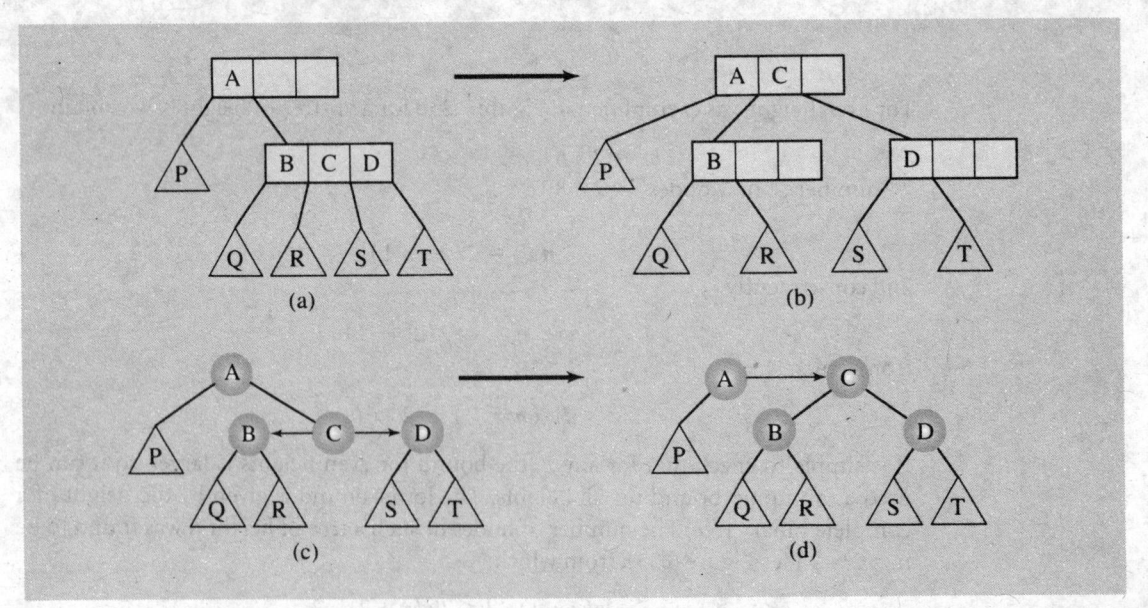

Resetting these three flags suggests the algorithm *flagFlipping*, which takes the following steps: If we visit a node n whose links are both horizontal, then reset the flag corresponding to the link from n's parent to n to horizontal and both flags in n to vertical.

If we have a situation as in Figure 7.25a, the split results in the 2–4 tree as in Figure 7.25b; applying *flagFlipping* to a vh-tree equivalent requires that only three bits are reset (Figures 7.25c and 7.25d).

FIGURE 7.25 (a–b) Split of a 4-node attached to a 3-node in a 2–4 tree and (c–d) a similar operation performed on one possible vh-tree equivalent to these two nodes.

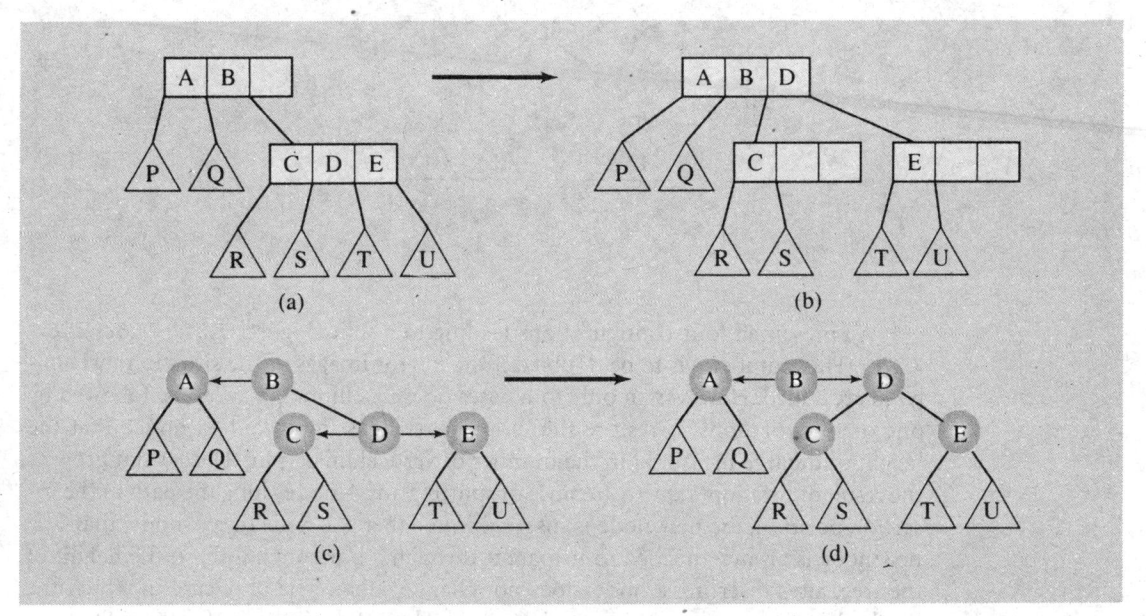

Figure 7.21 indicates that the same node of a 2–4 tree can have two equivalents in a vh-tree. Therefore, the situation in Figure 7.25a can be reflected not only by the tree in Figure 7.25c, but also by the tree in Figure 7.26a. If we proceed as before, by changing three flags as in Figure 7.25d, the tree in Figure 7.26b ends up with two consecutive horizontal links, which has no counterpart in any 2–4 tree. In this case, the three flag flips have to be followed by a rotation; namely, node B is rotated about node A, two flags are flipped, and the tree in Figure 7.26c is the same as in Figure 7.25d.

Figure 7.27a contains another way in which a 4-node is attached to a 3-node in a 2–4 tree before splitting. Figure 7.27b shows the tree after splitting. Applying *flagFlipping* to the tree in Figure 7.27c yields the tree in Figure 7.27d with two consecutive horizontal links. To restore the vh-tree property, two rotations and four flag flips are needed: Node C is rotated about node E, which is followed by two flag flips (Figure 7.27e), and then node C about node A, which is also followed by two flag flips. This all leads to the tree in Figure 7.27f.

FIGURE 7.26 Fixing a vh-tree that has consecutive horizontal links.

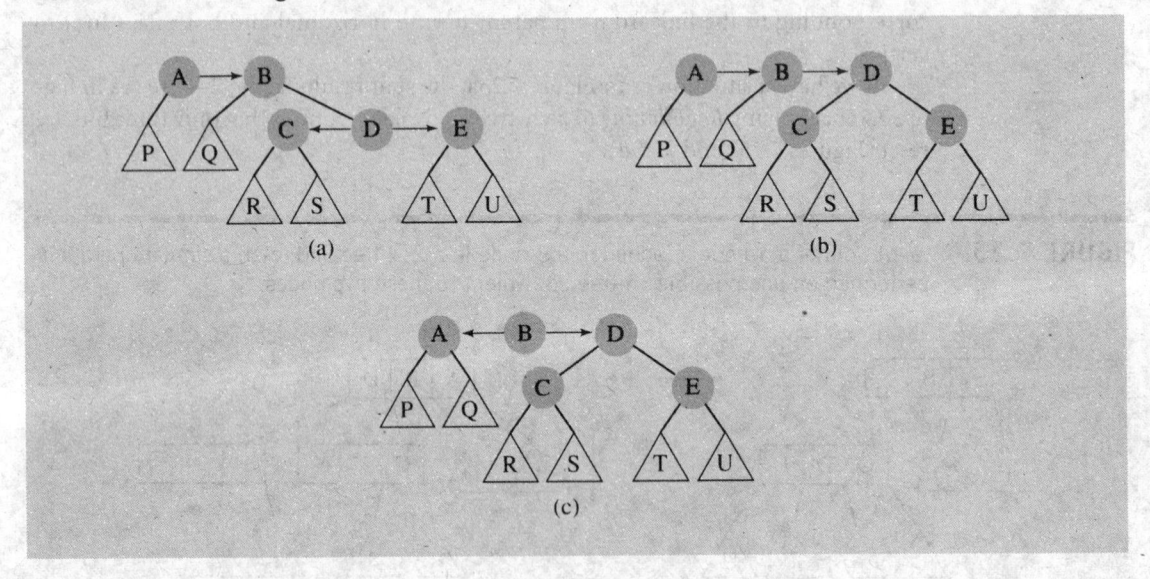

We presented four configurations leading to a split (Figures 7.24c, 7.25c, 7.26a, 7.27c). This number has to be doubled if the mirror images of the situation just analyzed are added. However, in only four cases does flag flipping have to be followed by one or two rotations to restore the vh-property. It is important to notice that the height of the tree measured in the number of vertical links (plus 1) does not grow as the result of rotation(s). Also, because of splitting any 4-node along the path to the insertion position, the new node is inserted into either a 2-node or a 3-node; that is, a new node is always attached to its parent through a horizontal link, so the height of the tree, after inserting a node, does not change either. The only case in which the height does grow is when the root is a 4-node. This is the ninth case for the split.

The vh-tree property can be distorted not only after a 4-node split, but also after including a new node in the tree, which leads to one or two rotations, as indicated at the end of the following insertion algorithm.

```
VHTreeInsert(K)
    create newNode and initialize it;
    if VHTree is empty
        root = newNode;
        return;
    for (p = root, prev = 0; p != 0;)
        if both p's flags are horizontal
            set them to vertical; //flag flipping
                mark prev's link connecting it with p as horizontal;
                if links connecting parent of prev with prev and prev with p are both horizontal
                    if both these links are left or both are right  // Figure 7.26b
                        rotate prev about its parent;
```

FIGURE 7.27 A 4-node attached to a 3-node in a 2–4 tree.

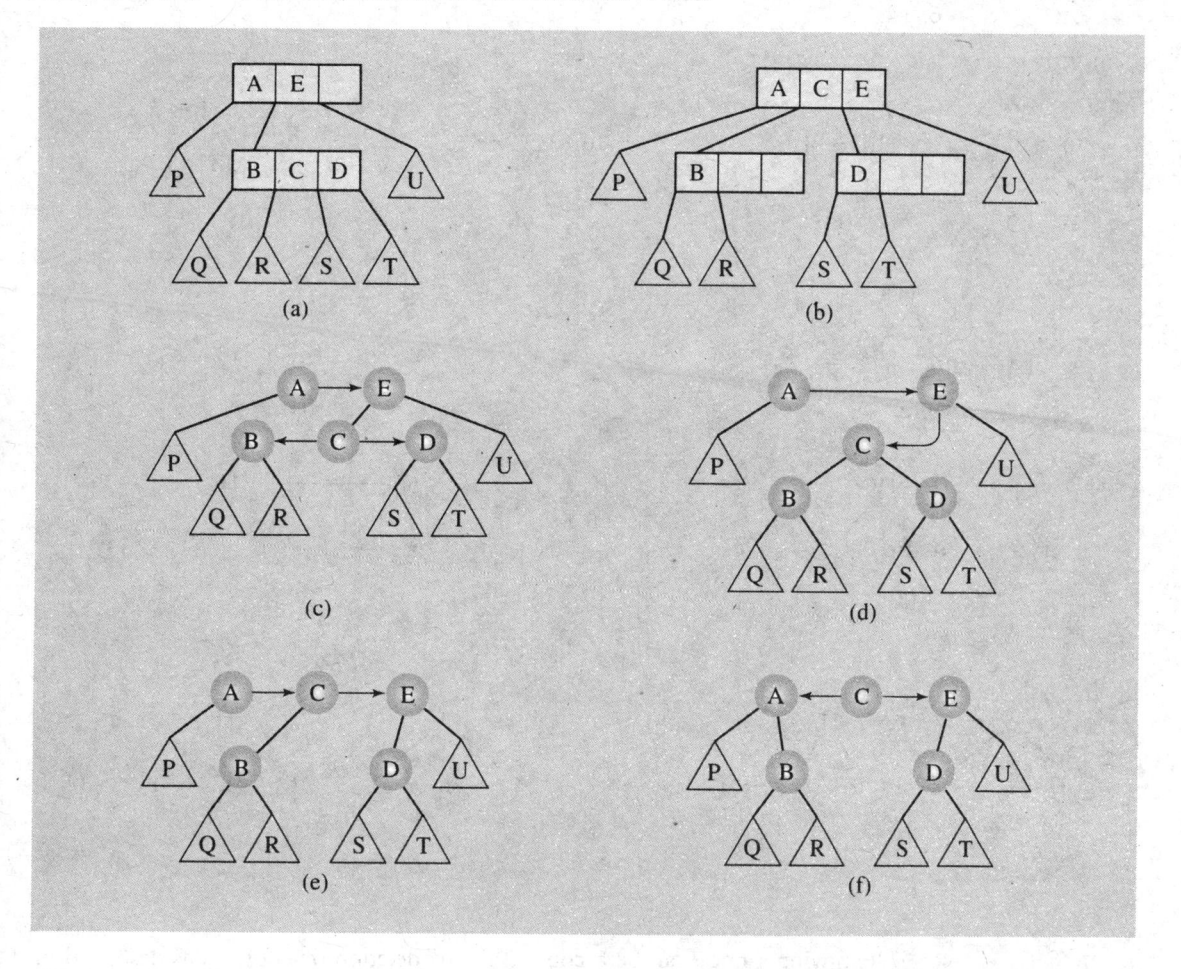

```
            else rotate p about prev and then p about its new parent;  // Figure 7.27d
        prev = p;
        if (p->key > K)
            p = p->left;
        else p = p->right;
attach newNode to prev;
mark prev's flag corresponding to its link to newNode to horizontal;
if link from prev's parent to prev is horizontal
    rotate prev about its parent or
    first rotate newNode about prev and then newNode about its new parent;
```

Figure 7.28 contains an example of inserting a sequence of numbers. Note that a double rotation has to be made in the tree in Figure 7.28h while 6 is being inserted. First 9 is rotated about 5 and then 9 is rotated about 11.

FIGURE 7.28 Building a vh-tree by inserting numbers in this sequence: 10, 11, 12, 13, 4, 5, 8, 9, 6, 14.

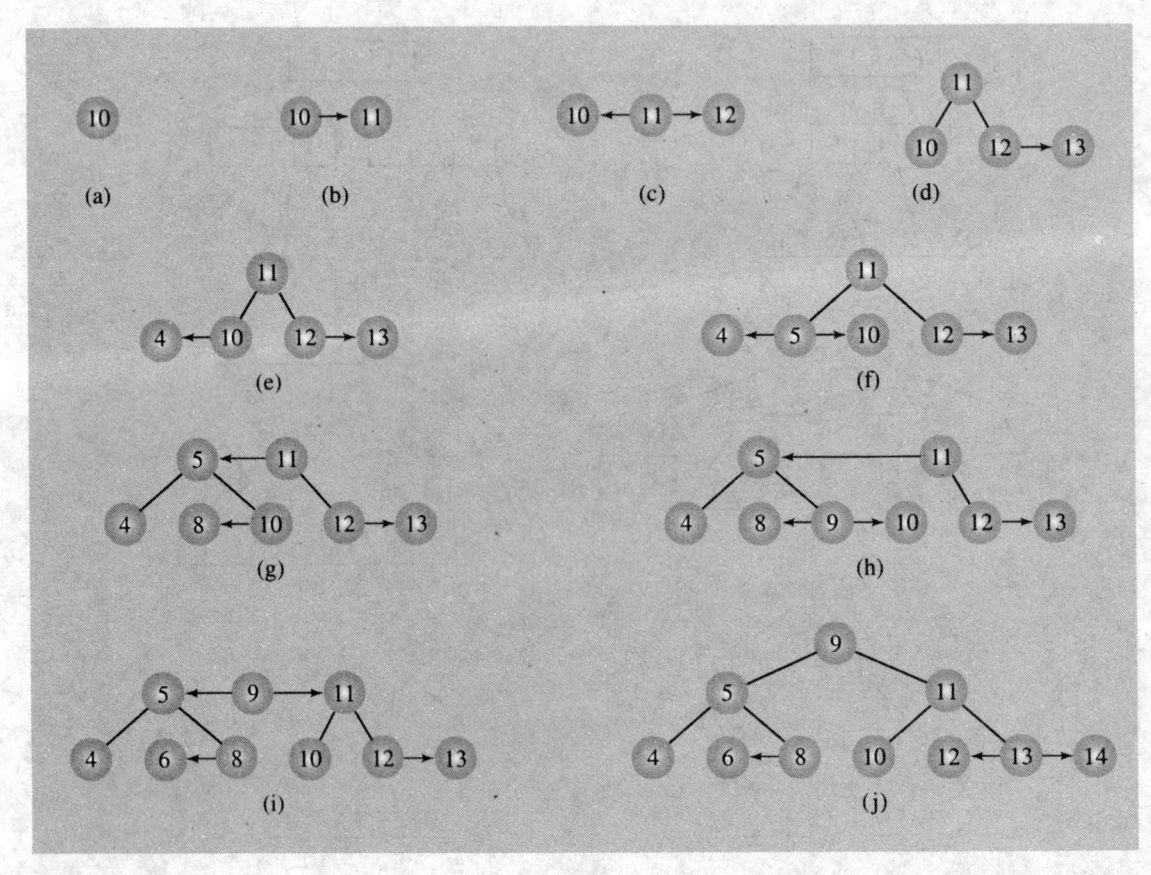

Removing a node can be accomplished by deletion by copying, as described in Section 6.6.2; that is, an immediate successor (or predecessor) is found in the tree, copied over the element to be removed, and the node that holds the original successor is removed from the tree. The successor is found by going one step to the right from the node that holds the element to be removed and then as far as possible to the left. The successor is on the last level of vertical links; that is, the successor may have one left descendant accessible through a horizontal link (in Figure 7.28h, a successor of 11, 12, has one such descendant, 13), or none (like 8, a successor of 5). In a plain binary search tree it is easy to remove such a successor. In the vh-tree, however, it may not be so. If the successor is connected to its parent with a horizontal link, it can simply be detached (like node 8 after copying 8 over 5 to remove number 5 from the tree in Figure 7.28h), but if the connection of the successor with no descendants with the parent is established through the vertical link, then removing this successor may violate the vh-tree property. For example, to remove 9 in the tree in Figure 7.28j, the successor 10 is found and copied over 9 and then node 10 is removed, but the path to the null left child of node 11 includes only one vertical node, whereas the paths to any other null node in the tree include two such links. One way to avoid the problem is to assure that when searching for the successor of a particular node, tree transformations are executed that make a vh-tree a valid vh-tree and cause the successor with no descendants to be connected to its parent with a horizontal link. To that end, a number of cases are distinguished with transformations corresponding to them. Figure 7.29 illustrates these cases and shows an arrow next to a link to indicate the currently scrutinized node and the next node to be checked afterwards.

- **Case 1.** Two 2-node siblings have a 2-node parent; the node and its descendants are merged into a 4-node (Figure 7.29a), which requires only two flag changes.
- **Case 2.** A 3-node with two 2-node descendants is transformed by splitting the 3-node into two 2-nodes and creating a 4-node from the three 2-nodes, as indicated in Figure 7.29b, at the cost of three flag changes.
- **Case 2a.** A 4-node with two 2-node descendants is split into a 2-node and a 3-node, and the three 2-nodes are merged into a 4-node (Figure 7.29c). This requires the same three flag changes as in Case 2.
- **Case 3.** When the end of a 3-node with an out-horizontal link is reached, the direction of the link is reversed through one rotation and two flag changes (Figure 7.29d).
- **Case 4.** A 2-node has a 3-node sibling (there can be any sized parent). Through one rotation—C about B—and two flag changes, the 2-node is expanded into a 3-node and the 3-node sibling is reduced to a 2-node (Figure 7.29e).
- **Case 5.** Similar to Case 4, except that the 3-node sibling has a different direction. The transformation is accomplished through two rotations—first, C about D and then C about B—and two flag changes (Figure 7.29f).
- **Case 5a.** A 2-node has a 4-node sibling (any parent). The 2-node is changed into a 3-node and the 4-node is turned into a 3-node with the same transformations as in Case 5 (Figure 7.29g).

Note that in all these cases we are concerned about changing the link that leads to a 2-node from vertical to horizontal (except Case 3, where the change is inside a 3-node). Nothing is done when the destination is a 3- or 4-node.

Required transformations are performed from the root until the successor of the node to be deleted is found. Because the node to be deleted must be found first, symmetrical cases to the cases already listed have to be included as well, so all in all, there are 15 cases: 1 requires no action, and the remaining 14 cases can be served with 10 different transformations. Examples of deletions are presented in Figure 7.30.

FIGURE 7.29 Deleting a node from a vh-tree.

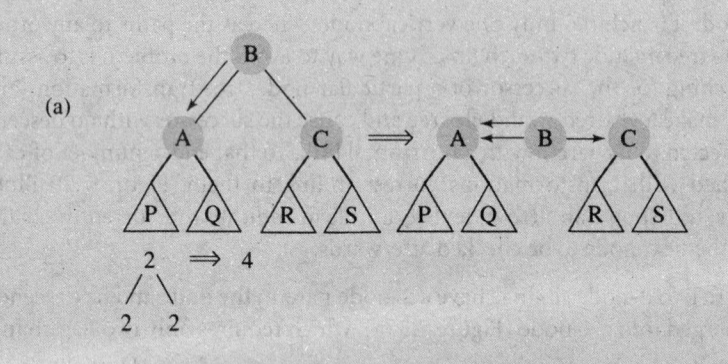

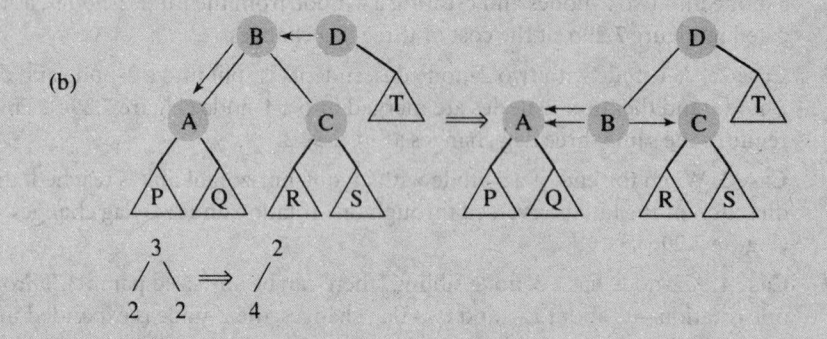

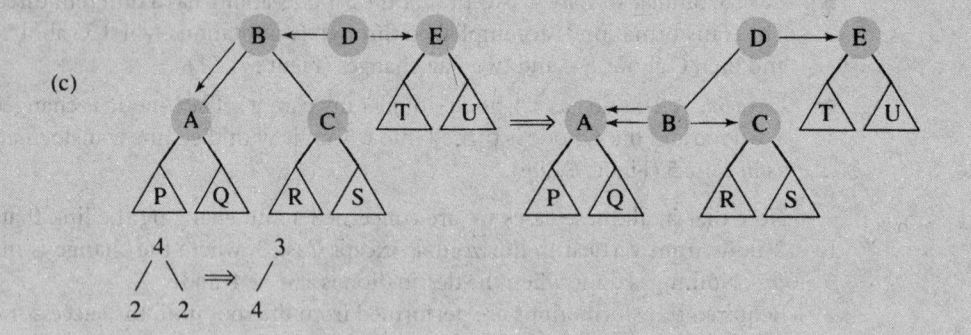

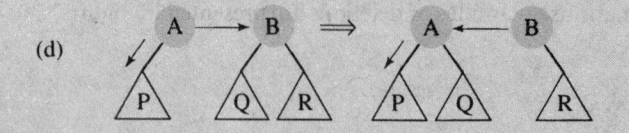

FIGURE 7.29 *(continued)*

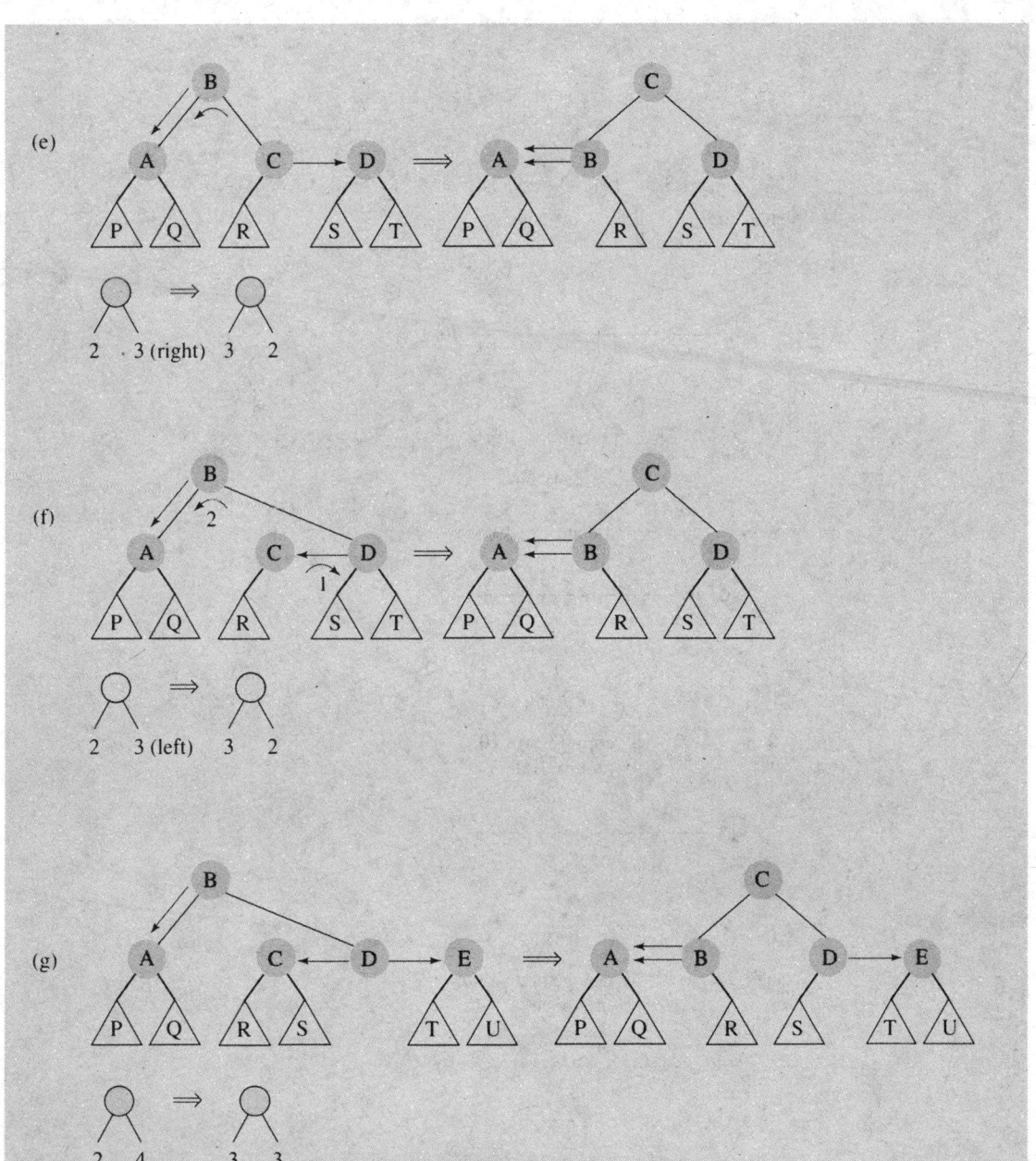

FIGURE 7.30 Examples of node deletions from a vh-tree.

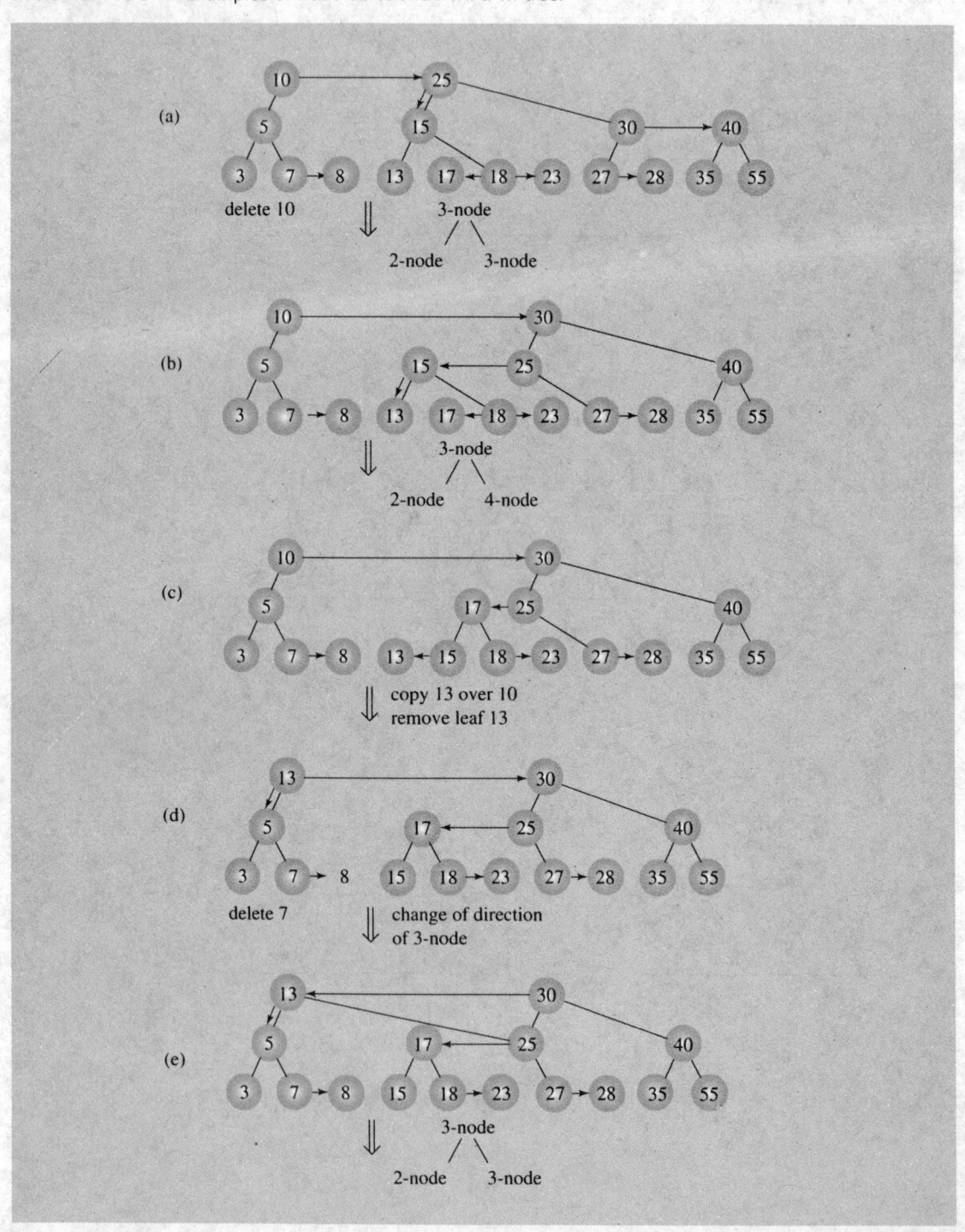

FIGURE 7.30 *(continued)*

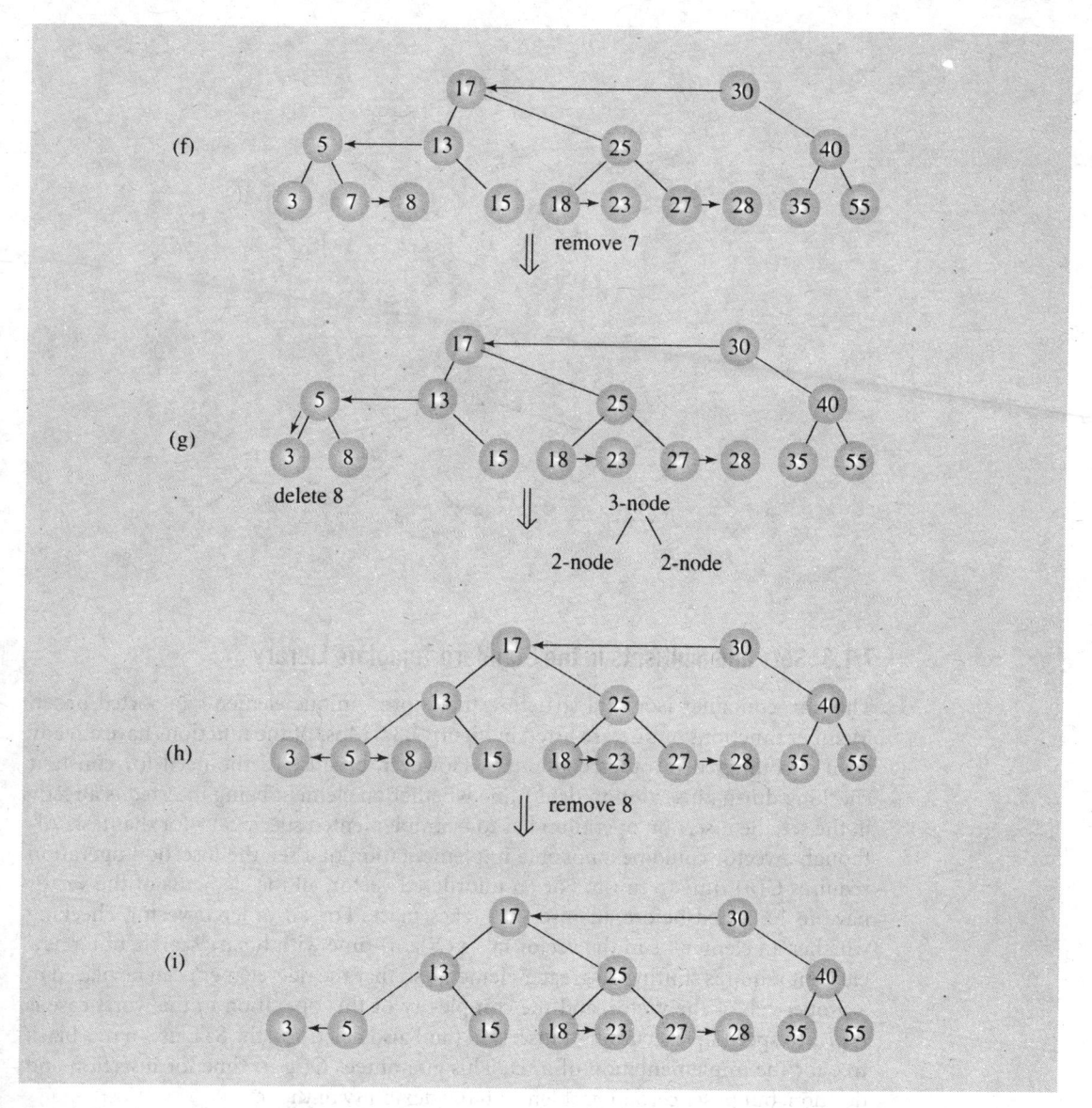

The vh-trees also include AVL trees. An AVL tree can be transformed into a vh-tree by converting the links connecting the roots of subtrees of even height with children of these roots of odd height into horizontal links. Figure 7.31 illustrates this conversion.

FIGURE 7.31 An example of converting an AVL tree (top part) into an equivalent vh-tree (bottom part).

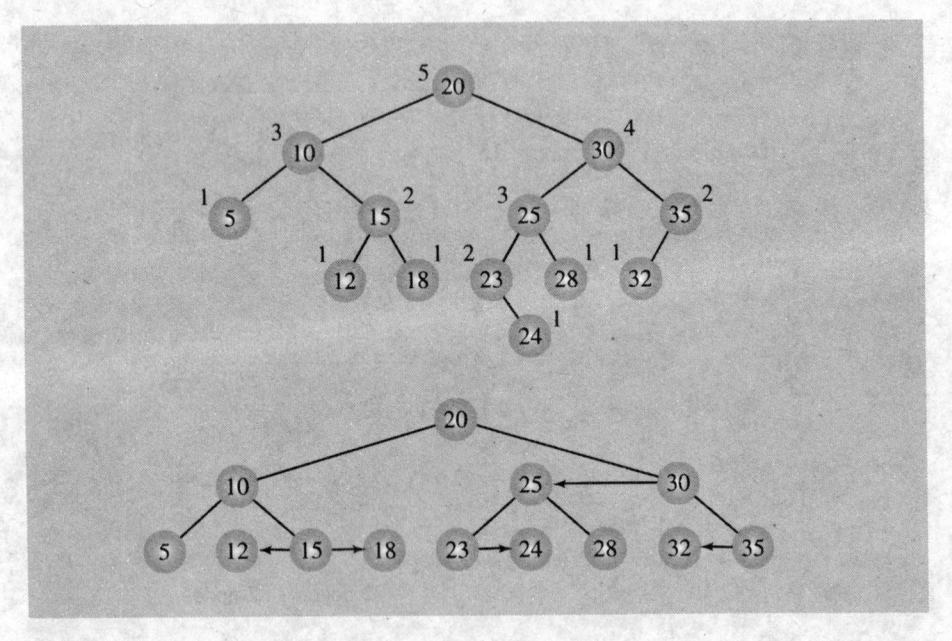

7.1.8 Sets and Multisets in the Standard Template Library

The set container is a data structure that stores unique elements in sorted order. Member functions of set are listed in Figure 7.32. Most of the functions have already been encountered in other containers. However, because of the need for constant checking during insertion to determine whether an element being inserted is already in the set, the insertion operation has to be implemented specifically for that task. Although a vector could be a possible implementation of a set, the insertion operation requires $O(n)$ time to finish. For an unordered vector, all the elements of the vector have to be tested before an insertion takes place. For an ordered vector, checking whether an element is in the vector takes $O(\lg n)$ time with binary search, but a new element requires shifting all greater elements so that the new element can be placed in a proper cell of the vector, and the complexity of this operation in the worst case is $O(n)$. To speed up execution of insertion (and also deletion), the STL uses a red-black tree for the implementation of a set. This guarantees $O(\lg n)$ time for insertion and deletion, but poses certain problems on the flexibility of the set.

A multiset uses the same member functions as a set (Figure 7.32) with two exceptions. First, set constructors are replaced with multiset constructors, but with the same parameters. Second, the member function pair<iterator, bool> insert(const T& el) is replaced with a member function iterator insert(const T& el), which returns an iterator referencing the newly inserted element. Because multiset allows for multiple copies of the same element, there is no need to check whether insertion is successful because it always is. The operation of some member functions for integer sets and multisets is illustrated in Figure 7.33. Sets and multisets are

FIGURE 7.32 Member functions of the container set.

Member Function	Operation
`iterator begin()` `const_iterator begin() const`	Return an iterator that references the first element in the set.
`void clear()`	Remove all the elements in the set.
`size_type count(const T& el) const`	Return the number of elements in the set equal to `el`.
`bool empty() const`	Return `true` if the set includes no elements and `false` otherwise.
`iterator end()` `const_iterator end() const`	Return an iterator that is past the last element of the set.
`pair<iterator, iterator>` `equal_range(const T& el) const`	Return a pair of iterators `<lower_bound(el), upper_bound(el)>` indicating a range of elements equal to `el`.
`void erase(iterator i)`	Remove the element referenced by iterator `i`.
`void erase(iterator first,` `iterator last)`	Remove the elements in the range indicated by iterators `first` and `last`.
`size_type erase(const T& el)`	Remove the elements equal to `el` and return as many as were deleted.
`iterator find(const T& el) const`	Return an iterator referencing the first element equal to `el`.
`pair<iterator,bool>` `insert(const T& el)`	Insert `el` into the set and return a pair `<position of el, true>` if `el` was inserted or `<position of el, false>` if `el` is already in the set.
`iterator insert(iterator i,` `const T& el)`	Insert `el` into the set before the element referenced by iterator `i`.
`void insert(iterator first,` `iterator last)`	Insert elements from the range indicated by iterators `first` and `last`.
`key_compare key_comp() const`	Return the comparison function for keys.
`iterator lower_bound(const T& el) const`	Return an iterator indicating the lower bound of the range of values equal to `el`.
`size_type max_size() const`	Return the maximum number of elements for the set.
`reverse_iterator rbegin()` `const_reverse_iterator rbegin() const`	Return an iterator referencing the last element in the set.
`reverse_iterator rend()` `const_reverse_iterator rend() const`	Return an iterator that is before the first element of the set.

Continues

FIGURE 7.32 *(continued)*

`set(comp = key_compare())`	Construct an empty set using a two-argument Boolean function `comp`.
`set(const set<T>& s)`	Copy constructor.
`set(iterator first, iterator last, comp = key_comp())`	Construct a set and insert elements from the range indicated by `first` and `last`.
`size_type size() const`	Return the number of elements in the set.
`void swap(set<T>& s)`	Swap the content of the set with the content of another set `s`.
`iterator upper_bound(const T& el) const`	Return an iterator indicating the upper bound of the range of values equal to `el`.
`value_compare value_comp() const`	Return the comparison function for values.

FIGURE 7.33 An example of application of the `set` and `multiset` member functions.

```
#include <iostream>
#include <set>
#include <iterator>

using namespace std;

template<class T>
void Union(const set<T>& st1, const set<T>& st2, set<T>& st3) {
    set<T> tmp(st2);
    if (&st1 != &st2)
        for (set<T>::iterator i = st1.begin(); i != st1.end(); i++)
            tmp.insert(*i);
    tmp.swap(st3);
}

int main() {
    ostream_iterator<int> out(cout," ");
    int a[] = {1,2,3,4,5};
    set<int> st1;
```

FIGURE 7.33 *(continued)*

```
set<int,greater<int> > st2;
st1.insert(6); st1.insert(7); st1.insert(8);   // st1 = (6 7 8)
st2.insert(6); st2.insert(7); st2.insert(8);   // st2 = (8 7 6)
set<int> st3(a,a+5);                            // st3 = (1 2 3 4 5)
set<int> st4(st3);                              // st4 = (1 2 3 4 5)
pair<set<int>::iterator,bool> pr;
pr = st1.insert(7);            // st1 = (6 7 8),   pr = (7 false)
pr = st1.insert(9);            // st1 = (6 7 8 9), pr = (9 true)
set<int>::iterator i1 = st1.begin(), i2 = st1.begin();
bool b1 = st1.key_comp()(*i1,*i1);              // b1 = false
bool b2 = st1.key_comp()(*i1,*++i2);            // b2 = true
bool b3 = st2.key_comp()(*i1,*i1);              // b3 = false
bool b4 = st2.key_comp()(*i1,*i2);              // b4 = false
st1.insert(2); st1.insert(4);
Union(st1,st3,st4); // st1 = (2 4 6 7 8 9) and st3 = (1 2 3 4 5) =>
                    // st4 = (1 2 3 4 5 6 7 8 9)

multiset<int> mst1;
multiset<int,greater<int> > mst2;
mst1.insert(6); mst1.insert(7); mst1.insert(8); // mst1 = (6 7 8)
mst2.insert(6); mst2.insert(7); mst2.insert(8); // mst2 = (8 7 6)
multiset<int> mst3(a,a+5);                      // mst3 = (1 2 3 4 5)
multiset<int> mst4(mst3);                       // mst4 = (1 2 3 4 5)
multiset<int>::iterator mpr = mst1.insert(7);   // mst1 = (6 7 7 8)
cout << *mpr << ' ';                            // 7
mpr = mst1.insert(9);                           // mst1 = (6 7 7 8 9)
cout << *mpr << ' ';                            // 9
multiset<int>::iterator i5 = mst1.begin(), i6 = mst1.begin();
i5++; i6++; i6++;                               // *i5 = 7, *i6 = 7
b1 = mst1.key_comp()(*i5,*i6);                  // b1 = false
return 0;
}
```

ordered with a relation `less<int>` or with a relation specified by the user, as for the set `st2` and multiset `mst2`. The relation can always be retrieved with the member function `key_comp()`.

A new number is inserted into a set if it is not already there. For example, an attempt to insert number 7 into the set `st1` = (6 7 8) is unsuccessful, which can be checked, because the pair `pr` of type `pair<set<int>::iterator,bool>` returned by function `insert()`,

```
pr = st1.insert(7);
```

equals <iterator referencing number 7, false>. The components of the pair can be accessed as pr.first and pr.second.

A more interesting situation arises for compound objects whose order is determined by the values of some of its data members. Consider the class Person defined in Section 1.8:

```
class Person {
public:
    . . . . . . . . . . .
    bool operator<(const Person& p) const {
        return strcmp(name,p.name) < 0;
    }
private:
    char *name;
    int age;
    friend class lesserAge;
};
```

With this definition and with an array of objects of type Person

```
Person p[] ={Person("Gregg",25),Person("Ann",30),Person("Bill",20),
             Person("Gregg",35),Person("Kay",30)};
```

we can declare and at the same time initialize two sets

```
set<Person> pSet1(p,p+5);
set<Person,lesserAge> pSet2(p,p+5);
```

By default, the first set is ordered with the operator < overloaded for class Person, and the second is ordered with the relation defined by the function object

```
class lesserAge {
public:
    int operator()(const Person& p1, const Person& p2) const {
        return p1.age < p2.age;
    }
};
```

so that

```
pSet1 = (("Ann,"30) ("Bill,"20) ("Gregg,"25) ("Kay,"30))
pSet2 = (("Bill,"20) ("Gregg,"25) ("Ann,"30) ("Gregg,"35))
```

The first set, pSet1, is ordered by name, and the second, pSet2, by age; therefore, in pSet1, each name appears only once so that the object ("Gregg,"35) is not included, whereas in pSet2, each age is unique so that object ("Kay",30) is not included. An interesting point to notice is that inequality suffices to accomplish this; that is, it excludes from a set repetitious entries depending on the criterion used to order the elements of the set. This is possible on account of implementation of the set. As already mentioned, a set is implemented as a red-black tree. An iterator in such a tree uses an inorder tree traversal to scan the nodes of the tree. Along the way, the inequality is used that is pertaining to a particular set to compare the element that we want to insert with the element in the current node t, for example, el < info(t) or lesserAge(el,info(t)). Note that if an element el is already in the tree in node nel then when scanning the tree to insert el again, we go one step to the right of nel because lesserAge(el,info(nel)) is false (no element is smaller than itself) and then all the way to the left reaching some node n. Node nel is the predecessor of n in inorder traversal so that if the order of the compared elements is now reversed, lesserAge(info(nel),el), then the result is false because info(nel) equals el, which means that el should not be inserted into the set. In this implementation, each node maintains a pointer to its parent so that if an iterator i references node n, then by ascending the tree through the parent link, the predecessor of n in inorder traversal referenced by --i can be reached.

If we declare now two multisets and initialize them,

```
multiset<Person> pSet3(p,p+5);
multiset<Person,lesserAge> pSet4(p,p+5);
```

then we create the multisets

pSet3 = (("Ann,"30) ("Bill,"20) ("Gregg,"25) ("Gregg,"35) ("Kay,"30))
pSet4 = (("Bill,"20) ("Gregg,"25) ("Ann,"30) ("Kay,"30) ("Gregg,"35))

With such declarations, we can output the range of all duplicates (e.g., all the "Gregg" objects in pSet1):

```
pair<multiset<Person>::iterator,multiset<Person>::iterator> mprP;
   mprP = pSet3.equal_range("Gregg");
   for (multiset<Person>::iterator i = mprP.first; i != mprP.second; i++)
       cout << *i;
```

which produces the output (Gregg,25) (Gregg,35), or the number of particular entries in multisets,

```
cout << pSet1.count("Gregg") << ' ' << pSet2.count(Person("",35));
```

which outputs numbers 2 2. Note that the member function count() requires an object of a particular class type, which means that count("Gregg") uses a constructor to build an object with which it is called, count(Person("Gregg",0)). This indicates that the age does not contribute to the result of the search in multiset pSet1 because only the name is included in the definition of operator <. Similarly, in pSet2, the name is irrelevant for the search, so it can be excluded in the request to find objects with a particular age.

It is important to notice that set-related algorithms cannot be used on sets. For example, the call

```
set_union(i1, i2, i3, i4, i5);
```

causes a compilation error because "l-value specifies const object." The reason for the message is that the set_union() algorithm takes nonrepetitive elements from the range indicated by iterators i1 and i2 and iterators i3 and i4 and copies them to a set starting from a position indicated by iterator i5. Therefore, the implementation of set_union() includes an assignment

```
*i5++ = *i1++;
```

To execute the assignment, the iterator i5 must not be constant, but the class set uses only constant iterators. The reason is that the set is implemented as a kind of binary search tree; therefore, a modification of information in a node of the tree can disturb the order of keys in the tree, which leads to improper output of many member functions of class set. Paradoxically, set-related algorithms are inapplicable to sets. To apply them, the set should be implemented differently (e.g., as an array, vector, list, or deque). Another solution is suggested in Figure 7.33 in a generic function Union(), which relies on the member functions insert() and swap() and can be applied to objects of any type class that support these two functions, including objects of type set.

7.1.9 Maps and Multimaps in the Standard Template Library

Maps are tables that can be indexed with any type of data. Hence, they are a generalization of arrays because arrays can be indexed only with constants and variables of ordinal types, such as characters and nonnegative integers, but not with strings or double numbers.

Maps use keys that are used as indexes and elements to be accessed through the keys. Like indexes in arrays, keys in maps are unique in that one key is associated with one element only. Thus, maps are also a generalization of sets. Like sets, maps are implemented as red-black trees. But unlike trees implementing sets that store elements only, trees implementing maps store pairs <key, element>. The pairs are ordered by an ordering function defined for keys, not for elements. Therefore, a particular element is found in the tree by locating a particular node using the key that is associated with this element and extracting the second element of the pair stored in this node. Unlike in sets, elements can now be modified because the tree is ordered by keys, not elements, which also means that keys in the tree cannot be modified. The possibility of modifying the elements accounts for two versions of many member functions listed in Figure 7.34, one for nonconstant iterators and another for constant iterators.

FIGURE 7.34   Member functions of the container map.

Member Function	Operation
iterator begin()	Return an iterator that references the first element in the map.
const_iterator begin() const	Return an iterator that references the first element in the map.
void clear()	Remove all the elements in the map.
size_type count(const K& key) const	Return the number of elements in the map with the key (0 or 1).
bool empty() const	Return true if the map includes no elements and false otherwise.
iterator end()	Return an iterator that is past the last element of the map.

FIGURE 7.34 (continued)

`const_iterator end() const`	Return a `const` iterator that is past the last element of the map.
`pair<iterator,iterator> equal_range(const K& key)`	Return a pair of iterators `<lower_bound(key), upper_bound(key)>` indicating a range of elements with the `key`.
`pair<const_iterator, const_iterator> equal_range(const K& key) const`	Return a pair of iterators `<lower_bound(key), upper_bound(key)>` indicating a range of elements with the `key`.
`void erase(iterator i)`	Remove the element referenced by iterator `i`.
`void erase(iterator first, iterator last)`	Remove the elements in the range indicated by iterators `first` and `last`.
`size_type erase(const K& key)`	Remove the elements with the `key` and return as many as were removed.
`iterator find(const K& key)`	Return an iterator referencing the first element with the `key`.
`const_iterator find(const K& key)const`	Return an iterator referencing the first element with the `key`.
`pair<iterator,bool> insert const pair <K,E>& (key,el))`	Insert pair `<key,el>` into the map and return a pair `<position of el, true>` if `el` was inserted or `<position of el, false>` if `el` is already in the map.
`iterator insert(iterator i, (const pair <K,E>& (key,el))`	Insert pair `<key,el>` into the map before the element referenced by iterator `i`.
`void insert(iterator first, iterator last)`	Insert pairs `<key,el>` from the range indicated by iterators `first` and `last`.
`key_compare key_comp() const`	Return the comparison function for keys.
`iterator lower_bound(const K& key)`	Return an iterator indicating the lower bound of the range of values with the `key`.
`const_iterator lower_bound(const K& key) const`	Return an iterator indicating the lower bound of the range of values with the `key`.
`map(comp = key_compare())`	Construct an empty map using a two-argument Boolean function `comp`.
`map(const map<K,E>& m)`	Copy constructor.
`map(iterator first, iterator last, comp = key_compare())`	Construct a map and insert elements from the range indicated by `first` and `last`.
`size_type max_size() const`	Return the maximum number of elements for the map.

Continues

FIGURE 7.34 (continued)

`T& operator[](const K& key)`	Return the element with the `key` if it is in the map; otherwise, insert it.
`reverse_iterator rbegin()`	Return an iterator referencing the last element in the map.
`const_reverse_iterator rbegin() const`	Return an iterator referencing the last element in the map.
`reverse_iterator rend()`	Return an iterator that is before the first element of the map.
`const_reverse_iterator rend() const`	Return an iterator that is before the first element of the map.
`size_type size() const`	Return the number of elements in the map.
`void swap(map<K,E>& m)`	Swap the content of the map with the content of another map `m`.
`iterator upper_bound(const K& key)`	Return an iterator indicating the upper bound of the range of values with the `key`.
`const_iterator upper_bound(const K& key) const`	Return an iterator indicating the upper bound of the range of values with the `key`.
`value_compare value_comp() const`	Return the comparison function for values.

An example program is shown in Figure 7.35. The map `cities` is indexed with objects of type `Person`. The map is initialized with three pairs `<Person object, string>`. The following assignment,

```
cities[Person("Kay",40)] = "New York";
```

uses as an index a new object, but the subscript operator `[]` is so defined for maps that it inserts the pair `<key, element>` if it is not in the map, as in the case of this assignment. The next assignment

```
cities["Jenny"] = "Newark";
```

implicitly uses a class constructor to generate an object `Person("Jenny",0)`, which is then used as a key. Because there is no entry in the map for this key, the pair `<Person("Jenny",0),"Newark">` is inserted into it.

New pairs can be explicitly inserted with the `insert()` member function, as also illustrated in the program in Figure 7.35. Two ways of creating a pair are shown. In the first `insert()` statement, a pair is generated with `value_type`, which is defined in the class map as another name for type `pair<key type, element type>`. In the second `insert()` statement, a pair is explicitly created. Both statements are attempts to insert a pair for an already existing key `Person("Kay",40)`; therefore, no insertion takes place, although the element (city name) is different than the one in the map

FIGURE 7.35 An example of application of the map member functions.

```cpp
#include <iostream>
#include <map>

using namespace std;

int main() {
    pair<Person,char*> p[] =
        {pair<Person,char*>(Person("Gregg",25),"Pittsburgh"),
         pair<Person,char*>(Person("Ann",30),"Boston"),
         pair<Person,char*>(Person("Bill",20),"Belmont")};
    map<Person,char*> cities(p,p+3);
    cities[Person("Kay",40)] = "New York";
    cities["Jenny"] = "Newark";
    cities.insert(map<Person,char*>::value_type(Person("Kay",40),
                                                "Detroit"));
    cities.insert(pair<Person,char*>(Person("Kay",40),"Austin"));
    map<Person,char*>::iterator i;
    for (i = cities.begin(); i != cities.end(); i++)
        cout << (*i).first << ' ' << (*i).second << endl;
    // output:
    //    (Ann,30) Boston
    //    (Bill,20) Belmont
    //    (Gregg,25) Pittsburgh
    //    (Jenny,0) Newark
    //    (Kay, 40) New York
    cities[p[1].first] = "Chicago";
    for (i = cities.begin(); i != cities.end(); i++)
        cout << (*i).first << ' ' << (*i).second << endl;
    // output:
    //    (Ann,30) Chicago
    //    (Bill,20) Belmont
    //    (Gregg,25) Pittsburgh
    //    (Jenny,0) Newark
    //    (Kay, 40) New York

    multimap<Person,char*> mCities(p,p+3);
    mCities.insert(pair<Person,char*>(Person("Kay",40),"Austin"));
    mCities.insert(pair<Person,char*>(Person("Kay",40),"Austin"));
    mCities.insert(pair<Person,char*>(Person("Kay",40),"Detroit"));
    multimap<Person,char*>::iterator mi;
    for (mi = mCities.begin(); mi != mCities.end(); mi++)
```

Continues

FIGURE 7.35 (continued)

```
        cout << (*mi).first << ' ' << (*mi).second << endl;
// output:
//    (Ann,30) Boston
//    (Bill,20) Belmont
//    (Gregg,25) Pittsburgh
//    (Kay, 40) Austin
//    (Kay, 40) Austin
//    (Kay, 40) Detroit
(*(mCities.find(Person("Kay",40)))).second = "New York";
for (mi = mCities.begin(); mi != mCities.end(); mi++)
        cout << (*mi).first << ' ' << (*mi).second << endl;
// output:
//    (Ann,30) Boston
//    (Bill,20) Belmont
//    (Gregg,25) Pittsburgh
//    (Kay, 40) New York
//    (Kay, 40) Austin
//    (Kay, 40) Detroit
    return 0;
}
```

with which this key is associated. To update an element in the map, an assignment has to be used, as illustrated in the program with the assignment

```
cities[p[1].first] = "Chicago";
```

Because the elements of the array p are pairs, a key that is the first element of the pair is accessed using dot notation, `p[1].first`. This way of accessing elements of pairs is also illustrated in the printing statements.

The program in Figure 7.35 is shown only to illustrate unconventional indexing. It would probably be more natural to include a city as another data member in each object. A more useful example concerns social security numbers and objects of type Person. If we wanted to create an array (or a vector or a deque) so that SSNs could be used as indexes, the array would need 1 billion cells because the largest SSN equals 999999999. But with maps, we can have only as many entries as the number of Person objects used in the program. For example, we can declare a map SSN

```
map<long,Person> SSN;
```

and then execute a number of assignments

```
SSN[123456789] = p[1].first;
SSN[987654321] = p[0].first;
```

```
SSN[222222222] = Person("Kay",40);
SSN[111111111] = "Jenny";
```

In this way, `SSN` has only four entries, although keys are very large numbers

SSN = ((111111111, ("Jenny,"0)), (123456789, ("Ann,"30)),
       (222222222, ("Kay,"40)), (987654321, ("Gregg,"25)))

Information is now very easily accessible and modifiable by using SSNs as the access keys.

A multimap is a map that allows for duplicate keys. The class `multimap` uses the same member functions as `map` with a few exceptions. Map constructors are replaced by multimap constructors with the same parameters, and the member function `pair<iterator,bool> insert(<key,el>)` is replaced by `iterator insert(<key,el>)`, which returns an iterator referencing the position in which the pair `<key,el>` was inserted. The subscript operator `[]` is not defined for multimaps. To modify an element in a multimap, the element can be found and then modified using the `find()` function, as shown in Figure 7.35. But this modifies only one entry, the first found by the function. To modify all the entries for a particular key, we can use a loop, as in

```
for (mi = mCities.lower_bound(Person("Kay",40));
     mi != mCities.upper_bound(Person("Kay",40)); mi++)
    (*mi).second = "New York";
```

7.2 TRIES

The preceding chapter showed that traversing a binary tree was guided by full key comparisons; each node contained a key that was compared to another key to find a proper path through the tree. The discussion of prefix B-trees indicated that this is not necessary and that only a portion of a key is required to determine the path. However, finding a proper prefix became an issue, and maintaining prefixes of an acceptable form and size made the process for insertion and deletion more complicated than in standard B-trees. A tree that uses parts of the key to navigate the search is called a *trie*. The name of the tree is appropriate, as it is a portion of the word re*trie*val with convoluted pronunciation: To distinguish a tree from a trie in speech, trie is pronounced "try."

Each key is a sequence of characters, and a trie is organized around these characters rather than entire keys. For simplicity, assume that all the keys are made out of five capital letters: A, E, I, P, R. Many words can be generated out of these five letters, but our examples will use only a handful of them.

Figure 7.36 shows a trie for words that are indicated in the vertical rectangles; this form was first used by E. Fredkin. These rectangles represent the leaves of the trie, which are nodes with actual keys. The internal nodes can be viewed as arrays of pointers to subtries. At each level *i*, the position of the array that corresponds to the *i*th letter of the key being processed is checked. If the pointer in this position is null, the key is not in the trie, which may mean a failure or a signal for key insertion. If

FIGURE 7.36 A trie of some words composed of the five letters A, E, I, R, and P. The sharp sign, #, indicates the end of a word, which can be a prefix of another word.

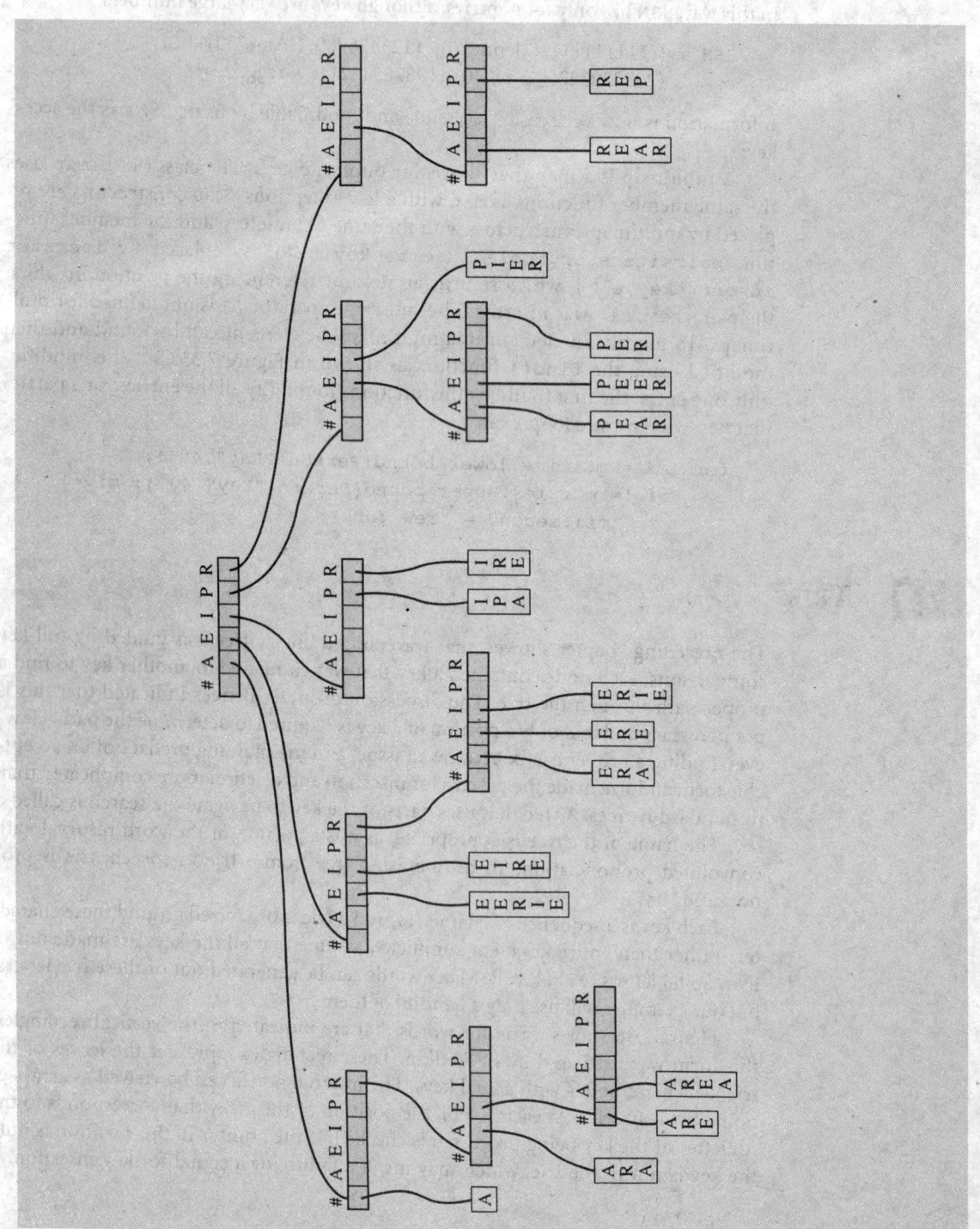

not, we continue processing until a leaf containing this key is found. For example, we check for the word "ERIE." At the first level of the trie, the pointer corresponding to the first letter of this word, "E," is checked. The pointer is not null, so we go to the second level of the trie, to the child of the root accessible from position "E"; now the pointer in the position indicated by the second letter, "R," is tested. It is not null either, so we descend down the trie one more level. At the third level, the third letter, "I," is used to access a pointer in this node. The pointer points to a leaf containing the word "ERIE." Thus, we conclude that the search is successful. If the desired word was "ERIIE," we would fail because we would access the same leaf as before, and obviously, the two words are different. If the word were "ERPIE," we would access the same node with a leaf that contains "ERIE," but this time "P" would be used to check the corresponding pointer in the node. Because the pointer is null, we would conclude that "ERPIE" is not in the trie.

There are at least two problems. First, how do we make a distinction between two words when one is a prefix of the other? For example, "ARE" is a prefix in "AREA." Thus, if we are looking for "ARE" in the trie, we must not follow the path leading to "AREA." To that end, a special character is used in each node guaranteed not to be used in any word, in this case, a sharp sign, "#." Now, while searching for "ARE" and after processing "A," "R," and "E," we find ourselves in a node at the fourth level of the trie, whose leaves are "ARE" and "AREA." Because we processed all letters of the key "ARE," we check the pointer corresponding to the end of words, "#," and because it is not empty, we conclude that the word is in the trie.

This last example points to another problem. Is it really necessary to store entire words in the trie? After we reached the fourth level when searching for "ARE" and the pointer for "#" is not null, do we have to go to the leaf to make a comparison between the key "ARE" and the contents of the leaf, also "ARE"? Not necessarily, and the example of prefix B-trees suggests the solution. The leaves may contain only the unprocessed suffixes of the words. It may even make the comparison faster in C++. If at each level of the trie the pointer w to a word is incremented to point to the next letter, then upon reaching a leaf, we need only check `strcmp(w, leaf->key)`. This chapter's case study adopts this approach.

This example restricted the number of letters used to five, but in a more realistic setting, all letters are used so that each node has 27 pointers (including "#"). The height of the trie is determined by the longest prefix, and for English words, the prefix should not be a long string. For most words, the matter is settled after several node visits, probably 5–7. This is true for 10,000 English words in the trie, and for 100,000. A corresponding perfectly balanced binary search tree for 10,000 words has a height $\lceil \lg 10{,}000 \rceil = 14$. Most words are stored on the lowest levels of this tree, so on the average, the search takes 13 node visits. (The average path length in a perfectly balanced tree of height h is $\lceil \lg h \rceil - 2$.) This is double the number of visits in the trie. For 100,000 words, the average number of visits in the tree increases by 3 because $\lceil \lg 100{,}000 \rceil = 17$; in the trie this number can increase by 1 or 2. Also, when making a comparison in the binary search tree, the comparison is made between the key searched for and the key in the current node, whereas in the trie only one character is used in each comparison except when comparing with a key in a leaf. Therefore, in situations where the speed of access is vital, such as in spell checkers, a trie is a very good choice.

Due to the fact that the trie has two types of nodes, inserting a key into a trie is a bit more complicated than inserting it into a binary search tree.

```
trieInsert(K)
    i = 0;
    p = the root;
    while not inserted
        if (K[i] == '\0')
            set the end-of-word marker in p to true;
        else if (p->ptrs[K[i]] == 0)
            create a leaf containing K and put its address in p->ptrs[K[i]];
        else if pointer p->ptrs[K[i]] points to a leaf
            K_L = key in leaf p->ptrs[K[i]]
            do create a nonleaf and put its address in p->ptrs[K[i]];
                p = the new nonleaf;
            while (K[i] == K_L[i++]);
            create a leaf containing K and put its address in p->ptrs[K[--i]];
            if (K_L[i] == '\0')
                set the end-of-word marker in p to true;
            else create a leaf containing K_L and put its address in p->ptrs[K_L[i]];
        else p = p->ptrs[K[i++]];
```

The inner do loop in this algorithm is needed when a prefix in the word K and in the word K_L is longer than the number of nodes in the path leading to the current node p. For example, before "REP" is inserted in the trie in Figure 7.36, the word "REAR" is stored in a leaf corresponding to the letter "R" of the root of the trie. If "REP" is now being inserted, it is not enough to replace this leaf by a nonleaf, because the second letters of both these words are the same letter, "E." Hence, one more nonleaf has to be created on the third level of the trie, and two leaves containing the words "REAR" and "REP" are attached to this nonleaf.

If we compare tries with binary search trees, we see that for tries, the order in which keys are inserted is irrelevant, whereas this order determines the shape of binary search trees. However, tries can be skewed by words or, rather, by the type of prefixes in words being inserted. The length of the longest identical prefix in two words determines the height of the trie. Therefore, the height of the trie is equal to the length of the longest prefix common to two words plus one (for a level to discriminate between the words with this prefix) plus one (for the level of leaves). The trie in Figure 7.36 has height five because the longest identical prefix, "ARE," is merely three letters long.

The main problem tries pose is the amount of space they require; a substantial amount of this space is basically wasted. Many nodes may have only a couple of non-null pointers, and yet the remaining 25 pointers must reside in memory. There is a burning need to decrease the amount of required space.

One way to reduce the size of a node is by storing only those pointers that are actually in use, as in Figure 7.37 (Briandais, 1959). However, the introduced flexibility concerning the size of each node somewhat complicates the implementation. Such tries can be implemented in the spirit of 2–4 tree implementation. All sibling nodes

FIGURE 7.37 The trie in Figure 7.36 with all unused pointer fields removed.

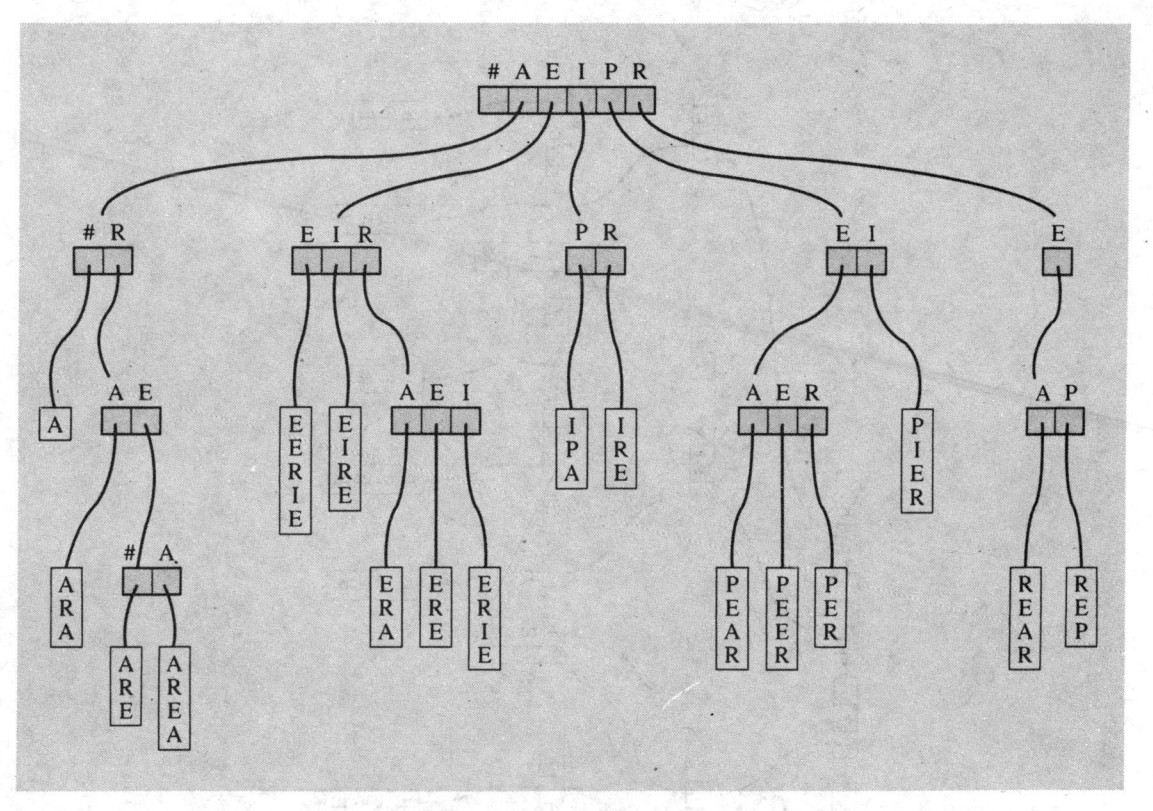

can be put on a linked list accessible from the parent node, as in Figure 7.38. One node of the previous trie corresponds now to a linked list. This means that random access of pointers stored in arrays is no longer possible, and the linked lists have to be scanned sequentially, although not exhaustively, because alphabetical order is most likely maintained. The space requirements are not insignificant either because each node now contains two pointers that may require two or four bytes, if not more, depending on the system.

Another way to reduce the space requirements is by changing the way words are tested (Rotwitt and Maine, 1971). A trie *a tergo* can be built in which the reverses of words are inserted. In our example, the number of nodes is about the same, but a trie *a tergo* representation for such words as "logged," "loggerhead," "loggia," and "logging" has leaves on the third level, not on the seventh, as in a forward trie. Admittedly, for some frequently used endings, such as "tion," "ism," and "ics," the problem reappears.

A variety of other orders can be considered, and checking every second character proved to be very useful (Bourne and Ford, 1961), but solving the problem of an optimal order cannot be solved in its generality, because the problem turns out to be extremely complex (Comer and Sethi, 1977).

FIGURE 7.38 The trie from Figure 7.37 implemented as a binary tree.

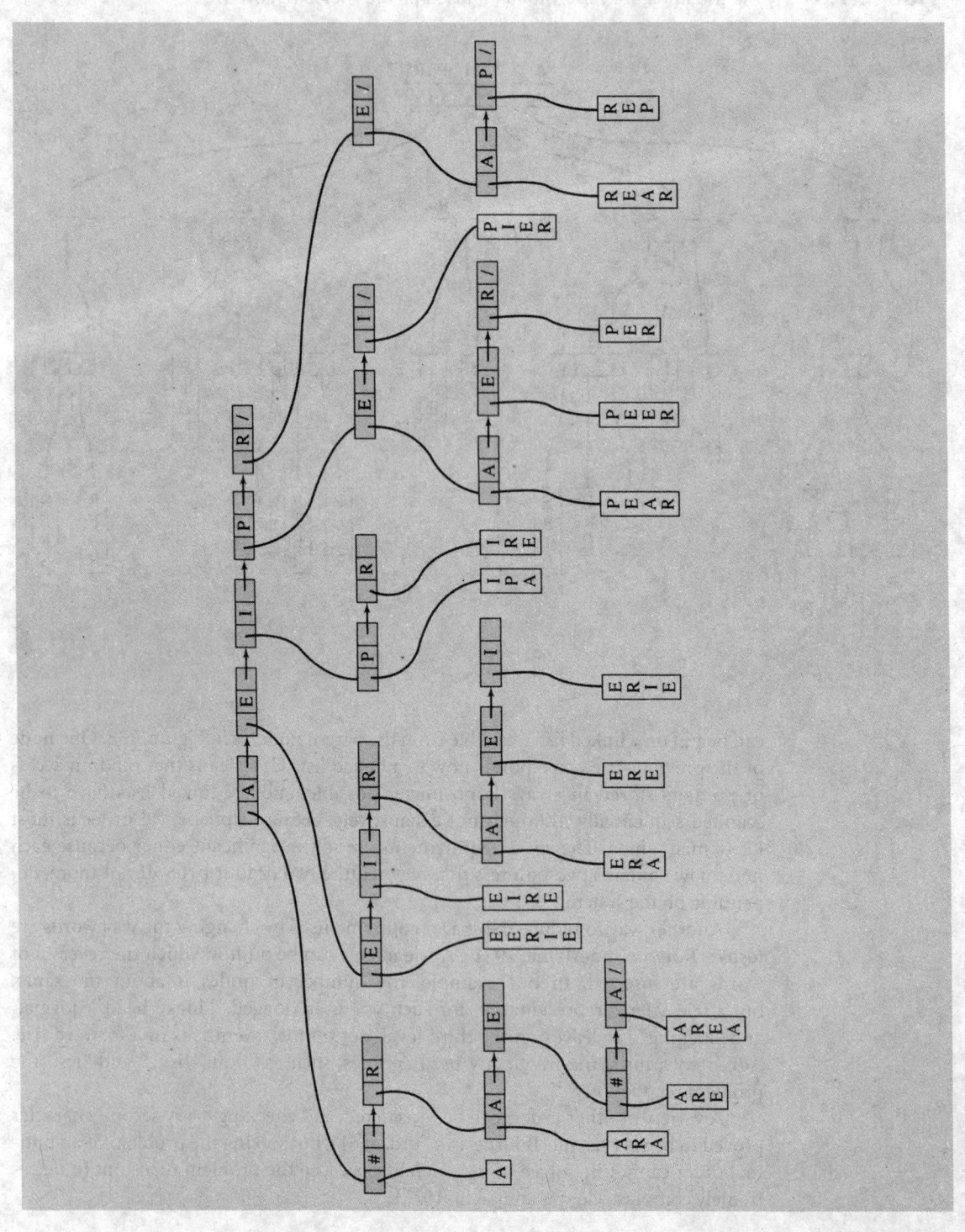

Another way to save space is to compress the trie. One method creates a large `cellArray` out of all the arrays in all nonleaf nodes by interleaving these arrays so that the pointers remain intact. The starting positions of these arrays are recorded in the encompassing `cellArray`. For example, the three nodes shown in Figure 7.39a containing pointers p_1 through p_7 to other nodes of the trie (including leaves) are put one by one into a `cellArray` in a nonconflicting way, as in Figure 7.39b. The problem is how to do that efficiently timewise and spacewise so that the algorithm is fast and the resulting array occupies substantially less space than all nonleaf nodes combined. In this example, all three nodes require $3 \cdot 6 = 18$ cells, and the `cellArray` has 11 cells, so the compression rate is $(18 - 11)/18$, 39%. However, if the cells are stored as in Figure 7.39c, the compression rate is $(18 - 10)/18$, 44%.

FIGURE 7.39 A part of a trie (a) before and (b) after compression using the `compressTrie()` algorithm and (c) after compressing it in an optimal way.

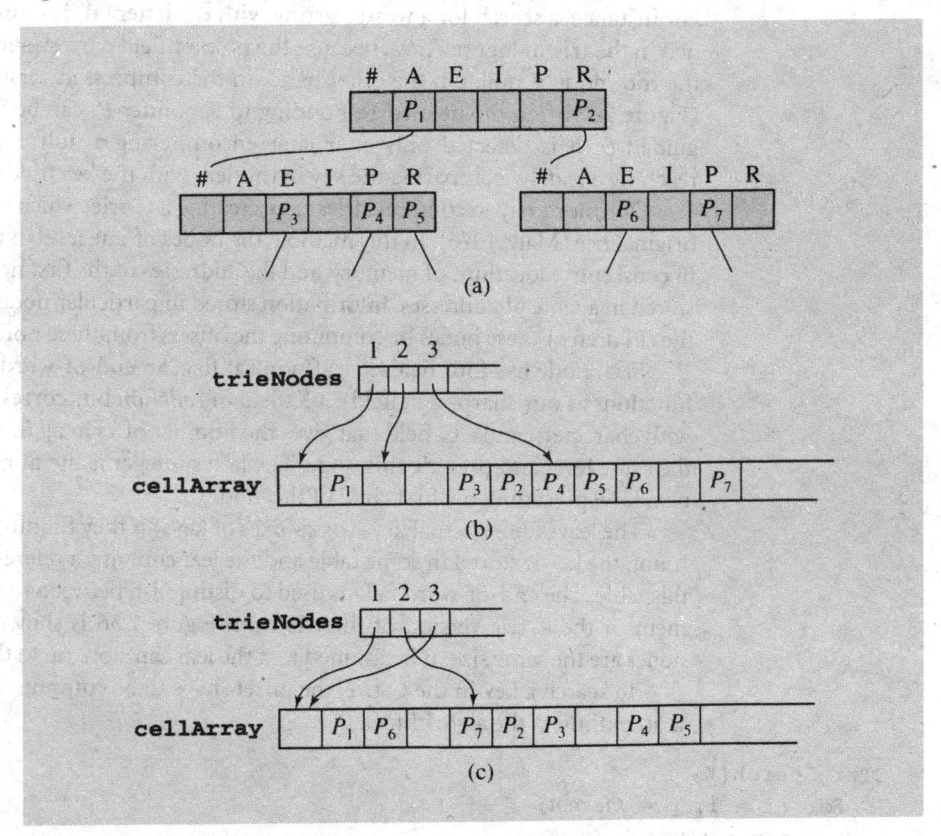

It turns out that the algorithm that compresses the trie is exponential in the number of nodes and inapplicable for large tries. Other algorithms may not render the optimal compression rate but are faster (cf. Al-Suwaiyel and Horowitz, 1984). One such algorithm is `compressTrie()`.

```
compressTrie()
    set to null all nodeNum*cellNum cells of cellArray;
    for each node
        for each position j of cellArray
            if after superimposing node on cellArray[j],...,cellArray[j+cellNum-1]
                no cell containing a pointer is superimposed on a cell with a pointer
                copy pointer cells from node to corresponding cells starting from cellArray[j];
                record j in trieNodes as the position of node in cellArray;
                break;
```

This is the algorithm that was applied to the trie in Figure 7.39a to render the arrays in Figure 7.39b. Searching such a compressed trie is similar to searching a regular trie. However, node accesses are mediated through the array trieNodes. If $node_1$ refers to $node_2$, the position of $node_2$ has to be found in this array, and then $node_2$ can be accessed in the cellArray.

The problem with using a compressed trie is that the search can lead us astray. For instance, a search for a word starting with the letter "P" is immediately discontinued in the trie in Figure 7.39a, because the pointer field corresponding to this letter in the root node is null. On the other hand, in the compressed version of the same trie (Figure 7.39b), in the field corresponding to P, pointer P_3 can be found. But the misguided path is detected only after later encountering a null pointer field or, after reaching a leaf, by comparing the key in this leaf with the key used in searching.

One more way to compress tries is by creating a C-trie, which is a bit version of the original trie (Maly, 1976). In this method, the nodes of one level of the C-trie are stored in consecutive locations of memory, and the addresses of the first nodes of each level are stored in a table of addresses. Information stored in particular nodes allows us to access the children of these nodes by computing the offsets from these nodes to their children.

Each node has four fields: a leaf/nonleaf flag, an end-of-word on/off field (which functions as our sharp-sign field), a K-field of *cellNum* bits corresponding to the cells with characters, and a C-field that gives the number of 1s in all the K-fields that are on the same level and precede this node. The latter integer is the number of nodes in the next level preceding the first child of this node.

The leaves store actual keys (or suffixes of keys) if they fit into the K-field+C-field. If not, the key is stored in some table and the leaf contains a reference to its position in this table. The end-of-word field is used to distinguish between these two cases. A fragment of the C-trie version of the trie from Figure 7.36 is shown in Figure 7.40. All nodes are the same size. It is assumed that the leaf can store up to three characters.

To search a key in the C-trie, the offsets have to be computed very carefully. Here is an outline of the algorithm:

```
CTrieSearch(K)
    for (i = 1, p = the root; ; i++)
        if p is a leaf
            if K is equal to the key(p)
                success;
            else failure;
        else if (K[i] == '\0')
            if end-of-word field is on
```

FIGURE 7.40 A fragment of the C-trie representation of the trie from Figure 7.36.

| 0 | 0 | 1 1 1 1 1 | 0 |

| 0 | 1 | 0 0 0 0 1 | 0 | 0 | 0 1 1 0 1 1 | 0 | 0 | 0 0 0 1 1 | 4 | 0 | 0 0 1 1 0 0 | 6 | 0 | 0 0 1 0 0 0 | 8 |

| 0 | 0 | 1 1 0 0 0 0 | 1 | 0 | | 1 | 0 | | 0 | 0 | 1 1 1 0 0 | 2 | 1 | 1 | I P A | 1 | 1 | I R E | 0 | 0 | 0 1 1 0 0 | 5 | ···

| E E R I E E I R E |

        success;
      else *failure;*
   else if *the bit corresponding to character* K[i] *is off*
      failure;
   else p = *address(the first node of level* i+1)
      +C-*field*(p)* *size(one node)*   // to skip all children of nodes
                                                  // in front of p on level i;
      +(*number of 1-bits in K-field*(p) *to the left of the bit* // to skip
       corresponding to K[i])* *size(one node)* // some children of p;

For example, to find "EERIE" in the C-trie in Figure 7.37, we first check in the root the bit corresponding to the first letter, "E." Because the bit is on and the root is not a leaf, we go to the second level. On the second level, the address of the node to be tested is determined by adding the address of the first node on this level to the length of one node, the first, in order to skip it. The bit of this nonleaf node corresponding to the second letter of our word, also an "E," is on, so we proceed to the third level. The address of the node to be tested is determined by adding the address of the first node of the third level to the size of one node (the first node of level three). We now access a leaf node with the end-of-word field set to 0. The table of words is accessed to make a comparison between the key looked for and the key in the table.

The compression is significant. One node of the original trie of 27 pointers of 2 bytes each occupies 54 bytes. One node of the C-trie requires 1 + 1 + 27 + 16 = 45 bits, which can be stored in 8 bytes. But it is not without a price. This algorithm requires putting nodes of one level tightly together, but storing one node at a time in memory by using new does not guarantee that the nodes are put in consecutive locations, especially in a multiuser environment. Therefore, the nodes from one level have to be generated first in temporary storage, and only then can a chunk of memory be requested that is large enough to accommodate all these nodes. This problem also indicates that the C-trie is ill-suited for dynamic updates. If the trie is generated only once, the C-trie is an excellent variation to be utilized. If, however, the trie needs to be frequently updated, this technique for trie compression should be abandoned.

7.3 Concluding Remarks

The survey of multiway trees in this chapter is by no means exhaustive. The number of different types of multiway trees is very large. My intention is to highlight the variety of uses to which these trees can be applied and show how the same type of tree can be applied to different areas. Of particular interest is a B-tree with all its variations. B^+-trees are commonly used in the implementation of indexes in today's relational databases. They allow very fast random access to the data, and they also allow fast sequential processing of the data.

The application of B-trees is not limited to processing information from secondary storage, although it was the original motivation in introducing these trees. A variant of B-trees, 2–4 trees, although unsuitable for processing information in secondary storage, turns out to be very useful in processing information in memory.

Also of particular use are tries, a different type of tree. With many variations, they have a vast scope of applications, and our case study illustrates one very useful application of tries.

7.4 Case Study: Spell Checker

An indispensable utility for any word processor is a spell checker, which allows the user to find as many spelling mistakes as possible. Depending on the sophistication of the spell checker, the user may even see possible corrections. Spell checkers are used mostly in an interactive environment; the user can invoke them at any time when using the word processor, make corrections on the fly, and exit even before processing the entire file. This requires writing a word processing program and, in addition to it, a spell checker module. This case study focuses on the use of tries. Therefore, the spell checker will be a stand-alone program to be used outside a word processor. It will process a text file in batch mode, not allowing word-by-word corrections after possible errors are detected.

The core of a spell checker is a data structure allowing efficient access to words in a dictionary. Such a dictionary most likely has thousands of words, so access has to be very fast to process a text file in a reasonable amount of time. Out of many possible data structures, the trie is chosen to store the dictionary words. The trie is first created after the spell checker is invoked, and afterward, the actual spell checking takes place.

For a large number of dictionary words, the size of the trie is very important because it should reside in main memory without recourse to virtual memory. But as we have already observed in this chapter, tries with fixed length nodes, as in Figure 7.36, are too wasteful. In most cases, only a fraction of the positions in each node is utilized, and the further away from the root, the smaller this fraction becomes (the root may be the only node with 26 children). Creating linked lists corresponding to all utilized letters for each node reduces wasted space, as in Figure 7.38. This approach has two disadvantages: The space required for the pointer fields can be substantial, and the linked lists force us to use sequential search. An improvement over the last solution is to reserve only as much space as required by the letters used by each node without resorting to the use of linked lists. We could use vectors, but to illustrate how flexible arrays are implemented, the program, when a need arises, substitutes larger arrays for existing arrays, copying the content of the old arrays to the new, and returning the old arrays to the operating system.

The key to the use of such pseudoflexible arrays is the implementation of a node. A node is an object that includes the following members: a leaf/nonleaf flag, an end-of-word flag, a pointer to a string, and a pointer to an array of pointers to structures of the same category. Figure 7.41 contains the trie utilizing the nodes of this structure. If a string attached to a certain node has to be extended, a new string is created that contains the contents of the old string with a new letter inserted into the proper position, a function performed by addCell(). The letters in each node are kept in alphabetical order.

The function insert() is an implementation of the algorithm trieInsert() discussed earlier in this chapter. Because the position of each letter may vary from one node to another, this position has to be determined each time, a function performed by position(). Should a letter be absent in a node, position() returns –1, which allows insert() to undertake the proper action.

Also, the discussion of tries in this chapter assumed that the leaves of the tries store full keys. This is not necessary because the prefixes of all words are implicitly stored in the trie and can be reconstructed by garnering all the letters on the path leading to the leaf. For example, to access the leaf with the word "ERIE," two nonleaves have to be passed through pointers corresponding to the letters "E" and "R." Therefore, it is enough to store the suffix "IE" in the leaf instead of the entire word "ERIE." By doing this, only 13 letters of suffixes of these words have to be retained in these leaves out of the 58 letters stored in all leaves of the trie in Figure 7.36, a substantial improvement.

We also included the function printTrie(), which prints the content of a trie sideways. The output generated by this function when applied to the trie in Figure 7.41 is as follows:

```
        >>REP|
        >>REA|R
    >>PI|ER
        >>PER|
        >>PEE|R
        >>PEA|R
    >>IR|E
    >>IP|A
        >>ERI|E
        >>ERE|
        >>ERA|
    >>EI|RE
    >>EE|RIE
           >>AREA|
           >>>ARE
        >>ARA|
    >>>A
```

360 ■ Chapter 7 Multiway Trees

FIGURE 7.41 An implementation of a trie that uses pseudoflexible arrays. The trie has the same words as the trie in Figure 7.36.

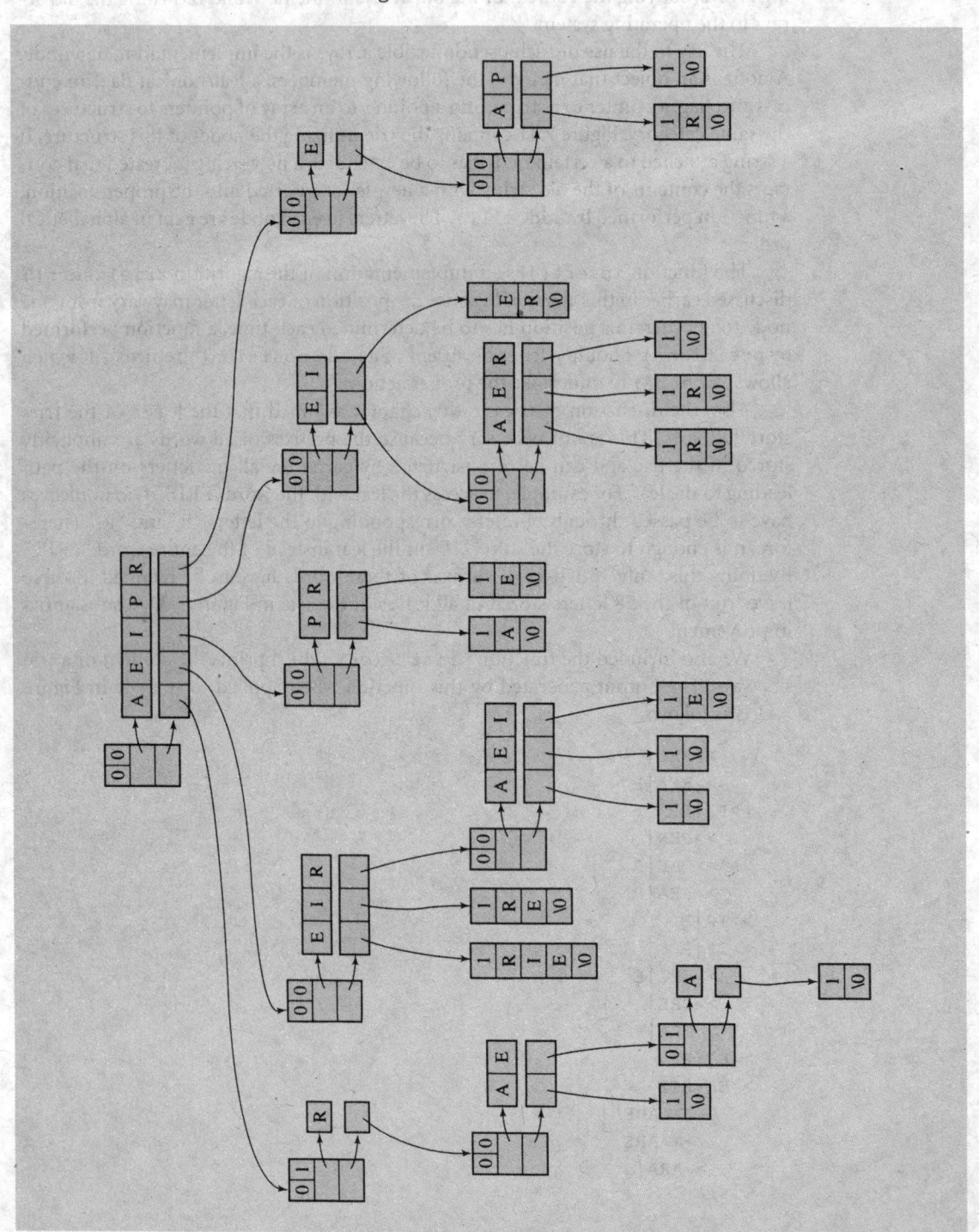

Three angle brackets indicate words for which the endOfWord flag has been set in the corresponding node. The words with two angle brackets have leaves in the trie. Sometimes these leaves contain only the character '\0'. The vertical bar separates the prefix reconstructed when scanning the trie from the suffix that was extracted from a leaf.

Spell checking works in a straightforward fashion by examining each word of a text file and printing out all misspelled words along with the line numbers where the misspelled words are found. Figure 7.42 contains the complete code of the spell checker.

FIGURE 7.42 Implementation of a spell checker using tries.

```
//*********************** trie.h ********************************
class Trie;

class TrieNonLeafNode {
public:
    TrieNonLeafNode() {
    }
    TrieNonLeafNode(char);
private:
    bool leaf, endOfWord;
    char *letters;
    TrieNonLeafNode **ptrs;
    friend class Trie;
};

class TrieLeafNode {
public:
    TrieLeafNode() {
    }
    TrieLeafNode(char*);
private:
    bool leaf;
    char *word;
    friend class Trie;
};

class Trie {
public:
    Trie() : notFound(-1) {
    }
    Trie(char*);
```

Continues

FIGURE 7.42 *(continued)*

```cpp
    void printTrie() {
        *prefix = '\0';
        printTrie(0,root,prefix);
    }
    void insert(char*);
    bool wordFound(char*);
private:
    TrieNonLeafNode *root;
    const int notFound;
    char prefix[80];
    int  position(TrieNonLeafNode*,char);
    void addCell(char,TrieNonLeafNode*,int);
    void createLeaf(char,char*,TrieNonLeafNode*);
    void printTrie(int,TrieNonLeafNode*,char*);
};

//*********************** trie.cpp *******************************

#include <iostream>
#include <cstring>
#include <cstdlib>
#include "trie.h"
using namespace std;

TrieLeafNode::TrieLeafNode(char *suffix) {
    leaf = true;
    word = new char[strlen(suffix)+1];
    if (word == 0) {
        cerr << "Out of memory2.\n";
        exit(-1);
    }
    strcpy(word,suffix);
}

TrieNonLeafNode::TrieNonLeafNode(char ch) {
    ptrs = new TrieNonLeafNode*;
    letters = new char[2];
    if (ptrs == 0 || letters == 0) {
        cerr << "Out of memory3.\n";
        exit(1);
    }
    leaf = false;
```

FIGURE 7.42 (continued)

```
        endOfWord = false;
        *ptrs = 0;
        *letters = ch;
        *(letters+1) = '\0';
    }

Trie::Trie(char* word) : notFound(-1) {
    root = new TrieNonLeafNode(*word); // initialize the root
    createLeaf(*word,word+1,root);     // to avoid later tests;
}

void Trie::printTrie(int depth, TrieNonLeafNode *p, char *prefix) {
    register int i;              // assumption: the root is not a leaf
    if (p->leaf) {               // and it is not null;
        TrieLeafNode *lf = (TrieLeafNode*) p;
        for (i = 1; i <= depth; i++)
            cout << "  ";
        cout << " >>" << prefix << "|" << lf->word << endl;
    }
    else {
        for (i = strlen(p->letters)-1; i >= 0; i--)
            if (p->ptrs[i] != 0) {                     // add the letter
                prefix[depth] = p->letters[i]; // corresponding to
                prefix[depth+1] = '\0';                // position i to prefix;
                printTrie(depth+1,p->ptrs[i],prefix);
            }
        if (p->endOfWord) {
            prefix[depth] = '\0';
            for (i = 1; i <= depth+1; i++)
                cout < "  ";
            cout << ">>>" << prefix << "\n";
        }
    }
}

int Trie::position(TrieNonLeafNode *p, char ch) {
    for (int i = 0; i < strlen(p->letters) && p->letters[i] != ch; i++);
    if (i < strlen(p->letters))
        return i;
    else return notFound;
}
```

Continues

FIGURE 7.42   *(continued)*

```
bool Trie::wordFound (char *word) {
    TrieNonLeafNode *p = root;
    TrieLeafNode *lf;
    int pos;
    while (true)
        if (p->leaf) {                              // node p is a leaf
            lf = (TrieLeafNode*) p;                 // where the matching
            if (strcmp(word,lf->word) == 0)         // suffix of word
                return true;                        // should be found;
            else return false;
        }
        else if (*word == '\0')                     // the end of word has
            if (p->endOfWord)                       // to correspond with
                return true;                        // the endOfWord marker
            else return false;                      // in node p set to true;
        else if ((pos = position(p,*word)) != notFound &&
                 p->ptrs[pos] != 0) {               // continue
            p = p->ptrs[pos];                       // path, if possible,
            word++;
        }
        else return false;                          // otherwise failure;
}

void Trie::addCell(char ch, TrieNonLeafNode *p, int stop) {
    int i, len = strlen(p->letters);
    char *s = p->letters;
    TrieNonLeafNode **tmp = p->ptrs;
    p->letters = new char[len+2];
    p->ptrs    = new TrieNonLeafNode*[len+1];
    if (p->letters == 0 || p->ptrs == 0) {
        cerr << "Out of memory1.\n";
        exit(1);
    }
    for (i = 0; i < len+1; i++)
        p->ptrs[i] = 0;
    if (stop < len)                 // if ch does not follow all letters in p,
        for (i = len; i >= stop+1; i--) {    // copy from tmp letters > ch;
            p->ptrs[i]    = tmp[i-1];
            p->letters[i] = s[i-1];
        }
```

FIGURE 7.42 (continued)

```
        p->letters[stop] = ch;
        for (i = stop-1; i >= 0; i--) {          // and letters < ch;
            p->ptrs[i]    = tmp[i];
            p->letters[i] = s[i];
        }
        p->letters[len+1] = '\0';
        delete [] s;
    }
}

void Trie::createLeaf(char ch, char *suffix, TrieNonLeafNode *p) {
    int pos = position(p,ch);
    if (pos == notFound) {
        for (pos = 0; pos < strlen(p->letters) &&
                      p->letters[pos] < ch; pos++);
        addCell(ch,p,pos);
    }
    p->ptrs[pos] = (TrieNonLeafNode*) new TrieLeafNode(suffix);
}

void Trie::insert (char *word) {
    TrieNonLeafNode *p = root;
    TrieLeafNode *lf;
    int offset, pos;
    char *hold = word;
    while (true) {
        if (*word == '\0') {                   // if the end of word reached,
            if (p->endOfWord)
                cout << "Duplicate entry1 " << hold << endl;
            else p->endOfWord = true;          // set endOfWord to true;
            return;
        }                                      // if position in p indicated
        pos = position(p,*word);
        if (pos == notFound) {                 // by the first letter of word
            createLeaf(*word,word+1,p);        // does not exist, create
            return;                            // a leaf and store in it the
        }                                      // unprocessed suffix of word;
        else if (pos != notFound &&            // if position *word is
                 p->ptrs[pos]->leaf) {         // occupied by a leaf,
            lf = (TrieLeafNode*) p->ptrs[pos]; // hold this leaf;
            if (strcmp(lf->word,word+1) == 0) {
```

Continues

FIGURE 7.42 *(continued)*

```
            cout << "Duplicate entry2 " << hold << endl;
            return;
    }
    offset = 0;
    // create as many non-leaves as the length of identical
    // prefix of word and the string in the leaf (for cell 'R',
    // leaf 'EP', and word 'REAR', two such nodes are created);
    do {
        pos = position(p,word[offset]);
        // word == "ABC", leaf = "ABCDEF" => leaf = "DEF";
        if (strlen(word) == offset+1) {
            p->ptrs[pos] = new TrieNonLeafNode(word[offset]);
            p->ptrs[pos]->endOfWord = true;
            createLeaf(lf->word[offset],lf->word + offset+1,
                                    p->ptrs[pos]);
            return;
        }
        // word == "ABCDE", leaf = "ABC" => leaf = "DEF";
        else if (strlen(lf->word) == offset) {
            p->ptrs[pos] = new TrieNonLeafNode(word[offset+1]);
            p->ptrs[pos]- >endOfWord = true;
            createLeaf(word[offset+1],word+offset+2,
                                    p->ptrs[pos]);
            return;
        }
        p->ptrs[pos] = new TrieNonLeafNode(word[offset+1]);
        p = p->ptrs[pos];
        offset++;
    } while (word[offset] == lf->word[offset-1]);
    offset--;
    // word = "ABCDEF", leaf = "ABCPQR" =>
    //    leaf('D') = "EF", leaf('P') = "QR";
    // check whether there is a suffix left:
    // word = "ABCD", leaf = "ABCPQR" =>
    //    leaf('D') = null, leaf('P') = "QR";
    char *s = "";
    if (strlen(word) > offset+2)
        s = word+offset+2;
    createLeaf(word[offset+1],s,p);
    // check whether there is a suffix left:
```

FIGURE 7.42   (continued)

```cpp
                // word = "ABCDEF", leaf = "ABCP" =>
                //    leaf('D') = "EF", leaf('P') = null;
                if (strlen(lf->word) > offset+1)
                     s = lf->word+offset+1;
                else s = "";
                createLeaf(lf->word[offset],s,p);
                delete [] lf->word;
                delete lf;
                return;
           }
           else {
                p = p->ptrs[pos];
                word++;
           }
       }
   }
}

//*********************** spellCheck.cpp **************************

#include <iostream>
#include <fstream>
#include <cstdlib>
#include <cstring>
#include <cctype>
#include "trie.h"
using namespace std;

char* strupr(char *s) {
    for (char *ss = s; *s = toupper(*s); s++);
    return ss;
}

int main(int argc, char* argv[])  {
    char fileName[25], s[80], ch;
    int i, lineNum = 1;
    ifstream dictionary("dictionary");
    if (dictionary.fail()) {
       cerr << "Cannot open 'dictionary'\n";
       exit(-1);
    }
```

Continues

FIGURE 7.42 (continued)

```cpp
    dictionary >> s;
    Trie trie(strupr(s));   // initialize root;
    while (dictionary >> s) // initialize trie;
        trie.insert(strupr(s));
    trie.printTrie();
    if (argc != 2) {
        cout << "Enter a file name: ";
        cin  >> fileName;
    }
    else strcpy(fileName,argv[1]);
    ifstream textFile(fileName);
    if (textFile.fail()) {
      cout << "Cannot open " << fileName << endl;
      exit(-1);
    }
    cout << "Misspelled words:\n";
    textFile.get(ch);
    while (!textFile.eof()) {
        while (true)
            if (!textFile.eof() && !isalpha(ch)) { // skip non-letters
                if (ch == '\n')
                    lineNum++;
                textFile.get(ch);
            }
            else break;
        if (textFile.eof())        // spaces at the end of textFile;
            break;
        for (i = 0; !textFile.eof() && isalpha(ch); i++) {
            s[i] = toupper(ch);
            textFile.get(ch);
        }
        s[i] = '\0';
        if (!trie.wordFound(s))
            cout << s << " on line " << lineNum << endl;
    }
    dictionary.close();
    textFile.close();
    return 0;
}
```

7.5 EXERCISES

1. What is the maximum number of nodes in a multiway tree of height h?

2. How many keys can a B-tree of order m and of height h hold?

3. Write a function that prints out the contents of a B-tree in ascending order.

4. The root of a B*-tree requires special attention because it has no sibling. A split does not render two nodes two-thirds full plus a new root with one key. Suggest some solutions to this problem.

5. Are B-trees immune to the order of the incoming data? Construct B-trees of order 3 (two keys per node) first for the sequence 1, 5, 3, 2, 4 and then for the sequence 1, 2, 3, 4, 5. Is it better to initialize B-trees with ordered data or with data in random order?

6. Draw all 10 different B-trees of order 3 that can store 15 keys and make a table that for each of these trees shows the number of nodes and the average number of visited nodes (Rosenberg and Snyder, 1981). What generalization can you make about them? Would this table indicate that (a) the smaller the number of nodes, the smaller the average number of visited nodes and (b) the smaller the average number of visited nodes, the smaller the number of nodes? What characteristics of the B-tree should we concentrate on to make them more efficient?

7. In all our considerations concerning B-trees, we assumed that the keys are unique. However, this does not have to be the case because multiple occurrences of the same key in a B-tree do not violate the B-tree property. If these keys refer to different objects in the data file (e.g., if the key is a name, and many people can have the same name), how would you implement such data file references?

8. What is the maximum height of a B$^+$-tree with n keys?

9. Occasionally, in a simple prefix B$^+$-tree, a separator can be as large as a key in a leaf. For example, if the last key in one leaf is "Herman" and the first key in the next leaf is "Hermann," then "Hermann" must be chosen as a separator in the parent of these leaves. Suggest a procedure to enforce the shorter separator.

10. Write a function that determines the shortest separator for two keys in a simple prefix B$^+$-tree.

11. Is it a good idea to use abbreviated forms of prefixes in the leaves of prefix B$^+$-trees?

12. If in two different positions, i and j, $i < j$, of a leaf in a bit-tree two D-bits are found such that $D_j = D_i$, what is the condition on at least one of the D-bits D_k for $i < k < j$?

13. If key K_i is deleted from a leaf of a bit-tree, then the D-bit between K_{i-1} and K_{i+1} has to be modified. What is the value of this D-bit if the values D_i and D_{i+1} are known? Make deletions in the leaf in Figure 7.17 to make an educated guess and then generalize this observation. In making a generalization, consider two cases: (a) $D_i < D_{i+1}$ and (b) $D_i > D_{i+1}$.

14. Write an algorithm that, for an R-tree, finds all entries in the leaves whose rectangles overlap a search rectangle R.

15. In the discussion of B-trees, which are comparable in efficiency to binary search trees, why are only B-trees of small order used and not B-trees of large order?
16. What is the worst case of inserting a key into a 2–4 tree?
17. What is the complexity of the `compressTrie()` algorithm in the worst case?
18. Can the leaves of the trie compressed with `compressTrie()` still have abbreviated versions of the words, namely, parts that are not included in the nonterminal nodes?
19. In the examples of tries analyzed in this chapter, we dealt with only 26 capital letters. A more realistic setting includes lowercase letters as well. However, some words require a capital letter at the beginning (names), and some require the entire word to be capitalized (acronyms). How can we solve this problem without including both lowercase and capital letters in the nodes?
20. A variant of a trie is a *digital tree,* which processes information on the level of bits. Because there are only two bits, only two outcomes are possible. Digital trees are binary. For example, to test whether the word "BOOK" is in the tree, we do not use the first letter, "B," in the root to determine to which of its children we should go, but the first bit, 0, of the first letter (ASCII(B) = 01000010), on the second level, the second bit, and so on before we get to the second letter. Is it a good idea to use a digital tree for a spell checking program, as was discussed in the case study?

7.6 PROGRAMMING ASSIGNMENTS

1. Extend our spell checking program to suggest the proper spelling of a misspelled word. Consider these types of misspellings: changing the order of letters (copmuter), omitting a letter (computr), adding a letter (compueter), dittography, i.e., repeating a letter (computter), and changing a letter (compurer). For example, if the letter i is exchanged with the letter $i + 1$, then the level i of the trie should be processed before level $i + 1$.

2. A *point quadtree* is a 4-way tree used to represent points on a plane (Samet 1989). A node contains a pair of coordinates (*latitude,longitude*) and pointers to four children that represent four quadrants, NW, NE, SW, and SE. These quadrants are generated by the intersection of the vertical and horizontal lines passing through point (*lat,lon*) of the plane. Write a program that accepts the names of cities and their geographical locations (*lat,lon*) and inserts them into the quadtree. Then, the program should give the names of all cities located within distance r from a location (*lat,lon*) or, alternatively, within distance r from a city C.

    Figure 7.43 contains an example. Locations on the map in Figure 7.43a are inserted into the quadtree in Figure 7.43b in the order indicated by the encircled numbers shown next to the city names. For instance, when inserting Pittsburgh into the quadtree, we check in which direction it is with respect to the root. The root stores the coordinates of Louisville, and Pittsburgh is NE from it; that is, it belongs to the second child of the root. But this child already stores a city, Washington. Therefore, we ask the same question concerning Pittsburgh with respect to the current node, the second child of the root: In which direction with respect to this city is Pittsburgh?

FIGURE 7.43 A map indicating (a) coordinates of some cities and (b) a quadtree containing the same cities.

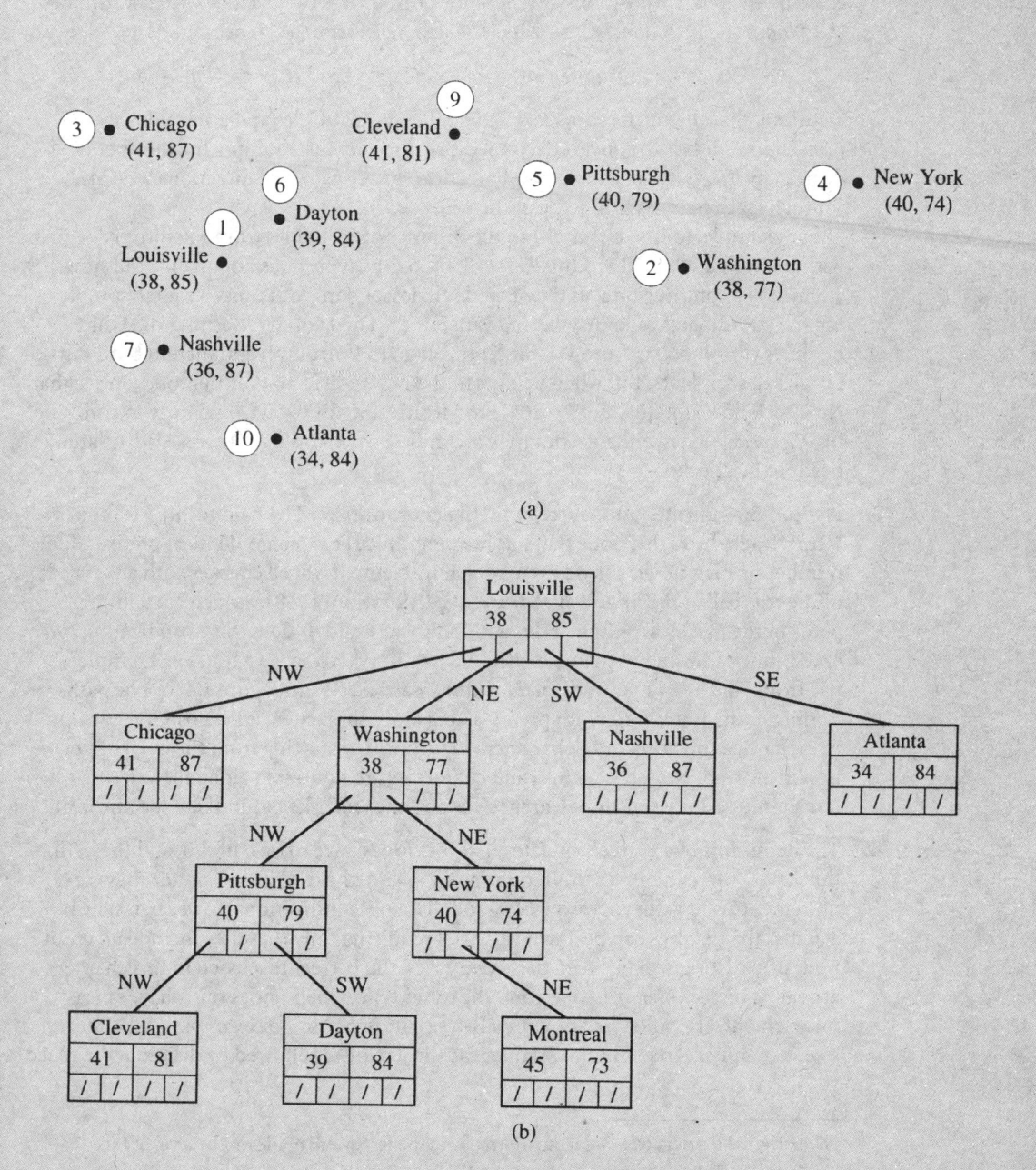

This time the answer is NW. Therefore, we go to the first child of the current node. The child is a null node, and therefore, the Pittsburgh node can be inserted here.

The problem is to not do an exhaustive search of the quadtree. So, if we are after cities within a radius r from a city C, then, for a particular node nd you find the distance between C and the city represented by nd. If the distance is within r, you have to continue to all four descendants of nd. If not, you continue to the descendants indicated by the relative positions. To measure a distance between cities with coordinates (lat_1, lon_1) and (lat_2, lon_2), the great circle distance formula can be used:

$$d = R \arccos(\sin(lat_1) \cdot \sin(lat_2) + \cos(lat_1) \cdot \cos(lat_2) \cdot \cos(lon_2 - lon_1))$$

assuming that the earth radius $R = 3{,}956$ miles and latitudes and longitudes are expressed in radians (to convert decimal degrees to radians, multiply the number of degrees by $\pi/180 = 0.017453293$ radians/degree). Also, for the directions west and south, negative angles should be used.

For example, to find cities within the distance of 200 miles from Pittsburgh, begin with the root and $d((38,85),(40,79)) = 350$, so Louisville does not qualify, but now you need to continue only in the SE and NE descendants of Louisville after comparing the coordinates of Louisville and Pittsburgh. Then you try Washington, which qualifies ($d = 175$), so, from Washington you go to Pittsburgh and then to both Pittsburgh's descendants. But when you get to the NE node from Washington, you see that New York does not qualify ($d = 264$), and from New York you would have to continue in SW and NW descendants, but they are null, so you stop right there. Also, Atlanta needs to be checked.

3. Figure 7.36 indicates one source of inefficiency for tries: The path to "REAR" and "REP" leads through a node that has just one child. For longer identical prefixes, the number of such nodes can be even longer. Implement a spell checker with a variation of the trie, called the *multiway Patricia tree* (Morrison, 1968),[3] which curtails the paths in the trie by avoiding nodes with only one child. It does this by indicating for each branch how many characters should be skipped to make a test. For example, a trie from Figure 7.44a is transformed into a Patricia tree in Figure 7.44b. The paths leading to the four words with prefix "LOGG" are shortened at the cost of recording in each node the number of characters to be omitted starting from the current position in a string. Now, because certain characters are not tested along the way, the final test should be between a key searched for and the *entire* key found in a specific leaf.

4. The definition of a B-tree stipulates that the nodes have to be half full, and the definition of a B*-tree increases this requirement to two-thirds. The reason for these requirements is to achieve reasonably good disk space utilization. However, it may be claimed that B-trees can perform very well requiring only that they include no empty nodes. To distinguish between these two cases, the B-trees discussed in this chapter are called *merge-at-half* B-trees, and the other type, when nodes must have at least one element, are called *free-at-empty* B-trees. It turns out, for example, that after a free-at-empty B-tree is built and then each insertion is followed by deletion, the space

---

[3] The original Patricia tree was a binary tree, and the tests were made on the level of bits.

FIGURE 7.44 (a) A trie with words having long identical prefixes and (b) a Patricia tree with the same words.

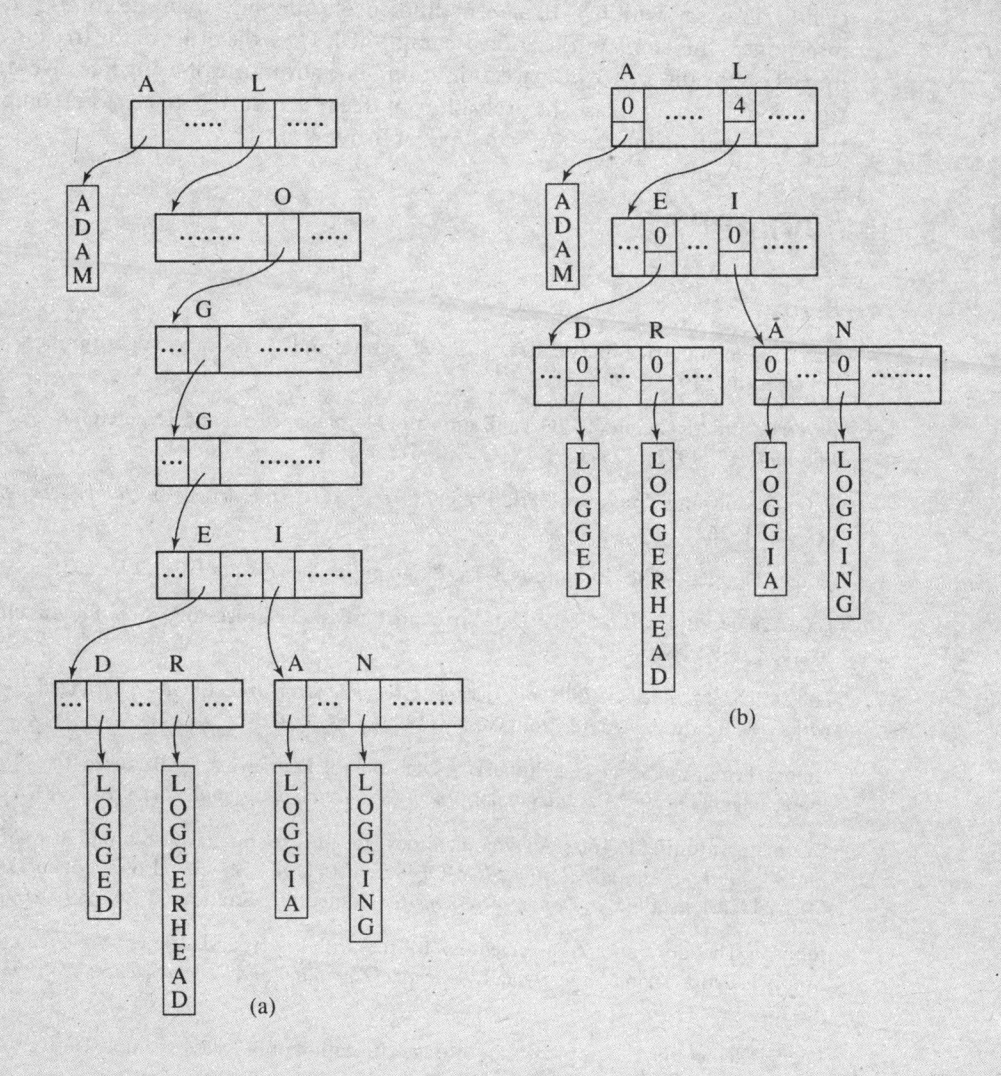

utilization is about 39% (Johnson and Shasha 1993), which is not bad considering the fact that this type of tree can have very small space utilization ($\frac{1}{m}$% for a tree of order m), whereas a merge-at-half B-tree has at least 50% utilization. Therefore, it may be expected that if the number of insertions outweighs the number of deletions, the gap between merge-at-half and free-at-empty B-trees will be bridged. Write a simulation program to check this contention. First build a large B-tree, and then run a simulation for this tree treating it as a merge-at-half B-tree and then as a free-at-empty B-tree for different ratios of number i of insertions to number d of deletions so that $\frac{i}{d} \geq 1$; that is, the number of insertions is not less than the number of deletions (the case when deletions outweigh insertions is not interesting, because eventually the tree disappears). Compare the space utilization for these different cases. For what ratio $\frac{i}{d}$ is the space utilization between these two types of B-trees sufficiently close (say, within 5–10 percent difference)? After how many deletions and insertions is this similar utilization accomplished? Does the order of the tree have an impact on the difference of space utilization? One advantage of using free-at-empty trees would be to decrease the probability of tree restructuring. In all cases, compare the tree restructuring rate for both types of B-trees.

BIBLIOGRAPHY

B-Trees

Bayer, R., "Symmetric Binary B-Trees: Data Structures and Maintenance Algorithms," *Acta Informatica* 1 (1972), 290–306.

Bayer, R., and McCreight, E., "Organization and Maintenance of Large Ordered Indexes," *Acta Informatica* 1 (1972), 173–189.

Bayer, Rudolf, and Unterauer, Karl, "Prefix B-Trees," *ACM Transactions on Database Systems* 2 (1977), 11–26.

Comer, Douglas, "The Ubiquitous B-Tree," *Computing Surveys* 11 (1979), 121–137.

Ferguson, David E., "Bit-Tree: A Data Structure for Fast File Processing," *Communications of the ACM* 35 (1992), No. 6, 114–120.

Folk, Michael J., Zoellick, Bill, and Riccardi, Greg, *File Structures: An Object-Oriented Approach with C++*, Reading, MA: Addison-Wesley (1998), Chs. 9, 10.

Guibas, Leo J., and Sedgewick, Robert, "A Dichromatic Framework for Balanced Trees," *Proceedings of the 19th Annual Symposium on Foundation of Computer Science* (1978), 8–21.

Guttman, Antonin, "R-Trees: A Dynamic Index Structure for Spatial Searching," *ACM SIGMOD '84 Proc. of Annual Meeting, SIGMOD Record* 14 (1984), 47–57 [also in Stonebraker, Michael (ed.), *Readings in Database Systems*, San Mateo, CA: Kaufmann, 1988, 599–609].

Johnson, Theodore, and Shasha, Dennis, "B-Trees with Inserts and Deletes: Why Free-at-Empty Is Better Than Merge-at-Half," *Journal of Computer and System Sciences* 47 (1993), 45–76.

Leung, Clement H. C., "Approximate Storage Utilization of B-Trees: A Simple Derivation and Generalizations," *Information Processing Letters* 19 (1984), 199–201.

McCreight, Edward M., "Pagination of B*-Trees with Variable-Length Records," *Communications of the ACM* 20 (1977), 670–674.

Rosenberg, Arnold L., and Snyder, Lawrence, "Time- and Space-Optimality in B-Trees," *ACM Transactions on Database Systems* 6 (1981), 174–193.

Sedgewick, Robert, *Algorithms in C*, Boston, MA: Addison-Wesley (1998), Ch. 13.

Sellis, Timos, Roussopoulos, Nick, and Faloutsos, Christos, "The R$^+$-Tree: A Dynamic Index for Multi-Dimensional Objects," *Proceedings of the 13th Conference on Very Large Databases* (1987), 507–518.

Stonebraker, M., Sellis, T., and Hanson, E., "Analysis of Rule Indexing Implementations in Data Base Systems," *Proceedings of the First International Conference on Expert Database Systems*, Charleston, SC (1986), 353–364.

Wedekind, H., "On the Selection of Access Paths in a Data Base System," in Klimbie, J. W., and Koffeman, K. L. (eds.), *Data Base Management*, Amsterdam: North-Holland (1974), 385–397.

Yao, Andrew Chi-Chih, "On Random 2–3 Trees," *Acta Informatica* 9 (1978), 159–170.

Tries

Bourne, Charles P., and Ford, Donald F., "A Study of Methods for Systematically Abbreviating English Words and Names," *Journal of the ACM* 8 (1961), 538–552.

Briandais, Rene de la, "File Searching Using Variable Length Keys," *Proceedings of the Western Joint Computer Conference* (1959), 295–298.

Comer, Douglas, and Sethi, Ravi, "The Complexity of Trie Index Construction," *Journal of the ACM* 24 (1977), 428–440.

Fredkin, Edward, "Trie Memory," *Communications of the ACM* 3 (1960), 490–499.

Maly, Kurt, "Compressed Tries," *Communications of the ACM* 19 (1976), 409–415.

Morrison, Donald R., "Patricia Trees," *Journal of the ACM* 15 (1968), 514–534.

Rotwitt, T., and de Maine, P. A. D., "Storage Optimization of Tree Structured Files Representing Descriptor Sets," *Proceedings of the ACM SIGFIDET Workshop on Data Description, Access and Control*, New York (1971), 207–217.

Al-Suwaiyel, M., and Horowitz, E., "Algorithms for Trie Compaction," *ACM Transactions on Database Systems* 9 (1984), 243–263.

Quadtrees

Finkel, R. A., and Bentley, J. L., "Quad Trees: A Data Structure for Retrieval on Composite Keys," *Acta Informatica* 4 (1974), 1–9.

Samet, Hanan, *The Design and Analysis of Spatial Data Structures*, Reading, MA: Addison-Wesley, 1989.

8 Graphs

In spite of the flexibility of trees and the many different tree applications, trees, by their nature, have one limitation, namely, they can only represent relations of a hierarchical type, such as relations between parent and child. Other relations are only represented indirectly, such as the relation of being a sibling. A generalization of a tree, a *graph*, is a data structure in which this limitation is lifted. Intuitively, a graph is a collection of vertices (or nodes) and the connections between them. Generally, no restriction is imposed on the number of vertices in the graph or on the number of connections one vertex can have to other vertices. Figure 8.1 contains examples of graphs. Graphs are versatile data structures that can represent a large number of different situations and events from diverse domains. Graph theory has grown into a sophisticated area of mathematics and computer science in the last 200 years since it was first studied. Many results are of theoretical interest, but in this chapter, some selected results of interest to computer scientists are presented. Before discussing different algorithms and their applications, several definitions need to be introduced.

A *simple graph* $G = (V, E)$ consists of a nonempty set V of *vertices* and a possibly empty set E of *edges*, each edge being a set of two vertices from V. The number of vertices and edges is denoted by $|V|$ and $|E|$, respectively. A *directed graph*, or a *digraph*, $G = (V, E)$ consists of a nonempty set V of vertices and a set E of edges (also called *arcs*), where each edge is a pair of vertices from V. The difference is that one edge of a simple graph is of the form $\{v_i, v_j\}$, and for such an edge, $\{v_i, v_j\} = \{v_j, v_i\}$. In a digraph, each edge is of the form (v_i, v_j), and in this case, $(v_i, v_j) \neq (v_j, v_i)$. Unless necessary, this distinction in notation will be disregarded, and an edge between vertices v_i and v_j will be referred to as $edge(v_i, v_j)$.

These definitions are restrictive in that they do not allow for two vertices to have more than one edge. A *multigraph* is a graph in which two vertices can be joined by multiple edges. Geometric interpretation is very simple (see Figure 8.1e). Formally, the definition is as follows: A multigraph $G = (V, E, f)$ is composed of a set of vertices V, a set of edges E, and a function $f: E \rightarrow \{\{v_i, v_j\} : v_i, v_j \in V \text{ and } v_i \neq v_j\}$. A *pseudograph* is

FIGURE 8.1 Examples of graphs: (a–d) simple graphs; (c) a complete graph K_4; (e) a multigraph; (f) a pseudograph; (g) a circuit in a digraph; (h) a cycle in the digraph.

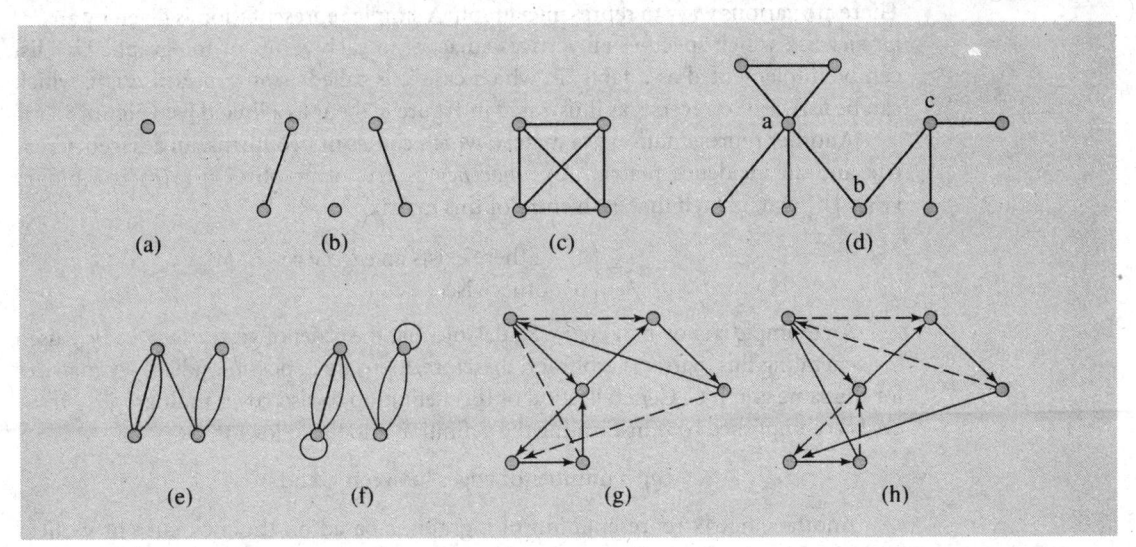

a multigraph with the condition $v_i \neq v_j$ removed, which allows for *loops* to occur; in a pseudograph, a vertex can be joined with itself by an edge (Figure 8.1f).

A *path* from v_1 to v_n is a sequence of edges $edge(v_1v_2), edge(v_2v_3), \ldots, edge(v_{n-1}v_n)$ and is denoted as path $v_1, v_2, v_3, \ldots, v_{n-1}, v_n$. If $v_1 = v_n$ and no edge is repeated, then the path is called a *circuit* (Figure 8.1g). If all vertices in a circuit are different, then it is called a *cycle* (Figure 8.1h).

A graph is called a *weighted graph* if each edge has an assigned number. Depending on the context in which such graphs are used, the number assigned to an edge is called its weight, cost, distance, length, or some other name.

A graph with *n* vertices is called *complete* and is denoted K_n if for each pair of distinct vertices there is exactly one edge connecting them; that is, each vertex can be connected to any other vertex (Figure 8.1c). The number of edges in such a graph $|E| =$

$$\binom{|V|}{2} = \frac{|V|!}{2!(|V|-2)!} = \frac{|V|(|V|-1)}{2} = O(|V|^2).$$

A *subgraph* G' of graph $G = (V,E)$ is a graph (V', E') such that $V' \subseteq V$ and $E' \subseteq E$. A subgraph *induced* by vertices V' is a graph (V', E') such that an edge $e \in E$ if $e \in E'$.

Two vertices v_i and v_j are called *adjacent* if the $edge(v_iv_j)$ is in E. Such an edge is called *incident with* the vertices v_i and v_j. The *degree* of a vertex v, $deg(v)$, is the number of edges incident with v. If $deg(v) = 0$, then v is called an *isolated vertex*. Part of the definition of a graph indicating that the set of edges E can be empty allows for a graph consisting only of isolated vertices.

8.1 Graph Representation

There are various ways to represent a graph. A simple representation is given by an *adjacency list*, which specifies all vertices adjacent to each vertex of the graph. This list can be implemented as a table, in which case it is called a *star representation*, which can be forward or reverse, as illustrated in Figure 8.2b, or as a linked list (Figure 8.2c).

Another representation is a matrix, which comes in two forms: an adjacency matrix and an incidence matrix. An *adjacency matrix* of graph $G = (V,E)$ is a binary $|V| \times |V|$ matrix such that each entry of this matrix

$$a_{ij} = \begin{cases} 1 & \text{if there exists an } edge(v_i v_j) \\ 0 & \text{otherwise} \end{cases}$$

An example is shown in Figure 8.2d. Note that the order of vertices $v_1, \ldots, v_{|V|}$ used for generating this matrix is arbitrary; therefore, there are $n!$ possible adjacency matrices for the same graph G. Generalization of this definition to also cover multigraphs can be easily accomplished by transforming the definition into the following form:

$$a_{ij} = \text{number of edges between } v_i \text{ and } v_j$$

Another matrix representation of a graph is based on the incidence of vertices and edges and is called an *incidence matrix*. An incidence matrix of graph $G = (V,E)$ is a $|V| \times |E|$ matrix such that

$$a_{ij} = \begin{cases} 1 & \text{if edge } e_j \text{ is incident with vertex } v_i \\ 0 & \text{otherwise} \end{cases}$$

Figure 8.2e contains an example of an incidence matrix. In an incidence matrix for a multigraph, some columns are the same, and a column with only one 1 indicates a loop.

Which representation is best? It depends on the problem at hand. If our task is to process vertices adjacent to a vertex v, then the adjacency list requires only $deg(v)$ steps, whereas the adjacency matrix requires $|V|$ steps. On the other hand, inserting or deleting a vertex adjacent to v requires linked list maintenance for an adjacency list (if such an implementation is used); for a matrix, it requires only changing 0 to 1 for insertion, or 1 to 0 for deletion, in one cell of the matrix.

FIGURE 8.2 Graph representations. (a) A graph represented as (b–c) an adjacency list, (d) an adjacency matrix, and (e) an incidence matrix.

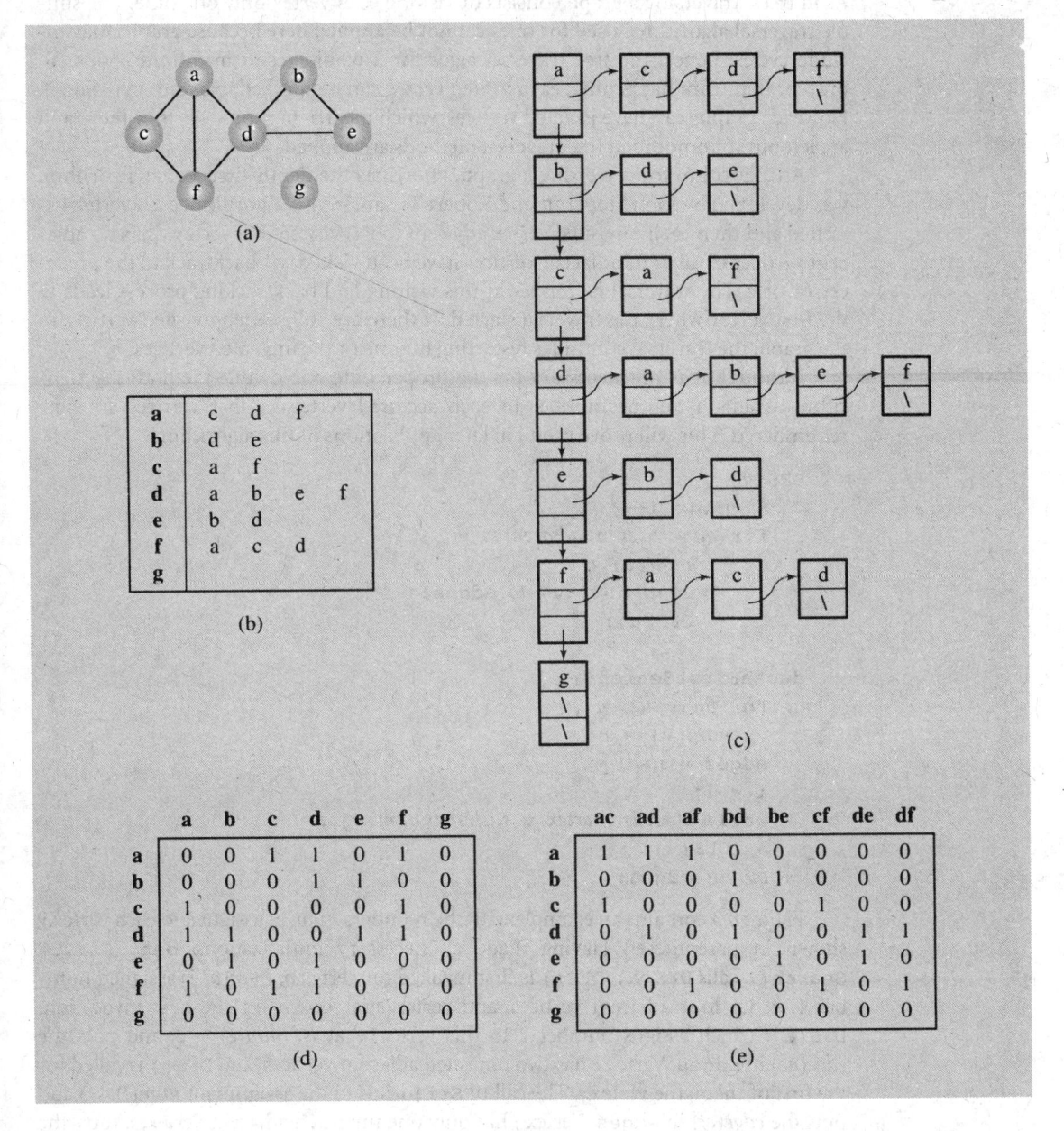

8.2 Graph Traversals

As in trees, traversing a graph consists of visiting each vertex only one time. The simple traversal algorithms used for trees cannot be applied here because graphs may include cycles; hence, the tree traversal algorithms would result in infinite loops. To prevent that from happening, each visited vertex can be marked to avoid revisiting it. However, graphs can have isolated vertices, which means that some parts of the graph are left out if unmodified tree traversal methods are applied.

An algorithm for traversing a graph, known as the depth-first search algorithm, was developed by John Hopcroft and Robert Tarjan. In this algorithm, each vertex v is visited and then each unvisited vertex adjacent to v is visited. If a vertex v has no adjacent vertices or all of its adjacent vertices have been visited, we backtrack to the predecessor of v. The traversal is finished if this visiting and backtracking process leads to the first vertex where the traversal started. If there are still some unvisited vertices in the graph, the traversal continues restarting for one of the unvisited vertices.

Although it is not necessary for the proper outcome of this method, the algorithm assigns a unique number to each accessed vertex so that vertices are now renumbered. This will prove useful in later applications of this algorithm.

```
DFS(v)
    num(v)= i++;
    for all vertices u adjacent to v
        if num(u) is 0
            attach edge(uv) to edges;
            DFS(u);

depthFirstSearch()
    for all vertices v
        num(v) = 0;
    edges = null;
    i = 1;
    while there is a vertex v such that num(v) is 0
        DFS(v);
    output edges;
```

Figure 8.3 contains an example with the numbers $num(v)$ assigned to each vertex v shown in parentheses. Having made all necessary initializations, depthFirstSearch() calls DFS(a). DFS() is first invoked for vertex a; $num(a)$ is assigned number 1. a has four adjacent vertices, and vertex e is chosen for the next invocation, DFS(e), which assigns number 2 to this vertex, that is, $num(e) = 2$, and puts the $edge(ae)$ in edges. Vertex e has two unvisited adjacent vertices, and DFS() is called for the first of them, the vertex f. The call DFS(f) leads to the assignment $num(f) = 3$ and puts the $edge(ef)$ in edges. Vertex f has only one unvisited adjacent vertex, i; thus, the fourth call, DFS(i), leads to the assignment $num(i) = 4$ and to the attaching of $edge(fi)$

FIGURE 8.3 An example of application of the `depthFirstSearch()` algorithm to a graph.

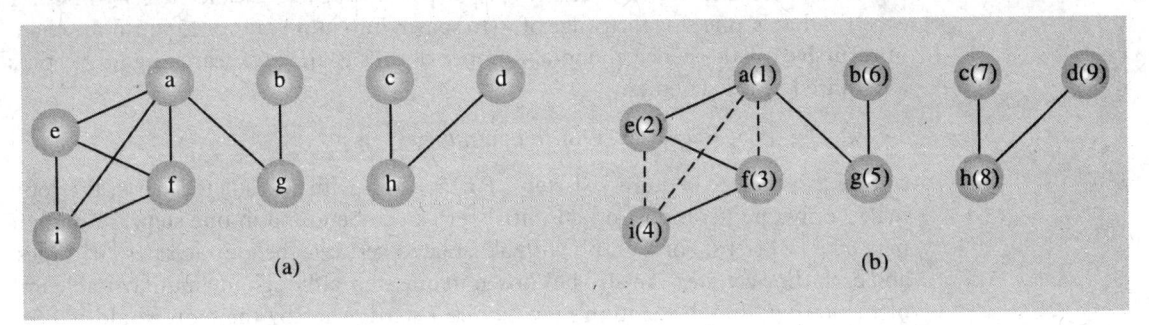

to edges. Vertex *i* has only visited adjacent vertices; hence, we return to call DFS(f) and then to DFS(e) in which vertex *i* is accessed only to learn that *num(i)* is not 0, whereby the *edge(ei)* is not included in edges. The rest of the execution can be seen easily in Figure 8.3b. Solid lines indicate edges included in the set edges.

Note that this algorithm guarantees generating a tree (or a forest, a set of trees) that includes or spans over all vertices of the original graph. A tree that meets this condition is called a *spanning tree*. The fact that a tree is generated is ascertained by the fact that the algorithm does not include in the resulting tree any edge that leads from the currently analyzed vertex to a vertex already analyzed. An edge is added to edges only if the condition in "if *num(u) is* 0" is true; that is, if vertex *u* reachable from vertex *v* has not been processed. As a result, certain edges in the original graph do not appear in the resulting tree. The edges included in this tree are called *forward edges* (or *tree edges*), and the edges not included in this tree are called *back edges* and are shown as dashed lines.

Figure 8.4 illustrates the execution of this algorithm for a digraph. Notice that the original graph results in three spanning trees, although we started with only two isolated subgraphs.

FIGURE 8.4 The `depthFirstSearch()` algorithm applied to a digraph.

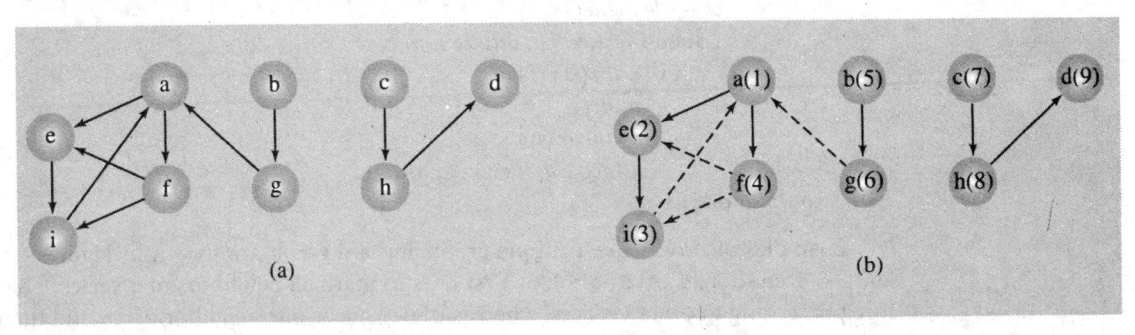

The complexity of depthFirstSearch() is $O(|V| + |E|)$ because (a) initializing $num(v)$ for each vertex v requires $|V|$ steps; (b) DFS(v) is called $deg(v)$ times for each v—that is, once for each edge of v (to spawn into more calls or to finish the chain of recursive calls)—hence, the total number of calls is $2|E|$; (c) searching for vertices as required by the statement

    while *there is a vertex* v *such that num*(v) *is* 0

can be assumed to require $|V|$ steps. For a graph with no isolated parts, the loop makes only one iteration, and an initial vertex can be found in one step, although it may take $|V|$ steps. For a graph with all isolated vertices, the loop iterates $|V|$ times, and each time a vertex can also be chosen in one step, although in an unfavorable implementation, the *i*th iteration may require *i* steps, whereby the loop would require $O(|V|^2)$ steps in total. For example, if an adjacency list is used, then for each v, the condition in the loop,

    for *all vertices* u *adjacent to* v

is checked $deg(v)$ times. However, if an adjacency matrix is used, then the same condition is used $|V|$ times, whereby the algorithm's complexity becomes $O(|V|^2)$.

As we shall see, many different algorithms are based on DFS(); however, some algorithms are more efficient if the underlying graph traversal is not depth first but breadth first. We have already encountered these two types of traversals in Chapter 6; recall that the depth-first algorithms rely on the use of a stack (explicitly, or implicitly, in recursion), and breadth-first traversal uses a queue as the basic data structure. Not surprisingly, this idea can also be extended to graphs, as shown in the following pseudocode:

```
breadthFirstSearch()
    for all vertices u
        num(u) = 0;
    edges = null;
    i = 1;
    while there is a vertex v such that num(v) == 0
        num(v)=i++;
        enqueue(v);
        while queue is not empty
            v = dequeue();
            for all vertices u adjacent to v
                if num(u) is 0
                    num(u) = i++;
                    enqueue(u);
                    attach edge(vu) to edges;
    output edges;
```

Examples of processing a simple graph and a digraph are shown in Figures 8.5 and 8.6. breadthFirstSearch() first tries to mark all neighbors of a vertex v before proceeding to other vertices, whereas DFS() picks one neighbor of a v and then proceeds to a neighbor of this neighbor before processing any other neighbors of v.

FIGURE 8.5 An example of application of the `breadthFirstSearch()` algorithm to a graph.

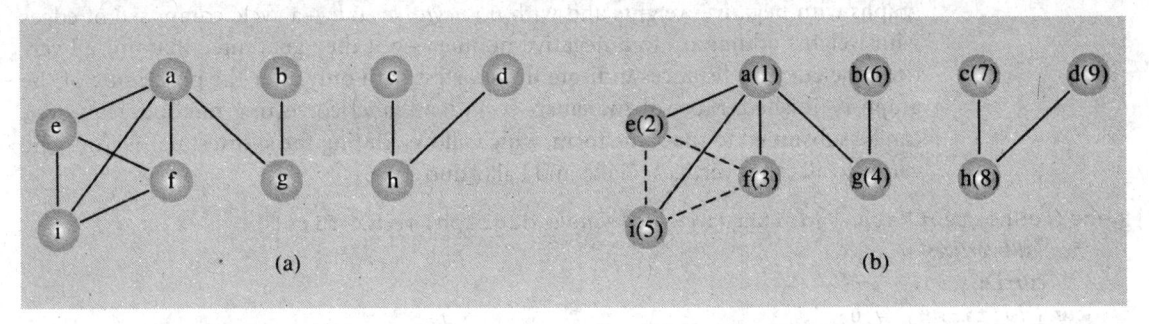

FIGURE 8.6 The `breadthFirstSearch()` algorithm applied to a digraph.

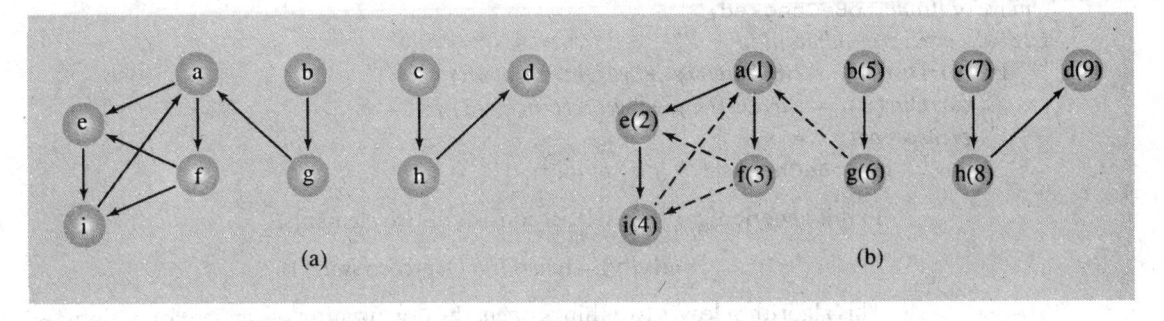

8.3 SHORTEST PATHS

Finding the shortest path is a classical problem in graph theory, and a large number of different solutions have been proposed. Edges are assigned certain weights representing, for example, distances between cities, times separating the execution of certain tasks, costs of transmitting information between locations, amounts of some substance transported from one place to another, and so on. When determining the shortest path from vertex v to vertex u, information about distances between intermediate vertices w has to be recorded. This information can be recorded as a label associated with these vertices, where the label is only the distance from v to w or the distance along with the predecessor of w in this path. The methods of finding the shortest path rely on these labels. Depending on how many times these labels are updated, the methods solving the shortest path problem are divided in two classes: label-setting methods and label-correcting methods.

For *label-setting methods*, in each pass through the vertices still to be processed, one vertex is set to a value that remains unchanged to the end of the execution. This, however, limits such methods to processing graphs with only positive weights. The second category includes *label-correcting methods*, which allow for the changing of *any* label during application of the method. The latter methods can be applied to graphs with negative weights and with no *negative cycle*—a cycle composed of edges with weights adding up to a negative number—but they guarantee that, for all vertices, the current distances indicate the shortest path only after the processing of the graph is finished. Most of the label-setting and label-correcting methods, however, can be subsumed to the same form, which allows finding the shortest paths from one vertex to all other vertices (Gallo and Pallottino, 1986):

```
genericShortestPathAlgorithm(weighted simple digraph, vertex first)
    for all vertices v
        currDist(v) = ∞;
    currDist(first) = 0;
    initialize toBeChecked;
    while toBeChecked is not empty
        v = a vertex in toBeChecked;
        remove v from toBeChecked;
        for all vertices u adjacent to v
            if currDist(u) > currDist(v) + weight(edge(vu))
                currDist(u) = currDist(v) + weight(edge(vu));
                predecessor(u) = v;
                add u to toBeChecked if it is not there;
```

In this generic algorithm, a label consists of two elements:

$$label(v) = (currDist(v), predecessor(v))$$

This algorithm leaves two things open: the organization of the set `toBeChecked` and the order of assigning new values to *v* in the assignment statement

v = a vertex in `toBeChecked`;

It should be clear that the organization of `toBeChecked` can determine the order of choosing new values for *v*, but it also determines the efficiency of the algorithm.

What distinguishes label-setting methods from label-correcting methods is the method of choosing the value for *v*, which is always a vertex in `toBeChecked` with the smallest current distance. One of the first label-setting algorithms was developed by Dijkstra.

In Dijkstra's algorithm, a number of paths $p_1, \ldots, p_n$ from a vertex *v* are tried, and each time, the shortest path is chosen among them, which may mean that the same path p_i can be continued by adding one more edge to it. But if p_i turns out to be longer than any other path that can be tried, p_i is abandoned and this other path is tried by resuming from where it was left and by adding one more edge to it. Because paths can lead to vertices with more than one outgoing edge, new paths for possible exploration are added for each outgoing edge. Each vertex is tried once, all paths leading from it are opened, and the vertex itself is put away and not used anymore. After all vertices are visited, the algorithm is finished. Dijkstra's algorithm is as follows:

```
DijkstraAlgorithm(weighted simple digraph, vertex first)
    for all vertices v
        currDist(v) = ∞;
    currDist(first) = 0;
    toBeChecked = all vertices;
    while toBeChecked is not empty
        v = a vertex in toBeChecked with minimal currDist(v);
        remove v from toBeChecked;
        for all vertices u adjacent to v and in toBeChecked
            if currDist(u) > currDist(v)+ weight(edge(vu))
                currDist(u) = currDist(v)+ weight(edge(vu));
                predecessor(u) = v;
```

Dijkstra's algorithm is obtained from the generic method by being more specific about which vertex is to be taken from toBeChecked so that the line

```
v = a vertex in toBeChecked;
```

is replaced by the line

```
v = a vertex in toBeChecked with minimal currDist(v);
```

and by extending the condition in the if statement whereby the current distance of vertices eliminated from toBeChecked is set permanently.[1] Note that the structure of toBeChecked is not specified, and the efficiency of the algorithms depends on the data type of toBeChecked, which determines how quickly a vertex with minimal distance can be retrieved.

Figure 8.7 contains an example. The table in this figure shows all iterations of the while loop. There are 10 iterations because there are 10 vertices. The table indicates the current distances determined up until the current iteration.

The list toBeChecked is initialized to $\{a\ b\ \ldots\ j\}$; the current distances of all vertices are initialized to a very large value, marked here as ∞; and in the first iteration, the current distances of d's neighbors are set to numbers equal to the weights of the edges from d. Now, there are two candidates for the next try, a and h, because d was excluded from toBeChecked. In the second iteration, h is chosen, because its current distance is minimal, and then the two vertices accessible from h, namely, e and i, acquire the current distances 6 and 10. Now, there are three candidates in toBeChecked for the next try, a, e, and i. a has the smallest current distance, so it is chosen in the third iteration. Eventually, in the tenth iteration, toBeChecked becomes empty and the execution of the algorithm completes.

The complexity of Dijkstra's algorithm is $O(|V|^2)$. The first for loop and the while loop are executed $|V|$ times. For each iteration of the while loop, (a) a vertex v in toBeChecked with minimal current distance has to be found, which requires $O(|V|)$ steps, and (b) the for loop iterates $deg(v)$ times, which is also $O(|V|)$. The efficiency can be improved by using a heap to store and order vertices and adjacency

---

[1]Dijkstra used six sets to ensure this condition, three for vertices and three for edges.

FIGURE 8.7 An execution of `DijkstraAlgorithm()`.

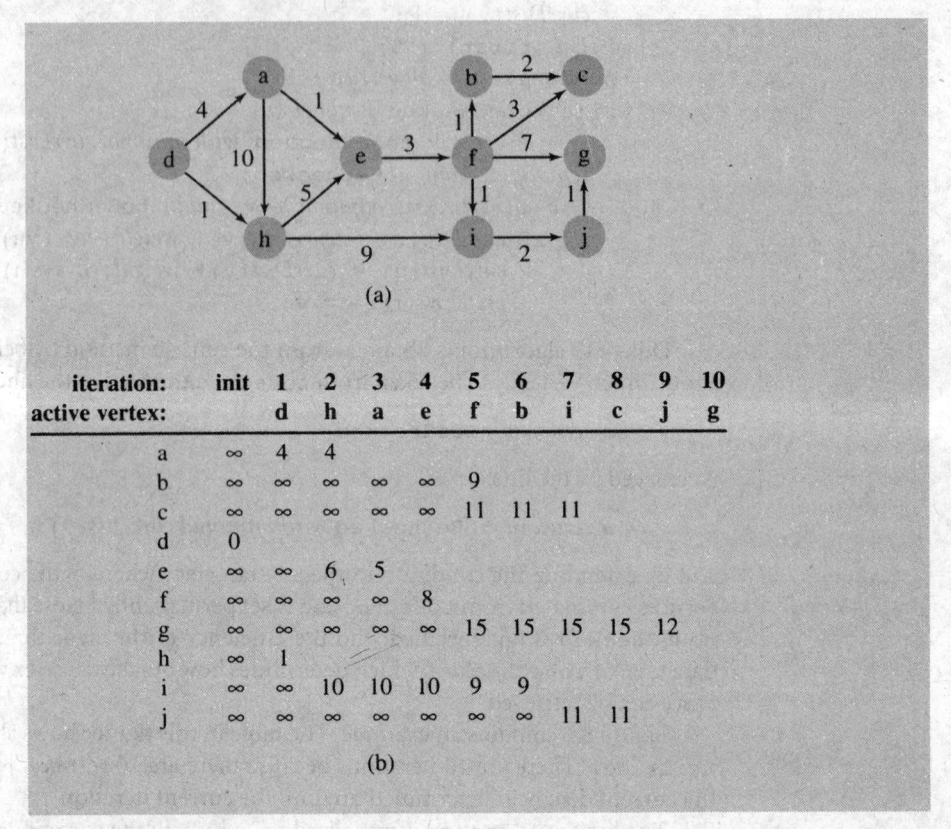

(a)

iteration: active vertex:	init	1 d	2 h	3 a	4 e	5 f	6 b	7 i	8 c	9 j	10 g
a	∞	4	4								
b	∞	∞	∞	∞	∞	9					
c	∞	∞	∞	∞	∞	11	11	11			
d	0										
e	∞	∞	6	5							
f	∞	∞	∞	∞	8						
g	∞	∞	∞	∞	∞	15	15	15	15	12	
h	∞	1									
i	∞	∞	10	10	10	9	9				
j	∞	∞	∞	∞	∞	∞	∞	11	11		

(b)

lists (Johnson, 1977). Using a heap turns the complexity of this algorithm into $O((|E| + |V|) \lg |V|)$; each time through the `while` loop, the cost of restoring the heap after removing a vertex is proportional to $O(\lg|V|)$. Also, in each iteration, only adjacent vertices are updated on an adjacency list so that the total updates for all vertices considered in all iterations are proportional to $|E|$, and each list update corresponds to the cost of $\lg|V|$ of the heap update.

Dijkstra's algorithm is not general enough in that it may fail when negative weights are used in graphs. To see why, change the weight of *edge(ah)* from 10 to –10. Note that the path *d, a, h, e* is now –1, whereas the path *d, a, e* as determined by the algorithm is 5. The reason for overlooking this less costly path is that the vertices with the current distance set from ∞ to a value are not checked anymore: First, successors of vertex *d* are scrutinized and *d* is removed from `toBeChecked`, then the vertex *h* is removed from `toBeChecked`, and only afterward is the vertex *a* considered as a candidate to be included in the path from *d* to other vertices. But now, the *edge(ah)* is not taken into consideration because the condition in the `for` loop prevents the algorithm from doing this. To overcome this limitation, a label-correcting method is needed.

One of the first label-correcting algorithms was devised by Lester Ford. Like Dijkstra's algorithm, it uses the same method of setting current distances, but Ford's method does not permanently determine the shortest distance for any vertex until it processes the entire graph. It is more powerful than Dijkstra's method in that it can process graphs with negative weights (but not graphs with negative cycles).

As required by the original form of the algorithm, all edges are monitored to find a possibility for an improvement of the current distance of vertices so that the algorithm can be presented in this pseudocode:

```
FordAlgorithm(weighted simple digraph, vertex first)
    for all vertices v
        currDist(v) = ∞;
    currDist(first) = 0;
    while there is an edge(vu) such that currDist(u) > currDist(v) + weight(edge(vu))
        currDist(u) = currDist(v) + weight(edge(vu));
```

To impose a certain order on monitoring the edges, an alphabetically ordered sequence of edges can be used so that the algorithm can repeatedly go through the entire sequence and adjust the current distance of any vertex, if needed. Figure 8.8 contains an example. The graph includes edges with negative weights. The table indicates iterations of the `while` loop and current distances updated in each iteration, where one iteration is defined as one pass through the edges. Note that a vertex can change its current distance during the same iteration. However, at the end, each vertex of the graph can be reached through the shortest path from the starting vertex (vertex c in the example in Figure 8.8).

FIGURE 8.8 FordAlgorithm() applied to a digraph with negative weights.

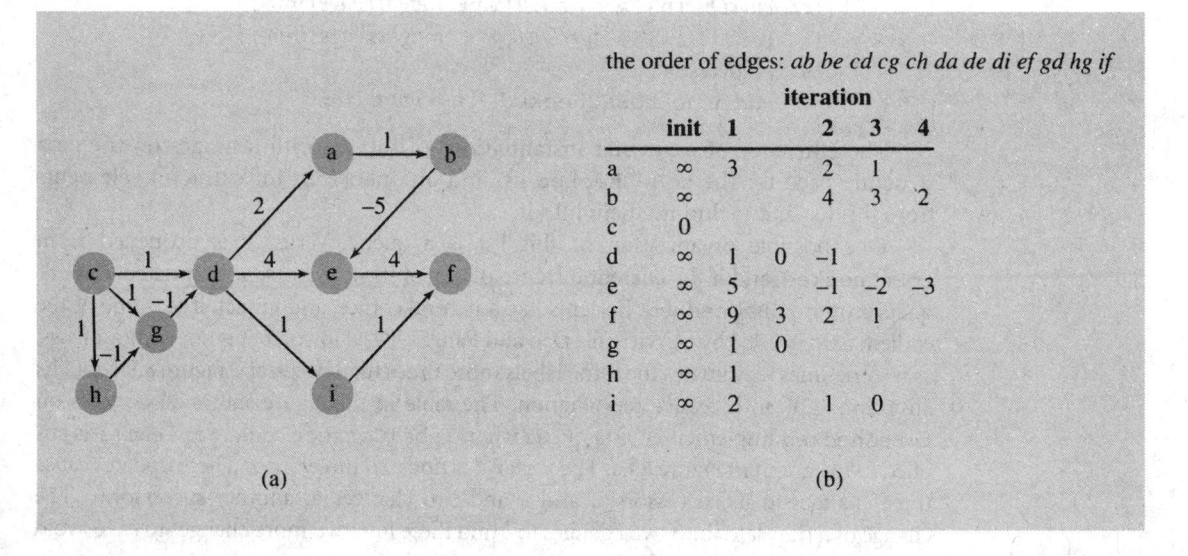

the order of edges: *ab be cd cg ch da de di ef gd hg if*

	init	1	2	3	4
a	∞	3	2	1	
b	∞		4	3	2
c	0				
d	∞	1	0	−1	
e	∞	5	−1	−2	−3
f	∞	9	3	2	1
g	∞	1	0		
h	∞	1			
i	∞	2	1	0	

(a)   (b)

The computational complexity of this algorithm is $O(|V||E|)$. There will be at most $|V|-1$ passes through the sequence of $|E|$ edges, because $|V|-1$ is the largest number of edges in any path. In the first pass, at least all one-edge paths are determined; in the second pass, all two-edge paths are determined; and so on. However, for graphs with irrational weights, this complexity is $O(2^{|V|})$ (Gallo and Pallottino 1986).

We have seen in the case of Dijkstra's algorithm that the efficiency of an algorithm can be improved by scanning edges and vertices in a certain order, which in turn depends on the data structure used to store them. The same holds true for label-correcting methods. In particular, FordAlgorithm() does not specify the order of checking edges. In the example illustrated in Figure 8.8, a simple solution is used in that all adjacency lists of all vertices were visited in each iteration. However, in this approach, all the edges are checked every time, which is not necessary, and more judicious organization of the list of vertices can limit the number of visits per vertex. Such an improvement is based on the genericShortestPathAlgorithm() by explicitly referring to the toBeChecked list, which in FordAlgorithm() is used only implicitly: It simply is the set of all vertices V and remains such for the entire run of the algorithm. This leads us to a general form of a label-correcting algorithm as expressed in this pseudocode:

```
labelCorrectingAlgorithm(weighted simple digraph, vertex first)
    for all vertices v
        currDist(v) = ∞;
    currDist(first) = 0;
    toBeChecked = {first};
    while toBeChecked is not empty
        v = a vertex in toBeChecked;
        remove v from toBeChecked;
        for all vertices u adjacent to v
            if currDist(u) > currDist(v)+ weight(edge(vu))
                currDist(u) = currDist(v)+ weight(edge(vu));
                predecessor(u) = v;
                add u to toBeChecked if it is not there;
```

The efficiency of particular instantiations of this algorithm hinges on the data structure used for the toBeChecked list and on operations for extracting elements from this list and including them into it.

One possible organization of this list is a queue: Vertex v is dequeued from toBeChecked, and if the current distance of any of its neighbors, u, is updated, u is enqueued onto toBeChecked. It seems like a natural choice, and in fact, it was one of the earliest, used in 1968 by C. Witzgall (Deo and Pang, 1984). However, it is not without flaws, as it sometimes reevaluates the same labels more times than necessary. Figure 8.9 contains an example of an excessive reevaluation. The table in this figure shows all changes on toBeChecked implemented as a queue when labelCorrectingAlgorithm() is applied to the graph in Figure 8.8a. The vertex d is updated three times. These updates cause three changes to its successors, a and i, and two changes to another successor, e. The change of a translates into two changes to b and these into two more changes to e. To avoid such repetitious updates, a doubly ended queue, or deque, can be used.

FIGURE 8.9 An execution of `labelCorrectingAlgorithm()`, which uses a queue.

									active vertex												
		c	d	g	h	a	e	i	d	g	b	f	a	e	i	d	b	f	a	i	e	b	f	e
queue			d	g	h	a	e	i	d	g	b	f	a	e	i	d	b	f	a	i	e	b	f	e
				g	h	a	e	i	d	g	b	f	a	e	i	d	b	f	a	i	e	b	f	e
					h	a	e	i	d	g	b	f	a	e	i	d	b	f	a	i	e	b	f	
						e	i	d	g	b	f	a	e		i	d	b		i	e				
							i	d	g	b	f		e	i	d									
													i	d										
a	∞	∞	3	3	3	3	3	3	2	2	2	2	2	2	2	1								
b	∞	∞	∞	∞	∞	∞	4	4	4	4	4	4	4	3	3	3	3	3	3	2				
c	0																								
d	∞		1	1	0	0	0	0	0	0	-1														
e	∞	∞	5	5	5	5	5	5	4	4	-1	-1	-1	-1	-1	-1	-2	-2	-2	-2	-2	-3			
f	∞	∞	∞	∞	∞	∞	9	3	3	3	3	3	3	3	2	2	2	2	2	1					
g	∞		1	1	1	0																			
h	∞		1																						
i	∞	∞	2	2	2	2	2	2	1	1	1	1	1	1	1	0								

The choice of a deque as a solution to this problem is attributed to D. D'Esopo (Pollack and Wiebenson, 1960) and was implemented by Pape. In this method, the vertices included in `toBeChecked` for the first time are put at the end of the list; otherwise, they are added at the front. The rationale for this procedure is that if a vertex v is included for the first time, then there is a good chance that the vertices accessible from v have not been processed yet, so they will be processed after processing v. On the other hand, if v has been processed at least once, then it is likely that the vertices reachable from v are still on the list waiting for processing; by putting v at the end of the list, these vertices may very likely be reprocessed due to the update of $currDist(v)$. Therefore, it is better to put v in front of their successors to avoid an unnecessary round of updates. Figure 8.10 shows changes in the deque during the execution of `labelCorrectingAlgorithm()` applied to the graph in Figure 8.8a. This time, the number of iterations is dramatically reduced. Although d is again evaluated three times, these evaluations are performed before processing its successors so that a and i are processed once and e twice. However, this algorithm has a problem of its own because in the worst case its performance is an exponential function of the number of vertices. (See Exercise 13 at the end of this chapter.) But in the average case, as Pape's experimental runs indicate, this implementation fares at least 60 percent better than the previous queue solution.

Instead of using a deque, which combines two queues, the two queues can be used separately. In this version of the algorithm, vertices stored for the first time are enqueued on $queue_1$, and on $queue_2$ otherwise. Vertices are dequeued from $queue_1$ if it is not empty, and from $queue_2$ otherwise (Gallo and Pallottino, 1988).

FIGURE 8.10 An execution of `labelCorrectingAlgorithm()`, which applies a deque.

	active vertex												
	c	d	g	d	h	g	d	a	e	i	b	e	f
deque	d	g	d	h	g	d	a	e	i	b	e	f	
		g	h	h	a	a	a	e	i	b	f	f	
		h	a	a	e	e	e	i	b	f			
			e	e	i	i	i						
			i	i									
a	∞	∞	3	3	2	2	2	1					
b	∞	∞	∞	∞	∞	∞	∞	∞	2				
c	0												
d	∞	1	1	0	0	0	−1						
e	∞	∞	5	5	4	4	4	3	3	3	3	−3	
f	∞	∞	∞	∞	∞	∞	∞	∞	∞	7	1		
g	∞	1	1	1	1	0							
h	∞	1											
i	∞	∞	2	2	1	1	1	0					

Another version of the label-correcting method is the *threshold algorithm*, which also uses two lists. Vertices are taken for processing from $list_1$. A vertex is added to the end of $list_1$ if its label is below the current threshold level, and to $list_2$ otherwise. If $list_1$ is empty, then the threshold level is changed to a value greater than a minimum label among the labels of the vertices in $list_2$, and then the vertices with the label values below the threshold are moved to $list_1$ (Glover, Glover, and Klingman, 1984).

Still another algorithm is a *small label first* method. In this method, a vertex is included at the front of a deque if its label is smaller than the label at the front of the deque; otherwise, it is put at the end of the deque (Bertsekas, 1993). To some extent, this method includes the main criterion of label-setting methods. The latter methods always retrieve the minimal element from the list; the small label first method puts a vertex with the label smaller than the label of the front vertex at the top. The approach can be carried to its logical conclusion by requiring each vertex to be included in the list according to its rank so that the deque turns into a priority queue and the resulting method becomes a label-correcting version of Dijkstra's algorithm.

8.3.1 All-to-All Shortest Path Problem

Although the task of finding all shortest paths from any vertex to any other vertex seems to be more complicated than the task of dealing with one source only, a method designed by Stephen Warshall and implemented by Robert W. Floyd and P. Z. Ingerman does it in a surprisingly simple way, provided an adjacency matrix is given that indicates all the edge weights of the graph (or digraph). The graph can include negative weights. The algorithm is as follows:

```
WFIalgorithm(matrix weight)
    for i = 1 to |V|
        for j = 1 to |V|
            for k = 1 to |V|
                if weight[j][k] > weight[j][i] + weight[i][k]
                    weight[j][k] = weight[j][i] + weight[i][k];
```

The outermost loop refers to vertices that may be on a path between the vertex with index j and the vertex with index k. For example, in the first iteration, when $i = 1$, all paths $v_j \ldots v_1 \ldots v_k$ are considered, and if there is currently no path from v_j to v_k and v_k is reachable from v_j, the path is established, with its weight equal to $p = weight(path(v_j v_1)) + weight(path(v_1 v_k))$, or the current weight of this path, $weight(path(v_j v_k))$, is changed to p if p is less than $weight(path(v_j v_k))$. As an example, consider the graph and the corresponding adjacency matrix in Figure 8.11. This figure also contains tables that show changes in the matrix for each value of i and the changes in paths as established by the algorithm. After the first iteration, the matrix and the graph remain the same, because a has no incoming edges (Figure 8.11a). They also remain the same in the last iteration, when $i = 5$; no change is introduced to the matrix because vertex e has no outgoing edges. A better path, one with a lower combined weight, is always chosen, if possible. For example, the direct one-edge path from b to e in Figure 8.11c is abandoned after a two-edge path from b to e is found with a lower weight, as in Figure 8.11d.

This algorithm also allows us to detect cycles if the diagonal is initialized to ∞ and not to zero. If any of the diagonal values are changed, then the graph contains a cycle. Also, if an initial value of ∞ between two vertices in the matrix is not changed to a finite value, it is an indication that one vertex cannot be reached from another.

The simplicity of the algorithm is reflected in the ease with which its complexity can be computed: All three `for` loops are executed $|V|$ times, so its complexity is $|V|^3$. This is a good efficiency for dense, nearly complete graphs, but in sparse graphs, there is no need to check for all possible connections between vertices. For sparse graphs, it may be more beneficial to use a one-to-all method $|V|$ times—that is, apply it to each vertex separately. This should be a label-setting algorithm, which as a rule has better complexity than a label-correcting algorithm. However, a label-setting algorithm cannot work with graphs with negative weights. To solve this problem, we have to modify the graph so that it does not have negative weights and it guarantees to have the same shortest paths as the original graph. Fortunately, such a modification is possible (Edmonds and Karp, 1972).

Observe first that, for any vertex v, the length of the shortest path to v is never greater than the length of the shortest path to any of its predecessors w plus the length of edge from w to v, or

$$dist(v) \leq dist(w) + weight(edge(wv))$$

FIGURE 8.11 An execution of `WFIalgorithm()`.

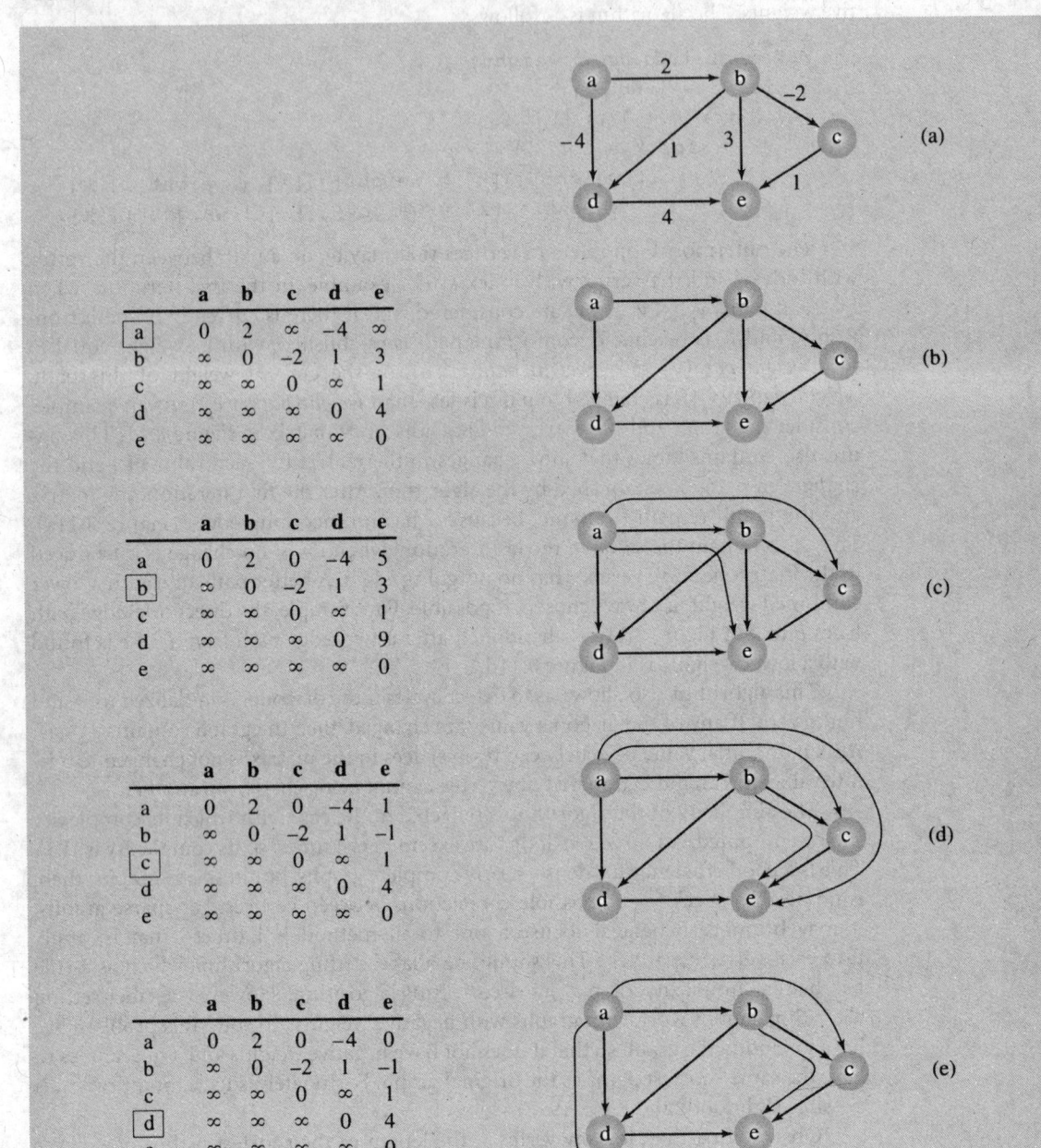

	a	b	c	d	e
☐ a	0	2	∞	−4	∞
b	∞	0	−2	1	3
c	∞	∞	0	∞	1
d	∞	∞	∞	0	4
e	∞	∞	∞	∞	0

	a	b	c	d	e
a	0	2	0	−4	5
☐ b	∞	0	−2	1	3
c	∞	∞	0	∞	1
d	∞	∞	∞	0	9
e	∞	∞	∞	∞	0

	a	b	c	d	e
a	0	2	0	−4	1
b	∞	0	−2	1	−1
☐ c	∞	∞	0	∞	1
d	∞	∞	∞	0	4
e	∞	∞	∞	∞	0

	a	b	c	d	e
a	0	2	0	−4	0
b	∞	0	−2	1	−1
c	∞	∞	0	∞	1
☐ d	∞	∞	∞	0	4
e	∞	∞	∞	∞	0

for any vertices v and w. This inequality is equivalent to the inequality

$$0 \leq weight'(edge(wv)) = weight(edge(vw)) + dist(w) - dist(v)$$

Hence, changing $weight(e)$ to $weight'(e)$ for all edges e renders a graph with nonnegative edge weights. Now note that the shortest path $v_1, v_2, \ldots, v_k$ is

$$\sum_{i=1}^{k-1} weight'(edge(v_i v_{i+1})) = \left(\sum_{i=1}^{k-1} weight(edge(v_i v_{i+1}))\right) + dist(v_1) - dist(v_k)$$

Therefore, if the length L' of the path from v_1 to v_k is found in terms of nonnegative weights, then the length L of the same path in the same graph using the original weights, some possibly negative, is $L = L' - dist(v_1) + dist(v_k)$.

But because the shortest paths have to be known to make such a transformation, the graph has to be preprocessed by one application of a label-correcting method. Only afterward are the weights modified, and then a label-setting method is applied $|V|$ times.

8.4 CYCLE DETECTION

Many algorithms rely on detecting cycles in graphs. We have just seen that, as a side effect, `WFIalgorithm()` allows for detecting cycles in graphs. However, it is a cubic algorithm, which in many situations is too inefficient. Therefore, other cycle detection methods have to be explored.

One such algorithm is obtained directly from `depthFirstSearch()`. For undirected graphs, small modifications in `DFS(v)` are needed to detect cycles and report them

```
cycleDetectionDFS(v)
    num(v) = i++;
    for all vertices u adjacent to v
        if num(u) is 0
            attach edge(uv) to edges;
            cycleDetectionDFS(u);
        else if edge(vu) is not in edges
            cycle detected;
```

For digraphs, the situation is a bit more complicated, because there may be edges between different spanning subtrees, called *side edges* (see $edge(ga)$ in Figure 8.4b). An edge (a back edge) indicates a cycle if it joins two vertices already included in the same spanning subtree. To consider only this case, a number higher than any number generated in subsequent searches is assigned to a vertex being currently visited after all its descendants have also been visited. In this way, if a vertex is about to be joined by an edge with a vertex having a lower number, we declare a cycle detection. The algorithm is now

```
digraphCycleDetectionDFS(v)
    num(v) = i++;
    for all vertices u adjacent to v
        if num(u) is 0
            pred(u) = v;
            digraphCycleDetectionDFS(u);
        else if num(u) is not ∞
            pred(u) = v;
            cycle detected;
    num(v) = ∞;
```

8.4.1 Union-Find Problem

Let us recall from a preceding section that depth-first search guaranteed generating a spanning tree in which no element of edges used by depthFirstSearch() led to a cycle with other elements of edges. This was due to the fact that if vertices *v* and *u* belonged to edges, then the *edge(vu)* was disregarded by depthFirstSearch(). A problem arises when depthFirstSearch() is modified so that it can detect whether a specific *edge(vu)* is part of a cycle (see Exercise 20). Should such a modified depth-first search be applied to each edge separately, then the total run would be $O(|E|(|E| + |V|))$, which could turn into $O(|V|^4)$ for dense graphs. Hence, a better method needs to be found.

The task is to determine if two vertices are in the same set. Two operations are needed to implement this task: finding the set to which a vertex *v* belongs and uniting two sets into one if vertex *v* belongs to one of them and *w* to another. This is known as the *union-find problem*.

The sets used to solve the union-find problem are implemented with circular linked lists; each list is identified by a vertex that is the root of the tree to which the vertices in the list belong. But first, all vertices are numbered with integers $0, \ldots, |V| - 1$, which are used as indexes in three arrays: root[ ] to store a vertex index identifying a set of vertices, next[ ] to indicate the next vertex on a list, and length[ ] to indicate the number of vertices in a list.

We use circular lists to be able to combine two lists right away, as illustrated in Figure 8.12. Lists L1 and L2 (Figure 8.12a) are merged into one by interchanging next pointers in both lists (Figure 8.12b or, the same list, Figure 8.12c). However, the vertices in L2 have to "know" to which list they belong; therefore, their root indicators have to be changed to the new root. Because it has to be done for all vertices of list L2, then L2 should be the shorter of the two lists. To determine the length of lists, the third array is used, length[ ], but only lengths for the identifying nodes (roots) have to be updated. Therefore, the lengths indicated for other vertices that were roots (and at the beginning each of them was) are disregarded.

FIGURE 8.12 Concatenating two circular linked lists.

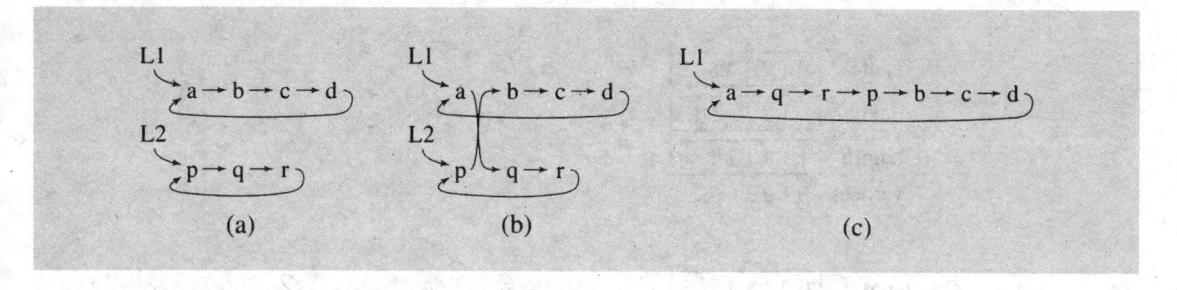

```
L1                    L1                    L1
  a→b→c→d               a  b→c→d              a→q→r→p→b→c→d
L2                    L2
  p→q→r                 p  q→r
   (a)                   (b)                    (c)
```

The union operation performs all the necessary tasks, so the find operation becomes trivial. By constantly updating the array root[ ], the set, to which a vertex j belongs, can be immediately identified, because it is a set whose identifying vertex is root[j]. Now, after the necessary initializations,

```
initialize()
    for i = 0 to |V| - 1
        root[i] = next[i] = i;
        length[i] = 1;
```

union() can be defined as follows:

```
union(edge(vu))
    if (root[u] == root[v])                              // disregard this edge,
        return;                                          // since v and u are in
    else if (length[root[v]] < length[root[u]])          // the same set; combine
        rt = root[v];                                    // two sets into one;
        length[root[u]] += length[rt];
        root[rt] = root[u];                              // update root of rt and
        for (j = next[rt]; j != rt; j = next[j])         // then other vertices
            root[j] = root[u];                           // in circular list;
        swap(next[rt],next[root[u]]);                    // merge two lists;
        add edge(vu) to spanningTree;
    else  // if length[root[v]] >= length[root[u]]
          // proceed as before, with v and u reversed;
```

An example of the application of union() to merge lists is shown in Figure 8.13. After initialization, there are $|V|$ unary sets or one-node linked lists, as in Figure 8.13a. After executing union() several times, smaller linked lists are merged into larger ones, and each time, the new situation is reflected in the three arrays, as shown in Figures 8.13b–d.

FIGURE 8.13 An example of application of `union()` to merge lists.

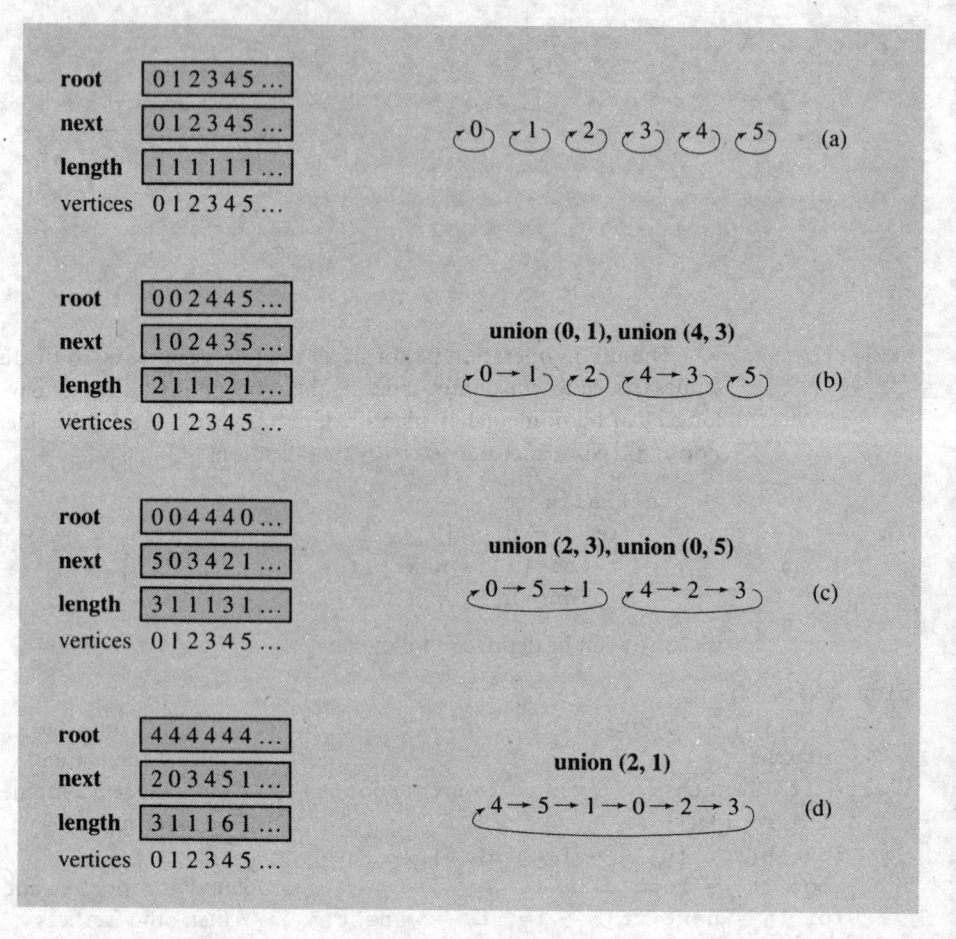

```
root     0 1 2 3 4 5 ...
next     0 1 2 3 4 5 ...                 ⌒0  ⌒1  ⌒2  ⌒3  ⌒4  ⌒5        (a)
length   1 1 1 1 1 1 ...
vertices 0 1 2 3 4 5 ...


root     0 0 2 4 4 5 ...
next     1 0 2 4 3 5 ...        union (0, 1), union (4, 3)
length   2 1 1 1 2 1 ...          ⌒0→1⌒  ⌒2⌒  ⌒4→3⌒  ⌒5⌒              (b)
vertices 0 1 2 3 4 5 ...


root     0 0 4 4 4 0 ...
next     5 0 3 4 2 1 ...        union (2, 3), union (0, 5)
length   3 1 1 1 3 1 ...          ⌒0→5→1⌒  ⌒4→2→3⌒                    (c)
vertices 0 1 2 3 4 5 ...


root     4 4 4 4 4 4 ...
next     2 0 3 4 5 1 ...             union (2, 1)
length   3 1 1 1 6 1 ...          ⌒4→5→1→0→2→3⌒                      (d)
vertices 0 1 2 3 4 5 ...
```

The complexity of `union()` depends on the number of vertices that have to be updated when merging two lists, specifically, on the number of vertices on the shorter list, because this number determines how many times the `for` loop in `union()` iterates. Because this number can be between 1 and $|V|/2$, the complexity of `union()` is given by $O(|V|)$.

8.5 SPANNING TREES

Consider the graph representing the airline connections between seven cities (Figure 8.14a). If the economic situation forces this airline to shut down as many connections as possible, which of them should be retained to make sure that it is still possible to reach any city from any other city, if only indirectly? One possibility is the graph in Figure 8.14b. City *a* can be reached from city *d* using the path *d, c, a*, but it is also pos-

FIGURE 8.14 A graph representing (a) the airline connections between seven cities and (b–d) three possible sets of connections.

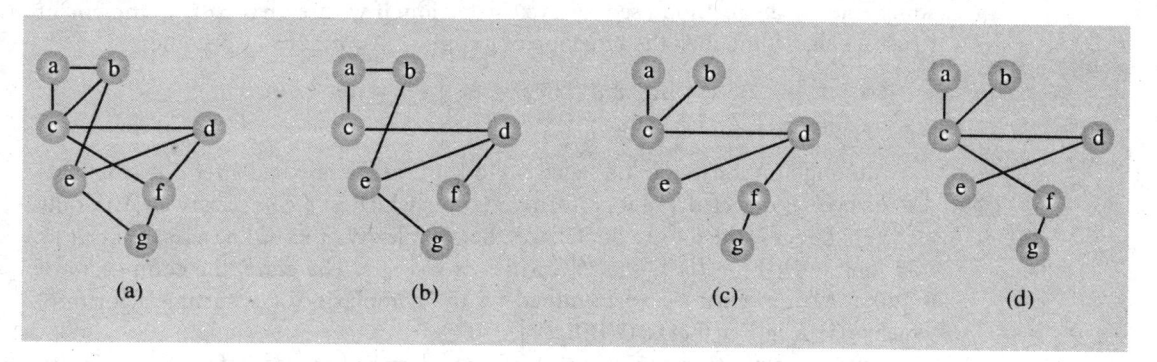

sible to use the path *d, e, b, a*. Because the number of retained connections is the issue, there is still the possibility we can reduce this number. It should be clear that the minimum number of such connections form a tree because alternate paths arise as a result of cycles in the graph. Hence, to create the minimum number of connections, a spanning tree should be created, and such a spanning tree is the by-product of `depthFirstSearch()`. Clearly, we can create different spanning trees (Figures 8.14c–d)—that is, we can decide to retain different sets of connections—but all these trees have six edges and we cannot do any better than that.

The solution to this problem is not optimal in that the distances between cities have not been taken into account. Because there are alternative six-edge connections between cities, the airline uses the cost of these connections to choose the best, guaranteeing the optimum cost. This can be achieved by having smallest distances for the six connections. This problem can now be phrased as finding a *minimum spanning tree*, which is a spanning tree in which the sum of the weights of its edges is minimal. The previous problem of finding a spanning tree in a simple graph is a case of the minimum spanning tree problem in that the weights for each edge are assumed to equal one. Therefore, each spanning tree is a minimum tree in a simple graph.

The minimum spanning tree problem has many solutions, and only two of them are presented here. (For a review of these methods, see Graham and Hell, 1985; see also ex. 26.)

One popular algorithm was devised by Joseph Kruskal. In this method, all edges are ordered by weight, and then each edge in this ordered sequence is checked to see whether it can be considered part of the tree under construction. It is added to the tree if no cycle arises after its inclusion. This simple algorithm can be summarized as follows:

```
KruskalAlgorithm(weighted connected undirected graph)
    tree = null;
    edges = sequence of all edges of graph sorted by weight;
    for (i = 1; i ≤ |E| and |tree| < |V| − 1; i++)
        if e_i from edges does not form a cycle with edges in tree
            add e_i to tree;
```

Figures 8.15ba–bf contain a step-by-step example of Kruskal's algorithm.

The complexity of this algorithm is determined by the complexity of the sorting method applied, which for an efficient sorting is $O(|E| \lg |E|)$. It also depends on the complexity of the method used for cycle detection. If we use union() to implement Kruskal's algorithm, then the for loop of KruskalAlgorithm() becomes

```
for (i = 1; i ≤ |E| and |tree| < |V| - 1; i++)
    union(e_i = edge(vu));
```

Although union() can be called up to $|E|$ times, it is exited after one (the first) test if a cycle is detected and it performs a union, which is of complexity $O(|V|)$, only for $|V| - 1$ edges added to tree. Hence, the complexity of KruskalAlgorithm()'s for loop is $O(|E| + (|V| - 1)|V|)$, which is $O(|V|^2)$. Therefore, the complexity of KruskalAlgorithm() is determined by the complexity of a sorting algorithm, which is $O(|E|\lg|E|)$, that is, $O(|E|\lg|V|)$.

Kruskal's algorithm requires that all the edges be ordered before beginning to build the spanning tree. This, however, is not necessary; it is possible to build a spanning tree by using any order of edges. A method was proposed by Dijkstra (1960) and independently by Robert Kalaba.

FIGURE 8.15 A spanning tree of graph (a) found, (ba–bf) with Kruskal's algorithm, (ca–cl) and with Dijkstra's method.

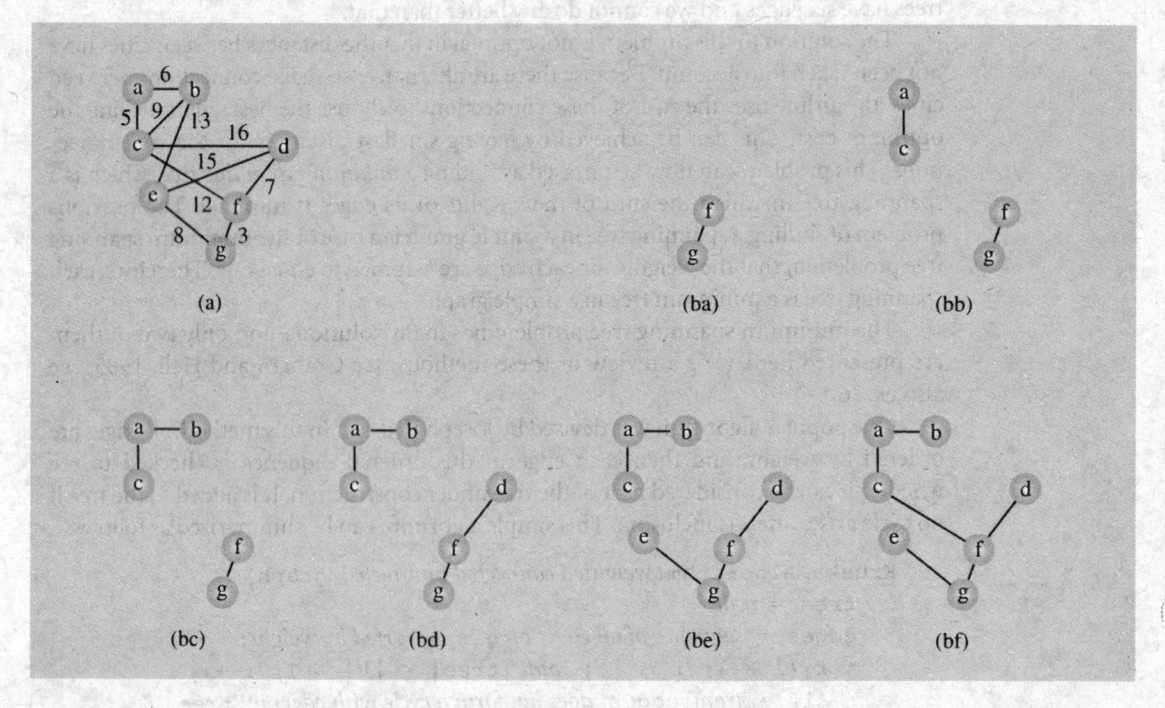

Section 8.5 Spanning Trees   399

FIGURE 8.15   *(continued)*

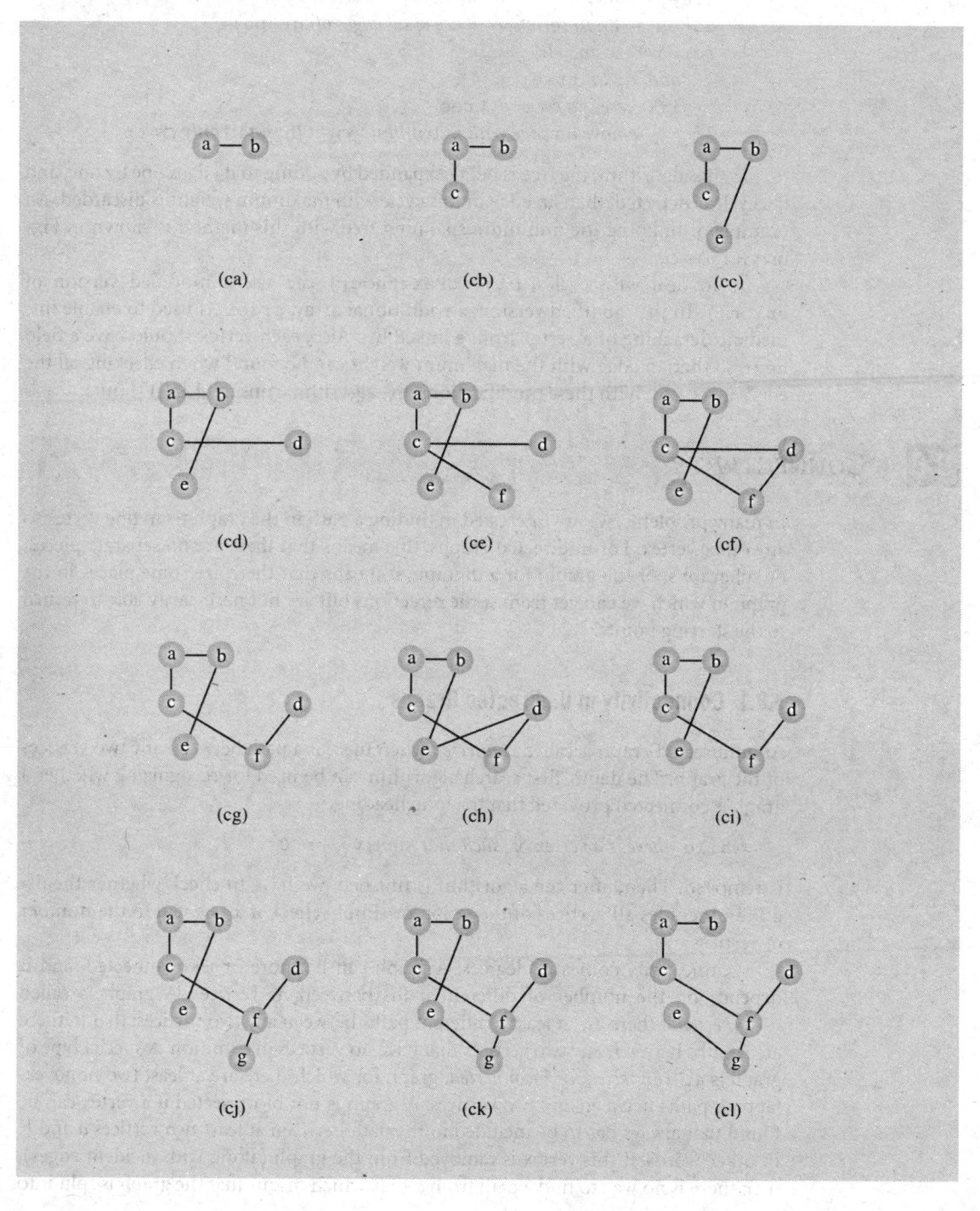

(ca)

(cb)

(cc)

(cd)

(ce)

(cf)

(cg)

(ch)

(ci)

(cj)

(ck)

(cl)

```
DijkstraMethod(weighted connected undirected graph)
    tree = null;
    edges = an unsorted sequence of all edges of graph;
    for j = 1 to |E|
        add e_j to tree;
        if there is a cycle in tree
            remove an edge with maximum weight from this only cycle;
```

In this algorithm, the tree is being expanded by adding to it edges one by one, and if a cycle is detected, then an edge in this cycle with maximum weight is discarded. An example of building the minimum spanning tree with this method is shown in Figures 8.15ca–cl.

To deal with cycles, `DijkstraMethod()` can use a modified version of `union()`. In the modified version, an additional array, `prior`, is used to enable immediate detaching of a vertex from a linked list. Also, each vertex should have a field `next` so that an edge with the maximum weight can be found when checking all the edges in a cycle. With these modifications, the algorithm runs in $O(|E||V|)$ time.

8.6 CONNECTIVITY

In many problems, we are interested in finding a path in the graph from one vertex to any other vertex. For undirected graphs, this means that there are no separate pieces, or subgraphs, of the graph; for a digraph, it means that there are some places in the graph to which we can get from some directions but are not necessarily able to return to the starting points.

8.6.1 Connectivity in Undirected Graphs

An undirected graph is called *connected* when there is a path between any two vertices of the graph. The depth-first search algorithm can be used for recognizing whether a graph is connected provided that the loop heading

```
while there is a vertex v such that num(v) == 0
```

is removed. Then, after the algorithm is finished, we have to check whether the list `edges` includes all vertices of the graph or simply check if `i` is equal to the number of vertices.

Connectivity comes in degrees: A graph can be more or less connected, and it depends on the number of different paths between its vertices. A graph is called *n-connected* if there are at least *n* different paths between any two vertices; that is, there are *n* paths between any two vertices that have no vertices in common. A special type of graph is a *2-connected*, or *biconnected*, graph for which there are at least two nonoverlapping paths between any two vertices. A graph is not biconnected if a vertex can be found that always has to be included in the path between at least two vertices *a* and *b*. In other words, if this vertex is removed from the graph (along with incident edges), then there is no way to find a path from *a* to *b*, which means that the graph is split into two separate subgraphs. Such vertices are called *articulation points*, or *cut-vertices*. Vertices *a* and *b* in Figure 8.1d are examples of articulation points. If an edge causes a graph to be split into two subgraphs, it is called a *bridge*, or *cut-edge*, as for example, the *edge(bc)* in Figure 8.1d. Connected subgraphs with no articulation points or bridges are called *blocks*, or—when they include at least two vertices—*biconnected components*. It is important to know how to decompose a graph into biconnected components.

Articulation points can be detected by extending the depth-first search algorithm. This algorithm creates a tree with forward edges (the graph edges included in the tree) and back edges (the edges not included). A vertex *v* in this tree is an articulation point if it has at least one subtree unconnected with any of its predecessors by a back edge; because it is a tree, certainly none of *v*'s predecessors is reachable from any of its successors by a forward link. For example, the graph in Figure 8.16a is transformed into a depth-first search tree (Figure 8.16c), and this tree has four articulation points, *b, d, h*, and *i*, because there is no back edge from any node below *d* to any node above it in the tree and no back edge from any vertex in the right subtree of *h* to any vertex above *h*. But vertex *g* cannot be an articulation point because its successor *h* is connected to a vertex above it. The four vertices divide the graph into the five blocks indicated in Figure 8.16c by dotted lines.

A special case for an articulation point is when a vertex is a root with more than one descendant. In Figure 8.16a, the vertex chosen for the root, *a*, has three incident edges, but only one of them becomes a forward edge in Figures 8.16b and 8.16c, because the other two are processed by depth-first search. Therefore, if this algorithm again recursively reaches *a*, there is no untried edge. If *a* were an articulation point, there would be at least one such untried edge, and this indicates that *a* is a cut vertex. So *a* is not an articulation point. To sum up, we say that a vertex *v* is an articulation point

1. if *v* is the root of the depth-first search tree and *v* has more than one descendant in this tree or
2. if at least one of *v*'s subtrees includes no vertex connected by a back edge with any of *v*'s predecessors.

To find articulation points, a parameter *pred(v)* is used, defined as $\min(num(v), num(u_1), \ldots, num(u_k))$, where $u_1, \ldots, u_k$ are vertices connected by a back edge with a descendant of *v* or with *v* itself. Because the higher a predecessor of *v* is, the lower its number is, choosing a minimum number means choosing the highest predecessor. For the tree in Figure 8.16c, $pred(c) = pred(d) = 1, pred(b) = 4$, and $pred(k) = 7$.

The algorithm uses a stack to store all currently processed edges. After an articulation point is identified, the edges corresponding to a block of the graph are output. The algorithm is given as follows:

```
blockDFS(v)
    pred(v) = num(v) = i++;
    for all vertices u adjacent to v
        if edge(uv) has not been processed
            push(edge(uv));
        if num(u) is 0
            blockDFS(u);
```

FIGURE 8.16 Finding blocks and articulation points using the `blockDFS()` algorithm.

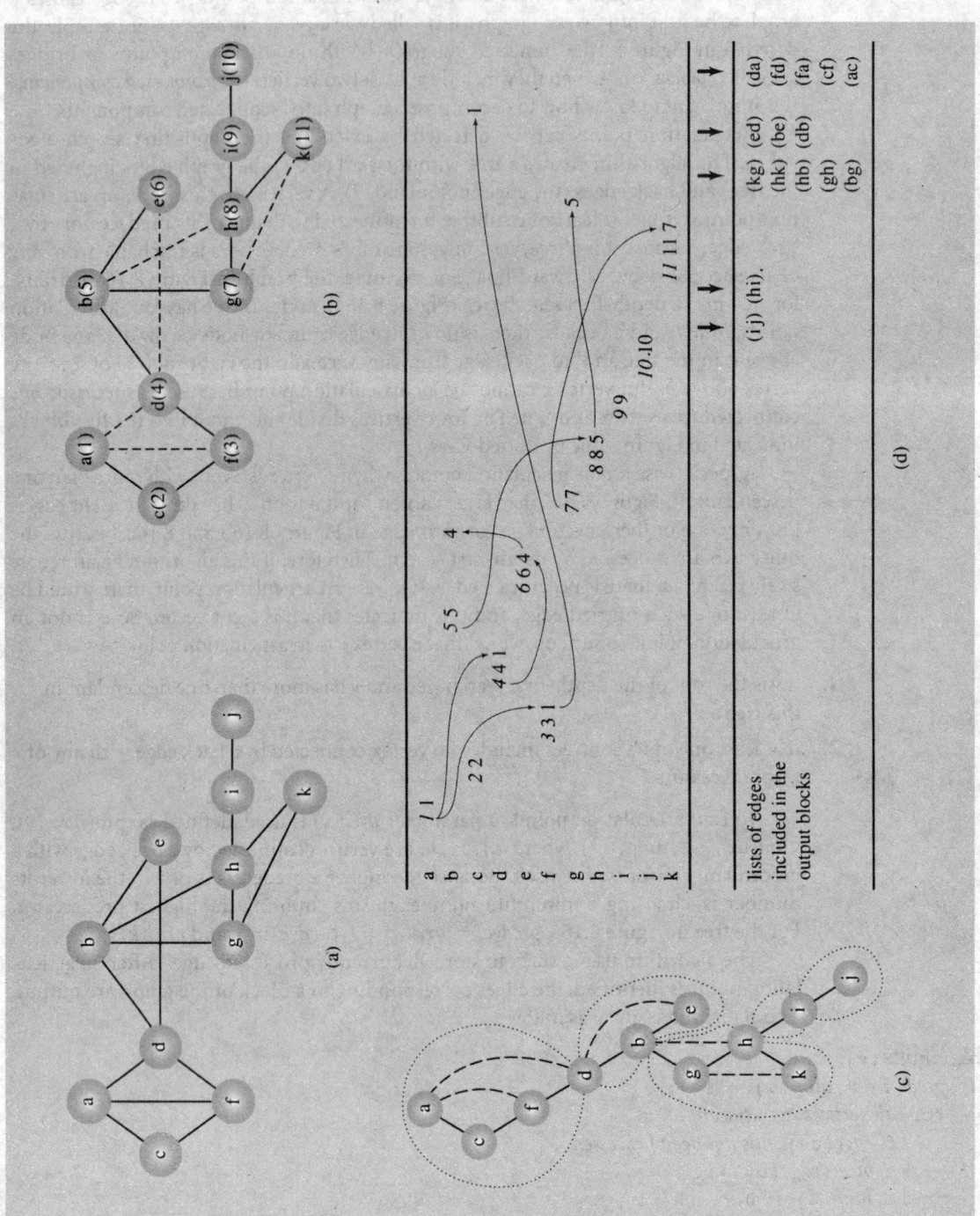

```
            if pred(u) ≥ num(v)              // if there is no edge from u to a
                e = pop();                   // vertex above v, output a block
                while e ≠ edge(vu)           // by popping all edges off the
                    output e;                // stack until edge(vu) is
                    e = pop();               // popped off;
                output e;                    // e == edge(vu);
            else pred(v) = min(pred(v),pred(u));  // take a predecessor higher up in
        else if u is not the parent of v     // tree;
            pred(v) = min(pred(v),num(u));   // update when back edge(vu) is
                                             // found;
blockSearch()
    for all vertices v
        num(v) = 0;
    i = 1;
    while there is a vertex v such that num(v) == 0
        blockDFS(v);
```

An example of the execution of this algorithm is shown in Figure 8.16d as applied to the graph in Figure 8.16a. The table lists all changes in *pred(v)* for vertices *v* processed by the algorithm, and the arrows show the source of the new values of *pred(v)*. For each vertex *v*, `blockDFS(v)` first assigns two numbers: *num(v)*, shown in italics, and *pred(v)*, which may change during the execution of `blockDFS(v)`. For example, *a* is processed first with *num(a)* and *pred(a)* set to 1. The *edge(ac)* is pushed onto the stack, and because *num(c)* is 0, the algorithm is invoked for *c*. At this point, *num(c)* and *pred(c)* are set to 2. Next, the algorithm is invoked for *f*, a descendant of *c*, so that *num(f)* and *pred(f)* are set to 3, and then it is invoked for *a*, a descendant of *f*. Because *num(a)* is not 0 and *a* is not *f*'s parent, *pred(f)* is set to 1 = min(*pred(f)*,*num(a)*) = min(3, 1).

This algorithm also outputs the edges in detected blocks, and these edges are shown in Figure 8.16d at the moment they were output after popping them off the stack.

8.6.2 Connectivity in Directed Graphs

For directed graphs, connectedness can be defined in two ways depending on whether or not the direction of the edges is taken into account. A directed graph is *weakly connected* if the undirected graph with the same vertices and the same edges is connected. A directed graph is *strongly connected* if for each pair of vertices there is a path between them in both directions. The entire digraph is not always strongly connected, but it may be composed of *strongly connected components* (SCC), which are defined as subsets of vertices of the graph such that each of these subsets induces a strongly connected digraph.

To determine SCCs, we also refer to depth-first search. Let vertex *v* be the first vertex of an SCC for which depth-first search is applied. Such a vertex is called the *root of the SCC*. Because each vertex *u* in this SCC is reachable from *v*, *num(v)* < *num(u)*, and only after all such vertices *u* have been visited, the depth-first search backtracks to *v*. In this case, which is recognized by the fact that *pred(v)* = *num(v)*, the SCC accessible from the root can be output.

The problem now is how to find all such roots of the digraph, which is analogous to finding articulation points in an undirected graph. To that end, the parameter *pred(v)* is also used, where *pred(v)* is the lower number chosen out of *num(v)* and *pred(u)*, where *u* is a vertex reachable from *v* and belonging to the same SCC as *v*. How can we determine whether two vertices belong to the same SCC before the SCC has been determined? The apparent circularity is solved by using a stack that stores all vertices belonging to the SCCs under construction. The topmost vertices on the stack belong to the currently analyzed SCC. Although construction is not finished, we at least know which vertices are already included in the SCC. The algorithm attributed to Tarjan is as follows:

```
strongDFS(v)
  pred(v) = num(v) = i++;
  push(v);
  for all vertices u adjacent to v
    if num(u) is 0
      strongDFS(u);
      pred(v) = min(pred(v),pred(u));            // take a predecessor higher up in
    else if num(u) < num(v) and u is on stack    // tree; update if back edge found
      pred(v) = min(pred(v),num(u));             // to vertex u is in the same SCC;
  if pred(v) == num(v)                           // if the root of a SCC is found,
    w = pop();                                   // output this SCC, i.e.,
    while w ≠ v                                  // pop all vertices off the stack
      output w;                                  // until v is popped off;
      w = pop();
    output w;                                    // w == v;

stronglyConnectedComponentSearch()
  for all vertices v
    num(v) = 0;
  i = 1;
  while there is a vertex v such that num(v) == 0
    strongDFS(v);
```

Figure 8.17 contains an example of execution of Tarjan's algorithm. The digraph in Figure 8.17a is processed by a series of calls to `strongDFS()`, which assigns to vertices *a* through *k* the numbers shown in parentheses in Figure 8.17b. During this process, five SCCs are detected: {*a,c,f*},{*b,d,e,g,h*},{*i*},{*j*}, and {*k*}. Figure 8.17c contains the depth-first search trees created by this process. Note that two trees are created so that the number of trees does not have to correspond to the number of SCCs, as the number of trees did not correspond to the number of blocks in the case for undirected graphs. Figure 8.17d indicates, in italics, numbers assigned to *num(v)* and all changes of parameter *pred(v)* for all vertices *v* in the graph. It also shows the SCC's output during the processing of the graph.

FIGURE 8.17 Finding strongly connected components with the `strongDFS()` algorithm.

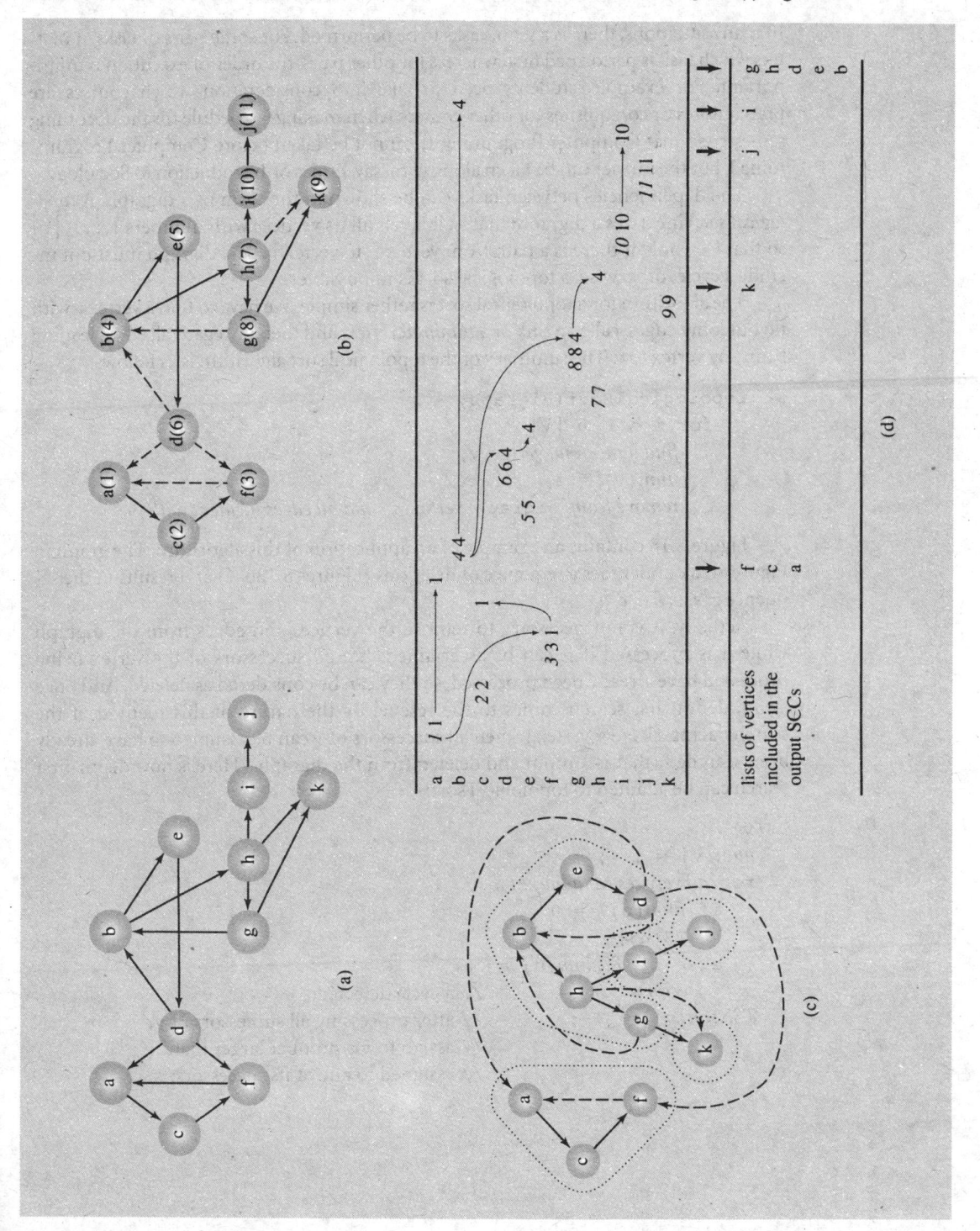

8.7 TOPOLOGICAL SORT

In many situations, there is a set of tasks to be performed. For some pairs of tasks, it matters which task is performed first, whereas for other pairs, the order of execution is unimportant. For example, students need to take into consideration which courses are prerequisites or corequisites for other courses when making a schedule for the upcoming semester so that Computer Programing II cannot be taken before Computer Programming I, but the former can be taken along with, say, Ethics or Introduction to Sociology.

The dependencies between tasks can be shown in the form of a digraph. A *topological sort* linearizes a digraph; that is, it labels all its vertices with numbers $1, \ldots, |V|$ so that $i < j$ only if there is a path from vertex v_i to vertex v_j. The digraph must not include a cycle; otherwise, a topological sort is impossible.

The algorithm for a topological sort is rather simple. We have to find a vertex v with no outgoing edges, called a *sink* or a *minimal vertex*, and then disregard all edges leading from any vertex to v. The summary of the topological sort algorithm is as follows:

```
topologicalSort(digraph)
    for i = 1 to |V|
        find a minimal vertex v;
        num(v) = i;
        remove from digraph vertex v and all edges incident with v;
```

Figure 8.18 contains an example of an application of this algorithm. The graph in Figure 8.18a undergoes a sequence of deletions (Figures 8.18b–f) and results in the sequence *g, e, b, f, d, c, a*.

Actually, it is not necessary to remove the vertices and edges from the digraph while it is processed if it can be ascertained that all successors of the vertex being processed have already been processed, so they can be considered as deleted. And once again, depth-first search comes to the rescue. By the nature of this method, if the search backtracks to a vertex v, then all successors of v can be assumed to have already been searched (that is, output and deleted from the digraph). Here is how depth-first search can be adapted to topological sort:

```
TS(v)
    num(v) = i++;
    for all vertices u adjacent to v
        if num(u) == 0
            TS(u);
        else if TSNum(u) == 0
            error;                  // a cycle detected;
    TSNum(v) = j++;                 // after processing all successors of v,
                                    // assign to v a number larger than
                                    // assigned to any of its successors;
```

FIGURE 8.18 Executing a topological sort.

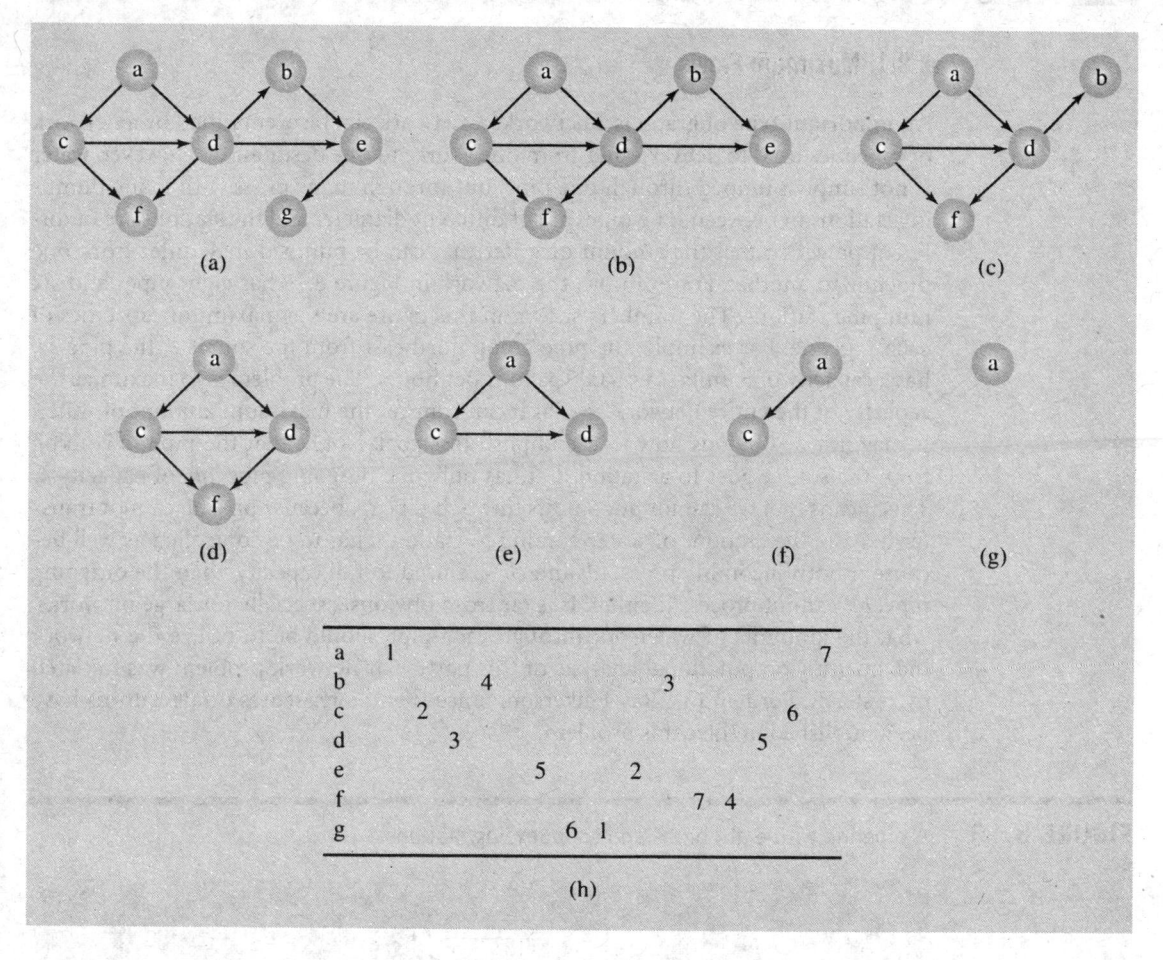

a	1					7
b			4		3	
c		2				6
d			3			5
e				5	2	
f						7 4
g				6	1	

(h)

```
topologicalSorting(digraph)
   for all vertices v
      num(v) = TSNum(v) = 0;
   i = j = 1;
   while there is a vertex v such that num(v) == 0
      TS(v);
   output vertices according to their TSNum's;
```

The table in Figure 8.18h indicates the order in which this algorithm assigns $num(v)$, the first number in each row, and $TSNum(v)$, the second number, for each vertex v of the graph in Figure 8.18a.

8.8 NETWORKS

8.8.1 Maximum Flows

An important type of graph is a network. A network can be exemplified by a network of pipelines used to deliver water from one source to one destination. However, water is not simply pumped through one pipe, but through many pipes with many pumping stations in between. The pipes are of different diameter and the stations are of different power so that the amount of water that can be pumped may differ from one pipeline to another. For example, the network in Figure 8.19 has eight pipes and six pumping stations. The numbers shown in this figure are the maximum capacities of each pipeline. For example, the pipe going northeast from the source s, the pipe sa, has a capacity of 5 units (say, 5,000 gallons per hour). The problem is to maximize the capacity of the entire network so that it can transfer the maximum amount of water. It may not be obvious how to accomplish this goal. Notice that the pipe sa coming from the source goes to a station that has only one outgoing pipe, ab, of capacity 4. This means that we cannot put 5 units through pipe sa, because pipe ab cannot transfer it. Also, the amount of water coming to station b has to be controlled as well because if both incoming pipes, ab and cb, are used to full capacity, then the outgoing pipe, bt, cannot process it either. It is far from obvious, especially for large networks, what the amounts of water put through each pipe should be to utilize the network maximally. Computational analysis of this particular network problem was initiated by Lester R. Ford and D. Ray Fulkerson. Since their work, scores of algorithms have been published to solve this problem.

FIGURE 8.19 A pipeline with eight pipes and six pumping stations.

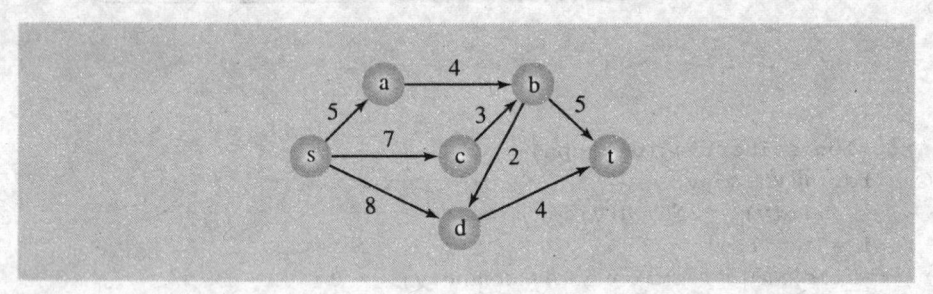

Before the problem is stated more formally, I would like to give some definitions. A *network* is a digraph with one vertex s, called the *source*, with no incoming edges, and one vertex t, called the *sink*, with no outgoing edges. (These definitions are chosen for their intuitiveness; however, in a more general case, both source and sink can be any two vertices.) With each edge e we associate a number $cap(e)$ called the *capacity* of the edge. A *flow* is a real function $f: E \rightarrow R$ that assigns a number to each edge of the network and meets these two conditions:

1. The flow through an edge e cannot be greater than its capacity, or $0 \le f(e) \le cap(e)$ (capacity constraint).
2. The total flow coming to a vertex v is the same as the total flow coming from it, or $\sum_u f(edge(uv)) = \sum_w f(edge(vw))$, where v is neither the source nor the sink (flow conservation).

The problem now is to maximize the flow f so that the sum $\sum_u f(edge(ut))$ has a maximum value for any possible function f. This is called a *maximum-flow* (or *max-flow*) *problem*.

An important concept used in the Ford-Fulkerson algorithm is the concept of cut. A *cut separating s and t* is a set of edges between vertices of set X and vertices of set $\overline{X}$; any vertex of the graph belongs to one of these sets, and source s is in X and sink t is in $\overline{X}$. For example, in Figure 8.19, if $X = \{s,a\}$, then $\overline{X} = \{b,c,d,t\}$, and the cut is the set of edges $\{(a,b),(s,c),(s,d)\}$. This means that if all edges belonging to this set are cut, then there is no way to get from s to t. Let us define the capacity of the cut as the sum of capacities of all its edges leading from a vertex in X to a vertex in $\overline{X}$; thus, $cap\{(a,b),(s,c),(s,d)\} = cap(a,b) + cap(s,c) + cap(s,d) = 19$. Now, it should be clear that the flow through the network cannot be greater than the capacity of any cut. This observation leads to the *max-flow min-cut theorem* (Ford and Fulkerson 1956):

Theorem. In any network, the maximal flow from s to t is equal to the minimal capacity of any cut.

This theorem states what is expressed in the simile of a chain being as strong as its weakest link. Although there may be cuts with great capacity, the cut with the smallest capacity determines the flow of the network. For example, although the capacity $cap\{(a,b),(s,c),(s,d)\} = 19$, two edges coming to t cannot transfer more than 9 units. Now we have to find a cut that has the smallest capacity among all possible cuts and transfer through each edge of this cut as many units as the capacity allows. To that end, a new concept is used.

A *flow-augmenting path* from s to t is a sequence of edges from s to t such that, for each edge in this path, the flow $f(e) < cap(e)$ on forward edges and $f(e) > 0$ on backward edges. It means that such a path is not optimally used yet, and it can transfer more units than it is currently transferring. If the flow for at least one edge of the path reaches its capacity, then obviously the flow cannot be augmented. Note that the path does not have to consist only of forward edges, so that examples of paths in Figure 8.19 are s, a, b, t, and s, d, b, t. Backward edges are what they are, backward; they push back some units of flow, decreasing the flow of the network. If they can be eliminated, then the overall flow in the network can be increased. Hence, the process of augmenting flows of paths is not finished until the flow for such edges is zero. Our task now is to find an augmenting path if it exists. There may be a very large number of paths from s to t, so finding an augmenting path is a nontrivial problem, and Ford and Fulkerson (1957) devised the first algorithm to accomplish it in a systematic manner.

The *labeling* phase of the algorithm consists of assigning to each vertex v a label, which is the pair

$$label(v) = (parent(v), slack(v))$$

where *parent(v)* is the vertex from which v is being accessed and *slack(v)* is the amount of flow that can be transferred from s to v. The forward and backward edges are treated differently. If a vertex u is accessed from v through a forward edge, then

$$label(u) = (v^+, \min(slack(v), slack(edge(vu))))$$

where

$$slack(edge(vu)) = cap(edge(vu)) - f(edge(vu))$$

which is the difference between the capacity of *edge(vu)* and the amount of flow currently carried by this edge. If the edge from v to u is backward (i.e., forward from u to v), then

$$label(u) = (v^-, \min(slack(v), f(edge(uv))))$$

and

$$slack(v) = \min(slack(parent(v)), slack(edge(parent(v)v)))$$

After a vertex is labeled, it is stored for later processing. In this process, only this *edge(vu)* is labeled, which allows for some more flow to be added. For forward edges, this is possible when $slack(edge(vu)) > 0$, and for backward edges when $f(edge(uv)) > 0$. However, finding one such path may not finish the entire process. The process is finished if we are stuck in the middle of the network unable to label any more edges. If we reach the sink t, the flows of the edges on the augmenting path that was just found are updated by increasing flows of forward edges and decreasing flows of backward edges, and the process restarts in the quest for another augmenting path. Here is a summary of the algorithm.

```
augmentPath(network with source s and sink t)
    for each edge e in the path from s to t
        if forward(e)
            f(e) += slack(t);
        else f(e) -= slack(t);

FordFulkersonAlgorithm(network with source s and sink t)
    set flow of all edges and vertices to 0;
    label(s) = (null,∞);
    labeled = {s};
    while labeled is not empty   // while not stuck;
        detach a vertex v from labeled;
        for all unlabeled vertices u adjacent to v
            if forward(edge(vu)) and slack(edge(vu)) > 0
                label(u) = (v⁺,min(slack(v),slack(edge(vu))))
            else if backward(edge(vu)) and f(edge(uv)) > 0
                label(u) = (v⁻,min(slack(v),f(edge(uv))));
            if u got labeled
                if u == t
                    augmentPath(network);
                    labeled = {s};   // look for another path;
                else include u in labeled;
```

Notice that this algorithm is noncommittal with respect to the way the network should be scanned. In exactly what order should vertices be included in `labeled` and detached from it? This question is left open, and we choose push and pop as implementations of these two operations, thereby processing the network in a depth-first fashion.

Figure 8.20 illustrates an example. Each edge has two numbers associated with it, the capacity and the current flow, and initially the flow is set to zero for each edge (8.20a). We begin by putting the vertex s in `labeled`. In the first iteration of the `while` loop, s is detached from `labeled`, and in the `for` loop, label $(s,2)$ is assigned to the first adjacent vertex, a; label $(s,4)$ to vertex c; and label $(s,1)$ to vertex e (Figure 8.20b), and all three vertices are pushed onto `labeled`. The `for` loop is exited, and because `labeled` is not empty, the `while` loop begins its second iteration. In this iteration, a vertex is popped off from `labeled`, which is e, and both unlabeled vertices incident to e, vertices d and f, are labeled and pushed onto `labeled`. Now, the third

FIGURE 8.20 An execution of `FordFulkersonAlgorithm()` using depth-first search.

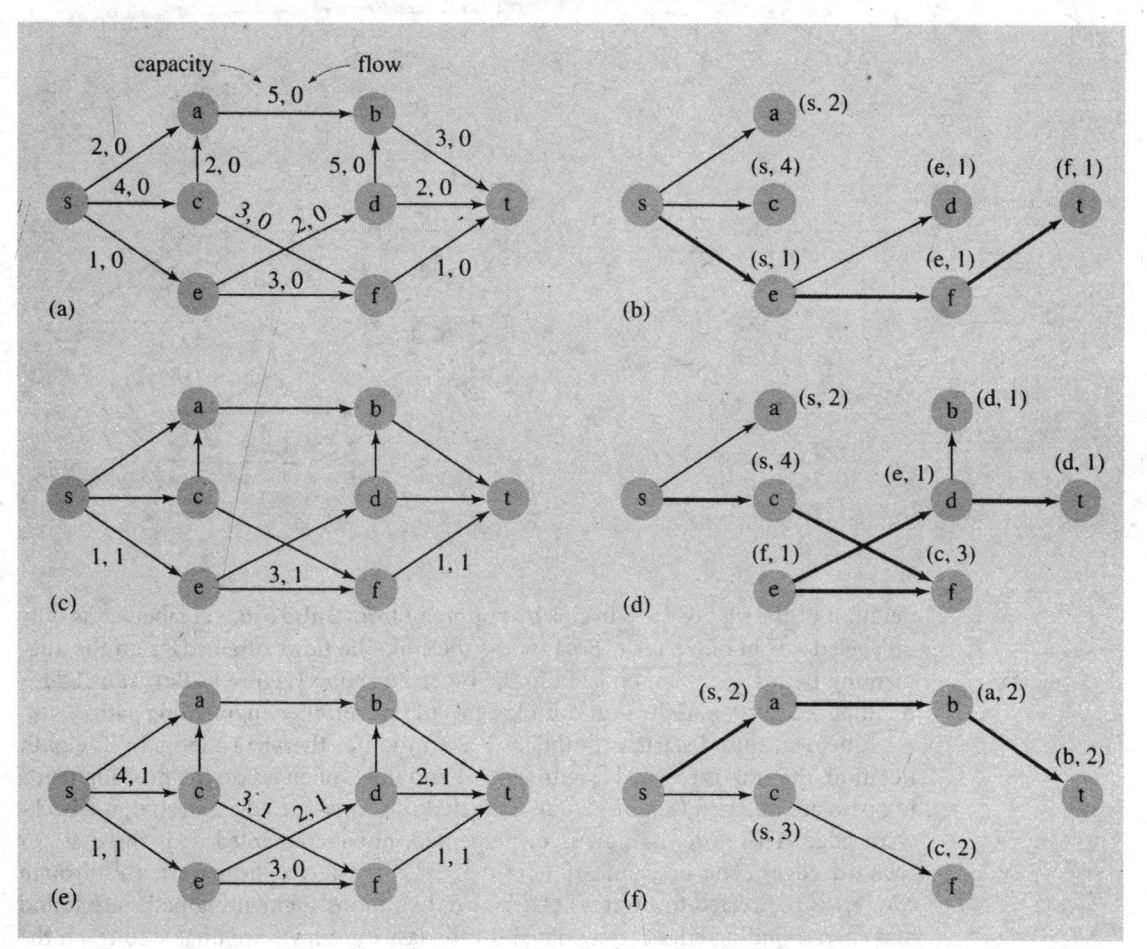

Continues

FIGURE 8.20 (continued)

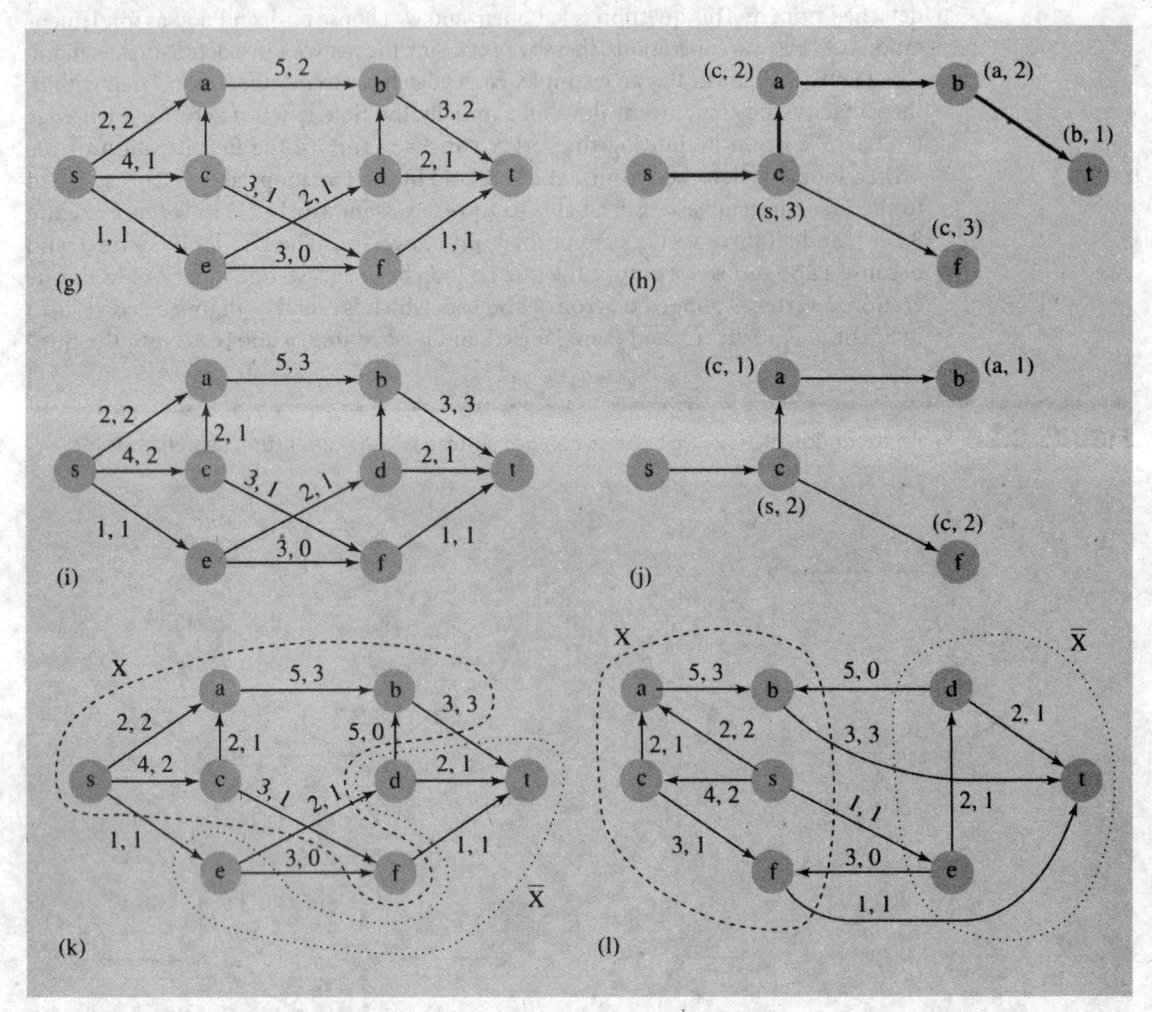

iteration of the `while` loop begins by popping *f* from `labeled` and labeling its only unlabeled neighbor, vertex *t*. Because *t* is the sink, the flows of all edges on the augmenting path *s, e, f, t* are updated in the inner `for` loop (Figure 8.20c), `labeled` is reinitialized to {*s*}, and the next round begins to find another augmenting path.

The next round starts with the fourth iteration of the `while` loop. In its eighth iteration, the sink is reached (Figure 8.20d) and flows of edges on the new augmenting path are updated (Figure 8.20e). Note that this time one edge, *edge(fe)*, is a backward edge. Therefore, its flow is decremented, not incremented as is the case for forward edges. The one unit of flow that was transferred from vertex *e* through *edge(ef)* is redirected to *edge(ed)*. Afterward, two more augmenting paths are found and corresponding edges are updated. In the last round, we are unable to reach the sink (Figure 8.20j), which means that all augmenting edges have been found and the maximum flow has been determined.

If after finishing execution of the algorithm all vertices labeled in the last round, including the source, are put in the set X and the unlabeled vertices in the set $\overline{X}$, then we have a min-cut (Figure 8.20k). For clarity, both sets are also shown in Figure 8.20l. Note that all the edges from X to $\overline{X}$ are used in full capacity, and all the edges from $\overline{X}$ to X do not transfer any flow at all.

The complexity of this implementation of the algorithm is not necessarily a function of the number of vertices and edges in the network. Consider the network in Figure 8.21. Using a depth-first implementation, we could choose the augmenting path s, a, b, t with flows of all three edges set to 1. The next augmenting path could be s, b, a, t with flows of two forward edges set to 1 and the flow of one backward $edge(ba)$ reset to 0. Next time, the augmenting path could be the same as the first, with flows of two edges set to 2 and with the vertical edge set to 1. It is clear that an augmenting path could be chosen $2 \cdot 10$ times, although there are only four vertices in the network.

FIGURE 8.21 An example of an inefficiency of `FordFulkersonAlgorithm()`.

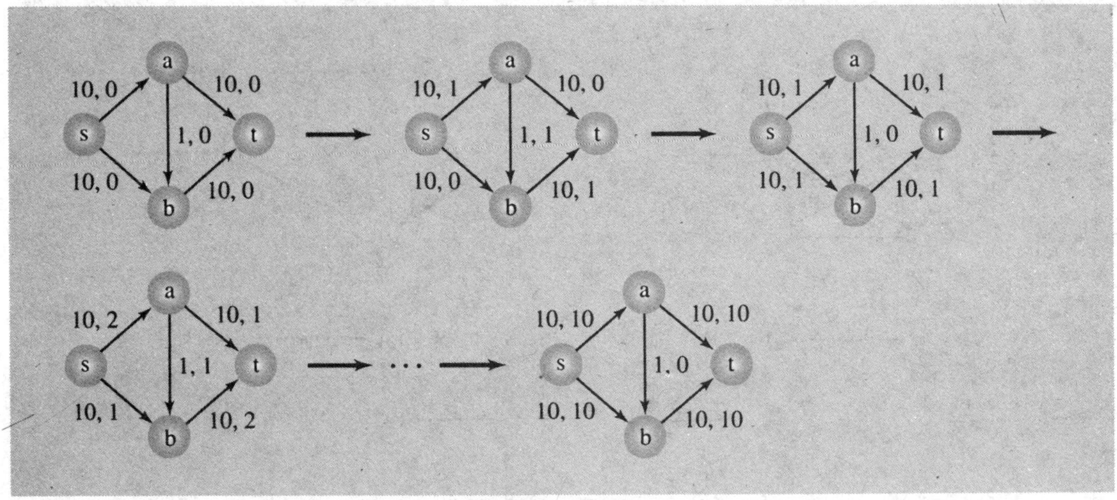

The problem with `FordFulkersonAlgorithm()` is that it uses the depth-first approach when searching for an augmenting path. But as already mentioned, this choice does not stem from the nature of this algorithm. The depth-first approach attempts to reach the sink as soon as possible. However, trying to find the shortest augmenting path gives better results. This leads to a breadth-first approach (Edmonds and Karp 1972). The breadth-first processing uses the same procedure as `FordFulkersonAlgorithm()` except that this time `labeled` is a queue. Figure 8.22 illustrates an example.

FIGURE 8.22 An execution of `FordFulkersonAlgorithm()` using breadth-first search.

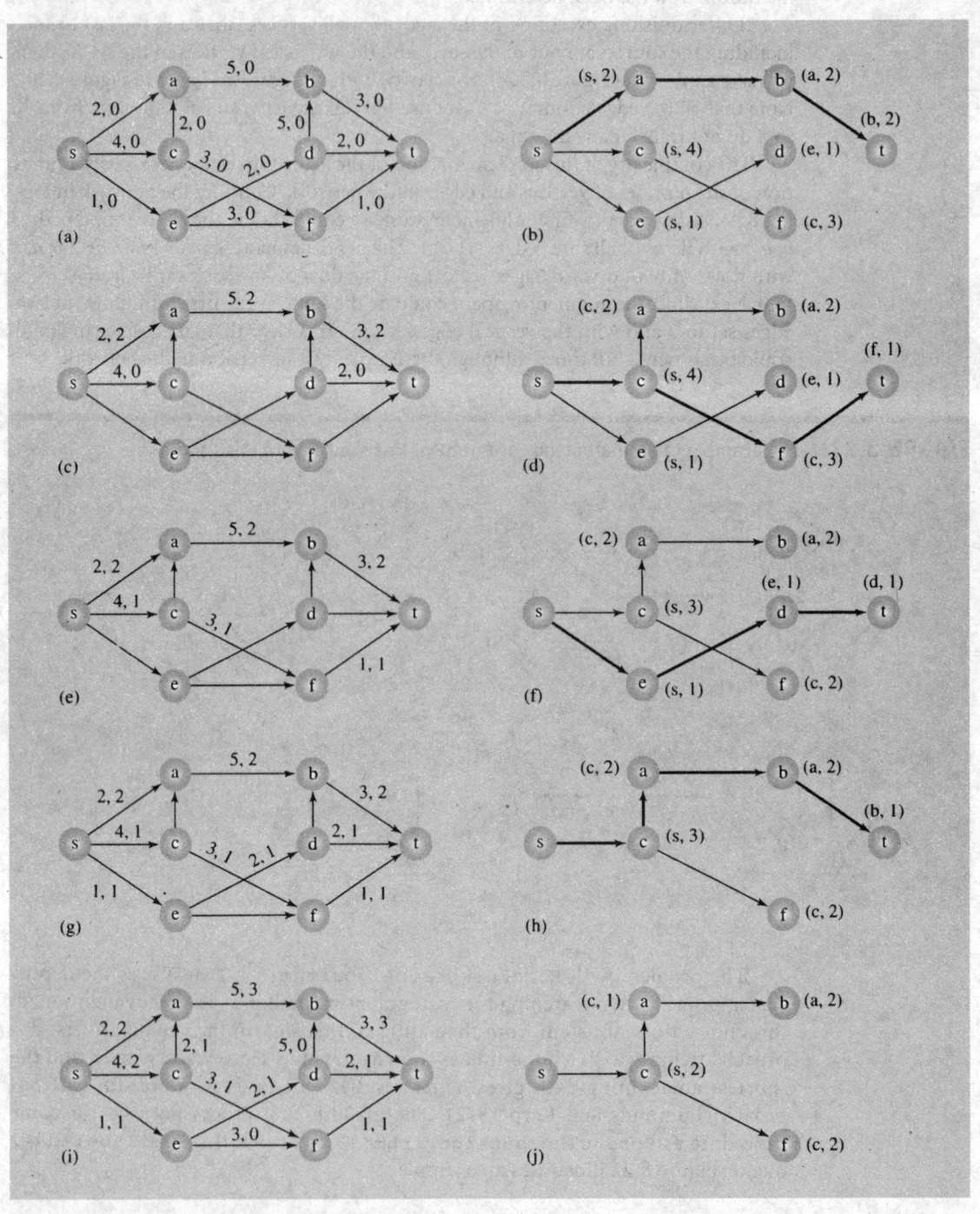

To determine one single augmenting path, the algorithm requires at most $2|E|$, or $O(|E|)$ steps, to check both sides of each edge. The shortest augmenting path in the network can have only one edge, and the longest path can have at most $|V| - 1$ edges. Therefore, there can be augmenting paths of lengths $1, 2, \ldots, |V| - 1$. The number of augmenting paths of a certain length is at most $|E|$. Therefore, to find all augmenting paths of all possible lengths, the algorithm needs to perform $O(|V||E|)$ steps. And because finding one such path is of order $O(|E|)$, the algorithm is of order $O(|V||E|^2)$.

Although the pure breadth-first search approach is better than the pure depth-first search implementation, it still is far from ideal. We will not fall into a loop of tiny increments of augmenting steps anymore, but there still seems to be a great deal of wasted effort. In breadth-first search, a large number of vertices are labeled to find the shortest path (shortest in a given iteration). Then all these labels are discarded to re-create them when looking for another augmenting path (*edge*(*sc*), *edge*(*se*), and *edge*(*cf*) in Figure 8.22b–d). Therefore, it is desirable to reduce this redundancy. Also, there is some merit to using the depth-first approach in that it attempts to aim at the goal, the sink, without expanding a number of paths at the same time and finally choosing only one and discarding the rest. Hence, the Solomonic solution appears to use both approaches, depth-first and breadth-first. Breadth-first search prepares the ground to prevent loops of small increments from happening (as in Figure 8.21) and to guarantee that depth-first search takes the shortest route. Only afterward, the depth-first search is launched to find the sink by aiming right at it. An algorithm based upon this principle was devised first by Efim A. Dinic (pronounced: dee-neetz).

In Dinic's algorithm, up to $|V| - 1$ passes (or phases) through the network are performed, and in each pass, all augmenting paths of the same length from the source to the sink are determined. Then, only some or all of these paths are augmented.

All augmenting paths form a *layered network* (also called a *level network*). Extracting layered networks from the underlying network starts from the lowest values. First, a layered network of a path of length one is found, if such a network exists. After the network is processed, a layered network of paths of length two is determined, if it exists, and so on. For example, the layered network with the shortest paths corresponding with the network in Figure 8.23a is shown in Figure 8.23b. In this network, all augmenting paths are of length three. A layered network with a single path of length one and layered networks with paths of length two do not exist. The layered network is created using breadth-first processing, and only forward edges that can carry more flow and backward edges that already carry some flow are included. Otherwise, even if an edge may lay on a short path from the source to the sink, it is not included. Note that the layered network is determined by breadth-first search that begins in the sink and ends in the source.

Now, because all the paths in a layered network are of the same length, it is possible to avoid redundant tests of edges that are part of augmenting paths. If in a current layered network there is no way to go from a vertex v to any of its neighbors, then in later tests in the same layered network there will be the same situation; hence, checking again all neighbors of v is not needed. Therefore, if such a dead-end vertex v is detected, all edges incident with v are marked as blocked so that there is no possibility to get to v from any direction. Also, all saturated edges are considered blocked. All blocked edges are shown in dashed lines in Figure 8.23.

FIGURE 8.23 An execution of `DinicAlgorithm()`.

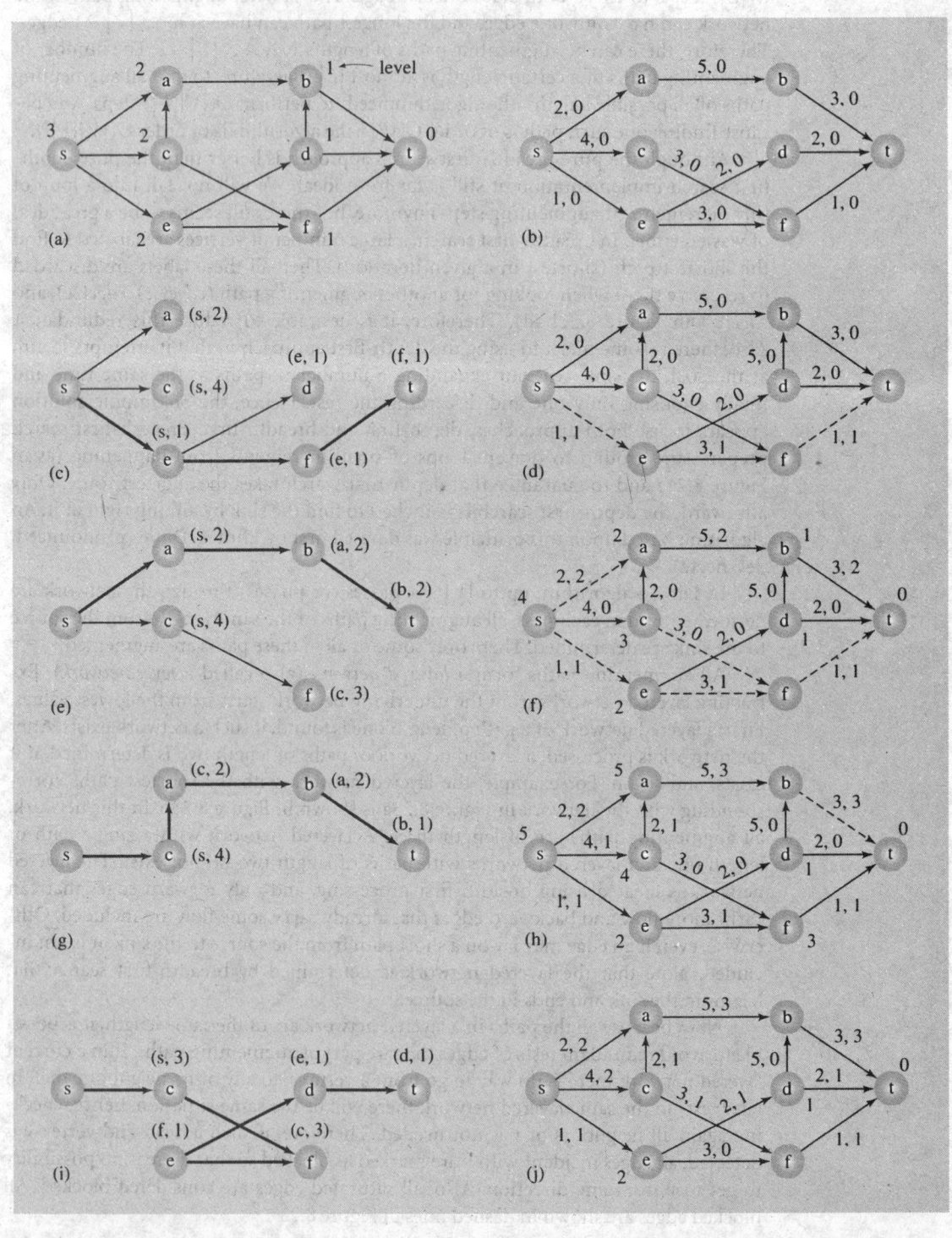

After a layered network is determined, the depth-first process finds as many augmenting paths as possible. Because all paths are of the same length, depth-first search does not go to the sink through some longer sequence of edges. After one such path is found, it is augmented and another augmenting path of the same length is looked for. For each such path, at least one edge becomes saturated so that eventually no augmenting path can be found. For example, in the layered network in Figure 8.23b that includes only augmenting paths three edges long, path *s, e, f, t* is found (Figure 8.23c), and all its edges are augmented (Figure 8.23d). Then only one more three-edge path is found, the path *s, a, b, t* (8.23e), because, for example, previous augmentation saturated *edge(ft)* so that the partial path *s, c, f* ends with a dead end. In addition, because no other vertex can be reached from *f*, all edges incident with *f* are blocked (Figure 8.23f) so that an attempt to find the third three-edge augmenting path only tests vertex *c*, but not vertex *f*, because *edge(cf)* is blocked.

If no more augmenting paths can be found, a higher level layered network is found, and augmenting paths for this network are searched for. The process stops when no layered network can be formed. For example, out of the network in Figure 8.23f, the layered network in Figure 8.23g is formed, which has only one four-edge path. To be sure, this is the only augmenting path for this network. After augmenting this path, the situation in the network is as in Figure 8.23h, and the last layered network is formed, which also has only one path, this time a path of five edges. The path is augmented (Figure 8.23j) and then no other layered network can be found. This algorithm can be summarized in the following pseudocode:

```
layerNetwork(network with source s and sink t)
    for all vertices u
        level(u) = -1;
    level(t) = 0;
    enqueue(t);
    while queue is not empty
        v = dequeue();
        for all vertices u adjacent to v such that level(u) == -1
            if forward(edge(uv)) and slack(edge(uv)) > 0 or
               backward(edge(uv)) and f(edge(vu)) > 0
                level(u) = level(v)+1;
                enqueue(u);
            if u == s
                return success;
    return failure;

processAugmentingPaths(network with source s and sink t)
    unblock all edges;
    labeled = {s};
    while labeled is not empty  // while not stuck;
        pop v from labeled;
        for all unlabeled vertices u adjacent to v such that
            edge(vu) is not blocked and level(v) == level(u) +1
            if forward(edge(vu)) and slack(edge(vu)) > 0
```

```
                        label(u) = (v⁺,min(slack(v),slack(edge(vu))))
                    else if backward(edge(vu)) and f(edge(uv)) > 0
                        label(u) = (v⁻,min(slack(v),f(edge(uv))));
                if u got labeled
                    if u == t
                        augmentPath(network);
                        block saturated edges;
                        labeled = {s};          // look for another path;
                    else push u onto labeled;
            if no neighbor of v has been labeled
                block all edges incident with v;

    DinicAlgorithm(network with source s sink t)
        set flows of all edges and vertices to 0;
        label(s) = (null,∞);
        while layerNetwork(network) is successful
            processAugmentingPaths(network);
```

What is the complexity of this algorithm? There are maximum $|V|-1$ layerings (phases) and up to $O(|E|)$ steps to layer the network. Hence, finding all the layered networks requires $O(|V||E|)$ steps. Moreover, there are $O(|E|)$ paths per phase (per one layered network) and, due to blocking, $O(|V|)$ steps to find one path, and because there are $O(|V|)$ layered networks, in the worst case, $O(|V|^2|E|)$ steps are required to find the augmenting paths. This estimation determines the efficiency of the algorithm, which is better than $O(|V||E|^2)$ for breadth-first `FordFulkersonAlgorithm()`. The improvement is in the number of steps to find one augmenting path, which is now $O(|V|)$, not $O(|E|)$, as before. The price for this improvement is the need to prepare the network by creating layered networks, which, as established, require additional $O(|V||E|)$ steps.

The difference in pseudocode for `FordFulkersonAlgorithm()` and `processAugmentingPaths()` is not large. The most important difference is in the amplified condition for expanding a path from a certain vertex v: Only the edges to adjacent vertices u that do not extend augmenting paths beyond the length of paths in the layered network are considered.

8.8.2 Maximum Flows of Minimum Cost

In the previous discussion, edges had two parameters, capacity and flow: how much flow they can carry and how much flow they are actually carrying. But although many different maximum flows through the network are possible, we choose the one dictated by the algorithm currently in use. For example, Figure 8.24 illustrates two possible maximum flows for the same network. Note that in the first case, the *edge(ab)* is not used at all; only in the second case are all the edges transferring some flow. The breadth-first algorithm leads to the first maximum flow and finishes our quest for maximum flow after identifying it. However, in many situations, this is not a good decision. If there are many possible maximum flows, it does not mean that any one of them is equally good.

FIGURE 8.24 Two possible maximum flows for the same network.

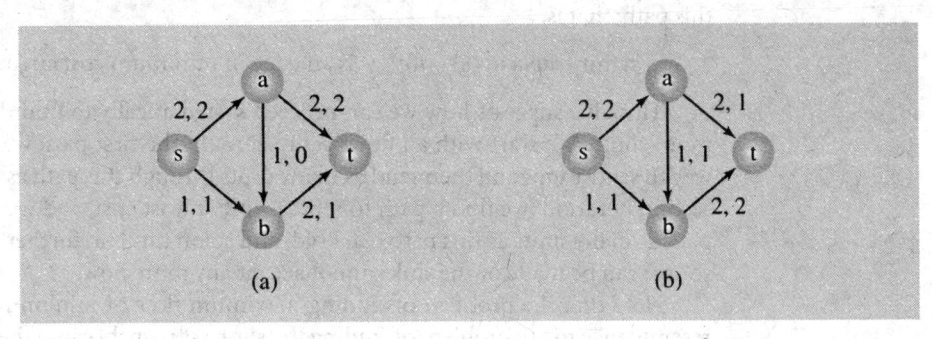

Consider the following example. If edges are roads between some locations, then it is not enough to know that a road has one or two lanes to choose a proper route. If the *distance(a,t)* is very long and *distance(a,b)* and *distance(b,t)* are relatively short, then it is better to consider the second maximum flow (Figure 8.24b) as a viable option rather than the first (Figure 8.24a). However, this may not be enough. The shorter way can have no pavement: It can be muddy, hilly, close to the avalanche areas, or sometimes blocked by boulders, among other disadvantages. Hence, using the distance as the sole criterion for choosing a road is insufficient. Taking the roundabout way may bring us to the destination faster and cheaper (to mention only time and gasoline burned).

We clearly need a third parameter for an edge: the *cost* of transferring one unit of flow through this edge. The problem now is how to find a maximum flow at minimum cost. More formally, if for each edge e, the *cost(e)* of sending one unit of flow is determined so that it costs $n \cdot cost(e)$ to transmit n units of flow over edge e, then we need to find a maximum flow f of minimum cost, or a flow such that

$$cost(f) = \min\{\sum_{e \in E} f(e) \cdot cost(e) : f \text{ is a maximum flow}\}$$

Finding all possible maximum flows and comparing their costs is not a feasible solution because the amount of work to find all such flows can be prohibitive. Algorithms are needed that find not only a maximum flow, but also the maximum flow at minimum cost.

One strategy is based on the following theorem, proven first by W. S. Jewell, R. G. Busacker, and P. J. Gowen, and implicitly used by M. Iri (Ford and Fulkerson 1962):

Theorem. If f is a minimal-cost flow with the flow value v and p is the minimum cost augmenting path sending a flow of value 1 from the source to the sink, then the flow $f + p$ is minimal and its flow value is $v + 1$.

The theorem should be intuitively clear. If we determined the cheapest way to send v units of flow through the network and afterward found a path that is the cheapest way for sending 1 unit of flow from the source to the sink, then we found the cheapest way to send $v + 1$ units using the route, which is a combination of the route already determined and the path just found. If this augmenting path allows for sending 1 unit for minimum cost, then it also allows for sending 2 units at minimum cost, and also 3 units, up to n units, where n is the maximum amount of units that can be sent through this path; that is,

$$n = \min\{capacity(e) - f(e) : e \text{ is an edge in minimum cost augmenting path}\}$$

This also suggests how we can proceed systematically to find the cheapest maximum route. We start with all flows set to zero. In the first pass, we find the cheapest way to send 1 unit and then send as many units through this path as possible. After the second iteration, we find a path to send 1 unit at least cost, and we send through this path as many units as this path can hold, and so on until no further dispatch from the source can be made or the sink cannot accept any more flow.

Note that the problem of finding maximum flow of minimum cost bears some resemblance to the problem of finding the shortest path, because the shortest path can be understood as the path with minimum cost. Hence, a procedure is needed to find the shortest path in the network so that as much flow as possible can be sent through this path. Therefore, a reference to an algorithm that solves the shortest path problem should not be surprising. We modify Dijkstra's algorithm used for solving the one-to-one shortest path problem (see Exercise 7 at the end of this chapter). Here is the algorithm:

```
modifiedDijkstraAlgorithm(network, s, t)
    for all vertices u
        f(u) = 0;
        cost(u) = ∞;
    set flows of all edges to 0;
    label(s) = (null,∞,0);
    labeled = null;
    while (true)
        v = a vertex not in labeled with minimal cost(v);
        if v == t
            if cost(t) == ∞ // no path from s to t can be found;
                return failure;
            else return success;
        add v to labeled;
        for all vertices u not in labeled and adjacent to v
            if forward(edge(vu)) and slack(edge(vu)) > 0 and cost(v) + cost(vu) < cost(u)
                label(u) = (v⁺,min(slack(v),slack(edge(vu))), cost(v) + cost(vu))
            else if backward(edge(vu)) and f(edge(uv)) > 0 and cost(v) − cost(uv) < cost(u)
                label(u) = (v⁻,min(slack(v),f(edge(uv)), cost(v) − cost(uv));

maxFlowMinCostAlgorithm(network with source s and sink t)
    while modifiedDijkstraAlgorithm(network,s,t) is successful
        augmentPath(network,s,t);
```

`modifiedDijkstraAlgorithm()` keeps track of three things at a time so that the label for each vertex is the triple

$$label(u) = (parent(u), flow(u), cost(u))$$

First, for each vertex u, it records the predecessor v, the vertex through which u is accessible from the source s. Second, it records the maximum amount of flow that can be pushed through the path from s to u and eventually to t. Third, it stores the cost of passing all the edges from the source to u. For forward $edge(vu)$, $cost(u)$ is the sum of the costs already accumulated in v plus the cost of pushing one unit of flow through $edge(vu)$. For backward $edge(vu)$, the unit cost of passing through this edge is subtracted from the $cost(v)$ and stored in $cost(u)$. Also, flows of edges included in augmented paths are updated; this task is performed by `augmentPath()` (see p. 410).

Figure 8.25 illustrates an example. In the first iteration of the `while` loop, `labeled` becomes $\{s\}$ and the three vertices adjacent to s are labeled, $label(a) = (s,2,6)$, $label(c) = (s,4,2)$, and $label(e) = (s,1,1)$. Then the vertex with the smallest cost is chosen, namely, vertex e. Now, `labeled` $= \{s,e\}$ and two vertices acquire new labels, $label(d) = (e,1,3)$ and $label(f) = (e,1,2)$. In the third iteration, vertex c is chosen, because its cost, 2, is minimal. Vertex a receives a new label, $(c,2,3)$, because the cost of accessing it from s through c is smaller than accessing it directly from s. Vertex f, which is adjacent to c, does not get a new label, because the cost of sending one unit of flow from s to f through c, 5, exceeds the cost of sending this unit through e, which is 2. In the fourth iteration, f is chosen, `labeled` becomes $\{s,e,c,f\}$, and $label(t) = (f,1,5)$. After the seventh iteration, the situation in the graph is as pictured in Figure 8.25b. The eighth iteration is exited right after the sink t is chosen, after which the path s, e, f, t is augmented (Figure 8.25c). The execution continues, `modifiedDijkstraAlgorithm()` is invoked four more times, and in the last invocation no other path can be found from s to t. Note that the same paths were found here as in Figure 8.20, although in a different order, which was due to the cost of these paths: 5 is the cost of the first detected path (Figure 8.25b), 6 is the cost of the second path (Figure 8.25d), 8 is the cost of the third (Figure 8.25f), and 9 is the cost of the fourth (Figure 8.25h). But the distribution of flows for particular edges allowing for the maximum flow is slightly different. In Figure 8.20k, $edge(sa)$ transmits 2 units of flow, $edge(sc)$ transmits 2 units, and $edge(ca)$ transmits 1 unit. In Figure 8.25i, the same three edges transmit 1, 3, and 2 units, respectively.

FIGURE 8.25 Finding a maximum flow of minimum cost.

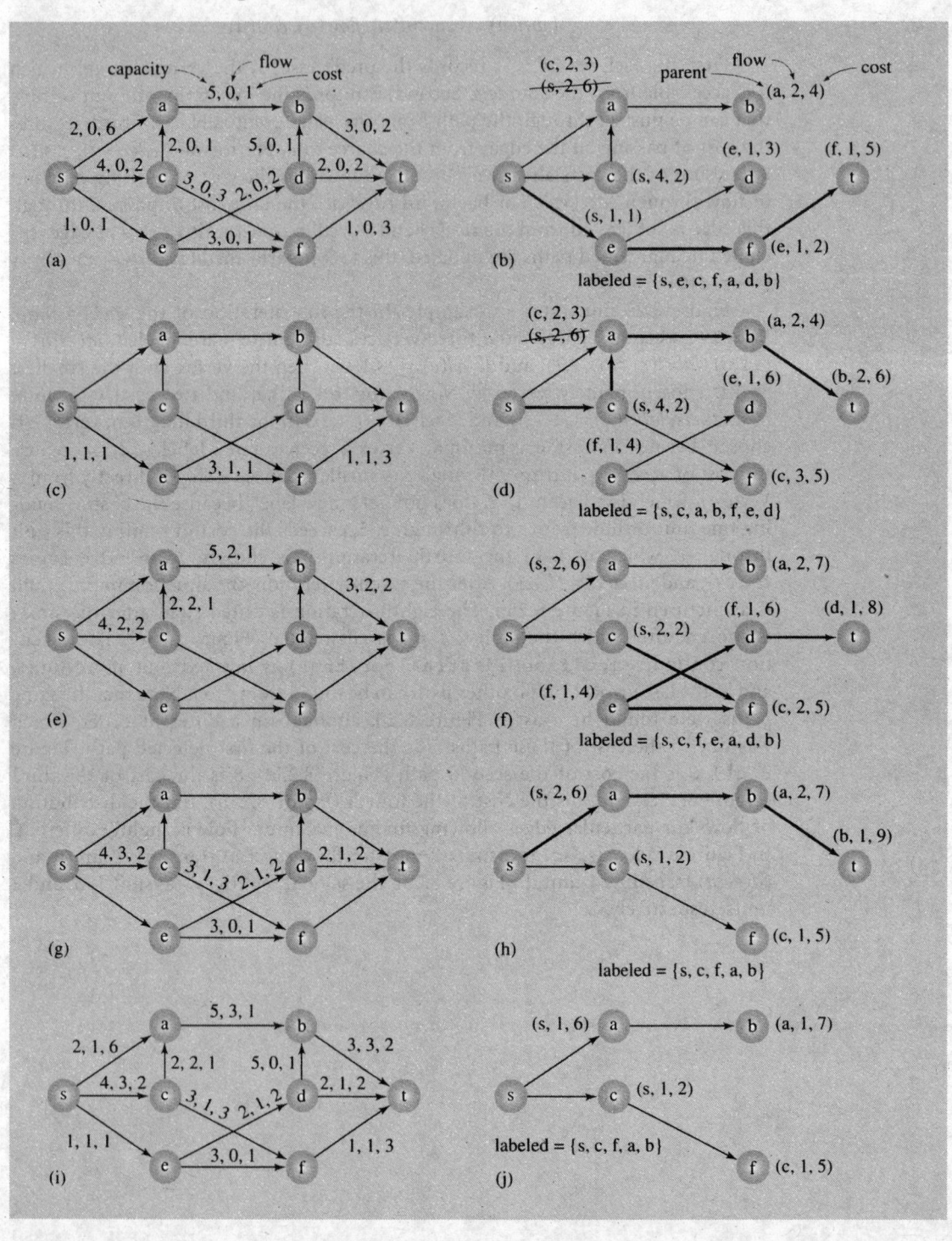

8.9 MATCHING

Suppose that there are five job openings a, b, c, d, and e and five applicants p, q, r, s, and t with qualifications shown in this table:

Applicants:	p	q	r	s	t
Jobs:	a b c	b d	a e	e	c d e

The problem is how to find a worker for each job; that is, how to match jobs with workers. There are many problems of this type. The job matching problem can be modeled with a bipartite graph. A *bipartite graph* is one in which the set of vertices V can be divided into two subsets V_1 and V_2 such that, for each $edge(vw)$, if vertex v is in one of the two sets V_1 or V_2, then w is in the other set. In this example, one set of vertices, V_1, represents applicants, the other set V_2, represents jobs, and edges represent jobs for which applicants are qualified (Figure 8.26). The task is to find a match between job and applicants so that one applicant is matched with one job. In a general case, there may not be enough applicants, or there may be no way to assign an applicant for each opening, even if the number of applicants exceeds the number of openings. Hence, the task now is to assign applicants to as many jobs as possible.

FIGURE 8.26 Matching five applicants with five jobs.

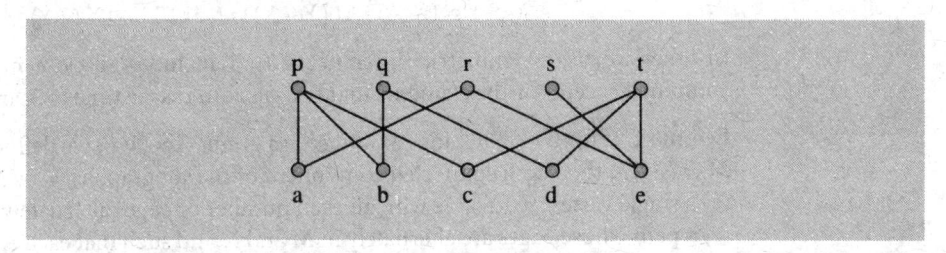

A *matching* M in a graph $G = (V,E)$ is a subset of edges, $M \subseteq E$, such that no two edges share the same vertex; that is, no two edges are adjacent. A *maximum matching* is a matching that contains a maximum number of edges so that the number of unmatched vertices (that is, vertices not incident with edges in M) is minimal. For example, in the graph in Figure 8.27, the sets $M_1 = \{edge(cd), edge(ef)\}$ and $M_2 = \{edge(cd), edge(ge), edge(fh)\}$ are matchings, but M_2 is a maximum matching, whereas M_1 is not. A *perfect matching* is a matching that pairs all the vertices of graph G. A matching $M = \{edge(pc), edge(qb), edge(ra), edge(se), edge(td)\}$ in Figure 8.26 is a perfect matching, but there is no perfect matching for the graph in Figure 8.27. A *matching problem* consists in finding a maximum matching for a certain graph G. The problem of finding a perfect matching is also called the *marriage problem*.

FIGURE 8.27 A graph with matchings $M_1 = \{edge(cd), edge(ef)\}$ and $M_2 = \{edge(cd), edge(ge), edge(fh)\}$.

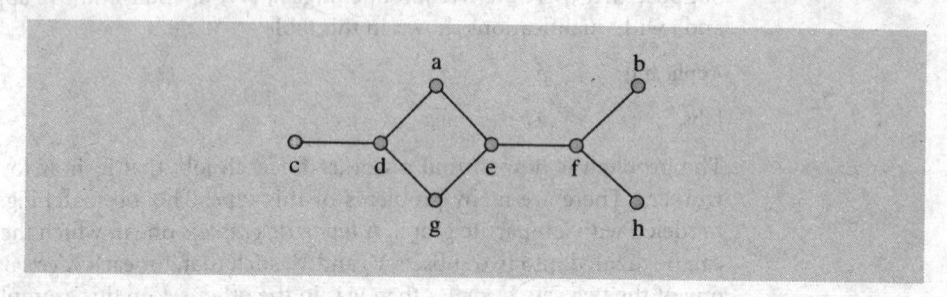

An *alternating path* for M is a sequence of edges $edge(v_1v_2)$, $edge(v_2v_3)$, ..., $edge(v_{k-1}v_k)$ that alternately belongs to M and to $E - M$ = set of edges that are not in M. An *augmenting path for M* is an alternating path where both end vertices are not incident with any edge in matching M. Thus, an augmenting path has an odd number of edges, $2k + 1$, k of them belonging to M and $k + 1$ not in M. If edges in M are replaced by edges not in M, then there is one more edge in M than before the interchange. Thus, the cardinality of the matching M is augmented by one.

A *symmetric difference* between two sets, $X \oplus Y$, is the set

$$X \oplus Y = (X - Y) \cup (Y - X) = (X \cup Y) - (X \cap Y)$$

In other words, a symmetric difference $X \oplus Y$ includes all elements from X and Y combined except for the elements that belong at the same time to X and Y.

Lemma 1. If for two matchings M and N in a graph $G = (V,E)$ we define a set of edges $M \oplus N \subseteq E$, then each connected component of the subgraph $G' = (V, M \oplus N)$ is either (a) a single vertex, (b) a cycle with an even number of edges alternately in M and N, or (c) a path whose edges are alternately in M and N and such that each end vertex of the path is matched only by one of the two matchings M and N (i.e., the whole path should be considered, not just part, to cover the entire connected component).

Proof. For each vertex v of G', $deg(v) \leq 2$, at most one edge of each matching can be incident with v; hence, each component of G' is either a single vertex, a path, or a cycle. If it is a cycle or a path, the edges must alternate between both matchings; otherwise, the definition of matching is violated. Thus, if it is a cycle, the number of edges must be even. If it is a path, then the degree of both end vertices is one so that they can be matched with only one of the matchings, not both. □

Figure 8.28 contains an example. A symmetric difference between matching $M = \{edge(ad), edge(bf), edge(gh), edge(ij)\}$ marked with dashed lines and matching $N = \{edge(ad), edge(cf), edge(gi), edge(hj)\}$ shown in dotted lines is the set $M \oplus N = \{edge(bf), edge(cf), edge(gh), edge(gi), edge(hj), edge(ij)\}$, which contains one path and a cycle (Figure 8.28b). The vertices of graph G that are not incident with any of the edges in $M \oplus N$ are isolated in the graph $G' = (V, M \oplus N)$.

Lemma 2. If M is a matching and P is an augmenting path for M, then $M \oplus P$ is a matching of cardinality $|M| + 1$.

Section 8.9 Matching   **425**

FIGURE 8.28   (a) Two matchings M and N in a graph $G = (V, E)$ and (b) the graph $G' = (V, M \oplus N)$.

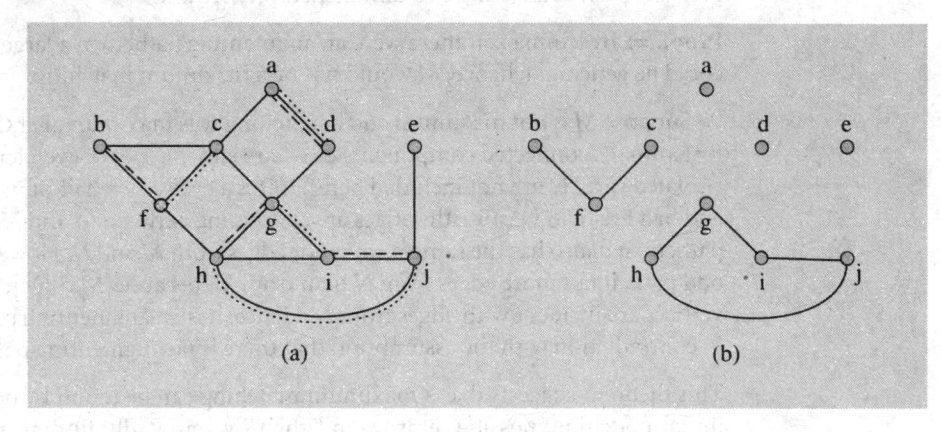

Proof. By definition of symmetric difference, $M \oplus P = (M - P) \cup (P - M)$. Except for the end vertices, all other vertices incident with edges in P are matched by edges in P. Hence, no edge in $M - P$ contains any vertex in P. Thus, edges in $M - P$ share no vertices with edges in $P - M$. Moreover, because P is a path with every other edge in $P - M$, then $P - M$ has no edges that share vertices. Hence, $(M - P) \cup (P - M)$ is a union of two nonoverlapping matchings and thus a matching. If $|P| = 2k + 1$, then $|M - P| = |M| - k$ because all edges in $M \cup P$ are excluded, and the number of edges in P but not in M, $|P - M| = k + 1$. Because $(M - P)$ and $(P - M)$ are not overlapping, $|(M - P) \cup (P - M)| = |M - P| + |P - M| = (|M| - k) + k + 1 = |M| + 1$. □

Figure 8.29 illustrates this lemma. For matching $M = \{edge(bf), edge(gh), edge(ij)\}$ shown with dashed lines, and augmenting path P for M, the path c, b, f, h, g, i, j, e, the resulting matching is $\{edge(bc), edge(ej), edge(fh), edge(gi)\}$, which includes all the edges from the path P that were originally excluded from M. So in effect the lemma finds a larger matching if in an augmenting path the roles of matched and unmatched edges are reversed.

FIGURE 8.29   (a) Augmenting path P and a matching M and (b) the matching $M \oplus P$.

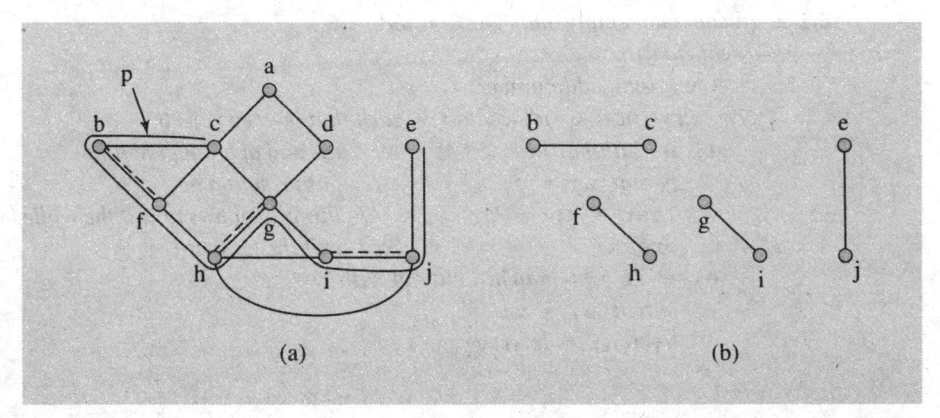

Theorem (Berge 1957). A matching M in a graph G is maximum if there is no augmenting path connecting two unmatched vertices in G.

Proof. $\Rightarrow$ By lemma 2, if there were an augmenting path, then a larger matching could be generated; hence, M would not be a maximum matching.

$\Leftarrow$ Suppose M is not maximum and a matching N is maximum. Let $G' = (V, M \oplus N)$. By lemma 1, connected components of G' are either cycles of even length or paths (isolated vertices are not included here). If it is a cycle, then half of its edges are in N and half are in M because the edges are alternating between M and N. If it is an even path, then it also has the same number of edges from M and N. However, if it is an odd path, it has more edges from N than from M, because $|N| > |M|$, and both end vertices are incident with edges from N. Hence, it is an augmenting path, which leads to contradiction with the assumption that there is no augmenting path. □

This theorem suggests that a maximum matching can be found by beginning with an initial matching, possibly empty, and then by repeatedly finding new augmenting paths and increasing the cardinality of matching until no such path can be found. This requires an algorithm to determine alternate paths. It is much easier to develop such an algorithm for bipartite graphs than for any other graphs; therefore, we start with a discussion of this simpler case.

To find an augmenting path, breadth-first search is modified to allow for always finding the shortest path. The procedure builds a tree, called a *Hungarian tree*, with an unmatched vertex in the root consisting of alternating paths, and a success is pronounced as soon as it finds another unmatched vertex than the one in the root (that is, as soon as it finds an augmenting path). The augmenting path allows for increasing the size of matching. After no such path can be found, the procedure is finished. The algorithm is as follows:

```
findMaximumMatching(bipartite graph)
    for all unmatched vertices v
        set level of all vertices to 0;
        set parent of all vertices to null;
        level(v) = 1;
        last = null;
        clear queue;
        enqueue(v);
        while queue is not empty and last is null
            v = dequeue();
            if level(v) is an odd number
                for all vertices u adjacent to v such that level(u) is 0
                    if u is unmatched              // the end of an augmenting
                        parent(u) = v;             // path is found;
                        last = u;                  // this also allows to exit the while loop;
                        break;                     // exit the for loop;
                    else if u is matched but not with v
                        parent(u) = v;
                        level(u) = level(v) + 1;
```

```
            enqueue(u);
    else // if level(v) is an even number
        enqueue(vertex u matched with v);
        parent(u) = v;
        level(u) = level(v) + 1;
if last is not null // augment matching by updating the augmenting path;
    for (u = last; u is not null; u = parent(parent(u)))
        matchedWith(u) = parent(u);
        matchedWith(parent(u)) = u;
```

An example is shown in Figure 8.30. For the current matching $M = \{(u_1, v_4), (u_2, v_2), (u_3, v_3), (u_5, v_5)\}$ (Figure 8.30a), we start from vertex u_4. First, three vertices adjacent to u_4 (namely, v_3, v_4, and v_5) are enqueued, all of them connected to u_4 with edges not in M. Then v_3 is dequeued, and because it is on an even level of the tree (Figure 8.30b), there is at most one successor to be considered, which is the vertex u_3 because $edge(u_3,v_3)$ is in M and u_3 is enqueued. Then successors of v_4 and v_5 are found—that is, u_1 and u_5, respectively—after which the vertex u_3 is considered. This vertex is on an odd level; hence, all vertices directly accessible from it through edges not in M are checked. There are three such vertices, v_2, v_4, and v_5, but only the first is not yet in the tree, so it is included now. Next, successors of u_1 are tested, but the only candidate, v_2, does not qualify because it is already in the tree. Finally, u_5 is checked, from which we arrive at an unmatched vertex v_6. This marks the end of an augmenting path; hence, the while loop is exited and then matching M is modified by including in M the edges in the newly found path that are not in M and excluding from M the edges of the path that are there. The path has one more edge in M than not in M, so after such modification the number of edges in M is increased by one. The new matching is shown in Figure 8.30c.

After finding and modifying an augmenting path, a search for another augmenting path begins. Because there are still two unmatched vertices, there still exists a possibility that a larger matching can be found. In the second iteration of the outer for loop, we begin with the vertex u_6, which eventually leads to the tree as in Figure 8.30d that includes an augmenting path, which in turn gives a matching as in Figure 8.30e. There are no unmatched vertices left; thus, the maximum matching just found is also a perfect matching.

Complexity of the algorithm is found as follows. Each alternating path increases the cardinality of matching by one, and because the maximum number of edges in matching M is $|V|/2$, the number of iterations of the outer for loop is at most $|V|/2$. Moreover, finding one augmenting path requires $O(|E|)$ steps so that the total cost of finding a maximum matching is $O(|V||E|)$.

8.9.1 Stable Matching Problem

In the example of matching applicants with jobs, any successful maximum matching was acceptable because it did not matter to applicants what job they got and it did not matter to the employers whom they hired. But usually this is not the case. Applicants have their preferences, and so do employers. In the *stable matching problem*, also called the *stable marriage problem*, there are two nonoverlapping sets U and W of the same cardinality. Each element of U has a ranking list of elements of W, and each

FIGURE 8.30 Application of the `findMaximumMatching()` algorithm. Matched vertices are connected with solid lines.

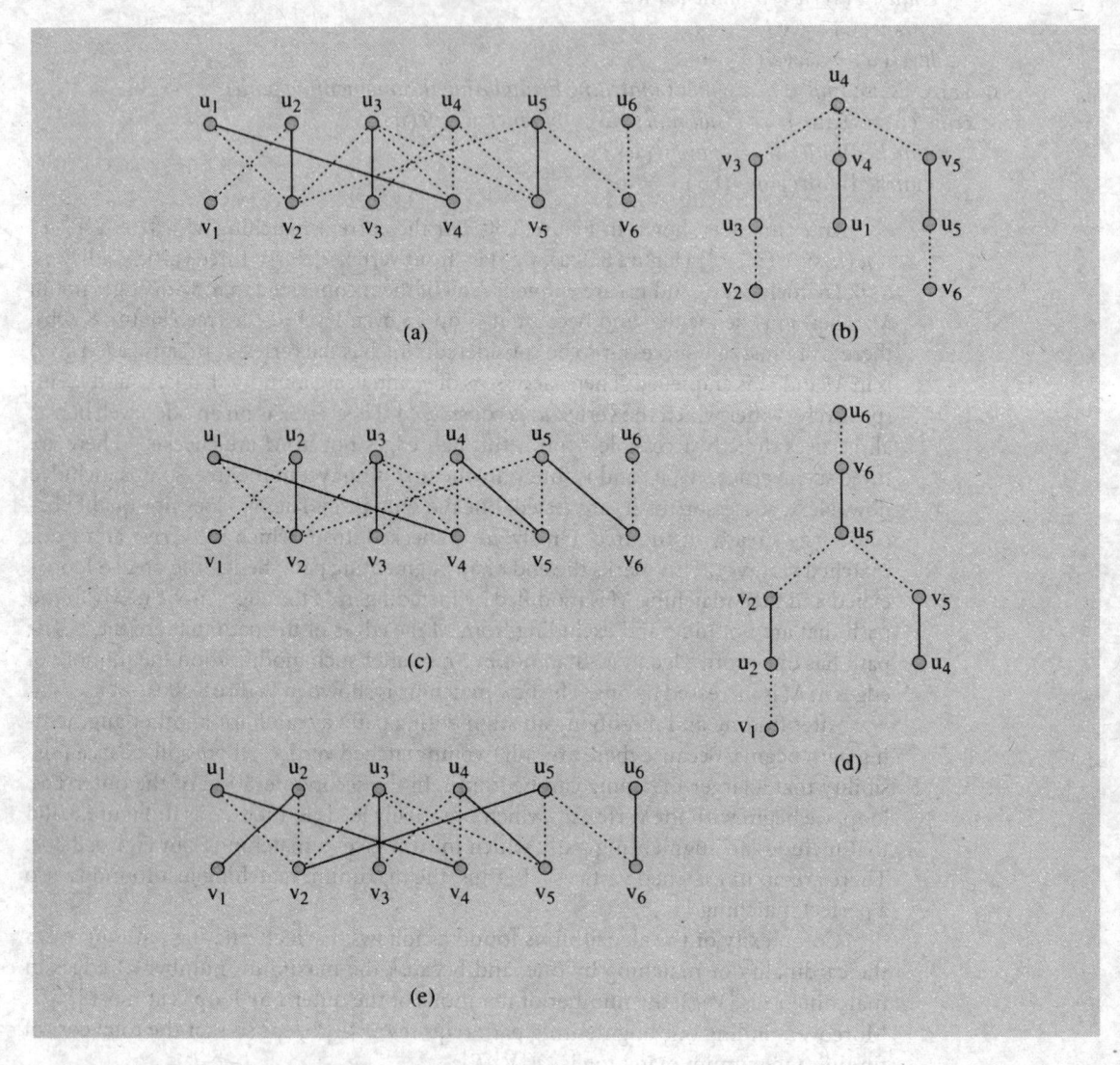

element of W has a preference list consisting of elements of U. Ideally, the elements should be matched with their highest preferences, but because of possible conflicts between different lists (for example, the same w can be first on two ranking lists), a matching should be created that is stable. A matching is *unstable* if two such elements, u and w, rank each other higher than the elements with which they are currently matched; otherwise, the matching is *stable*. Consider sets $U = \{u_1, u_2, u_3, u_4\}$ and $W = \{w_1, w_2, w_3, w_4\}$ and ranking lists

$u_1: w_2 > w_1 > w_3 > w_4$ $w_1: u_3 > u_2 > u_1 > u_4$
$u_2: w_3 > w_2 > w_1 > w_4$ $w_2: u_1 > u_3 > u_4 > u_2$
$u_3: w_3 > w_4 > w_1 > w_2$ $w_3: u_4 > u_2 > u_3 > u_1$
$u_4: w_2 > w_3 > w_4 > w_1$ $w_4: u_2 > u_1 > u_3 > u_4$

The matching $(u_1, w_1), (u_2, w_2), (u_3, w_4), (u_4, w_3)$ is unstable because there are two elements, u_1 and w_2, that prefer each other over the elements with which they are currently matched: u_1 prefers w_2 over w_1 and w_2 prefers u_1 over u_2.

A classical algorithm to find a stable matching was designed by Gale and Shapley (1962), who also show that a stable matching always exists.

```
stableMatching(graph = (U∪W,M))            // U∩W = null, |U| = |W|, M = null;
    while there is an unmatched element u∈U
        w = the highest remaining choice from W on u's list;
        if w is unmatched
            matchedWith(u) = w;
            matchedWith(w) = u;            // include edge(uw) in matching M;
        else if w is matched and w ranks u higher that its current match
            matchedWith(matchedWith(w)) = null;  // remove edge(matchedWith(w), w) from M;
            matchedWith(u) = w;            // include edge(uw) in M;
            matchedWith(w) = u;
```

Because the list of choices for each $u \in U$ decreases in each iteration, each list is of length $|W| = |U|$ and there are $|U|$ such lists, one for each u, the algorithm executes $O(|U|^2)$ iterations: $|U|$ times in the best case and $|U|^2$ in the worst case.

Consider an application of this algorithm to the set U and W defined before with the specified rankings. In the first iteration, u_1 is chosen and matched immediately with the unmatched w_2 that is highest on u_1's ranking list. In the second iteration, u_2 is successfully matched with its highest choice, w_3. In the third iteration, an attempt is made to match u_3 with its preference, w_3, but w_3 is already matched and w_3 prefers its current match, u_2, more than u_3, so nothing happens. In the fourth iteration, u_3 is matched with its second preference, w_4, which is currently unmatched. In the fifth iteration, a match is tried for u_4 and w_2, but unsuccessfully, because w_2 is matched already with u_1, and u_1 is ranked by w_2 higher than u_4. In the sixth iteration, a successful attempt is made to match u_4 with its second choice, w_3: w_3 is matched with u_2, but it prefers u_4 over u_2, so u_2 becomes unmatched and u_4 is matched with w_3. Now, u_2 has to be matched. The summary of all steps is given in the following table:

iteration	u	w	matched pairs
1	u_1	w_2	(u_1, w_2)
2	u_2	w_3	$(u_1, w_2), (u_2, w_3)$
3	u_3	w_3	$(u_1, w_2), (u_2, w_3)$
4	u_3	w_4	$(u_1, w_2), (u_2, w_3), (u_3, w_4)$
5	u_4	w_2	$(u_1, w_2), (u_2, w_3), (u_3, w_4)$
6	u_4	w_3	$(u_1, w_2), (u_3, w_4), (u_4, w_3)$
7	u_2	w_2	$(u_1, w_2), (u_3, w_4), (u_4, w_3)$
8	u_2	w_1	$(u_1, w_2), (u_2, w_1), (u_3, w_4), (u_4, w_3)$

Note that an asymmetry is implied in this algorithm concerning whose rankings are more important. The algorithm is working in favor of elements of set U. When the roles of sets U and W are reversed, then w's immediately have their preferred choices and the resulting stable matching is

$$(u_1, w_2), (u_2, w_4), (u_3, w_1), (u_4, w_3)$$

and u_2 and u_3 are matched with w's—w_4 and w_1, respectively—that are lower on their ranking lists than the w's chosen before—w_1 and w_4, respectively.

8.9.2 Assignment Problem

The problem of finding a suitable matching becomes more complicated in a weighted graph. In such a graph, we are interested in finding a matching with the maximum total weight. The problem is called an *assignment problem*. The assignment problem for complete bipartite graphs with two sets of vertices of the same size is called an *optimal assignment problem*.

An $O(|V|)^3$ algorithm is due to Kuhn (1955) and Munkres (1957) (Bondy and Murty 1976; Thulasiraman and Swamy 1992). For a bipartite graph $G = (V,E)$, $V = U \cup W$, we define a labeling function $f: U \cup W \to R$ such that a label $f(v)$ is a number assigned to each vertex v such that for all vertices v, u, $f(u) + f(v) \geq weight(edge(uv))$. Create a set $H = \{edge(uv) \in E: f(u) + f(v) = weight(edge(uv))\}$ and then an *equality subgraph* $G_f = (V, H)$. The Kuhn-Munkres algorithm is based on the theorem stating that if for a labeling function f and an equality subgraph G_f, graph G contains a perfect matching, then this matching is optimal: for any matching M in G, $\Sigma f(u) + \Sigma f(v) \geq weight(M)$, for any perfect matching M_p, $\Sigma f(u) + \Sigma f(v) = weight(M_p)$; that is, $weight(M) \leq \Sigma f(u) + \Sigma f(v) = weight(M_p)$.

The algorithm expands the equality subgraph G_f until a perfect matching can be found in it, which will also be an optimal matching for graph G.

```
optimalAssignment()
    G_f = equality subgraph for some vertex labeling f;
    M  = matching in G_f;
    S  = {some unmatched vertex u};  // beginning of an augmenting path P;
    T  = null;
    while M is not a perfect matching
        Γ(S) = {w: ∃u∈S: edge(uw)∈G_f};  // vertices adjacent in G_f to the vertices in S;
        if Γ(S) == T
            d = min{(f(u) + f(w) - weight(edge(uw))): u∈S, w∉T};
            for each vertex v
                if v ∈ S
                    f(v) = f(v) - d;
                else if v ∈ T
                    f(v) = f(v) + d;
            construct new equality subgraph G_f and new matching M;
        else  // if T ⊂ Γ(S)
            w = a vertex from Γ(S) - T;
```

```
     if w is unmatched // the end of the augmenting path P;
         P = augmenting path just found;
         M = M ⊕ P;
         S = {some unmatched vertex u};
         T = null;
     else S = S ∪ {matching neighbor of w in M};
         T = T ∪ {w};
```

For an example, see Figure 8.31. A complete bipartite graph $G = (\{u_1, \ldots, u_4\} \cup \{w_1, \ldots, w_4\}, E)$ has weights defined by the matrix in Figure 8.31a.

FIGURE 8.31 An example of application of the `optimalAssignment()` algorithm.

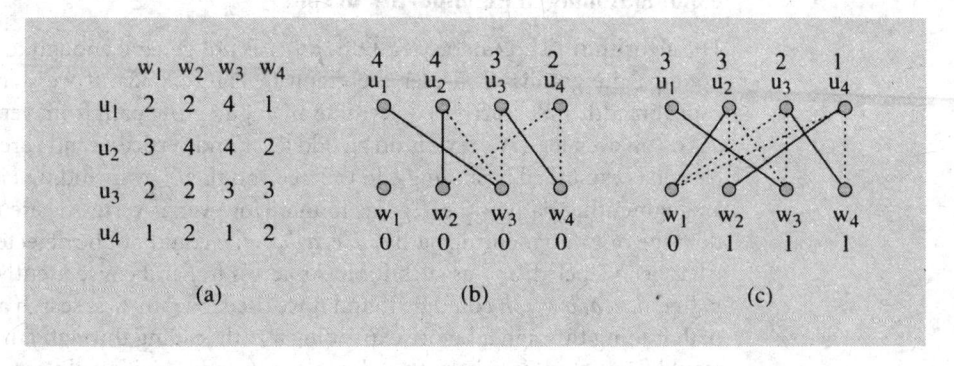

	w_1	w_2	w_3	w_4
u_1	2	2	4	1
u_2	3	4	4	2
u_3	2	2	3	3
u_4	1	2	1	2

(a)

(b) labels: u_1: 4, u_2: 4, u_3: 3, u_4: 2; w_1: 0, w_2: 0, w_3: 0, w_4: 0

(c) labels: u_1: 3, u_2: 3, u_3: 2, u_4: 1; w_1: 0, w_2: 1, w_3: 1, w_4: 1

0. For an initial labeling, we choose the function f such that $f(u) = \max\{weight(edge(uw)): w \in W\}$, that is, the maximum weight in the weight matrix in the row for vertex u, and $f(w) = 0$, so that for the graph G, the initial labeling is as in Figure 8.31b. We choose a matching as in Figure 8.31b and set S to $\{u_4\}$ and T to null.

1. In the first iteration of the `while` loop, $\Gamma(S) = \{w_2, w_4\}$, because both w_2 and w_4 are neighbors of u_4, which is the only element of S. Because T ⊂ $\Gamma(S)$—that is, ∅ ⊂ $\{w_2, w_4\}$—the outer `else` clause is executed, whereby $w = w_2$ (we simply choose the first element in $\Gamma(S)$ and not in T), and because w_2 is matched, the inner `else` clause is executed, in which we extend S to $\{u_2, u_4\}$, because u_2 is both matched and adjacent to w_2, and extend T to $\{w_2\}$.

All the iterations are summarized in the following table.

Iteration	$\Gamma(S)$	w	S	T
0	∅		$\{u_4\}$	∅
1	$\{w_2, w_4\}$	w_2	$\{u_2, u_4\}$	$\{w_2\}$
2	$\{w_2, w_3, w_4\}$	w_3	$\{u_1, u_2, u_4\}$	$\{w_2, w_3\}$
3	$\{w_2, w_3, w_4\}$	w_4	$\{u_1, u_2, u_3, u_4\}$	$\{w_2, w_3, w_4\}$
4	$\{w_2, w_3, w_4\}$			

In the fourth iteration, the condition of the outer `if` statement becomes true because sets T and $\Gamma(S)$ are now equal, so the distance $d = \min\{(f(u) + f(w) - weight(edge(uw)): u \in S, w \notin T\}$ is computed. Because w_1 is the only vertex not in T = $\{w_2, w_3, w_4\}$, $d = \min\{(f(u) + f(w_1) - weight(edge(uw_1)): u \in S = \{u_1, u_2, u_3, u_4\}\} = \min\{(4+0-2), (4+0-3), (3+0-2), (2+0-1)\} = 1$. With this distance, the labels of vertices in graph G are updated to become labels in Figure 8.31c. The labels of all four vertices in S are decremented by $d = 1$, and all three vertices in T are incremented by the same value. Next, an equality subgraph is created that includes all the edges, as in Figure 8.31c, and then the matching is found that includes edges drawn with solid lines. This is a perfect matching, and hence, an optimal assignment, which concludes the execution of the algorithm.

8.9.3 Matching in Nonbipartite Graphs

The algorithm `findMaximumMatching()` is not general enough to properly process nonbipartite graphs. Consider the graph in Figure 8.32a. If we start building a tree using breadth-first search to determine an augmenting path from vertex c, then vertex d is on an even level, vertex e is on an odd level, and vertices a and f are on an even level. Next, a is expanded by adding b to the tree and then f by including i in the tree so that an augmenting path c, d, e, f, g, i is found. However, if vertex i were not in the graph, then the only augmenting path c, d, e, a, b, g, f, h could not be detected because vertex g has been labeled, and as such it blocks access to f and consequently to vertex h. The path c, d, e, a, b, g, f, h could be found if we used a depth-first search and expanded the path leading through a before expanding a path leading through f, because the search would first determine the path c, d, e, a, b, g, f, and then it would access h from f. However, if h was not in the graph, then the very same depth-first search would miss the path c, d, e, f, g, i because first the path c, d, e, a, b, g, f with vertices g and f would be expanded so that the detection of path c, d, e, f, g, i is ruled out.

FIGURE 8.32 Application of the `findMaximumMatching()` algorithm to a nonbipartite graph.

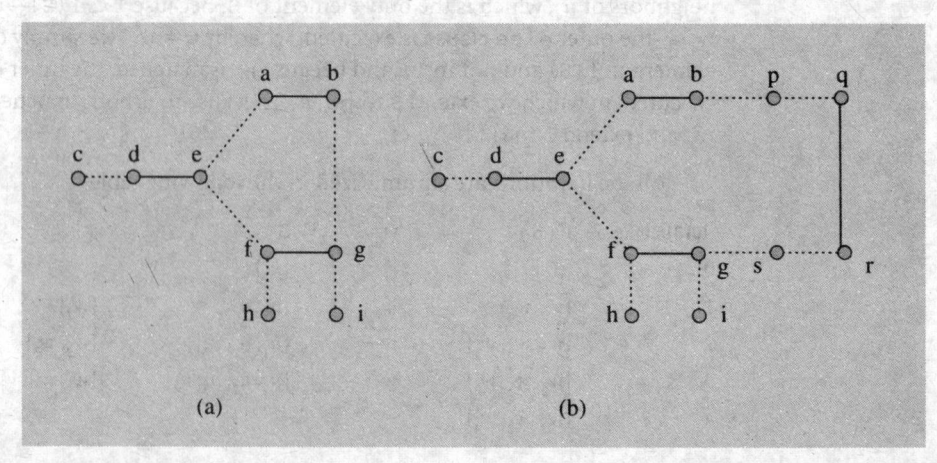

A source of the problem is the presence of cycles with an odd number of edges. But it is not just the odd number of edges in a cycle that causes the problem. Consider the graph in Figure 8.32b. The cycle e, a, b, p, q, r, s, g, f, e has nine edges, but `findMaximumMatching()` is successful here, as the reader can easily determine (both depth-first search and breadth-first search first find path c, d, e, a, b, p and then path h, f, g, i). The problem arises in a special type of cycle with an odd number of edges, which are called blossoms. The technique of determining augmenting paths for graphs with blossoms is due to Jack Edmonds. But first some definitions.

A *blossom* is an alternating cycle $v_1, v_2, \ldots, v_{2k-1} v_1$ such that $edge(v_1 v_2)$ and $edge(v_{2k-1} v_1)$ are not in matching. In such a cycle, the vertex v_1 is called the *base* of the blossom. An even length alternating path is called a *stem*; a path of length zero that has only one vertex is also a stem. A blossom with a stem whose edge in matching is incident with the base of the blossom is called a *flower*. For example, in Figure 8.32a, path c, d, e and path e are stems, and cycle e, a, b, g, f, e is a blossom with the base e.

The problems with blossoms arise if a prospective augmenting path leads to a blossom through the base. Depending on which edge is chosen to continue the path, we may or may not obtain an augmenting path. However, if the blossom is entered through any other vertex v than the base, the problem does not arise because we can choose only one of the two edges of v. Hence, an idea is to prevent a blossom from possibly harmful effects by detecting the fact that a blossom is being entered through its base. The next step is to temporarily remove the blossom from the graph by putting in place of its base a vertex that represents such a blossom and to attach to this vertex all edges connected to the blossom. The search for an augmenting path continues, and if an augmenting path that includes a vertex representing a blossom is found, the blossom is expanded and the path through it is determined by going backward from the edge that leads to the blossom to one of the edges incident with the base.

The first problem is how to recognize that a blossom has been entered through the base. Consider the Hungarian tree in Figure 8.33a, which is generated using breadth-first search in the graph in Figure 8.32a. Now, if we try to find neighbors of b, then only g qualifies because $edge(ab)$ is in matching, and thus only edges not in matching can be included starting from b. Such edges would lead to vertices on an even level of the tree. But g has already been labeled and it is located on an odd level. This marks a blossom detection. If a labeled vertex is reached through different paths, one of them requiring this vertex to be on an even level and another on a odd level, then we know that we are in the middle of a blossom entered through its base. Now we trace the paths from g and b back in the tree until a common root is found. This common root, vertex e in our example, is the base of the detected blossom. The blossom is now replaced by a vertex A, which leads to a transformed graph, as in Figure 8.33b. The search for an augmenting path restarts from vertex A and continues until such a path is found, namely, path c, d, A, h. Now we expand the blossom represented by A and trace the augmenting path through the blossom. We do that by starting from $edge(hA)$, which is now $edge(hf)$. Because it is an edge not in matching, then from f only $edge(fg)$ can be chosen so that the augmenting path can be alternating. Moving through vertices f, g, b, a, e, we determine the part of the augmenting path, c, d, A, h, which corresponds to A (Figure 8.33c) so that the full augmenting path is c, d, e, a, b, g, f, h. After the path is processed, we obtain a new matching, as in Figure 8.33d.

FIGURE 8.33 Processing a graph with a blossom.

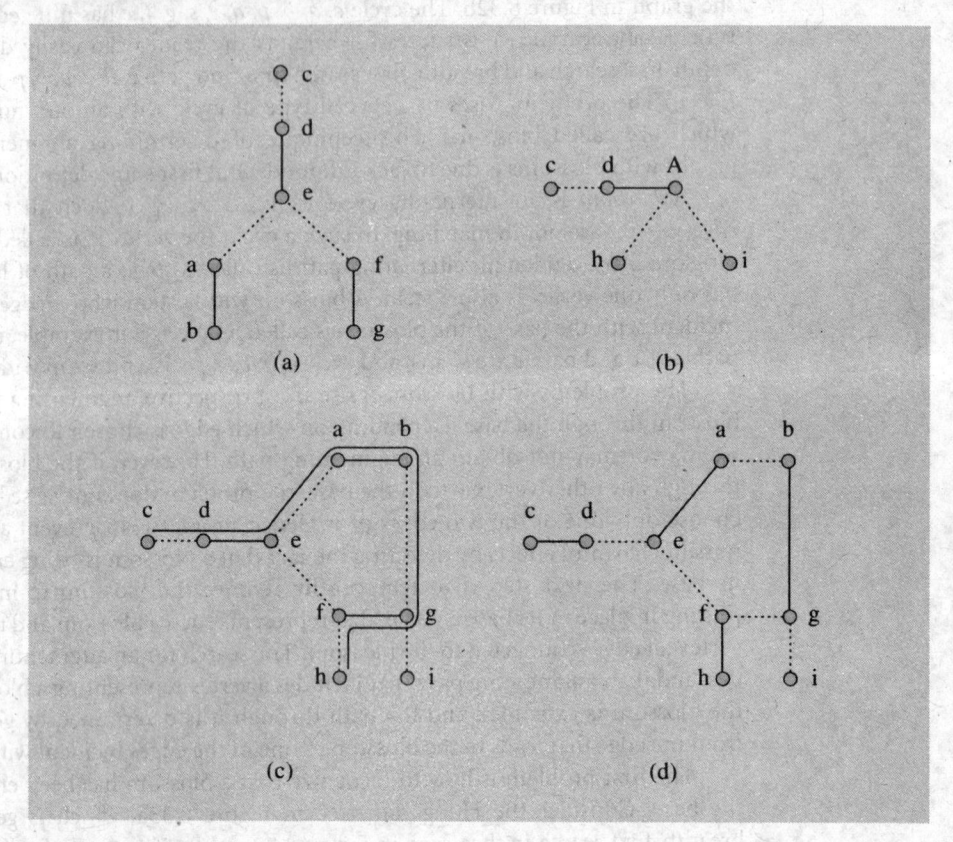

8.10 EULERIAN AND HAMILTONIAN GRAPHS

8.10.1 Eulerian Graphs

An *Eulerian trail* in a graph is a path that includes all edges of the graph only once. An *Eulerian cycle* is a cycle that is also an Eulerian trail. A graph that has an Eulerian cycle is called an *Eulerian graph*. A theorem proven by Euler (pronounced: oiler) says that a graph is Eulerian if every vertex of the graph is incident to an even number of edges. Also, a graph contains an Eulerian trail if it has exactly two vertices incident with an odd number of edges.

The oldest algorithm that allows us to find an Eulerian cycle if this is possible is due to Fleury (1883). The algorithm takes great care in not traversing a bridge—that is, an edge whose removal would disconnect the graphs G_1 and G_2—because if traversal of G_1 is not completed before traversing such an edge to pass to G_2, it would not be possible to return to G_1. As Fleury himself phrases it, the algorithm consists in "taking

Section 8.10 Eulerian and Hamiltonian Graphs

an isolating path (= a bridge) only when there is no other path to take." Only after the entire subgraph G_1 has been traversed can the path lead through such an edge. Fleury's algorithm is as follows:

```
FleuryAlgorithm(undirected graph)
    v = a starting vertex;      // any vertex;
    path = v;
    untraversed = graph;
    while v has untraversed edges
        if edge(vu) is the only one untraversed edge
            e = edge(vu);
            remove v from untraversed;
        else e = edge(vu) which is not a bridge in untraversed;
        path = path + u;
        remove e from untraversed;
        v = u;
    if untraversed has no edges
        success;
    else failure;
```

Note that for cases when a vertex has more than one untraversed edge, a connectivity checking algorithm should be applied.

An example of finding an Eulerian cycle is shown in Figure 8.34. It is critical that before an edge is chosen, a test is made to determine whether the edge is a bridge in the untraversed subgraph. For example, if in the graph in Figure 8.34a the traversal begins in vertex b to reach vertex a through vertices $e, f, b,$ and c, thereby using the path b, e, f, b, c, a, then we need to be careful which untraversed edge is chosen in a: $edge(ab), edge(ad),$ or $edge(ae)$ (Figure 8.34b). If we choose $edge(ab)$, then the remaining three untraversed edges are unreachable, because in the yet untraversed subgraph $untraversed = (\{a,b,d,e\}, \{edge(ab), edge(ad), edge(ae), edge(de)\})$, $edge(ab)$ is a bridge because it disconnects two subgraphs of $untraversed$, $(\{a,d,e\}, \{edge(ad), edge(ae), edge(de)\})$ and $(\{b\}, \emptyset)$.

FIGURE 8.34 Finding an Eulerian cycle.

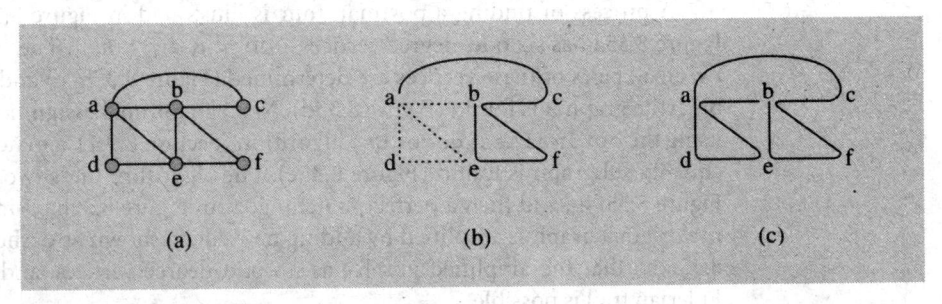

The Chinese Postman Problem

The *Chinese postman problem* is stated as follows: A postman picks up mail at the post office, delivers the mail to houses in a certain area, and returns to the post office (Kwan, 1962). The walk should have a shortest distance when traversing each street at least once. The problem can be modeled with a graph G whose edges represent streets and their lengths and vertices represent street corners in which we want to find a minimum closed walk. Let us observe first that if the graph G is Eulerian, then each Eulerian cycle gives a solution; however, if the graph G is not Eulerian, then it can be so amplified that it becomes an Eulerian graph G^* in which every edge e appears as many times as the number of times it is used in the postman's walk. If so, we want to construct such a graph G^* in which the sum of distances of the added edges is minimal. First, odd-degree vertices are grouped into pairs (u, w) and a path of new edges is added to an already existing path between both vertices of each pair, thereby forming the graph G^*. The problem consists now in so grouping the odd-degree vertices that the total distance of the added paths is minimum. The following algorithm for solving this problem is due to Edmonds and Johnson (Edmonds 1965; Edmonds and Johnson 1973; see Gibbons 1985).

```
ChinesePostmanTour(G = (V, E))
    ODD = set of all odd-degree vertices of G;
    if ODD is not empty
        E* = E;
        G* = (V, E*);
        find the shortest paths between all pairs of odd-degree vertices;
        construct a complete bipartite graph H = (U∪W, E'), ODD == (v_1,...,v_{2k}), such that
            U = (u_1,...,u_{2k}) and u_i is a copy of v_i;
            W = (w_1,...,w_{2k}) and w_i is a copy of v_i;
            dist(edge(u_i w_i)) = -∞;
            dist(edge(u_i w_j)) = -dist(edge(v_i v_j)) for i≠j;
        find optimal assignment M in H;
        for each edge(u_i w_j) ∈ M such that v_i is still an odd-degree vertex
            E* = E*∪{edge(uw) ∈ path(u_i w_j): path(u_i w_j) is minimum};
    find Eulerian path in G*;
```

Note that the number of odd-degree vertices is even (Exercise 44).

A process of finding a postman tour is illustrated in Figure 8.35. The graph in Figure 8.35a has six odd-degree vertices, ODD = {c, d, f, g, h, j}. The shortest paths between all pairs of these vertices are determined (Figure 8.35b–c) and then a complete bipartite graph H is found (Figure 8.35d). Next, an optimal assignment M is found. By using the `optimalAssignment()` algorithm (Section 8.9.1), a matching in an initial equality subgraph is found (Figure 8.35e). The algorithm finds two matchings, as in Figure 8.35f–g, and then a perfect matching, as in Figure 8.35h. Using this matching, the original graph is amplified by adding new edges, shown as dashed lines in Figure 8.35i, so that the amplified graph has no odd-degree vertices, and thus finding an Eulerian trail is possible.

Section 8.10 Eulerian and Hamiltonian Graphs   ■   437

FIGURE 8.35   Solving the Chinese postman problem.

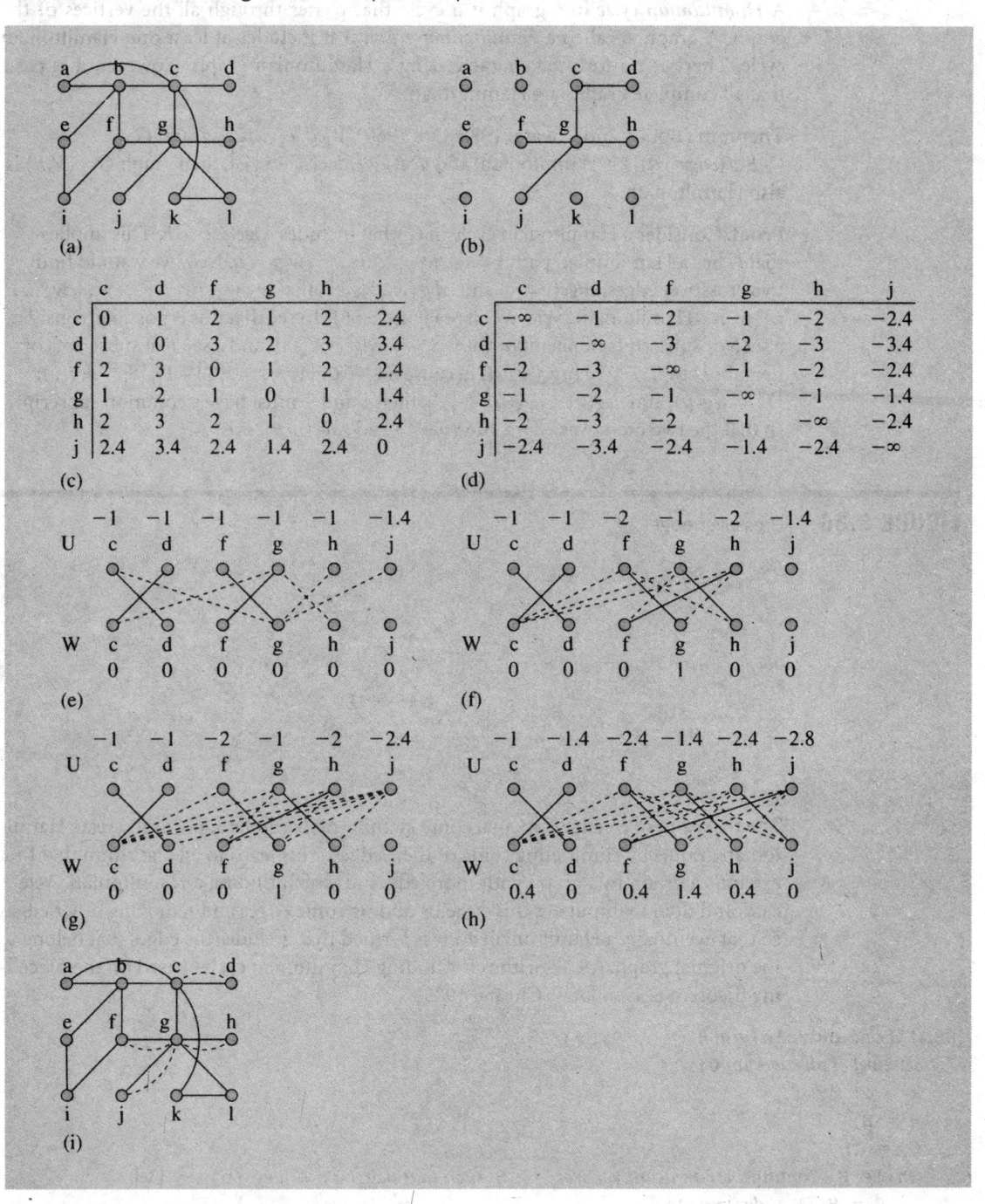

(a)

(b)

(c)

	c	d	f	g	h	j
c	0	1	2	1	2	2.4
d	1	0	3	2	3	3.4
f	2	3	0	1	2	2.4
g	1	2	1	0	1	1.4
h	2	3	2	1	0	2.4
j	2.4	3.4	2.4	1.4	2.4	0

(d)

	c	d	f	g	h	j
c	$-\infty$	-1	-2	-1	-2	-2.4
d	-1	$-\infty$	-3	-2	-3	-3.4
f	-2	-3	$-\infty$	-1	-2	-2.4
g	-1	-2	-1	$-\infty$	-1	-1.4
h	-2	-3	-2	-1	$-\infty$	-2.4
j	-2.4	-3.4	-2.4	-1.4	-2.4	$-\infty$

(e)

```
    -1   -1   -1   -1   -1   -1.4
U    c    d    f    g    h    j
W    c    d    f    g    h    j
     0    0    0    0    0    0
```

(f)

```
    -1   -1   -2   -1   -2   -1.4
U    c    d    f    g    h    j
W    c    d    f    g    h    j
     0    0    0    1    0    0
```

(g)

```
    -1   -1   -2   -1   -2   -2.4
U    c    d    f    g    h    j
W    c    d    f    g    h    j
     0    0    0    1    0    0
```

(h)

```
    -1   -1.4  -2.4  -1.4  -2.4  -2.8
U    c    d    f    g    h    j
W    c    d    f    g    h    j
    0.4   0   0.4   1.4   0.4   0
```

(i)

8.10.2 Hamiltonian Graphs

A *Hamiltonian cycle* in a graph is a cycle that passes through all the vertices of the graph. A graph is called a *Hamiltonian graph* if it includes at least one Hamiltonian cycle. There is no formula characterizing a Hamiltonian graph. However, it is clear that all complete graphs are Hamiltonian.

Theorem (Bondy and Chvátal 1976; Ore 1960). If $edge(vu) \notin E$, graph $G^* = (V, E \cup \{edge(vu)\})$ is Hamiltonian, and $deg(v) + deg(u) \geq |V|$, then graph $G = (V,E)$ is also Hamiltonian.

Proof. Consider a Hamiltonian cycle in G^* that includes $edge(vu) \notin E$. This implies that G has a Hamiltonian path $v = w_1, w_2, \ldots, w_{|V|-1}, w_{|V|} = u$. Now we want to find two crossover edges, $edge(vw_{i+1})$ and $edge(w_j u)$, such that $w_1, w_{i+1}, w_{i+2}, \ldots, w_{|V|}, w_j, \ldots, w_2, w_1$ is a Hamiltonian cycle in G (see Figure 8.36). To see that this is possible, consider a set S of subscripts of neighbors of v, $S = \{j: edge(vw_{j+1})\}$, and a set T of subscripts of neighbors of u, $T = \{j: edge(w_j u)\}$. Because $S \cup T \subseteq \{1, 2, \ldots, |V|-1\}$, $|S| = deg(v)$, $|T| = deg(u)$, and $deg(v) + deg(u) \geq |V|$, then S and T must have a common subscript so that the two crossover edges, $edge(vw_{i+1})$ and $edge(w_j u)$, exist. □

FIGURE 8.36 Crossover edges.

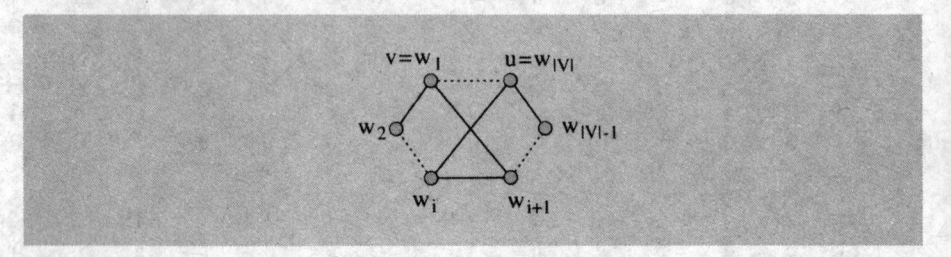

The theorem, in essence, says that some Hamiltonian graphs allow us to create Hamiltonian graphs by eliminating some of their edges. This leads to an algorithm that first expands a graph to a graph with more edges in which finding a Hamiltonian cycle is easy, and then manipulates this cycle by adding some edges and removing other edges so that eventually a Hamiltonian cycle is formed that includes the edges that belong to the original graph. An algorithm for finding Hamiltonian cycles based on the preceding theorem is as follows (Chvátal 1985):

```
HamiltonianCycle(graph G = (V,E))
    set label of all edges to 0;
    k = 1;
    H = E;
    G_H = G;
    while G_H contains nonadjacent vertices v, u such that deg_H(v) + deg_H(u) ≥ |V|
        H = H ∪ {edge(vu)};
        G_H = (V, H);
```

```
    label(edge(vu)) = k++;
if there exists a Hamiltonian cycle C
    while (k = max{label(edge(pq)): edge(pq)∈C}) > 0
        C = a cycle due to a crossover with each edge labeled by a number < k;
```

Figure 8.37 contains an example. In the first phase, the `while` loop is executed to create graph G_H based on graph G in Figure 8.37a. In each iteration, two nonadjacent vertices are connected with an edge if the total number of their neighbors is not less than the number of all vertices in the graph. We first look at all the vertices not adjacent to a. For vertex c, $deg_H(a) + deg_H(c) = 6 \geq |V| = 6$, the *edge(ac)* labeled with number 1 is included in H. Next, vertex e is considered, and because the degree of a just increased by acquiring a new neighbor, b, $deg_H(a) + deg_H(e) = 6$, so the *edge(ae)* labeled with 2 is included in H. The next vertex, for which we try to establish new neighbors, is b of degree 2, for which there are three nonadjacent vertices, d, e, and f with degrees 3, 3, and 3, respectively; therefore, the sum of b's degree and a degree of any of the three vertices does not reach 6, and no edge is now included in H. In the next iterations of the `while` loop, all possible neighbors of vertices c, d, e, and f are tested, which results in graph H as in Figure 8.37b with new edges shown as dashed lines with their labels.

In the second phase of `HamiltonianCycle()`, a Hamiltonian cycle in H is found, a, c, e, f, d, b, a. In this cycle, an edge with the highest label is found, *edge(ef)* (Figure 8.37c). The vertices in the cycle are so ordered that the vertices in this edge are on the extreme ends. Then by moving left to right in this sequence of vertices, we try to find crossover edges by checking edges from two neighbor vertices to the vertices at the ends of the sequence so that the edges cross each other. The first possibility is vertices d and b with *edge(bf)* and *edge(de)*, but this pair is rejected because the label of *edge(bf)* is greater than the largest label of the current cycle, 6. After this, the vertices b and a and the edges connecting them to the ends of the sequence *edge(af)* and *edge(be)* are checked; the edges are acceptable (their labels are 0 and 5), so the old cycle f, d, b, a, c, e, f is transformed into a new cycle f, a, c, e, b, d, f. This is shown beneath the diagram in Figure 8.37d with two new edges crossing each other and also in a sequence and in the diagram in Figure 8.37d.

In the new cycle, *edge(be)* has the highest label, 5, so the cycle is presented with the vertices of this edge, b and e, shown as the extremes of the sequence b, d, f, a, c, e (Figure 8.37e). To find crossover edges, we first investigate the pair of crossover edges, *edge(bf)* and *edge(de)*, but the label of *edge(bf)*, 7, is greater than the largest label of the current Hamiltonian cycle, 5, so the pair is discarded. Next, we try the pair *edge(ab)* and *edge(ef)*, but because of the magnitude of label of *edge(ef)*, 6, the pair is not acceptable. The next possibility is the pair *edge(bc)* and *edge(ae)*, which is acceptable, so a new cycle is formed, b, c, e, a, f, d, b (Figure 8.37e). In this cycle, a pair of crossover edges is found, *edge(ab)* and *edge(de)*, and a new cycle is formed, b, a f, d, e, c (Figure 8.37f), which includes edges only with labels equal to 0 (that is, only edges from graph G), which marks the end of execution of the algorithm with the last cycle being Hamiltonian and built only from edges in G.

440 ■ Chapter 8 Graphs

FIGURE 8.37 Finding a Hamiltonian cycle.

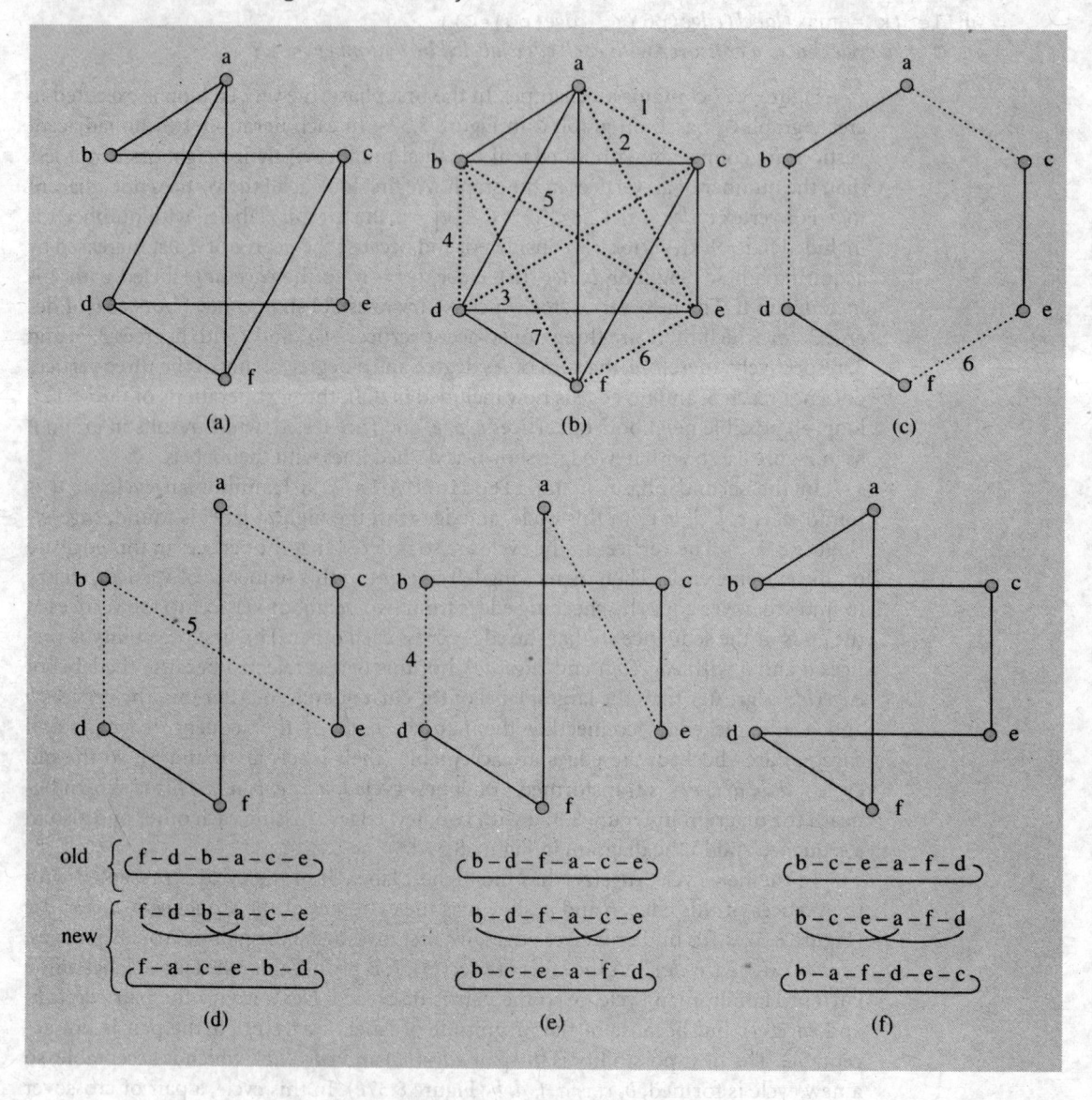

(a) (b) (c)

(d) (e) (f)

old { f–d–b–a–c–e   b–d–f–a–c–e   b–c–e–a–f–d

new { f–d–b–a–c–e   b–d–f–a–c–e   b–c–e–a–f–d
      f–a–c–e–b–d   b–c–e–a–f–d   b–a–f–d–e–c

The Traveling Salesman Problem

The *traveling salesman problem* (TSP) consists in finding a minimum tour; that is, in visiting once each city from a set of cities and then returning home so that the total distance traveled by the salesman is minimal. If distances between each pair of n cities are known, then there are $(n-1)!$ possible routes (the number of permutations of the vertices starting with a vertex v_1) or tours (or $\frac{(n-1)!}{2}$ if two tours traveled in opposite directions are equated). The problem is then in finding a minimum Hamiltonian cycle.

Most versions of TSP rely on the triangle inequality, $dist(v_i v_j) \leq dist(v_i v_k) + dist(v_k v_j)$. One possibility is to add to an already constructed path $v_1, \ldots, v_j$ a city v_{j+1} that is closest to v_j (a greedy algorithm). The problem with this solution is that the last $edge(v_n v_1)$ may be as long as the total distance for the remaining edges.

One approach is to use a minimum spanning tree. Define the length of the tree to be the sum of lengths of all the edges in the tree. Because removing one edge from the tour results in a spanning tree, then the minimum salesman tour cannot be shorter than the length of the minimum spanning tree mst, $length(minTour) \geq length(mst)$. Also, a depth-first search of the tree traverses each edge twice (when going down and then when backtracking) to visit all vertices (cities), whereby the length of the minimum salesman tour is at most twice the length of the minimum spanning tree, $2length(mst) \geq length(minTour)$. But a path that includes each edge twice goes through some vertices twice, too. Each vertex, however, should be included only once in the path. Therefore, if vertex v has already been included in such a path, then its second occurrence in a subpath $\ldots w\ v\ u \ldots$ is eliminated and the subpath is contracted to $\ldots w\ u \ldots$ whereby the length of the path is shortened due to the triangle inequality. For example, the minimum spanning tree for the complete graph that connects the cities a through h in Figure 8.38a is given in Figure 8.38b, and depth-first search renders the path in Figure 8.38c. By repeatedly applying the triangle inequality (Figure 8.38c–i), the path is transformed into the path in Figure 8.38i in which each

FIGURE 8.38 Using a minimum spanning tree to find a minimum salesman tour.

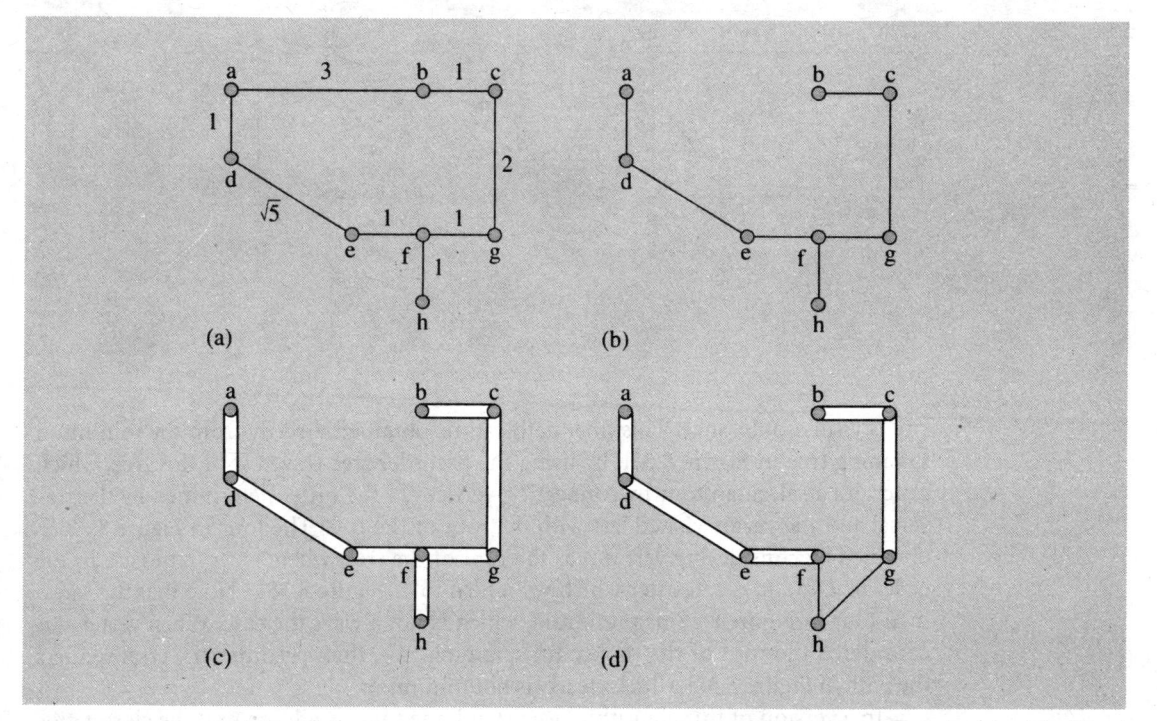

Continues

FIGURE 8.38   (continued)

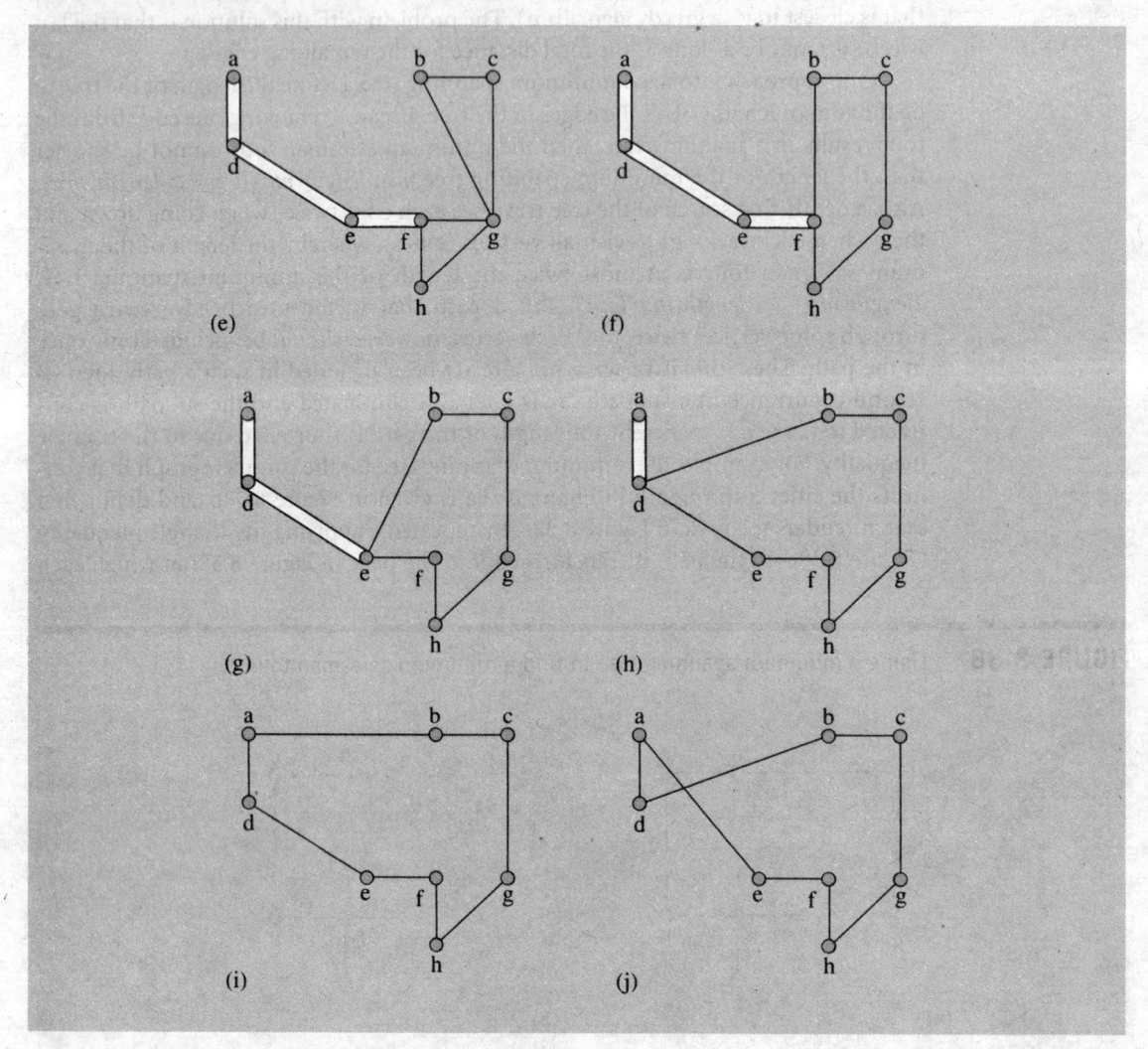

city is visited only once. This final path can be obtained directly from the minimum spanning tree in Figure 8.38b by using the preorder tree traversal of this tree, which generates a salesman tour by connecting vertices in the order determined by the traversal and the vertex visited last with the root of the tree. The tour in Figure 8.38i is obtained by considering vertex *a* as the root of the tree, whereby the cities are in the order *a, d, e, f, h, g, c, b*, after which we return to *a* (Figure 8.38i). Note that the salesman tour in Figure 8.38i is minimum, which is not always the case. When vertex *d* is considered the root of the minimum spanning tree, then preorder traversal renders the path in Figure 8.38j, which clearly is not minimum.

In a version of this algorithm, we extend one tour by adding to it the closest city. Because the tour is kept in one piece, it bears resemblance to the Jarník-Prim method.

Section 8.10 Eulerian and Hamiltonian Graphs

```
nearestAdditionAlgorithm(cities V)
    tour = {edge(vv)} for some v;
    while tour has less than |V| edges
        v_i = a vertex not on the tour closest to it;
        v_p = a vertex on the tour closest to v_i  (edge(v_p v_i) ∉ tour);
        v_q = a vertex on the tour such that edge(v_p v_q) ∈ tour;
        tour = tour ∪ {edge(v_p v_i), edge(v_i v_q)} − {edge(v_p v_q)};
```

In this algorithm, $edge(v_p v_q)$ is one of two edges that connects the city v_p on the tour to one of its two neighbors v_q on the tour. An example application of the algorithm is presented in Figure 8.39.

It may appear that the cost of execution of this algorithm is rather high. To find v_i and v_p in one iteration, all combinations should be tried, which is $\left(\sum_{i=1}^{|V|-1} i(|V|-i)\right) = \frac{(|V|-1)|V|(|V|+1)}{6} = O(|V|^3)$. However, a speedup is possible by carefully structuring the data. After the first vertex v is determined and used to initialize the tour, distances from each other vertex u to v are found, and two fields are properly set up for u: the field $distance = distance(uv)$ and $distanceTo = v$; at the same time, a vertex v_{min} with the minimum distance is determined. Then, in each iteration, $v_p = v_{min}$ from the previous iteration. Next, each vertex u not on the tour is checked to

FIGURE 8.39 Applying the nearest insertion algorithm to the cities in Figure 8.38a.

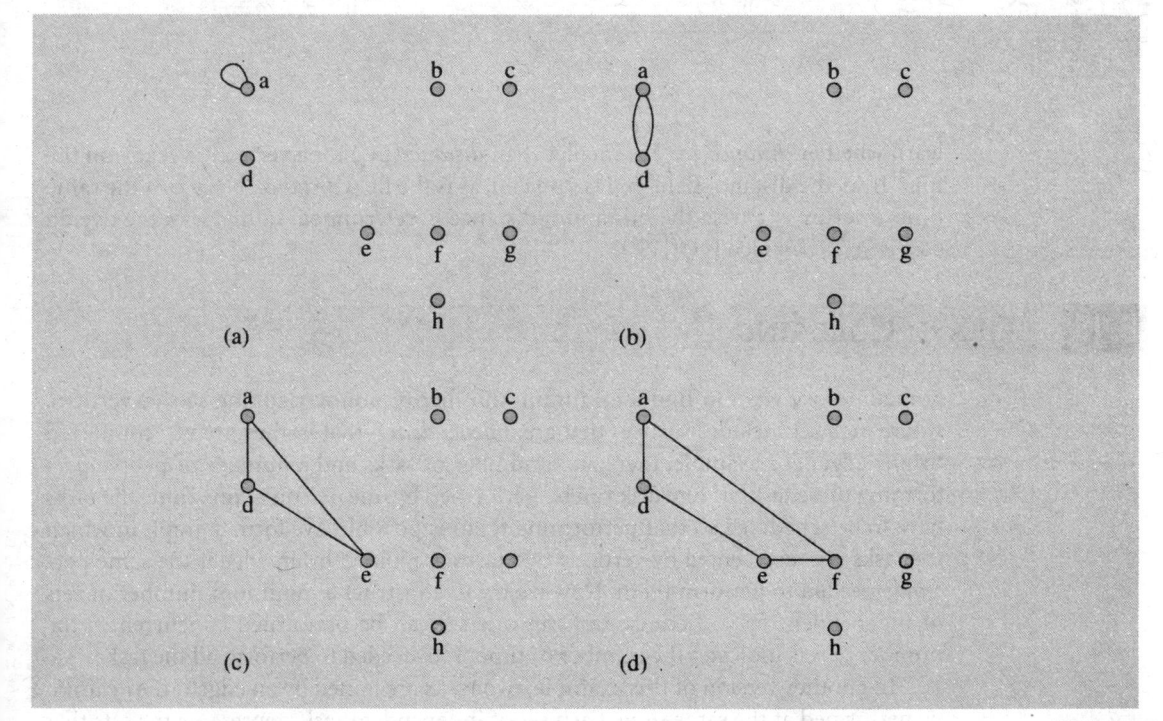

Continues

FIGURE 8.39 *(continued)*

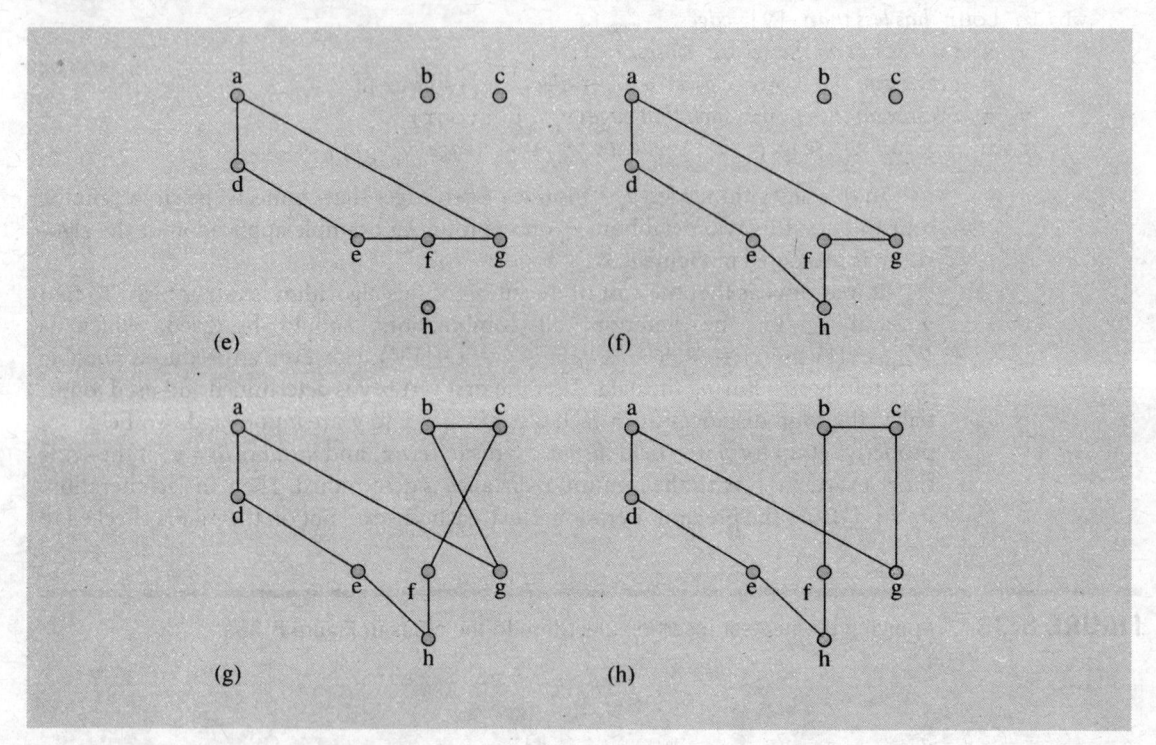

learn whether *distance*(uv_p) is smaller than *distance*(uv_r) for a vertex v_r already on the tour. If so, the distance field in *u* is updated, as is the field *distanceTo* = v_p. At the same time, a vertex v_{min} with the minimum distance is determined. In this way, the overall cost is $\sum_{i=1}^{|V|-1} i$, which is $O(|V|^2)$.

8.11 GRAPH COLORING

Sometimes we want to find a minimum number of nonoverlapping sets of vertices, where each set includes vertices that are *independent*—that is, they are not connected by any edge. For example, there are a number of tasks and a number of people performing these tasks. If one task can be performed by one person at one time, the tasks have to be scheduled so that performing them is possible. We form a graph in which the tasks are represented by vertices; two tasks are joined by an edge if the same person is needed to perform them. Now we try to construct a minimum number of sets of independent tasks. Because tasks in one set can be performed concurrently, the number of sets indicates the number of time slots needed to perform all the tasks.

In another version of this example, two tasks are joined by an edge if they cannot be performed at the same time. Each set of independent tasks represents the sets that can be performed concurrently, but this time the minimum number of sets indicates the minimum number of people needed to perform the tasks. Generally, we join by an edge two vertices when they are not allowed to be members of the same class. The problem can be rephrased by saying that we assign colors to vertices of the graph so that two vertices joined by an edge have a different color, and the problem amounts to *coloring* the graph with the minimum number of colors. More formally, if we have a set of colors C then we wish to find a function $f:V \rightarrow C$ such that if there is an $edge(vw)$, then $f(v) \neq f(w)$, and also C is of minimum cardinality. The minimum number of colors used to color the graph G is called the *chromatic number* of G and is denoted $\chi(G)$. A graph for which $k = \chi(G)$ is called *k-colorable*.

There may be more than one minimum set of colors C. No general formula exists for the chromatic number of any arbitrary graph. For some special cases, however, the formula is rather easy to determine: for a complete graph K_n, $\chi(K_n) = n$; for a cycle C_{2n} with an even number of edges, $\chi(C_{2n}) = 2$; for a cycle C_{2n+1} with an odd number of edges, $\chi(C_{2n+1}) = 3$; and for a bipartite graph G, $\chi(G) \leq 2$.

Determining a chromatic number of a graph is an NP-complete problem. Therefore, methods should be used that can approximate the exact graph coloring reasonably well—that is, methods that allow for coloring a graph with the number of colors that is not much larger than the chromatic number.

One general approach, called *sequential coloring*, is to establish the sequence of vertices and a sequence of colors before coloring them, and then color the next vertex with the lowest number possible.

```
sequentialColoringAlgorithm(graph = (V, E))
    put vertices in a certain order  v_p_1, v_p_2, ..., v_p_v;
    put colors in a certain order  c_1, c_2, ..., c_k;
    for i = 1 to |V|
        j = the smallest index of color that does not appear in any neighbor of v_pi;
        color(v_p_i) = c_j;
```

The algorithm is not specific about the criteria by which vertices are ordered (the order of colors is immaterial). One possibility is to use an ordering according to indices already assigned to the vertices before the algorithm is invoked, as in Figure 8.40b, which gives a $O(|V|^2)$ algorithm. The algorithm, however, may result in a number of colors that is vastly different from the chromatic number for a particular graph.

Theorem (Welsh and Powell 1967). For the sequential coloring algorithm, the number of colors needed to color the graph, $\chi(G) \leq \max_i \min(i, deg(v_{p_i}) + 1)$.

Proof. When coloring the ith vertex, at most $\min(i - 1, deg(v_{p_i}))$ of its neighbors already have colors; therefore, its color is at most $\min(i, deg(v_{p_i}) + 1)$. Taking the maximum value over all vertices renders the upper bound. □

For the graph in Figure 8.40a, $\chi(G) \leq \max_i \min(i, deg(v_{p_i}) + 1) = \max(\min(1, 4), \min(2, 4), \min(3, 3), \min(4, 3), \min(5, 3), \min(6, 5), \min(7, 6), \min(8, 4)) = \max(1, 2, 3, 3, 3, 5, 6, 4) = 6$.

The theorem suggests that the sequence of vertices should be organized so that vertices with high degrees should be placed at the beginning of the sequence so that min(position in sequence, $deg(v)$) = position in sequence, and the vertices with low degree should be placed at the end of the sequence so that min(position in sequence,

FIGURE 8.40 (a) A graph used for coloring; (b) colors assigned to vertices with the sequential coloring algorithm that orders vertices by index number; (c) vertices are put in the largest first sequence; (d) graph coloring obtained with the Brélaz algorithm.

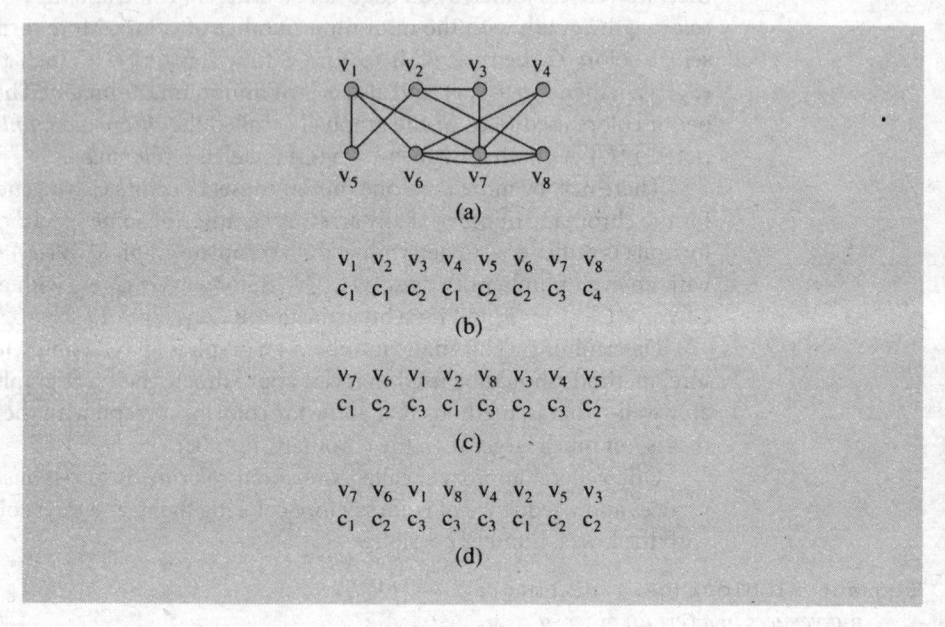

$$
\begin{array}{cccccccc}
v_1 & v_2 & v_3 & v_4 & v_5 & v_6 & v_7 & v_8 \\
c_1 & c_1 & c_2 & c_1 & c_2 & c_2 & c_3 & c_4
\end{array}
$$
(b)

$$
\begin{array}{cccccccc}
v_7 & v_6 & v_1 & v_2 & v_8 & v_3 & v_4 & v_5 \\
c_1 & c_2 & c_3 & c_1 & c_3 & c_2 & c_3 & c_2
\end{array}
$$
(c)

$$
\begin{array}{cccccccc}
v_7 & v_6 & v_1 & v_8 & v_4 & v_2 & v_5 & v_3 \\
c_1 & c_2 & c_3 & c_3 & c_3 & c_1 & c_2 & c_2
\end{array}
$$
(d)

$deg(v)) = deg(v)$. This leads to the *largest first* version of the algorithm in which the vertices are ordered in descending order according to their degrees. In this way, the vertices from Figure 8.40a are ordered in the sequence $v_7, v_6, v_1, v_2, v_8, v_3, v_4, v_5$, where the vertex v_7, with the largest number of neighbors, is colored first, as shown in Figure 8.40c. This ordering also gives a better estimate of the chromatic number, because now $\chi(G) \leq \max(\min(1, deg(v_7) + 1), \min(2, deg(v_6) + 1), \min(3, deg(v_1) + 1), \min(4, deg(v_2) + 1), \min(5, deg(v_8) + 1), \min(6, deg(v_3) + 1), \min(7, deg(v_4) + 1), \min(8, deg(v_5) + 1)) = \max(1, 2, 3, 4, 4, 3, 3, 3) = 4$.

The largest first approach is guided by the first principle, and so they use only one criterion to generate a sequence of vertices to be colored. However, this restriction can be lifted so that two or more criteria can be used at the same time. This is particularly important in breaking ties. In our example, if two vertices have the same degree, a vertex with the smaller index is chosen. In an algorithm proposed by Brélaz (1979), the primary criterion relies on the *saturation degree* of a vertex v, which is the number of different colors used to color neighbors of v. Should a tie occur, it is broken by choosing a vertex with the largest *uncolored degree*, which is the number of uncolored vertices adjacent to v.

```
BrelazColoringAlgorithm(graph)
    for each vertex v
        saturationDeg(v) = 0;
        uncoloredDeg(v) = deg(v);
    put colors in a certain order c_1, c_2, ..., c_k;
    while not all vertices are processed
        v = a vertex with highest saturation degree or,
            in case of a tie, vertex with maximum uncolored degree;
        j = the smallest index of color that does not appear in any neighbor of v;
        for each uncolored vertex u adjacent to v
            if no vertex adjacent to u is assigned color c_j
                saturationDeg(u)++;
            uncoloredDeg(u)--;
        color(v) = c_j;
```

For an example, see Figure 8.40d. First, v_7 is chosen and assigned color c_1 because v_7 has the highest degree. Next, saturation degrees of vertices $v_1, v_3, v_4, v_6,$ and v_8 are set to one because they are vertices adjacent to v_7. From among these five vertices, v_6 is selected because it has the largest number of uncolored neighbors. Then, saturation degrees of v_1 and v_8 are increased to two, and because both saturation and uncolored degrees of the two vertices are equal, we choose v_1 as having a lower index. The remaining color assignments are shown in Figure 8.40d.

The `while` loop is executed $|V|$ times; v is found in $O(|V|)$ steps and the `for` loop takes $deg(v)$ steps, which is also $O(|V|)$; therefore, the algorithm runs in $O(|V|^2)$ time.

8.12 NP-COMPLETE PROBLEMS IN GRAPH THEORY

In this section, NP-completeness of some problems in graph theory is presented.

8.12.1 The Clique Problem

A *clique* in a graph G is a complete subgraph of G. The clique problem is to determine whether G contains a clique K_m for some integer m. The problem is NP, because we can guess a set of m vertices and check in polynomial time whether a subgraph with these vertices is a clique. To show that the problem is NP-complete, we reduce the 3-satisfiability problem (see Section 2.10) to the clique problem. We perform reduction by showing that for a Boolean expression BE in CNF with three variables we can construct such a graph that the expression is satisfiable if there is a clique of size m in the graph. Let m be the number of alternatives in BE, that is,

$$BE = A_1 \land A_2 \land \ldots \land A_m$$

and each $A_i = (p \lor q \lor r)$ for $p \in \{x, \neg x\}$, $q \in \{y, \neg y\}$, and $r \in \{z, \neg z\}$, where $x, y,$ and z are Boolean variables.

We construct a graph whose vertices represent all the variables and their negations found in BE. Two vertices are joined by an edge if variables they represent are in different alternatives and the variables are not complementary—that is, one is not a negation of the other. For example, for the expression

$$BE = (x \vee y \vee \neg z) \wedge (x \vee \neg y \vee \neg z) \wedge (w \vee \neg x \vee \neg y)$$

a corresponding graph is in Figure 8.41. With this construction, an edge between two vertices represents a possibility of both variables represented by the vertices to be true at the same time. An m-clique represents a possibility of one variable from each alternative to be true, which renders the entire BE true. In Figure 8.41, each triangle represents a 3-clique. In this way, if BE is satisfiable, then an m-clique can be found. It is also clear that if an m-clique exists, then BE is satisfiable. This shows that the satisfiability problem is reduced to the clique problem, and the latter is NP-complete because the former has already been shown to be NP-complete.

FIGURE 8.41 A graph corresponding to the Boolean expression
$(x \vee y \vee \neg z) \wedge (x \vee \neg y \vee \neg z) \wedge (w \vee \neg x \vee \neg y)$.

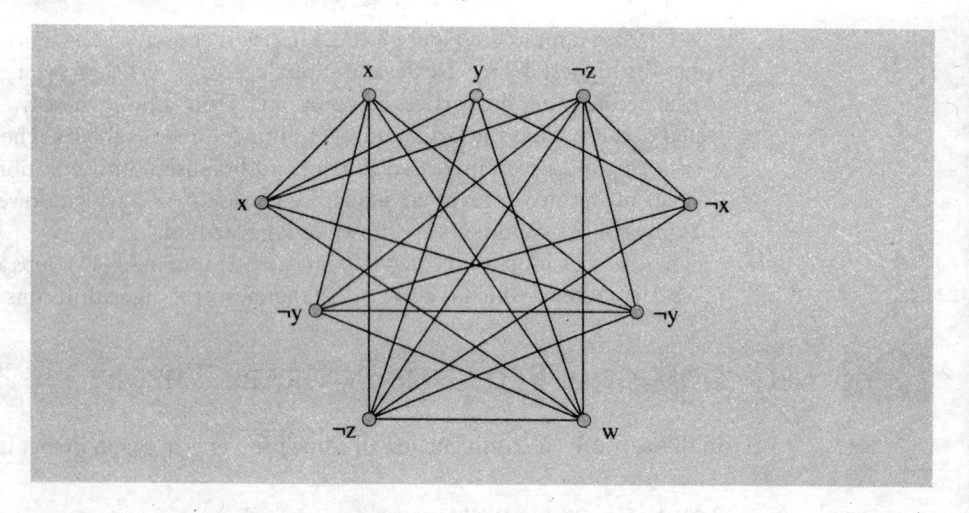

8.12.2 The 3-Colorability Problem

The 3-colorability problem is a question of whether a graph can be properly colored with three colors. We prove that the problem is NP-complete by reducing to it the 3-satisfiability problem. The 3-colorability problem is NP because we can guess a coloring of vertices with three colors and check in quadratic time that the coloring is correct (for each of the $|V|$ vertices check the color of up to $|V| - 1$ of its neighbors). To reduce the 3-satisfiability problem to the 3-colorability problem, we utilize an auxiliary 9-subgraph. A 9-subgraph takes 3 vertices, $v_1, v_2,$ and v_3, from an existing graph and adds 6 new vertices and 10 edges, as in Figure 8.42a. Consider the set $\{f, t, n\}$ (fuchsia/false, turquoise/true, nasturtium/neutral) of three colors used to color a graph. The reader can easily check the validity of the following lemma.

Lemma. 1) If all three vertices, $v_1, v_2,$ and v_3, of a 9-subgraph are colored with f, then vertex v_4 must also be colored with f to have the 9-subgraph colored correctly. 2) If only colors t and f can be used to color vertices $v_1, v_2,$ and v_3 of a 9-subgraph, and at least one is colored with t, then vertex v_4 can be colored with t. □

FIGURE 8.42 (a) A 9-subgraph; (b) a graph corresponding to the Boolean expression $(\neg w \vee x \vee y) \wedge (\neg w \vee \neg y \vee z) \wedge (w \vee \neg y \vee \neg z)$.

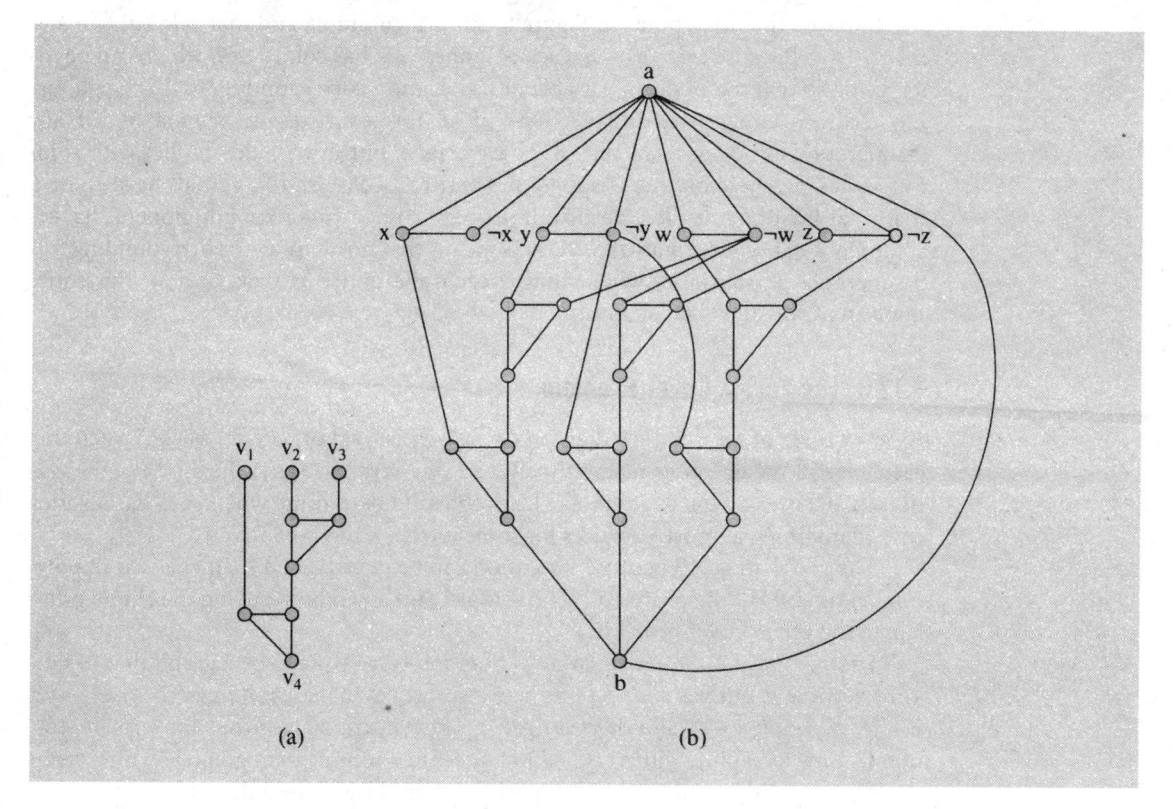

Now, for a given Boolean expression *BE* consisting of *k* alternatives we construct a graph in the following fashion. The graph has two special vertices, *a* and *b*, and *edge(ab)*. Moreover, the graph includes one vertex for each variable used in *BE* and the negation of this variable. For each pair of vertices *x* and $\neg x$, the graph includes *edge(ax)*, *edge(a(¬x))*, and *edge(x(¬x))*. Next, for each alternative *p*, *q*, or *r* included in *BE*, the graph has a 9-subgraph whose vertices v_1, v_2, and v_3 correspond to the three Boolean variables or their negations *p*, *q*, and *r* in this alternative. Finally, for each 9-subgraph, the graph includes $edge(v_4 b)$. A graph corresponding to the Boolean expression

$$(\neg w \vee x \vee y) \wedge (\neg w \vee \neg y \vee z) \wedge (w \vee \neg y \vee \neg z)$$

is presented in Figure 8.42b.

We now claim that if a Boolean expression *BE* is satisfiable, then the graph corresponding to it is 3-colorable. For each variable *x* in *BE*, we set *color(x) = t* and *color(¬x) = f* when *x* is true, and *color(x) = f* and *color(¬x) = t* otherwise. A Boolean expression is satisfiable if each alternative A_i in *BE* is satisfiable, which takes place when at least one variable *x* or its negation $\neg x$ in A_i is true. Because, except for *b* (whose color is about to be determined), each neighbor of *a* has color *t* or *f*, and because at least one of the three vertices v_1, v_2, and v_3 of each 9-subgraph has color t, each 9-subgraph is 3-colorable, and $color(v_4) = t$; by setting $color(a) = n$ and $color(b) = f$, the entire graph is 3-colorable.

Suppose that a graph as in Figure 8.42b is 3-colorable and that $color(a) = n$ and $color(b) = f$. Because $color(a) = n$, each neighbor of a has color f or t, which can be interpreted so that the Boolean variable or its negation corresponding to this neighboring vertex is either true or false. Only if all three vertices, v_1, v_2, and v_3, of any 9-subgraph have color f can vertex v_4 have color f, but this would conflict with color f of vertex b. Therefore, no 9-subgraph's vertices v_1, v_2, and v_3 can all have color f; that is, at least one of these vertices must have color t (the remaining one(s) having color f, not n, because $color(a) = n$). This means that no alternative corresponding to a 9-subgraph can be false, which means each alternative is true, and so the entire Boolean expression is satisfiable.

8.12.3 The Vertex Cover Problem

A *vertex cover* of an undirected graph $G = (V, E)$ is a set of vertices $W \subseteq V$ such that each edge in the graph is incident to at least one vertex from W. In this way, the vertices in W cover all the edges in E. The problem to determine whether G has a vertex cover containing at most k vertices for some integer k is NP-complete.

The problem is NP because a solution can be guessed and then checked in polynomial time. That the problem is NP-complete is shown by reducing the clique problem to the vertex cover problem.

First, define a *complement graph* $\overline{G}$ of graph $G = (V, E)$ to be a graph that has the same vertices V, but has connections between vertices that are not in G; that is, $\overline{G} = (V, \overline{E} = \{edge(uv): u, v \in V \text{ and } edge(uv) \notin E\})$. The reduction algorithm converts in polynomial time a graph G with a $(|V| - k)$-clique into a complement graph $\overline{G}$ with a vertex cover of size k. If $C = (V_C, E_C)$ is a clique in G, then vertices from the set $V - V_C$ cover all the edges in $\overline{G}$, because $\overline{G}$ has no edges with both endpoints in V_C. Consequently, $V - V_C$ is a vertex cover in $\overline{G}$ (see Figure 8.43a for a graph with a clique and 8.43b for a complement graph with a vertex cover). Suppose now that $\overline{G}$ has a vertex cover W; that is, an edge is in $\overline{E}$ if at least one endpoint of the edge is in W. Now, if none of the endpoints of an edge is in W, the edge is in graph G—that is, the latter endpoints are in $V - W$, and

FIGURE 8.43 (a) A graph with a clique; (b) a complement graph.

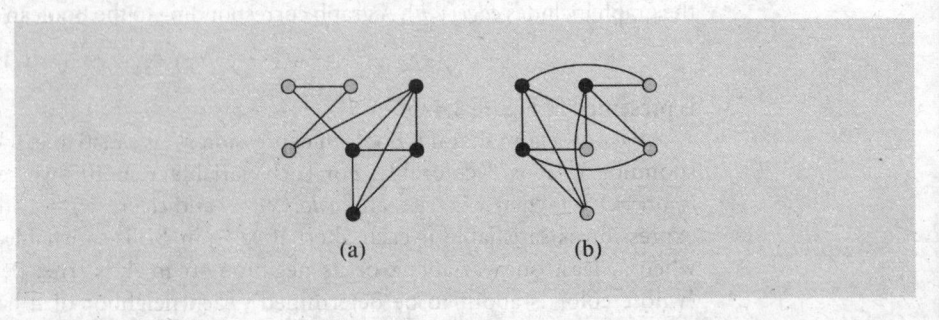

thus $V_C = V - W$ generates a clique. This proves that the positive answer to the clique problem is, through conversion, a positive answer to a vertex cover problem, and thus the latter is an NP-complete problem because the former is.

8.12.4 The Hamiltonian Cycle Problem

The contention that finding a Hamiltonian cycle in a simple graph G is an NP-complete problem is shown by reducing the vertex cover problem to the Hamiltonian cycle problem. First, we introduce an auxiliary concept of a 12-subgraph that is depicted in Figure 8.44a. The reduction algorithm converts each *edge(vu)* of graph G into a 12-subgraph so that one side of the subgraph, with vertices a and b, corresponds to a vertex v of G, and the other side, with vertices c and d, corresponds to vertex u. After entering one side of a 12-subgraph, for instance, at a, we can go through all the 12 vertices in order a, c, d, b and exit the 12-subgraph on the same side, at b. Also, we can go directly from a to b and if there is a Hamiltonian cycle in the entire graph, the vertices c and b are traversed during another visit of the 12-subgraph. Note that any other path through the 12-subgraph renders building a Hamiltonian cycle of the entire graph impossible.

Provided that we have a graph G, we build a graph G_H as follows. Create vertices $u_1, \ldots, u_k$, where k is the parameter corresponding to the vertex cover problem for graph G. Then, for each edge of G, a 12-subgraph is created; the 12-subgraphs associated with vertex v are connected together on the sides corresponding to v. Each endpoint of such a string of 12-subgraphs is connected to vertices $u_1, \ldots, u_k$. The result of transforming graph G for $k = 3$ in Figure 8.44b is the graph G_H in Figure 8.44c. To avoid clutter, the figure shows only some complete connections between endpoints of strings of 12-subgraphs and vertices u_1, u_2, and u_3, indicating only the existence of remaining connections. The claim is that there is a vertex cover of size k in graph G if there is a Hamiltonian cycle in graph G_H. Assume that $W = \{v_1, \ldots, v_k\}$ is a vertex cover in G. Then there is a Hamiltonian cycle in G_H formed in the following way. Beginning with u_1, go through the sides of 12-subgraphs that correspond to v_1. For a particular 12-subgraph, go through all of its 12 vertices if the other side of the 12-subgraph does not correspond to a vertex in the cover W; otherwise, go straight through the 12-subgraph. In the latter case, six vertices corresponding to a vertex w are not currently traversed, but they are traversed when processing the part of the Hamiltonian cycle corresponding to w. After the end of the string of 12-subgraphs is reached, go to u_2, and from here process the string of 12-subgraphs corresponding to v_2, and so on. For the last vertex u_k, process v_k and end the path at u_1, thereby creating a Hamiltonian cycle. Figure 8.44c presents with a thick line the part of the Hamiltonian cycle corresponding to v_1 that begins at u_1 and ends at u_2. Because the cover $W = \{v_1, v_2, v_6\}$, the processing continues for v_2 at u_2 and ends at u_3, and then for v_6 at u_3 and ends at u_1.

Conversely, if G_H has a Hamiltonian cycle, it includes subpaths through k 12-subgraph strings that correspond to k vertices in G_C that form a cover.

Consider now this version of the traveling salesman problem. In a graph with distances assigned to each edge we try to determine whether there is a cycle with total distance with the combined distance not greater than an integer k. That the problem is NP-complete can be straightforwardly shown by reducing it to the Hamiltonian path problem.

FIGURE 8.44 (a) A 12-subgraph; (b) a graph G and (c) its transformation, graph G_H.

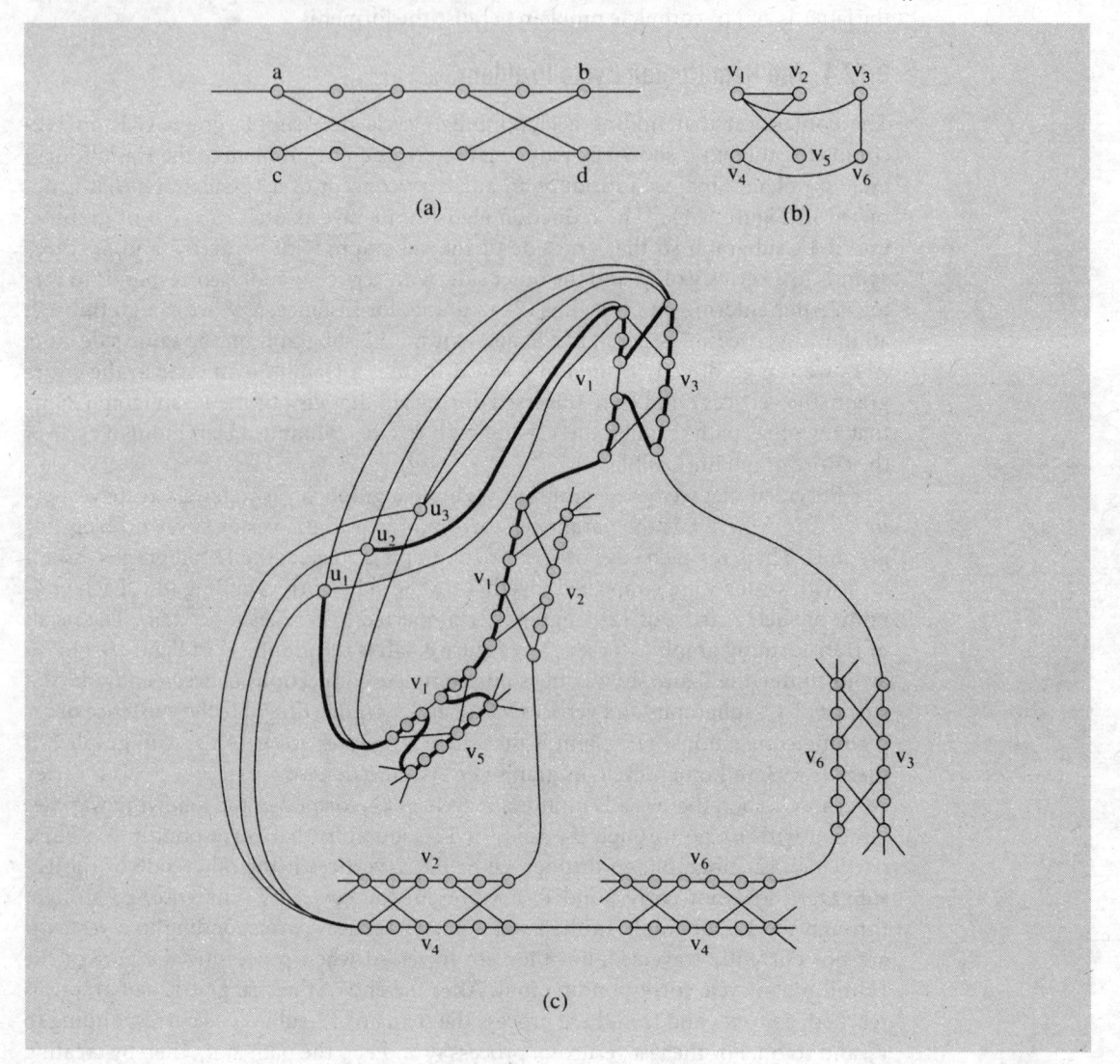

8.13 CASE STUDY: DISTINCT REPRESENTATIVES

Let there be a set of committees, $C = \{C_1, \ldots, C_n\}$, each committee having at least one person. The problem is to determine, if possible, representatives from each committee so that the committee is represented by one person and each person can represent only one committee. For example, if there are three committees, $C_1 = \{M_5, M_1\}$, $C_2 = \{M_2, M_4, M_3\}$, and $C_3 = \{M_3, M_5\}$, then one possible representation is: member M_1 represents committee C_1, M_2 represents C_2, and M_5 represents C_3. However, if we have these three committees, $C_4 = C_5 = \{M_6, M_7\}$, and $C_6 = \{M_7\}$, then no distinct representation can be created, because there are only two members in all three committees combined. The latter observation has been proven by P. Hall in the *system of distinct representatives* theorem, which can be phrased in the following way:

Theorem. A nonempty collection of finite nonempty sets $C_1, \ldots, C_n$ has a system of distinct representatives if for any $i \leq n$, the union $C_{k_1} \cup \ldots \cup C_{k_i}$ has at least i elements.

The problem can be solved by creating a network and trying to find a maximum flow in this network. For example, the network in Figure 8.45a can represent the membership of the three committees, C_1, C_2, and C_3. There is a dummy source vertex connected to nodes representing committees, the committee vertices are connected to vertices representing their members, and the member vertices are all connected to a dummy sink vertex. We assume that each edge e's capacity $cap(e) = 1$. A system of distinct representatives is found if the maximum flow in the network equals the number of committees. The paths determined by a particular maximum flow algorithm determine the representatives. For example, member M_1 represents the committee C_1 if a path s, C_1, M_1, t is determined.

FIGURE 8.45 (a) A network representing membership of three committees, C_1, C_2, and C_3, and (b) the first augmenting path found in this network.

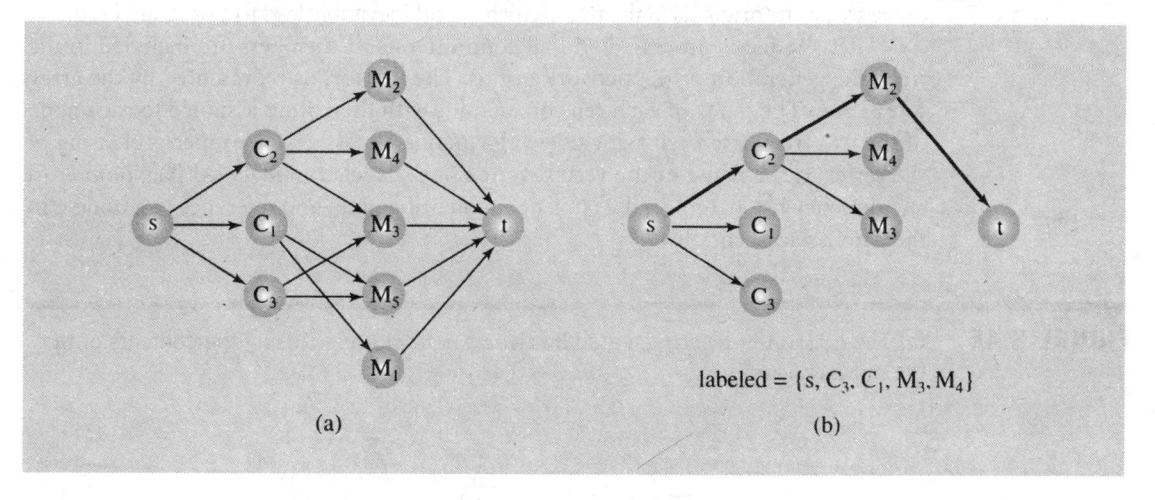

The implementation has two main stages. First, a network is created using a set of committees and members stored in a file. Then, the network is processed to find augmenting paths corresponding to members representing committees. The first stage is specific to the system of distinct representatives. The second stage can be used for finding the maximum flow of any network because it assumes that the network has been created before it begins.

When reading committees and members from a file, we assume that the name of a committee is always followed by a colon and then by a list of members separated by commas and ended with a semicolon. An example is the following file `committees`, which includes information corresponding to Figure 8.45:

```
C2: M2, M4, M3;
C1: M5, M1;
C3: M3, M5;
```

In preparation for creating a network, two trees are generated: a `committeeTree` and a `memberTree`. The information stored in each node includes the name of the committee or member, an `idNum` assigned by the program using a running counter `numOfVertices`, and an adjacency list to be included later in the network. Figure 8.46a shows `committeeTree` corresponding to the committees of the example file. The adjacency lists are implemented using the STL class `list`. The adjacency lists are shown in simplified form, with member `idNum` only and with forward link only (the STL lists are implemented as doubly linked lists; see Section 3.7). A fuller form of the lists is shown in Figure 8.46b, but only with the forward links. The names of the members are shown above the nodes of these adjacency lists. A separate adjacency list, `sourceList`, is built for the source vertex. Note that the member `adjacent` in a `NetTreeNode` is not of type `list<Vertex>` but a pointer type `list<Vertex>*`. Because these lists are later transferred to the array `vertices`, a nonpointer type would use the assignment that copies the lists for the tree to the array, which is inefficient. More important, the program uses pointers to vertices stored in lists in the trees (`twin` in `Vertex` and `corrVer` in `VertexArrayRec`) when creating the trees, and the copies of the vertices, of course, would have different addresses than the originals in the trees, which would inevitably lead to a program crash.

After the file is processed and all committees and members are included in the trees, the generation of the network begins. The network is represented by the array `vertices`. The index of each cell corresponds with the `idNum` assigned to each node of the two trees. Each cell `i` includes information necessary for proper processing of the vertex `i`: the name of the vertex, vertex slack, labeled/nonlabeled flag, pointer to an adjacency list, parent in the current augmenting path, and reference to a node `i` in the parent's adjacency list.

FIGURE 8.46 (a) The `committeeTree` created by `readCommittees()` using the contents of the file `committees`.

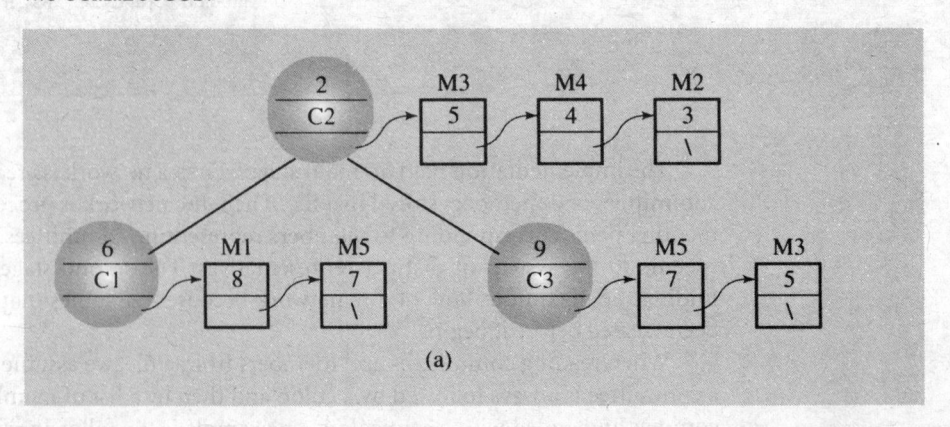

FIGURE 8.46 (b) the network representation created by `FordFulkersonMaxFlow()`.

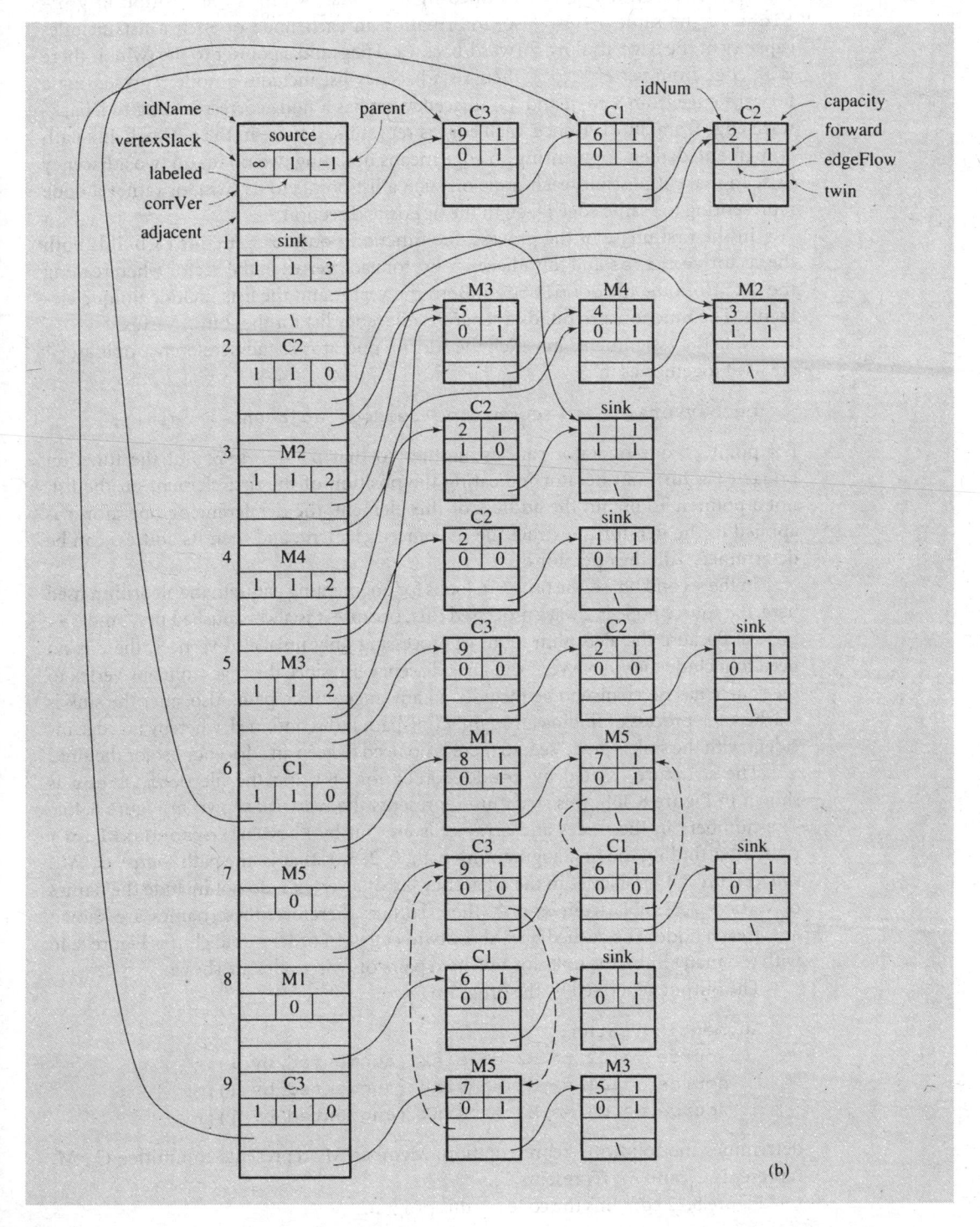

An adjacency list of a vertex in position `i` represents edges incident with this vertex. Each node on the list is identified by its `idNum`, which is the position in `vertices` of the same vertex. Also, information in each node of such a list includes capacity of the edge, its flow, forward/backward flag, and a pointer to the twin. If there is an edge from vertex `i` to `j`, then `i`'s adjacency list includes a node representing a forward edge from `i` to `j`, and `j`'s adjacency list has a node corresponding to a backward edge from `j` to `i`. Hence, each edge is represented twice in the network. If a path is augmented, then augmenting an edge means updating two nodes on two adjacency lists. To make it possible, each node on such a list points to its twin, or rather a node representing the same edge taken in the opposite direction.

In the first phase of the process, the function `readCommittees()` builds both the vector `vertices` and the adjacency list for each vertex in the vector when reading the data from the file `committees`. Both the vector and the lists include unique elements. The function also builds a separate adjacency list for the source vertex.

Note the assignments in which the address operator `&` and dereference operator `*` are used together, as in

```
memberVerAddr = &*committeeTreeNode.adjacent->begin();
```

For pointers, one operator cancels another so that `p == &*p`. But the function `begin()` returns an iterator indicating the position of the first element on the list, not a pointer. To obtain the address of this element, the dereferencing operator `*` is applied to the iterator to extract the referenced element, and then its address can be determined with the operator `&`.

In the second phase, the program looks for augmenting paths. In the algorithm used here, the source node is always processed first, because it is always pushed first onto `labeled`. Because the algorithm requires processing only unlabeled vertices, there is no need to include the source vertex in any adjacency list, since the edge from any vertex to the source has no chance to be included in any augmenting path. Also, after the sink is reached, the process of finding an augmenting path is discontinued, whereby no edge incident with the sink is processed, so there is no need to keep an adjacency list for the sink.

The structure created by `readCommittees()` using the file `committees` is shown in Figure 8.46b; this structure represents the network shown in Figure 8.45a. The numbers in the nodes and array cells are put by `FordFulkersonMaxFlow()` right after finding the first augmenting path, 0, 2, 3, 1, that is, the path *source*, C_2, M_2, *sink* (Figure 8.45b). Nodes in the adjacency list of a vertex `i` do not include the names of vertices accessible from `i`, only their `idNum`; therefore, these names are shown above each node. The dashed lines show twin edges. In order not to clutter Figure 8.46 with too many links, the links for only two pairs of twin nodes are shown.

The output generated by the program

```
Augmenting paths:
    source => C2 => M2 => sink (augmented by 1);
    source => C1 => M5 => sink (augmented by 1);
    source => C3 => M3 => sink (augmented by 1);
```

determines the following representation: Member M_2 represents committee C_2, M_5 represents C_1, and M_3 represents C_3.

Figure 8.47 contains the code for this program.

FIGURE 8.47 An implementation of the distinct representatives problem.

```cpp
#include <iostream>
#include <fstream>
#include <cctype>
#include <crstdlib>
#include <string>
#include <limits>
#include <list>
#include <stack>
#include <iterator>
using namespace std;
#include "genBST.h"

class VertexArrayRec;
class LocalTree;
class Network;

class Vertex {
public:
    Vertex() {
    }
    Vertex(int id, int c, int ef, bool f, Vertex *t = 0) {
        idNum = id; capacity = c; edgeFlow = ef; forward = f; twin = t;
    }
    bool operator== (const Vertex& v) const {
        return idNum == v.idNum;
    }
    bool operator!= (const Vertex& v) const { // required
        return idNum != v.idNum;
    }
    bool operator< (const Vertex& v) const {  // by the compiler;
        return idNum < v.idNum;
    }
    bool operator> (const Vertex& v) const {
        return idNum > v.idNum;
    }
private:
    int idNum, capacity, edgeFlow;
    bool forward;     // direction;
    Vertex *twin;     // edge in the opposite direction;
    friend class Network;
```

Continues

FIGURE 8.47   *(continued)*

```
        friend ostream& operator<< (ostream&, const Vertex&);
};

class NetTreeNode {
public:
    NetTreeNode(forwardArrayRec **v = 0) {
        verticesPtr = v;
        adjacent = new list<Vertex>;
    }
    bool operator<  (const NetTreeNode& tr) const {
        return strcmp(idName,tr.idName) < 0;
    }
    bool operator== (const NetTreeNode& tr) const {
        return strcmp(idName,tr.idName) == 0;
    }
private:
    int idNum;
    char *idName;
    VertexArrayRec **verticesPtr;
    list<Vertex> *adjacent;
    friend class Network;
    friend class LocalTree;
    friend ostream& operator<< (ostream&,const NetTreeNode&);
};

class VertexArrayRec {
public:
    VertexArrayRec() {
        adjacent = 0;
    }
private:
    char *idName;
    int vertexSlack;
    bool labeled;
    int parent;
    Vertex *corrVer;          // corresponding vertex: vertex on parent's
    list<Vertex> *adjacent;   // list of adjacent vertices with the same
    friend class Network;     // idNum as the cell's index;
    friend class LocalTree;
    friend ostream& operator<< (ostream&,const Network&);
};
```

FIGURE 8.47   *(continued)*

```
// define new visit() to be used by inorder() from genBST.h;
class LocalTree : public BST<NetTreeNode> {
    void visit(BSTNode<NetTreeNode>* p) {
        (*(p->key.verticesPtr))[p->key.idNum].idName   = p->key.idName;
        (*(p->key.verticesPtr))[p->key.idNum].adjacent
                                                = p->key.adjacent;
    }
};

class Network {
public:
    Network() : sink(1), source(0), none(-1), numOfVertices(2) {
        verticesPtr = new VertexArrayRec*;
    }
    void readCommittees(char *committees);
    void FordFulkersonMaxFlow();
private:
    const int sink, source, none;
    int numOfVertices;
    VertexArrayRec *vertices;
    VertexArrayRec **verticesPtr; // used by visit() in LocalTree to
                                  // update vertices;
    int edgeSlack(Vertex *u) const {
        return u->capacity - u->edgeFlow;
    }
    int min(int n, int m) const {
        return n < m ? n : m;
    }
    bool Labeled(Vertex *v) const {
        return vertices[v->idNum].labeled;
    }
    void label(Vertex*,int);
    void augmentPath();
    friend class LocalTree;
    friend ostream& operator<< (ostream&,const Network&);
};

ostream& operator<< (ostream& out, const NetTreeNode& tr) {
    out << tr.idNum << ' ' << tr.idName << ' ';
    return out;
}
```

Continues

FIGURE 8.47 *(continued)*

```cpp
ostream& operator<< (ostream& out, const Vertex& vr) {
    out << vr.idNum  << ' ' << vr.capacity << ' ' << vr.edgeFlow << ' '
        << vr.forward << "| ";
    return out;
}

ostream& operator<< (ostream& out, const Network& net) {
    ostream_iterator<Vertex> output(out," ");
    for (int i = 0; i < net.numOfVertices; i++) {
        out << i << ": "
            << net.vertices[i].idName << '|'
            << net.vertices[i].vertexSlack << '|'
            << net.vertices[i].labeled << '|'
            << net.vertices[i].parent << '|'
            << /* net.vertices[i].corrVer << */ "-> ";
        if (net.vertices[i].adjacent != 0)
            copy (net.vertices[i].adjacent->begin(),
                  net.vertices[i].adjacent->end(),output);
        out << endl;
    }
    return out;
}

void Network::readCommittees(char *fileName) {
    char i, name[80], *s;
    LocalTree committeeTree, memberTree;
    Vertex memberVer(0,1,0,false), commVer(0,1,0,true);
    Vertex *commVerAddr, *memberVerAddr;
    NetTreeNode committeeTreeNode(verticesPtr),
                memberTreeNode(verticesPtr), *member;
    list<Vertex> *sourceList = new list<Vertex>;
    ifstream fIn(fileName);
    if (fIn.fail()) {
        cerr << "Cannot open " << fileName << endl;
        exit(-1);
    }
    while (!fIn.eof()) {
        fIn >> name[0]; // skip leading spaces;
        if (fIn.eof())  // spaces at the end of file;
            break;
        for (i = 0; name[i] != ':'; )
            name[++i] = fIn.get();
```

FIGURE 8.47 (continued)

```
            for (i--; isspace(name[i]); i--); // discard trailing spaces;
            name[i+1] = '\0';
            s = strdup(name);
            committeeTreeNode.idNum = commVer.idNum = numOfVertices++;
            committeeTreeNode.idName = s;
            for (bool lastMember = false; lastMember == false; ) {
                fIn >> name[0]; // skip leading spaces;
                for (i = 0; name[i] != ',' && name[i] != ';'; )
                    name[++i] = fIn.get();
                if (name[i] == ';')
                    lastMember = true;
                for (i--; isspace(name[i]); i--); // discard trailing spaces;
                name[i+1] = '\0';
                s = strdup(name);
                memberTreeNode.idName = s;
                commVer.forward = false;
                if ((member = memberTree.search(memberTreeNode)) == 0) {
                    memberVer.idNum = memberTreeNode.idNum =
                                    numOfVertices++;
                    memberTreeNode.adjacent->push_front(Vertex(sink,1,0,true));
                    memberTreeNode.adjacent->push_front(commVer);
                    commVerAddr = &*memberTreeNode.adjacent->begin();
                    memberTree.insert(memberTreeNode);
                    memberTreeNode.adjacent = new list<Vertex>;
                }
                else {
                    memberVer.idNum = member->idNum;
                    member->adjacent->push_front(commVer);
                    commVerAddr = &*member->adjacent->begin();
                }
                memberVer.forward = true;
                committeeTreeNode.adjacent->push_front(memberVer);
                memberVerAddr = &*committeeTreeNode.adjacent->begin();
                memberVerAddr->twin = commVerAddr;
                commVerAddr->twin = memberVerAddr;
            }
            commVer.forward = true;
            sourceList->push_front(commVer);
            committeeTree.insert(committeeTreeNode);
            committeeTreeNode.adjacent = new list<Vertex>;
    }
```

Continues

FIGURE 8.47 *(continued)*

```
    fIn.close();
    cout << "\nCommittee tree:\n"; committeeTree.printTree();
    cout << "\nMember tree:\n";         memberTree.printTree();
    vertices = *verticesPtr = new VertexArrayRec[numOfVertices];
    if (vertices == 0) {
        cerr << "Not enough memory\n";
        exit(-1);
    }
    vertices[source].idName  = "source";
    vertices[sink].idName    = "sink";
    vertices[source].adjacent = sourceList;
    vertices[source].parent = none;
    committeeTree.inorder(); // transfer data from both trees
    memberTree.inorder();    // to array vertices[];
}

void Network::label(Vertex *u, int v) {
    vertices[u->idNum].labeled = true;
    if (u->forward)
         vertices[u->idNum].vertexSlack =
             min(vertices[v].vertexSlack,edgeSlack(u));
    else vertices[u->idNum].vertexSlack =
             min(vertices[v].vertexSlack,u->edgeFlow);
    vertices[u->idNum].parent  = v;
    vertices[u->idNum].corrVer = u;
}

void Network::augmentPath() {
    register int i, sinkSlack = vertices[sink].vertexSlack;
    Stack<char*> path;
    for (i = sink; i != source; i = vertices[i].parent) {
        path.push(vertices[i].idName);
        if (vertices[i].corrVer->forward)
             vertices[i].corrVer->edgeFlow += sinkSlack;
        else vertices[i].corrVer->edgeFlow -= sinkSlack;
        if (vertices[i].parent != source && i != sink)
             vertices[i].corrVer->twin->edgeFlow =
                 vertices[i].corrVer->edgeFlow;
    }
    for (i = 0; i < numOfVertices; i++)
        vertices[i].labeled = false;
    cout << "  source";
```

FIGURE 8.47   *(continued)*

```
        while (!path.empty())
            cout << " => " << path.pop();
        cout << " (augmented by " << sinkSlack << ");\n";
    }

    void Network::FordFulkersonMaxFlow() {
        Stack<int> labeled;
        Vertex *u;
        list<Vertex>::iterator it;
        for (int i = 0; i < numOfVertices; i++) {
            vertices[i].labeled = false;
            vertices[i].vertexSlack = 0;
            vertices[i].parent = none;
        }
        vertices[source].vertexSlack = INT_MAX;
        labeled.push(source);
        cout << "Augmenting paths:\n";
        while (!labeled.empty()) {    // while not stuck;
            int v = labeled.pop();
            for (it = vertices[v].adjacent->begin(), u = &*it;
                 it != vertices[v].adjacent->end(); it++, u = &*it)
                if (!Labeled(u)) {
                    if (u->forward && edgeSlack(u) > 0)
                        label(u,v);
                    else if (!u->forward && u->edgeFlow > 0)
                        label(u,v);
                    if (Labeled(u))
                        if (u->idNum == sink) {
                            augmentPath();
                            while (!labeled.empty())
                                labeled.pop();    // clear the stack;
                            labeled.push(source);// look for another path;
                            break;
                        }
                        else {
                            labeled.push(u->idNum);
                            vertices[u->idNum].labeled = true;
                        }
                }
        }
    }
```

Continues

FIGURE 8.47 *(continued)*

```
int main(int argc, char* argv[]) {
    char fileName[80];
    Network net;
    if (argc != 2) {
        cout << "Enter a file name: ";
        cin.getline(fileName,80);
    }
    else strcpy(fileName,argv[1]);
    net.readCommittees(fileName);
    cout << net;
    net.FordFulkersonMaxFlow();
    cout << net;
    return 0;
}
```

8.14 EXERCISES

1. Look carefully at the definition of a graph. In one respect, graphs are more specific than trees. What is it?

2. What is the relationship between the sum of the degrees of all vertices and the number of edges of graph $G = (V,E)$?

3. What is the complexity of `breadthFirstSearch()`?

4. Show that a simple graph is connected if it has a spanning tree.

5. Show that a tree with n vertices has $n - 1$ edges.

6. How can `DijkstraAlgorithm()` be applied to undirected graphs?

7. How can `DijkstraAlgorithm()` be modified to become an algorithm for finding the shortest path from vertex a to b?

8. The last clause from `genericShortestPathAlgorithm()`

    add u to `toBeChecked` *if it is not there;*

    is not included in `DijkstraAlgorithm()`. Can this omission cause any trouble?

9. Modify `FordAlgorithm()` so that it does not fall into an infinite loop if applied to a graph with negative cycles.

10. For what digraph does the `while` loop of `FordAlgorithm()` iterate only one time? Two times?

11. Can `FordAlgorithm()` be applied to undirected graphs?

12. Make necessary changes in `FordAlgorithm()` to adapt it to solving the all-to-one shortest path problem and apply the new algorithm to vertex f in the graph in Figure 8.8. Using the same order of edges, produce a table similar to the table shown in this figure.

13. The D'Esopo-Pape algorithm is exponential in the worst case. Consider the following method to construct pathological graphs of n vertices (Kershenbaum, 1981), each vertex identified by a number $1, \ldots, n$:

```
KershenbaumAlgorithm()
    construct a two-vertex graph with vertices 1 and 2, and edge(1,2) = 1;
    for k = 3 to n
        add vertex k;
        for i = 2 to r-1
            add edge(k,i) with weight(edge(k,i)) = weight(edge(1,i));
            weight(edge(1,i)) = weight(1,i) + 2^{k-3} + 1;
        add edge(1,k) with weight(edge(1,k)) = 1;
```

The vertices adjacent to vertex 1 are put in ascending order and the remaining adjacency lists are in descending order. Using this algorithm, construct a five-vertex graph and execute the D'Esopo-Pape algorithm showing all changes in the deque and all edge updates. What generalization can you make about applying Pape's method to such graphs?

14. What do you need to change in genericShortestPathAlgorithm() in order to convert it to Dijkstra's one-to-all algorithm?

15. Enhance WFIalgorithm() to indicate the shortest paths, in addition to their lengths.

16. WFIalgorithm() finishes execution gracefully even in the presence of a negative cycle. How do we know that the graph contains such a cycle?

17. The original implementation of WFIalgorithm() given by Floyd is as follows:

```
WFIalgorithm2(matrix weight)
    for i = 1 to |V|
        for j = 1 to |V|
            if weight[j,i] < ∞
                for k = 1 to |V|
                    if weight[i,k] < ∞
                        if (weight[j][k] > weight[j][i] + weight[i][k])
                            weight[j][k] = weight[j][i] + weight[i][k];
```

Is there any advantage to this longer implementation?

18. One method of finding shortest paths from all vertices to all other vertices requires us to transform the graph so that it does not include negative weights. We may be tempted to do it by simply finding the smallest negative weight k and adding $-k$ to the weights of all edges. Why is this method inapplicable?

19. For which edges does ≤ in the inequality

$$dist(v) \leq dist(w) + weight(edge(wv)) \text{ for any vertex } w$$

become <?

20. Modify cycleDetectionDFS() so that it could determine whether a particular edge is part of a cycle in an undirected graph.

21. Our implementation of union() requires three arrays. Is it possible to use only two of them and still have the same information concerning roots, next vertices, and lengths? Consider using negative numbers.

22. When would KruskalAlgorithm() require $|E|$ iterations?

23. How can the second minimum spanning tree be found?

24. Is the minimum spanning tree unique?

25. How can the algorithms for finding the minimum spanning tree be used to find the maximum spanning tree?

26. Apply the following two algorithms to find the minimum spanning tree to the graph in Figure 8.15a.

    a. Probably the first algorithm for finding the minimum spanning tree was devised in 1926 by Otakar Borůvka (pronounced: boh-roof-ka). In this method, we start with $|V|$ one-vertex trees, and for each vertex v, we look for an *edge(vw)* of minimum weight among all edges outgoing from v and create small trees by including these edges. Then, we look for edges of minimal weight that can connect the resulting trees to larger trees. The process is finished when one tree is created. Here is a pseudocode for this algorithm:

```
BorůvkaAlgorithnm(weighted connected undirected graph)
    make each vertex the root of a one-node tree;
    while there is more than one tree
        for each tree t
            e = minimum weight edge(vu) where v is included in t and u is not;
            create a tree by combining t and the tree that includes u
                if such a tree does not exist yet;
```

    b. Another algorithm was discovered by Vojtech Jarník (pronounced: yar-neek) in 1936 and later rediscovered by Robert Prim. In this method, all of the edges are also initially ordered, but a candidate for inclusion in the spanning tree is an edge that not only does not lead to cycles in the tree, but also is incident to a vertex already in the tree:

```
JarnikPrimAlgorithm(weighted connected undirected graph)
    tree = null;
    edges = sequence of all edges of graph sorted by weight;
    for i = 1 to |V| - 1
        for j = 1 to |edges|
            if e_j from edges does not form a cycle with edges in tree and
                is incident to a vertex in tree
                    add e_j to tree;
                    break;
```

27. The algorithm `blockSearch()`, when used for undirected graphs, relies on the following observation: In a depth-first search tree created for an undirected graph, each back edge connects a successor to a predecessor (and not, for instance, a sibling to a sibling). Show the validity of this observation.

28. What is the complexity of `blockSearch()`?

29. Blocks in undirected graphs are defined in terms of edges, and the algorithm `blockDFS()` stores edges on the stack to output blocks. On the other hand, SCCs in digraphs are defined in terms of vertices, and the algorithm `strongDFS()` stores vertices on the stack to output SCC. Why?

30. Consider a possible implementation of `topologicalSort()` by using in it the following routine:

    ```
    minimalVertex(digraph)
        v = a vertex of digraph;
        while v has a successor
            v = successor(v);
        return v;
    ```

    What is the disadvantage of using this implementation?

31. A *tournament* is a digraph in which there is exactly one edge between every two vertices.

    a. How many edges does a tournament have?
    b. How many different tournaments of *n* edges can be created?
    c. Can each tournament be topologically sorted?
    d. How many minimal vertices can a tournament have?
    e. A *transitive tournament* is a tournament that has *edge(vw)* if it has *edge(vu)* and *edge(uw)*. Can such a tournament have a cycle?

32. Does considering loops and parallel edges complicate the analysis of networks? How about multiple sources and sinks?

33. `FordFulkersonAlgorithm()` assumes that it terminates. Do you think such an assumption is safe?

34. `FordFulkersonAlgorithm()` executed in a depth-first fashion has some redundancy. First, all outgoing edges are pushed onto the stack and then the last is popped off to be followed by the algorithm. For example, in the network in Figure 8.20a, first, all three edges coming out of vertex *s* are pushed, and only afterward is the last of them, *edge(se)*, followed. Modify `FordFulkersonAlgorithm()` so that the first edge coming out of a certain vertex is immediately followed, and the second is followed only if the first does not lead to the sink. Consider using recursion.

35. Find the capacity of the cut determined by the set $X = \{s,d\}$ in the graph in Figure 8.19.

36. What is the complexity of `DinicAlgorithm()` in a network where all edges have a capacity of one?

37. Why does `DinicAlgorithm()` start from the sink to determine a layered network?

38. Apply to Figure 8.38a the following approximation algorithms (Johnson and Papadimitriou 1985; Rosenkrantz et al. 1977) to solve the traveling salesman problem.

    a. The *nearest neighbor algorithm* (*next best method*) begins with an arbitrary vertex *v* and then finds a vertex *w* not on the tour that is closest to the vertex *u* last added and includes in the tour *edge(uw)* and *edge(wv)* after deleting *edge(vu)*.

b. The *nearest insertion algorithm* is obtained from `nearestAdditionAlgorithm()` by finding two vertices v_q and v_r in the tour that minimize the expression.

$$edge(v_q v_i) + edge(v_i v_r) - edge(v_q v_r)$$

In this way, a new vertex v_i is inserted in the best place in the existing tour, which may not be next to v_p.

c. The *cheapest insertion algorithm* is obtained from `nearestAdditionAlgorithm()` by including in tour a new vertex v_i that minimizes the length of the new tour.

d. The *farthest insertion algorithm* is just like the nearest insertion algorithms except that it requires that v_i is farthest from tour, not closest.

e. The *nearest merger algorithm*, which corresponds to the Borůvka algorithm:

```
nearestMergerAlgorithm(cities V)
    create |V| tours such that tour_i = {edge(v_i v_i)};
    while there are at least two tours
        find two closest tours tour_i and tour_j
        (the minimum distance between city v_s ∈ tour_i and v_t ∈ tour_j is the smallest for all tours);
        find edge(v_k v_l) ∈ tour_i and edge(v_p v_q) ∈ tour_j such that they minimize
            dist(edge(v_k v_p)) + dist(edge(v_l v_q)) - dist(edge(v_k v_l)) - dist(edge(v_p v_q));
        tour_i = tour_i ∪ tour_j ∪ {edge(v_k v_p), edge(v_l v_q)} - {edge(v_k v_l), edge(v_p v_q)};
        remove tour_j;
```

39. Consider a bipartite graph $G = (\{u_1, u_2 \ldots u_k\} \cup \{w_1, w_2 \ldots w_k\}, \{(edge(u_i w_j): i \neq j\}$. How many colors are needed to color vertices of G with `sequentialColoringAlgorithm()` if vertices are colored in this order

    a. $u_1, \ldots, u_k, w_1, \ldots, w_k$?
    b. $u_1, w_1, u_2, w_2, \ldots, u_k, w_k$?

40. What is the vertex cover for a matching? For a bipartite graph?

41. Show a coloring of the 9-subgraph from Figure 8.42a in which one vertex x, y, or z is t, the color of the other two vertices is f, and $color(p) = f$.

42. Show that the 2-colorability problem can be solved in polynomial time.

43. Show that the number of odd-degree vertices in simple graphs is even.

44. The member function `readCommittees()` in the case study uses two trees, `committeeTree` and `memberTree`, to generate adjacency lists and then initialize the array `vertices`. However, one tree would be sufficient. What do you think is the reason for using two trees, not one?

45. What would be the output of the program in the case study if linked lists in Figure 8.46b were in the reverse order?

8.15 PROGRAMMING ASSIGNMENTS

1. All algorithms discussed in this chapter for determining the minimum spanning tree have one thing in common: They start building the tree from the beginning and they add new edges to the structure, which eventually becomes such a tree. However, we can go in the opposite direction and build this tree by successively removing edges to break cycles in the graph until no circuit is left. In this way, the graph turns into the tree. The edges chosen for removal should be the edges of maximum weight among those that can break any cycle in the tree (for example, Dijkstra's method). This algorithm somewhat resembles the Kruskal method, but because it works in the opposite direction, it can be called a Kruskal method *à rebours*. Use this approach to find the minimum spanning tree for the graph of distances between at least a dozen cities.

2. Write a graphics demonstration program to show the difference between Kruskal's method and Jarník-Prim's algorithm. Randomly generate 50 vertices and display them in the left half of the screen. Then, randomly generate 200 edges and display them. Make sure that the graph is connected. After the graph is ready, create the minimum spanning tree using Kruskal's method and display each edge included in the tree. (Use a different color than the one used during graph generation.) Then, display the same graph in the right half of the screen, create the minimum spanning tree using Jarník-Prim's algorithm, and display all edges being included in the tree.

3. An important problem in database management is preventing deadlocks between transactions. A transaction is a sequence of operations on records in the database. In large databases, many transactions can be executed at the same time. This can lead to inconsistencies if the order of executing operations is not monitored. However, this monitoring may cause transactions to block each other, thereby causing a deadlock. To detect a deadlock, a wait-for graph is constructed to show which transaction waits for which. Use a binary locking mechanism to implement a wait-for graph. In this mechanism, if a record R is accessed by a transaction T, then T puts a lock on R and this record cannot be processed by any other transaction before T is finished. Release all locks put on by a transaction T when T finishes. The input is composed of the following commands: $read(T,A)$, $write(T,A)$, $end(T)$. For example, if input is

$$read(T_1,A_1), read(T_2,A_2), read(T_1,A_2), write(T_1,A_2), end(T_1) \ldots$$

then T_1 is suspended when attempting to execute the step $read(T_1,A_2)$, and $edge(T_1,T_2)$ is created, because T_1 waits for T_2 to finish. If T_1 does not have to wait, resume its execution. After each graph update, check for a cycle in the graph. If a cycle is detected, interrupt execution of the youngest transaction T and put its steps at the end of the input.

Note that some records might have been modified by such a transaction, so they should be restored to their state before T started. But such a modification could have been used by another transaction, which should also be interrupted. In this program, do not address the problem of restoring the values of records (the problem of rolling back transactions and of cascading this rolling back). Concentrate on updating and monitoring the wait-for graph. Note that if a transaction is finished, its vertex should be removed from the graph, which may be what other transactions are waiting for.

4. Write a rudimentary spreadsheet program. Display a grid of cells with columns A through H and rows 1 through 20. Accept input in the first row of the screen. The commands are of the form *column row entry*, where *entry* is a number, a cell address preceded by a plus (e.g., +A5), a string, or a function preceded by an at sign, @. The functions are: *max, min, avg,* and *sum*. During execution of your program, build and modify the graph reflecting the situation in the spreadsheet. Show the proper values in proper cells. If a value in a cell is updated, then the values in all cells depending on it should also be modified. For example, after the following sequence of entries:

```
A1 10
B1 20
A2 30
D1 +A1
C1 @sum(A1..B2)
D1 +C1
```

both cells C1 and D1 should display the number 60.

Consider using a modification of the interpreter from Chapter 5 as an enhancement of this spreadsheet so that arithmetic expressions could also be used to enter values, such as

```
C3 2*A1
C4 @max(A1..B2) - (A2 + B2)
```

BIBLIOGRAPHY

Ahuja, Ravindra K., Magnanti, Thomas L., and Orlin, James B., *Network Flows; Theory, Algorithms, and Applications,* Englewood Cliffs, NJ: Prentice Hall, 1993.

Berge, Claude, "Two Theorems in Graph Theory," *Proceedings of the National Academy of Sciences of the USA* 43 (1957), 842–844.

Bertsekas, Dimitri P., "A Simple and Fast Label Correcting Algorithm for Shortest Paths," *Networks* 23 (1993), 703–709.

Bondy, J. A., and Chvátal, V., "A Method in Graph Theory," *Discrete Mathematics* 15 (1976), 111–135.

Bondy, John A., and Murty, U. S. R., *Graph Theory with Applications,* New York: Elsevier, 1976.

Brélaz, Daniel, "New Methods to Color the Vertices of a Graph," *Communications of the ACM* 22 (1979), 251–256.

Chvátal, V., "Hamiltonian Cycles," in Lawler, E. L., Lenstra, J. K., Rinnoy, Kan, A. H. G., and Shmoys, D. B. (eds.), *The Traveling Salesman Problem,* New York: Wiley (1985), 403–429.

Deo, Narsingh, and Pang, Chi-yin, "Shortest Path Algorithms: Taxonomy and Annotation," *Networks* 14 (1984), 275–323.

Dijkstra, E. W., "A Note on Two Problems in Connection with Graphs," *Numerische Mathematik* 1 (1959), 269–271.

Dijkstra, E. W., "Some Theorems on Spanning Subtrees of a Graph," *Indagationes Mathematicae* 28 (1960), 196–199.

Dinic, E. A., "Algorithm for Solution of a Problem of Maximum Flow in a Network with Power Estimation" [Mistranslation of: with Polynomial Bound], *Soviet Mathematics Doklady* 11 (1970), 1277–1280.

Edmonds, Jack, "Paths, Trees, and Flowers," *Canadian Journal of Mathematics* 17 (1963), 449–467.

Edmonds, Jack, "The Chinese Postman Problem," *Operations Research* 13 (1965), Suppl. 1, B–73.

Edmonds, Jack, and Johnson, Elias L., "Matching, Euler Tours and the Chinese Postman," *Mathematical Programming* 5 (1973), 88–124.

Edmonds, Jack, and Karp, Richard M., "Theoretical Improvement in Algorithmic Efficiency for Network Flow Problems," *Journal of the ACM* 19 (1972), 248–264.

Fleury, "Deux Problèmes de Géométrie de Situation," *Journal de Mathématiques Élémentaires* (1883), 257–261.

Floyd, Robert W., "Algorithm 97: Shortest Path," *Communications of the ACM* 5 (1962), 345.

Ford, L. R., and Fulkerson, D. R., "Maximal Flow Through a Network," *Canadian Journal of Mathematics* 8 (1956), 399–404.

Ford, L. R., and Fulkerson, D. R., "A Simple Algorithm for Finding Maximal Network Flows and an Application to the Hitchcock Problem," *Canadian Journal of Mathematics* 9 (1957), 210–218.

Ford, L. R., and Fulkerson, D. R., *Flows in Networks*, Princeton, NJ: Princeton University Press, 1962.

Gale, D., and Shapley, L. S., "College Admissions and the Stability of Marriage," *American Mathematical Monthly* 69 (1962), 9–15.

Gallo, Giorgio, and Pallottino, Stefano, "Shortest Path Methods: A Unified Approach," *Mathematical Programming Study* 26 (1986), 38–64.

Gallo, Giorgio, and Pallottino, Stefano, "Shortest Path Methods," *Annals of Operations Research* 7 (1988), 3–79.

Gibbons, Alan, *Algorithmic Graph Theory*, New York: Cambridge University Press, 1985.

Glover, Fred, Glover, Randy, and Klingman, Darwin, "Computational Study of an Improved Shortest Path Algorithm," *Networks* 14 (1984), 25–36.

Gould, Ronald, *Graph Theory*, Menlo Park, CA: Benjamin Cummings, 1988.

Graham, R. L., and Hell, Pavol, "On the History of the Minimum Spanning Tree Problem," *Annals of the History of Computing* 7 (1985), 43–57.

Hall, Philip, "On Representatives of Subsets," *Journal of the London Mathematical Society* 10 (1935), 26–30.

Ingerman, P. Z., "Algorithm 141: Path Matrix," *Communications of the ACM* 5 (1962), 556.

Johnson, Donald B., "Efficient Algorithms for Shortest Paths in Sparse Networks," *Journal of the ACM* 24 (1977), 1–13.

Johnson, D. S., and Papadimitriou, C. H., "Performance Guarantees for Heuristics," in Lawler, E. L., Lenstra, J. K., Rinnoy, Kan A. H. G., and Shmoys, D. B. (eds.), *The Traveling Salesman Problem*, New York: Wiley (1985), 145–180.

Kalaba, Robert, "On Some Communication Network Problems," *Combinatorial Analysis*, Providence, RI: American Mathematical Society (1960), 261–280.

Kershenbaum, Aaron, "A Note on Finding Shortest Path Trees," *Networks* 11 (1981), 399–400.

Kruskal, Joseph B., "On the Shortest Spanning Tree of a Graph and the Traveling Salesman Problem," *Proceedings of the American Mathematical Society* 7 (1956), 48–50.

Kuhn, Harold W., "The Hungarian Method for the Assignment Problem," *Naval Research Logistics Quarterly* 2 (1955), 83–97.

Kwan, Mei-ko, "Graphic Programming Using Odd or Even Points," *Chinese Mathematics* 1 (1962), 273–277, translation of a paper published in *Acta Mathematica Sinica* 10 (1960), 263–266.

Munkres, James, "Algorithms for the Assignment Problem and Transportation Problems," *Journal of the Society of Industrial and Applied Mathematics* 5 (1957), 32–38.

Ore, Oystein, "Note on Hamilton Circuits," *American Mathematical Monthly* 67 (1960), 55.

Papadimitriou, Christos H., and Steiglitz, Kenneth, *Combinatorial Optimization: Algorithms and Complexity*, Englewood Cliffs, NJ: Prentice Hall, 1982.

Pape, U., "Implementation and Efficiency of Moore-Algorithms for the Shortest Route Problem," *Mathematical Programming* 7 (1974), 212–222.

Pollack, Maurice, and Wiebenson, Walter, "Solutions of the Shortest-Route Problem—A Review," *Operations Research* 8 (1960), 224–230.

Prim, Robert C., "Shortest Connection Networks and Some Generalizations," *Bell System Technical Journal* 36 (1957), 1389–1401.

Rosenkrantz, Daniel J., Stearns, Richard E., and Lewis, Philip M., "An Analysis of Several Heuristics for the Traveling Salesman Problem," *SIAM Journal on Computing* 6 (1977), 563–581.

Tarjan, Robert E., *Data Structures and Network Algorithms*, Philadelphia: Society for Industrial and Applied Mathematics, 1983.

Thulasiraman, K., and Swamy, M. N. S., *Graphs: Theory and Algorithms*, New York: Wiley, 1992.

Warshall, Stephen, "A Theorem on Boolean Matrices," *Journal of the ACM* 9 (1962), 11–12.

Welsh, D. J. A., and Powell, M. B., "An Upper Bound for the Chromatic Number of a Graph and Its Application to Timetabling Problems," *Computer Journal* 10 (1967), 85–86.

9 Sorting

The efficiency of data handling can often be substantially increased if the data are sorted according to some criteria of order. For example, it would be practically impossible to find a name in the telephone directory if the names were not alphabetically ordered. The same can be said about dictionaries, book indexes, payrolls, bank accounts, student lists, and other alphabetically organized materials. The convenience of using sorted data is unquestionable and must be addressed in computer science as well. Although a computer can grapple with an unordered telephone book more easily and quickly than a human can, it is extremely inefficient to have the computer process such an unordered data set. It is often necessary to sort data before processing.

The first step is to choose the criteria that will be used to order data. This choice will vary from application to application and must be defined by the user. Very often, the sorting criteria are natural, as in the case of numbers. A set of numbers can be sorted in ascending or descending order. The set of five positive integers (5, 8, 1, 2, 20) can be sorted in ascending order resulting in the set (1, 2, 5, 8, 20) or in descending order resulting in the set (20, 8, 5, 2, 1). Names in the phone book are ordered alphabetically by last name, which is the natural order. For alphabetic and nonalphabetic characters, the American Standard Code for Information Interchange (ASCII) code is commonly used, although other choices such as Extended Binary Coded Decimal Interchange Code (EBCDIC) are possible. Once a criterion is selected, the second step is how to put a set of data in order using that criterion.

The final ordering of data can be obtained in a variety of ways, and only some of them can be considered meaningful and efficient. To decide which method is best, certain criteria of efficiency have to be established and a method for quantitatively comparing different algorithms must be chosen.

To make the comparison machine-independent, certain critical properties of sorting algorithms should be defined when comparing alternative methods. Two such properties are the number of comparisons and the number of data movements. The choice of these two properties should not be surprising. To sort a set of data, the data have to be compared and moved as necessary; the efficiency of these two operations depends on the size of the data set.

Because determining the precise number of comparisons is not always necessary or possible, an approximate value can be computed. For this reason, the number of comparisons and movements is approximated with big-O notation by giving the order of magnitude of these numbers. But the order of magnitude can vary depending on the initial ordering of data. How much time, for example, does the machine spend on data ordering if the data are already ordered? Does it recognize this initial ordering immediately or is it completely unaware of that fact? Hence, the efficiency measure also indicates the "intelligence" of the algorithm. For this reason, the number of comparisons and movements is computed (if possible) for the following three cases: best case (often, data already in order), worst case (usually, data in reverse order), and average case (data in random order). Some sorting methods perform the same operations regardless of the initial ordering of data. It is easy to measure the performance of such algorithms, but the performance itself is usually not very good. Many other methods are more flexible, and their performance measures for all three cases differ.

The number of comparisons and the number of movements do not have to coincide. An algorithm can be very efficient on the former and perform poorly on the latter, or vice versa. Therefore, practical reasons must aid in the choice of which algorithm to use. For example, if only simple keys are compared, such as integers or characters, then the comparisons are relatively fast and inexpensive. If strings or arrays of numbers are compared, then the cost of comparisons goes up substantially, and the weight of the comparison measure becomes more important. If, on the other hand, the data items moved are large, such as structures, then the movement measure may stand out as the determining factor in efficiency considerations. All theoretically established measures have to be used with discretion, and theoretical considerations should be balanced with practical applications. After all, the practical applications serve as a rubber stamp for theory decisions.

Sorting algorithms, whose number can be counted in the dozens, are of different levels of complexity. A simple method can be only 20 percent less efficient than a more elaborate one. If sorting is used in the program once in a while and only for small sets of data, then using a sophisticated and slightly more efficient algorithm may not be desirable; the same operation can be performed using a simpler method and simpler code. But if thousands of items are to be sorted, then a gain of 20 percent must not be neglected. Simple algorithms often perform better with a small amount of data than their more complex counterparts whose effectiveness may become obvious only when data samples become very large.

9.1 ELEMENTARY SORTING ALGORITHMS

9.1.1 Insertion Sort

An *insertion sort* starts by considering the two first elements of the array `data`, which are `data[0]` and `data[1]`. If they are out of order, an interchange takes place. Then, the third element, `data[2]`, is considered and inserted into its proper place. If `data[2]` is less than `data[0]` and `data[1]`, these two elements are shifted by one position; `data[0]` is placed at position 1, `data[1]` at position 2, and `data[2]` at position 0. If `data[2]` is less than `data[1]` and not less than `data[0]`, then only `data[1]` is moved to position 2 and its place is taken by `data[2]`. If, finally, `data[2]` is not less than both its predecessors, it stays in its current position. Each element `data[i]` is inserted into its proper location *j* such that $0 \le j \le i$, and all elements greater than `data[i]` are moved by one position.

An outline of the insertion sort algorithm is as follows:

```
insertionsort(data[],n)
    for i = 1 to n-1
        move all elements data[j] greater than data[i] by one position;
        place data[i] in its proper position;
```

Note that sorting is restricted only to a fraction of the array in each iteration, and only in the last pass is the whole array considered. Figure 9.1 shows what changes are made to the array [5 2 3 8 1] when `insertionsort()` executes.

FIGURE 9.1 The array [5 2 3 8 1] sorted by insertion sort.

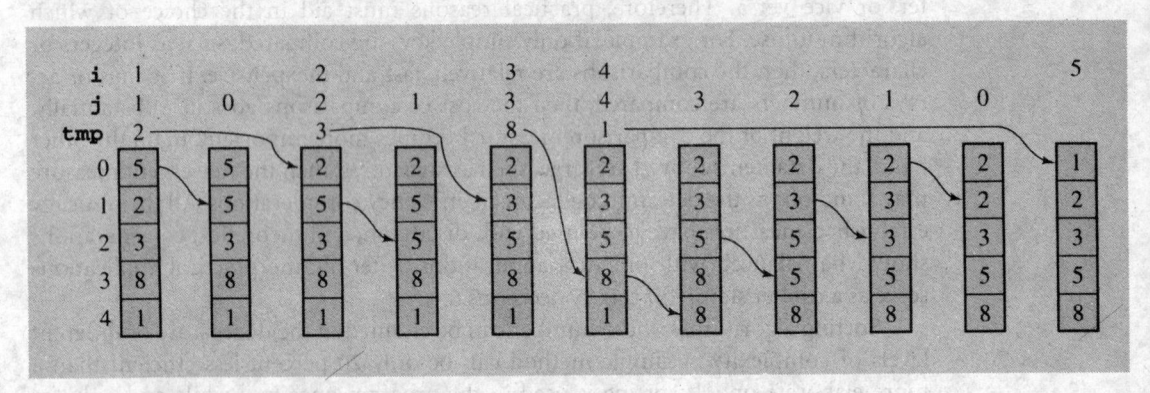

Because an array having only one element is already ordered, the algorithm starts sorting from the second position, position 1. Then for each element `tmp = data[i]`, all elements greater than `tmp` are copied to the next position, and `tmp` is put in its proper place.

An implementation of insertion sort is:

```cpp
template<class T>
void insertionsort(T data[], int n) {
    for (int i = 1,j; i < n; i++) {
        T tmp = data[i];
        for (j = i; j > 0 && tmp < data[j-1]; j--)
            data[j] = data[j-1];
        data[j] = tmp;
    }
}
```

An advantage of using insertion sort is that it sorts the array only when it is really necessary. If the array is already in order, no substantial moves are performed; only the variable `tmp` is initialized, and the value stored in it is moved back to the same position. The algorithm recognizes that part of the array is already sorted and stops execution accordingly. But it recognizes only this, and the fact that elements may already be in their proper positions is overlooked. Therefore, they can be moved from these positions and then later moved back. This happens to numbers 2 and 3 in the example in Figure 9.1. Another disadvantage is that if an item is being inserted, all elements greater than the one being inserted have to be moved. Insertion is not localized and may require moving a significant number of elements. Considering that an element can be moved from its final position only to be placed there again later, the number of redundant moves can slow down execution substantially.

To find the number of movements and comparisons performed by `insertionsort()`, observe first that the outer `for` loop always performs $n - 1$ iterations. However, the number of elements greater than `data[i]` to be moved by one position is not always the same.

The best case is when the data are already in order. Only one comparison is made for each position i, so there are $n - 1$ comparisons, which is $O(n)$, and $2(n - 1)$ moves, all of them redundant.

The worst case is when the data are in reverse order. In this case, for each i, the item `data[i]` is less than every item `data[0]`,...,`data[i-1]`, and each of them is moved by one position. For each iteration i of the outer `for` loop, there are i comparisons, and the total number of comparisons for all iterations of this loop is

$$\sum_{i=1}^{n-1} i = 1 + 2 + \cdots + (n-1) = \frac{n(n-1)}{2} = O(n^2)$$

The number of times the assignment in the inner `for` loop is executed can be computed using the same formula. The number of times `tmp` is loaded and unloaded in the outer `for` loop is added to that, resulting in the total number of moves:

$$\frac{n(n-1)}{2} + 2(n-1) = \frac{n^2 + 3n - 4}{2} = O(n^2)$$

Only extreme cases have been taken into consideration. What happens if the data are in random order? Is the sorting time closer to the time of the best case, $O(n)$, or to the worst case, $O(n^2)$? Or is it somewhere in between? The answer is not immediately evident, and requires certain introductory computations. The outer `for` loop always executes $n - 1$ times, but it is also necessary to determine the number of iterations for the inner loop.

For every iteration i of the outer `for` loop, the number of comparisons depends on how far away the item `data[i]` is from its proper position in the currently sorted subarray `data[0...i]`. If it is already in this position, only one test is performed that compares `data[i]` and `data[i-1]`. If it is one position away from its proper place, two comparisons are performed: `data[i]` is compared with `data[i-1]` and then with `data[i-2]`. Generally, if it is j positions away from its proper location, `data[i]` is compared with $j + 1$ other elements. This means that, in iteration i of the outer `for` loop, there are either $1, 2, \ldots,$ or i comparisons.

Under the assumption of equal probability of occupying array cells, the average number of comparisons of data[i] with other elements during the iteration i of the outer for loop can be computed by adding all the possible numbers of times such tests are performed and dividing the sum by the number of such possibilities. The result is

$$\frac{1+2+\ldots+i}{i} = \frac{\frac{1}{2}i(i+1)}{i} = \frac{i+1}{2}$$

To obtain the average number of all comparisons, the computed figure has to be added for all i's (for all iterations of the outer for loop) from 1 to $n-1$. The result is

$$\sum_{i=1}^{n-1}\frac{i+1}{2} = \frac{1}{2}\sum_{i=1}^{n-1}i + \sum_{i=1}^{n-1}\frac{1}{2} = \frac{\frac{1}{2}n(n-1)}{2} + \frac{1}{2}(n-1) = \frac{n^2+n-2}{4}$$

which is $O(n^2)$ and approximately one-half of the number of comparisons in the worst case.

By similar reasoning, we can establish that, in iteration i of the outer for loop, data[i] can be moved either $0, 1, \ldots$, or $i-1$ times; that is

$$\frac{0+1+\ldots+(i-1)}{i} = \frac{\frac{1}{2}i(i-1)}{i} = \frac{i-1}{2}$$

times plus two unconditional movements (to tmp and from tmp). Hence, in all the iterations of the outer for loop we have, on the average,

$$\sum_{i=1}^{n-1}\left(\frac{i-1}{2}+2\right) = \frac{1}{2}\sum_{i=1}^{n-1}i + \sum_{i=1}^{n-1}\frac{3}{2} = \frac{\frac{1}{2}n(n-1)}{2} + \frac{3}{2}(n-1) = \frac{n^2+5n-6}{4}$$

movements, which is also $O(n^2)$.

This answers the question: Is the number of movements and comparisons for a randomly ordered array closer to the best or to the worst case? Unfortunately, it is closer to the latter, which means that, on the average, when the size of an array is doubled, the sorting effort quadruples.

9.1.2 Selection Sort

Selection sort is an attempt to localize the exchanges of array elements by finding a misplaced element first and putting it in its final place. The element with the lowest value is selected and exchanged with the element in the first position. Then, the smallest value among the remaining elements data[1], ..., data[n-1] is found and put in the second position. This selection and placement by finding, in each pass i, the lowest value among the elements data[i], ..., data[n-1] and swapping it with data[i] are continued until all elements are in their proper positions. The following pseudocode reflects the simplicity of the algorithm:

```
selectionsort(data[],n)
    for i = 0 to n-2
        select the smallest element among data[i],...,data[n-1];
        swap it with data[i];
```

It is rather obvious that n-2 should be the last value for i, because if all elements but the last have been already considered and placed in their proper positions, then the nth element (occupying position n-1) has to be the largest. An example is shown in Figure 9.2. Here is a C++ implementation of selection sort:

```cpp
template<class T>
void selectionsort(T data[], int n) {
    for (int i = 0,j,least; i < n-1; i++) {
        for (j = i+1, least = i; j < n; j++)
            if (data[j] < data[least])
                least = j;
        swap(data[least],data[i]);
    }
}
```

FIGURE 9.2. The array [5 2 3 8 1] sorted by selection sort.

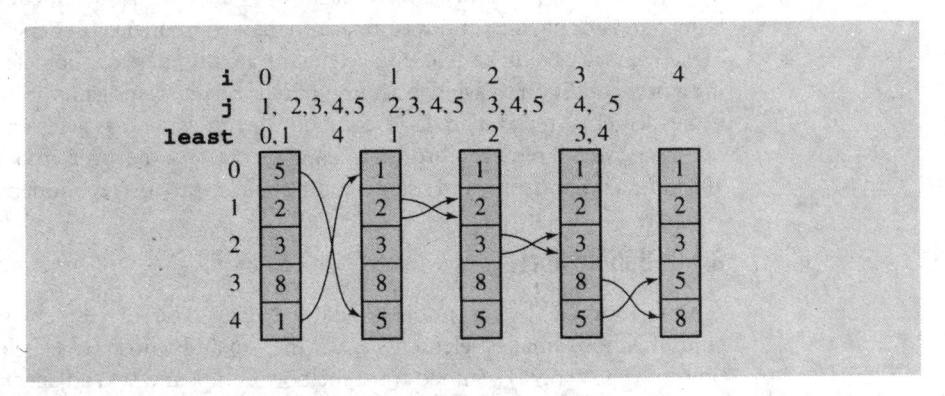

where the function swap() exchanges elements data[least] and data[i] (see the end of Section 1.2). Note that least is not the smallest element but its position.

The analysis of the performance of the function selectionsort() is simplified by the presence of two for loops with lower and upper bounds. The outer loop executes n − 1 times, and for each i between 0 and n − 2, the inner loop iterates j = (n − 1) − i times. Because comparisons of keys are done in the inner loop, there are

$$\sum_{i=0}^{n-2}(n-1-i) = (n-1) + \cdots + 1 = \frac{n(n-1)}{2} = O(n^2)$$

comparisons. This number stays the same for all cases. There can be some savings only in the number of swaps. Note that if the assignment in the if statement is executed, only the index j is moved, not the item located currently at position j. Array elements are swapped unconditionally in the outer loop as many times as this loop executes, which is n-1. Thus, in all cases, items are moved the same number of times, $3 \cdot (n-1)$.

The best thing about this sort is the required number of assignments, which can hardly be beaten by any other algorithm. However, it might seem somewhat unsatisfactory that the total number of exchanges, $3 \cdot (n-1)$, is the same for all cases. Obviously, no exchange is needed if an item is in its final position. The algorithm disregards that and swaps such an item with itself, making three redundant moves. The problem can be alleviated by making swap() a conditional operation. The condition preceding the swap() should indicate that no item less than data[least] has been found among elements data[i+1], ..., data[n-1]. The last line of selectionsort() might be replaced by the lines:

```
if (data[i] != data[least])
    swap (data[least], data[i]);
```

This increases the number of array element comparisons by $n-1$, but this increase can be avoided by noting that there is no need to compare items. We proceed as we did in the case of the if statement of selectionsort() by comparing the indexes and not the items. The last line of selectionsort() can be replaced by:

```
if (i != least)
    swap (data[least], data[i]);
```

Is such an improvement worth the price of introducing a new condition in the procedure and adding $n-1$ index comparisons as a consequence? It depends on what types of elements are being sorted. If the elements are numbers or characters, then interposing a new condition to avoid execution of redundant swaps gains little in efficiency. But if the elements in data are large compound entities such as arrays or structures, then one swap (which requires three assignments) may take the same amount of time as, say, 100 index comparisons, and using a conditional swap() is recommended.

9.1.3 Bubble Sort

A bubble sort can be best understood if the array to be sorted is envisaged as a vertical column whose smallest elements are at the top and whose largest elements are at the bottom. The array is scanned from the bottom up, and two adjacent elements are interchanged if they are found to be out of order with respect to each other. First, items data[n-1] and data[n-2] are compared and swapped if they are out of order. Next, data[n-2] and data[n-3] are compared, and their order is changed if necessary, and so on up to data[1] and data[0]. In this way, the smallest element is bubbled up to the top of the array.

However, this is only the first pass through the array. The array is scanned again comparing consecutive items and interchanging them when needed, but this time, the last comparison is done for data[2] and data[1] because the smallest element is already in its proper position, namely, position 0. The second pass bubbles the second smallest element of the array up to the second position, position 1. The procedure continues until the last pass when only one comparison, data[n-1] with data[n-2], and possibly one interchange are performed.

A pseudocode of the algorithm is as follows:

```
bubblesort(data[],n)
    for i = 0 to n−2
        for j = n−1 down to i+1
            swap elements in position j and j-1 if they are out of order;
```

Section 9.1 Elementary Sorting Algorithms ■ 481

Figure 9.3 illustrates the changes performed in the integer array [5 2 3 8 1] during the execution of `bubblesort()`. Here is an implementation of bubble sort:

```
template<class T>
void bubblesort(T data[], int n) {
    for (int i = 0; i < n-1; i++)
        for (int j = n-1; j > i; --j)
            if (data[j] < data[j-1])
                swap(data[j],data[j-1]);
}
```

FIGURE 9.3 The array [5 2 3 8 1] sorted by bubble sort.

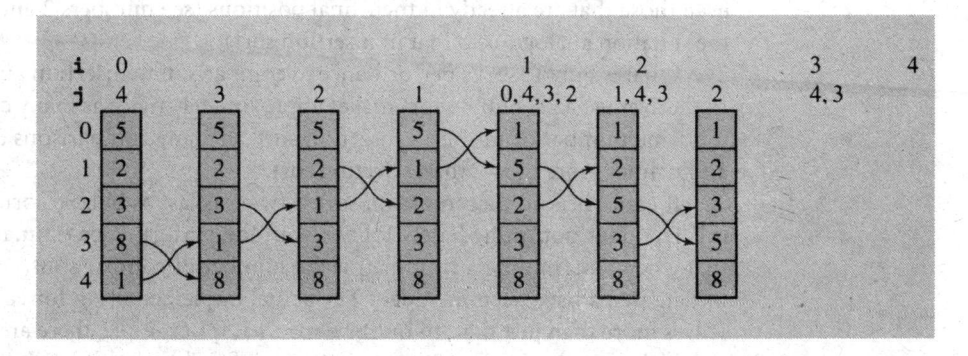

The number of comparisons is the same in each case (best, average, and worst) and equals the total number of iterations of the inner `for` loop

$$\sum_{i=0}^{n-2}(n-1-i) = \frac{n(n-1)}{2} = O(n^2)$$

comparisons. This formula also computes the number of swaps in the worst case when the array is in reverse order. In this case, $3\frac{n(n-1)}{2}$ moves have to be made.

The best case, when all elements are already ordered, requires no swaps. If an i-cell array is in random order, then the number of swaps can be any number between zero and $i-1$; that is, there can be either no swap at all (all items are in ascending order), one swap, two swaps, ..., or $i-1$ swaps. The array processed by the inner `for` loop is `data[i],...,data[n-1]`, and the number of swaps in this subarray—if its elements are randomly ordered—is either zero, one, two, ..., or $n-1-i$. After averaging the sum of all these possible numbers of swaps by the number of these possibilities, the average number of swaps is obtained, which is

$$\frac{0+1+2+\cdots+(n-1-i)}{n-i} = \frac{n-i-1}{2}$$

If all these averages for all the subarrays processed by `bubblesort()` are added (that is, if such figures are summed over all iterations i of the outer `for` loop), the result is

$$\sum_{i=0}^{n-2}\frac{n-i-1}{2} = \frac{1}{2}\sum_{i=0}^{n-2}(n-1) - \frac{1}{2}\sum_{i=0}^{n-2}i$$
$$= \frac{(n-1)^2}{2} - \frac{(n-1)(n-2)}{4} = \frac{n(n-1)}{4}$$

swaps, which is equal to $\frac{3}{4}n(n-1)$ moves.

The main disadvantage of bubble sort is that it still painstakingly bubbles items step by step up toward the top of the array. It looks at two adjacent array elements at a time and swaps them if they are not in order. If an element has to be moved from the bottom to the top, it is exchanged with every element in the array. It does not skip them as selection sort did. In addition, the algorithm concentrates only on the item that is being bubbled up. Therefore, all elements that distort the order are moved, even those that are already in their final positions (see numbers 2 and 3 in Figure 9.3, the situation analogous to that in insertion sort).

What is bubble sort's performance in comparison to insertion and selection sort? In the average case, bubble sort makes approximately twice as many comparisons and the same number of moves as insertion sort, as many comparisons as selection sort, and n times more moves than selection sort.

It could be said that insertion sort is twice as fast as bubble sort. In fact it is, but this fact does not immediately follow from the performance estimates. The point is that when determining a formula for the number of comparisons, only comparisons of data items have been included. The actual implementation for each algorithm involves more than just that. In `bubblesort()`, for example, there are two loops, both of which compare indexes: `i` and `n-1` in the first loop, `j` and `i` in the second. All in all, there are $\frac{n(n-1)}{2}$ such comparisons, and this number should not be treated too lightly. It becomes negligible if the data items are large structures. But if `data` consists of integers, then comparing the data takes a similar amount of time as comparing indexes. A more thorough treatment of the problem of efficiency should focus on more than just data comparison and exchange. It should also include the overhead necessary for implementation of the algorithm.

9.2 DECISION TREES

The three sorting methods analyzed in previous sections were not very efficient. This leads to several questions: Can any better level of efficiency for a sorting algorithm be expected? Can algorithms, at least theoretically, be more efficient by executing faster? If so, when can we be satisfied with an algorithm and be sure that the sorting speed is unlikely to be increased? We need a quantitative measurement to estimate a *lower bound* of sorting speed.

This section focuses on the comparisons of two elements and not the element interchange. The questions are: On the average, how many comparisons have to be made to sort n elements? Or what is the best estimate of the number of item comparisons if an array is assumed to be ordered randomly?

Every sorting algorithm can be expressed in terms of a binary tree in which the arcs carry the labels Y(es) or N(o). Nonterminal nodes of the tree contain conditions or queries for labels, and the leaves have all possible orderings of the array to which the algorithm is applied. This type of tree is called a *decision tree*. Because the initial ordering cannot be predicted, all possibilities have to be listed in the tree in order for the sorting procedure to grapple with any array and any possible initial order of data. This initial order determines which path is taken by the algorithm and what sequence of comparisons is actually chosen. Note that different trees have to be drawn for arrays of different length.

Figure 9.4 illustrates decision trees for insertion sort and bubble sort for an array [a b c]. The tree for insertion sort has six leaves, and the tree for bubble sort has eight leaves. How many leaves does a tree for an n-element array have? Such an array can be ordered in $n!$ different ways, as many ways as the possible permutations of the array elements, and all of these orderings have to be stored in the leaves of the decision tree. Thus, the tree for insertion sort has six leaves because $n = 3$, and $3! = 6$.

FIGURE 9.4 Decision trees for (a) insertion sort and (b) bubble sort as applied to the array [a b c].

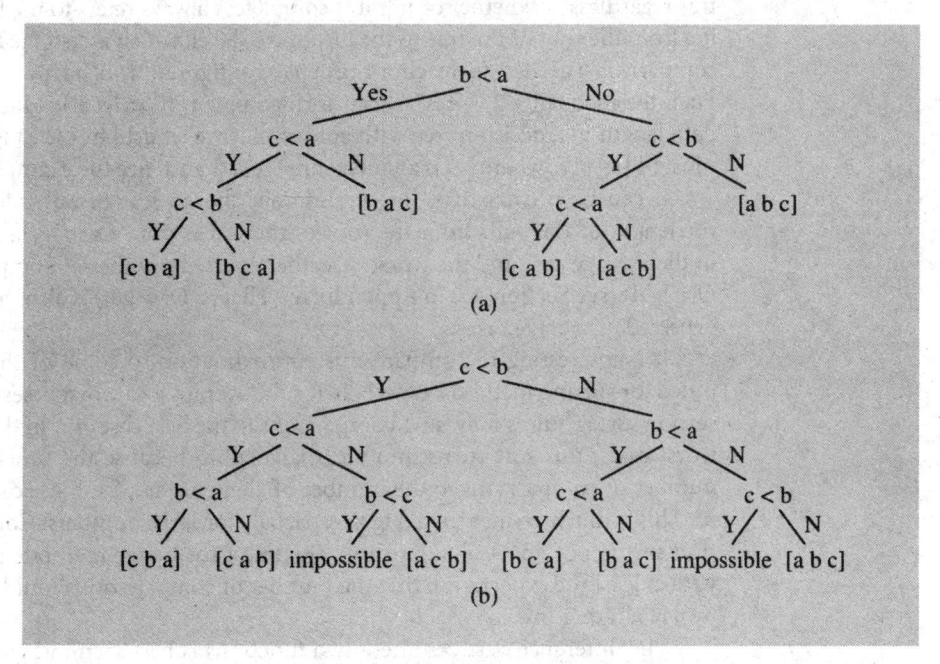

```
                              b < a
                    Yes   ─────────   No
                  c < a                       c < b
                Y /   \ N                   Y /   \ N
            c < b    [b a c]             c < a   [a b c]
           Y /  \ N                     Y /  \ N
       [c b a] [b c a]              [c a b] [a c b]
                              (a)

                              c < b
                      Y ─────────── N
                  c < a                       b < a
                Y /   \ N                   Y /   \ N
            b < a     b < c               c < a    c < b
           Y / \ N   Y / \ N             Y / \ N  Y / \ N
      [c b a][c a b] impossible [a c b] [b c a][b a c] impossible [a b c]
                              (b)
```

But as the example of the decision tree for bubble sort indicates, the number of leaves does not have to equal $n!$. In fact, it is never less than $n!$, which means that it can be greater than $n!$. This is a consequence of the fact that a decision tree can have leaves corresponding to failures, not only to possible orderings. The failure nodes are reached by an inconsistent sequence of operations. Also, the total number of leaves can be greater than $n!$ because some orderings (permutations) can occur in more than one leaf, because the comparisons may be repeated.

One of the interesting properties of decision trees is the average number of arcs traversed from the root to reach a leaf. Because one arc represents one comparison, the average number of arcs reflects the average number of key comparisons when executing a sorting algorithm.

As already established in Chapter 6, an i-level complete decision tree has 2^{i-1} leaves, $2^{i-1}-1$ nonterminal nodes (for $i \geq 1$) and 2^i-1 total nodes. Because all non-complete trees with the same number of i levels have fewer nodes than that, $k + m \leq 2^i - 1$, where m is the number of leaves and k is the number of nonleaves. Also, $k \leq 2^{i-1}-1$ and $m \leq 2^{i-1}$ (Section 6.1 and Figure 6.5). The latter inequality is used as an approximation for m. Hence, in an i-level decision tree, there are at most 2^{i-1} leaves.

Now, a question arises: What is a relationship between the number of leaves of a decision tree and the number of all possible orderings of an n-element array? There are $n!$ possible orderings, and each one of them is represented by a leaf in a decision tree. But the tree also has some extra nodes due to repetitions and failures. Therefore, $n! \leq m \leq 2^{i-1}$, or $2^{i-1} \geq n!$. This inequality answers the following question: How many comparisons are performed when using a certain sorting algorithm for an n-element array in the worst case? Or rather, what is the lowest or the best figure expected in the worst case? Note that this analysis pertains to the worst case. We assume that i is a level of a tree regardless of whether or not it is complete; i always refers to the longest path leading from the root of the tree to the lowest tree level, which is also the largest number of comparisons needed to reach an ordered configuration of array stored in the root. First, the inequality $2^{i-1} \geq n!$ is transformed into $i-1 \geq \lg(n!)$, which means that the path length in a decision tree with at least $n!$ leaves must be at least $\lg(n!)$, or rather, it must be $\lceil \lg(n!) \rceil$, where $\lceil x \rceil$ is an integer not less than x. See the example in Figure 9.5.

It can be proven that, for a randomly chosen leaf of an m-leaf decision tree, the length of the path from the root to the leaf is not less than $\lg m$ and that, both in the average case and the worst case, the required number of comparisons, $\lg(n!)$, is $O(n \lg n)$ (see Section A.2 in Appendix A). That is, $O(n \lg n)$ is also the best that can be expected in average cases.

It is interesting to compare this approximation to some of the numbers computed for sorting methods, especially for the average and worst cases. For example, insertion sort requires only $n-1$ comparisons in the best case, but in the average and the worst cases, this sort turns into an n^2 algorithm because the functions relating the number of comparisons to the number of elements are, for these cases, the big-Os of n^2. This is much greater than $n \lg n$, especially for large numbers. Consequently, insertion sort is not an ideal algorithm. The quest for better methods can be continued with at least the expectation that the number of comparisons should be approximated by $n \lg n$ rather than by n^2.

The difference between these two functions is best seen in Figure 9.6 if the performance of the algorithms analyzed so far is compared with the expected performance $n \lg n$ in the average case. The numbers in the table in Figure 9.6 show that if 100 items are sorted, the desired algorithm is four times faster than insertion sort and eight times faster than selection sort and bubble sort. For 1,000 items, it is 25 and 50 times faster, respectively. For 10,000, the difference in performance differs by factors of 188 and 376, respectively. This can only serve to encourage the search for an algorithm embodying the performance of the function $n \lg n$.

FIGURE 9.5 Examples of decision trees for an array of three elements.

These are some possible decision trees for an array of three elements. These trees must have at least $3! = 6$ leaves. For the sake of the example, it is assumed that each tree has one extra leaf (a repetition or a failure). In the worst and average cases, the number of comparisons is $i - 1 \geq \lceil \lg(n!) \rceil$. In this example, $n = 3$, so $i - 1 \geq \lceil \lg 3! \rceil = \lceil \lg 6 \rceil \approx \lceil 2.59 \rceil = 3$. And, in fact, only for the best balanced tree (a), the nonrounded length of the average path is less than three.

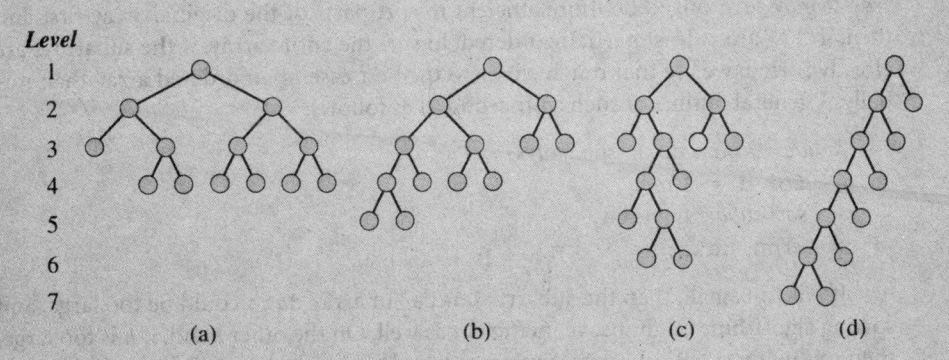

Level
1
2
3
4
5
6
7
  (a)   (b)   (c)   (d)

These are the sums of the lengths of the paths from the root to all leaves in trees (a) – (d) and the average path lengths:

(a) $2 + 3 + 3 + 3 + 3 + 3 + 3 = 20$; average $= \frac{20}{7} \approx 2.86$

(b) $4 + 4 + 3 + 3 + 3 + 2 + 2 = 21$; average $= \frac{21}{7} = 3$

(c) $2 + 4 + 5 + 5 + 3 + 2 + 2 = 23$; average $= \frac{23}{7} \approx 3.29$

(d) $6 + 6 + 5 + 4 + 3 + 2 + 1 = 27$; average $= \frac{27}{7} \approx 3.86$

FIGURE 9.6 Number of comparisons performed by the simple sorting method and by an algorithm whose efficiency is estimated by the function $n \lg n$.

sort type	n	100	1,000	10,000
insertion	$\frac{n(n-1)}{4}$	2,475	249,750	24,997,500
selection, bubble	$\frac{n(n-1)}{2}$	4,950	499,500	49,995,000
expected	$n \lg n$	664	9,966	132,877

9.3 EFFICIENT SORTING ALGORITHMS

9.3.1 Shell Sort

The $O(n^2)$ limit for a sorting method is much too large and must be broken to improve efficiency and decrease run time. How can this be done? The problem is that the time required for ordering an array by the three sorting algorithms usually grows faster than the size of the array. In fact, it is customarily a quadratic function of that size. It may turn out to be more efficient to sort parts of the original array first and then, if they are at least partially ordered, to sort the entire array. If the subarrays are already sorted, we are that much closer to the best case of an ordered array than initially. A general outline of such a procedure is as follows:

    divide data *into* h *subarrays;*
        for i = 1 *to* h
            sort subarray data$_i$;
    sort array data;

If h is too small, then the subarrays data$_i$ of array data could be too large, and sorting algorithms might prove inefficient as well. On the other hand, if h is too large, then too many small subarrays are created, and although they are sorted, it does not substantially change the overall order of data. Lastly, if only one such partition of data is done, the gain on the execution time may be rather modest. To solve that problem, several different subdivisions are used, and for every subdivision, the same procedure is applied separately, as in:

    determine numbers $h_t \ldots h_1$ *of ways of dividing array* data *into subarrays;*
    for (h=h$_t$; t > 1; t--, h=h$_t$)
        divide data *into* h *subarrays;*
        for i = 1 *to* h
            sort subarray data$_i$;
    sort array data;

This idea is the basis of the *diminishing increment sort,* also known as *Shell sort* and named after Donald L. Shell who designed this technique. Note that this pseudocode does not identify a specific sorting method for ordering the subarrays; it can be any simple method. Usually, however, Shell sort uses insertion sort.

The heart of Shell sort is an ingenious division of the array data into several subarrays. The trick is that elements spaced further apart are compared first, then the elements closer to each other are compared, and so on, until adjacent elements are compared on the last pass. The original array is logically subdivided into subarrays by picking every h_tth element as part of one subarray. Therefore, there are h_t subarrays, and for every $h = 1, \ldots, h_t$,

$$\text{data}_{h_t h}[i] = \text{data}[h_t \cdot i + (h-1)]$$

For example, if $h_t = 3$, the array data is subdivided into three subarrays data$_1$, data$_2$, and data$_3$ so that

Section 9.3 Efficient Sorting Algorithms ■ 487

$data_{31}[0] = data[0], data_{31}[1] = data[3], \ldots, data_{31}[i] = data[3*i], \ldots$
$data_{32}[0] = data[1], data_{32}[1] = data[4], \ldots, data_{32}[i] = data[3*i+1], \ldots$
$data_{33}[0] = data[2], data_{33}[1] = data[5], \ldots, data_{33}[i] = data[3*i+2], \ldots$

and these subarrays are sorted separately. After that, new subarrays are created with an $h_{t-1} < h_t$, and insertion sort is applied to them. The process is repeated until no subdivisions can be made. If $h_t = 5$, the process of extracting subarrays and sorting them is called a 5-sort.

Figure 9.7 shows the elements of the array data that are five positions apart and are logically inserted into a separate array—"logically" because physically they still occupy the same positions in data. For each value of increment h_t, there are h_t subarrays, and each of them is sorted separately. As the value of the increment decreases, the number of subarrays decreases accordingly, and their sizes grow. Much of data's disorder has been removed in the earlier iterations, so on the last pass, the array is much closer to its final form than before all the intermediate h-sorts.

FIGURE 9.7 The array [10 8 6 20 4 3 22 1 0 15 16] sorted by Shell sort.

data before 5-sort	10	8	6	20	4	3	22	1	0	15	16
Five subarrays before sorting	10	—	—	—	—	3	—	—	—	—	16
		8	—	—	—	—	22	—	—	—	—
			6	—	—	—	—	1	—	—	—
				20	—	—	—	—	0	—	—
					4	—	—	—	—	15	—
Five subarrays after sorting	3	—	—	—	—	10	—	—	—	—	16
		8	—	—	—	—	22	—	—	—	—
			1	—	—	—	—	6	—	—	—
				0	—	—	—	—	20	—	—
					4	—	—	—	—	15	—
data after 5-sort and before 3-sort	3	8	1	0	4	10	22	6	20	15	16
Three subarrays before sorting	3	—	—	0	—	—	22	—	—	15	—
		8	—	—	4	—	—	6	—	—	16
			1	—	—	10	—	—	20	—	—
Three subarrays after sorting	0	—	—	3	—	—	15	—	—	22	—
		4	—	—	6	—	—	8	—	—	16
			1	—	—	10	—	—	20	—	—
data after 3-sort and before 1-sort	0	4	1	3	6	10	15	8	20	22	16
data after 1-sort	0	1	3	4	6	8	10	15	16	20	22

One problem still has to be addressed, namely, choosing the optimal value of the increment. In the example in Figure 9.7, the value of 5 is chosen to begin with, then 3, and 1 is used for the final sort. But why these values? Unfortunately, no convincing answer can be given. In fact, any decreasing sequence of increments can be used as long as the last one, h_1, is equal to 1. Donald Knuth has shown that even if there are only two increments, $(\frac{16n}{\pi})^{\frac{1}{3}}$ and 1, Shell sort is more efficient than insertion sort because it takes $O(n^{\frac{5}{3}})$ time instead of $O(n^2)$. But the efficiency of Shell sort can be improved by using a larger number of increments. It is imprudent, however, to use sequences of increments such as 1, 2, 4, 8, . . . or 1, 3, 6, 9, . . . because the mixing effect of data is lost.

For example, when using 4-sort and 2-sort, a subarray, $data_{2,i}$, for $i = 1, 2$, consists of elements of two arrays, $data_{4,i}$ and $data_{4,j}$, where $j = i + 2$, and only those. It is much better if elements of $data_{4,i}$ do not meet together again in the same array because a faster reduction in the number of exchange inversions is achieved if they are sent to different arrays when performing the 2-sort. Using only powers of 2 for the increments, as in Shell's original algorithm, the items in the even and odd positions of the array do not interact until the last pass, when the increment equals 1. This is where the mixing effect (or lack thereof) comes into play. But there is no formal proof indicating which sequence of increments is optimal. Extensive empirical studies along with some theoretical considerations suggest that it is a good idea to choose increments satisfying the conditions

$$h_1 = 1$$
$$h_{i+1} = 3h_i + 1$$

and stop with h_t for which $h_{t+2} \geq n$. For $n = 10,000$, this gives the sequence

$$1, 4, 13, 40, 121, 364, 1093, 3280$$

Experimental data have been approximated by the exponential function, the estimate, $1.21n^{\frac{5}{4}}$, and the logarithmic function $.39n \ln^2 n - 2.33n \ln n = O(n \ln^2 n)$. The first form fits the results of the tests better. $1.21n^{1.25} = O(n^{1.25})$ is much better than $O(n^2)$ for insertion sort, but it is still much greater than the expected $O(n \lg n)$ performance.

Figure 9.8 contains a function to sort the array $data$ using Shell sort. Note that before sorting starts, increments are computed and stored in the array $increments$.

The core of Shell sort is to divide an array into subarrays by taking elements h positions apart. Three features of this algorithm vary from one implementation to another:

1. The sequence of increments
2. A simple sorting algorithm applied in all passes except the last
3. A simple sorting algorithm applied only in the last pass, for 1-sort

In our implementation, as in Shell's, insertion sort is applied in all h-sorts, but other sorting algorithms can be used. For example, Dobosiewicz uses bubble sort for the last pass and insertion sort for other passes. Incerpi and Sedgewick use two iterations of cocktail shaker sort and a version of bubble sort in each h-sort and finish with insertion sort, obtaining what they call a *shakersort*. All these versions perform better

FIGURE 9.8   Implementation of Shell sort.

```
template<class T>
void ShellSort(T data[], int arrSize) {
    register int i, j, hCnt, h;
    int increments[20], k;
 // create an appropriate number of increments h
    for (h = 1, i = 0; h < arrSize; i++) {
        increments[i] = h;
        h = 3*h + 1;
    }
 // loop on the number of different increments h
    for (i--; i >= 0; i--) {
        h = increments[i];
      // loop on the number of subarrays h-sorted in ith pass
        for (hCnt = h; hCnt<2*h; hCnt++) {
          // insertion sort for subarray containing every hth element of
            for (j = hCnt; j < arrSize; ) {    // array data
                T tmp = data[j];
                k = j;
                while (k-h >= 0 && tmp < data[k-h]) {
                    data[k] = data[k-h];
                    k -= h;
                }
                data[k] = tmp;
                j += h;
            }
        }
    }
}
```

than simple sorting methods, although there are some differences in performance among versions. Analytical results concerning the complexity of these sorts are not available. All results regarding complexity are of an empirical nature.

9.3.2 Heap Sort

Selection sort makes $O(n^2)$ comparisons and is very inefficient, especially for large n. But it performs relatively few moves. If the comparison part of the algorithm is improved, the end results can be promising.

Heap sort was invented by John Williams and uses the approach inherent to selection sort. Selection sort finds among the n elements the one that precedes all other

$n - 1$ elements, then the least element among those $n - 1$ items, and so forth, until the array is sorted. To have the array sorted in ascending order, heap sort puts the largest element at the end of the array, then the second largest in front of it, and so on. Heap sort starts from the end of the array by finding the largest elements, whereas selection sort starts from the beginning using the smallest elements. The final order in both cases is indeed the same.

Heap sort uses a heap as described in Section 6.9. A heap is a binary tree with the following two properties:

1. The value of each node is not less than the values stored in each of its children.
2. The tree is perfectly balanced and the leaves in the last level are all in the leftmost positions.

A tree has the heap property if it satisfies condition 1. Both conditions are useful for sorting, although this is not immediately apparent for the second condition. The goal is to use only the array being sorted without using additional storage for the array elements; by condition 2, all elements are located in consecutive positions in the array starting from position 0, with no unused position inside the array. In other words, condition 2 reflects the packing of an array with no gaps.

Elements in a heap are not perfectly ordered. It is known only that the largest element is in the root node and that, for each other node, all its descendants are not greater than the element in this node. Heap sort thus starts from the heap, puts the largest element at the end of the array, and restores the heap that now has one less element. From the new heap, the largest element is removed and put in its final position, and then the heap property is restored for the remaining elements. Thus, in each round, one element of the array ends up in its final position, and the heap becomes smaller by this one element. The process ends with exhausting all elements from the heap and is summarized in the following pseudocode:

```
heapsort(data[],n)
    transform data into a heap;
        for i = downto 2
        swap the root with the element in position i;
        restore the heap property for the tree data[0],...,data[i-1];
```

In the first phase of heap sort, an array is transformed into a heap. In this process, we use a bottom-up method devised by Floyd and described in Section 6.9.2. All steps leading to the transformation of the array [2 8 6 1 10 15 3 12 11] into a heap are illustrated in Figure 9.9 (see also Figure 6.58).

The second phase begins after the heap has been built (Figures 9.9g and 9.10a). At that point, the largest element, number 15, is moved to the end of the array. Its place is taken by 8, thus violating the heap property. The property has to be restored, but this time it is done for the tree without the largest element, 15. Because it is already in its proper position, it does not need to be considered anymore and is removed (pruned) from the tree (indicated by the dashed lines in Figure 9.10). Now, the largest element among `data[0], . . . , data[n-2]` is looked for. To that end, the function `moveDown()` from Section 6.9 (Figure 6.56) is called to construct a heap out of all the

FIGURE 9.9 Transforming the array [2 8 6 1 10 15 3 12 11] into a heap.

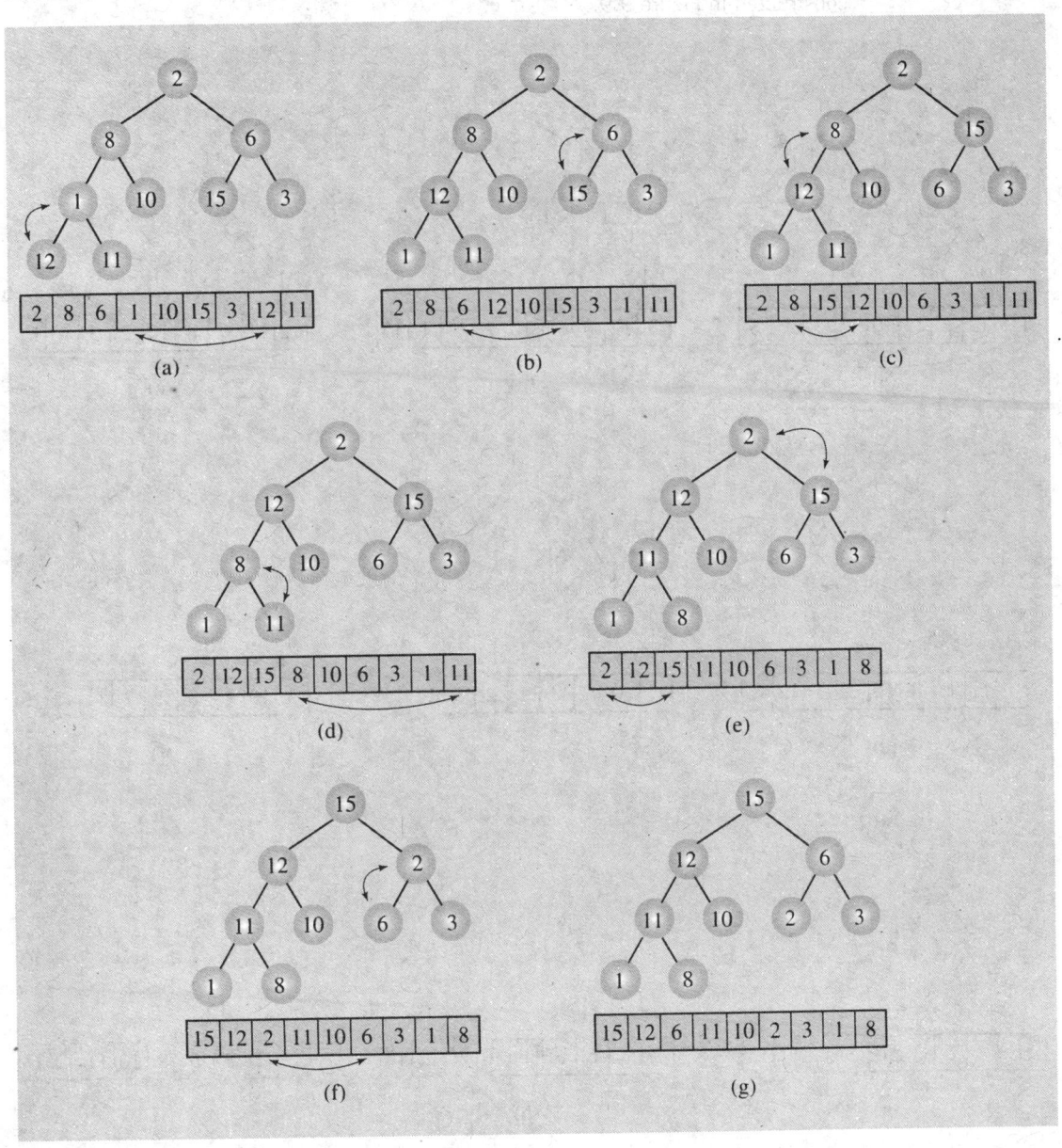

elements of `data` except the last, `data[n-1]`, which results in the heap in Figure 9.10c. Number 12 is sifted up and then swapped with 1, giving the tree in Figure 9.10d. The function `moveDown()` is called again to select 11 (Figure 9.10e), and the element is swapped with the last element of the current subarray, which is 3 (Figure

FIGURE 9.10 Execution of heap sort on the array [15 12 6 11 10 2 3 1 8], which is the heap constructed in Figure 9.9.

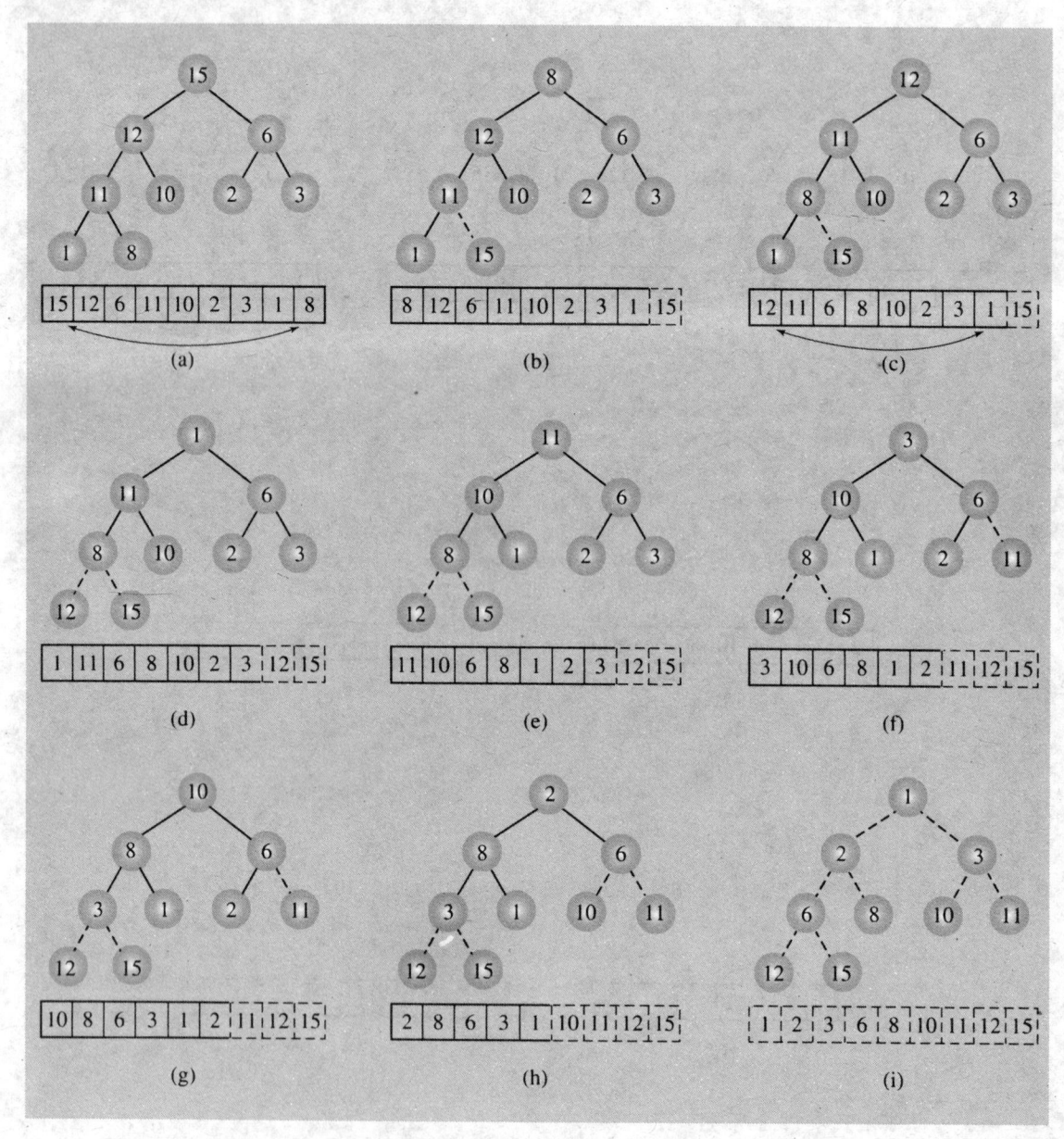

9.10f). Now 10 is selected (Figure 9.10g) and exchanged with 2 (Figure 9.10h). The reader can easily construct trees and heaps for the next passes through the loop of `heapsort()`. After the last pass, the array is in ascending order and the tree is ordered accordingly. An implementation of `heapsort()` is as follows:

```
template<class T>
void heapsort(T data[], int size) {
    for (int i = size/2 - 1; i >= 0; --i)   // create the heap;
        moveDown (data,i,size-1);
    for (int i = size-1; i >= 1; --i) {
        swap(data[0],data[i]); // move the largest item to data[i];
        moveDown(data,0,i-1);  // restore the heap property;
    }
}
```

Heap sort might be considered inefficient because the movement of data seems to be very extensive. First, all effort is applied to moving the largest element to the leftmost side of the array in order to move it to the furthest right. But therein lies its efficiency. In the first phase, to create the heap, heapsort() uses moveDown(), which performs $O(n)$ steps (see Section 6.9.2).

In the second phase, heapsort() exchanges $n-1$ times the root with the element in position i and also restores the heap $n-1$ times, which in the worst case causes moveDown() to iterate $\lg i$ times to bring the root down to the level of the leaves. Thus, the total number of moves in all executions of moveDown() in the second phase of heapsort() is $\sum_{i=1}^{n-1} \lg i$, which is $O(n \lg n)$. In the worst case, heapsort() requires $O(n)$ steps in the first phase, and in the second phase, $n-1$ swaps and $O(n \lg n)$ operations to restore the heap property, which gives $O(n) + O(n \lg n) + (n-1) = O(n \lg n)$ exchanges for the whole process in the worst case.

For the best case, when the array contains identical elements, moveDown() is called $\frac{n}{2}$ times in the first phase, but no moves are performed. In the second phase, heapsort() makes one swap to move the root element to the end of the array, resulting in only $n-1$ moves. Also, in the best case, n comparisons are made in the first phase and $2(n-1)$ in the second. Hence, the total number of comparisons in the best case is $O(n)$. However, if the array has distinct elements, then in the best case the number of comparisons equals $n \lg n - O(n)$ (Ding and Weiss, 1992).

9.3.3 Quicksort

Shell sort approached the problem of sorting by dividing the original array into subarrays, sorting them separately, and then dividing them again to sort the new subarrays until the whole array is sorted. The goal was to reduce the original problem to subproblems that can be solved more easily and quickly. The same reasoning was a guiding principle for C. A. R. Hoare, who invented an algorithm appropriately called *quicksort*.

The original array is divided into two subarrays, the first of which contains elements less than or equal to a chosen key called the *bound* or *pivot*. The second subarray includes elements equal to or greater than the bound. The two subarrays can be sorted separately, but before this is done, the partition process is repeated for both subarrays. As a result, two new bounds are chosen, one for each subarray. The four subarrays are created because each subarray obtained in the first phase is now divided into two segments. This process of partitioning is carried down until there are only one-cell arrays that do not need to be sorted at all. By dividing the task of sorting a large array into two simpler tasks and then dividing those tasks into even simpler tasks, it turns out that in the process of getting prepared to sort, the data have already been sorted. Because the sorting has been somewhat dissipated in the preparation process, this process is the core of quicksort.

Quicksort is recursive in nature because it is applied to both subarrays of an array at each level of partitioning. This technique is summarized in the following pseudocode:

```
quicksort(array[])
    if length(array) > 1
        choose bound; // partition array into subarray₁ and subarray₂
        while there are elements left in array
            include element either in subarray₁ = {el: el ≤ bound}
                or in subarray₂ = {el: el ≥ bound};
        quicksort(subarray₁);
        quicksort(subarray₂);
```

To partition an array, two operations have to be performed: A bound has to be found and the array has to be scanned to place the elements in the proper subarrays. However, choosing a good bound is not a trivial task. The problem is that the subarrays should be approximately the same length. If an array contains the numbers 1 through 100 (in any order) and 2 is chosen as a bound, then an imbalance results: The first subarray contains only 1 number after partitioning, whereas the second has 99 numbers.

A number of different strategies for selecting a bound have been developed. One of the simplest consists of choosing the first element of an array. That approach can suffice for some applications; however, because many arrays to be sorted already have many elements in their proper positions, a more cautious approach is to choose the element located in the middle of the array. This approach is incorporated in the implementation in Figure 9.11.

Another task is scanning the array and dividing the elements between its two subarrays. The pseudocode is vague about how this can be accomplished. In particular, it does not decide where to place an element equal to the bound. It only says that elements are placed in the first subarray if they are less than or the same as the bound and in the second if they are greater than or the same as the bound. The reason is that the difference between the lengths of the two subarrays should be minimal. Therefore, elements equal to the bound should be so divided between the two subarrays as to make this difference in size minimal. The details of handling this depend on a particular implementation, and one such implementation is given in Figure 9.11. In this implementation, `quicksort(data[],n)` preprocesses the array to be sorted by locating the largest element in the array and exchanging it with the last element of the array. Having the largest element at the end of the array prevents the index `lower` from running off the end of the array. This could happen in the first inner `while` loop if the bound were the largest element in the array. The index `lower` would be constantly incremented, eventually causing an abnormal program termination. Without this preprocessing, the first inner `while` loop would have to be

```
while (lower < last && data[lower] < bound)
```

The first test, however, would be necessary only in extreme cases, but it would be executed in each iteration of this `while` loop.

FIGURE 9.11 Implementation of quicksort.

```cpp
template<class T>
void quicksort(T data[], int first, int last) {
    int lower = first+1, upper = last;
    swap(data[first],data[(first+last)/2]);
    T bound = data[first];
    while (lower <= upper) {
        while (data[lower] < bound)
            lower++;
        while (bound < data[upper])
            upper--;
        if (lower < upper)
            swap(data[lower++],data[upper--]);
        else lower++;
    }
    swap(data[upper],data[first]);
    if (first < upper-1)
        quicksort (data,first,upper-1);
    if (upper+1 < last)
        quicksort (data,upper+1,last);
}

template<class T>
void quicksort(T data[], int n) {
    int i, max;
    if (n < 2)
        return;
    for (i = 1, max = 0; i < n; i++)   // find the largest
        if (data[max] < data[i])       // element and put it
            max = i;                    // at the end of data[];
    swap(data[n-1],data[max]);  // largest el is now in its
    quicksort(data,0,n-2);       // final position;
}
```

In this implementation, the main property of the bound is used, namely, that it is a boundary item. Hence, as befits the boundary item, it is placed on the borderline between the two subarrays obtained as a result of one call to `quicksort()`. In this way, the bound is located in its final position and can be excluded from further processing. To ensure that the bound is not moved around, it is stashed in the first position, and after partitioning is done, it is moved to its proper position, which is the rightmost position of the first subarray.

Figure 9.12 contains an example of partitioning the array [8 5 4 7 6 1 6 3 8 12 10]. In the first partitioning, the largest element in the array is located and exchanged with

FIGURE 9.12 Partitioning the array [8 5 4 7 6 1 6 3 8 12 10] with `quicksort()`.

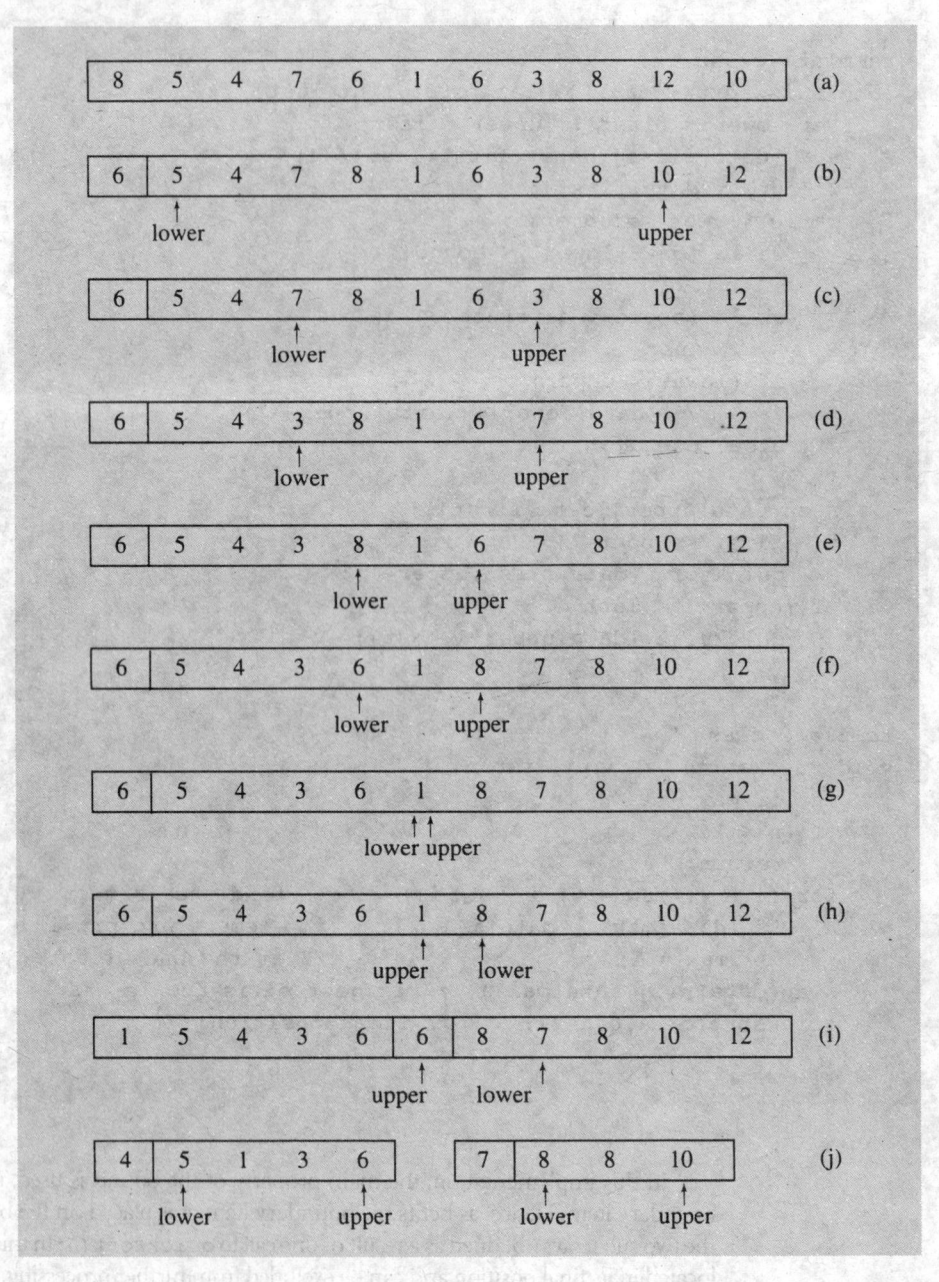

8	5	4	7	6	1	6	3	8	12	10	(a)

| 6 | 5 | 4 | 7 | 8 | 1 | 6 | 3 | 8 | 10 | 12 | (b) |

↑ lower   ↑ upper

| 6 | 5 | 4 | 7 | 8 | 1 | 6 | 3 | 8 | 10 | 12 | (c) |

         ↑ lower   ↑ upper

| 6 | 5 | 4 | 3 | 8 | 1 | 6 | 7 | 8 | 10 | 12 | (d) |

         ↑ lower   ↑ upper

| 6 | 5 | 4 | 3 | 8 | 1 | 6 | 7 | 8 | 10 | 12 | (e) |

              ↑ lower ↑ upper

| 6 | 5 | 4 | 3 | 6 | 1 | 8 | 7 | 8 | 10 | 12 | (f) |

              ↑ lower ↑ upper

| 6 | 5 | 4 | 3 | 6 | 1 | 8 | 7 | 8 | 10 | 12 | (g) |

                 ↑↑ lower upper

| 6 | 5 | 4 | 3 | 6 | 1 | 8 | 7 | 8 | 10 | 12 | (h) |

              ↑ upper ↑ lower

| 1 | 5 | 4 | 3 | 6 | 6 | 8 | 7 | 8 | 10 | 12 | (i) |

              ↑ upper ↑ lower

| 4 | 5 | 1 | 3 | 6 | | 7 | 8 | 8 | 10 | | (j) |

↑ lower ↑ upper   ↑ lower ↑ upper

the last element, resulting in the array [8 5 4 7 6 1 6 3 8 10 12]. Because the last element is already in its final position, it does not have to be processed anymore. Therefore, in the first partitioning, `lower` = 1, `upper` = 9, and the first element of the array, 8, is exchanged with the bound, 6 in position 4, so that the array is [6 5 4 7 8 1 6 3 8 10 12]

(Figure 9.12b). In the first iteration of the outer while loop, the inner while loop moves lower to position 3 with 7, which is greater than the bound. The second inner while loop moves upper to position 7 with 3, which is less than the bound (Figure 9.12c). Next the elements in these two cells are exchanged, giving the array [6 5 4 3 8 1 6 7 8 10 12] (Figure 9.12d). Then lower is incremented to 4 and upper is decremented to 6 (Figure 9.12e). This concludes the first iteration of the outer while loop.

In its second iteration, neither of the two inner while loops modifies any of the two indexes because lower indicates a position occupied by 8, which is greater than the bound, and upper indicates a position occupied by 6, which is equal to the bound. The two numbers are exchanged (Figure 9.12f), and then both indexes are updated to 5 (Figure 9.12g).

In the third iteration of the outer while loop, lower is moved to the next position containing 8, which is greater than the bound, and upper stays in the same position because 1 in this position is less than the bound (Figure 9.12h). But at that point, lower and upper cross each other, so no swapping takes place, and after a redundant increment of lower to 7, the outer while loop is exited. At that point, upper is the index of the rightmost element of the first subarray (with the element not exceeding the bound), so the element in this position is exchanged with the bound (Figure 9.12i). In this way, the bound is placed in its final position and can be excluded from subsequent processing. Therefore, the two subarrays that are processed next are the left subarray, with elements to the left of the bound, and the right subarray, with elements to its right (Figure 9.12j). Then partitioning is performed for these two subarrays separately, and then for subarrays of these subarrays, until subarrays have less than two elements. The entire process is summarized in Figure 9.13, in which all the changes in all current arrays are indicated.

The worst case occurs if in each invocation of quicksort(), the smallest (or largest) element of the array is chosen for the bound. This is the case if we try to sort the array [5 3 1 2 4 6 8]. The first bound is 1, and the array is broken into an empty array and the array [3 5 2 4 6] (the largest number, 8, does not participate in partitioning). The new bound is 2, and again only one nonempty array, [5 3 4 6], is obtained as the result of partitioning. The next bound and array returned by partition are 3 and [5 4 6], then 4 and [5 6], and finally 5 and [6]. The algorithm thus operates on arrays of size $n-1, n-2, \ldots, 2$. The partitions require $n - 2 + n - 3 + \cdots + 1$ comparisons, and for each partition, only the bound is placed in the proper position. This results in a run time equal to $O(n^2)$, which is hardly a desirable result, especially for large arrays or files.

The best case is when the bound divides an array into two subarrays of approximately length $\frac{n}{2}$. If the bounds for both subarrays are well chosen, the partitions produce four new subarrays, each of them with approximately $\frac{n}{4}$ cells. If, again, the bounds for all four subarrays divide them evenly, the partitions give eight subarrays, each with approximately $\frac{n}{8}$ elements. Therefore, the number of comparisons performed for all partitions is approximately equal to

$$n + 2\frac{n}{2} + 4\frac{n}{4} + 8\frac{n}{8} + \ldots + n\frac{n}{n} = n(\lg n + 1)$$

which is $O(n \lg n)$. This is due to the fact that parameters in the terms of this sum (and also the denominators) form a geometric sequence so that $n = 2^k$ for $k = \lg n$ (assuming that n is a power of 2).

FIGURE 9.13 Sorting the array [8 5 4 7 6 1 6 3 8 12 10] with `quicksort()`.

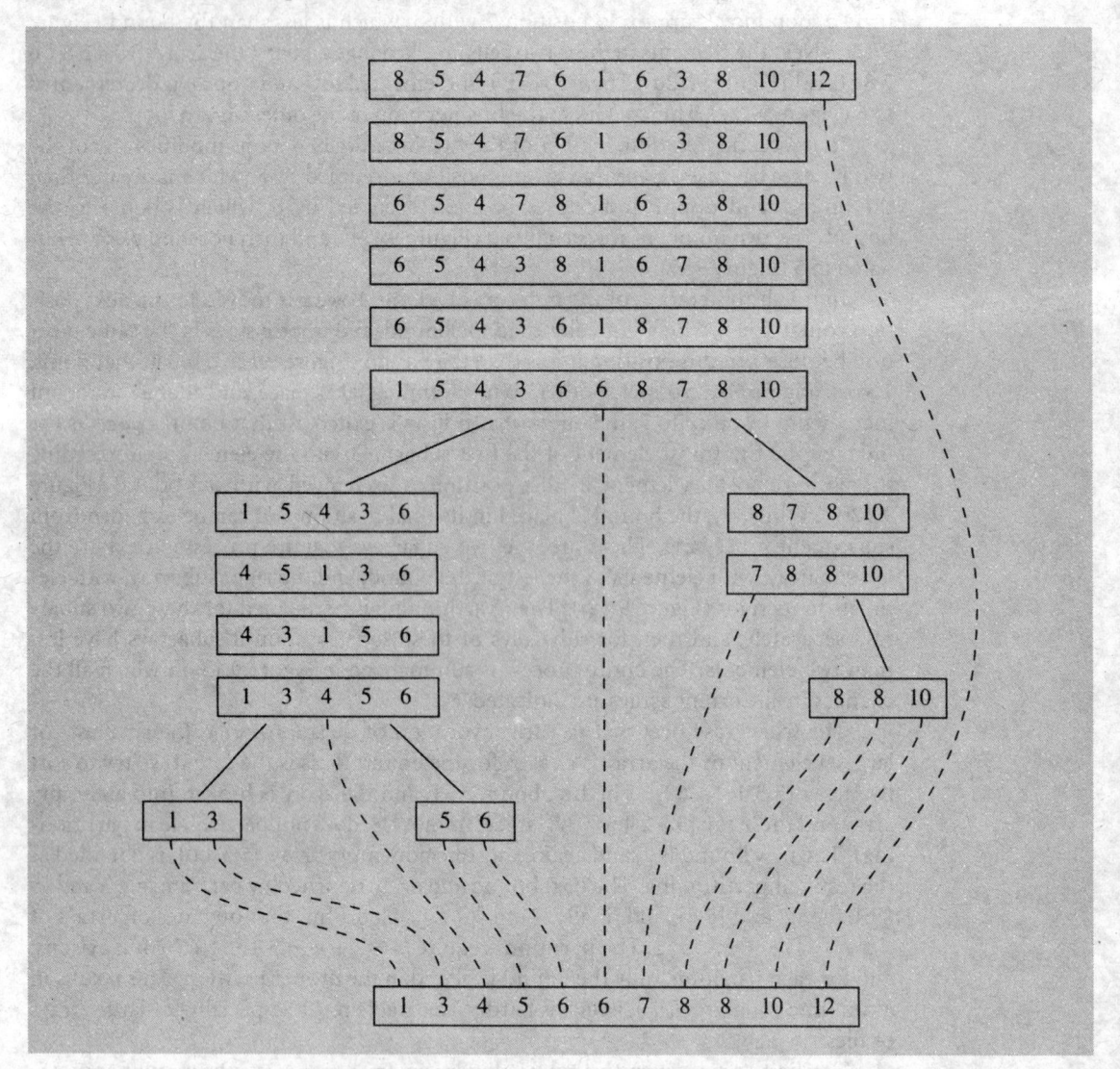

Now we can answer the question asked before: Is the average case, when the array is ordered randomly, closer to the best case, $n \lg n$, or to the worst, $O(n^2)$? Some calculations show that the average case requires only $O(n \lg n)$ comparisons (see Appendix A.3), which is the desired result. The validity of this figure can be strengthened by referring to the tree obtained after disregarding the bottom rectangle in Figure 9.13. This tree indicates how important it is to keep the tree balanced, for the smaller the number of levels, the quicker the sorting process. In the extreme case, the tree can be turned into a linked list in which every nonleaf node has only one child. That rather rare phenomenon is possible and prevents us from calling quicksort the ideal sort. But quicksort seems to be closest to such an ideal because, as analytic studies indicate, it outperforms other efficient sorting methods by at least a factor of 2.

How can the worst case be avoided? The partition procedure should produce arrays of approximately the same size, which can be achieved if a good bound is chosen. This is the crux of the matter: How can the best bound be found? Only two methods will be mentioned. The first method randomly generates a number between `first` and `last`. This number is used as an index of the bound, which is then interchanged with the first element of the array. In this method, the partition process proceeds as before. Good random number generators may slow down the execution time as they themselves often use sophisticated and time-consuming techniques. Thus, this method is not highly recommended.

The second method chooses a median of three elements: the first, middle, and last. For the array [1 5 4 7 8 6 6 3 8 12 10], 6 is chosen from the set [1 6 10], and for the first generated subarray, the bound 4 is chosen from the set [1 4 6]. Obviously, there is the possibility that all three elements are always the smallest (or the largest) in the array, but it does not seem very likely.

Is quicksort the best sorting algorithm? It certainly is—usually. It is not bulletproof, however, and some problems have already been addressed in this section. First, everything hinges on which element of the file or array is chosen for the bound. Ideally, it should be the median element of the array. An algorithm to choose a bound should be flexible enough to handle all possible orderings of the data to be sorted. Because some cases always slip by these algorithms, from time to time quicksort can be expected to be anything but quick.

Second, it is inappropriate to use quicksort for small arrays. For arrays with fewer than 30 items, insertion sort is more efficient than quicksort (Cook and Kim, 1980). In this case the initial pseudocode can be changed to

```
quicksort2 (array)
    if length(array) > 30
        partition array into subarray₁ and subarray₂;
        quicksort2(subarray₁);
        quicksort2(subarray₂);
    else insertionsort(array);
```

and the implementations changed accordingly. However, the table in Figure 9.18 later in this chapter indicates that the improvement is not significant.

9.3.4 Mergesort

The problem with quicksort is that its complexity in the worst case is $O(n^2)$ because it is difficult to control the partitioning process. Different methods of choosing a bound attempt to make the behavior of this process fairly regular; however, there is no guarantee that partitioning results in arrays of approximately the same size. Another strategy is to make partitioning as simple as possible and concentrate on merging the two sorted arrays. This strategy is characteristic of *mergesort*. It was one of the first sorting algorithms used on a computer and was developed by John von Neumann.

The key process in mergesort is merging sorted halves of an array into one sorted array. However, these halves have to be sorted first, which is accomplished by merging the already sorted halves of these halves. This process of dividing arrays into two halves stops when the array has fewer than two elements. The algorithm is recursive in nature and can be summarized in the following pseudocode:

```
mergesort(data)
    if data have at least two elements
        mergesort(left half of data);
        mergesort(right half of data);
        merge(both halves into a sorted list);
```

Merging two subarrays into one is a relatively simple task, as indicated in this pseudocode:

```
merge(array1, array2, array3)
    i1, i2, i3 are properly initialized;
    while both array2 and array3 contain elements
        if array2[i2] < array3[i3]
            array1[i1++] = array2[i2++];
        else array1[i1++] = array3[i3++];
    load into array1 the remaining elements of either array2 or array3;
```

For example, if array2 = [1 4 6 8 10] and array3 = [2 3 5 22], then the resulting array1 = [1 2 3 4 5 6 8 10 22].

The pseudocode for merge() suggests that array1, array2, and array3 are physically separate entities. However, for the proper execution of mergesort(), array1 is a concatenation of array2 and array3 so that array1 before the execution of merge() is [1 4 6 8 10 2 3 5 22]. In this situation, merge() leads to erroneous results, because after the second iteration of the while loop, array2 is [1 2 6 8 10] and array1 is [1 2 6 8 10 2 3 5 22]. Therefore, a temporary array has to be used during the merging process. At the end of the merging process, the contents of this temporary array are transferred to array1. Because array2 and array3 are subarrays of array1, they do not need to be passed as parameters to merge(). Instead, indexes for the beginning and the end of array1 are passed, because array1 can be a part of another array. The new pseudocode is

```
merge (array1, first, last)
    mid = (first + last) / 2;
    i1 = 0;
    i2 = first;
    i3 = mid + 1;
    while both left and right subarrays of array1 contain elements
        if array1[i2] < array1[i3]
            temp[i1++] = array1[i2++];
        else temp[i1++] = array1[i3++];
    load into temp the remaining elements of array1;
    load to array1 the content of temp.
```

The entire array1 is copied to temp and then temp is copied back to array1, so the number of movements in each execution of merge() is always the same and is equal to 2 · (last − first + 1). The number of comparisons depends on the ordering in array1. If array1 is in order or if the elements in the right half precede the elements in the left half, the number of comparisons is (first + last)/2. The worst case is when the last element of one half precedes only the last element of the other half, as in [1 6 10 12] and [5 9 11 13]. In this case, the number of comparisons is last − first. For an *n*-element array, the number of comparisons is *n* − 1.

The pseudocode for mergesort() is now

```
mergesort (data, first, last)
    if first < last
        mid = (first + last) / 2;
        mergesort(data, first, mid);
        mergesort(data, mid+1, last);
        merge(data, first, last);
```

Figure 9.14 illustrates an example using this sorting algorithm. This pseudocode can be used to analyze the computing time for mergesort. For an *n*-element array, the number of movements is computed by the following recurrence relation:

$$M(1) = 0$$
$$M(n) = 2M\left(\frac{n}{2}\right) + 2n$$

FIGURE 9.14 The array [1 8 6 4 10 5 3 2 22] sorted by mergesort.

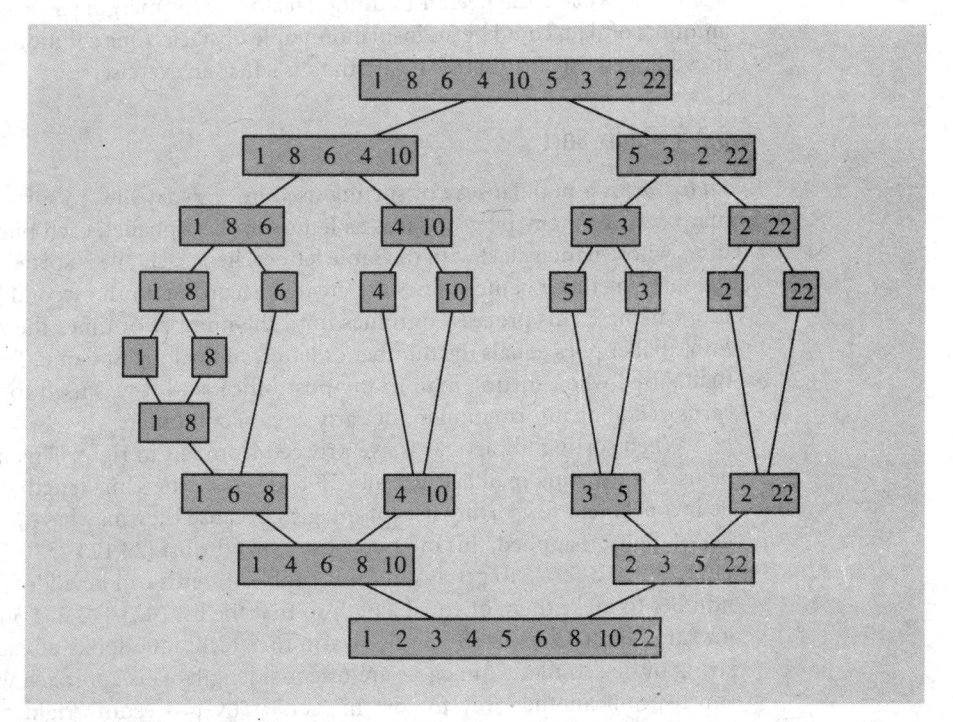

$M(n)$ can be computed in the following way:

$$M(n) = 2\left(2M\left(\frac{n}{4}\right) + 2\left(\frac{n}{2}\right)\right) + 2n = 4M\left(\frac{n}{4}\right) + 4n$$

$$= 4\left(2M\left(\frac{n}{8}\right) + 2\left(\frac{n}{4}\right)\right) + 4n = 8M\left(\frac{n}{8}\right) + 6n$$

$$\vdots$$

$$= 2^i M\left(\frac{n}{2^i}\right) + 2in$$

Choosing $i = \lg n$ so that $n = 2^i$ allows us to infer

$$M(n) = 2^i M\left(\frac{n}{2^i}\right) + 2in = nM(1) + 2n \lg n = 2n \lg n = O(n \lg n)$$

The number of comparisons in the worst case is given by a similar relation:

$$C(1) = 0$$

$$C(n) = 2C\left(\frac{n}{2}\right) + n - 1$$

which also results in $C(n)$ being $O(n \lg n)$.

Mergesort can be made more efficient by replacing recursion with iteration (see the exercises at the end of this chapter) or by applying insertion sort to small portions of an array, a technique that was suggested for quicksort. However, mergesort has one serious drawback: the need for additional storage for merging arrays, which for large amounts of data could be an insurmountable obstacle. One solution to this drawback uses a linked list; analysis of this method is left as an exercise.

9.3.5 Radix Sort

Radix sort is a popular way of sorting used in everyday life. To sort library cards, we may create as many piles of cards as letters in the alphabet, each pile containing authors whose names start with the same letter. Then, each pile is sorted separately using the same method; namely, piles are created according to the second letter of the authors' names. This process continues until the number of times the piles are divided into smaller piles equals the number of letters of the longest name. This method is actually used when sorting mail in the post office, and it was used to sort 80-column cards of coding information in the early days of computers.

When sorting library cards, we proceed from left to right. This method can also be used for sorting mail because all ZIP codes have the same length. However, it may be inconvenient for sorting lists of integers because they may have an unequal number of digits. If applied, this method would sort the list [23 123 234 567 3] into the list [123 23 234 3 567]. To get around this problem, zeros can be added in front of each number to make them of equal length so that the list [023 123 234 567 003] is sorted into the list [003 023 123 234 567]. Another technique looks at each number as a string of bits so that all integers are of equal length. This approach will be discussed shortly. Still another way to sort integers is by proceeding right to left, and this method is discussed now.

When sorting integers, 10 piles numbered 0 through 9 are created, and initially, integers are put in a given pile according to their rightmost digit so that 93 is put in pile 3. Then, piles are combined and the process is repeated, this time with the second rightmost digit; in this case, 93 ends up on pile 9. The process ends after the leftmost digit of the longest number is processed. The algorithm can be summarized in the following pseudocode:

```
radixsort()
    for d = 1 to the position of the leftmost digit of longest number
        distribute all numbers among piles 0 through 9 according to the dth digit;
        put all integers on one list;
```

The key to obtaining a proper outcome is the way the 10 piles are implemented and then combined. For example, if these piles are implemented as stacks, then the integers 93, 63, 64, 94 are put on piles 3 and 4 (other piles being empty):

```
pile 3: 63 93
pile 4: 94 64
```

These piles are then combined into the list 63, 93, 94, 64. When sorting them according to the second rightmost digit, the piles are as follows:

```
pile 6: 64 63
pile 9: 94 93
```

and the resulting list is 64, 63, 94, 93. The processing is finished, but the result is an improperly sorted list.

However, if piles are organized as queues, the relative order of elements on the list is retained. When integers are sorted according to the digit in position d, then within each pile, integers are sorted with regard to the part of the integer extending from digit 1 to $d-1$. For example, if after the third pass, pile 5 contains the integers 12534, 554, 3590, then this pile is ordered with respect to the two rightmost digits of each number. Figure 9.15 illustrates another example of radix sort.

Here is an implementation of radix sort:

```
void radixsort(long data[], int n) {
    register int i, j, k, factor;
    const int radix = 10;
    const int digits = 10; // the maximum number of digits for a long
    Queue<long> queues[radix];    // integer;
    for (i = 0, factor = 1; i < digits; factor *= radix, i++) {
        for (j = 0; j < n; j++)
            queues[(data[j] / factor) % radix].enqueue(data[j]);
        for (j = k = 0; j < radix; j++)
            while (!queues[j].empty())
                data[k++] = queues[j].dequeue();
    }
}
```

This algorithm does not rely on data comparison as did the previous sorting methods. For each integer from data, two operations are performed: division by a

FIGURE 9.15 Sorting the list 10, 1234, 9, 7234, 67, 9181, 733, 197, 7, 3 with radix sort.

data = [10 1234 9 7234 67 9181 733 197 7 3]

```
                                                      7
                              3      7234            197
              10      9181   733     1234             67                  9
piles:         0        1     2       3       4       5       6       7       8       9
                                           pass 1
```

data = [10 9181 733 3 1234 7234 67 197 7 9]

```
              9
              7             7234
              3             1234
              0     10       733                     67             9181    197
piles:        0     1         2       3       4       5      6       7       8       9
                                           pass 2
```

data = [3 7 9 10 733 1234 7234 67 9181 197]

```
piles:        67
              10
               9
               7    197    7234                                    773
               3    9181   1234
               0      1      2       3       4       5      6       7       8       9
                                           pass 3
```

data = [3 7 9 10 67 9181 197 1234 7234 733]

```
piles:       733
             197
              67
              10
               9
               7
               3    1234                                          7234           9181
               0      1      2       3       4       5      6       7       8       9
                                           pass 4
```

data = [3 7 9 10 67 197 733 1234 7234 9181]

factor to disregard digits following digit *i* being processed in the current pass and division modulo radix (equal to 10) to disregard all digits preceding *i* for a total of 2*n*digits = $O(n)$ operations. The operation div can be used, which combines both / and %. In each pass, all integers are moved to piles and then back to data for a total of 2*n*digits = $O(n)$ moves. The algorithm requires additional space for piles, which if implemented as linked lists, is equal to *kn* words depending on the size *k* of the pointers. Our implementation uses only for loops with counters; therefore, it requires the same amount of passes for each case: best, average, and worst. The body of the only while loop is always executed *n* times to dequeue integers from all queues.

The foregoing discussion treated integers as combinations of digits. But as already mentioned, they can be regarded as combinations of bits. This time, division and division modulo are not appropriate, because for each pass, one bit for each number has to be extracted. In this case, only two queues are required.

An implementation is as follows:

```
void bitRadixsort(long data[], int n) {
    register int i, j, k, mask = 1;
    const int bits = 31;
    Queue<long> queues[2];
    for (i = 0; i < bits; i++) {
        for (j = 0; j < n; j++)
            queues[data[j] & mask ? 1 : 0].enqueue(data[j]);
        mask <<= 1;
        k = 0;
        while (!queues[0].empty())
            data[k++] = queues[0].dequeue();
        while (!queues[1].empty())
            data[k++] = queues[1].dequeue();
    }
}
```

Division is replaced here by the bitwise-and operation &. The variable mask has one bit set to 1 and the rest are set to 0. After each iteration, this 1 is shifted to the left. If data[j] & mask has a nonzero value, then data[j] is put in queues[1]; otherwise, it is put in queues[0]. Bitwise and is much faster than integer division, but in this example, 31 passes are needed, instead of 10 when the largest integer is 10 digits long. This means $62n$ data movements as opposed to $20n$.

Quicker operations cannot outweigh a larger number of moves: bitRadixsort() is much slower than radixsort() because the queues are implemented as linked lists, and for each item included in a particular queue, a new node has to be created and attached to the queue. For each item copied back to the original array, the node has to be detached from the queue and disposed of using delete. Although theoretically obtained performance $O(n)$ is truly impressive, it does not include operations on queues, and it hinges upon the efficiency of the queue implementation.

A better implementation is an array of size n for each queue, which requires creating these queues only once. The efficiency of the algorithm depends only on the number of exchanges (copying to and from queues). However, if radix r is a large number and a large amount of data has to be sorted, then this solution requires r queues of size n, and the number $(r + 1) \cdot n$ (original array included) may be unrealistically large.

A better solution uses one integer array queues of size n representing linked lists of indexes of numbers belonging to particular queues. Cell i of the array queueHeads contains an index of the first number in data that belongs to this queue, whose dth digit is i. queueTails[] contains a position in data of the last number whose dth digit is i. Figure 9.16 illustrates the situation after the first pass, for $d = 1$. queueHeads[4] is 1, which means that the number in position 1 in data, 1234, is the first number found in data with 4 as the last digit. Cell queues[1] contains 3, which is an index of the next number in data with 4 as the last digit, 7234. Finally, queues[3] is −1 to indicate the end of the numbers meeting this condition.

FIGURE 9.16 An implementation of radix sort.

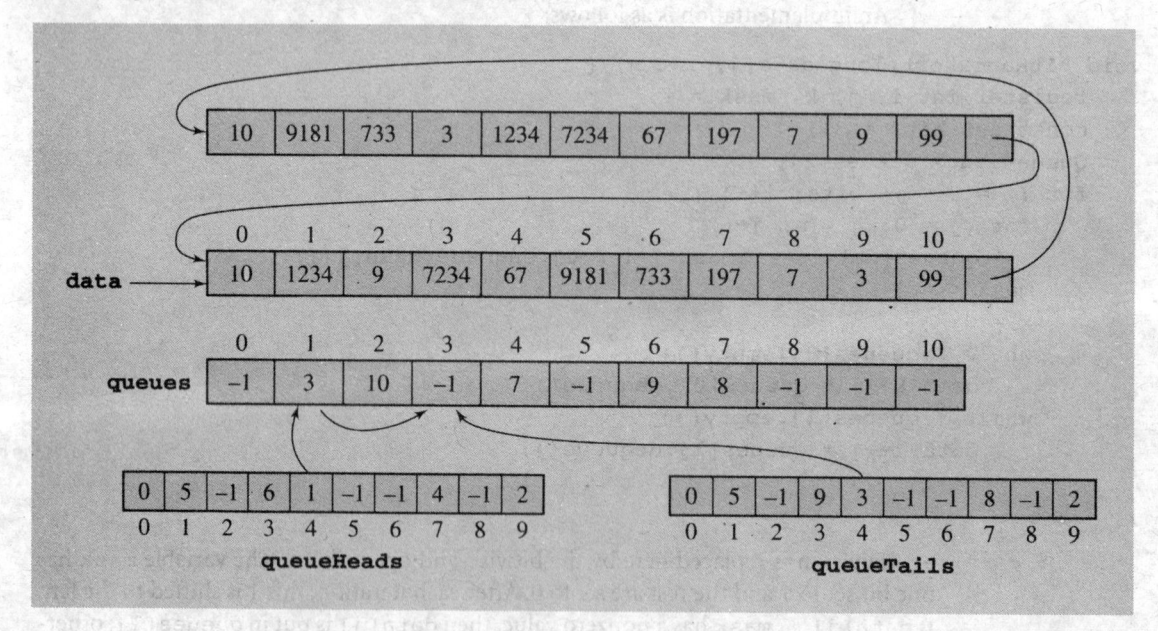

	10	9181	733	3	1234	7234	67	197	7	9	99	

	0	1	2	3	4	5	6	7	8	9	10
data	10	1234	9	7234	67	9181	733	197	7	3	99

	0	1	2	3	4	5	6	7	8	9	10
queues	−1	3	10	−1	7	−1	9	8	−1	−1	−1

0	5	−1	6	1	−1	−1	4	−1	2
0	1	2	3	4	5	6	7	8	9

queueHeads

0	5	−1	9	3	−1	−1	8	−1	2
0	1	2	3	4	5	6	7	8	9

queueTails

The next stage orders data according to information gathered in queues. It copies all the data from the original array to some temporary storage and then back to this array. To avoid the second copy, two arrays can be used, constituting a two-element circular linked list. After copying, the pointer to the list is moved to the next node, and the array in this node is treated as storage of numbers to be sorted. The improvement is significant because the new implementation runs at least two times faster than the implementation that uses queues (see Figure 9.18 later in this chapter).

9.4 SORTING IN THE STANDARD TEMPLATE LIBRARY

The STL provides many sorting functions, particularly in the library <algorithm>. The functions are implementing some of the sorting algorithms discussed in this chapter: quicksort, heap sort, and mergesort. A program in Figure 9.17 demonstrates these functions. For descriptions of the functions, see also Appendix B.

The first set of functions are partial sorting functions. The first version picks k = middle − first smallest elements from a container and puts them in order in the range [first, middle). For example,

   partial_sort(v1.begin(),v1.begin()+3,v1.end());

takes the three smallest integers from the entire vector v1 and puts them in the first three cells of the vector. The remaining integers are put in cells four through seven. In this way, v1 = [1,4,3,6,7,2,5] is transformed into v1 = [1,2,3,6,7,4,5]. This version

FIGURE 9.17 Demonstration of sorting functions.

```cpp
#include <iostream>
#include <vector>
#include <algorithm>
#include <functional> // greater<>

using namespace std;

class Person {
public:
    Person(char *n = "", int a = 0) {
        name = strdup(n);
        age = a;
    }
    bool operator==(const Person& p) const {
        return strcmp(name,p.name) == 0;
    }
    bool operator<(const Person& p) const {
        return strcmp(name,p.name) < 0;
    }
private:
    char *name;
    int age;
    friend ostream& operator<< (ostream& out, const Person& p) {
        out << "(" << p.name << "," << p.age << ")";
        return out;
    }
};

bool f1(int n) {
    return n < 5;
}

template<class T>
void printVector(char *s, const vector<T>& v) {
    cout << s << " = (";
    if (v.size() == 0) {
        cout << ")\n";
        return;
    }
    for (vector<T>::const_iterator i = v.begin(); i! = v.end()-1; i++)
        cout << *i << ',';
```

Continues

FIGURE 9.17 *(continued)*

```
        cout << *i << ")\n";
}

int main() {
    int a[] = {1,4,3,6,7,2,5};
    vector<int> v1(a,a+7), v2(a,a+7), v3(6,9), v4(6,9);
    vector<int>::iterator i1, i2, i3, i4;
    partial_sort(v1.begin(),v1.begin()+3,v1.end());
    printVector("v1",v1);                       // v1 = (1,2,3,6,7,4,5)
    partial_sort(v2.begin()+1,v2.begin()+4,v2.end(),greater<int>());
    printVector("v2",v2);                       // v2 = (1,7,6,5,3,2,4)
    i3 = partial_sort_copy(v2.begin(),v2.begin()+4,v3.begin(),v3.end());
    printVector("v3",v3);                       // v3 = (1,5,6,7,9,9)
    cout << *(i3-1) << ' ' << *i3 << endl;      // 7 9
    i4 = partial_sort_copy(v1.begin(),v1.begin()+4,v4.begin(),v4.end(),
                            greater<int>());
    printVector("v4",v4);                       // v4 = (6,3,2,1,9,9)
    cout << *(i4-1) << ' ' << *i4 << endl;      // 1 9
    i1 = partition(v1.begin(),v1.end(),f1);     // v1 = (1,2,3,4,7,6,5)
    printVector("v1",v1);
    cout << *(i1-1) << ' ' << *i1 << endl;      // 4 7
    i2 = partition(v2.begin(),v2.end(),bind2nd(less<int>(),5));
    printVector("v2",v2);                       // v2 = (1,4,2,3,5,6,7)
    cout << *(i2-1) << ' ' << *i2 << endl;      // 3 5
    sort(v1.begin(),v1.end());                  // v1 = (1,2,3,4,5,6,7)
    sort(v1.begin(),v1.end(),greater<int>());   // v1 = (7,6,5,4,3,2,1)

    vector<Person> pv1, pv2;
    for (int i = 0; i < 20; i++) {
        pv1.push_back(Person("Josie",60 - i));
        pv2.push_back(Person("Josie",60 - i));
    }
    sort(pv1.begin(),pv1.end());            // pv1 = ((Josie,41)...(Josie,60))
    stable_sort(pv2.begin(),pv2.end());// pv2 = ((Josie,60)...(Josie,41))

    vector<int> heap1, heap2, heap3(a,a+7), heap4(a,a+7);
    for (i = 1; i <= 7; i++) {
        heap1.push_back(i);
        push_heap(heap1.begin(),heap1.end());
        printVector("heap1",heap1);
    }
```

FIGURE 9.17 (continued)

```
    // heap1 = (1)
    // heap1 = (2,1)
    // heap1 = (3,1,2)
    // heap1 = (4,3,2,1)
    // heap1 = (5,4,2,1,3)
    // heap1 = (6,4,5,1,3,2)
    // heap1 = (7,4,6,1,3,2,5)
    sort_heap(heap1.begin(),heap1.end());  // heap1 = (1,2,3,4,5,6,7)
    for (i = 1; i <= 7; i++) {
        heap2.push_back(i);
        push_heap(heap2.begin(),heap2.end(),greater<int>());
        printVector("heap2",heap2);
    }
    // heap2 = (1)
    // heap2 = (1,2)
    // heap2 = (1,2,3)
    // heap2 = (1,2,3,4)
    // heap2 = (1,2,3,4,5)
    // heap2 = (1,2,3,4,5,6)
    // heap2 = (1,2,3,4,5,6,7)
    sort_heap(heap2.begin(),heap2.end(),greater<int>());
    printVector("heap2",heap2);              // heap2 = (7,6,5,4,3,2,1)
    make_heap(heap3.begin(),heap3.end());    // heap3 = (7,6,5,1,4,2,3)
    sort_heap(heap3.begin(),heap3.end());    // heap3 = (1,2,3,4,5,6,7)
    make_heap(heap4.begin(),heap4.end(),greater<int>());
    printVector("heap4",heap4);              // heap4 = (1,4,2,6,7,3,5)
    sort_heap(heap4.begin(),heap4.end(),greater<int>());
    printVector("heap4",heap4);              // heap4 = (7,6,5,4,3,2,1)
    return 0;
}
```

orders elements in ascending order. An ordering relation can be changed if it is provided as the fourth parameter to the function call. For example,

```
partial_sort(v2.begin()+1,v2.begin()+4,v2.end(),greater<int>());
```

picks the largest three integers from the vector v2 and puts them in positions two to four in descending order. The order of the remaining integers outside this range is not specified. This call transforms v2 = [1,4,3,6,7,2,5] into v2 = [1,7,6,5,3,2,4].

The third version of partial sorting takes k = last1 - first1 or last2 - first2, whichever is smaller, first elements from the range [first1,last1), and writes them over elements in the range [first2,last2). The call

```
i3 = partial_sort_copy(v2.begin(),v2.begin()+4,v3.begin(),v3.end());
```

takes the first four integers in vector v2 = [1,7,6,5], and puts them in ascending order in the first four cells of v3 so that v3 = [9,9,9,9,9,9] is transformed into v3 = [1,5,6,7,9,9]. As an extra, an iterator is returned that refers to the first position after the last copied number. The fourth version of the partial sorting function is similar to the third, but it allows for providing a relation with which elements are sorted. The program in Figure 9.17 also demonstrates the partition function, which orders two ranges with respect to one another by putting in the first range numbers for which a one-argument Boolean function is true; numbers for which the function is false are in the second range. The call

```
i1 = partition(v1.begin(),v1.end(),f1);
```

uses an explicitly defined function f1() and transforms v1 = [·1,2,3,6,7,4,5] into v1 = [1,2,3,4,7,6,5] by putting all numbers less than 5 in front of numbers that are not less than 5. The call

```
i2 = partition(v2.begin(),v2.end(),bind2nd(less<int>(),5));
```

accomplished the same for v2 with the built-in functional bind2nd, which binds the second argument of the operator < to 5, effectively generating a function that works just like f1().

Probably the most useful is the function sort() that implements quicksort. The call

```
sort(v1.begin(),v1.end());
```

transforms v1 = [1,2,3,6,7,4,5] into fully ordered v1 = [1,2,3,4,5,6,7]. The second version of sort() allows for specifying any ordering relation.

The STL also provides stable sorting functions. A sorting algorithm is said to be *stable* if equal keys remain in the same relative order in output as they are initially (that is, if data[i] equals data[j] for $i < j$ and the ith element ends up in the kth position and the jth element in the mth position, then $k < m$). To see the difference between sorting and stable sorting, consider a vector of objects of type Person. The definition of operator< orders the Person objects by name without taking age into account. Therefore, two objects with the same name but with different ages are considered equal, just as in the definition of operator==, although this operator is not used in sorting; only the less than operator is. After creating two equal vectors, pv1 and pv2, both equal to [("Josie,"60) ... ("Josie,"41)], we see that sort() transforms this vector into [("Josie,"41) ... ("Josie,"60)], but stable_sort() leaves it intact by retaining the relative order of equal objects. The feat is possible by using mergesort in the stable sorting routine. Note that, for small numbers of objects, sort() is also stable because, for a small number of elements, insertion sort is used, not quicksort (see quicksort2() at the end of Section 9.3.3).

9.5 CONCLUDING REMARKS

Figure 9.18 compares the run times for different sorting algorithms and different numbers of integers being sorted. They were all run on a PC. At each stage, the number of integers has been doubled to see the factors by which the run times raise. These factors are included in each column except for the first three columns and are shown along

FIGURE 9.18 Comparison of run times for different sorting algorithms and different numbers of integers to be sorted.

	10,000			20,000		
	Ascending	Random	Descending	Ascending	Random	Descending
insertionsort	.00	3.18	6.54	.00 —	13.51 4.2	26.69 4.1
selectionsort	5.65	5.17	5.55	20.82 3.7	22.03 4.3	22.41 4.0
bubblesort	5.22	10.88	15.60	20.87 4.0	45.09 4.1	1 m 2.51 4.0
Shellsort	.00	.05	.00	.00 —	.11 2.2	.05 —
heapsort	.06	.06	.05	.11 1.8	.11 1.8	.11 2.2
mergesort	.00	.05	.06	.11 —	.11 1.8	.11 1.8
quicksort	.00	.05	.00	.00 —	.05 1.0	.05 —
quicksort2	.00	.00	.00	.05 —	.06 —	.06 —
radixsort	.44	.44	.44	.82 1.9	.88 2.0	.88 2.0
bitRadixsort	.99	1.09	1.10	2.03 2.1	2.14 2.0	2.14 1.9
radixsort2	.17	.11	.17	.33 1.9	.33 3.1	.33 1.9
bitRadixsort2	.21	.22	.22	.44 .21	.44 2.0	.44 2.0

	40,000			80,000		
	Ascending	Random	Descending	Ascending	Random	Descending
insertionsort	.00 —	53.66 4.0	1 m 54.35 4.3	.05 —	3 m 46.62 4.2	7 m 40.94 4.0
selectionsort	1 m 24.47 4.2	1 m 31.01 4.1	1 m 30.96 4.1	5 m 55.64 4.2	6 m 6.30 4.0	6 m 21.13 4.2
bubblesort	1 m 27.55 4.2	2 m 59.40 4.0	4 m 15.68 4.1	6 m 6.91 4.2	12 m 27.15 4.2	17 m 14.02 4.0
Shellsort	0.11 —	.28 2.5	.16 2.7	.22 2.0	.66 2.4	.33 2.1
heapsort	.25 2.3	.27 1.6	.28 2.5	.55 2.2	.66 2.4	.49 1.7
mergesort	.16 1.5	.22 2.0	.22 2.0	.49 2.2	.49 2.2	.44 2.0
quicksort	.06 —	.11 2.2	.05 1.0	.11 1.8	.28 2.5	.17 2.8
quicksort2	.05 1.0	.11 1.8	.06 1.0	.11 2.2	.27 2.5	.11 1.8
radixsort	1.70 2.1	1.76 2.0	1.76 2.0	3.46 2.0	3.52 2.1	3.52 2.0
bitRadixsort	4.07 2.0	4.17 1.9	4.23 2.0	8.41 2.1	8.40 2.0	8.45 2.0
radixsort2	.66 2.0	.60 1.8	.65 2.0	1.38 2.1	1.32 2.2	1.43 2.2
bitRadixsort2	.94 2.1	.99 2.2	.99 2.3	1.97 2.1	2.08 2.1	1.98 2.0

with the run times. The factors are rounded to one decimal place, whereas run times (in seconds) are rounded to two decimal places. For example, heap sort required 0.25 seconds to sort an array of 40,000 integers in ascending order and 0.55 seconds to sort 8,000 integers, also in ascending order. Doubling the number of data is associated with the increase of run time by a factor of 0.55/0.25 = 2.2, and the number 2.2 follows 0.55 in the fourth column.

The table in Figure 9.18 indicates that the run time for elementary sorting methods, which are squared algorithms, grows approximately by a factor of 4 after the amount of data is doubled, whereas the same factor for nonelementary methods, whose complexity is $O(n \lg n)$, is approximately 2. This is also true for the four implementations of radix sort, whose complexity equals $2n$digits or $2n$bits. The table also shows that quicksort is the fastest algorithm among all sorting methods; most of the time, it runs at least twice as fast as any other algorithm.

9.6 CASE STUDY: ADDING POLYNOMIALS

Adding polynomials is a common algebraic operation and is usually a simple calculation. It is a known rule that, to add two terms, they must contain the same variables raised to the same powers, and the resulting term retains these variables and powers, except that its coefficient is computed by simply adding coefficients of both terms. For example,

$$3x^2y^3 + 5x^2y^3 = 8x^2y^3$$

but $3x^2y^3$ and $5x^2z^3$ or $3x^2y^3$ and $5x^2y^2$ cannot be conveniently added because the first pair of terms has different variables, and the variables in the second pair are raised to different powers. We would like to write a program that computes the sum of two polynomials entered by the user. For example, if

$$3x^2y^3 + 5x^2w^3 - 8x^2w^3z^4 + 3$$

and

$$-2x^2w^3 + 9y - 4xw - x^2y^3 + 8x^2w^3z^4 - 4$$

are entered, the program should output

$$-4wx + 3w^3x^2 + 2x^2y^3 + 9y - 1$$

To be more exact, the input and output for this problem should be as follows:

```
Enter two polynomials ended with a semicolon:
3x2y3 + 5x2w3 - 8x2w3z4 + 3;
- 2x2w3 + 9y - 4xw - x2y3 + 8x2w3z4 - 4;
The result is:
- 4wx + 3w3x2 + 2x2y3 + 9y - 1
```

It has to be observed that the order of variables in a term is irrelevant; for example, x^2y^3 and y^3x^2 represent exactly the same term. Therefore, before any addition is performed, the program should order all of the variables in each term to make the terms homogeneous and add them properly. Thus, there are two major tasks to be implemented: ordering variables in each term of both polynomials and adding the polynomials. But before we embark on the problem of implementing the algorithms, we have to decide how to represent polynomials in C++. Out of many possibilities, a linked list representation is chosen with each node on the list representing one term. A term contains information about a coefficient, variables, and exponents included in the term. Because each variable and its exponent belong together, they are also kept together in an object of type `Variable`. Also, because the number of variables can vary from one term to another, a vector of the `Variable` objects is used to store in one node information concerning variables in one term. A polynomial is simply a linked list of such nodes. For example,

$$-x^2y + y - 4x^2y^3 + 8x^2w^3z^4$$

is represented by the list in Figure 9.19a, which in C++ looks like the list in Figure 9.19b.

FIGURE 9.19 A linked list representation of the expression $-x^2y + y - 4x^2y^3 + 8x^2w^3z^4$.

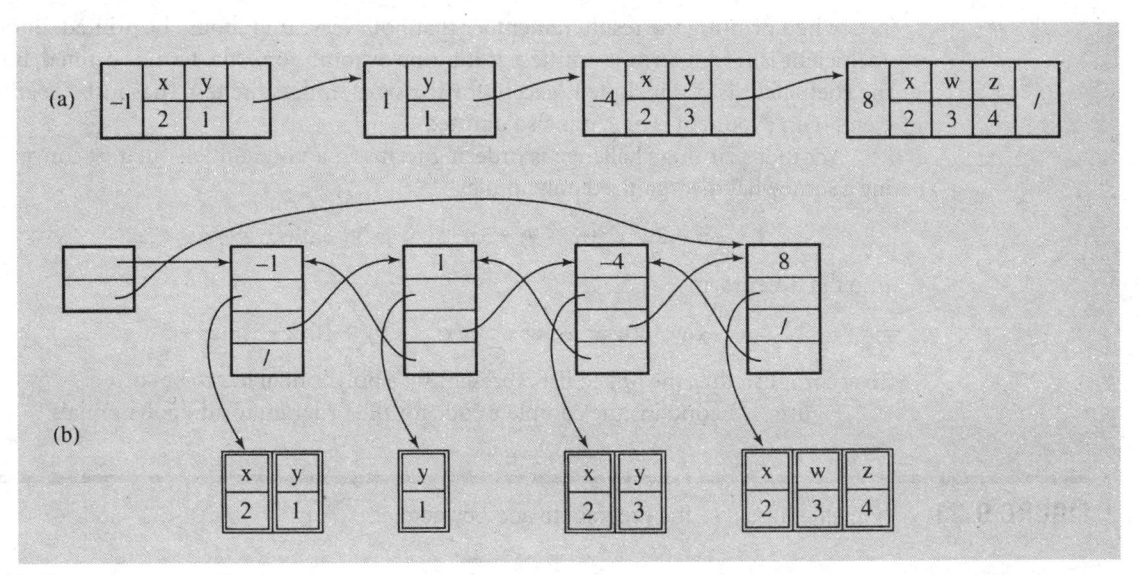

The first operation to perform on these polynomials is to order variables in their terms. After a polynomial is entered, each term is sorted separately by sorting vectors accessible from the nodes of the list.

The second task is to add the polynomials. The process begins by creating a linked list consisting of copies of nodes of the polynomials to be added. In this way, the two polynomials are not affected and can be used in other operations.

Now, addition is reduced to simplification. In the linked list, all equal terms (equal except for the coefficients) have to be collapsed together, and redundant nodes must be eliminated. For example, if the list being processed is as in Figure 9.20a, then the resulting list in Figure 9.20b results from the simplification operation.

FIGURE 9.20

Transforming (a) a list representing the expression $-x^2y^3 + 3x^2y^3 + y^2z + 2x^2y^3 - 2y^2z$ into (b) a list representing a simplified version of this expression, $4x^2y^3 - y^2z$.

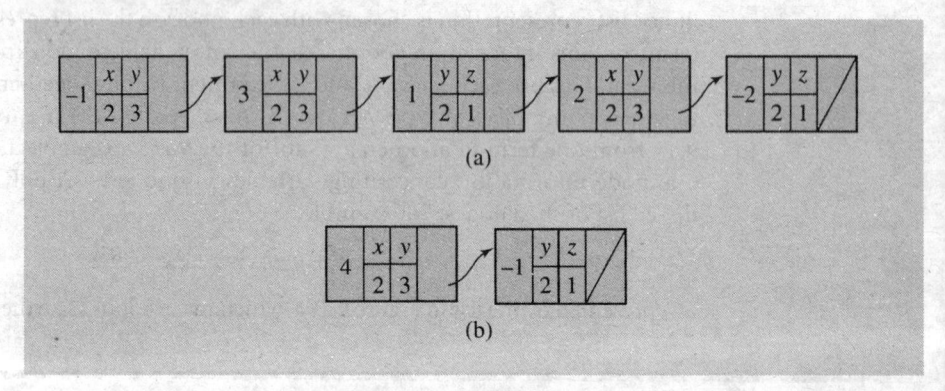

When printing the result, remember that not everything should be printed. If the coefficient is zero, a term is omitted. If it is one or minus one, the term is printed, but the coefficient is not included (except for the sign), unless the term has just a coefficient. If an exponent is one, it is also omitted.

Another printing challenge is ordering terms in a polynomial—that is, converting a somewhat disorganized polynomial

$$-z^2 - 2w^2x^3 + 5 + 9y - 5z - 4wx - x^2y^3 + 3w^2x^3z^4 + 10yz$$

into the tidier form

$$-4wx - 2w^2x^3 + 3w^2x^3z^4 - x^2y^3 + 9y + 10yz - 5z - z^2 + 5$$

To accomplish this, the linked list representing a polynomial has to be sorted.

Figure 9.21 contains the complete code for the program to add polynomials.

FIGURE 9.21

Implementation of the program to add polynomials.

```
#include <iostream>
#include <cctype>
#include <cstdlib>
#include <vector>
#include <list>
#include <algorithm>

using namespace std;

class Variable {
```

FIGURE 9.21 (continued)

```
public:
    char id;
    int exp;
    Variable() { // required by <vector>;
    }
    Variable(char c, int i) {
        id = c; exp = i;
    }
    bool operator== (const Variable& v) const {
        return id  == v.id  && exp == v.exp;
    }
    bool operator< (const Variable& v) const { // used by sort();
        return id < v.id;
    }
};

class Term {
public:
    Term() {
        coeff = 0;
    }
    int coeff;
    vector<Variable> vars;
    bool operator== (const Term&) const;
    bool operator!= (const Term& term) const { // required by <list>
        return !(*this == term);
    }
    bool operator< (const Term&) const;
    bool operator> (const Term& term) const {  // required by <list>
        return *this != term && (*this < term);
    }
    int min(int n, int m) const {
        return (n < m) ? n : m;
    }
};

class Polynomial {
public:
    Polynomial() {
    }
    Polynomial operator+ (const Polynomial&) const;
```

Continues

FIGURE 9.21   *(continued)*

```
        void error(char *s) {
            cerr << s << endl; exit(1);
        }
private:
    list<Term> terms;
    friend istream& operator>> (istream&, Polynomial&);
    friend ostream& operator<< (ostream&, const Polynomial&);
};

// two terms are equal if all variables are the same and
// corresponding variables are raised to the same powers;
// the first cell of the node containing a term is excluded
// from comparison, since it stores coefficient of the term;

bool Term::operator== (const Term& term) const {
    for (int i = 0; i < min(vars.size(),term.vars.size()) &&
                    vars[i] == term.vars[i]; i++);
    return i == vars.size() && vars.size() == term.vars.size();
}

bool Term::operator< (const Term& term2) const { // used by sort();
    if (vars.size() == 0)
        return false;            // *this is just a coefficient;
    else if (term2.vars.size() == 0)
        return true;             // term2 is just a coefficient;
    for (int i = 0; i < min(vars.size(),term2.vars.size()); i++)
        if (vars[i].id < term2.vars[i].id)
            return true;         // *this precedes term2;
        else if (term2.vars[i].id < vars[i].id)
            return false;        // term2 precedes *this;
        else if (vars[i].exp < term2.vars[i].exp)
            return true;         // *this precedes term2;
        else if (term2.vars[i].exp < vars[i].exp)
            return false;        // term2 precedes *this;
    return ((int)vars.size()-(int)term2.vars.size() < 0) ? true : false;
}

Polynomial Polynomial::operator+ (const Polynomial& polyn2) const {
    Polynomial result;
    list<Term>::iterator p1, p2;
    bool erased;
```

FIGURE 9.21   *(continued)*

```
        for (p1 = terms.begin(); p1 != terms.end(); p1++) // create a new
            result.terms.push_back(*p1);                  // polyn from
                                                          // copies of *this
        for (p1 = polyn2.terms.begin(); p1 != polyn2.terms.end(); p1++)
            result.terms.push_back(*p1);                  // and polyn2;
        for (p1 = result.terms.begin(); p1 != result.terms.end(); ) {
            for (p2 = p1, p2++, erased = false; p2 != result.terms.end();
                 p2++)
                if (*p1 == *p2) {          // if two terms are equal (except
                    p1->coeff += p2->coeff; // for the coefficient), add the
                    result.terms.erase(p2); // two coefficients and erase
                    if (p1->coeff == 0)    // a redundant term; if the
                        result.terms.erase(p1); // coefficient in retained
                    erased = true;         // term is zero, break;
                                           // erase the term as well;

                }
            if (erased)         // restart processing from the beginning
                p1 = result.terms.begin();  // if any node was erased;
            else p1++;
        }
        result.terms.sort();
        return result;
    }

    istream& operator>> (istream& in, Polynomial& polyn) {
        char ch, sign, coeffUsed, id;
        int exp;
        Term term;
        in >> ch;
        while (true) {
            coeffUsed = 0;
            if (!isalnum(ch) && ch != ';' && ch != '-' && ch != '+')
                polyn.error("Wrong character entered2");
            sign = 1;
            while (ch == '-' || ch == '+') { // first get sign(s) of Term
                if (ch == '-')
                    sign *= -1;
                ch = in.get();
                if (isspace(ch))
                    in >> ch;
```

FIGURE 9.21 *(continued)*

```
        }
        if (isdigit(ch)) {                  // and then its coefficient;
            in.putback(ch);
            in >> term.coeff;
            ch = in.get();
            term.coeff *= sign;
            coeffUsed = 1;
        }
        else term.coeff = sign;
        for (int i = 0; isalnum(ch); i++) { // process this term:
            id = ch;                        // get a variable name
            ch = in.get();
            if (isdigit(ch)) {              // and an exponent (if any);
                in.putback(ch);
                in >> exp >> ch;
            }
            else exp = 1;
            term.vars.push_back(Variable(id,exp));
        }
        polyn.terms.push_back(term);// and include it in the linked list;
        term.vars.resize(0);
        if (isspace(ch))
            in >> ch;
        if (ch == ';')                      // finish if a semicolon is entered;
            if (coeffUsed || i > 0)
                break;
            else polyn.error("Term is missing");  // e.g., 2x - ; or
                                                  // just ';'
        else if (ch != '-' && ch != '+')          // e.g., 2x  4y;
            polyn.error("wrong character entered");
    }
    for (list<Term>::iterator i = polyn.terms.begin();
                              i != polyn.terms.end(); i++)
        if (i->vars.size() > 1)
            sort(i->vars.begin(),i->vars.end());
    return in;
}

ostream& operator<< (ostream& out, const Polynomial& polyn) {
    int afterFirstTerm = 0, i;
```

FIGURE 9.21 *(continued)*

```
    for (list<Term>::iterator pol = polyn.terms.begin();
                          pol != polyn.terms.end(); pol++) {
        out.put(' ');
        if (pol->coeff < 0)                 // put '-' before polynomial
            out.put('-');                   // and between terms (if
                                            // needed);
        else if (afterFirstTerm)            // don't put '+' in front of
            out.put('+');                   // polynomial;
        afterFirstTerm++;
        if (abs(pol->coeff) != 1)           // print a coefficient
            out << ' ' << abs(pol->coeff);  // if it is not 1 nor -1, òr
        else if (pol->vars.size() == 0)     // the term has only a
                                            // coefficient
            out << " 1";
        else out.put(' ');
        for (i = 1; i <= pol->vars.size(); i++) {
            out << pol->vars[i-1].id;       // print a variable name
            if (pol->vars[i-1].exp != 1)    // and an exponent, only
                out << pol->vars[i-1].exp;  // if it is not 1;
        }
    }
    out << endl;
    return out;
}

int main() {
    Polynomial polyn1, polyn2;
    cout << "Enter two polynomials, each ended with a semicolon:\n";
    cin  >> polyn1 >> polyn2;
    cout << "The result is:\n" << polyn1 + polyn2;
    return 0;
}
```

9.7 EXERCISES

1. Many operations can be performed faster on sorted than on unsorted data. For which of the following operations is this the case?
   a. checking whether one word is an anagram of another word, e.g., *plum* and *lump*
   b. finding an item with a minimum value
   c. computing an average of values
   d. finding the middle value (the median)
   e. finding the value that appears most frequently in the data

2. The function `bubblesort()` is inefficient because it continues execution after an array is sorted by performing unnecessary comparisons. Therefore, the number of comparisons in the best and worst cases is the same. The implementation can be improved by making a provision for the case when the array is already sorted. Modify `bubblesort()` by adding a flag to the outer `for` loop indicating whether it is necessary to make the next pass. Set the flag to true every time an interchange occurs, which indicates that there is a need to scan the array again.

3. Will `bubblesort()` work properly if the inner loop

   ```
   for (int j = n-1; j > i; --j)
   ```

   is replaced by

   ```
   for (int j = n-1; j > 0; --j)
   ```

   What is the complexity of the new version?

4. In our implementation of bubble sort, a sorted array was scanned bottom-up to bubble up the smallest element. What modifications are needed to make it work top-down to bubble down the largest element?

5. A *cocktail shaker sort* designed by Donald Knuth is a modification of bubble sort in which the direction of bubbling changes in each iteration: In one iteration, the smallest element is bubbled up; in the next, the largest is bubbled down; in the next, the second smallest is bubbled up; and so forth. Implement this new algorithm and explore its complexity.

6. Insertion sort goes sequentially through the array when making comparisons to find a proper place for an element currently processed. Consider using binary search instead and give a complexity of the resulting insertion sort.

7. Draw decision trees for all the elementary sorting algorithms as applied to the array [a b c d].

8. Which of the algorithms discussed in this chapter is easily adaptable to singly linked lists? To doubly linked lists?

9. What exactly are the smallest and largest numbers of movements and comparisons to sort four elements using `heapsort()`, `quicksort()`, and `mergesort()`?

10. Implement and test `mergesort()`.

11. Show that for mergesort the number of comparisons $C(n) = n \lg n - 2^{\lg n} + 1$.

12. Implement and analyze the complexity of the following nonrecursive version of mergesort. First, merge subarrays of length 1 into $\frac{n}{2}$ two-cell subarrays, possibly one

of them being a one-cell array. The resulting arrays are then merged into $\frac{n}{4}$ four-cell subarrays possibly, with one smaller array, having one, two, or three cells, and so on, until the entire array is ordered. Note that this is a bottom-up approach to the mergesort implementation, as opposed to the top-down approach discussed in this chapter.

13. `mergesort()` merges the subarrays of an array that is already in order. Another top-down version of mergesort alleviates this problem by merging only *runs*, subarrays with ordered elements. Merging is applied only after two runs are determined. For example, in the array [6 7 8 3 4 10 11 12 13 2], runs [6 7 8] and [3 4] are first merged to become [3 4 6 7 8], then runs [10 11 12 13] and [2] are merged to become [2 10 11 12 13], and finally, runs [3 4 6 7 8] and [2 10 11 12 13] are merged to become [2 3 4 6 7 8 10 11 12 13]. Implement this algorithm and investigate its complexity. A mergesort that takes advantage of a partial ordering of data (that is, uses the runs) is called a *natural sort*. A version that disregards the runs by always dividing arrays into (almost) even sections is referred to as *straight merging*.

14. To avoid doubling the workspace needed when arrays are sorted with mergesort, it may be better to use a linked list of data instead of an array. In what situations is this approach better? Implement this technique and discuss its complexity.

15. Which sorting algorithms are stable?

16. Consider a *slow sorting* algorithm, which applies selection sort to every *i*th element of an *n*-element array, where *i* takes on values $n/2, n/3, \ldots, n/n$ (Julstrom 1992). First, selection sort is applied to two elements of the array, the first and the middle elements, then to three elements, separated by the distance $n/3$, and so on, and finally to every element. Compute the complexity of this algorithm.

9.8 PROGRAMMING ASSIGNMENTS

1. At the end of the section discussing quicksort, two techniques for choosing the bound were mentioned—using a randomly chosen element from the file and using a median element of the first, middle, and last elements of the array. Implement these two versions of quicksort, apply them to large arrays, and compare their run times.

2. The board of education maintains a database of substitute teachers in the area. If a temporary replacement is necessary for a certain subject, an available teacher is sent to the school that requests him or her. Write a menu-driven program maintaining this database.
    The file `substitutes` lists the first and last names of the substitute teachers, an indication of whether or not they are currently available (Y(es) or N(o)), and a list of numbers that represents the subjects they can teach. An example of `substitutes` is:

```
Hilliard Roy         Y 0 4 5
Ennis John           N 2 3
Nelson William       Y 1 2 4 5
Baird Lyle           Y 1 3 4 5
Geiger Melissa       N 3 5
Kessel Warren        Y 3 4 5
Scherer Vincent      Y 4 5
Hester Gary          N 0 1 2 4
Linke Jerome         Y 0 1
Thornton Richard     N 2 3 5
```

Create a `class teacher` with three data members: index, left child, and right child. Declare an array `subjects` with the same number of cells as the number of subjects, each cell storing a pointer to `class Teacher`, which is really a pointer to the root of a binary search tree of teachers teaching a given subject. Also, declare an array `names` with each cell holding one entry from the file.

First, prepare the array `names` to create binary search trees. To do that, load all entries from `substitutes` to `names` and sort `names` using one of the algorithms discussed in this chapter. Afterward, create a binary tree using the function `balance()` from Section 6.7: Go through the array `names`, and for each subject associated with each name, create a node in the tree corresponding to this subject. The index field of such a node indicates a location in `names` of a teacher teaching this subject. (Note that `insert()` in `balance()` should be able to go to a proper tree.) Figure 9.22 shows the ordered array `names` and trees accessible from `subjects` as created by `balance()` for our sample file.

FIGURE 9.22 Data structures used by the board of education for substitute teachers.

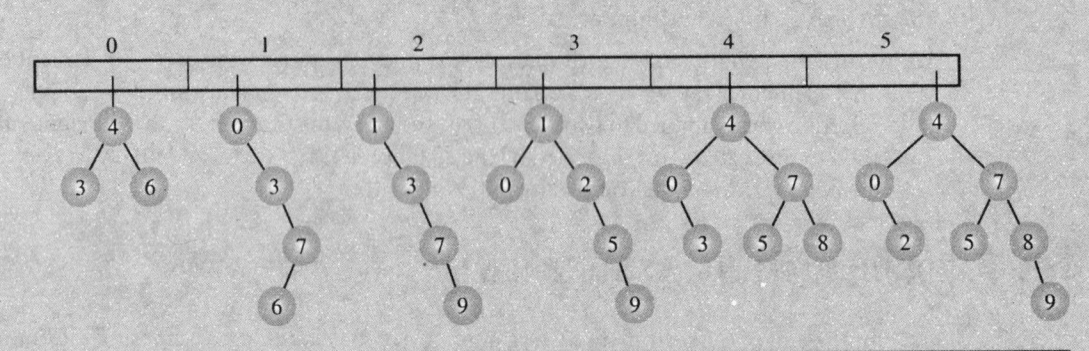

Baird	Ennis	Geiger	Hester	Hilliard	Kessel	Linke	Nelson	Scherer	Thornton
Lyle	John	Melissa	Gary	Roy	Warren	Jerome	William	Vincent	Richard
Y	N	N	N	Y	Y	Y	Y	Y	N
1 3 4 5	2 3	3 5	0 1 2 4	0 4 5	3 4 5	0 1	1 2 4 5	4 5	2 3 5

Allow the user to reserve teachers if so requested. If the program is finished and an exit option is chosen, load all entries from `names` back to `substitutes`, this time with updated availability information.

3. Implement different versions of Shell sort by mixing simple sorts used in h-sorts, 1-sort, and different sequences of increments. Run each version with (at least) any of the following sequences:

   a. $h_1 = 1, h_{i+1} = 3h_i + 1$ and stop with h_t for which $h_{t+2} \geq n$ (Knuth)
   b. $2^k - 1$ (Hibbard)

c. $2^k + 1$ (Papernov and Stasevich)
   d. Fibonacci numbers
   e. $\frac{n}{2}$ is the first increment and then $h_i = .75h_{i+1}$ (Dobosiewicz)

   Run all these versions for at least five sets of data of sizes 1,000, 5,000, 10,000, 50,000, and 100,000. Tabulate and plot the results and try to approximate them with some formula expressing the complexity of these versions.

4. Extend the program from the case study to include polynomial multiplication.

5. Extend the program from the case study to include polynomial differentiation. For the rules, see the exercises in Chapter 5.

BIBLIOGRAPHY

Sorting Algorithms

Flores, Ivan, *Computer Sorting*, Englewood Cliffs, NJ: Prentice Hall, 1969.

Knuth, Donald E., *The Art of Computer Programming, Vol. 3: Sorting and Searching*, Reading, MA: Addison-Wesley, 1998.

Lorin, Harold, *Sorting and Sort Systems*, Reading, MA: Addison-Wesley, 1975.

McLuckie, Keith, and Barber, Angus, *Sorting Routines for Microcomputers*, Basingstoke: United Kingdom Macmillan, 1986.

Mehlhorn, Kurt, *Data Structures and Algorithms, Vol. 1: Sorting and Searching*, Berlin: Springer, 1984.

Reynolds, Carl W., "Sorts of Sorts," *Computer Language* (March 1988), 49–62.

Rich, R., *Internal Sorting Methods Illustrated with PL/1 Programs*, Englewood Cliffs, NJ: Prentice Hall, 1972.

Shell Sort

Dobosiewicz, W., "An Efficient Variation of Bubble Sort," *Information Processing Letters* 11 (1980), 5–6.

Gale, David, and Karp, Richard M., "A Phenomenon in the Theory of Sorting," *Journal of Computer and System Sciences* 6 (1972), 103–115.

Incerpi, Janet, and Sedgewick, Robert, "Practical Variations of Shellsort," *Information Processing Letters* 26 (1987/88), 37–43.

Poonen, Bjorn, "The Worst Case in Shellsort and Related Algorithms," *Journal of Algorithms* 15 (1993), 101–124.

Pratt, Vaughan R., *Shellsort and Sorting Networks*, New York: Garland, 1979.

Shell, Donald L., "A High-Speed Sorting Procedure," *Communications of the ACM* 2 (1959), 30–32.

Weiss, Mark A., and Sedgewick, Robert, "Tight Lower Bounds for Shellsort," *Journal of Algorithms* 11 (1990), 242–251.

Heap Sort

Carlsson, Svente, "Average-Case Results on Heapsort," *BIT* 27 (1987), 2–17.

Ding, Yuzheng, and Weiss, Mark A., "Best Case Lower Bounds for Heapsort," *Computing* 49 (1992), 1–9.

Wegener, Ingo, "Bottom-up-Heap Sort, A New Variant of Heap Sort Beating on Average quick Sort," in Rovan, B. (ed.), *Mathematical Foundations of Computer Science,* Berlin: Springer (1990), 516–522.

Williams, John W. J., "Algorithm 232: Heapsort," *Communications of the ACM* 7 (1964), 347–348.

Quicksort

Cook, Curtis R., and Kim, Do Jin, "Best Sorting Algorithm for Nearly Sorted Lists," *Communications of the ACM* 23 (1980), 620–624.

Dromey, R. G., "Exploiting Partial Order with Quicksort," *Software Practice and Experience* 14 (1984), 509–518.

Frazer, William D., and McKellar, Archie C., "Samplesort: A Sampling Approach to Minimal Storage Tree Sorting," *Journal of the ACM* 17 (1970), 496–507.

Hoare, Charles A. R., "Algorithm 63: Quicksort," *Communications of the ACM* 4 (1961), 321.

Hoare, Charles A. R., "Quicksort," *Computer Journal* 2 (1962), 10–15.

Huang, B. C., and Knuth, Donald, "A One-Way, Stackless Quicksort Algorithm," *BIT* 26 (1986), 127–130.

Motzkin, Dalia, "Meansort," *Communications of the ACM* 26 (1983), 250–251.

Sedgewick, Robert, *Quicksort,* New York: Garland, 1980.

Mergesort

Dvorak, S., and Durian, B., "Unstable Linear Time $O(1)$ Space Merging," *The Computer Journal* 31 (1988), 279–283.

Huang, B. C., and Langston, M. A., "Practical In-Place Merging," *Communications of the ACM* 31 (1988), 348–352.

Knuth, Donald, "Von Neumann's First Computer Program," *Computing Surveys* 2 (1970), 247–260.

Slow Sorting

Julstrom, A., "Slow Sorting: A Whimsical Inquiry," *SIGCSE Bulletin* 24 (1992), No. 3, 11–13.

Decision Trees

Moret, B. M. E., "Decision Trees and Algorithms," *Computing Surveys* 14 (1982), 593–623.

Hashing 10

The main operation used by the searching methods described in the preceding chapters was comparison of keys. In a sequential search, the table that stores the elements is searched successively to determine which cell of the table to check, and the key comparison determines whether an element has been found. In a binary search, the table that stores the elements is divided successively into halves to determine which cell of the table to check, and again, the key comparison determines whether an element has been found. Similarly, the decision to continue the search in a binary search tree in a particular direction is accomplished by comparing keys.

A different approach to searching calculates the position of the key in the table based on the value of the key. The value of the key is the only indication of the position. When the key is known, the position in the table can be accessed directly, without making any other preliminary tests, as required in a binary search or when searching a tree. This means that the search time is reduced from $O(n)$, as in a sequential search, or from $O(\lg n)$, as in a binary search, to 1 or at least $O(1)$; regardless of the number of elements being searched, the run time is always the same. But this is just an ideal, and in real applications, this ideal can only be approximated.

We need to find a function h that can transform a particular key K, be it a string, number, record, or the like, into an index in the table used for storing items of the same type as K. The function h is called a *hash function*. If h transforms different keys into different numbers, it is called a *perfect hash function*. To create a perfect hash function, which is always the goal, the table has to contain at least the same number of positions as the number of elements being hashed. But the number of elements is not always known ahead of time. For example, a compiler keeps all variables used in a program in a symbol table. Real programs use only a fraction of the vast number of possible variable names, so a table size of 1,000 cells is usually adequate.

But even if this table can accommodate all the variables in the program, how can we design a function h that allows the compiler to immediately access the position associated with each variable? All the letters of the variable name can be added together and the sum can be used as an index. In this case, the table needs 3,782 cells (for a variable K made out of 31 letters "z," $h(K) = 31 \cdot 122 = 3{,}782$). But even with this size, the function h does not return unique values. For example, $h(\text{"abc"}) = h(\text{"acb"})$. This problem is called *collision*, and is discussed later. The worth of a hash function depends on how well it avoids collisions. Avoiding collisions can be achieved by making the function more sophisticated, but this sophistication should not go too far because the computational cost in determining $h(K)$ can be prohibitive, and less sophisticated methods may be faster.

10.1 HASH FUNCTIONS

The number of hash functions that can be used to assign positions to n items in a table of m positions (for $n \leq m$) is equal to m^n. The number of perfect hash functions is the same as the number of different placements of these items in the table and is equal to $\frac{m!}{(m-n)!}$. For example, for 50 elements and a 100-cell array, there are $100^{50} = 10^{100}$ hash functions, out of which "only" 10^{94} (one in a million) are perfect. Most of these functions are too unwieldy for practical applications and cannot be represented by a concise formula. However, even among functions that can be expressed with a formula, the number of possibilities is vast. This section discusses some specific types of hash functions.

10.1.1 Division

A hash function must guarantee that the number it returns is a valid index to one of the table cells. The simplest way to accomplish this is to use division modulo *TSize* = *sizeof*(*table*), as in $h(K) = K$ mod *TSize*, if K is a number. It is best if *TSize* is a prime number; otherwise, $h(K) = (K$ mod $p)$ mod *TSize* for some prime $p >$ *TSize* can be used. However, nonprime divisors may work equally well as prime divisors provided they do not have prime factors less than 20 (Lum et al. 1971). The division method is usually the preferred choice for the hash function if very little is known about the keys.

10.1.2 Folding

In this method, the key is divided into several parts (which conveys the true meaning of the word *hash*). These parts are combined or folded together and are often transformed in a certain way to create the target address. There are two types of folding: *shift folding* and *boundary folding*.

The key is divided into several parts and these parts are then processed using a simple operation such as addition to combine them in a certain way. In shift folding, they are put underneath one another and then processed. For example, a social security number (SSN) 123-45-6789 can be divided into three parts, 123, 456, 789, and then these parts can be added. The resulting number, 1,368, can be divided modulo *TSize* or, if the size of the table is 1,000, the first three digits can be used for the address. To be sure, the division can be done in many different ways. Another possibility is to divide the same number 123-45-6789 into five parts (say, 12, 34, 56, 78, and 9), add them, and divide the result modulo *TSize*.

With boundary folding, the key is seen as being written on a piece of paper that is folded on the borders between different parts of the key. In this way, every other part will be put in the reverse order. Consider the same three parts of the SSN: 123, 456, and 789. The first part, 123, is taken in the same order, then the piece of paper with the second part is folded underneath it so that 123 is aligned with 654, which is the second part, 456, in reverse order. When the folding continues, 789 is aligned with the two previous parts. The result is 123 + 654 + 789 = 1,566.

In both versions, the key is usually divided into even parts of some fixed size plus some remainder and then added. This process is simple and fast, especially when bit patterns are used instead of numerical values. A bit-oriented version of shift folding is obtained by applying the exclusive-or operation, ^.

In the case of strings, one approach processes all characters of the string by "xor'ing" them together and using the result for the address. For example, for the string "abcd," $h(\text{"abcd"}) = \text{"a"}\char`^\text{"b"}\char`^\text{"c"}\char`^\text{"d."}$ However, this simple method results in addresses between the numbers 0 and 127. For better result, chunks of strings are "xor'ed" together rather than single characters. These chunks are composed of the number of characters equal to the number of bytes in an integer. An integer in a C++ implementation for the IBM PC computer is 2 bytes long, $h(\text{"abcd"}) = \text{"ab"} \cdot \text{xor} \text{ "cd"}$ (most likely divided modulo $TSize$). Such a function is used in the case study in this chapter.

10.1.3 Mid-Square Function

In the mid-square method, the key is *squared* and the middle or *mid* part of the result is used as the address. If the key is a string, it has to be preprocessed to produce a number by using, for instance, folding. In a mid-square hash function, the entire key participates in generating the address so that there is a better chance that different addresses are generated for different keys. For example, if the key is 3,121, then $3,121^2$ = 9,740,641, and for the 1,000-cell table, $h(3,121) = 406$, which is the middle part of $3,121^2$. In practice, it is more efficient to choose a power of 2 for the size of the table and extract the middle part of the bit representation of the square of a key. If we assume that the size of the table is 1,024, then, in this example, the binary representation of $3,121^2$ is the bit string 100101001010000101100001, with the middle part shown in italics. This middle part, the binary number 0101000010, is equal to 322. This part can easily be extracted by using a mask and a shift operation.

10.1.4 Extraction

In the extraction method, only a part of the key is used to compute the address. For the social security number 123-45-6789, this method might use the first four digits, 1,234; the last four, 6,789; the first two combined with the last two, 1,289; or some other combination. Each time, only a portion of the key is used. If this portion is carefully chosen, it can be sufficient for hashing, provided the omitted portion distinguishes the keys only in an insignificant way. For example, in some university settings, all international students' ID numbers start with 999. Therefore, the first three digits can be safely omitted in a hash function that uses student IDs for computing table positions. Similarly, the starting digits of the ISBN code are the same for all books published by the same publisher (e.g., 0534 for Brooks/Cole Publishing Company). Therefore, they should be excluded from the computation of addresses if a data table contains only books from one publisher.

10.1.5 Radix Transformation

Using the radix transformation, the key K is transformed into another number base; K is expressed in a numerical system using a different radix. If K is the decimal number 345, then its value in base 9 (nonal) is 423. This value is then divided modulo *TSize*, and the resulting number is used as the address of the location to which K should be hashed. Collisions, however, cannot be avoided. If *TSize* = 100, then although 345 and 245 (decimal) are not hashed to the same location, 345 and 264 are because 264 decimal is 323 in the nonal system, and both 423 and 323 return 23 when divided modulo 100.

10.2 COLLISION RESOLUTION

Note that straightforward hashing is not without its problems, because for almost all hash functions, more than one key can be assigned to the same position. For example, if the hash function h_1 applied to names returns the ASCII value of the first letter of each name (i.e., $h_1(name) = name[0]$), then all names starting with the same letter are hashed to the same position. This problem can be solved by finding a function that distributes names more uniformly in the table. For example, the function h_2 could add the first two letters (i.e., $h_2(name) = name[0] + name[1]$), which is better than h_1. But even if all the letters are considered (i.e., $h_3(name) = name[0] + \cdots + name[\mathtt{strlen}(name) - 1]$), the possibility of hashing different names to the same location still exists. The function h_3 is the best of the three because it distributes the names most uniformly for the three defined functions, but it also tacitly assumes that the size of the table has been increased. If the table has only 26 positions, which is the number of different values returned by h_1, there is no improvement using h_3 instead of h_1. Therefore, one more factor can contribute to avoiding conflicts between hashed keys, namely, the size of the table. Increasing this size may lead to better hashing, but not always! These two factors—hash function and table size—may minimize the number of collisions, but they cannot completely eliminate them. The problem of collision has to be dealt with in a way that always guarantees a solution.

There are scores of strategies that attempt to avoid hashing multiple keys to the same location. Only a handful of these methods are discussed in this chapter.

10.2.1 Open Addressing

In the open addressing method, when a key collides with another key, the collision is resolved by finding an available table entry other than the position (address) to which the colliding key is originally hashed. If position $h(K)$ is occupied, then the positions in the probing sequence

$$norm(h(K) + p(1)), norm(h(K) + p(2)), \ldots, norm(h(K) + p(i)), \ldots$$

are tried until either an available cell is found or the same positions are tried repeatedly or the table is full. Function p is a *probing function*, i is a *probe*, and *norm* is a *normalization function*, most likely, division modulo the size of the table.

The simplest method is *linear probing*, for which $p(i) = i$, and for the ith probe, the position to be tried is $(h(K) + i)$ mod *TSize*. In linear probing, the position in which a key can be stored is found by sequentially searching all positions starting from the position calculated by the hash function until an empty cell is found. If the end of the table is reached and no empty cell has been found, the search is continued from the beginning of the table and stops—in the extreme case—in the cell preceding the one from which the search started. Linear probing, however, has a tendency to create clusters in the table. Figure 10.1 contains an example where a key K_i is hashed to the position i. In Figure 10.1a, three keys—$A_5, A_2,$ and A_3—have been hashed to their home positions. Then B_5 arrives (Figure 10.1b), whose home position is occupied by A_5. Because the next position is available, B_5 is stored there. Next, A_9 is stored with no problem, but B_2 is stored in position 4, two positions from its home address. A large cluster has already been formed. Next, B_9 arrives. Position 9 is not available, and because it is the last cell of the table, the search starts from the beginning of the table, whose first slot can now host B_9. The next key, C_2, ends up in position 7, five positions from its home address.

FIGURE 10.1 Resolving collisions with the linear probing method. Subscripts indicate the home positions of the keys being hashed.

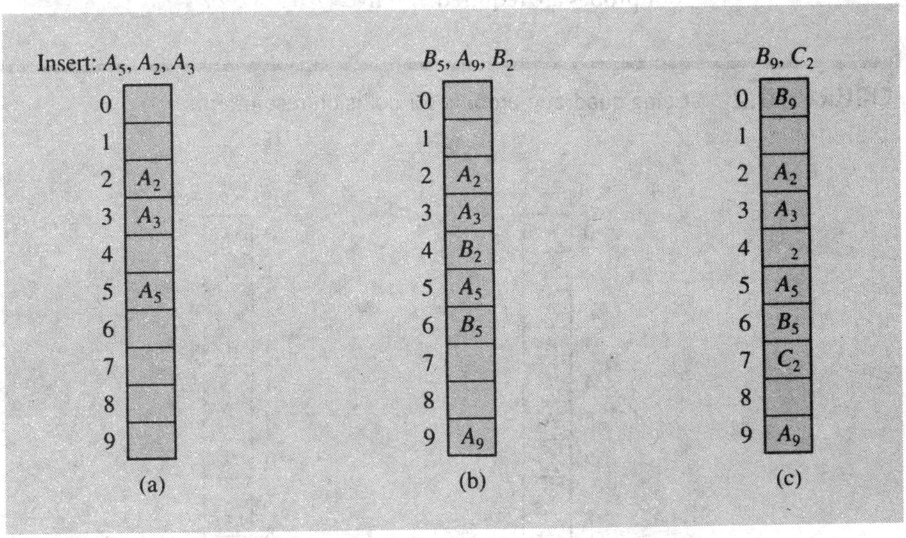

In this example, the empty cells following clusters have a much greater chance to be filled than other positions. This probability is equal to $(sizeof(cluster) + 1)/TSize$. Other empty cells have only $1/TSize$ chance of being filled. If a cluster is created, it has a tendency to grow, and the larger a cluster becomes, the larger is the likelihood that it will become even larger. This fact undermines the performance of the hash table for storing and retrieving data. The problem at hand is how to avoid cluster buildup. An answer can be found in a more careful choice of the probing function p.

One such choice is a quadratic function so that the resulting formula is

$$p(i) = h(K) + (-1)^{i-1}((i+1)/2)^2 \text{ for } i = 1, 2, \ldots, TSize - 1$$

This rather cumbersome formula can be expressed in a simpler form as a sequence of probes:

$$h(K) + i^2, h(K) - i^2 \text{ for } i = 1, 2, \ldots, (TSize - 1)/2$$

Including the first attempt to hash K, this results in the sequence:

$$h(K), h(K) + 1, h(K) - 1, h(K) + 4, h(K) - 4, \ldots, h(K) + (TSize - 1)^2/4,$$
$$h(K) - (TSize - 1)^2/4$$

all divided modulo $TSize$. The size of the table should not be an even number, because only the even positions or only the odd positions are tried depending on the value of $h(K)$. Ideally, the table size should be a prime $4j + 3$ of an integer j, which guarantees the inclusion of all positions in the probing sequence (Radke 1970). For example, if $j = 4$, then $TSize = 19$, and assuming that $h(K) = 9$ for some K, the resulting sequence of probes is[1]

$$9, 10, 8, 13, 5, 18, 0, 6, 12, 15, 3, 7, 11, 1, 17, 16, 2, 14, 4$$

The table from Figure 10.1 would have the same keys in a different configuration, as in Figure 10.2. It still takes two probes to locate B_2 in some location, but for C_2, only four probes are required, not five.

FIGURE 10.2 Using quadratic probing for collision resolution.

	Insert: A_5, A_2, A_3		B_5, A_9, B_2		B_9, C_2
0		0		0	B_9
1		1	B_2	1	B_2
2	A_2	2	A_2	2	A_2
3	A_3	3	A_3	3	A_3
4		4		4	
5	A_5	5	A_5	5	A_5
6		6	B_5	6	B_5
7		7		7	
8		8		8	C_2
9		9	A_9	9	A_9
	(a)		(b)		(c)

---

[1] Special care should be taken for negative numbers. When implementing these formulas, the operator % means division modulo a modulus. However, this operator is usually implemented as the *remainder* of division. For example, –6 % 23 is equal to –6, and not to 17, as expected. Therefore, when using the operator % for the implementation of division modulo, the modulus (the right operand of %) should be added to the result when the result is negative. Therefore, (–6 % 23) + 23 returns 17.

Note that the formula determining the sequence of probes chosen for quadratic probing is not $h(K) + i^2$, for $i = 1, 2, \ldots, TSize - 1$, because the first half of the sequence

$$h(K) + 1, h(K) + 4, h(K) + 9, \ldots, h(K) + (TSize - 1)^2$$

covers only half of the table, and the second half of the sequence repeats the first half in the reverse order. For example, if $TSize = 19$, and $h(K) = 9$, then the sequence is

$$9, 10, 13, 18, 6, 15, 7, 1, 16, 14, 14, 16, 1, 7, 15, 6, 18, 13, 10$$

This is not an accident. The probes that render the same address are of the form

$$i = TSize/2 + 1 \text{ and } j = TSize/2 - 1$$

and they are probes for which

$$i^2 \bmod TSize = j^2 \bmod TSize$$

that is,

$$(i^2 - j^2) \bmod TSize$$

In this case,

$$(i^2 - j^2) = (TSize/2 + 1)^2 - (TSize/2 - 1)^2$$
$$= (TSize^2/4 + TSize + 1 - TSize^2/4 + TSize - 1)$$
$$= 2TSize$$

and to be sure, $2TSize \bmod TSize = 0$.

Although using quadratic probing gives much better results than linear probing, the problem of cluster buildup is not avoided altogether, because for keys hashed to the same location, the same probe sequence is used. Such clusters are called *secondary clusters*. These secondary clusters, however, are less harmful than primary clusters.

Another possibility is to have p be a random number generator (Morris 1968), which eliminates the need to take special care about the table size. This approach prevents the formation of secondary clusters, but it causes a problem with repeating the same probing sequence for the same keys. If the random number generator is initialized at the first invocation, then different probing sequences are generated for the same key K. Consequently, K is hashed more than once to the table, and even then it might not be found when searched. Therefore, the random number generator should be initialized to the same seed for the same key before beginning the generation of the probing sequence. This can be achieved in C++ by using the `srand()` function with a parameter that depends on the key; for example, $p(i) = \text{srand}(sizeof(K)) \cdot i$ or $\text{srand}(K[0]) + i$. To avoid relying on `srand()`, a random number generator can be written that assures that each invocation generates a unique number between 0 and $TSize - 1$. The following algorithm was developed by Robert Morris for tables with $TSize = 2^n$ for some integer n:

```
generateNumber()
    static int r = 1;
    r = 5*r;
    r = mask out n + 2 low-order bits of r;
    return r/4;
```

The problem of secondary clustering is best addressed with *double hashing*. This method utilizes two hash functions, one for accessing the primary position of a key, h, and a second function, h_p, for resolving conflicts. The probing sequence becomes

$$h(K), h(K) + h_p(K), \ldots, h(K) + i \cdot h_p(K), \ldots$$

(all divided modulo *TSize*). The table size should be a prime number so that each position in the table can be included in the sequence. Experiments indicate that secondary clustering is generally eliminated because the sequence depends on the values of h_p, which, in turn, depend on the key. Therefore, if the key K_1 is hashed to the position j, the probing sequence is

$$j, j + h_p(K_1), j + 2 \cdot h_p(K_1), \ldots$$

(all divided modulo *TSize*). If another key K_2 is hashed to $j + h_p(K_1)$, then the next position tried is $j + h_p(K_1) + h_p(K_2)$, not $j + 2 \cdot h_p(K_1)$, which avoids secondary clustering if h_p is carefully chosen. Also, even if K_1 and K_2 are hashed primarily to the same position j, the probing sequences can be different for each. This, however, depends on the choice of the second hash function, h_p, which may render the same sequences for both keys. This is the case for function $h_p(K) = \texttt{strlen}(K)$, when both keys are of the same length.

Using two hash functions can be time-consuming, especially for sophisticated functions. Therefore, the second hash function can be defined in terms of the first, as in $h_p(K) = i \cdot h(K) + 1$. The probing sequence for K_1 is

$$j, 2j + 1, 5j + 2, \ldots$$

(modulo *TSize*). If K_2 is hashed to $2j + 1$, then the probing sequence for K_2 is

$$2j + 1, 4j + 3, 10j + 11, \ldots$$

which does not conflict with the former sequence. Thus, it does not lead to cluster buildup.

How efficient are all these methods? Obviously, it depends on the size of the table and on the number of elements already in the table. The inefficiency of these methods is especially evident for *unsuccessful searches*, searching for elements not in the table. The more elements in the table, the more likely it is that clusters will form (primary or secondary) and the more likely it is that these clusters are large.

Consider the case when linear probing is used for collision resolution. If K is not in the table, then starting from the position $h(K)$, all consecutively occupied cells are checked; the longer the cluster, the longer it takes to determine that K, in fact, is not in the table. In the extreme case, when the table is full, we have to check all the cells starting from $h(K)$ and ending with $(h(K) - 1)$ mod *TSize*. Therefore, the search time increases with the number of elements in the table.

There are formulas that approximate the number of times for successful and unsuccessful searches for different hashing methods. These formulas were developed by Donald Knuth and are considered by Thomas Standish to be "among the prettiest in computer science." Figure 10.3 contains these formulas. Figure 10.4 contains a table showing the number of searches for different percentages of occupied cells. This table indicates that the formulas from Figure 10.3 provide only approximations of the number of searches. This is particularly evident for the higher percentages. For example, if 90 percent of the cells are occupied, then linear probing requires 50 trials to determine that the key being searched for is not in the table. However, for the full table of 10 cells, this number is 10, not 50.

FIGURE 10.3 Formulas approximating, for different hashing methods, the average numbers of trials for successful and unsuccessful searches (Knuth, 1998).

	linear probing	quadratic probing[a]	double hashing
successful search	$\frac{1}{2}\left(1 + \frac{1}{1-LF}\right)$	$1 - \ln(1-LF) - \frac{LF}{2}$	$\frac{1}{LF} \ln \frac{1}{1-LF}$
unsuccessful search	$\frac{1}{2}\left(1 + \frac{1}{(1-LF)^2}\right)$	$\frac{1}{1-LF} - LF - \ln(1-LF)$	$\frac{1}{1-LF}$

Loading factor $LF = \frac{\text{number of elements in the table}}{\text{table size}}$

[a] the formulas given in this column approximate any open addressing method which causes secondary clusters to arise, and quadratic probing is only one of them.

FIGURE 10.4 The average numbers of successful searches and unsuccessful searches for different collision resolution methods.

LF	Linear Probing		Quadratic Probing		Double Hashing	
	Successful	Unsuccessful	Successful	Unsuccessful	Successful	Unsuccessful
0.05	1.0	1.1	1.0	1.1	1.0	1.1
0.10	1.1	1.1	1.1	1.1	1.1	1.1
0.15	1.1	1.2	1.1	1.2	1.1	1.2
0.20	1.1	1.3	1.1	1.3	1.1	1.2
0.25	1.2	1.4	1.2	1.4	1.2	1.3
0.30	1.2	1.5	1.2	1.5	1.2	1.4
0.35	1.3	1.7	1.3	1.6	1.2	1.5
0.40	1.3	1.9	1.3	1.8	1.3	1.7
0.45	1.4	2.2	1.4	2.0	1.3	1.8
0.50	1.5	2.5	1.4	2.2	1.4	2.0
0.55	1.6	3.0	1.5	2.5	1.5	2.2
0.60	1.8	3.6	1.6	2.8	1.5	2.5
0.65	1.9	4.6	1.7	3.3	1.6	2.9
0.70	2.2	6.1	1.9	3.8	1.7	3.3
0.75	2.5	8.5	2.0	4.6	1.8	4.0
0.80	3.0	13.0	2.2	5.8	2.0	5.0
0.85	3.8	22.7	2.5	7.7	2.2	6.7
0.90	5.5	50.5	2.9	11.4	2.6	10.0
0.95	10.5	200.5	3.5	22.0	3.2	20.0

For the lower percentages, the approximations computed by these formulas are closer to the real values. The table in Figure 10.4 indicates that if the table is 65 percent full, then linear probing requires, on average, fewer than two trials to find an element in the table. Because this number is usually an acceptable limit for a hash function, linear probing requires 35 percent of the spaces in the table to be unoccupied to keep performance at an acceptable level. This may be considered too wasteful, especially for very large tables or files. This percentage is lower for a quadratic probing (25 percent) and for double hashing (20 percent), but it may still be considered large. Double hashing requires one cell out of five to be empty, which is a relatively high fraction. But all these problems can be solved by allowing more than one item to be stored in a given position or in an area associated with one position.

10.2.2 Chaining

Keys do not have to be stored in the table itself. In *chaining*, each position of the table is associated with a linked list or *chain* of structures whose `info` fields store keys or references to keys. This method is called *separate chaining*, and a table of references (pointers) is called a *scatter table*. In this method, the table can never overflow, because the linked lists are extended only upon the arrival of new keys, as illustrated in Figure 10.5. For short linked lists, this is a very fast method, but increasing the length of these lists can significantly degrade retrieval performance. Performance can be improved by maintaining an order on all these lists so that, for unsuccessful searches, an exhaustive search is not required in most cases or by using self-organizing linked lists (Pagli, 1985).

FIGURE 10.5 In chaining, colliding keys are put on the same linked list.

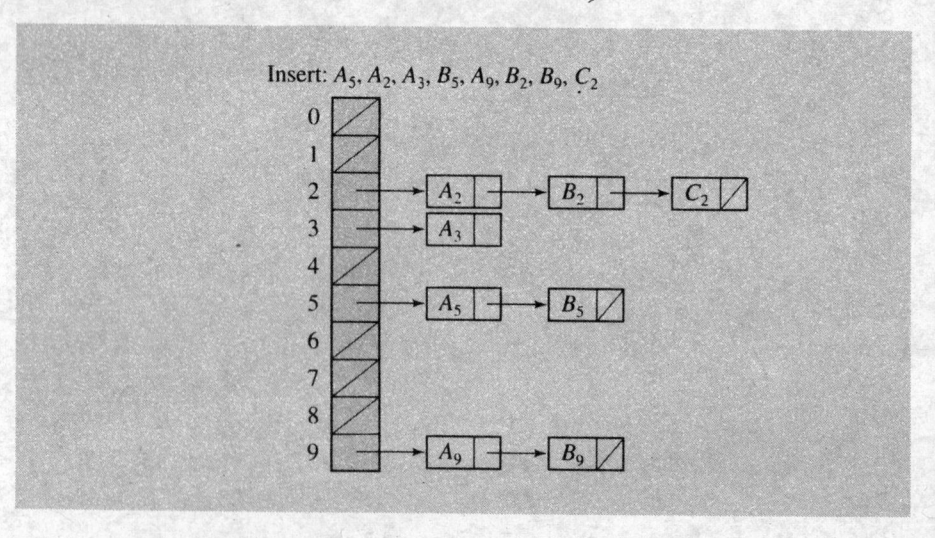

Insert: $A_5, A_2, A_3, B_5, A_9, B_2, B_9, C_2$

This method requires additional space for maintaining pointers. The table stores only pointers, and each node requires one pointer field. Therefore, for n keys, $n +$ *TSize* pointers are needed, which for large n can be a very demanding requirement.

A version of chaining called *coalesced hashing* (or *coalesced chaining*) combines linear probing with chaining. In this method, the first available position is found for a key colliding with another key, and the index of this position is stored with the key already in the table. In this way, a sequential search down the table can be avoided by directly accessing the next element on the linked list. Each position *pos* of the table stores an object with two members: `info` for a key and `next` with the index of the next key that is hashed to *pos*. Available positions can be marked by, say, –2 in `next`; –1 can be used to indicate the end of a chain. This method requires *TSize* · *sizeof*(`next`) more space for the table in addition to the space required for the keys. This is less than for chaining, but the table size limits the number of keys that can be hashed into the table.

An overflow area known as a *cellar* can be allocated to store keys for which there is no room in the table. This area should be located dynamically if implemented as a list of arrays.

Figure 10.6 illustrates an example where coalesced hashing puts a colliding key in the last position of the table. In Figure 10.6a, no collision occurs. In Figure 10.6b, B_5 is put in the last cell of the table, which is found occupied by A_9 when it arrives. Hence, A_9 is attached to the list accessible from position 9. In Figure 10.6c, two new colliding keys are added to the corresponding lists.

FIGURE 10.6 Coalesced hashing puts a colliding key in the last available position of the table.

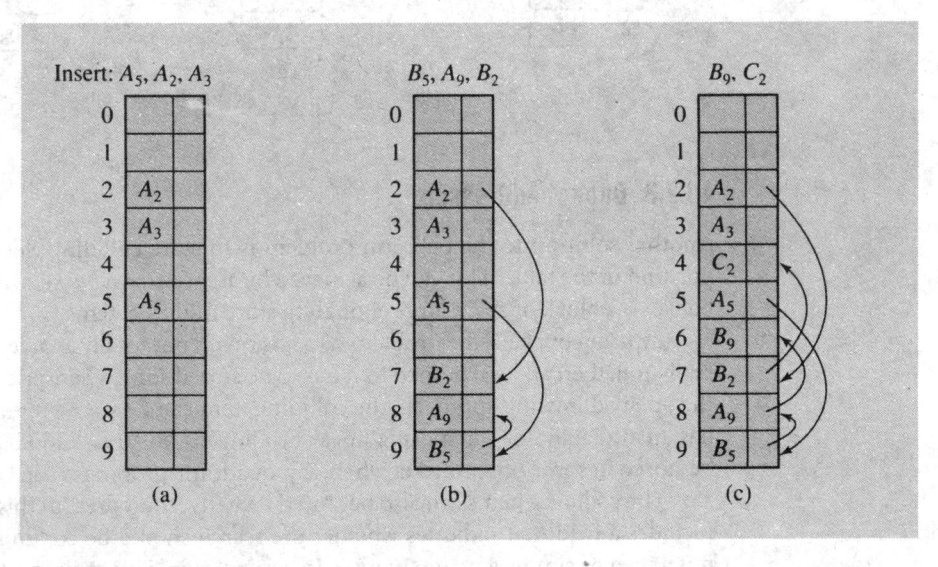

Figure 10.7 illustrates coalesced hashing that uses a cellar. Noncolliding keys are stored in their home positions, as in Figure 10.7a. Colliding keys are put in the last available slot of the cellar and added to the list starting from their home position, as in Figure 10.7b. In Figure 10.7c, the cellar is full, so an available cell is taken from the table when C_2 arrives.

FIGURE 10.7 Coalesced hashing that uses a cellar.

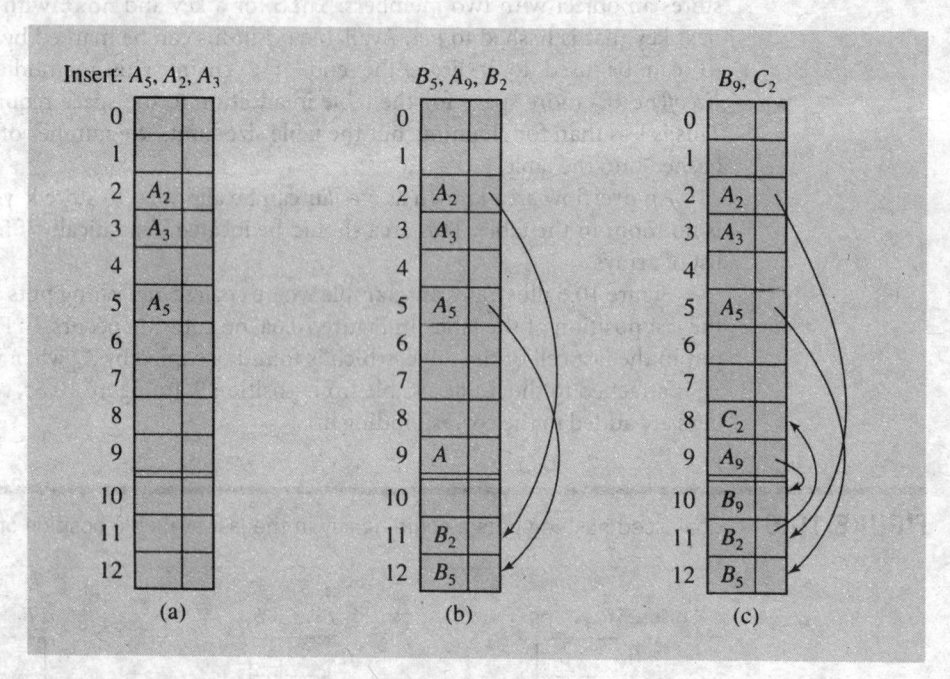

10.2.3 Bucket Addressing

Another solution to the collision problem is to store colliding elements in the same position in the table. This can be achieved by associating a *bucket* with each address. A bucket is a block of space large enough to store multiple items.

By using buckets, the problem of collisions is not totally avoided. If a bucket is already full, then an item hashed to it has to be stored somewhere else. By incorporating the open addressing approach, the colliding item can be stored in the next bucket if it has an available slot when using linear probing, as illustrated in Figure 10.8, or it can be stored in some other bucket when, say, quadratic probing is used.

The colliding items can also be stored in an overflow area. In this case, each bucket includes a field that indicates whether the search should be continued in this area or not. It can be simply a yes/no marker. In conjunction with chaining, this marker can be the number indicating the position in which the beginning of the linked list associated with this bucket can be found in the overflow area (see Figure 10.9).

FIGURE 10.8 Collision resolution with buckets and linear probing method.

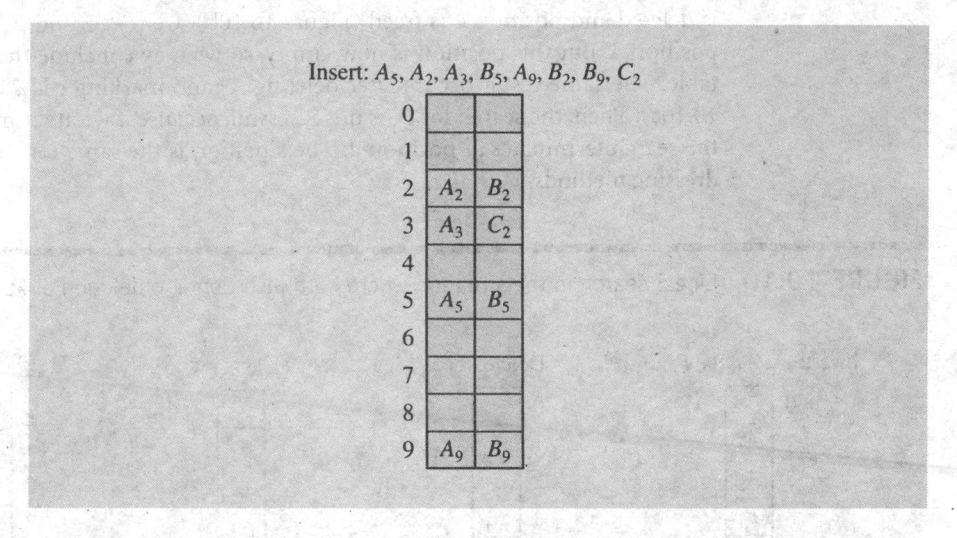

Insert: $A_5, A_2, A_3, B_5, A_9, B_2, B_9, C_2$

0		
1		
2	A_2	B_2
3	A_3	C_2
4		
5	A_5	B_5
6		
7		
8		
9	A_9	B_9

FIGURE 10.9 Collision resolution with buckets and overflow area.

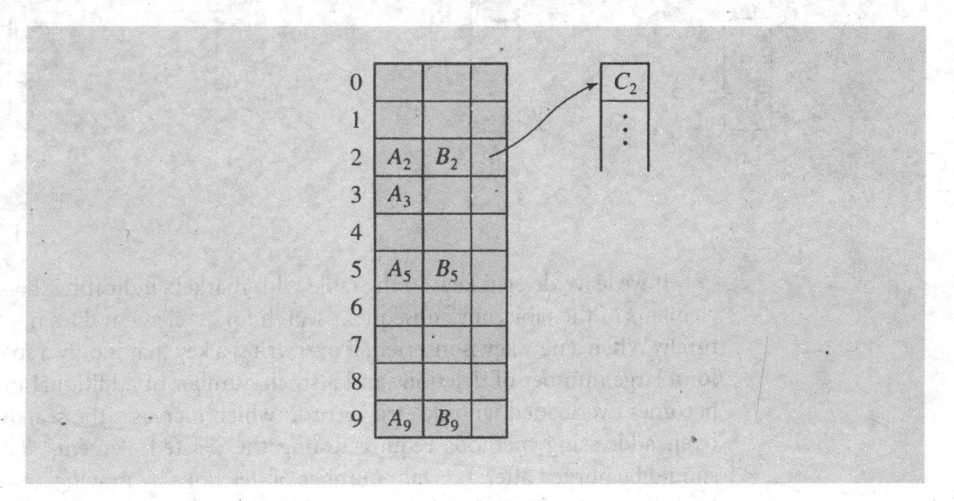

0				C_2
1				⋮
2	A_2	B_2		
3	A_3			
4				
5	A_5	B_5		
6				
7				
8				
9	A_9	B_9		

10.3 DELETION

How can we remove data from a hash table? With a chaining method, deleting an element leads to the deletion of a node from a linked list holding the element. For other methods, a deletion operation may require a more careful treatment of collision resolution, except for the rare occurrence when a perfect hash function is used.

Consider the table in Figure 10.10a in which the keys are stored using linear probing. The keys have been entered in the following order: A_1, A_4, A_2, B_4, B_1. After A_4 is deleted and position 4 is freed (Figure 10.10b), we try to find B_4 by first checking position 4. But this position is now empty, so we may conclude that B_4 is not in the table. The same result occurs after deleting A_2 and marking cell 2 as empty (Figure 10.10c). Then, the search for B_1 is unsuccessful, because if we are using linear probing, the search terminates at position 2. The situation is the same for the other open addressing methods.

FIGURE 10.10 Linear search in the situation where both insertion and deletion of keys are permitted.

	Insert: A_1, A_4, A_2, B_4, B_1	Delete: A_4	Delete: A_2	
0				
1	A_1	A_1	A_1	A_1
2	A_2	A_2		B_1
3	B_1	B_1	B_1	
4	A_4			B_4
5	B_4	B_4	B_4	
6				
7				
8				
9				
	(a)	(b)	(c)	(d)

If we leave deleted keys in the table with markers indicating that they are not valid elements of the table, any subsequent search for an element does not terminate prematurely. When a new key is inserted, it overwrites a key that is only a space filler. However, for a large number of deletions and a small number of additional insertions, the table becomes overloaded with deleted records, which increases the search time because the open addressing methods require testing the deleted elements. Therefore, the table should be purged after a certain number of deletions by moving undeleted elements to the cells occupied by deleted elements. Cells with deleted elements that are not overwritten by this procedure are marked as free. Figure 10.10d illustrates this situation.

10.4 PERFECT HASH FUNCTIONS

All the cases discussed so far assume that the body of data is not precisely known. Therefore, the hash function only rarely turned out to be an ideal hash function in the sense that it immediately hashed a key to its proper position and avoided any collisions. In most cases, some collision resolution technique had to be included, because sooner or later, a key would arrive that conflicted with another key in the table. Also, the number of keys is rarely known in advance, so the table had to be large enough to accommodate all the arriving data. Moreover, the table size contributed to the number of collisions: A larger table has a smaller number of collisions (provided the hash function took table size into consideration). All this was caused by the fact that the body of data to be hashed in the table was not precisely known ahead of time. Therefore, a hash function was first devised and then the data were processed.

In many situations, however, the body of data is fixed, and a hash function can be devised after the data are known. Such a function may really be a perfect hash function if it hashes items on the first attempt. In addition, if such a function requires only as many cells in the table as the number of data so that no empty cell remains after hashing is completed, it is called a *minimal perfect hash function*. Wasting time for collision resolution and wasting space for unused table cells are avoided in a minimal perfect hash function.

Processing a fixed body of data is not an uncommon situation. Consider the following examples: a table of reserved words used by assemblers or compilers, files on unerasable optical disks, dictionaries, and lexical databases.

Algorithms for choosing a perfect hash function usually require tedious work due to the fact that perfect hash functions are rare. As already indicated for 50 elements and a 100-cell array, only one in 1 million is perfect. Other functions lead to collisions.

10.4.1 Cichelli's Method

One algorithm to construct a minimal perfect hash function was developed by Richard J. Cichelli. It is used to hash a relatively small number of reserved words. The function is of the form

$$h(word) = (length(word) + g(firstletter(word)) + g(lastletter(word))) \bmod TSize$$

where g is the function to be constructed. The function g assigns values to letters so that the resulting function h returns unique hash values for all words in a predefined set of words. The values assigned by g to particular letters do not have to be unique. The algorithm has three parts: computation of the letter occurrences, ordering the words, and searching. The last step is the heart of this algorithm and uses an auxiliary function `try()`. Cichelli's algorithm for constructing g and h is as follows:

```
choose a value for Max;
compute the number of occurrences of each first and last letter in the set of all words;
order all words in accordance to the frequency of occurrence of the first and the last letters;
search(wordList)
    if wordList is empty
        halt;
    word = first word from wordList;
    wordList = wordList with the first word detached;
    if the first and the last letters of word are assigned g-values
        try(word,-1,-1);   // -1 signifies 'value already assigned'
        if success
            search(wordList);
```

            put word *at the beginning of* wordList *and detach its hash value*;
        else if *neither the first nor the last letter has a g-value*
            for *each* n,m *in* {0,..., Max}
                try(word,n,m);
                if *success*
                    search(wordList);
                put word *at the beginning of* wordList *and detach its hash value*;
        else if *either the first or the last letter has a g-value*
            for *each* n *in* {0,..., Max}
                try(word,-1,n) *or* try(word,n,-1);
                if *success*
                    search(wordList);
                put word *at the beginning of* wordList *and detach its hash value*;

try(word,firstLetterValue,lastLetterValue)
    if h(word) *has not been claimed*
        reserve h(word);
        assign firstLetterValue *and/or* lastLetterValue *as g-values of firstletter*(word)
            and/or lastletter(word) *if they are not* -1;
        return *success*;
    return *failure*;

We can use this algorithm to build a hash function for the names of the nine Muses: Calliope, Clio, Erato, Euterpe, Melpomene, Polyhymnia, Terpsichore, Thalia, and Urania. A simple count of the letters renders the number of times a given letter occurs as a first and last letter (case sensitivity is disregarded): E (6), A (3), C (2), O (2), T (2), M (1), P (1), and U (1). According to these frequencies, the words can be put in the following order: Euterpe (E occurs six times as the first and the last letter), Calliope, Erato, Terpsichore, Melpomene, Thalia, Clio, Polyhymnia, and Urania.

Now the procedure search() is applied. Figure 10.11 contains a summary of its execution, in which Max = 4. First, the word Euterpe is tried. E is assigned the g-value of 0, whereby h(Euterpe) = 7, which is put on the list of reserved hash values. Everything goes well until Urania is tried. All five possible g-values for U result in an already reserved hash value. The procedure backtracks to the preceding step, when Polyhymnia was tried. Its current hash value is detached from the list, and the g-value of 1 is tried for P, which causes a failure, but 2 for P gives 3 for the hash value, so the algorithm can continue. Urania is tried again five times, then the fifth attempt is successful. All the names have been assigned unique hash values and the search process is finished. If the g-values for each letter are A = C = E = O = M = T = 0, P = 2, and U = 4, then h is the minimal perfect hash function for the nine Muses.

The searching process in Cichelli's algorithm is exponential because it uses an exhaustive search, and thus, it is inapplicable to a large number of words. Also, it does not guarantee that a perfect hash function can be found. For a small number of words, however, it usually gives good results. This program often needs to be run only once, and the resulting hash function can be incorporated into another program. Cichelli applied his method to the Pascal reserved words. The result was a hash func-

FIGURE 10.11 Subsequent invocations of the searching procedure with *Max* = 4 in Cichelli's algorithm assign the indicated values to letters and to the list of reserved hash values. The asterisks indicate failures.

			reserved hash values
Euterpe	$E = 0$	$h = 7$	{7}
Calliope	$C = 0$	$h = 8$	{7 8}
Erato	$O = 0$	$h = 5$	{5 7 8}
Terpsichore	$T = 0$	$h = 2$	{2 5 7 8}
Melpomene	$M = 0$	$h = 0$	{0 2 5 7 8}
Thalia	$A = 0$	$h = 6$	{0 2 5 6 7 8}
Clio		$h = 4$	{0 2 4 5 6 7 8}
Polyhymnia	$P = 0$	$h = 1$	{0 1 2 4 5 6 7 8}
Urania	$U = 0$	$h = 6*$	{0 1 2 4 5 6 7 8}
Urania	$U = 1$	$h = 7*$	{0 1 2 4 5 6 7 8}
Urania	$U = 2$	$h = 8*$	{0 1 2 4 5 6 7 8}
Urania	$U = 3$	$h = 0*$	{0 1 2 4 5 6 7 8}
Urania	$U = 4$	$h = 1*$	{0 1 2 4 5 6 7 8}
Polyhymnia	$P = 1$	$h = 2*$	{0 2 4 5 6 7 8}
Polyhymnia	$P = 2$	$h = 3$	{0 2 3 4 5 6 7 8}
Urania	$U = 0$	$h = 6*$	{0 2 3 4 5 6 7 8}
Urania	$U = 1$	$h = 7*$	{0 2 3 4 5 6 7 8}
Urania	$U = 2$	$h = 8*$	{0 2 3 4 5 6 7 8}
Urania	$U = 3$	$h = 0*$	{0 2 3 4 5 6 7 8}
Urania	$U = 4$	$h = 1$	{0 1 2 3 4 5 6 7 8}

tion that reduced the run time of a Pascal cross-reference program by 10 percent after it replaced the binary search used previously.

There have been many successful attempts to extend Cichelli's technique and overcome its shortcomings. One technique modified the terms involved in the definition of the hash function. For example, other terms, the alphabetical positions of the second to last letter in the word, are added to the function definition (Sebesta and Taylor, 1986), or the following definition is used (Haggard and Karplus, 1986):

$$h(word) = length(word) + g_1(firstletter(word)) + \cdots + g_{length(word)}(lastletter(word))$$

Cichelli's method can also be modified by partitioning the body of data into separate buckets for which minimal perfect hash functions are found. The partitioning is performed by a grouping function, *gr*, which for each word indicates the bucket to which it belongs. Then a general hash function is generated whose form is

$$h(word) = bucket_{gr(word)} + h_{gr(word)}(word)$$

(e.g., Lewis and Cook, 1986). The problem with this approach is that it is difficult to find a generally applicable grouping function tuned to finding minimal perfect hash functions.

Both these ways—modifying hash function and partitioning—are not entirely successful if the same Cichelli's algorithm is used. Although Cichelli ends his paper with the adage: "When all else fails, try brute force," the attempts to modify his approach included devising a more efficient searching algorithm to circumvent the need for brute force. One such approach is incorporated in the FHCD algorithm.

10.4.2 The FHCD Algorithm

An extension of Cichelli's approach was proposed by Thomas Sager. The FHCD algorithm is an extension of Sager's method and it is discussed in this section. The FHCD algorithm (Fox et al., 1992) searches for a minimal perfect hash function of the form

$$h(word) = h_0(word) + g(h_1(word)) + g(h_2(word))$$

(modulo *TSize*), where g is the function to be determined by the algorithm. To define the functions h_i, three tables—T_0, T_1, and T_2—of random numbers are defined, one for each function h_i. Each word is equal to a string of characters $c_1 c_2 \ldots c_m$ corresponding to a triple $(h_0(word), h_1(word), h_2(word))$ whose elements are calculated according to the formulas

$$h_0 = (T_0(c_1) + \cdots + T_0(c_m)) \bmod n$$
$$h_1 = (T_1(c_1) + \cdots + T_1(c_m)) \bmod r$$
$$h_2 = ((T_2(c_1) + \cdots + T_2(c_m)) \bmod r) + r$$

where n is the number of all words in the body of data, r is a parameter usually equal to $n/2$ or less, and $T_i(c_j)$ is the number generated in table T_i for c_j. The function g is found in three steps: *mapping, ordering,* and *searching.*

In the mapping step, n triples $(h_0(word), h_1(word), h_2(word))$ are created. The randomness of functions h_i usually guarantees the uniqueness of these triples; should they not be unique, new tables T_i are generated. Next, a *dependency graph* is built. It is a bipartite graph with half of its vertices corresponding to the h_1 values and labeled 0 through $r - 1$ and the other half to the h_2 values and labeled r through $2r - 1$. Each word corresponds to an edge of the graph between the vertices $h_1(word)$ and $h_2(word)$. The mapping step is expected to take $O(n)$ time.

As an example, we again use the set of names of the nine Muses. To generate three tables T_i, the standard function rand() is used, and with these tables, a set of nine triples is computed, as shown in Figure 10.12a. Figure 10.12b contains a corresponding dependency graph with $r = 3$. Note that some vertices cannot be connected to any other vertices, and some pairs of vertices can be connected with more than one arc.

The ordering step rearranges all the vertices so that they can be partitioned into a series of levels. When a sequence $v_1, \ldots, v_t$ of vertices is established, then a level $K(v_i)$ of keys is defined as a set of all the edges that connect v_i with those v_js for which $j \leq i$. The sequence is initiated with a vertex of maximum degree. Then, for each successive position i of the sequence, a vertex v_i is selected from among the vertices having at least one connection to the vertices $v_1, \ldots, v_{i-1}$, which has maximal degree. When no such vertex can be found, any vertex of maximal degree is chosen from among the unselected vertices. Figure 10.12c contains an example.

In the last step, searching, hash values are assigned to keys level by level. The g-value for the first vertex is chosen randomly among the numbers $0, \ldots, n - 1$. For the

Section 10.4 Perfect Hash Functions   543

FIGURE 10.12 Applying the FHCD algorithm to the names of the nine Muses.

(a)

Value of:	h_0	h_1	h_2
Calliope	(0	1	5)
Clio	(7	1	4)
Erato	(3	2	5)
Euterpe	(6	2	3)
Melpomene	(3	1	5)
Polyhymnia	(8	2	4)
Terpsichore	(8	0	5)
Thalia	(8	2	3)
Urania	(0	2	4)

(b)

(c)

Level	Node	Arcs
0	2	
1	5	Erato
2	1	Calliope, Melpomene
3	4	Clio, Polyhymnia, Urania
4	3	Euterpe, Thalia
5	0	Terpsichore

(d)

Level	Vertex	g-value		
0	2	2		
1	5	6	h(Erato)	$= (3 + 2 + 6) \% 9 = 2$
2	1	4	h(Calliope)	$= (0 + 4 + 6) \% 9 = 1$
2	1	4	h(Melpomene)	$= (3 + 4 + 6) \% 9 = 4$
3	4	2	h(Clio)	$= (7 + 6 + 2) \% 9 = 6$
3	4	2	h(Polyhymnia)	$= (8 + 6 + 2) \% 9 = 7$
3	4	2	h(Urania)	$= (0 + 6 + 2) \% 9 = 8$
4	3	4	h(Euterpe)	$= (6 + 2 + 4) \% 9 = 3$
4	3	4	h(Terpsichore)	$= (8 + 2 + 4) \% 9 = 5$
4	3	4	h(Thalia)	$= (8 + 4 + 6) \% 9 = 0$

(e)

Function g	
0	4
1	4
2	2
3	4
4	2
5	6

other vertices, because of their construction and ordering, we have the following relation: If $v_i < r$, then $v_i = h_1$. Thus, each word in $K(v_i)$ has the same value $g(h_1(word)) = g(v_i)$. Also, $g(h_2(word))$ has already been defined, because it is equal to some v_j that has already been processed. Analogical reasoning can be applied to the case when $v_i > r$ and then $v_i = h_2$. For each word, either $g(h_1(word))$ or $g(h_2(word))$ is known. The second g-value is found randomly for each level so that the values obtained from the formula of the minimal perfect hash function h indicate the positions in the hash table that are available. Because the first choice of a random number will not always fit all words on a given level to the hash table, both random numbers may need to be tried.

The searching step for the nine Muses starts with randomly choosing $g(v_1)$. Let $g(2) = 2$, where $v_1 = 2$. The next vertex is $v_2 = 5$ so that $K(v_2) = \{\text{Erato}\}$. According to Figure 10.12a, $h_0(\text{Erato}) = 3$, and because the edge *Erato* connects v_1 and v_2, either $h_1(\text{Erato})$ or $h_2(\text{Erato})$ must be equal to v_1. We can see that $h_1(\text{Erato}) = 2 = v_1$; hence, $g(h_1(\text{Erato})) = g(v_1) = 2$. A value for $g(v_2) = g(h_2(\text{Erato})) = 6$ is chosen randomly. From this, $h(\text{Erato}) = (h_0(\text{Erato}) + g(h_1(\text{Erato})) + g(h_2(\text{Erato}))) \bmod TSize = (3 + 2 + 6) \bmod 9 = 2$. This means that position 2 of the hash table is no longer available. The new g-value, $g(5) = 6$, is retained for later use.

Now, $v_3 = 1$ is tried, with $K(v_3) = \{\text{Calliope, Melpomene}\}$. The h_0-values for both words are retrieved from the table of triples, and the $g(h_2)$-values are equal to 6 for both words, because $h_2 = v_2$ for both of them. Now we must find a random $g(h_1)$-value such that the hash function h computed for both words renders two numbers different from 2, because position two is already occupied. Assume that this number is 4. As a result, $h(\text{Calliope}) = 1$ and $h(\text{Melpomene}) = 4$. Figure 10.12d contains a summary of all the steps. Figure 10.12e shows the values of the function g. Through these values of g, the function h becomes a minimal perfect hash function. However, because g is given in tabular form and not with a neat formula, it has to be stored as a table to be used every time function h is needed, which may not be a trivial task. The function $g : \{0, \ldots, 2r - 1\} \rightarrow \{0, \ldots, n - 1\}$, and the size of g's domain increases with r. The parameter r is approximately $n/2$, which for large databases means that the table storing all values for g is not of a negligible size. This table has to be kept in main memory to make computations of the hash function efficient.

10.5 HASH FUNCTIONS FOR EXTENDIBLE FILES

All the methods discussed so far work on tables of fixed sizes. This is a reasonable assumption for arrays, but for files it may be too restrictive. After all, file sizes change dynamically by adding new elements or deleting old ones. Some hashing techniques can be used in this situation, such as coalesced hashing or hashing with chaining, but some of them may be inadequate. New techniques have been developed that specifically take into account the variable size of the table or file. We can distinguish two classes of such techniques: directory and directoryless.

In the directory schemes, key access is mediated by the access to a directory or an index of keys in the structure. There are several techniques and modifications to those techniques in the category of the directory schemes. We mention only a few: *expandable hashing* (Knott, 1971), *dynamic hashing* (Larson, 1978), and *extendible hashing* (Fagin et al., 1979). All three methods distribute keys among buckets in a similar fashion. The main difference is the structure of the index (directory). In expandable hashing and dynamic hashing, a binary tree is used as an index of buckets. On the other hand, in extendible hashing, a directory of records is kept in a table.

One directoryless technique is *virtual hashing,* defined as "any hashing which may dynamically change its hashing function" (Litwin, 1978). This change of hashing function compensates for the lack of a directory. An example of this approach is linear hashing (Litwin, 1980). In the following pages, one method from each category is discussed.

10.5.1 Extendible Hashing

Assume that a hashing technique is applied to a dynamically changing file composed of buckets, and each bucket can hold only a fixed number of items. Extendible hashing accesses the data stored in buckets indirectly through an index that is dynamically adjusted to reflect changes in the file. The characteristic feature of extendible hashing is the organization of the index, which is an expandable table.

A hash function applied to a certain key indicates a position in the index and not in the file (or table of keys). Values returned by such a hash function are called *pseudokeys*. In this way, the file requires no reorganization when data are added to it or deleted from it, because these changes are indicated in the index. Only one hash function h can be used, but depending on the size of the index, only a portion of the address $h(K)$ is utilized. A simple way to achieve this effect is by looking at the address $h(K)$ as a string of bits from which only the i leftmost bits can be used. The number i is called the *depth* of the directory. In Figure 10.13a, the depth is equal to two.

FIGURE 10.13 An example of extendible hashing.

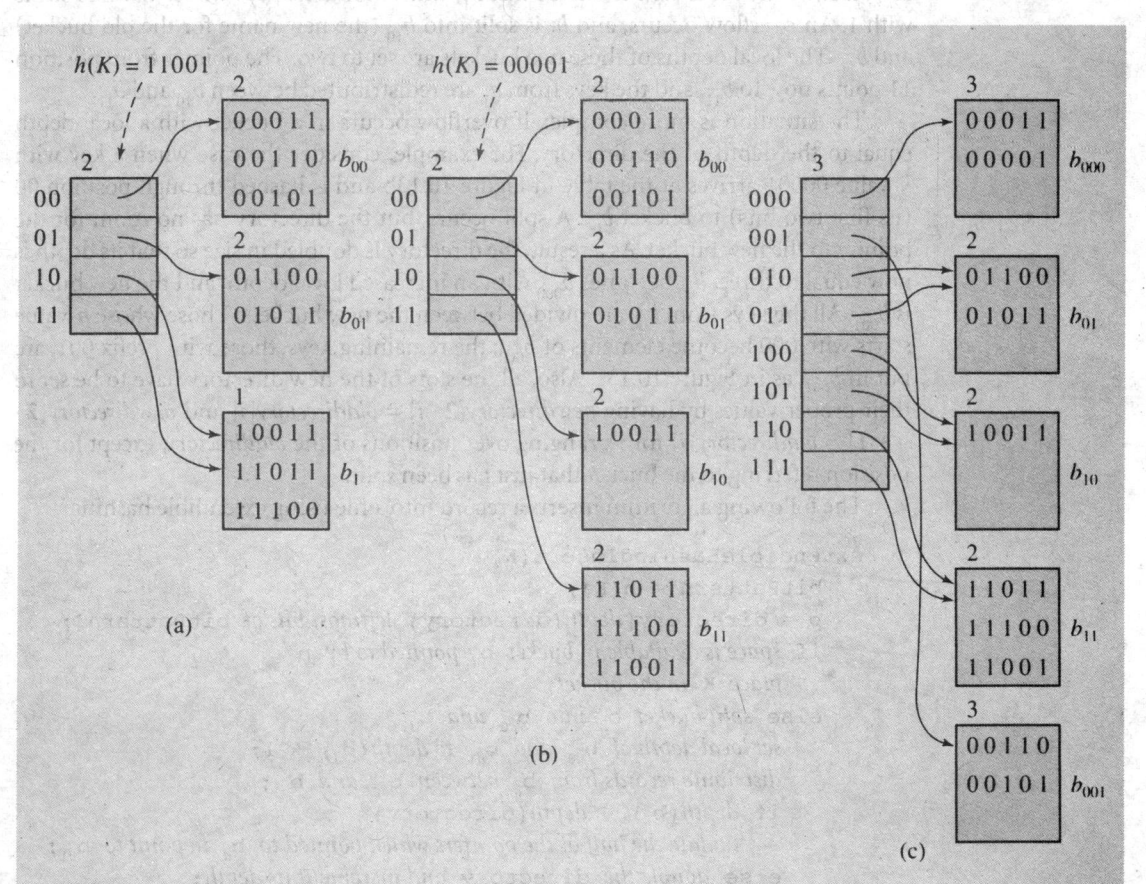

As an example, assume that the hash function h generates patterns of five bits. If this pattern is the string 01011 and the depth is two, then the two leftmost bits, 01, are considered to be the position in the directory containing the pointer to a bucket in which the key can be found or into which it is to be inserted. In Figure 10.13, the values of h are shown in the buckets, but these values represent only the keys that are actually stored in these buckets.

Each bucket has a *local depth* associated with it that indicates the number of leftmost bits in $h(K)$. The leftmost bits are the same for all keys in the bucket. In Figure 10.13, the local depths are shown on top of each bucket. For example, the bucket b_{00} holds all keys for which $h(K)$ starts with 00. More important, the local depth indicates whether the bucket can be accessed from only one location in the directory or from at least two. In the first case, when the local depth is equal to the depth of directory, it is necessary to change the size of the directory after the bucket is split in the case of overflow. When the local depth is smaller than the directory depth, splitting the bucket only requires changing half of the pointers pointing to this bucket so that they point to the newly created one. Figure 10.13b illustrates this case. After a key with h-value 11001 arrives, its two first bits (because depth = 2) direct it to the fourth position of the directory, from which it is sent to the bucket b_1, which contains keys whose h-value starts with 1. An overflow occurs, and b_1 is split into b_{10} (the new name for the old bucket) and b_{11}. The local depths of these two buckets are set to two. The pointer from position 11 points now to b_{11}, and the keys from b_1 are redistributed between b_{10} and b_{11}.

The situation is more complex if overflow occurs in a bucket with a local depth equal to the depth of the directory. For example, consider the case when a key with h-value 00001 arrives at the table in Figure 10.13b and is hashed through position 00 (its first two bits) to bucket b_{00}. A split occurs, but the directory has no room for the pointer to the new bucket. As a result, the directory is doubled in size so that its depth is now equal to three, b_{00} becomes b_{000} with an increased local depth, and the new bucket is b_{001}. All the keys from b_{00} are divided between the new buckets: Those whose h-value starts with 000 become elements of b_{000}; the remaining keys, those with prefix 001, are put in b_{001}, as in Figure 10.13c. Also, all the slots of the new directory have to be set to their proper values by having *newdirectory*$[2 \cdot i]$ = *olddirectory*$[i]$ and *newdirectory*$[2 \cdot i + 1]$ = *olddirectory*$[i]$ for i's ranging over positions of the *olddirectory*, except for the position referring to the bucket that just has been split.

The following algorithm inserts a record into a file using extendible hashing:

```
extendibleHashingInsert(K)
    bitPattern = h(K);
    p = directory[depth(directory) leftmost bits of bitPattern];
    if space is available in bucket b_d pointed to by p
        place K in the bucket;
    else split bucket b_d into b_d0 and b_d1;
        set local depth of b_d0 and b_d1 to depth(b_d) + 1;
        distribute records from b_d between b_d0 and b_d1;
        if depth(b_d) < depth(directory)
            update the half of the pointers which pointed to b_d to point to b_d1;
        else double the directory and increment its depth;
            set directory entries to proper pointers;
```

An important advantage of using extendible hashing is that it avoids a reorganization of the file if the directory overflows. Only the directory is affected. Because the directory in most cases is kept in main memory, the cost of expanding and updating it is very small. However, for large files of small buckets, the size of the directory can become so large that it may be put in virtual memory or explicitly in a file, which may slow down the process of using the directory. Also, the size of the directory does not grow uniformly, because it is doubled if a bucket with a local depth equal to the depth of the directory is split. This means that for large directories there will be many redundant entries in the directory. To rectify the problem of an overgrown directory, David Lomet proposed using extendible hashing until the directory becomes too large to fit into main memory. Afterward, the buckets are doubled instead of the directory, and the bits in the bit pattern $h(K)$ that come after the first $depth$ bits are used to distinguish between different parts of the bucket. For example, if $depth = 3$ and a bucket b_{10} has been quadrupled, its parts are distinguished with bit strings 00, 01, 10, and 11. Now, if $h(K) = 101 01 101$, the key K is searched for in the second portion, 01, of b_{101}.

10.5.2 Linear Hashing

Extendible hashing allows the file to expand without reorganizing it, but it requires storage space for an index. In the method developed by Witold Litwin, no index is necessary because new buckets generated by splitting existing buckets are always added in the same linear way, so there is no need to retain indexes. To this end, a pointer *split* indicates which bucket is to be split next. After the bucket pointed to by *split* is divided, the keys in this bucket are distributed between this bucket and the newly created bucket, which is added to the end of the table. Figure 10.14 contains a sequence of initial splits in which *TSize* = 3. Initially, the pointer *split* is zero. If the

FIGURE 10.14 Splitting buckets in the linear hashing technique.

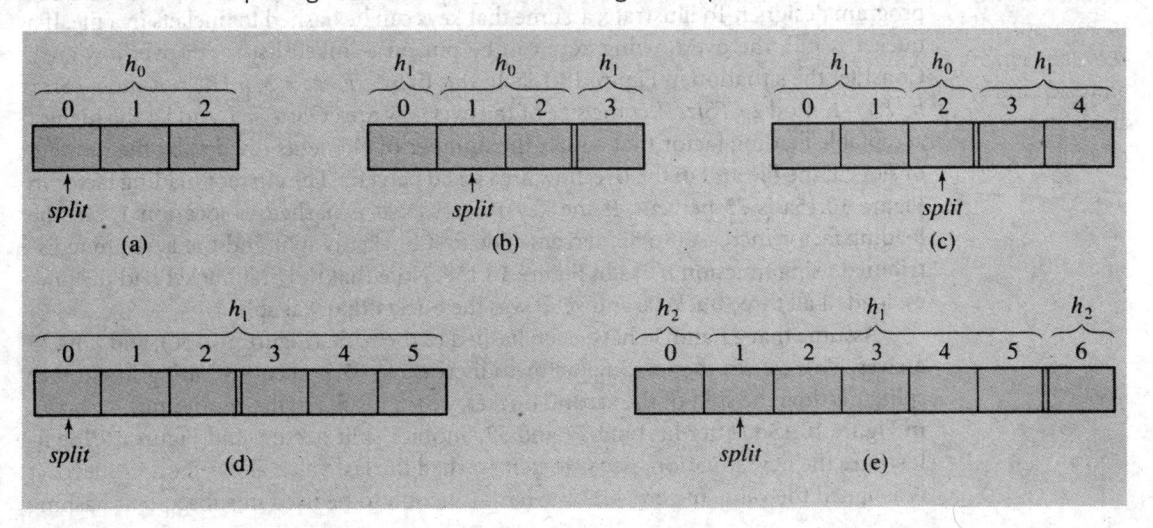

loading factor exceeds a certain level, a new bucket is created, keys from bucket zero are distributed between bucket zero and bucket three, and *split* is incremented. How is this distribution performed? If only one hash function is used, then it hashes keys from bucket zero to bucket zero before and after splitting. This means that one function is not sufficient.

At each level of splitting, linear hashing maintains two hash functions, h_{level} and $h_{level+1}$, such that $h_{level}(K) = K \bmod (TSize \cdot 2^{level})$. The first hash function, h_{level}, hashes keys to buckets that have not yet been split on the current level. The second function, $h_{level+1}$ is used for hashing keys to already split buckets. The algorithm for linear hashing is as follows:

initialize: `split = 0; level = 0;`

```
linearHashingInsert(K)
    if h_level(K) < split       // bucket h_level(K) has been split;
        hashAddress = h_level+1(K);
    else hashAddress = h_level(K);
    insert K in a corresponding bucket or an overflow area if possible;
    while the loading factor is high or K not inserted
        create a new bucket with index split + TSize * 2^level;
        redistribute keys from bucket split between buckets split and split + TSize * 2^level;
        split++;
        if split == TSize * 2^level   // all buckets on the current
                                      // level have been split;
            level++;                  // proceed to the next level;
            split = 0;
        try to insert K if not inserted yet;
```

It may still be unclear when to split a bucket. Most likely, as the algorithm assumes, a threshold value of the loading factor is used to decide whether to split a bucket. This threshold has to be known in advance, and its magnitude is chosen by the program designer. To illustrate, assume that keys can be hashed to buckets in a file. If a bucket is full, the overflowing keys can be put on a linked list in an overflow area. Consider the situation in Figure 10.15a. In this figure, $TSize = 3$, $h_0(K) = K \bmod TSize$, $h_1(K) = K \bmod 2 \cdot TSize$. Let the size of the overflow area $OSize = 3$, and let the highest acceptable loading factor that equals the number of elements divided by the number of slots in the file and in the overflow area be 80 percent. The current loading factor in Figure 10.15a is 75 percent. If the key 10 arrives, it is hashed to location 1, but the loading factor increases to 83 percent. The first bucket is split and the keys are redistributed using function h_1, as in Figure 10.15b. Note that the first bucket had the lowest load of all three buckets, and yet it was the bucket that was split.

Assume that 21 and 36 have been hashed to the table (Figure 10.15c), and now 25 arrives. This causes the loading factor to increase to 87 percent, resulting in another split, this time the split of the second bucket, which results in the configuration shown in Figure 10.15d. After hashing 27 and 37, another split occurs, and Figure 10.15e illustrates the new situation. Because *split* reached the last value allowed on this level, it is assigned the value of zero, and the hash function to be used in subsequent hashing

Section 10.5 Hash Functions for Extendible Files ■ 549

FIGURE 10.15 Inserting keys to buckets and overflow areas with the linear hashing technique.

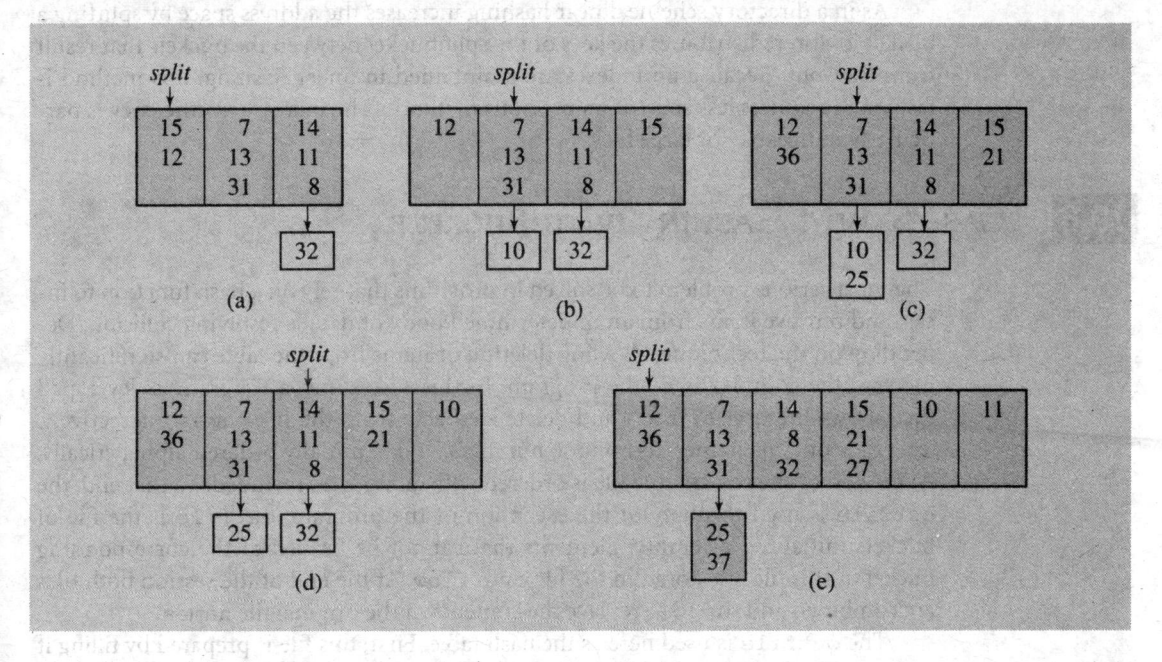

is h_1, the same as before, and a new function, h_2, is defined as K mod $4 \cdot TSize$. All of these steps are summarized in the following table:

K	h(K)	Number of Items	Number of Cells	Loading Factor	Split	Hash Functions	
		9	9 + 3	9/12 = 75%	0	K mod 3	K mod 6
10	1	10	9 + 3	10/12 = 83%	0	K mod 3	K mod 6
		10	9 + 3	10/15 = 67%	1	K mod 3	K mod 6
21	3	11	12 + 3	11/15 = 73%	1	K mod 3	K mod 6
36	0	12	12 + 3	12/15 = 80%	1	K mod 3	K mod 6
25	1	13	12 + 3	13/15 = 87%	1	K mod 3	K mod 6
		13	12 + 3	13/18 = 72%	2	K mod 3	K mod 6
27	3	14	15 + 3	14/18 = 78%	2	K mod 3	K mod 6
37	1	15	15 + 3	15/18 = 83%	2	K mod 3	K mod 6
		15	18 + 3	15/21 = 71%	0	K mod 6	K mod 12

Note that linear hashing requires the use of some overflow area because the order of splitting is predetermined. In the case of files, this may mean more than one file access. This area can be explicit and different from buckets, but it can be introduced somewhat in the spirit of coalesced hashing by utilizing empty space in the buckets

(Mullin, 1981). In a directory scheme, on the other hand, an overflow area is not necessary, although it can be used.

As in a directory scheme, linear hashing increases the address space by splitting a bucket. It also redistributes the keys of the split bucket between the buckets that result from the split. Because no indexes are maintained in linear hashing, this method is faster and requires less space than previous methods. The increase in efficiency is particularly noticeable for large files.

10.6 CASE STUDY: HASHING WITH BUCKETS

The most serious problem to be solved in programs that rely on a hash function to insert and retrieve items from an undetermined body of data is resolving collision. Depending on the technique, allowing deletion of items from the table can significantly increase the complexity of the program. In this case study, a program is developed that allows the user to insert and delete elements from the file names interactively. This file contains names and phone numbers, and is initially ordered alphabetically. At the end of the session, the file is ordered with all updates included. To that end, the outfile is used throughout the execution of the program. outfile is the file of buckets initialized as empty. Elements that cannot be hashed to the corresponding bucket in this file are stored in the file overflow. At the end of the session both files are combined and sorted to replace the contents of the original file names.

The outfile is used here as the hash table. First, this file is prepared by filling it with tableSize * bucketSize empty records (one record is simply a certain number of bytes). Next, all entries of names are transferred to outfile to buckets indicated by the hash function. This transfer is performed by the function insertion(), which includes the hashed item in the bucket indicated by the hash function or in overflow if the bucket is full. In the latter case, overflow is searched from the beginning, and if a position occupied by a deleted record is found, the overflow item replaces it. If the end of overflow is reached, the item is put at the end of this file.

After initializing outfile, a menu is displayed and the user chooses to insert a new record, delete an old one, or exit. For insertion, the same function is used as before. No duplicates are allowed. When the user wants to delete an item, the hash function is used to access the corresponding bucket, and the linear search of positions in the bucket is performed until the item is found, in which case the deletion marker "#" is written over the first character of the item in the bucket. However, if the item is not found and the end of the bucket is reached, the search continues sequentially in overflow until either the item is found and marked as deleted or the end of the file is encountered.

If the user chooses to exit, the undeleted entries of overflow are transferred to outfile, and all undeleted entries of outfile are sorted using an external sort. To that end, quicksort is applied both to outfile and to an array pointers[ ], which contains the addresses of entries in outfile. For comparison, the entries in outfile can be accessed, but the elements in pointers[ ] are moved, not the elements of outfile.

After this indirect sorting is accomplished, the data in outfile have to be put in alphabetical order. This is accomplished by transferring entries from outfile to sorted using the order indicated in pointers[ ]; that is, by going down the array and retrieving the entry in outfile through the address stored in the currently accessed cell. After that, names is deleted and sorted is renamed names.

Section 10.6 Case Study: Hashing with Buckets ■ 551

Here is an example. If the contents of the original file are

```
Adam 123-4567      Brenda 345-5352     Brendon 983-7373
Charles 987-1122   Jeremiah 789-4563   Katherine 823-1573
Patrick 757-4532   Raymond 090-9383    Thorsten 929-6632
```

the hashing ge:nerates the outfile:

```
Katherine 823-1573   |*******************||
Adam 123-4567        |Brenda 345-5352    ||
Raymond 090-9383     |Thorsten 929-6632  ||
```

and the file overflow:

```
Brendon 983-7373    |Charles 987-1122   ||
Jeremiah 789-4563   |Patrick 757-4532   ||
```

(The vertical bars are *not* included in the file; one bar divides the records in the same bucket, and two bars separate different buckets.)

After inserting Carol 654-6543 and deleting Brenda 345-5352 and Jeremiah 789-4563, the file's contents are:

```
outfile:
Katherine 823-1573   |Carol 654-6543     ||
Adam 123-4567        |#renda 345-5352    ||
Raymond 090-9383     |Thorsten 929-6632  ||
```

and overflow:

```
Brendon 983-7373    |Charles 987-1122   ||
#eremiah 789-4563   |Patrick 757-4532   ||
```

A subsequent insertion of Maggie 733-0983 and deletion of Brendon 983-7373 changes only overflow:

```
#rendon 983-7373    |Charles 987-1122   ||
Maggie 733-0983     |Patrick 757-4532   ||
```

After the user chooses to exit, undeleted records from overflow are transferred to outfile, which now includes:

```
Katherine 823-1573   |Carol 654-6543     ||
Adam 123-4567        |#renda 345-5352    ||
Raymond 090-9383     |Thorsten 929-6632  ||
Charles 987-1122     |Maggie 733-0983    ||
Patrick 757-4532     |
```

This file is sorted and the outcome is:

```
Adam 123-4567        |Carol 654-6543      ||
Charles 987-1122     |Katherine 823-1573  ||
Maggie 733-0983      |Patrick 757-4532    ||
Raymond 090-9383     |Thorsten 929-6632   ||
```

Figure 10.16 contains the code for this program.

FIGURE 10.16 Implementation of hashing using buckets.

```cpp
#include <iostream>
#include <fstream>
#include <cstring>
#include <cctype>
#include <iomanip>
#include <cstdio> // remove(), rename();
using namespace std;
const int bucketSize = 2, tableSize = 3, strLen = 20;
const int recordLen = strLen;

class File {
public:
    File() : empty('*'), delMarker('#') {
    }
    void processFile(char*);
private:
    const char empty, delMarker;
    long *pointers;
    fstream outfile, overflow, sorted;
    int  hash(char*);
    void swap(long& i, long& j) {
        long tmp = i; i = j; j = tmp;
    }
    void getName(char*);
    void insert(char line[]) {
        getName(line); insertion(line);
    }
    void insertion(char*);
    void excise(char*);
    void partition(int,int,int&);
    void QSort(int,int);
    void sortFile();
    void combineFiles();
};

unsigned long File::hash(char *s) {
    unsigned long xor = 0, pack;
    int i, j, slength; // exclude trailing blanks;
    for (slength = strlen(s); isspace(s[slength-1]); slength—);
    for (i = 0; i < slength; ) {
        for (pack = 0, j = 0; ; j++, i++) {
            pack |= (unsigned long) s[i];        // include s[i] in the
```

FIGURE 10.16 (continued)

```
                if (j == 3 || i == slength - 1) { // rightmost byte of pack;
                    i++;
                    break;
                }
                pack <<= 8;
        }                   // xor at one time 8 bytes from s;
        xor ^= pack;  // last iteration may put less
    }                       // than 8 bytes in pack;
    return (xor % tableSize) * bucketSize * recordLen;
}// return byte position of home bucket for s;

void File::getName(char line[]) {
    cout << "Enter a name: ";
    cin.getline(line,recordLen+1);
    for (int i = strlen(line); i < recordLen; i++)
        line[i] = ' ';
    line[recordLen] = '\0';
}

void File::insertion(char line[]) {
    int address = hash(line), counter = 0;
    char name[recordLen+1];
    bool done = false, inserted = false;
    outfile.clear();
    outfile.seekg(address,ios::beg);
    while (!done && outfile.get(name,recordLen+1)) {
        if (name[0] == empty || name[0] == delMarker) {
            outfile.clear();
            outfile.seekg(address+counter*recordLen,ios::beg);
            outfile << line << setw(strlen(line)-recordLen);
            done = inserted = true;
        }
        else if (!strcmp(name,line)) {
            cout << line << " is already in the file\n";
            return;
        }
        else counter++;
        if (counter == bucketSize)
            done = true;
        else outfile.seekg(address+counter*recordLen,ios::beg);
    }
```

Continues

FIGURE 10.16 *(continued)*

```
    if (!inserted) {
        done = false;
        counter = 0;
        overflow.clear();
        overflow.seekg(0,ios::beg);
        while (!done && overflow.get(name,recordLen+1)) {
            if (name[0] == delMarker)
                done = true;
            else if (!strcmp(name,line)) {
                cout << line << " is already in the file\n";
                return;
            }
            else counter++;
        }
        overflow.clear();
        if (done)
            overflow.seekg(counter*recordLen,ios::beg);
        else overflow.seekg(0,ios::end);
        overflow << line << setw(strlen(line)-recordLen);
    }
}

void File::excise(char line[]) {
    getName(line);
    int address = hash(line), counter = 0;
    bool done = false, removed = false;
    char name2[recordLen+1];
    outfile.clear();
    outfile.seekg(address,ios::beg);
    while (!done && outfile.get(name2,recordLen+1)) {
        if (!strcmp(line,name2)) {
            outfile.clear();
            outfile.seekg(address+counter*recordLen,ios::beg);
            outfile.put(delMarker);
            done = removed = true;
        }
        else counter++;
        if (counter == bucketSize)
            done = true;
        else outfile.seekg(address+counter*recordLen,ios::beg);
    }
```

FIGURE 10.16 *(continued)*

```
        if (!removed) {
            done = false;
            counter = 0;
            overflow.clear();
            overflow.seekg(0,ios::beg);
            while (!done && overflow.get(name2,recordLen+1)) {
                if (!strcmp(line,name2)) {
                    overflow.clear();
                    overflow.seekg(counter*recordLen,ios::beg);
                    overflow.put(delMarker);
                    done = removed = true;
                }
                else counter++;
                overflow.seekg(counter*recordLen,ios::beg);
            }
        }
        if (!removed)
            cout << line << " is not in database\n";
}

void File::partition (int low, int high, int& pivotLoc) {
    char rec[recordLen+1], pivot[recordLen+1];
    register int i, lastSmall;
    swap(pointers[low],pointers[(low+high)/2]);
    outfile.seekg(pointers[low]*recordLen,ios::beg);
    outfile.clear();
    outfile.get(pivot,recordLen+1);
    for (lastSmall = low, i = low+1; i <= high; i++) {
        outfile.clear();
        outfile.seekg(pointers[i]*recordLen,ios::beg);
        outfile.get(rec,recordLen+1);
        if (strcmp(rec,pivot) < 0) {
            lastSmall++;
            swap(pointers[lastSmall],pointers[i]);
        }
    }
    swap(pointers[low],pointers[lastSmall]);
    pivotLoc = lastSmall;
}
```

Continues

FIGURE 10.16 *(continued)*

```
void File::QSort(int low, int high) {
    int pivotLoc;
    if (low < high) {
        partition(low, high, pivotLoc);
        QSort(low, pivotLoc-1);
        QSort(pivotLoc+1, high);
    }
}

void File::sortFile() {
    char rec[recordLen+1];
    QSort(1,pointers[0]);    // pointers[0] contains the # of elements;
    rec[recordLen] = '\0';   // put data from outfile in sorted order
    for (int i = 1; i <= pointers[0]; i++) {         // in file sorted;
        outfile.clear();
        outfile.seekg(pointers[i]*recordLen,ios::beg);
        outfile.get(rec,recordLen+1);
        sorted << rec << setw(strlen(rec)-recordLen);
    }
}

// data from overflow file and outfile are all stored in outfile and
// prepared for external sort by loading positions of the data to an
// array;

void File::combineFiles() {
    int counter = bucketSize*tableSize;
    char rec[recordLen+1];
    outfile.clear();
    overflow.clear();
    outfile.seekg(0,ios::end);
    overflow.seekg(0,ios::beg);
    while (overflow.get(rec,recordLen+1)) { // transfer from
        if (rec[0] != delMarker) {          // overflow to outfile only
            counter++;                      // valid (non-removed) items;
            outfile << rec << setw(strlen(rec)-recordLen);
        }
    }
    pointers = new long[counter+1];   // load to array pointers positions
    int arrCnt = 1;                   // of valid data stored in output file;
    for (int i = 0; i < counter; i++) {
```

FIGURE 10.16 (continued)

```
        outfile.clear();
        outfile.seekg(i*recordLen,ios::beg);
        outfile.get(rec,recordLen+1);
        if (rec[0] != empty && rec[0] != delMarker)
            pointers[arrCnt++] = i;
    }
    pointers[0] = --arrCnt; // store the number of data in position 0;
}

void File::processFile(char *fileName) {
    ifstream fIn(fileName);
    if (fIn.fail()) {
        cerr << "Cannot open " << fileName << endl;
        return;
    }
    char command[strLen+1] = " ";
    outfile.open("outfile",ios::in|ios::out|ios::trunc);
    sorted.open("sorted",ios::in|ios::out|ios::trunc);
    overflow.open("overflow",ios::in|ios::out|ios::trunc);
    for (int i = 1; i <= tableSize*bucketSize*recordLen; i++)
                                          // initialize
        outfile << empty;                 // outfile;
    char line[recordLen+1];
    while (fIn.get(line,recordLen+1)) // load infile to outfile;
        insertion(line);
    while (strcmp(command,"exit")) {
        cout << "Enter a command (insert, remove, or exit): ";
        cin.getline(command,strLen+1);
        if (!strcmp(command,"insert"))
            insert(line);
        else if (!strcmp(command,"remove"))
            excise(line);
        else if (strcmp(command,"exit"))
            cout << "Wrong command entered, please retry.\n";
    }
    combineFiles();
    sortFile();
    outfile.close();
    sorted.close();
    overflow.close();
    fIn.close();
```

Continues

FIGURE 10.16 (continued)

```
    remove(fileName);
    rename("sorted",fileName);
}
int main(int argc, char* argv[]) {
    char fileName[30];
    if (argc != 2) {
        cout << "Enter a file name: ";
        cin.getline(fileName,30);
    }
    else strcpy(fileName,argv[1]);
    File fClass;
    fClass.processFile(fileName);
    return 0;
}
```

The hash function used in the program appears overly complicated. The function hash() applies the exclusive-or function to the chunks of four characters of a string. For example, the hash value corresponding to the string "ABCDEFGHIJ" is the number "ABCD"^"EFGH"^"IJ"—that is, in hexadecimal notation, 0x41424344^0x45464748^0x0000494a. It may appear that the same outcome can be generated with the function

```
unsigned long File::hash2(char *s) {
    unsigned long xor, remainder;
    for (xor = 0; strlen(s) >= 4; s += 4)
         xor^ = *reinterpret_cast<unsigned long*>(s);
    if (strlen(s) != 0) {
         strcpy(reinterpret_cast<char*>(&remainder),s);
         xor ^= remainder;
    }
    return (xor % tableSize) * bucketSize * recordLen;
}
```

The problem with this simpler function is that it may return different values for the same string. The outcome depends on the way numbers are stored on a particular system, which in turn depends on the "endianness" supported by the system. If a system is big-endian, it stores the most significant byte in the lowest address; that is, numbers are stored "big-end-first." In a little-endian system, the most significant byte is in the highest address. For example, number 0x12345678 is stored as 0x12345678 in the big-endian system—first the contents of the highest order byte, 12, then the contents of the lower order byte, 34, and so on. On the other hand, the same number is stored as 0x78563412 in the little-endian system—first the contents of the lowest order byte, 78, then the contents of the higher order byte, 56, and so on. Consequently, on the big-endian systems, the statement

```
xor ^= *reinterpret_cast<unsigned long*>(s);
```

xors with `xor` the substring "ABCD" of string `s` = "ABCDEFGHIJ" because the cast forces the system to treat the first four characters in `s` as representing a long number without changing the order of the characters. On a little-endian system, the same four characters are read in reverse order, lowest order byte first. Therefore, to prevent this system dependence, one character of `s` is processed at a time and included in `xor`, after which the contents of `xor` are shifted to the left by eight bits to make room for another character. This is a way of simulating a big-endian reading.

10.7 EXERCISES

1. What is the minimum number of keys that are hashed to their home positions using the linear probing technique? Show an example using a 5-cell array.

2. Consider the following hashing algorithm (Bell and Kaman 1970). Let Q and R be the quotient and remainder obtained by dividing K by $TSize$, and let the probing sequence be created by the following recurrence formula:

$$h_i(K) = \begin{cases} R & \text{if } i = 0 \\ (h_{i-1}(K) + Q) \bmod TSize & \text{otherwise} \end{cases}$$

   What is the desirable value of $TSize$? What condition should be imposed on Q?

3. Is there any advantage to using binary search trees instead of linked lists in the separate chaining method?

4. In Cichelli's method for constructing the minimal hash function, why are all words first ordered according to the occurrence of the first and the last letters? The subsequent searching algorithm does not make any reference to this order.

5. Trace the execution of the searching algorithm used in Cichelli's technique with $Max = 3$. (See the illustration of such a trace for $Max = 4$ in Figure 10.11.)

6. In which case does Cichelli's method not guarantee to generate a minimal perfect hash function?

7. Apply the FHCD algorithm to the nine Muses with $r = n/2 = 4$ and then with $r = 2$. What is the impact of the value of r on the execution of this algorithm?

8. Strictly speaking, the hash function used in extendible hashing also dynamically changes. In what sense is this true?

9. Consider an implementation of extendible hashing that allows buckets to be pointed to by only one pointer. The directory contains null pointers so that all pointers in the directory are unique except the null pointers. What keys are stored in the buckets? What are the advantages and disadvantages of this implementation?

10. How would the directory used in extendible hashing be updated after splitting if the last $depth$ bits of $h(K)$ are considered an index to the directory, not the first $depth$ bits?

11. List the similarities and differences between extendible hashing and B$^+$-trees.

12. What is the impact of the uniform distribution of keys over the buckets in extendible hashing on the frequency of splitting?

13. Apply the linear hashing method to hash numbers 12, 24, 36, 48, 60, 72, and 84 to an initially empty table with three buckets and with three cells in the overflow area. What problem can you observe? Can this problem bring the algorithm to a halt?

14. Outline an algorithm to delete a key from a table when the linear hashing method is used for inserting keys.

15. The function `hash()` applied in the case study uses the exclusive or (xor) operation to fold all the characters in a string. Would it be a good idea to replace it by bitwise-and or bitwise-or?

10.8 PROGRAMMING ASSIGNMENTS

1. As discussed in this chapter, the linear probing technique used for collision resolution has a rapidly deteriorating performance if a relatively small percentage of the cells are available. This problem can be solved using another technique for resolving collisions, and also by finding a better hash function, ideally, a perfect hash function. Write a program that evaluates the efficiency of various hashing functions combined with the linear probing method. Have your program write a table similar to the one in Figure 10.4, which gives the averages for successful and unsuccessful trials of locating items in the table. Use functions for operating on strings and a large text file whose words will be hashed to the table. Here are some examples of such functions (all values are divided modulo *TSize*):

    a. FirstLetter(s) + SecondLetter(s) + · · · + LastLetter(s)

    b. FirstLetter(s) + LastLetter(s) + length(s) (Cichelli)

    c. ```
       for (i = 1, index = 0; i < strlen(s); i++)
           index = (26 * index + s[i] - ' '); (Ramakrishna)
       ```

2. Another way of improving the performance of hashing is to allow reorganization of the hash table during insertions. Write a program that compares the performance of linear probing with the following self-organization hashing methods:

    a. *Last-come-first-served hashing* places a new element in its home position, and in case of a collision, the element that occupies this position is inserted in another position using a regular linear probing method to make room for the arriving element (Poblete and Munro, 1989).

    b. *Robin Hood hashing* checks the number of positions two colliding keys are away from their home positions and continues searching for an open position for the key closer to its home position (Celis et al., 1985).

3. Write a program that inserts records into a file and retrieves and deletes them using either extendible hashing or the linear hashing technique.

4. Extend the program presented in the case study by creating a linked list of overflowing records associated with each bucket of the intermediate file `outfile`. Note that if a bucket has no empty cells, the search continues in the overflow area. In the extreme case, it may mean that the bucket holds only deleted items, and new items are inserted in the overflow area. Therefore, it may be advantageous to have a purging function that, after a certain number of deletions, is automatically invoked. This function transfers items from the overflow area to the main file, which are hashed to buckets with deleted items. Write such a function.
       ```

BIBLIOGRAPHY

Bell, James R., and Kaman, Charles H., "The Linear Quotient Hash Code," *Communications of the ACM* 13 (1970), 675–677.

Celis, P., Larson P., and Munro J. I., "Robin Hood Hashing," *Proceedings of the 26th IEEE Symposium on the Foundations of Computer Science* (1985), 281–288.

Cichelli, Richard J., "Minimal Perfect Hash Function Made Simple," *Communications of the ACM* 23 (1980), 17–19.

Czech, Zbigniew J., and Majewski, Bohdan S., "A Linear Time Algorithm for Finding Minimal Perfect Hash Functions," *Computer Journal* 36 (1993), 579–587.

Enbody, R. J., and Dy, H. C., "Dynamic Hashing Schemes," *Computing Surveys* 20 (1988), 85–113.

Fagin, Ronald, Nievergelt, Jurg, Pippenger, Nicholas, and Strong, H. Raymond, "Extendible Hashing—A Fast Access Method for Dynamic Files," *ACM Transactions on Database Systems* 4 (1979), 315–344.

Fox, Edward A., Heath, Lenwood S., Chen, Qi F., and Daoud, Amjad M., "Practical Minimal Perfect Hash Functions for Large Databases," *Communications of the ACM* 35 (1992), 105–121.

Haggard, G., and Karplus, K., "Finding Minimal Perfect Hash Functions," *SIGCSE Bulletin* 18 (1986), No. 1, 191–193.

Knott, G. D., "Expandable Open Addressing Hash Table Storage and Retrieval," *Proceedings of the ACM SIGFIDET Workshop on Data Description, Access, and Control* (1971), 186–206.

Knuth, Donald, *The Art of Computer Programming*, Vol. 3, Reading, MA: Addison-Wesley, 1998.

Larson, Per A., "Dynamic Hashing," *BIT* 18 (1978), 184–201.

Larson, Per A., "Dynamic Hash Tables," *Communications of the ACM* 31 (1988), 446–457.

Lewis, Ted G., and Cook, Curtis R., "Hashing for Dynamic and Static Internal Tables," *IEEE Computer* (October 1986), 45–56.

Litwin, Witold, "Virtual Hashing: A Dynamically Changing Hashing," *Proceedings of the Fourth Conference of Very Large Databases* (1978), 517–523.

Litwin, Witold, "Linear Hashing: A New Tool for File and Table Addressing," *Proceedings of the Sixth Conference of Very Large Databases* (1980), 212–223.

Lomet, David B., "Bounded Index Exponential Hashing," *ACM Transactions on Database Systems* 8 (1983), 136–165.

Lum, V. Y., Yuen, P. S. T., and Dood, M., "Key-to-Address Transformation Techniques: A Fundamental Performance Study on Large Existing Formatted Files," *Communications of the ACM* 14 (1971), 228–239.

Morris, Robert, "Scatter Storage Techniques," *Communications of the ACM* 11 (1968), 38–44.

Mullin, James K., "Tightly Controlled Linear Hashing Without Separate Overflow Storage," *BIT* 21 (1981), 390–400.

Pagli, L., "Self-Adjusting Hash Tables," *Information Processing Letters* 21 (1985), 23–25.

Poblete, Patricio V., and Munro, J. Ian, "Last-Come-First-Served Hashing," *Journal of Algorithms* 10 (1989), 228–248.

Radke, Charles E., "The Use of the Quadratic Search Residue," *Communications of the ACM* 13 (1970), 103–105.

Sager, Thomas J., "A Polynomial Time Generator for Minimal Perfect Hash Functions," *Communications of the ACM* 28 (1985), 523–532.

Sebesta, Robert W., and Taylor, Mark A., "Fast Identification of Ada and Modula-2 Reserved Words," *Journal of Pascal, Ada, and Modula-2* (March/April 1986), 36–39.

Tharp, Alan L., *File Organization and Processing*, New York: Wiley, 1988.

Vitter, Jeffrey S., and Chen, Wen C., *Design and Analysis of Coalesced Hashing*, New York: Oxford University Press, 1987.

11 Data Compression

Transfer of information is essential for the proper functioning of any structure on any level and any type of organization. The faster an exchange of information occurs, the smoother the structure functions. Improvement of the rate of transfer can be achieved by improving the medium through which data are transferred or by changing the data themselves so that the same information can be transmitted within a shorter time interval.

Information can be represented in a form that exhibits some redundancy. For example, in a database, it is enough to say about a person that he is "M" or she is "F," instead of spelling out the whole words, "male" and "female," or to use 1 and 2 to represent the same information. The number one hundred twenty-eight can be stored as 80 (hexadecimal), 128, 1000000 (binary), CXXVIII, $\rho\kappa\eta$ (the Greek language used letters as digits), or | | | . . . | (128 bars). If numbers are stored as the sequences of digits representing them, then 80 is the shortest form. Numbers are represented in binary form in computers.

11.1 CONDITIONS FOR DATA COMPRESSION

When transferring information, the choice of the data representation determines how fast the transfer is performed. A judicious choice can improve the throughput of a transmission channel without changing the channel itself. There are many different methods of *data compression* (or *compaction*) that reduce the size of the representation without affecting the information itself.

Assume that there are n different symbols used to code messages. For a binary code, $n = 2$; for Morse code, $n = 3$: the dot, the dash, and the blank separating the sequences of dots and dashes that represent letters. Assume also that all symbols m_i forming a set M have been independently chosen and are known to have probabilities of occurrence $P(m_i)$, and the symbols are coded with strings of 0s and 1s. Then $P(m_1) + \cdots + P(m_n) = 1$. The information content of the set M, called the *entropy* of the source M, is defined by

$$L_{ave} = P(m_1)L(m_1) + \cdots + P(m_n)L(m_n) \tag{11.1}$$

where $L(m_i) = -\lg(P(m_i))$, which is the minimum length of a codeword for symbol m_i. Claude E. Shannon established in 1948 that Equation 11.1 gives the best possible average length of a codeword when the source symbols and the probabilities of their use are known. No data compression algorithm can be better than L_{ave}, and the closer it is to this number, the better is its compression rate.

For example, if there are three symbols, m_1, m_2, and m_3, with the probabilities .25, .25, and .5, respectively, then the lengths of the codewords assigned to them are:

$$-\lg(P(m_1)) = -\lg(P(m_2)) = -\lg(.25) = \lg\left(\frac{1}{.25}\right) = \lg(4) = 2 \text{ and}$$

$$-\lg(P(m_3)) = \lg(2) = 1$$

and the average length of a codeword is

$$L_{ave} = P(m_1) \cdot 2 + P(m_2) \cdot 2 + P(m_3) \cdot 1 = 1.5$$

Various data compression techniques attempt to minimize the average codeword length by devising an optimal code (that is, an assignment of codewords to symbols) that depends on the probability P with which a symbol is being used. If a symbol is issued infrequently, it can be assigned a long codeword. For frequently issued symbols, very short encodings are more to the point.

Some restrictions need to be imposed on the prospective codes:

1. Each codeword corresponds to exactly one symbol.
2. Decoding should not require any look ahead; after reading each symbol it should be possible to determine whether the end of a string encoding a symbol of the original message has been reached. A code meeting this requirement is called a code with the *prefix property*, and it means that no codeword is a prefix of another codeword. Therefore, no special punctuation is required to separate two codewords in a coded message.

The second requirement can be illustrated by three different encodings of three symbols, as given in the following table:

Symbol	$code_1$	$code_2$	$code_3$
A	1	1	11
B	2	22	12
C	12	12	21

The first code does not allow us to make a distinction between AB and C, because both are coded as 12. The second code does not have this ambiguity, but it requires a look ahead, as in 1222: The first 1 can be decoded as A. The following 2 may indicate that A was improperly chosen, and 12 should have been decoded as C. It may be that A is a proper choice if the third symbol is 2. Because 2 is found, AB is chosen as the tentatively decoded string, but the fourth symbol is another 2. Hence, the first turn was wrong, and A has been ill-chosen. The proper decoding is CB. All these problems arise because both $code_1$ and $code_2$ violate the prefix property. Only $code_3$ can be unambiguously decoded as read.

For an optimal code, two more stipulations are specified.

3. The length of the codeword for a given symbol m_i should not exceed the length of the codeword of a less probable symbol m_j; that is, if $P(m_i) \leq P(m_j)$, then $L(m_i) \geq L(m_j)$ for $1 \leq i, j \leq n$.

4. In an optimal encoding system, there should not be any unused short codewords either as stand-alone encodings or as prefixes for longer codewords, because this would mean that longer codewords were created unnecessarily. For example, the sequence of codewords 01, 000, 001, 100, 101 for a certain set of five symbols is not optimal because the codeword 11 is not used anywhere; this encoding can be turned into an optimal sequence 01, 10, 11, 000, 001.

In the following sections, several data compression methods are presented. To compare the efficiency of these methods when applied to the same data, the same measure is used. This measure is the *compression rate* (also called the *fraction of data reduction*), and it is defined as the ratio

$$\frac{\text{length(input)} - \text{length(output)}}{\text{length(input)}} \tag{11.2}$$

It is expressed as a percentage indicating the amount of redundancy removed from the input.

11.2 HUFFMAN CODING

The construction of an optimal code was developed by David Huffman, who utilized a tree structure in this construction: a binary tree for a binary code. The algorithm is surprisingly simple and can be summarized as follows:

```
Huffman()
    for each symbol create a tree with a single root node and order all trees
        according to the probability of symbol occurrence;
    while more than one tree is left
        take the two trees t₁, t₂ with the lowest probabilities p₁, p₂ (p₁ ≤ p₂)
            and create a tree with t₁ and t₂ as its children and with
            the probability in the new root equal to p₁ + p₂;
    associate 0 with each left branch and 1 with each right branch;
    create a unique codeword for each symbol by traversing the tree from the root
        to the leaf containing the probability corresponding to this
        symbol and by putting all encountered 0s and 1s together;
```

The resulting tree has a probability of 1 in its root.

It should be noted that the algorithm is not deterministic in the sense of producing a unique tree because, for trees with equal probabilities in the roots, the algorithm does not prescribe their positions with respect to each other either at the beginning or during execution. If t_1 with probability p_1 is in the sequence of trees and the new tree t_2 is created with $p_2 = p_1$, should t_2 be positioned to the left of t_1 or to the right? Also, if there are three trees, $t_1, t_2,$ and t_3, with the same lowest probability in the entire sequence, which two trees should be chosen to create a new tree? There are three possibilities for choosing two trees. As a result, different trees can be obtained depending on where the trees with equal probabilities are placed in the sequence with respect to each other. Regardless of the shape of the tree, the average length of the codeword remains the same.

To assess the compression efficiency of the Huffman algorithm, a definition of the *weighted path length* is used, which is the same as Equation 11.1 except that $L(m_i)$ is interpreted as the number of 0s and 1s in the codeword assigned to symbol m_i by this algorithm.

Figure 11.1 contains an example for the five letters A, B, C, D, and E with probabilities .39, .21, .19, .12, and .09, respectively. The tree in Figures 11.1a and 11.1b are different in the way in which the two nodes containing a probability of .21 have been chosen to be combined with tree .19 to create a tree of .40. Regardless of the choice, the lengths of the codewords associated with the five letters A through E are the same, namely, 2, 2, 2, 3, and 3, respectively. However, the codewords assigned to them are slightly different, as shown in Figures 11.1c and 11.1d, which present abbreviated (and more commonly used) versions of the way the trees in Figures 11.1a and 11.1b were created. The average length for the latter two trees is

$$L_{\text{Huf}} = .39 \cdot 2 + .21 \cdot 2 + .19 \cdot 2 + .12 \cdot 3 + .09 \cdot 3 = 2.21$$

which is very close to 2.09 (only 5 percent off), the average length computed according to Equation 11.1:

$$L_{\text{ave}} = .39 \cdot 1.238 + .21 \cdot 2.252 + .19 \cdot 2.396 + .12 \cdot 3.059 + .09 \cdot 3.474 = 2.09$$

Corresponding letters in Figures 11.1a and 11.1b have been assigned codewords of the same length. Obviously, the average length for both trees is the same. But each way of building a Huffman tree, starting from the same data, should result in the same average length, regardless of the shape of the tree. Figure 11.2 shows two Huffman trees for the letters P, Q, R, S, and T with the probabilities .1, .1, .1, .2, and .5, respectively. Depending on how the lowest probabilities are chosen, different codewords are assigned to these letters with different lengths, at least for some of them. However, the average length remains the same and is equal to 2.0.

The Huffman algorithm can be implemented in a variety of ways, at least as many as the number of ways a priority queue can be implemented. The priority queue is the natural data structure in the context of the Huffman algorithm because it requires removing the two smallest probabilities and inserting the new probability in the proper position.

One way to implement this algorithm is to use a singly linked list of pointers to trees, which reflects closely what Figure 11.1a illustrates. The linked list is initially ordered according to the probabilities stored in the trees, all of them consisting of just a root. Then, repeatedly, the two trees with the smallest probabilities are chosen; the tree with the smaller probability is replaced by a newly created tree, and the node with the pointer to the tree with the higher probability is removed from the linked list. From trees having the same probability in their roots, the first tree encountered is chosen.

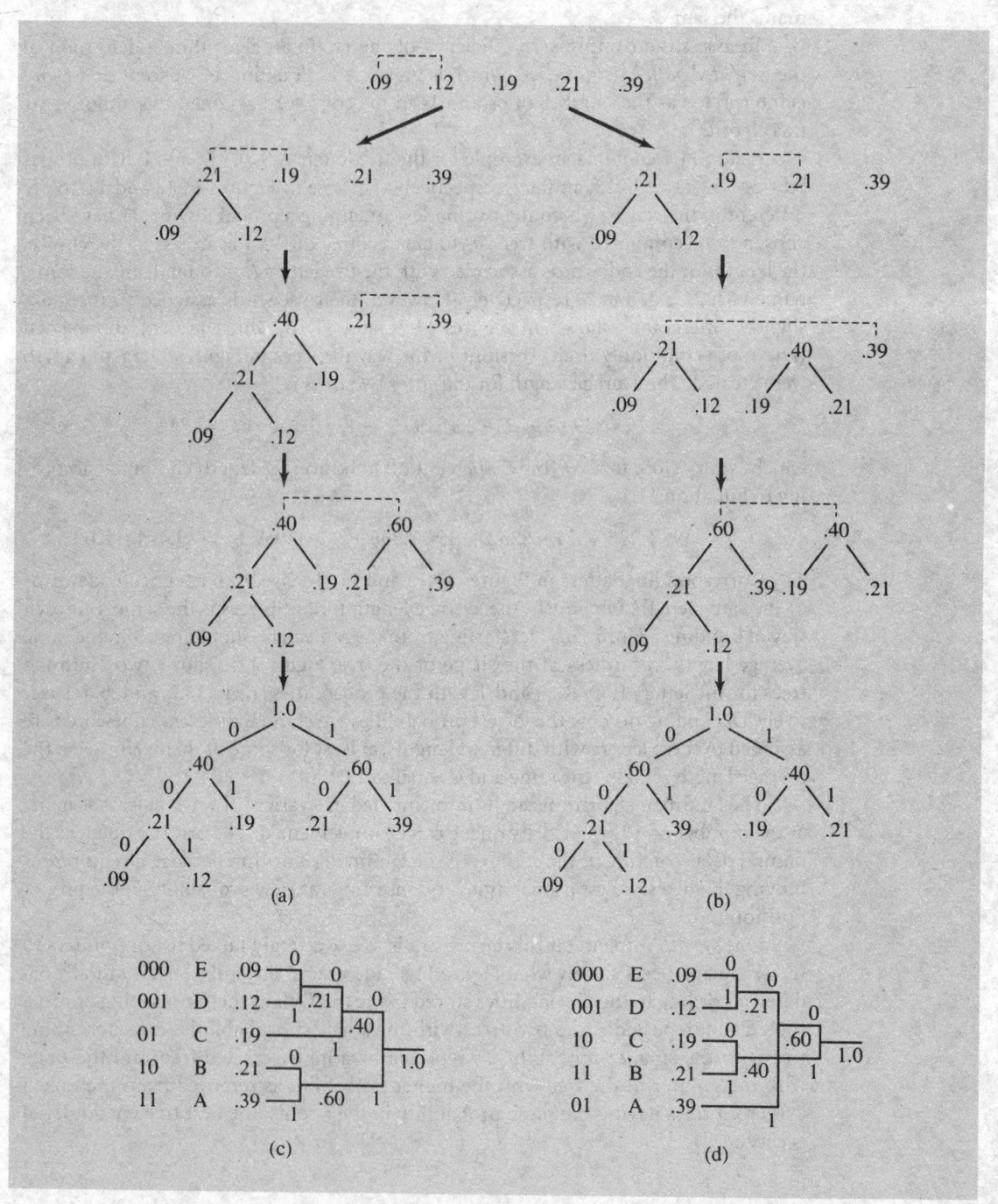

FIGURE 11.1 Two Huffman trees created for five letters A, B, C, D, and E with probabilities .39, .21, .19, .12, and .09.

FIGURE 11.2 Two Huffman trees generated for letters P, Q, R, S, and T with probabilities .1, .1, .1, .2, and .5.

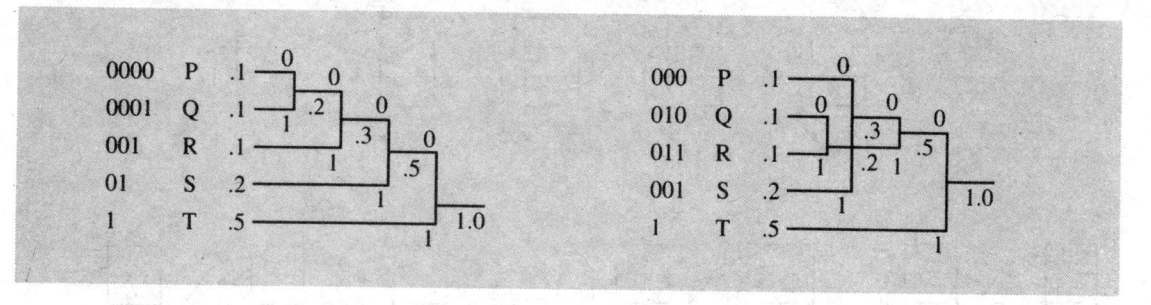

In another implementation, all probability nodes are first ordered, and that order is maintained throughout the operation. From such an ordered list, the first two trees are always removed to create a new tree from them, which is inserted close to the end of the list. To that end, a doubly linked list of pointers to trees with immediate access to the beginning and to the end of this list can be used. Figure 11.3 contains a trace of the execution of this algorithm for the letters A, B, C, D, and E with the same probabilities as in Figure 11.1. Codewords assigned to these letters are also indicated in Figure 11.3. Note that they are different from the codewords in Figure 11.1, although their lengths are the same.

The two preceding algorithms built Huffman trees bottom-up by starting with a sequence of trees and collapsing them together to a gradually smaller number of trees and, eventually, to one tree. However, this tree can be built top-down, starting from the highest probability. But only the probabilities to be placed in the leaves are known. The highest probability, to be put in the root, is known if lower probabilities, in the root's children, have been determined; the latter are known if still lower probabilities have been computed, and so on. Therefore, creating nonterminal nodes has to be deferred until the probabilities to be stored in them are found. It is very convenient to use the following recursive algorithm to implement a Huffman tree:

```
createHuffmanTree(prob)
    declare the probabilities p₁, p₂, and the Huffman tree Htree;
    if only two probabilities are left in prob
        return a tree with p₁, p₂ in the leaves and p₁ + p₂ in the root;
    else remove the two smallest probabilities from prob and assign them to p₁ and p₂;
        insert p₁ + p₂ to prob;
        Htree = CreateHuffmanTree(prob);
        in Htree make the leaf with p₁ + p₂ the parent of two leaves with p₁ and p₂;
        return Htree;
```

Figure 11.4 contains a summary of the trace of the execution of this algorithm for the letters A, B, C, D, and E with the probabilities as shown in Figure 11.1. Indentation indicates consecutive calls to `createHuffmanTree()`.

One implementation of a priority queue is a min heap, which can also be used to implement this algorithm. In this heap, each nonterminal node has a smaller probability

FIGURE 11.3 Using a doubly linked list to create the Huffman tree for the letters from Figure 11.1.

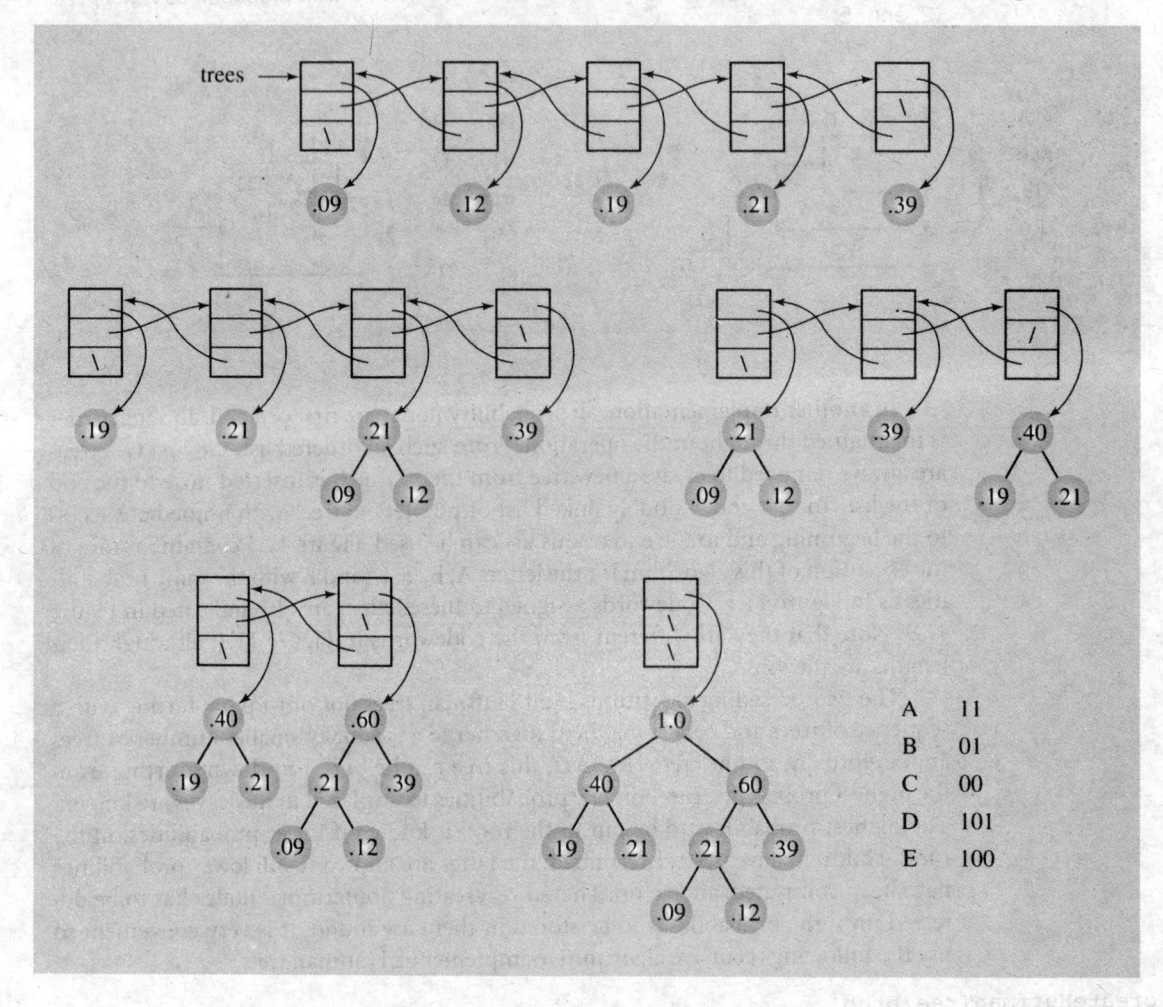

A	11
B	01
C	00
D	101
E	100

than the probabilities in its children, and because the smallest probability is in the root, that one is simple to remove. But after it is removed, the root is empty. Therefore, the largest element is put in the root and the heap property is restored. Then the second element can be removed from the root and replaced with a new element, which represents the sum of the probability of the root and the probability previously removed.

Afterward, the heap property has to be restored again. After one such sequence of operations, the heap has one less node: Two probabilities from the previous heap have been removed and a new one has been added. But it is not enough to create the Huffman tree: The new probability is a parent of the probabilities just removed, and this information must be retained. To that end, three arrays can be used: *indexes* containing the indexes of the original probabilities and the probabilities created during the process of creating the Huffman tree; *probabilities*, an array of the original and newly created probabilities; and *parents*, an array of indexes indicating the position of the

FIGURE 11.4 Top-down construction of a Huffman tree using recursive implementation.

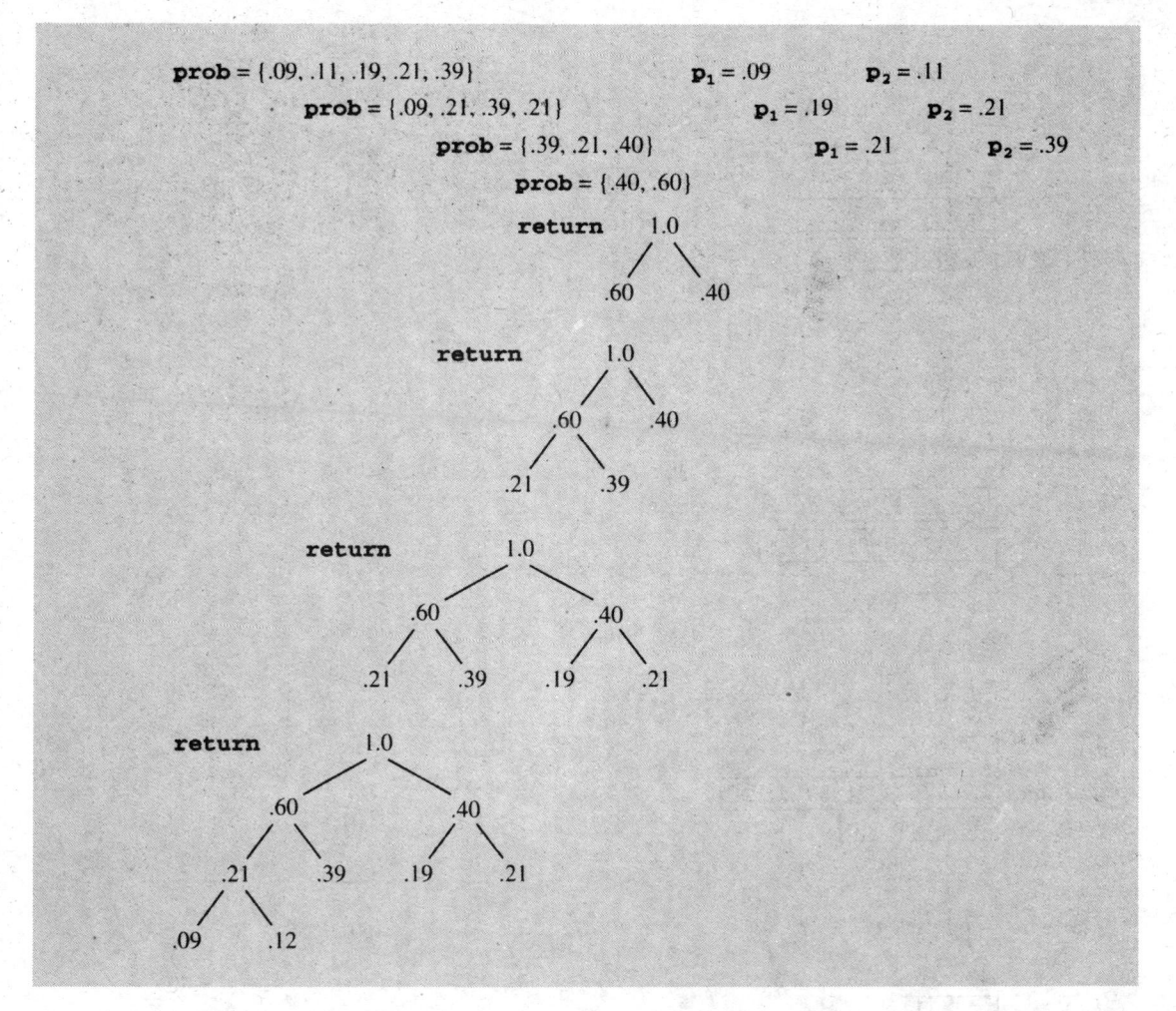

$\mathbf{prob} = \{.09, .11, .19, .21, .39\}$ $p_1 = .09$ $p_2 = .11$
$\mathbf{prob} = \{.09, .21, .39, .21\}$ $p_1 = .19$ $p_2 = .21$
$\mathbf{prob} = \{.39, .21, .40\}$ $p_1 = .21$ $p_2 = .39$
$\mathbf{prob} = \{.40, .60\}$

return 1.0
   .60   .40

return 1.0
   .60   .40
.21   .39

return 1.0
  .60   .40
.21 .39 .19 .21

return 1.0
 .60   .40
.21 .39 .19 .21
.09 .12

parents of the elements stored in *probabilities*. A positive number in *parents* indicates the left child, and a negative number indicates the right child. Codewords are created by accumulating 0s and 1s when going from leaves to the root using the array *parents*, which functions as an array of pointers. It is important to note that in this particular implementation probabilities are sorted indirectly: The heap is actually made up of indexes to probabilities, and all exchanges take place in *indexes*.

Figure 11.5 illustrates an example of using a heap to implement the Huffman algorithm. The heaps in steps (a), (e), (i), and (m) in Figure 11.5 are ready for processing. First, the highest probability is put in the root, as in steps (b), (f), (j), and (n) of Figure 11.5. Next, the heap is restored, as in steps (c), (g), (k), and (o), and the root probability is set to the sum of the two smallest probabilities, as in steps (d), (h), (l), and (p). Processing is complete when there is only one node in the heap.

FIGURE 11.5 Huffman algorithm implemented with a heap.

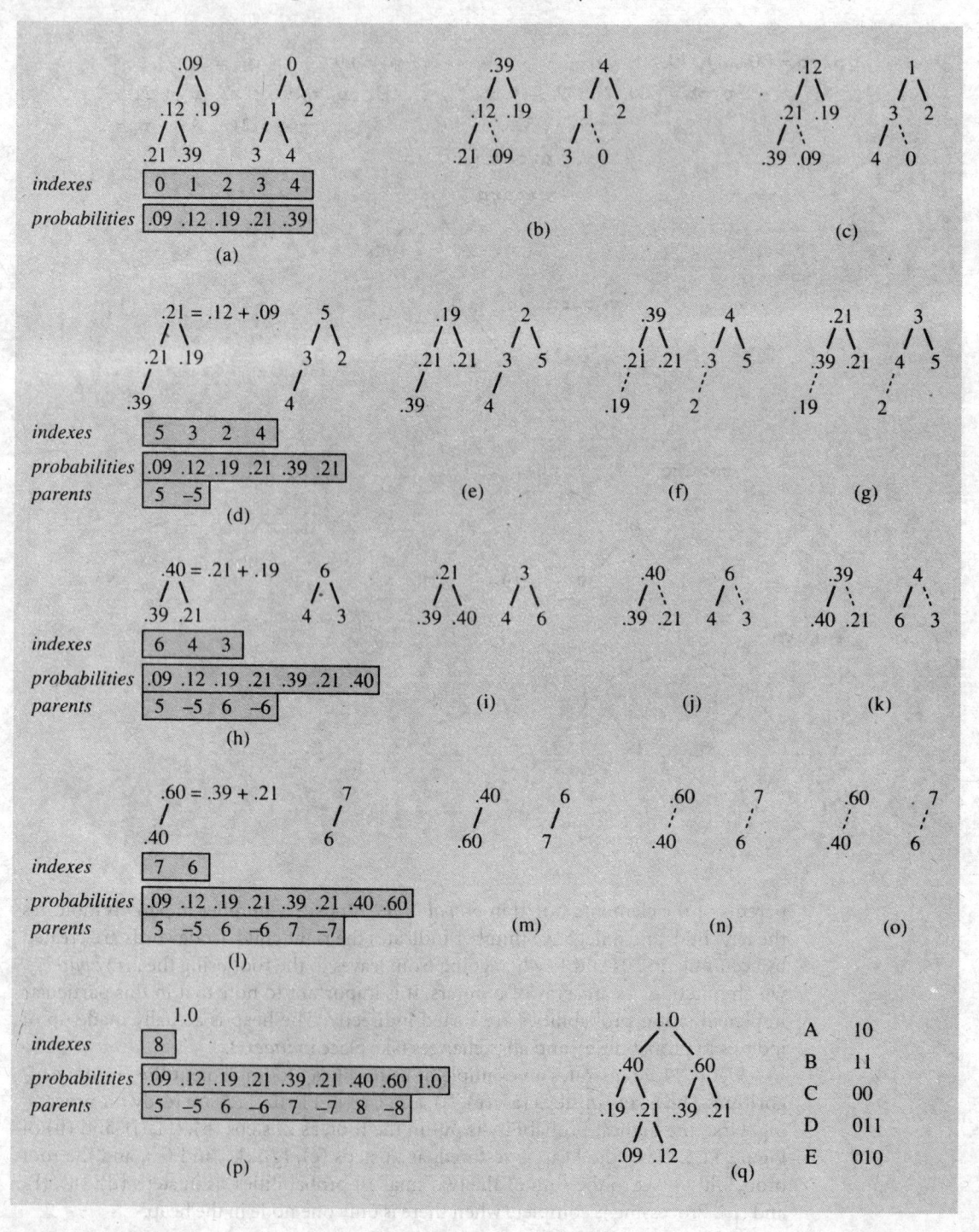

Using the Huffman tree, a table can be constructed that gives the equivalents for each symbol in terms of 1s and 0s encountered along the path leading to each of the leaves of the tree. For our example, the tree from Figure 11.3 will be used, and the resulting table is

A	11
B	01
C	00
D	101
E	100

The coding process transmits coded equivalents of the symbols to be sent. For example, instead of sending ABAAD, the sequence 11011111101 is dispatched with the average number of bits per one letter equal to 11/5 = 2.2, almost the same as 2.09, the value specified by the formula for L_{ave}. To decode this message, the conversion table has to be known to the message receiver. Using this table, a Huffman tree can be constructed with the same paths as the tree used for coding, but its leaves would (for the sake of efficiency) store the symbols instead of their probabilities. In this way, upon reaching a leaf, the symbol can be retrieved directly from it. Using this tree, each symbol can be decoded uniquely. For example, if 1001101 is received, then we try to reach a leaf of the tree using the path indicated by leading 1s and 0s. In this case, 1 takes us to the right, 0 to the left, and another 0 again to the left, whereby we end up in a leaf containing E. After reaching this leaf, decoding continues by starting from the root of the tree and trying to reach a leaf using the remaining 0s and 1s. Because 100 has been processed, 1101 has to be decoded. Now, 1 takes us to the right and another 1 again to the right, which is a leaf with A. We start again from the root, and the sequence 01 is decoded as B. The entire message is now decoded as EAB.

At this point, a question can be asked: Why send 11011111101 instead of ABAAD? This is supposed to be data compression, but the coded message is twice as long as the original. Where is the advantage? Note precisely the way in which messages are sent. A, B, C, D, and E are single letters, and letters, being characters, require 1 byte (8 bits) to be sent, using the extended ASCII code. Therefore, the message ABAAD requires 5 bytes (40 bits). On the other hand, 0s and 1s in the coded version can be sent as single bits. Therefore, if 11011111101 is regarded not as a sequence of the characters "0" and "1," but as a sequence of bits, then only 11 bits are needed to send the message, about one-fourth of what is required to send the message in its original form, ABAAD.

This example raises one problem: Both the encoder and the decoder have to use the same coding, the same Huffman tree. Otherwise, the decoding will be unsuccessful. How can the encoder let the decoder know which particular code has been used? There are at least three possibilities:

1. Both the encoder and the decoder agree beforehand on a particular Huffman tree and both use it for sending any message.
2. The encoder constructs the Huffman tree afresh every time a new message is sent and sends the conversion table along with the message. The decoder either uses the table to decode the message or reconstructs the corresponding Huffman tree and then performs the translation.
3. The decoder constructs the Huffman tree during transmission and decoding.

The second strategy is more versatile, but its advantages are visible only when large files are encoded and decoded. For our simple example, ABAAD, sending both the table of codewords and the coded message 11011111101 is hardly perceived as data compression. However, if a file contains a message of 10,000 characters using the characters A through E, then the space saved is significant. Using the probabilities indicated earlier for these letters, we project that there are approximately 3,900 As, 2,100 Bs, 1,900 Cs, 1,200 Ds, and 900 Es. Hence, the number of bits needed to code this file is

$$3{,}900 \cdot 2 + 2{,}100 \cdot 2 + 1{,}900 \cdot 2 + 1{,}200 \cdot 3 + 900 \cdot 3 = 22{,}100 \text{ bits} = 2{,}762.5 \text{ bytes}$$

which is approximately one-fourth the 10,000 bytes required for sending the original file. Even if the conversion table is added to the file, this proportion is only minimally affected.

However, even with this approach, there may be some room for improvement. As indicated, an ideal compression algorithm should give the same average codeword length as computed from Equation 11.1. The symbols from Figure 11.1 have been assigned codewords whose average length is 2.21, approximately 5 percent worse than the ideal 2.09. Sometimes, however, the difference is larger. Consider, for example, three symbols, X, Y, and Z, with probabilities .1, .1, and .8. Figure 11.6a shows a Huff-

FIGURE 11.6 Improving the average length of the codeword by applying the Huffman algorithm to (b) pairs of letters instead of (a) single letters.

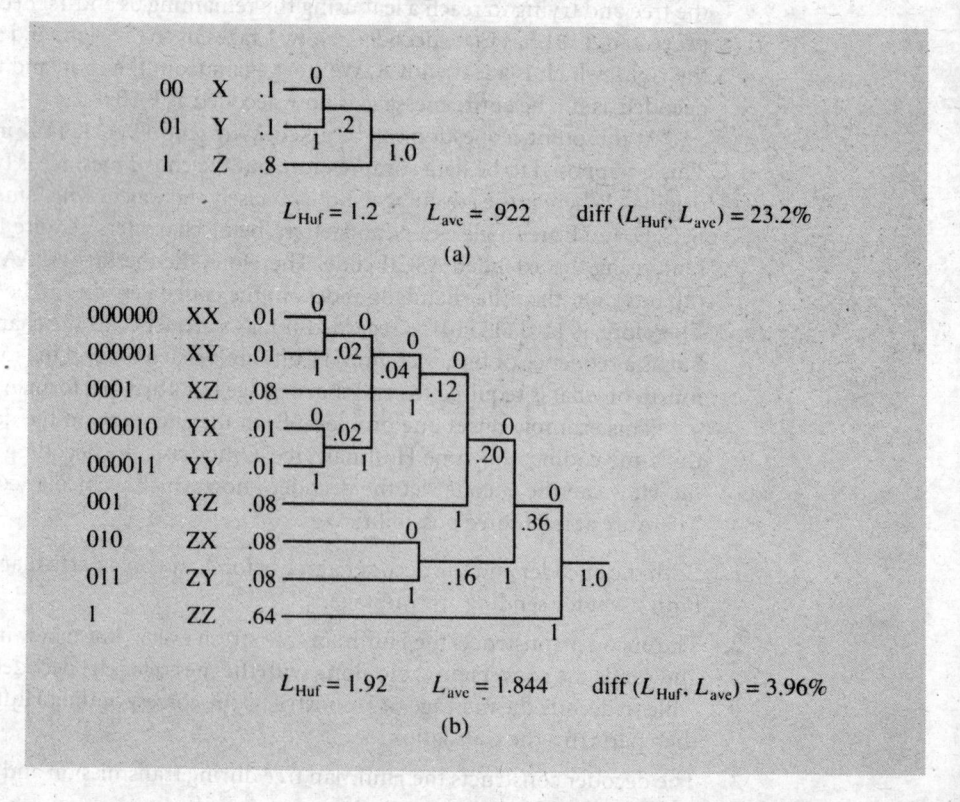

```
00   X  .1 ─┐0
01   Y  .1 ─┘1 .2 ─┐0
1    Z  .8 ────────┘1 1.0
```

$L_{Huf} = 1.2 \qquad L_{ave} = .922 \qquad \text{diff}(L_{Huf}, L_{ave}) = 23.2\%$

(a)

```
000000  XX  .01 ─┐0
000001  XY  .01 ─┘1 .02 ─┐0
0001    XZ  .08 ─────────┘1 .04 ─┐0
000010  YX  .01 ─┐0              │
000011  YY  .01 ─┘1 .02 ─┐1      .12 ─┐0
001     YZ  .08 ─────────┘               │
010     ZX  .08 ─────────┐0             .20 ─┐0
011     ZY  .08 ─────────┘1 .16 ─┐1           │
1       ZZ  .64 ──────────────────┘          .36 ─┐0
                                                     1.0
```

$L_{Huf} = 1.92 \qquad L_{ave} = 1.844 \qquad \text{diff}(L_{Huf}, L_{ave}) = 3.96\%$

(b)

man tree for these symbols, with codewords assigned to them. The average length, according to this tree, is

$$L_{Huf} = 2 \cdot .1 + 2 \cdot .1 + 1 \cdot .8 = 1.2$$

and the best expected average, L_{ave}, is .922. Therefore, there is a possibility we can improve the Huffman coding by approximately 23.2 percent, ignoring the fact that, at this point, a full 23.2 percent improvement is not possible because the average is below 1. How is this possible? As already stated, all Huffman trees result in the same average weighted path length. Therefore, no improvement can be expected if only the symbols X, Y, and Z are used to construct this tree.

On the other hand, if all possible pairs of symbols are used for building a Huffman tree, the data rate can be reduced. Figure 11.6b illustrates this procedure. Out of three symbols, X, Y, and Z, nine pairs are created whose probabilities are computed by multiplying the probability of both symbols. For example, because the probability for both X and Y is .1, the probability of pair XY is .01 = .1 · .1. The average L_{Huf} is 1.92 and the expected average L_{ave} is 1.84 (twice the previous L_{ave}), with the difference between these averages being 4 percent. This represents a 19.2 percent improvement at the cost of including a larger conversion table (nine entries instead of three) as part of the message to be sent. If the message is large and the number of symbols used in the message is relatively small, then the increase in the size of the table is insignificant. However, for a large number of symbols, the size of the table may be much too large to notice any improvement. For 26 English letters, the number of pairs is 676, which is considered relatively small. But if all printable characters have to be distinguished in an English text, from the blank character (ASCII code 32) to the tilde (ASCII code 126), plus the carriage return character, then there are $(126 - 32 + 1) + 1 = 96$ characters and 9,216 pairs of characters. Many of these pairs are not likely to occur at all (e.g., XQ or KZ), but even if 50 percent of them are found, the resulting table containing these pairs along with codewords associated with them may be too large to be useful.

Using pairs of symbols is still a good idea, even if the number of symbols is large. For example, a Huffman tree can be constructed for all symbols and for all pairs of symbols that occur at least five times. The efficiency of the variations of Huffman encoding can be measured by comparing the size of compressed files. Experiments were performed on an English text, a PL/1 program file, and a digitized photographic image (Rubin 1976). When only single characters were used, the compression rates were approximately 40 percent, 60 percent, and 50 percent, respectively. When single characters were used along with the 100 most frequent groups (not only two characters long), the compression rates were 49 percent, 73 percent, and 52 percent. When the 512 most frequent groups were used, the compression rates were around 55 percent, 71 percent, and 62 percent.

11.2.1 Adaptive Huffman Coding

The foregoing discussion assumed that the probabilities of messages are known in advance. A natural question is: How do we know them?

One solution computes the number of occurrences of each symbol expected in messages in some fairly large sample of texts of, say, 10 million characters. For messages in natural languages such as English, such samples may include some literary works, newspaper articles, and a portion of an encyclopedia. After each character's frequency has been determined, a conversion table can be constructed for use by both the sending and receiving ends of the data transfer. This eliminates the need to include such a table every time a file is transmitted.

However, this method may not be useful for sending some specialized files, even if written in English. A computer science paper includes a much higher percentage of digits and parentheses, especially if it includes extensive illustrations in LISP or C++ code, than a paper on the prose of Jane Austen. In such circumstances, it is more judicious to use the text to be sent to determine the needed frequencies, which also requires enclosing the table as overhead in the file being sent. A preliminary pass through this file is required before an actual conversion table can be constructed; however, the file to be preprocessed may be very large, and preprocessing slows down the entire transmission process. Also, the file to be sent may not be known in its entirety when it is being sent, and yet compression is necessary; for example, when a text is being typed and sent line by line, there is no way to know the contents of the whole file at the time of sending. In such a situation, adaptive compression is a viable solution.

An adaptive Huffman encoding technique was devised first by Robert G. Gallager and then improved by Donald Knuth. The algorithm is based on the following *sibling property*: If each node has a sibling (except for the root) and the breadth-first right-to-left tree traversal generates a list of nodes with nonincreasing frequency counters, it can be proven that a tree with the sibling property is a Huffman tree (Faller, 1974; Gallager, 1978).

In adaptive Huffman coding, the Huffman tree includes a counter for each symbol, and the counter is updated every time a corresponding input symbol is being coded. Checking whether the sibling property is retained assures that the Huffman tree under construction is still a Huffman tree. If the sibling property is violated, the tree has to be restructured to restore this property. Here is how this is accomplished.

First, it is assumed that the algorithm maintains a doubly linked list nodes that contains the nodes of the tree ordered by breadth-first right-to-left tree traversal. A *block$_i$* is a part of the list where each node has frequency i, and the first node in each block is called a *leader*. For example, Figure 11.7 shows the Huffman tree and also the list nodes = (**7** **4** **3** **2** 2 **1** 1 1 1 **0**) that has six blocks—*block$_7$*, *block$_4$*, *block$_3$*, *block$_2$*, *block$_1$*, and *block$_0$*—with leaders shown with counters in boldface.

All unused symbols are kept in one node with a frequency of 0, and each symbol encountered in the input has its own node in the tree. Initially, the tree has just one 0-node that includes all symbols. If an input symbol did not yet appear in the input, the 0-node is split in two, with the new 0-node containing all symbols except the newly encountered one and the node referring to this new symbol with counter set to 1; both nodes become children of the one parent whose counter is also set to 1. If an input symbol already has a node p in the tree, its counter is incremented. However, such an increment may endanger the sibling property, so this property has to be restored by exchanging the node p with the leader of the block to which p currently belongs, except when this leader is p's parent. This node is found by going in nodes from p toward the beginning of this list. If p belongs to block$_i$ before increment, it is swapped with the leader of this block, whereby it is included in block$_{i+1}$. Then the counter increment is done for p's possibly new parent, which may also lead to a tree transformation to restore the sibling property. This process is continued until the root is reached. In this way, the counters are updated on the *new* path from p to the root rather than on its old path. For each symbol, the codeword is is-

FIGURE 11.7 Doubly-linked list nodes formed by breadth-first right-to-left tree traversal.

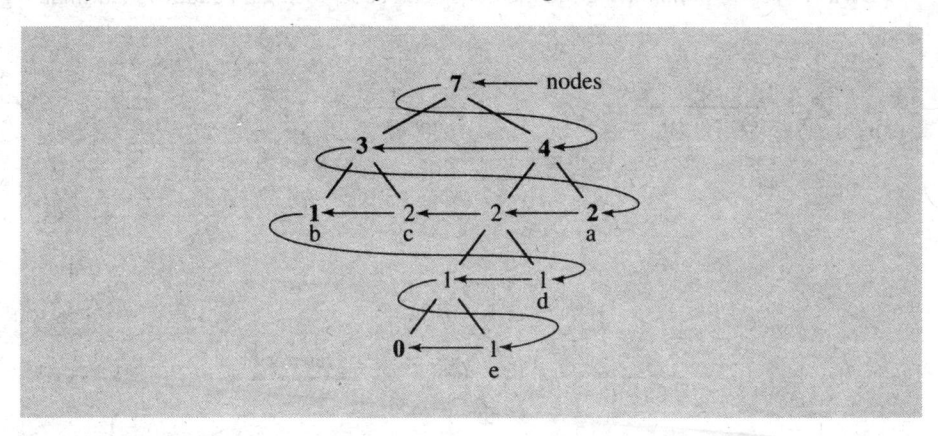

sued, which is obtained by scanning the Huffman tree from the root to the node corresponding to this symbol *before* any transformation in the tree takes place.

Two different types of codewords are transmitted during this process. If a symbol being coded has already appeared, then the normal coding procedure is applied: The Huffman tree is scanned from the root to the node holding this symbol to determine its codeword. If a symbol appears in the input for the first time, it is in the 0-node, but just sending the Huffman codeword of the 0-node does not suffice. Therefore, along with the codeword allowing us to reach the 0-node, the codeword is sent that indicates the position of the encountered symbol. For the sake of simplicity, we assume that position n is coded as n 1s followed by a 0. Zero is used to separate the 1s from those belonging to the next codeword. For example, when the letter c is coded for the first time, its codeword, 001110, is a combination of the codeword for the 0-node, 00, and the codeword 1110 indicating that c can be found in the third position in the list of unused symbols associated with 0-node. These two codewords (or rather, parts of one codeword) are marked in Figure 11.8 by underlining them separately. After a symbol is removed from the list in 0-node, its place is taken by the last symbol of this list. This also indicates that the encoder and receiver have to agree on the alphabet being used and its ordering. The algorithm is shown in this pseudocode:

```
FGKDynamicHuffmanEncoding(symbol s)
    p = leaf that contains symbol s;
    c = Huffman codeword for s;
if p is the 0-node
    c = c concatenated with the number of 1s representing position of s in 0-node and with 0;
    write the last symbol in 0-node over s in this node;
    create a new node q for symbol s and set its counter to 1;
    p = new node to become the parent of both 0-node and node q;
    counter(p) = 1;
    include the two new nodes to nodes;
else increment counter(p);
while p is not the root
```

FIGURE 11.8 Transmitting the message "aafcccbd" using an adaptive Huffman algorithm.

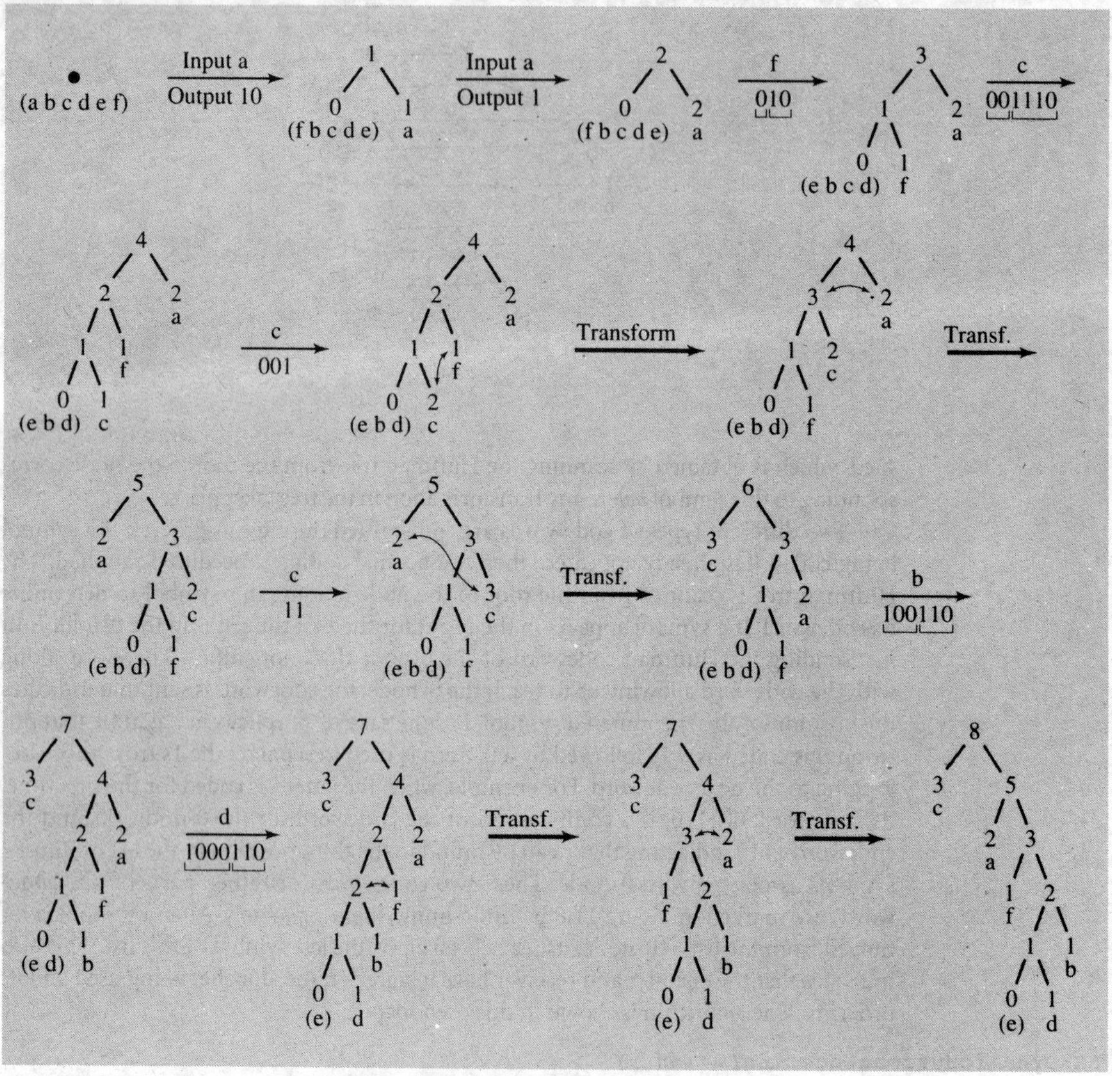

```
if p violates the sibling property
    if the leader of the block_i that still includes p is not parent(p)
        swap p with the leader;
    p = parent(p);
    increment counter(p);
return codeword c;
```

A step-by-step example for string *aafcccbd* is shown in Figure 11.8.

1. Initially, the tree includes only the 0-node with all the source letters (a, b, c, d, e, f). After the first input letter, a, only the codeword for the position occupied by a in the 0-node is output. Because it is the first position, one 1 is output, followed by a 0. The last letter in the 0-node is placed in the first position, and a separate node is created for the letter a. The node, with the frequency count set to 1, becomes a child of another new node that is also the parent of the 0-node.

2. After the second input letter, also an a, 1 is output, which is the Huffman codeword for the leaf that includes a. The frequency count of a is incremented to 2, which violates the sibling property, but because the leader of the block is the parent of node p (that is, of node a), no swap takes place; only p is updated to point to its parent, and then p's frequency count is incremented.

3. The third input letter, f, is a letter output for the first time; thus, the Huffman codeword for the 0-node, 0, is generated first, followed by the number of 1s corresponding to the position occupied by f in the 0-node, followed by 0: 10. The letter e is put in place of letter f in the 0-node, a new leaf for f is created, and a new node becomes the parent of the 0-node and the leaf just created. Node p, which is the parent of leaf f, does not violate the sibling property, so p is updated, p = *parent*(p), thereby becoming the root that is incremented.

4. The fourth input letter is c, which appears for the first time in the input. The Huffman codeword for the 0-node is generated, followed by three 1s and a 0 because c is the third letter in the 0-node. After that, d is put in place of c in the 0-node, and c is put in a newly created leaf; p is updated twice, allowing for incrementing counters of two nodes, left child of the root and the root itself.

5. The letter c is the next input letter; thus, first the Huffman codeword for its leaf is given, 001. Next, because the sibling property is violated, the node p (that is, the leaf c), is swapped with the leader f of $block_1$ that still includes this leaf. Then, p = *parent*(p), and the new parent p of the c node is incremented, which leads to another violation of the sibling property and to an exchange of node p with the leader of $block_2$, namely, with the node a. Next, p = *parent*(p), node p is incremented, but because it is the root, the process of updating the tree is finished.

6. The sixth input letter is c, which has a leaf in the tree, so first, the Huffman codeword, 11, of the leaf is generated and the counter of node c is incremented. The node p, which is the node c, violates the sibling property, so p is swapped with the leader, node a, of $block_3$. Now, p = *parent*(p), p's counter is incremented, and because p is the root, the tree transformation is concluded for this input letter. The remaining steps can be traced in Figure 11.8.

It is left to the reader to make appropriate modifications to this pseudocode to obtain an `FGKDynamicHuffmanDecoding`(*codeword* c) algorithm.

It is possible to design a Huffman coding that does not require any initial knowledge of the set of symbols used by the encoder (Cormack and Horspool, 1984). The Huffman tree is initialized to a special escape character. If a new symbol is to be sent, it is preceded by the escape character (or its current codeword in the tree) and followed by the symbol itself. The receiver can now know this symbol so that if its codeword arrives later on, it can be properly decoded. The symbol is inserted in the tree by making the leaf L with the lowest frequency a nonleaf so that L has two children, one pertaining to the symbol previously in L and one to the new symbol.

Adaptive Huffman coding surpasses simple Huffman coding in two respects: It requires only one pass through the input, and it adds only an alphabet to the output. Both versions are relatively fast, and more important, they can be applied to any kind of file, not only to text files. In particular, they can compress object or executable files. The problem with executable files, however, is that they generally use larger character sets than source code files, and the distribution of these characters is more uniform than in text files. Therefore, the Huffman trees are large, the codewords are of similar length, and the output file is not much smaller than the original; it is compressed merely by 10–20 percent.

11.3 RUN-LENGTH ENCODING

A *run* is defined as a sequence of identical characters. For example, the string $s = $ "aaabba" has three runs: a run of three "a"s followed by runs of two "b"s and one "a." The run-length encoding technique takes advantage of the presence of runs and represents them in an abbreviated, compressed form.

If runs are of the same characters, as in the string $s = $ "nnnn***r%%%%%%%," then instead of transmitting this string, information about runs can be transferred. Each run is coded by the pair (n, ch), where ch is a character and n is the integer representing the number of consecutive characters ch in the run. The string s is coded as 4n3*1r7%. However, a problem arises if one of the characters being transferred is a digit, as in 11111111111544444, which is represented as 1111554 (for eleven 1s, one 5, and five 4s). Therefore, for each run, instead of the number n, a character can be used whose ASCII value is n. For example, the run of 43 consecutive letters "c" is represented as +c ("+" has ASCII code 43), and the run of 49 1s is coded as 11 ("1" has ASCII code 49).

This technique is efficient only when at least two-character runs are transmitted, because for one-character runs, the codeword is twice as long as the character. Therefore, the technique should be applied only to runs of at least two characters. This requires using a marker indicating that what is being transmitted is either a run in an abbreviated form or a literal character. Three characters are needed to represent a run: a compression marker cm, a literal character ch, and a counter n, which make up a triple $\langle cm, ch, n \rangle$. The problem of choosing the compression marker is especially delicate, because it should not be confused with a literal character being transmitted. If a regular text file is transmitted, then the character '~'+1 can be chosen. If there is no restriction on the characters transmitted, then whenever the compression marker itself occurs in the input file, we transmit the compression marker twice. The decoder discards one such marker upon receiving two of them in a row and retains just one as part of the data being received, for example, \\ to print just one backslash. Because for each literal marker two of them must be sent, an infrequently used marker should be chosen. In addition, runs of markers are not sent in compressed form.

Compressing runs results in a sequence of three characters, so this technique should be applied to runs of at least four characters. The maximum length of a run that can be represented by the triple $\langle cm, ch, n \rangle$ is 255 for 8-bit ASCII if the number n

represents the number of characters in the run. But because only runs of four or more characters are encoded, n can represent the number of actual characters in the run minus 4; for example, if $n = 1$, then there are five characters in the run. In this case, the longest run representable by one triple has 259 characters.

Run-length encoding is only modestly efficient for text files in which only the blank character has a tendency to be repeated. In this case, a predecessor of this technique can be applied, *null suppression*, which compresses only runs of blanks and eliminates the need to identify the character being compressed. As a result, pairs $\langle cm, n \rangle$ are used for runs of three or more blanks. This simple technique is used in the IBM 3780 BISYNC transmission protocol where throughput gain is between 30 and 50 percent.

Run-length encoding is very useful when applied to files that are almost guaranteed to have many runs of at least four characters. One example is relational databases. All records in the same relational database file have to be of equal length. Records (rows, tuples) are collections of fields, which may be—and most often are—longer than the information stored in them. Therefore, they have to be padded with some character, thereby creating a large collection of runs whose only purpose is to fill up free space in each field of every record.

Another candidate for compression using run-length encoding is fax images, which are composed of combinations of black and white pixels. For low resolution, there are about 1.5 million pixels per page.

A serious drawback of run-length encoding is that it relies entirely on the occurrences of runs. In particular, this method taken by itself is unable to recognize the high frequency of the occurrence of certain symbols that call for short codewords. For example, AAAABBBB can be compressed, because it is composed of two runs, but ABABABAB cannot, although both messages are made up of the same letters. On the other hand, ABABABAB is compressed by Huffman encoding into the same number of codewords as AAAABBBB without taking into consideration the presence of runs. Therefore, it seems appropriate to combine both methods as in this chapter's case study.

11.4 ZIV-LEMPEL CODE

The problem with some of the methods discussed thus far is that they require some knowledge about the data before encoding takes place. A "pure form" of the Huffman encoder has to know the frequencies of symbol occurrences before codewords are assigned to the symbols. Some versions of the adaptive Huffman encoding can circumvent this limitation, not by relying on previous knowledge of the source characteristics, but by building this knowledge in the course of data transmission. Such a method is called a *universal coding scheme,* and Ziv-Lempel code is an example of a universal data compression code.

In a version of the Ziv-Lempel method called LZ77, a buffer of symbols is maintained. The first l_1 positions hold the l_1 most recently encoded symbols from the input, and the remaining l_2 positions contain the l_2 symbols about to be encoded. In each iteration, starting from one of the first l_1 positions, the buffer is searched for a substring matching a prefix of a string located in the second portion of the buffer. If such a match is found, a codeword is transmitted; the codeword is a triple composed of the position in which the match was found, the length of the match, and the first mismatching symbol. Then, the entire content of the buffer is shifted to the left by the length of match plus one. Some symbols are shifted out. Some new symbols from the input are shifted in. To initiate this process, the first l_1 positions are filled with l_1 copies of the first symbol of the input.

As an example, consider the case when $l_1 = l_2 = 4$, and the input is the string "aababacbaacbaadaaa...." Positions in the buffer are indexed with the numbers 0–7. The initial situation is shown at the top of Figure 11.9. The first symbol of the input is "a," and positions 0 through 3 are filled up with "a"s. The first four symbols of the input, "aaba," are placed in the remaining positions. The longest prefix matching any substring that begins in any position between 0 and 3 is "aa." Therefore, the generated codeword is a triple ⟨2, 2, b⟩, or simply 22b: The match starts in position two, it is two symbols long, and the symbol following this match is "b." Next, a left shift occurs, three "a"s are shifted out, and the string "bac" is shifted in. The longest match also starts in position two and is three symbols long, namely, "aba," with "c" following it. The issued codeword is 23c. Figure 11.9 illustrates a few more steps.

FIGURE 11.9 Encoding the string "aababacbaacbaadaaa..." with LZ77.

Input	Buffer	Code Transmitted
aababacbaacbaadaa...	aaaa	a
aababacbaacbaadaa...	aaaaaaba	22b
abacbaacbaadaaa...	aaababac	23c
baacbaadaaa...	abacbaac	12a
cbaadaaa...	cbaacbaa	03a
daaa...	cbaadaaa	30d
aaa...	...	

The numbers l_1 and l_2 are chosen in this example so that only 2 bits are needed for each. Because each symbol requires 1 byte (8 bits), one codeword can be stored in 12 bits. Therefore, l_1 and l_2 should be powers of 2, so that no binary number is unused. If l_1 is 5, then 3 bits are needed to code all possible positions 0 through 4, and the 3-bit combinations corresponding to the numbers 5, 6, and 7 are not used.

A more frequently applied version of Ziv-Lempel algorithm called LZW uses a table of codewords created during data transmission. A simple algorithm for encoding can be presented as follows (Miller and Wegman, 1985; Welch, 1984):

```
LZWcompress()
    enter all letters to the table;
    initialize string s to the first letter from input;
    while any input left
```

```
        read character c;
        if s+c is in the table
            s = s+c;
        else output codeword(s);
            enter s+c to the table;
            s = c;
      output codeword(s);
```

String s is always at least one character long. After reading a new character, the concatenation of string s and character c is checked in the table. A new character is read if the concatenation s+c is in the table. If it is not, the codeword for s is output, the concatenation s+c is stored in the table, and s is initialized to c. Figure 11.10 shows a trace of the execution of this procedure applied to the input "aababacbaacbaadaaa....". The figure contains the generated output; the strings included in the table in full form and in abbreviated form, represented by a number and a character.

FIGURE 11.10 LZW applied to the string "aababacbaacbaadaaa...."

Encoder		Table	
Input	Output	Index (Codeword)	Full String	Abbreviated String
		1	a	a
		2	b	b
		3	c	c
a		4	d	d
a	1	5	aa	1a
b	1	6	ab	1b
ab	2	7	ba	2a
a	6	8	aba	6a
c	1	9	ac	1c
ba	3	10	cb	3b
ac	7	11	baa	7a
baa	9	12	acb	9b
d	11	13	baad	11d
aa	4	14	da	4a
a	5	15	aaa	5a
...				

A crucial component of efficiency is the organization of the table. Clearly, for more realistic examples, hundreds and thousands of entries can be expected in this table so that an efficient searching method has to be used. A second concern is the size of the table, which grows particularly when new long strings are entered in it. The problem of size is addressed by storing in the table codewords for the prefix and the last characters of strings. For example, if "ba" is assigned the codeword 7, then "baa" can be stored in the table as a number of its prefix, "ba," and the last character, "a," that is, as 7a. In this way, all table entries have the same length. The problem of searching is addressed by using a hash function.

For decoding, the same table is created by updating it for each incoming code except the first. For each codeword, a corresponding prefix and a character are retrieved from the table. Because the prefix is also a codeword (except for single characters), it requires another table lookup, as the entire string is decoded. This is clearly a recursive procedure that may be implemented with an explicit stack. This is necessary because the decoding process applied to prefixes yields a string in the reverse order. The decoding procedure can be summarized as follows:

```
LZWdecompress()
    enter all letters to the table;
    read priorcodeword and output one character corresponding to it;
    while codewords are still left
        read codeword;
        if codeword is not in the table    // special case: c+s+c+s+c, also if s is null;
            enter in table string(priorcodeword) + firstchar(string(priorcodeword));
            output string(priorcodeword) + firstchar(string(priorcodeword));
        else enter in table string(priorcodeword) + firstchar(string(codeword));
            output string(codeword);
        priorcodeword = codeword;
```

This relatively simple algorithm has to consider a special case, when a codeword being processed has no corresponding entry in the table. This situation arises when the string being decoded contains a substring "cScSc," where "c" is a single character, and "cS" is already in the table.

All of the discussed compression algorithms are widely used. UNIX has three compression programs: *pack* uses the Huffman algorithm, *compact* is based on the adaptive Huffman method, and *compress* uses LZW coding. According to system manuals, *pack* compresses text files by 25–40 percent, *compact* by 40 percent, and *compress* by 40–50 percent. The rate of compression is better for Ziv-Lempel coding. It is also faster.

11.5 CASE STUDY: HUFFMAN METHOD WITH RUN-LENGTH ENCODING

As indicated in the discussion of run-length encoding, this method is suitable for files that are almost guaranteed to have many runs of at least four symbols; otherwise, no compression is achieved. The Huffman algorithm, on the other hand, can be applied

Section 11.5 Case Study: Huffman Method with Run-Length Encoding ■ 585

to files with any runs, including runs of one to three symbols long. This method can be applied to single symbols, such as letters, but also to pairs of symbols, to triples, and to a collection of variable length sequences of symbols. Incorporating run-length encoding in the Huffman method works exceedingly well for files with many long runs and moderately well for files with a small number of runs and a large number of different symbols.

For files with no runs, this method is reduced to plain Huffman encoding. In this approach, a file to be compressed is scanned first to determine all runs, including one-, two-, and three-symbols long runs. Runs composed of the same symbols but of different lengths are treated as different "super-symbols," which are used to create a Huffman tree. For example, if the message to be compressed is AAABAACCAABA, then the super-symbols included in the Huffman tree are AAA, B, AA, CC, and A, and not symbols A, B, and C. In this way, the number of codewords to be created grows from three for the symbols to five for the super-symbols. The conversion table becomes larger, but the codewords assigned to the runs are much shorter than in straight run-length encoding. In run-length encoding, this codeword is always 3 bytes long (24 bits). In Huffman code, it may be even 1 bit long.

First, an input file is scanned and all super-symbols are collected in the vector `data` by the function `garnerData()` and sorted according to the frequency of occurrence. Figure 11.11a illustrates the positions of the data in the sorted vector. Next, the sorted data are stored in the output file to be used by the decoder to create

FIGURE 11.11 (a) Contents of the array `data` after the message AAABAACCAABA has been processed. (b) Huffman tree generated from these data.

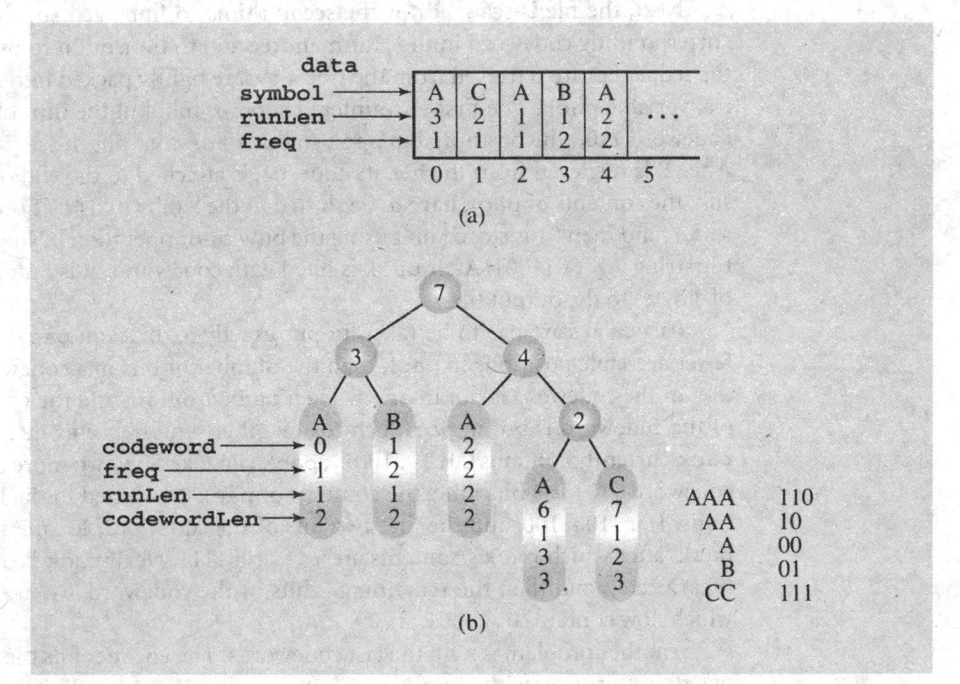

the same Huffman tree that the encoder is about to create. `createHuffmanTree()` generates the tree of Huffman codewords using information collected in `data`. To that end, a doubly linked list of single node trees similar to the list in Figure 11.3 is created first. Then, repeatedly, the two trees with the lowest frequencies are combined to create one tree, which eventually results in one Huffman tree, as in Figure 11.11b.

After the tree has been created, the positions of all nodes, in particular the leaves, can be determined whereby the codewords of all symbols in the leaves can be generated. Each node in this tree has seven data members, but only five of them are shown, just for leaves. The codewords are stored as numbers that represent binary sequences of 0s and 1s. For example, the codeword for CC is 7, 111 in binary. However, these numbers are always the same length, and 7 is stored as 3 bits set to 1 preceded by 29 bits set to 0, 0 . . . 0111. It is, therefore, unclear how many bits out of 32 are included in the sequence representing the codeword for a certain symbol. Is it 111, 0111, 00111, or some other sequence? The codeword field for single As is 0. Is the codeword for A 0, 00, 000, or some more 0s? To avoid ambiguity, the `codewordLen` field stores the number of bits included in the codeword for a given symbol. `codewordLen` for A is 2 and `codeword` is 0, so the codeword sequence representing A is 00.

After the Huffman tree is generated and the leaves are filled with relevant information, the process of coding information in the input file can be initiated. Because searching for particular symbols directly in the tree is too time-consuming, an array `chars[]` of linked lists corresponding to each ASCII symbol is created. The nodes of the linked lists are simply leaves of the tree linked through right pointers, and each list has as many nodes as the number of different run lengths of a given symbol in the input file. It gives immediate access to a particular linked list, but some linked lists may be long if there are many run lengths of a given symbol.

Next, the file is scanned for the second time to find each super-symbol and its corresponding codeword in the Huffman tree and to transmit it to the output file. As the sequences are retrieved from the tree, they are tightly packed into a 4-byte numerical variable `pack`. The first encountered super-symbol in the input file is AAA with codeword 110, which is stored in `pack` so that `pack` contains the sequence 0 . . . 0110. After B is retrieved from the file, its code, 01, is attached to the end of `pack`. As a result, the contents of `pack` have to be shifted to the left by two positions to make room for 01, and then 01 is stored in it using the bitwise-or operation |. Now, `pack` contains the string 0 . . . 011001. After `pack` is filled with codewords, it is output as a sequence of 4 bytes to the output file.

Particular care has to be taken to put exactly 32 bytes in `pack`. When there are fewer available positions in `pack` than the number of bits in a codeword, only a portion of the codeword is put in `pack`. Then `pack` is output and the remaining portion of the codeword is put in `pack` before any other symbol is encoded. For example, if `pack` currently contains 001 . . . 10011, `pack` can take only two more bits. Because the codeword 1101 is 4 bits long, the contents of `pack` are shifted to the left by two positions, 1 . . . 1001100, and the first two bits of the codeword, 11, are put at the end of `pack`, after which `pack`'s contents are 1 . . . 1001111. Next, `pack` is output as 4 bytes (characters), and then the remaining 2 bits of the codeword, 01, are put into `pack`, which now contains 0 . . . 001.

Another problem is with the last codewords. The encoder fills the output file with bytes (in this case, with chunks of 4 bytes), each containing 8 bits. What happens if there are no symbols left, but there is still room in `pack`? The decoder has to know that some bits at the end of the file should not be decoded. If they are, some spurious characters will be added to the decoded file. In this implementation, the problem is solved by transmitting at the beginning of the encoded file the number of characters to be decoded. The decoder decodes only this number of codewords. Even if some bits are left in the encoded file, they are not included in the decoding process. This is a problem that arises in our example. The message AAABAACCAABA is encoded as the sequence of codewords 110, 01, 10, 111, 10, 01, 00, and the contents of `pack` are 00000000000000001100110111100100. If the encoding process is finished, the contents are shifted to the left by the number of unused bits, whereby `pack` becomes 11001101111001000000000000000000 and is output as a sequence of 4 bytes, 11001101, 1100100, 00000000, and 00000000 or, in more readable decimal notation, as 205, 228, 0, and 0. The last 16 bits do not represent any codewords, and if it is not indicated, they are decoded as eight As, whose codeword is 00. To prevent this, the output file includes the number of encoded characters, namely, 12: A, A, A, B, A, A, C, C, A, A, B, and A. The output file also includes the number of all symbols in the Huffman tree. For this example, it is the number 5, because five different super-symbols can be found in the input file and in the Huffman tree: AAA, B, AA, CC, and A. Therefore, the structure of the output file is as follows: the number of super-symbols, `dataSize`, number of characters, contents of `data` (symbols, run lengths, and frequencies), and codewords of all super-symbols found in the input file.

The decoder is much simpler than the encoder because it uses information supplied by the encoder in the header of the encoded message. The decoder re-creates first the array `data[]` in `inputFrequencies()`, then reconstructs the Huffman tree with the same `createHuffmanTree()` and `createCodewords()` decoder used, and finally, in `decode()`, scans the tree in the order determined by the stream of bits in the compressed file to find in its leaves the encrypted symbols.

As expected, this implementation gives particularly good results for database files, with a compression rate of 60 percent. The compression rate for LISP files is 50 percent (runs of parentheses); for text files, 40 percent; and for executable files, merely 13 percent.

Figure 11.12 contains the complete code for the encoder.

FIGURE 11.12 Implementation of Huffman method with run-length encoding.

```
//********************   HuffmanCoding.h   ********************

#include <vector>
#include <algorithm>

class HuffmanNode {
public:
    char symbol;
    unsigned long codeword, freq;
```

Continues

FIGURE 11.12 *(continued)*

```cpp
        unsigned int runLen, codewordLen;
        HuffmanNode *left, *right;
        HuffmanNode() {
            left = right = 0;
        }
        HuffmanNode(char s, unsigned long f, unsigned int r,
            HuffmanNode *lt = 0, HuffmanNode *rt = 0) {
            symbol = s; freq = f; runLen = r; left = lt; right = rt;
        }
    };

    class ListNode {
    public:
        HuffmanNode *tree;
        ListNode *next, *prev;
        ListNode() {
            next = prev = 0;
        }
        ListNode(ListNode *p, ListNode *n) {
            prev = p; next = n;
        }
    };

    class DataRec {
    public:
        char symbol;
        unsigned int runLen;
        unsigned long freq;
        DataRec() {
        }
        bool operator== (const DataRec& dr) const { // used by find();
            return symbol == dr.symbol && runLen == dr.runLen;
        }
        bool operator< (const DataRec& dr) const {  // used by sort();
            return freq < dr.freq;
        }
    };

    class HuffmanCoding {
    public:
        HuffmanCoding() : mask(0xff), bytes(4), bits(8), ASCII(256) {
            chars = new HuffmanNode*[ASCII+1];
        }
```

FIGURE 11.12 *(continued)*

```
    void compress(char*,ifstream&);
    void decompress(char*,ifstream&);
private:
    const unsigned int bytes, bits, ASCII;
    unsigned int dataSize;
    const unsigned long mask;
    unsigned long charCnt;
    ofstream fOut;
    HuffmanNode *HuffmanTree, **chars;
    vector<DataRec> data;
    void error(char *s) {
        cerr << s << endl; exit(1);
    }
    void output(unsigned long pack);
    void garnerData(ifstream&);
    void outputFrequencies(ifstream&);
    void read2ByteNum(unsigned int&,ifstream&);
    void read4ByteNum(unsigned long&,ifstream&);
    void inputFrequencies(ifstream&);
    void createHuffmanTree();
    void createCodewords(HuffmanNode*,unsigned long,int);
    void transformTreeToArrayOfLists(HuffmanNode*);
    void encode(ifstream&);
    void decode(ifstream&);
};

void HuffmanCoding::output(unsigned long pack) {
    char *s = new char[bytes];
    for (int i = bytes - 1; i >= 0; i--) {
        s[i] = pack & mask;
        pack >>= bits;
    }
    for (i = 0; i < bytes; i++)
        fOut.put(s[i]);
}

void HuffmanCoding::garnerData(ifstream& fIn) {
    char ch, ch2;
    DataRec r;
    vector<DataRec>::iterator i;
    r.freq = 1;
    for (fIn.get(ch); !fIn.eof(); ch = ch2) {
```

Continues

FIGURE 11.12 *(continued)*

```
            for (r.runLen = 1, fIn.get(ch2); !fIn.eof() && ch2 == ch;
r.runLen++)
            fIn.get(ch2);
        r.symbol = ch;
        if ((i = find(data.begin(),data.end(),r)) == data.end())
            data.push_back(r);
        else i->freq++;
    }
    sort(data.begin(),data.end());
}

void HuffmanCoding::outputFrequencies(ifstream& fIn) {
    unsigned long temp4;
    char ch = data.size();
    unsigned int temp2 = data.size();
    temp2 >>= bits;
    fOut.put(char(temp2)).put(ch);
    fIn.clear();
    output((unsigned long)fIn.tellg());
    for (int j = 0; j < data.size(); j++) {
        fOut.put(data[j].symbol);
        ch = temp2 = data[j].runLen;
        temp2 >>= bits;
        fOut.put(char(temp2)).put(ch);
        temp4 = data[j].freq;
        output(temp4);
    }
}

void HuffmanCoding::read2ByteNum(unsigned int& num, ifstream& fIn) {
    num = fIn.get();
    num <<= bits;
    num |= fIn.get();
}

void HuffmanCoding::read4ByteNum(unsigned long& num, ifstream& fIn) {
    num = (unsigned long) fIn.get();
    for (int i = 1; i < 4; i++) {
        num <<= bits;
        num |= (unsigned long) fIn.get();
    }
}
```

FIGURE 11.12 *(continued)*

```cpp
void HuffmanCoding::inputFrequencies(ifstream& fIn) {
    DataRec r;
    read2ByteNum(dataSize,fIn);
    read4ByteNum(charCnt,fIn);
    data.reserve(dataSize);
    for (int j = 0; !fIn.eof() && j < dataSize; j++) {
        r.symbol = fIn.get();
        read2ByteNum(r.runLen,fIn);
        read4ByteNum(r.freq,fIn);
        data.push_back(r);
    }
}

void HuffmanCoding::createHuffmanTree() {
    ListNode *p, *newNode, *head, *tail;
    unsigned long newFreq;
    head = tail = new ListNode;              // initialize list pointers;
    head->tree = new
                 HuffmanNode(data[0].symbol,data[0].freq,data[0].runLen);
    for (int i = 1; i < data.size(); i++) {  // create the rest of the
                                             // list;
        tail->next = new ListNode(tail,0);
        tail = tail->next;
        tail->tree =
            new HuffmanNode(data[i].symbol,data[i].freq,data[i].runLen);
    }
    while (head != tail) {                   // create one Huffman tree;
        newFreq = head->tree->freq + head->next->tree->freq; // two
                                             // lowest frequencies
        for (p = tail; p != 0 && p->tree->freq > newFreq; p = p->prev);
        newNode = new ListNode(p,p->next);
        p->next = newNode;
        if (p == tail)
            tail = newNode;
        else newNode->next->prev = newNode;
        newNode->tree =
            new HuffmanNode('\0',newFreq,0,head->tree,head->next->tree);
        head = head->next->next;
        delete head->prev->prev;
        delete head->prev;
        head->prev = 0;
    }
}
```

Continues

FIGURE 11.12 (continued)

```
        HuffmanTree = head->tree;
        delete head;
}

void HuffmanCoding::createCodewords(HuffmanNode *p, unsigned long
codeword, int level) {
    if (p->left == 0 && p->right == 0) {          // if p is a leaf,
        p->codeword    = codeword;                // store codeword
        p->codewordLen = level;                   // and its length,
    }
    else {                                        // otherwise add 0
        createCodewords(p->left,  codeword<<1,    level+1); // for left
                                                            // branch
        createCodewords(p->right,(codeword<<1)+1,level+1); // and 1 for
                                                            // right;
    }
}

void HuffmanCoding::transformTreeToArrayOfLists(HuffmanNode *p) {
    if (p->left == 0 && p->right == 0) {          // if p is a leaf,
        p->right = chars[(unsigned char)p->symbol]; // include it in
        chars[(unsigned char)p->symbol] = p;      // a list associated
    }                                             // with symbol found in p;
    else {
        transformTreeToArrayOfLists(p->left);
        transformTreeToArrayOfLists(p->right);
    }
}

void HuffmanCoding::encode(ifstream& fIn) {
    unsigned long packCnt = 0, hold, maxPack = bytes*bits, pack = 0;
    char ch, ch2;
    int bitsLeft, runLength;
    for (fIn.get(ch); !fIn.eof(); ) {
        for (runLength = 1, fIn.get(ch2); !fIn.eof() && ch2 == ch;
             runLength++)
            fIn.get(ch2);
        for (HuffmanNode *p = chars[(unsigned char) ch];
             p != 0 && runLength != p->runLen; p = p->right)
            ;
        if (p == 0)
            error("A problem in encode()");
        if (p->codewordLen < maxPack - packCnt) {  // if enough room in
```

FIGURE 11.12 *(continued)*

```
            pack = (pack << p->codewordLen) | p->codeword; // pack to
                                                          // store new
            packCnt += p->codewordLen;                    // codeword, shift
        }                                                 // its content to the
                                                          // left and attach
                                                          // new codeword;
        else {                                            // otherwise move
            bitsLeft = maxPack - packCnt;                 // pack's content to
            pack <<= bitsLeft;                            // the left by the
            if (bitsLeft != p->codewordLen) {             // number of left
                hold = p->codeword;                       // spaces and if new
                hold >>= p->codewordLen - bitsLeft;       // codeword is
                                                          // longer than room
                pack |= hold;                             // left, transfer
            }                                             // only as many bits as
                                                          // can fit in pack;
            else pack |= p->codeword;                     // if new codeword
                                                          // exactly fits in
                                                          // pack, transfer it;
            output(pack);                                 // output pack as
                                                          // four chars;
            if (bitsLeft != p->codewordLen) {             // transfer
                pack = p->codeword;                       // unprocessed bits
                packCnt = maxPack - (p->codewordLen - bitsLeft); // of
                packCnt = p->codewordLen - bitsLeft;      // new codeword to
                                                          // pack;
            }
            else packCnt = 0;
        }
        ch = ch2;
    }
    if (packCnt != 0) {
        pack <<= maxPack - packCnt; // transfer leftover codewords and
                                    // some 0s
        output(pack);
    }
}

void HuffmanCoding::compress(char *inFileName, ifstream& fIn) {
    char outFileName[30];
    strcpy(outFileName,inFileName);
    if (strchr(outFileName,'.'))            // if there is an extension
```

Continues

FIGURE 11.12 *(continued)*

```cpp
            strcpy(strchr(outFileName,'.')+1,"z");// overwrite it with 'z'
    else strcat(outFileName,".z");            // else add extension '.z';
    fOut.open(outFileName,ios::out|ios::binary);
    garnerData(fIn);
    outputFrequencies(fIn);
    createHuffmanTree();
    createCodewords(HuffmanTree,0,0);
    for (int i = 0; i <= ASCII; i++)
        chars[i] = 0;
    transformTreeToArrayOfLists(HuffmanTree);
    fIn.clear();       // clear especially the eof flag;
    fIn.seekg(0,ios::beg);
    encode(fIn);
    fIn.clear();
    cout.precision(2);
    cout << "Compression rate = " <<
            100.0*(fIn.tellg()-fOut.tellp())/fIn.tellg() << "%\n"
         << "Compression rate without table = " <<
            100.0*(fIn.tellg()-
                fOut.tellp()+data.size()*(2+4))/fIn.tellg();
    fOut.close();
}

void HuffmanCoding::decode(ifstream& fIn) {
    unsigned long chars;
    char ch, bitCnt = 1, mask = 1;
    mask <<= bits - 1;          // change 00000001 to 100000000;
    for (chars = 0, fIn.get(ch); !fIn.eof() && chars < charCnt; ) {
        for (HuffmanNode *p = HuffmanTree; ; ) {
            if (p->left == 0 && p->right == 0) {
                for (int j = 0; j < p->runLen; j++)
                    fOut.put(p->symbol);
                chars += p->runLen;
                break;
            }
            else if ((ch & mask) == 0)
                p = p->left;
            else p = p->right;
            if (bitCnt++ == bits) { // read next character from fIn
                fIn.get(ch);         // if all bits in ch are checked;
                bitCnt = 1;
            }                        // otherwise move all bits in ch
            else ch <<= 1;           // to the left by one position;
```

FIGURE 11.12 *(continued)*

```
        }
    }
}

void HuffmanCoding::decompress(char *inFileName, ifstream& fIn) {
    char outFileName[30];
    strcpy(outFileName,inFileName);
    if (strchr(outFileName,'.'))               // if there is an extension
        strcpy(strchr(outFileName,'.')+1,"dec");// overwrite it with 'z'
    else strcat(outFileName,".dec");           // else add extension '.z';
    fOut.open(outFileName,ios::out|ios::binary);
    inputFrequencies(fIn);
    createHuffmanTree();
    createCodewords(HuffmanTree,0,0);
    decode(fIn);
    fOut.close();
}

//*********************** HuffmanEncoder.cpp **********************

#include <iostream>
#include <fstream>
#include <cstring>
using namespace std;
#include "HuffmanCoding.h"

int main(int argc, char* argv[]) {
    char fileName[30];
    HuffmanCoding Huffman;
    if (argc != 2) {
        cout << "Enter a file name: ";
        cin  >> fileName;
    }
    else strcpy(fileName,argv[1]);
    ifstream fIn(fileName,ios::binary);
    if (fIn.fail()) {
        cerr << "Cannot open " << fileName << endl;
        return 0;
    }
    Huffman.decompress(fileName,fIn);
    fIn.close();
    return 0;
}
```

11.6 EXERCISES

1. For which probabilities $P(m_i)$ of n symbols is the average length maximal? When is it minimal?

2. Find L_{ave} for the letters X, Y, and Z and their probabilities .05, .05, and .9 and compare it to L_{Huf} computed for single letters and pairs of letters, as in Figure 11.6. Does L_{Huf} satisfactorily approximate L_{ave}? How can we remedy the problem?

3. Assess the complexity of all the implementations of the Huffman algorithm suggested in this chapter.

4. What are the lengths of the Huffman codewords of the least probable messages with respect to each other?

5. In the adaptive Huffman algorithm, first the codeword for an encountered symbol is issued and then the conversion table is updated. Could the table be updated first and then the new codeword for this symbol be issued? Why or why not?

6. The functions `createCodewords()` and `transformTreeToArrayOfLists()` used in the case study seem to be vulnerable because the first thing they both do is access the pointer `left` of node `p`, which would be dangerous if `p` were null; therefore, the body of both functions should apparently be preceded by the condition `if (p != 0)`. Explain why this is not necessary.

7. What problem arises if, in run-length encoding, triples of the form $\langle cm, n, ch \rangle$ are used instead of triples of the form $\langle cm, ch, n \rangle$?

8. In Figure 11.9, $l_1 = l_2 = 4 = 2^2$. In what respect does the choice of $l_1 = l_2 = 16 = 2^4$ simplify the implementation of the LZ77 algorithm?

9. In which situation does the LZ77 algorithm perform best? Worst?

10. Describe the process of decoding using LZ77. What string is coded by this sequence of codewords: b, 31a, 23b, 30c, 21a, 32b?

11. Using LZW with the table initialized with the letters a, b, c, decode the string coded as 1 2 4 3 1 4 9 5 8 12 2.

11.7 PROGRAMMING ASSIGNMENTS

1. A large number of messages with very low probabilities in a long series of messages require a large number of very long codewords (Hankamer 1979). Instead, one codeword can be assigned to all these messages, and if needed, this codeword is sent along with the message. Write a program for coding and decoding this approach by adapting the Huffman algorithm.

2. Write an encoder and a decoder that use the run-length encoding technique.

3. Write an encoder and a decoder using run-length encoding to transmit voice, with the voice simulated by a certain function f. Voice is generated continuously, but it is measured at $t_0, t_1, \ldots$, where $t_i - t_{i-1} = \delta$, for some time interval δ. If $|f(t_i) - f(t_{i-1})| < \epsilon$ for some tolerance ϵ, then the numbers $f(t_i)$ and $f(t_{i-1})$ are treated as equal. Therefore, for runs of such equal values, a compressed version can be transmitted in the form of a triple $\langle cm, f(t_i), n \rangle$ with cm being a negative number. In Figure 11.13, circles represent the numbers included in a run indicated by the first preceding bullet; in this example, two runs are sent. What is a potential danger of this technique, known also as the *zero-order predictor*? How can this be solved? Try your program on the functions $\frac{\sin n}{n}$ and $\ln n$.

FIGURE 11.13 A function representing voice frequency.

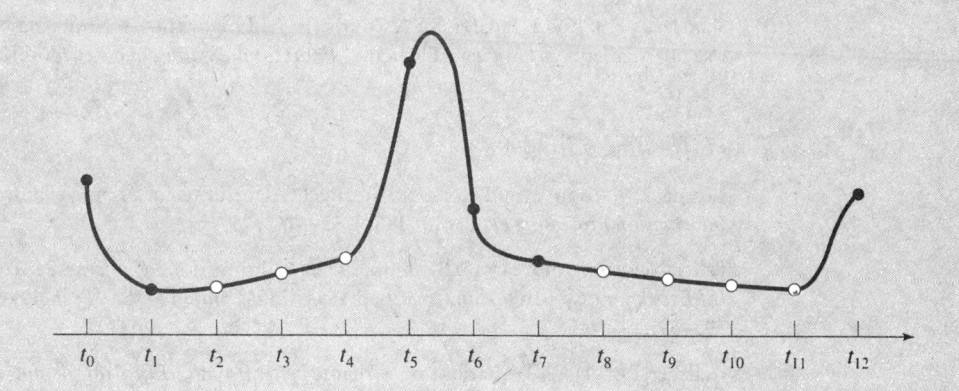

4. Static dictionary techniques are characterized by using a predefined dictionary of patterns encoded with unique codewords. After a dictionary is established, the problem of using it most efficiently still remains. For example, for a dictionary = {*ability, ility, pec, re, res, spect, tab*}, the word *respectability* can be broken down in two ways: *res, pec, tab, ility* and *re, spect, ability*; that is, the first division requires four codewords for this word, whereas the second requires only three. The algorithm parses the word or words and determines which of the two choices will be made. Of course, for a large dictionary, there may be more than two possible parsings of the same word or phrase. By far the most frequently used technique is a *greedy algorithm* that finds the longest match in the dictionary. For our example, the match *res* is longer than *re*; therefore, the word *respectability* is divided into four components with the greedy strategy. An optimal parsing can be found by adapting a shortest path algorithm. (Bell, Cleary, and Witten 1990; Schuegraf and Heaps 1974). Write a program that for a dictionary of patterns compresses a text file. For each string s, create a digraph with $length(s)$ nodes. Edges are labeled with the dictionary patterns, and their codeword lengths are edges' costs. Two nodes i and j are connected with an edge if the dictionary contains a pattern $s[i] \ldots s[j-1]$. The shortest path represents the shortest sequences of codewords for patterns found in the path.

BIBLIOGRAPHY

Data Compression Methods

Bell, Timothy C., Cleary, J. G., and Witten, Ian H., *Text Compression*, Englewood Cliffs, NJ: Prentice Hall, 1990.

Drozdek, Adam, *Elements of Data Compression*, Pacific Grove, CA: Brooks/Cole, 2002.

Lelever, Debra A. and Hirschberg, Daniel S., "Data Compression," *ACM Computing Surveys* 19 (1987), 261–296.

Rubin, Frank, "Experiments in Text File Compression," *Communications of the ACM* 19 (1976), 617–623.

Salomon, David, *Data Compression: The Complete Reference*, New York: Springer, 2000.

Schuegraf, E. J., and Heaps, H. S., "A Comparison of Algorithms for Data-Base Compression by Use of Fragments as Language Elements," *Information Storage and Retrieval* 10 (1974), 309–319.

Huffman Coding

Cormack, Gordon V., and Horspool, R. Nigel, "Algorithms for Adaptive Huffman Codes," *Information Processing Letters* 18 (1984), 159–165.

Faller, Newton, "An Adaptive System for Data Compression," *Conference Record of the Seventh IEEE Asilomar Conference on Circuits, Systems, and Computers*, San Francisco: IEEE (1974), 593–597.

Gallager, Robert G., "Variations on a Theme of Huffman," *IEEE Transactions on Information Theory* IT-24 (1978), 668–674.

Hankamer, M., "A Modified Huffman Procedure with Reduced Memory Requirement," *IEEE Transactions on Communication* COM-27 (1979), 930–932.

Huffman, David A., "A Method for the Construction of Minimum-Redundancy Codes," *Proceedings of the Institute of Radio Engineers* 40 (1952), 1098–1101.

Knuth, Donald E., "Dynamic Huffman Coding," *Journal of Algorithms* 6 (1985), 163–180.

Run-Length Encoding

Pountain, Dick, "Run-Length Encoding," *Byte* 12 (1987), No. 6, 317–320.

Ziv-Lempel Code

Miller, Victor S., and Wegman, Mark N., "Variations on a Theme by Ziv and Lempel," in Apostolico, A., and Galil, Z. (eds.), *Combinatorial Algorithms on Words*, Berlin: Springer (1985), 131–140.

Welch, Terry A., "A Technique for High-Performance Data Compression," *Computer* 17 (1984), 6, 8–19.

Ziv, Jacob, and Lempel, Abraham, "A Universal Algorithm for Sequential Data Compression," *IEEE Transactions on Information Theory* IT-23 (1977), 337–343.

Memory Management 12

The preceding chapters rarely looked behind the scenes to see how programs are actually executed and how variables of different types are stored. The reason is that this book emphasizes data structures rather than the inner workings of the computer. The latter belongs more to a book about operating systems or assembly language programming than to a discussion of data structures.

But at least in one case, such a reference was inescapable, namely, when discussing recursion in Chapter 5. Using recursion was explained in terms of the run-time stack and how a computer actually works. We also alluded to this level when discussing dynamic memory allocation. It is hard to discuss dynamic memory allocation without a keen awareness of the structure of computer memory and the realization that without `new`, pointers may point to unallocated memory locations. In addition, `delete` should be used to avoid exhausting the computer's memory resources. Managing memory in C++ is the responsibility of the programmer, and memory may become clogged with unreachable locations that were not deallocated with `delete` if the programmer is not sufficiently careful. The most efficient and elegant program structure may be disempowered if too much memory is allocated.

The *heap* is the region of main memory from which portions of memory are dynamically allocated upon request of a program. (This heap has nothing to do with the special tree structure called a heap in Section 6.9.) In languages such as FORTRAN, COBOL, or BASIC, the compiler determines how much memory is needed to run programs. In languages that allow dynamic memory allocation, the amount of memory required cannot always be determined prior to the program run. To that end, the heap is used. If a C program requests memory by issuing `malloc()` or `calloc()` and a C++ or Pascal program does it by issuing a call to `new`, a certain amount of bytes is allocated from the heap, and the address to the first byte of this portion is returned. Also, in these languages, unused memory has to be specifically released by the programmer through `dispose()` in Pascal, `free()` in C, and `delete` in C++. In some languages, there is no need to explicitly release memory. Unused memory is simply abandoned and then automatically reclaimed by the operating system. Automatic storage reclamation is a luxury that is not part of every language environment. It emerged with LISP and it is part of functional languages, but logic languages and most object-oriented languages, including Smalltalk, Prolog, Modula-3, Eiffel, and Java, also have automatic storage reclamation.

The maintenance of free memory blocks, assigning specific memory blocks to the user programs if necessary and cleaning memory from unneeded blocks to return them to the memory pool, is performed by a part of the operating system called a *memory manager*. The memory manager also performs other functions, such as scheduling access to shared data, moving code and data between main and secondary memory, and keeping one process away from another. This is particularly important in the multiprogramming system, where many different processes can reside in memory at the same time and the CPU serves for a brief amount of time each of the processes in turn. The processes are put in memory in free spaces and removed if either space is needed for other processes to be served or they've completed execution.

One problem that a well-designed memory manager has to solve is that of the configuration of available memory. When returning memory with `delete`, the programmer has no control over this configuration. In particular, after many allocations and deallocations, the heap is divided into small pieces of available memory sandwiched between chunks of memory in use. If a request comes to allocate n bytes of memory, the request may not be met if there is not enough contiguous memory in the heap, although the total of available memory may far surpass n. This phenomenon is called *external fragmentation*. Changing memory configuration and, in particular, putting available memory in one part of the heap and allocated memory in another solves this problem. Another problem is *internal fragmentation*, when allocated memory chunks are larger than requested. External fragmentation amounts to the presence of wasted space between allocated segments of memory; internal fragmentation amounts to the presence of unused memory inside the segments.

12.1 THE SEQUENTIAL-FIT METHODS

A simple organization of memory could require a linked list of all memory blocks, which is updated after a block is either requested or returned. The blocks on such linked lists can be organized in a variety of ways, according to the block sizes or the block addresses. Whenever a block is requested, a decision has to be made concerning which block to allocate and how to treat the portion of the block exceeding the requested size.

For reasons of efficiency, doubly linked lists of blocks are maintained with links residing in the blocks. Each available block of memory uses a portion of itself for two links. Also, both available and reserved blocks have two fields to indicate their status (available or reserved) and their size.

In the sequential-fit methods, all available memory blocks are linked, and the list is searched to find a block whose size is larger than or the same as the requested size. A simple policy for handling returned blocks of memory is to coalesce them with neighboring blocks and reflect this fact by properly adjusting the links in the linked list.

The order of searching the list for such a block determines the division of these methods into several categories. The *first-fit* algorithm allocates the first block of memory large enough to meet the request. The *best-fit* algorithm allocates a block that is closest in size to the request. The *worst-fit* method finds the largest block on the list so that, after returning its portion equal to the requested size, the remaining part is large enough to be used in later requests. The *next-fit* method allocates the next available block that is sufficiently large.

Figure 12.1a contains a memory configuration after several requests and returns of memory blocks. Figure 12.1b illustrates which portion of memory would be allocated by which sequential-fit method to satisfy a request for 8KB of memory.

FIGURE 12.1 Memory allocation using sequential-fit methods.

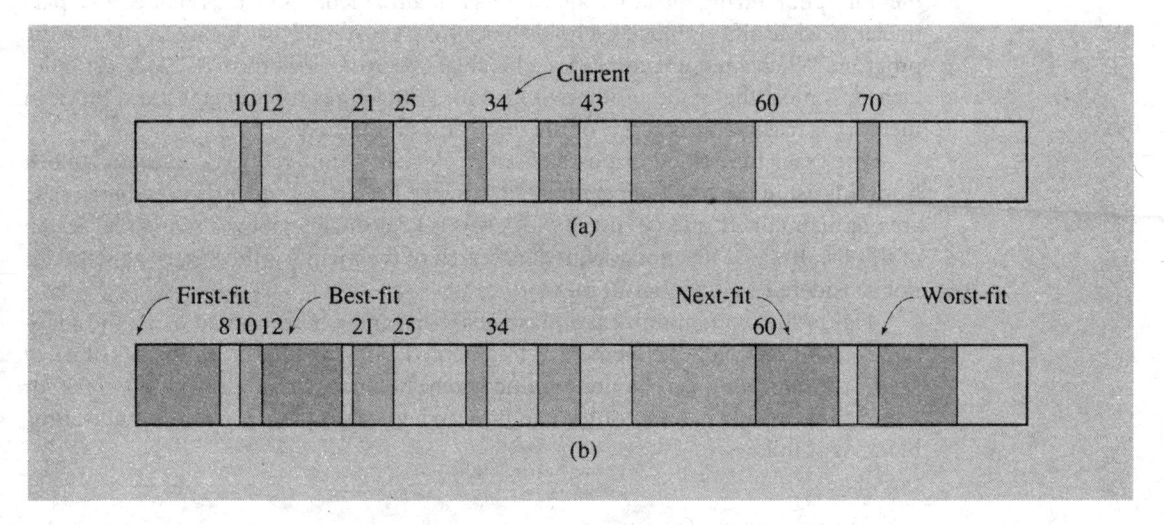

The most efficient method is the first-fit procedure. The next-fit method is of comparable speed but causes more extensive external fragmentation because it scans the list of blocks starting from the current position and reaches the end of the list much earlier than the first-fit method. But the best-fit algorithm is even worse in that respect because it searches for the closest match with respect to size. The parts of blocks remaining after returning the required size are small and practically unusable. The worst-fit algorithm attempts to prevent this type of fragmentation by avoiding, or at least delaying, the creation of small blocks.

The way the blocks are organized on the list determines how fast the search for an available block succeeds or fails. For example, to optimize the best-fit and the worst-fit methods, the blocks should be arranged by size. For other methods, the address ordering is adequate.

12.2 THE NONSEQUENTIAL-FIT METHODS

The sequential-fit methods being what they are, may become inefficient for large memory. In the case of large memory, a nonsequential search is desirable. One strategy divides memory into an arbitrary number of lists, each list holding blocks of the same size (Ross, 1967). Larger blocks are split into smaller blocks to satisfy requests, and new lists may be created. Because the number of such lists can become large, they can be organized as a tree.

Another approach is based on the observation that the number of sizes requested by a program is limited, although the sizes may differ from one program to another. Therefore, the lists of blocks of different sizes can be kept short if it can be determined which sizes are the most popular. This leads to an *adaptive exact-fit* technique that dynamically creates and adjusts storage block lists that fit the requests exactly (Oldehoeft and Allan, 1985).

In adaptive exact-fit, a size-list of block lists of a particular size returned to the memory pool during the last T allocations is maintained. A block b is added to a particular block list if this block list holds blocks of b's size and b has been returned by the program. When a request comes for a block of b's size, a block from its block list is detached to meet the request; otherwise, a more time-consuming search for a block in memory is triggered using one of the sequential-fit methods.

The exact-fit method disposes of entire block lists if no request comes for a block from this list in the last T allocations. In this way, lists of infrequently used block sizes are not maintained, and the list of block lists is kept small to allow a sequential search of this list. Because it is not a sequential search of the memory, the exact-fit method is not considered a sequential-fit method.

Figure 12.2 contains an example of a size-list and a heap created using the adaptive exact-fit method. The memory is fragmented, but if a request comes for a block of size 7, the allocation can be done immediately, because the size-list has an entry for size 7; thus, memory does not have to be searched. A simple algorithm for allocating blocks is as follows:

```
t = 0;
allocate (reqSize)
    t++;
    if a block list b1 with reqSize blocks is on sizeList
        lastref(b1) = t;
        b = head of blocks(b1);
        if b was the only block accessible from b1
            detach b1 from sizeList;
    else b = search-memory-for-a-block-of(reqSize);
    dispose of all block lists on sizeList for which t - lastref(b1) < T;
    return b;
```

A procedure for returning blocks is even simpler.

This algorithm highlights the problem of memory fragmentation. The algorithm must be expanded to deal with this problem successfully. One solution is to write a function to compact memory after a certain number of allocations and deallocations. A noncompacting approach may consist in liquidating the size-list and building it anew after some predetermined period. The authors of this method claim that fragmentation problems "failed to materialize," but that can be attributed to the configurations of their tests. Such problems certainly materialize in sequential-fit methods and in another nonsequential-fit strategy to be discussed in the next section.

FIGURE 12.2 An example configuration of a size-list and heap created by the adaptive exact-fit method.

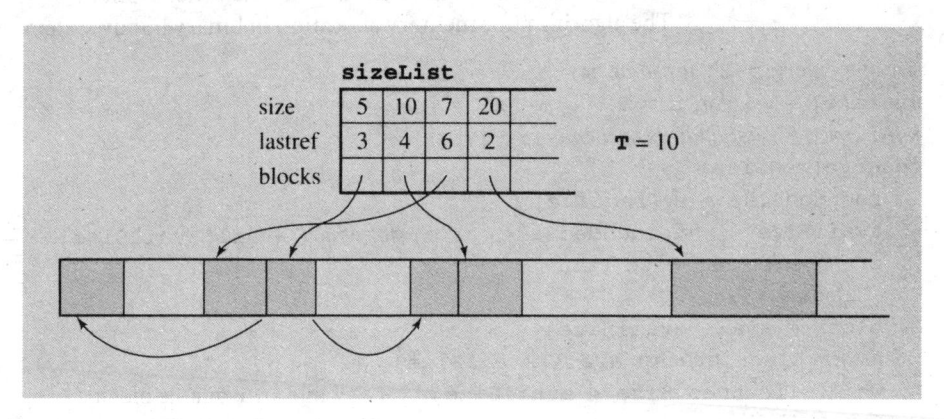

```
              sizeList
     size    | 5 | 10 | 7 | 20 |
     lastref | 3 |  4 | 6 |  2 |      T = 10
     blocks  |   |    |   |    |
```

12.2.1 Buddy Systems

Nonsequential memory management methods known as *buddy systems* do not just assign memory in sequential slices, but divide it into two buddies that are merged whenever possible. In the buddy system, two buddies are never free. A block can have either a buddy used by the program or none.

The classic buddy system is the *binary buddy system* (Knowlton, 1965). The binary buddy system assumes that storage consists of 2^m locations for some integer m, with addresses $0, \ldots, 2^m - 1$, and that these locations can be organized into blocks whose lengths can only be powers of 2. There is also an array `avail[]` such that, for each $i = 0, \ldots, m$, `avail[i]` is the head of a doubly linked list of blocks of the same size, 2^i.

The name of this method is derived from the fact that each block of memory (except the entire memory) is coupled with a buddy *of the same size* that participates with the block in reserving and returning chunks of memory. The buddy of a block of length 2^i is determined by complementing bit $i + 1$ in the address of this block. This is strictly related to the lengths of blocks, which can only be powers of 2. In particular, all blocks of size 2^i have 0s in the i rightmost positions and differ only in the remaining bits. For example, if memory has only eight locations, then the possible addresses of blocks of size one are {000, 001, 010, 011, 100, 101, 110, 111}, addresses of blocks of size two are {000, 010, 100, 110}, of size four {000, 100}, and of size eight {000}. Note that in the second set of addresses, the last bit is 0, and the addresses refer to blocks of size 2^1. The addresses in the third set have two ending 0s, because the size of the blocks is 2^2. Now, for the second set, there are two pairs of blocks and their buddies: {(000, 010), (100, 110)}; for the third set, there is only one pair: (000, 100). Hence, the difference between the address of a block of size 2^i and the address of its buddy is only in bit $i + 1$.

If a request arrives to allocate a memory block of size s, then the buddy system returns a memory block whose size is greater than or equal to s. There are many candidates for such blocks, so the list of such blocks is checked in `avail[]` whose size k is the smallest among all $k \geq s$. This list of blocks can be found in location `avail[k]`. If the list is empty, then the next list of blocks is checked in position $k + 1$, then in position $k + 2$, and so on. The search continues until a nonempty list is found (or the end of avail[ ] is reached), and then a block is detached from it.

The algorithm for memory allocation in binary buddy systems is as follows:

```
size of memory = 2^m for some m;
avail[i] = -1 for i = 0,...,m-1;
avail[m] = first address in memory;
reserve(reqSize)
   roundedSize = ⌈lg(reqSize)⌉;
   availSize = min(roundedSize,...,m) for which avail[availSize] > -1;
   if no such availSize exists
      failure;
   block = avail[availSize];
   detach block from list avail[availSize];
   while (rounded Size < availSize)        // while an available block
      availSize--;                          // is too large - split it;
      block = left half of block;
      insert buddy of block in list avail[availSize];
   return block;
```

Each free block of the buddy system should include four fields indicating its status, its size, and its two neighbors in the list. On the other hand, reserved blocks include only a status field and a size field. Figure 12.3a illustrates the structure of a free block in the buddy system. The block is marked as free with the status field set to 0. The size is specified as 2^5 locations. No predecessor is specified, so this block is pointed to by avail[5]. The size of its successor is also 2^5 locations. Figure 12.3b illustrates a reserved block whose status field is set to 1.

FIGURE 12.3 Block structure in the binary buddy system.

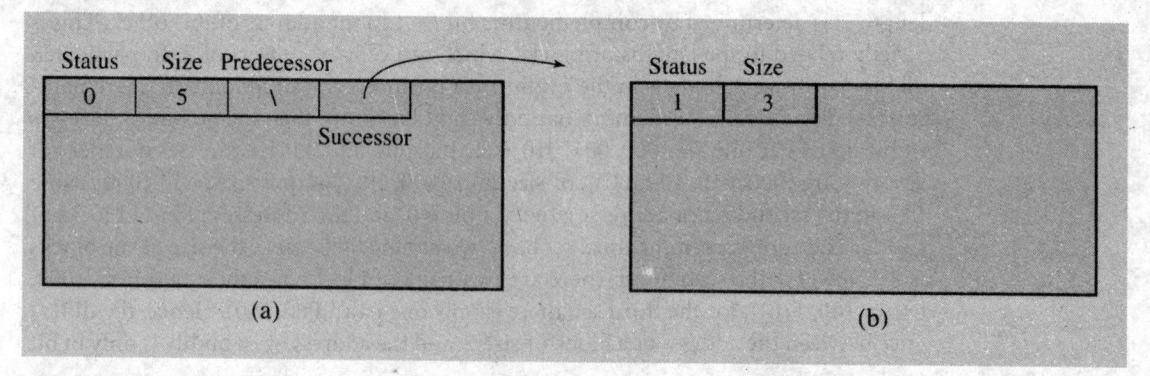

Figure 12.4 contains an example of reserving three blocks, assuming that the memory in use is of size $2^7 = 128$ locations. First, the entire memory is free (Figure 12.4a). Then, 18 locations are requested, so roundedSize = ⌈lg(18)⌉ = 5. But avail-

Size = 7, so the memory is split into two buddies, each of size 2^6. The second buddy is marked as available by setting the status field and including it the list avail[6] (Figure 12.4b). availSize is still greater than roundedSize, so another iteration of the while loop of reserve() is executed. The first block is split into two and the second buddy is included in the list avail[5] (Figure 12.4c). The first buddy is marked as reserved and returned to the caller of reserve() for use. Note that only a portion of the returned block is really needed; however, the entire block is marked as reserved.

Next, a block of 14 locations is requested; now, roundedSize = $\lceil \lg(14) \rceil$ = 4, availSize = 5, and the block pointed to by avail[5] is claimed (Figure 12.4d). This block is too large because roundedSize < availSize, so the block is divided into two buddies. The first buddy is marked as reserved and returned, and the second is included in a list (Figure 12.4e). Finally, a block of 16 locations is requested. After two iterations of the while loop of reserve(), the configuration pictured in Figure 12.4g emerges; there are two available blocks of 16 locations and both are linked up in list avail[4].

To be sure, blocks of memory are not only claimed, but also returned; hence, they have to be included in the pool of available blocks. Before they are included, the status of each block's buddy is checked. If the buddy is available, the block is combined with its buddy to create a block twice as large as before the combination. If the buddy of the new block is available, it is also combined with its buddy, resulting in a still larger block of memory. This process continues until the entire memory is combined into one block or a buddy is not available. This coalescing creates blocks of available memory as large as possible. The algorithm of including a block in the pool of available blocks is as follows:

```
include(block)
   blockSize = size(block);
   buddy = address(block) with bit blockSize+1 set to its complement;
   while status(buddy) is 0                    // buddy has not
         and size(buddy) == blockSize          // been claimed;
         and blockSize != lg(size of memory)   // buddy exists;
      detach buddy from list avail[blockSize];
      block = block plus body;    // coalesce block and its buddy;
      set status(block) to 0;
      blockSize++;
      buddy = address(now extended block) with bit blockSize+1 set to its complement;
   include block in list avail[blockSize];
```

Figure 12.5 illustrates this process. A block previously claimed is now released (Figure 12.5a), and because the buddy of this block is free, it is combined with the block, resulting in a double-sized block, which is included in the list avail[5] (Figure 12.5b). Releasing another block allows the memory manager to combine this block with its buddy and the resulting block with its buddy (Figure 12.5c). Note that the free portion of the leftmost block (marked with the darker screen) did not participate in this coalescing process and is still considered occupied. Also, the two rightmost blocks in Figure 12.5c, although adjacent, were not combined because they are not buddies. Buddies in the binary buddy method have to be of the same size.

FIGURE 12.4 Reserving three blocks of memory using the binary buddy system.

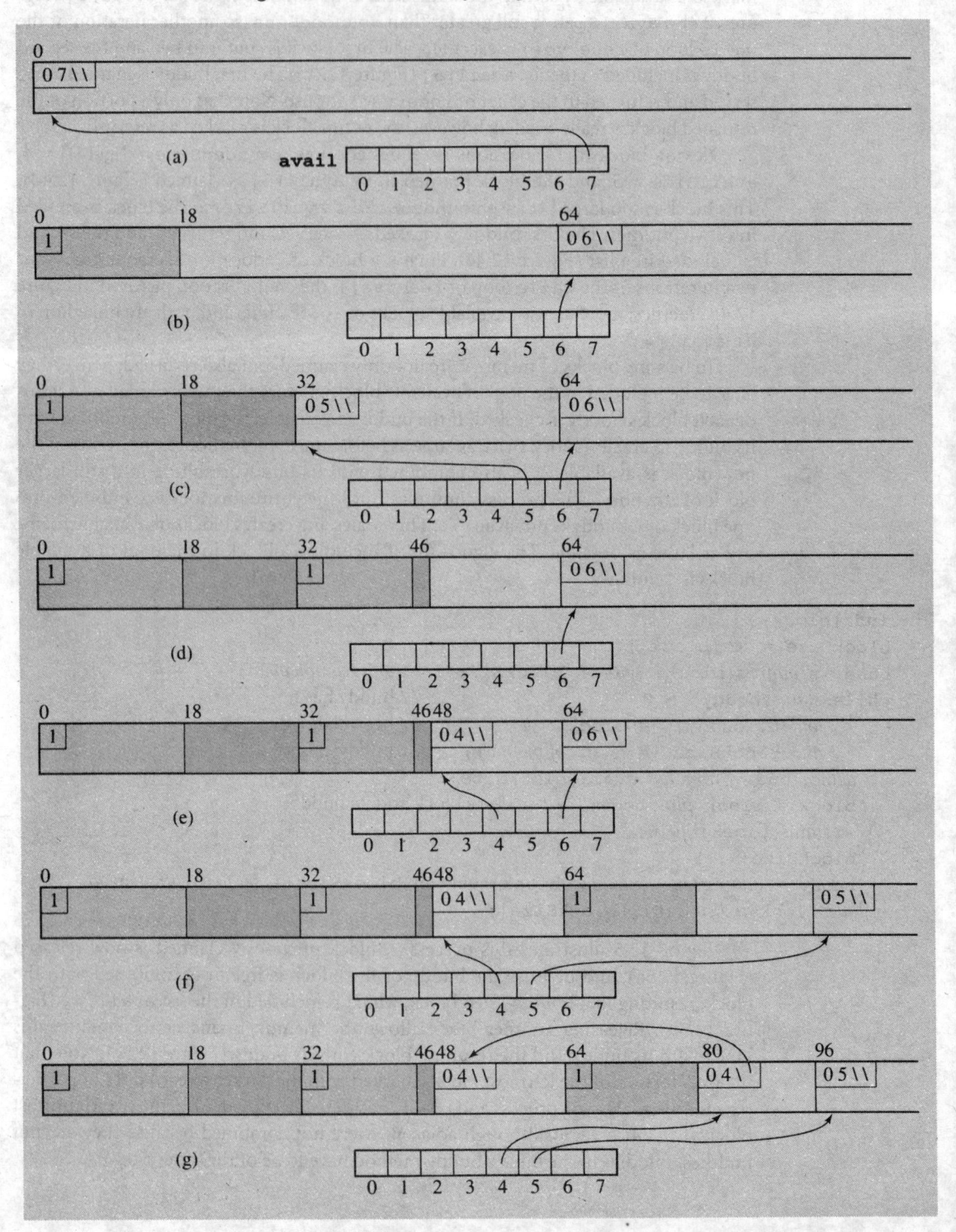

FIGURE 12.5 (a) Returning a block to the pool of blocks, (b) resulting in coalescing one block with its buddy. (c) Returning another block leads to two coalescings.

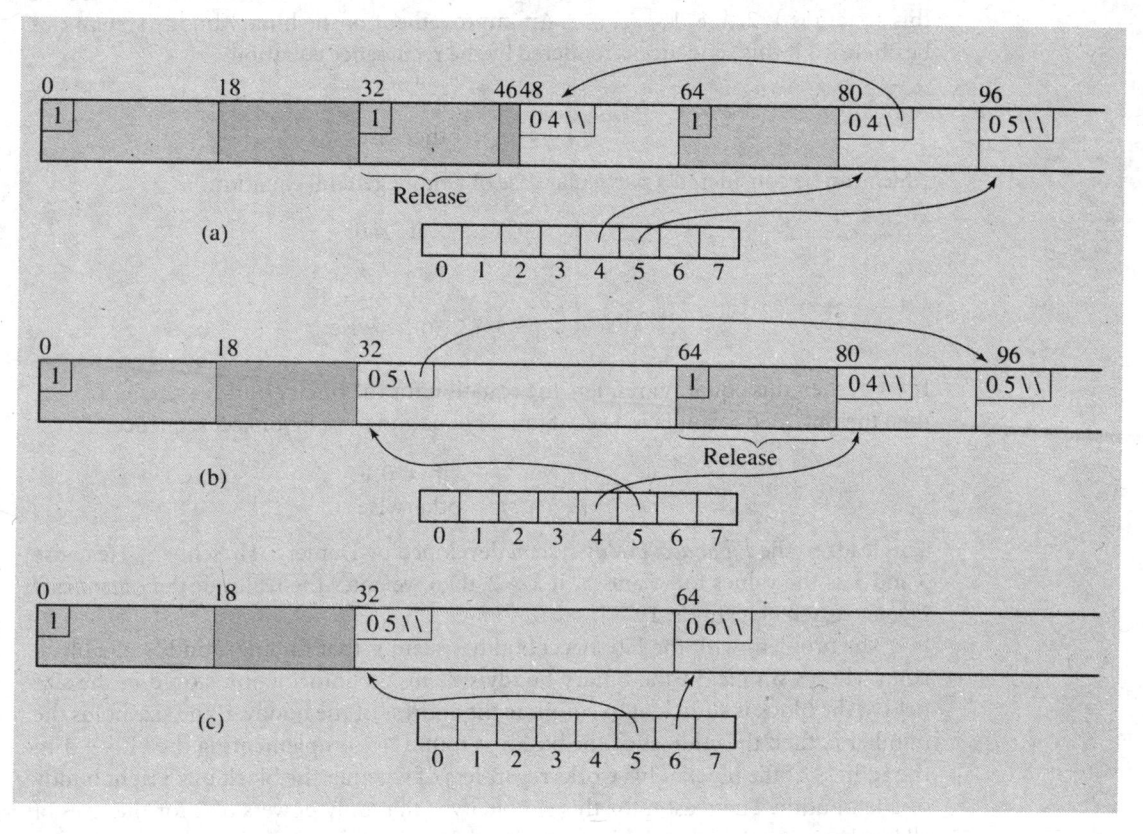

The binary buddy system, although relatively efficient in terms of speed, may be inefficient in terms of space. Figure 12.4d shows that the two leftmost blocks amount to a size of 48 locations, but only 32 of them are in use, because the user really needs 18 + 14 locations. This means that one-third of these two blocks is wasted. This can get even worse if the number of locations requested is always slightly more than a power of 2. In this case, approximately 50 percent of memory is not in actual use. This is a problem with internal fragmentation that results from the need to round all requests to the nearest larger power of 2.

Also, there may be a problem with external fragmentation; a request may be refused although the amount of available space is sufficient to meet it. For example, for the configuration of memory in Figure 12.4g, a request for 50 locations is refused because there is no block available with a size of 64 locations or more. A request for 33 locations is treated similarly and for the same reason, although there are 33 available consecutive locations. But one of these locations belongs to another block, which puts it out of reach.

These problems are brought about by the fact that the binary buddy system uses a simple division of blocks into two even parts, which results in the division of memory not sufficiently tuned to incoming requests. The sequence of block sizes possible in this system is 1, 2, 4, 8, 16, ..., 2^m. An improvement of the binary buddy system can be obtained if this sequence is rendered by the recurrence equation

$$s_i = \begin{cases} 1 & \text{if } i = 0 \\ s_{i-1} + s_{i-1} & \text{otherwise} \end{cases}$$

which can be considered a particular case of a more general equation:

$$s_i = \begin{cases} c_1 & \text{if } i = 0, \\ \vdots & \vdots \\ c_k & \text{if } i = k-1 \\ s_{i-1} + s_{i-2} & \text{otherwise} \end{cases}$$

If $k = 1$, then this equation renders the equation for the binary buddy system. If $k = 2$, then the obtained formula is a very familiar equation for a Fibonacci sequence:

$$s_i = \begin{cases} 1 & \text{if } i = 0, 1 \\ s_{i-1} + s_{i-1} & \text{otherwise} \end{cases}$$

This leads to the *Fibonacci buddy system* developed by Daniel S. Hirschberg. He chose 3 and 5 as the values for s_0 and s_1. If $k > 2$, then we enter the realm of the *generalized Fibonacci systems* (Hinds, 1975).

The problem with the Fibonacci buddy system is that finding a buddy of a block is not always simple. In the binary buddy system, the information stored in the size field of the block is sufficient to compute the address of the buddy. If the size holds the number k, then the address of the buddy is found by complementing the bit $k + 1$ in the address of the block. This works regardless of whether the block has a right buddy or a left buddy. The reason for this simplicity is that only powers of 2 for the sizes of all blocks are used, and each block and its buddy are of the same size.

In the Fibonacci system, this approach is inapplicable, yet it is necessary to know whether a returned block has a right or a left buddy in order to combine the two. Not surprisingly, finding the buddy of a block may be rather demanding in terms of time or space. To this end, Hirschberg used a table that could have nearly 1,000 entries if buffers of up to 17,717 locations are allowed. His method can be simplified if a proper flag is included in each block, but a binary Left/Right flag may be insufficient. If block b_1 marked as Left is coalesced with its buddy, block b_2, then the question is: How do you find the buddy of the resulting block, b_3? An elegant solution uses two binary flags instead of one: a buddy-bit and a memory-bit (Cranston and Thomas, 1975). If a block b_1 is split into blocks b_{left} and b_{right}, then buddy-bit(b_{left}) = 0, buddy-bit(b_{right}) = 1, memory-bit(b_{left}) = buddy-bit(b_1), and finally, memory-bit(b_{right}) = memory-bit(b_1) (see Figure 12.6a). The last two assignments preserve some information about predecessors: Memory-bit(b_{left}) indicates whether its parent is a left or right buddy, and memory-bit(b_{right}) is a bit of information to indicate the same status for one of the predecessors of its parent. Note that the coalescing process is an exact reversal of splitting (see Figure 12.6b).

FIGURE 12.6
(a) Splitting a block of size $Fib(k)$ into two buddies using the buddy-bit and the memory-bit. (b) Coalescing two buddies utilizing information stored in buddy- and memory-bits.

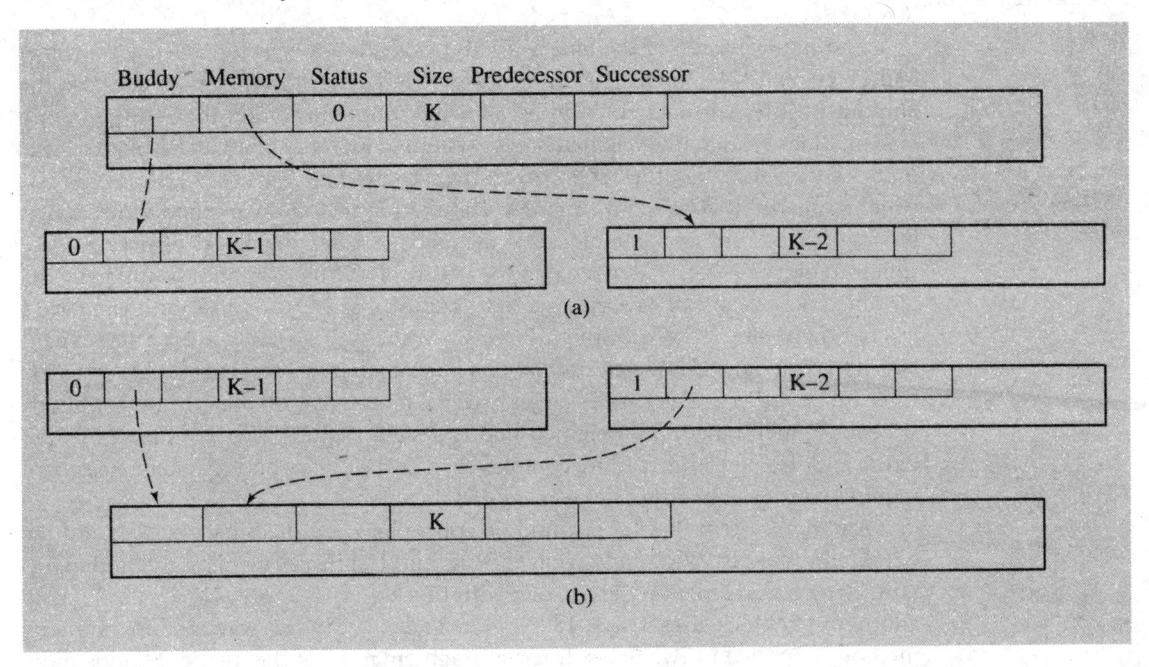

The algorithms for reserving blocks and for returning them are in many respects similar to the algorithms used for the binary buddy system. An algorithm for reserving blocks is as follows:

```
avail[i] = -1 for i = 0,...,m-1;
avail[m] = first address in memory;

reserveFib(reqSize)
    availSize = the position of the first Fibonacci number greater than reqSize
                for which avail[availSize] > -1;
    if no such availSize exists
        failure;
    block = avail[availSize];
    detach block from list avail[availSize];
    while Fib(availSize-1) > reqSize   // while an available block is
                                        // too large - split it; choose
        if reqSize ≤ Fib(availSize-2)  // smaller of the buddies if it's
            insert block's larger part in avail[availSize-1];  // large enough;
            block = block's smaller part;
        else insert block's smaller part in avail[availSize-2];
            block = block's larger part;
```

```
      availSize = size(block);
      set flags(block);
      set flags(block's buddy);
   return block;
```

Another extension of the binary buddy system is a *weighted buddy system* (Shen and Peterson 1974). Its goal, as in the case of Fibonacci systems, is to decrease the amount of internal fragmentation by allowing more block sizes than in the binary system. Block sizes in the weighted buddy system in memory of 2^m unary blocks are 2^k for $0 \leq k \leq m$, and $3 \cdot 2^k$, for $0 \leq k \leq m - 2$; the sizes are 1, 2, 3, 4, 6, 8, 12, 16, 24, 32, ..., which is nearly twice as many different sizes than in the binary method. If necessary, blocks of size 2^k are split into blocks $3 \cdot 2^{k-2}$ and 2^{k-2}, and the blocks of size $3 \cdot 2^k$ are split into blocks 2^{k+1} and 2^k. Note that the buddy of a 2^k block cannot be uniquely determined because it can have either a right buddy of size 2^{k+1} or $3 \cdot 2^k$ or it can have a left buddy of size 2^{k-1}. To distinguish between these three cases, a 2-bit flag *type* is added to each block. However, simulations indicate that the weighted buddy system is three times slower and generates larger external fragmentation than the binary buddy system. As mentioned, the weighted buddy system requires two additional bits per block, and the algorithm is more complex than in the binary buddy system, because it requires considering more cases when coalescing blocks.

A buddy system that takes a middle course between the binary system and the weighted system is a *dual buddy system* (Page and Hagins, 1986). This method maintains two separate memory areas, one with block sizes 1, 2, 4, 8, 16, ..., 2^i, ... and another with block sizes 3, 6, 9, 18, 36, ..., $3 \cdot 2^j$, ... In this way, the binary buddy method is applied in two areas. Internal fragmentation of the dual method is more or less halfway between that of the binary and weighted methods. External fragmentation in the dual buddy system is almost the same as that of the binary buddy system.

To conclude this discussion, observe that internal fragmentation is often inversely proportional to external fragmentation because internal fragmentation is avoided if allocated blocks are as close in size to the requested blocks as possible. But this means that some small splinter blocks are generated that are of little use. These small blocks can be compacted together to form a large block with sequential-fit methods, but compaction does not square very well with the buddy system approach. In fact, the variant buddy system, which is an elaboration of the weighted buddy method, attempts to compact memory, but the complexity of the algorithm undermines its usefulness (Bromley, 1980).

12.3 GARBAGE COLLECTION

As mentioned at the beginning of this chapter, some languages have automatic storage reclamation in their environment so that no explicit return on unused memory cells must be done by any program. The program can allocate memory through the function new, but there is no need to return the allocated memory block to the operating system if the block is not needed any longer. The block is simply abandoned, and it will be reclaimed by a method called a *garbage collector* that is automatically invoked to collect unused memory cells when the program is idle or when memory resources are exhausted.

The garbage collector views the heap as a collection of memory cells, or nodes, each cell composed of several fields. Depending on the garbage collector, the fields can be different. For example, in LISP, a cell has two pointers, *head* and *tail* (or in LISP terminology, *car* and *cdr*), to other cells, except for atomic cells that have no pointers. The cells include headers with such elements as an atom/nonatom flag and a marked/unmarked flag. Data that are included may be stored in yet another field of a cell or in the portion of atomic cells used for pointers in nonatomic cells. Moreover, if variable sized cells are used, the header includes the number of bytes in the data field. Using more than two pointer fields is also possible. Pointers to all linked structures currently utilized by the program are stored in a *root set,* which contains all *root pointers.* The garbage collector's task is to determine those parts of memory that are accessible from any of these pointers and parts that are not currently in use and can be returned to the free memory pool.

Garbage collection methods usually include two phases, which may be implemented as distinct passes or can be integrated:

1. The *marking* phase—to identify all currently used cells.
2. The *reclamation* phase—when all unmarked cells are returned to the memory pool; this phase can also include heap compaction.

12.3.1 Mark-and-Sweep

A classical method of collecting garbage is the *mark-and-sweep* technique, which clearly distinguishes the two phases. First, memory cells currently in use are marked by traversing each linked structure, and then the memory is swept to glean unused (garbage) cells and put them together in a memory pool.

Marking

A simple marking procedure looks very much like preorder tree traversal. If a node is not marked, then it is marked, and if it is not an atomic node, marking continues for its *head* and for its *tail:*

```
marking (node)
    if node is not marked
        mark node;
        if node is not an atom
            marking(head(node));
            marking(tail(node));
```

This procedure is called for each element of the root set. The problem with this succinct and elegant algorithm is that it may cause the run-time stack to overflow, which is a very real prospect considering the fact that the list being marked can be very long. Therefore, an explicit stack can be used so that there is no need to store on the run-time stack the data necessary to properly resume execution after returning from recursive calls. Here is an example of an algorithm that uses an explicit stack:

```
markingWithStack (node)
    push(node);
    while stack is not empty
        node = pop();
        while node is an unmarked nonatom
            mark node;
            push(tail(node));
            node = head(node);
        if node is an unmarked atom
            mark node;
```

The problem of an overflow is not avoided altogether. If the stack is implemented as an array, the array may turn out to be too small. If it is implemented as a linked list, it may be impossible to use, because the stack requires memory resources that have just been used up and in the restoration of which the stack was supposed to participate. There are two ways to avoid this predicament: by using a stack of limited size and invoking some operations in case of stack overflow or by trying not to use any stack at all.

A useful algorithm that requires no explicit stack was developed by Schorr and Waite. The basic idea is to, in a sense, incorporate the stack in the list being processed. This technique belongs in the same category as the stackless tree traversal techniques discussed in Section 6.4.3. In the Schorr and Waite marking method, some links are temporarily reversed when traversing the list to "remember" the path back, and their original setting is restored after marking all cells accessible from a position in which the reversal has been performed. When a marked node or an atom is encountered, the algorithm returns to the preceding node. However, it can return to a node through the *head* field or through the *tail* field. In the former case, the *tail* path has to be explored, and the algorithm has to use a marker to indicate whether both *head* and *tail* paths have been checked or only the *head* path has been checked. To that end, the algorithm uses one additional bit called a *tag* bit. If the *head* of a cell is accessed, then the tag bit remains zero so that, upon return to this cell, the path accessible from *tail* will be followed, in which case the tag bit is set to one and reset to zero upon return. The summary of the algorithms is as follows:

```
invertLink (p1, p2, p3)
    tmp = p3;
    p3 = p1;
    p1 = p2;
    p2 = tmp;

SWmarking (curr)
    prev = null;
    while (1)
        mark curr;
        if head(curr) is marked or atom
            if head(curr) is an unmarked atom
```

```
                    mark head(curr);
              while tail(curr) is marked or atom
                if tail(curr) is an unmarked atom
                    mark tail(curr);
                while prev is not null and tag(prev) is 1 // go back
                    tag(prev) = 0;
                    invertLink(curr,prev,tail(prev));
                if prev is not null
                    invertLink(curr,prev,head(prev));
                else finished;
              tag(curr) = 1;
              invertLink(prev,curr,tail(curr));
           else invertLink(prev,curr,head(curr));
```

Figure 12.7 illustrates an example. Each part of this figure shows changes in the list after the indicated operations have been performed. Note that atom nodes do not require a tag bit. Figure 12.7a contains the list before marking. Each nonatomic node has four parts: a marking bit, a tag bit, and *head* and *tail* fields. The marking and tag bits are initialized to 0. There is one more bit not shown in this figure, an atom/nonatom flag.

Here is a description of each iteration of the `while` loop and the figure number that contains the structure of the list after that iteration.

Iteration 1: Execute `invertLink(prev,curr,`*head*`(curr))` (Figure 12.7b).

Iteration 2: Execute another `invertLink(prev,curr,`*head*`(curr))` (Figure 12.7c).

Iteration 3: Execute still another `invertLink(prev,curr,`*head*`(curr))` (Figure 12.7d).

Iteration 4: Mark *tail*(curr) and execute `invertLink(curr,prev,`*head*`(prev))` (Figure 12.7e), execute another `invertLink(curr,prev,`*head*`(prev))` (Figure 12.7f), set *tag*(curr) to 1, and execute `invertLink(prev,curr,`*tail*`(curr))` (Figure 12.7g).

Iteration 5: Mark *tail*(curr) to 1, set *tag*(prev) to 0, and execute `invertLink(curr,prev,`*tail*`(prev))` (Figure 12.7h). Execute `invertLink(curr,prev,`*head*`(prev))` (Figure 12.7i), set *tag*(curr) to 1, and execute `invertLink(prev,curr,`*tail*`(curr))` (Figure 12.7j).

Iteration 6: Set *tag*(prev) to 0 and execute `invertLink(curr,prev,`*tail*`(prev))` (Figure 12.7k). The algorithm completes and `prev` becomes *null*.

Note that the algorithm has no problem with cycles in lists. `SWmarking()` is slower than `markingWithStack()`, because it requires two visits per cell, pointer maintenance, and an additional bit. Hence, disposing of a stack does not seem to be the best solution. Other approaches attempt to combine a stack with some form of overflow handling. Schorr and Waite proposed such a solution by resorting to their link inversion technique if a fixed-length stack becomes full. Other techniques are more discriminating about what information should be stored on the stack. For example, `markingWithStack()` unnecessarily pushes onto the stack the nodes that have empty *tail* fields—nodes whose processing is finished after the *head* path is finished.

FIGURE 12.7 An example of execution of the Schorr and Waite algorithm for marking used memory cells.

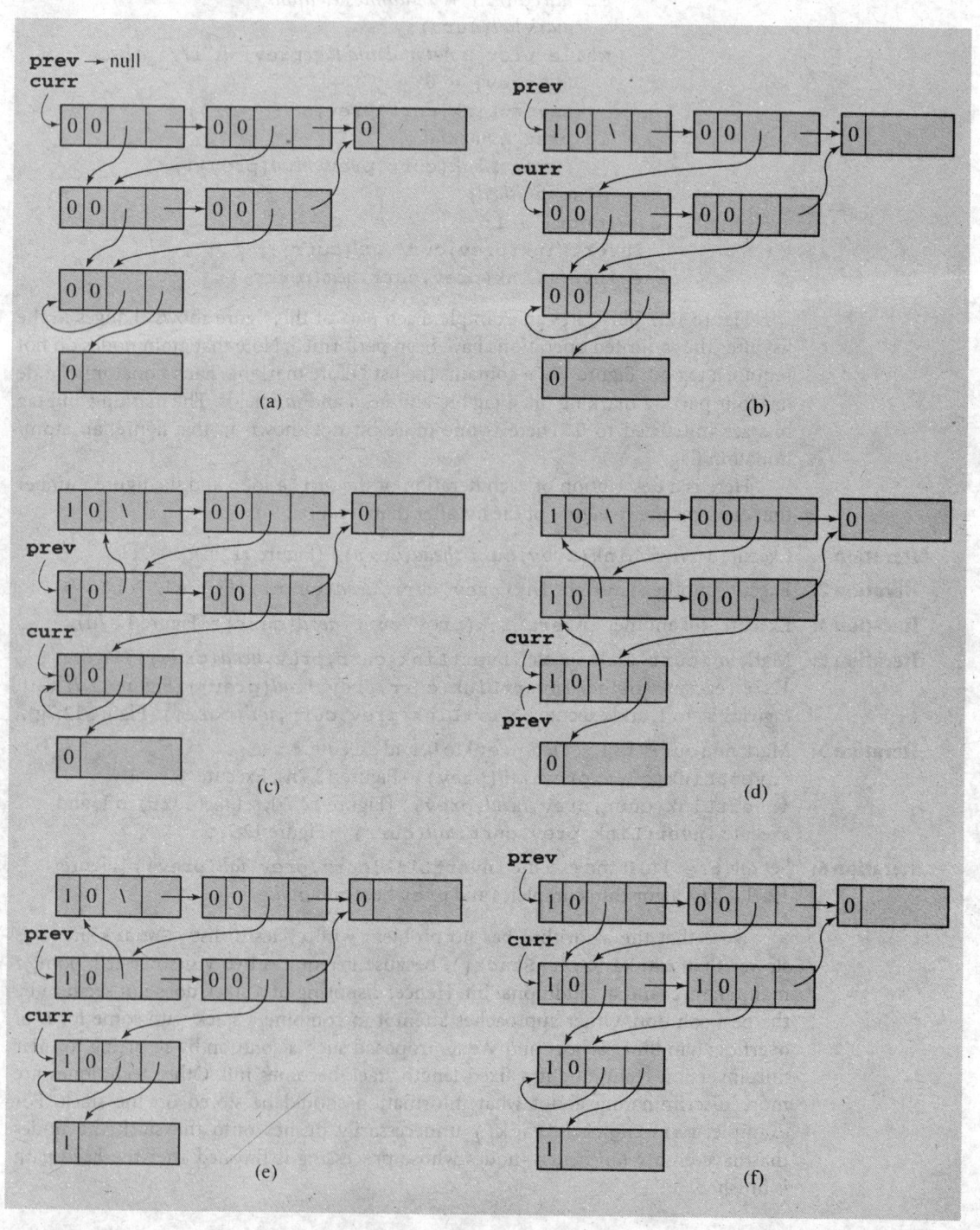

FIGURE 12.7 *(continued)*

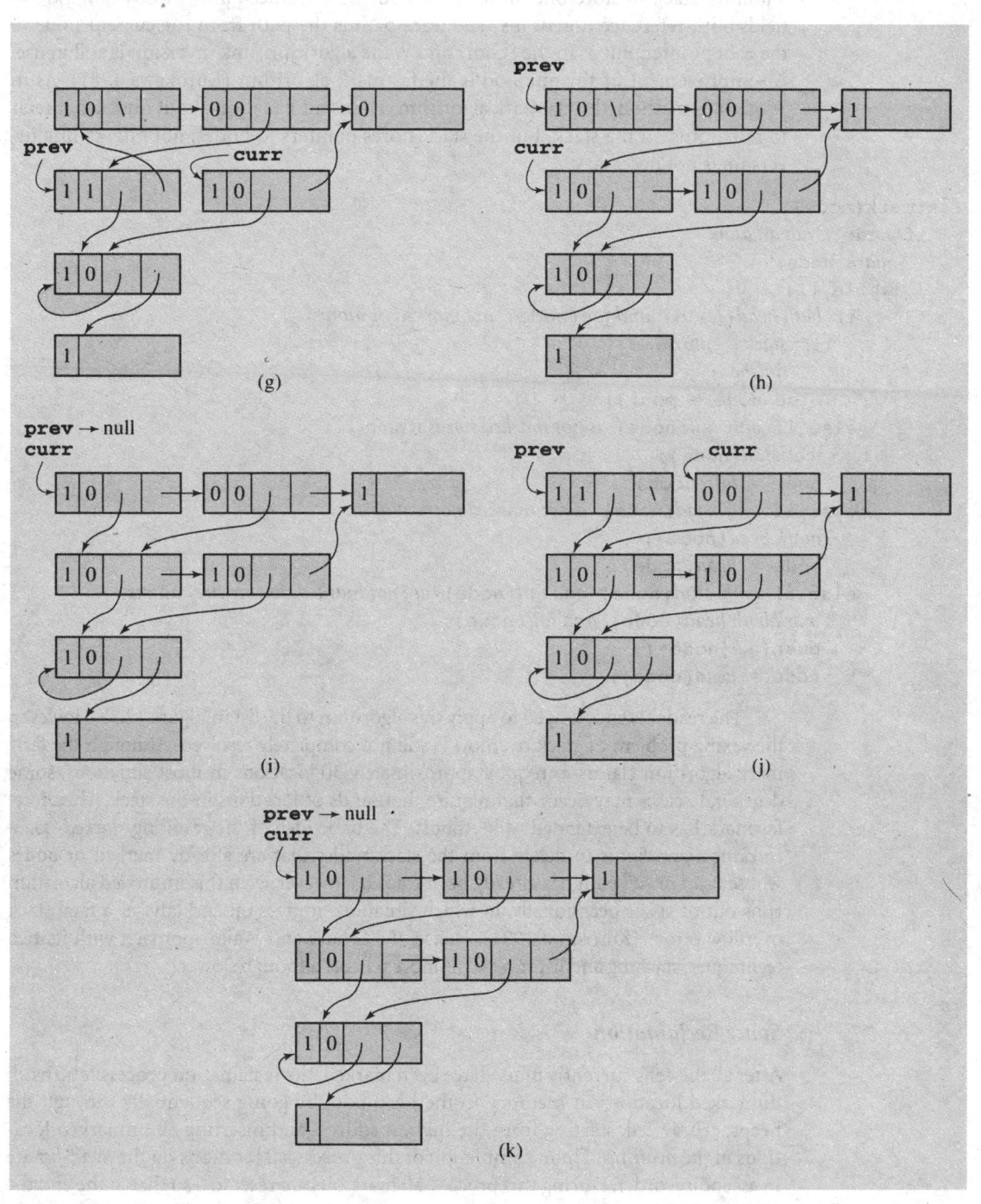

The method devised by Wegbreit requires no tag bit and uses a bit stack instead of a pointer stack to store one bit for each node on the trace path whose *head* and *tail* fields both reference nonatoms. The trace path is the path from the current node to the root pointer. But as in the Schorr and Waite algorithm, link inversion is still in use. An improvement of this method is the fastmark algorithm (Kurokawa 1981). As in Wegbreit's method, the fastmark algorithm retains information about nodes that refer to nonatoms on the stack. But the stack stores pointers to nodes, not bits, so link inversion is not necessary.

```
fastmark(node)
    if node is not an atom
        mark node;
        while (1)
            if both head(node) and tail(node) are marked or atoms
                if stack is empty
                    finished;
                else node = pop();
            else if only tail(node) is not marked nor is it atom
                mark tail(node);
                node = tail(node);
            else if only head(node) is not marked nor is it atom
                mark head(node);
                node = head(node);
            else if both head(node) and tail(node) are not marked nor are they atoms
                mark both head(node) and tail(node);
                push(tail(node));
                node = head(node);
```

The reader is encouraged to apply this algorithm to the list in Figure 12.7a. However, the vexing problem of stack overflow is still not completely resolved. Although the fastmark algorithm claims to require approximately 30 locations in most situations, some degenerate cases may occur that require thousands of locations in the stack. Therefore, fastmark has to be extended to be robust. The basic idea of the resulting *stacked-node-checking algorithm* is to delete from the stack nodes that are already marked or nodes whose *head* or *tail* path has already been traced. However, even this improved algorithm runs out of space occasionally, in which situation "it gives up and advises a fatal stack overflow error" (Kurokawa 1981). Hence, the Schorr and Waite approach with its two techniques, stacking and list reversal, is more reliable although slower.

Space Reclamation

After all the cells currently in use have been marked, the reclamation process returns all unmarked locations in memory to the heap pool by going sequentially through the heaps, cell by cell, starting from the highest address and inserting all unmarked locations in the *avail-list*. Upon completion of this process, all locations on the *avail-list* are in ascending order. During this process, all mark bits are reset to 0 so that at the end the mark bits of all used and unused locations are 0. This simple algorithm is as follows:

```
sweep()
    for each location from the last to the first
        if mark(location) is 0
            insert location in front of availList;
        else set mark(location) to 0;
```

The sweep() algorithm makes a pass through the entire memory. If we add a pass required for marking and the subsequent maintenance of the availList containing locations sparsely scattered throughout the heap, this rather undesirable situation calls for improvement.

Compaction

After the reclamation process is complete, the available locations are interspersed with the cells being used by the program. This requires *compaction*. If all available cells are in contiguous order, then there is no need to maintain the availList. Also, if garbage collection is used for reclaiming cells of variable cells, then having all available cells in sequence is highly desirable. Compaction is also necessary when garbage collection processes virtual memory. In this way, responses to memory requests can be performed with a minimal number of accesses. Another situation in which compacting is beneficial is when the run-time stack and a heap are used at the same time. C++ is an example of a language implemented in this way. The heap and the stack are in opposite sides of memory and they grow toward one another. If occupied memory cells on the heap can be kept away from the stack, then the stack has more room for expansion.

A simple *two-pointer algorithm* for heap compaction uses an approach similar to the one utilized in partitioning in quicksort: Two pointers scan the heap starting from opposite sides of memory. After the first pointer finds an unmarked cell and the second finds a marked cell, the contents of the marked cell are moved to the unmarked cell and its new location is recorded in the old location. This process continues after the pointers cross. Then, the compacted part is scanned to readjust the *head* and *tail* pointers. If the pointers of the copied cells refer to locations beyond the compacted area, the old locations are accessed to retrieve the new address. Here is the algorithm:

```
compact()
    lo = the bottom of heap;
    hi = the top of heap;
    while (lo < hi)     // scan the entire heap;
        while *lo (the cell pointed to by lo) is marked
            lo++;
        while *hi is not marked
            hi--;
        unmark cell *hi;
        *lo = *hi;
        tail(*hi--) = lo++;   // leave forwarding address;
    lo = the bottom of heap;
    while (lo <= hi)    // scan only the compacted area;
```

```
            if *lo is not atom and head(*lo) > hi
                head(*lo) = tail(head(*lo));
            if *lo is not atom and tail(*lo) > hi
                tail(*lo) = tail(tail(*lo));
            lo++;
```

Figure 12.8 illustrates this process in the case of two available spots in front of cell A to which cells B and C can be moved. Figure 12.8a illustrates the situation in the heap before compaction. In Figure 12.8b, cells B and C have been moved into these spots with the *tail* fields of the old cells indicating the new positions. Figure 12.8c illustrates the compacted part of the heap after checking the *head* and *tail* fields of all cells and updating them in case they referred to positions beyond the compacted area.

This simple algorithm is inefficient in that it requires one pass through the heap to mark cells, one pass to move marked cells into contiguous locations, and one pass through the compacted area to update pointers; two and a half heap passes are required. One way to reduce the number of passes is to integrate marking and sweeping, which opens up a new category of methods.

FIGURE 12.8   An example of heap compaction.

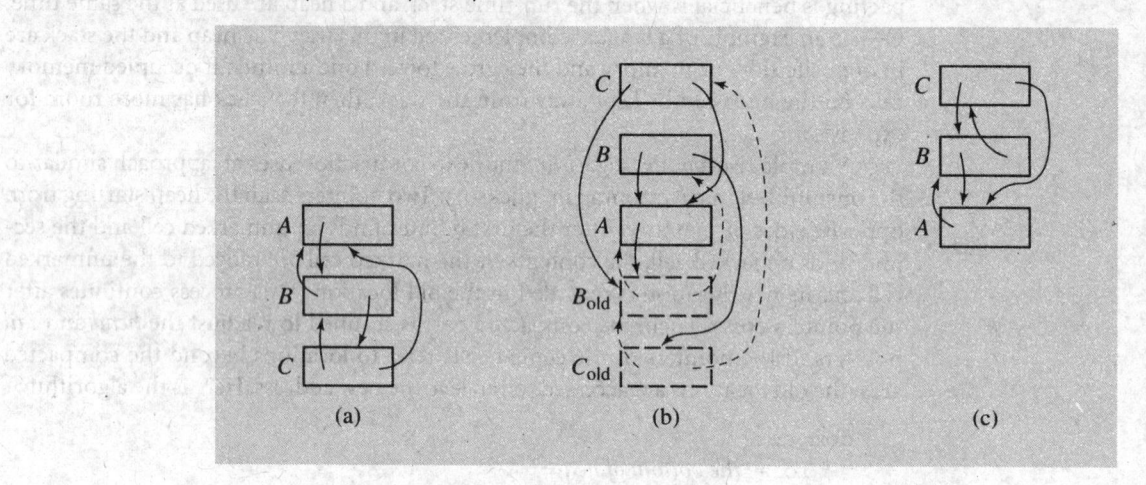

12.3.2 Copying Methods

Copying algorithms are cleaner than the previous methods in that they do not touch garbage. They process only the cells accessible from the root pointers and put them together; the unprocessed cells are available. An example of a copying method is the *stop-and-copy algorithm,* which divides the heap into two *semispaces,* one of which is only used for allocating memory (Fenichel and Yochelson, 1969). After the allocation pointer reaches the end of the semispace, all the cells being used are copied to the second semispace, which becomes an active space, and the program resumes execution (see Figure 12.9).

FIGURE 12.9 (a) A situation in heap before copying the contents of cells in use from semispace$_1$ to semispace$_2$ and (b) the situation right after copying. All used cells are packed contiguously.

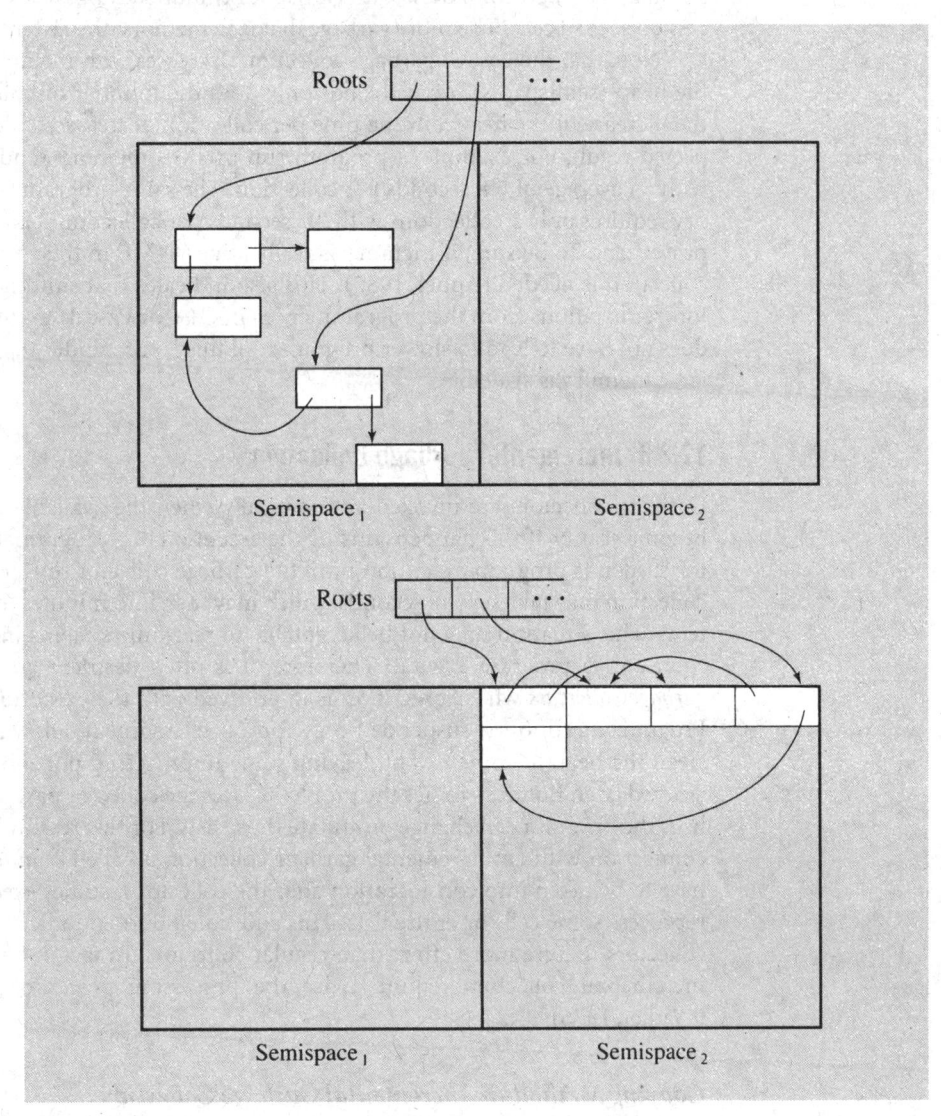

Lists can be copied using breadth-first traversal (Cheney, 1970). If lists were just binary trees with no cross-references, the algorithm would be the same as the breadth-first tree traversal discussed in Section 6.4.1. However, lists can have cycles, and cells on one list can point to cells on another. In the latter case, this algorithm would produce multiple copies of the same cell. In the former case it would fall into an infinite loop. The problem can be easily solved, as in `compact()`, by retaining a forward address in the cell being copied. This allows the copying procedure to refer to a cell after it has already been copied. This algorithm requires no marking phase and no stack. The breadth-first traversal also allows it to combine two tasks: copying lists and updating pointers. The algorithm deals with garbage only indirectly because it does not really access unneeded cells. The more garbage that is in memory, the faster the algorithm.

Note that the cost of garbage collection decreases with the increase in the size of the heap (semispaces). Actually, not only does the number of collections drop with the increase of the heap, but the time per collection decreases, which is a more unexpected result. For example, a program run in 4MB memory requires 34 collections with an average of 6.8 seconds per collection. The same program run in 16MB memory requires only 3 collections with 2.7 seconds per collection—a very significant improvement. To be sure, if memory is really large (64MB in this example), no garbage collection is needed (Appel, 1987). This also indicates that shifting the responsibility for free locations from the programmer (as in C++ or Pascal) to the garbage collector does not have to lead to slower programs. All this is true under the assumption that a large memory is available.

12.3.3 Incremental Garbage Collection

Garbage collectors are invoked automatically when the available memory resources become scanty. If this happens during the execution of a program, the garbage collector suspends program execution until the garbage collector finishes its task. Garbage collection may take several seconds, which may turn into minutes in time-sharing systems. This situation may not be acceptable in real-time systems in which the fast response of a program is vital. Therefore, it is often desirable to create *incremental garbage collectors* whose execution is interleaved with the execution of the program. Program execution is suspended only for a brief moment, allowing the collector to clean the heap to some extent, leaving some unprocessed portion of the heap to be cleaned later. But therein lies the problem. After the collector partially processes some lists, the program can change or mutate those lists. For this reason, a program used in connection with an incremental garbage collection is called a *mutator*. Such changes have to be taken into consideration after the collector resumes execution, possibly to reprocess some cells or entire lists. This additional burden indicates that incremental collectors require more effort than regular collectors. In fact, it has been shown that incremental collectors require twice the processing power of regular collectors (Wadler, 1976).

Copying Methods in Incremental Garbage Collection

An incremental algorithm based on the stop-and-copy technique has been devised by Henry Baker (1978). As in stop-and-copy, the Baker algorithm also uses two semispaces, called *fromspace* and *tospace*, which are both active to ensure proper cooperation between the mutator and the collector. The basic idea is to allocate cells in tospace starting from its top, and to always copy the same number, k, of cells from fromspace to tospace upon request. In this way, the collector can perform its task without incurring any undue interruption of the mutator's work. After all reachable cells have been copied to tospace, the roles of the semispaces are interchanged.

The collector maintains two pointers. The first pointer is *scan*, which points to a cell whose *head* and *tail* lists should be copied to tospace if they still are in fromspace. Because these lists may be larger than k, they may not be processed at one time. Up to k cells accessible by breadth-first traversal are copied from fromspace, and the copies are put at the end of the queue. This queue is simply accessible by the second pointer, *bottom*, which points to the beginning of the free space in tospace. The collector can process *tail* of the current cell during the same time slice, but it may wait until the next turn. Figure 12.10 contains an example. If a request comes to allocate a cell whose *head* points to P and *tail* to Q (as in LISP's *cons*(P, Q)), with both P and Q residing in tospace, then a new cell is allocated in the upper part of tospace with both its pointer fields properly initialized. Assuming that $k = 2$, two cells are copied from the *head* list of the cell pointed to by *scan*, and the *tail* is processed when the next request comes. As in stop-and-copy, Baker's algorithm retains a forwarding address in the original cell in fromspace to its copy in tospace, just in case later allocations refer to this original.

Special care must be taken when the *head* and/or *tail* of a cell being allocated refer to a cell in fromspace that is either already copied or still in fromspace. Because the cells at the top of tospace are not processed by the collector, retaining a pointer in any of them to fromspace cells leads to fatal consequences after fromspace becomes tospace because the latter cells are now considered available and filled with new contents. The mutator could at one point use the pointer to the original, and at a later point could use a copy, leading to inconsistencies. Hence, the mutator is preceded by a *read barrier* that precludes utilizing references to cells in fromspace. In the case of a reference to fromspace, we have to check whether this cell has a forwarding address, an address in its *tail* to a location in tospace. If the answer is yes, the forwarding address is used in the allocation; otherwise, the cell referred to in the current allocation has to be copied *before* the actual allocation takes place. For example, if the *head* of a cell to be allocated is to point to P, a cell in fromspace that has already been copied, as illustrated in Figure 12.11a, P's new address is stored in *head* (see Figure 12.11b). If the *tail* of the new cell is to point to Q, which is still untouched in fromspace, Q is copied to tospace (along with one descendant, because $k = 2$) and only afterward is the *tail* of the new cell initialized to the copy of Q.

Baker's algorithm lends itself to various modifications and improvements. For example, to avoid constant condition tests when allocating new cells, an indirection field is included in every cell. If a cell is in tospace, the indirection field points to itself; otherwise, it points to its copy in tospace (Brooks, 1984). Tests are avoided, but indirection pointers have to be maintained for every cell instead. Another way to solve this problem is by utilizing hardware facilities, if available. For example, memory protection facilities can prevent the mutator's access to cells not processed by the collector: All pages of the heap with unprocessed cells are read-protected (Ellis, Li, and Appel, 1988). If the mutator attempts to access such a page, the access is trapped and an exception raised, forcing the collector to process this page so that the mutator can resume execution. But this method can undermine the incremental collection, because after the semispaces change roles, traps are invoked frequently, and each trap requires that an entire page of the heap be processed. Some additional provisions may be needed such as not requiring a scan of the entire page in case of a trap. On the other hand, if the heap is not accessed too frequently, this is not a problem.

FIGURE 12.10 A situation in memory (a) before and (b) after allocating a cell with *head* and *tail* pointers referring to cells *P* and *Q* in tospace according to the Baker algorithm.

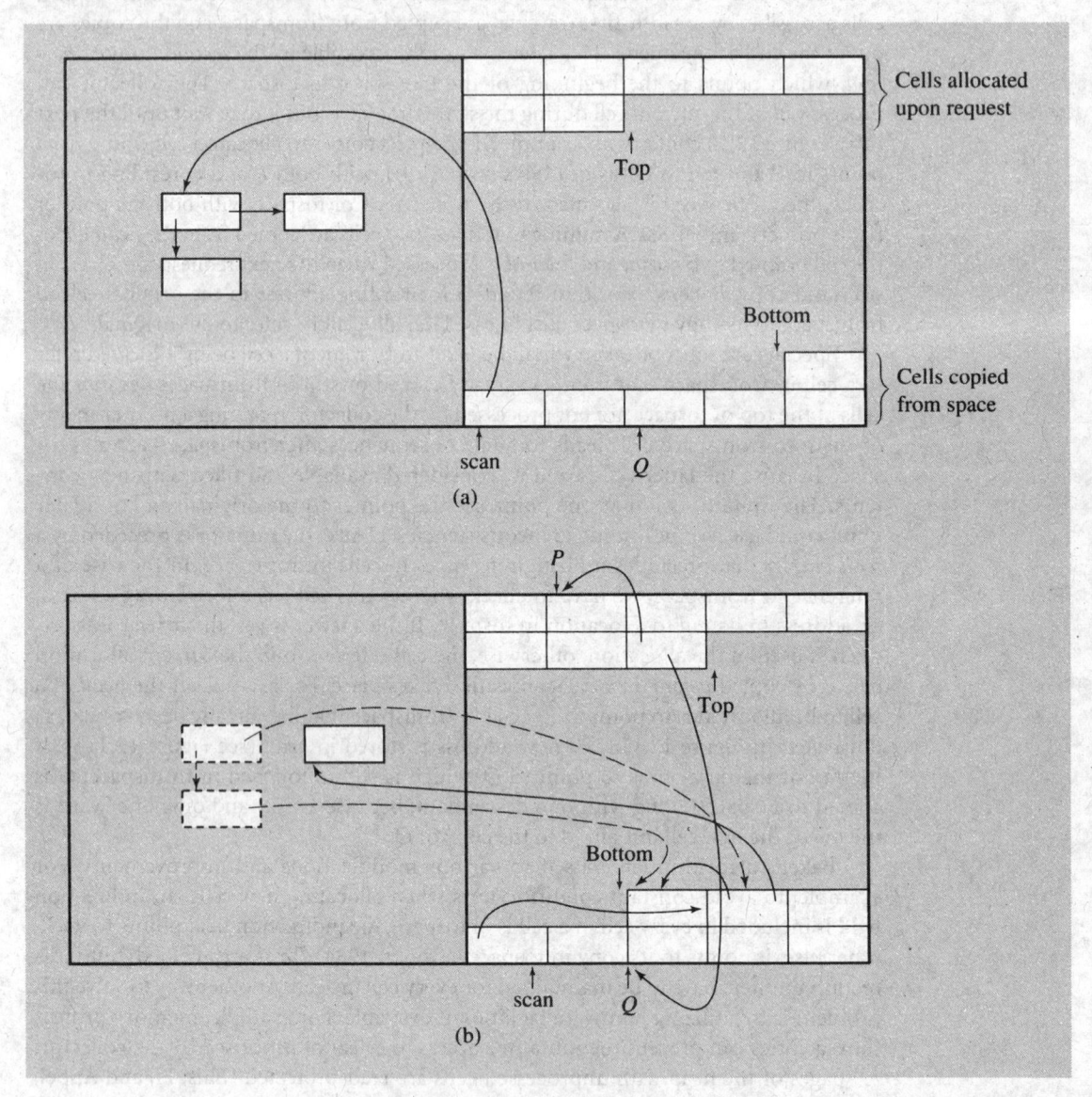

An interesting modification to Baker's algorithm is a technique based on the observation that most allocated cells are needed for a very short time; only some of them are used for longer timespans. This leads to a *generational garbage collection* technique that divides all allocated cells into at least two generations and focuses its attention on the youngest generation, which generates most of the garbage. Such cells do not need

FIGURE 12.11 Changes performed by the Baker algorithm when addresses P and Q refer to cells in fromspace, P to an already copied cell, Q to a cell still in fromspace.

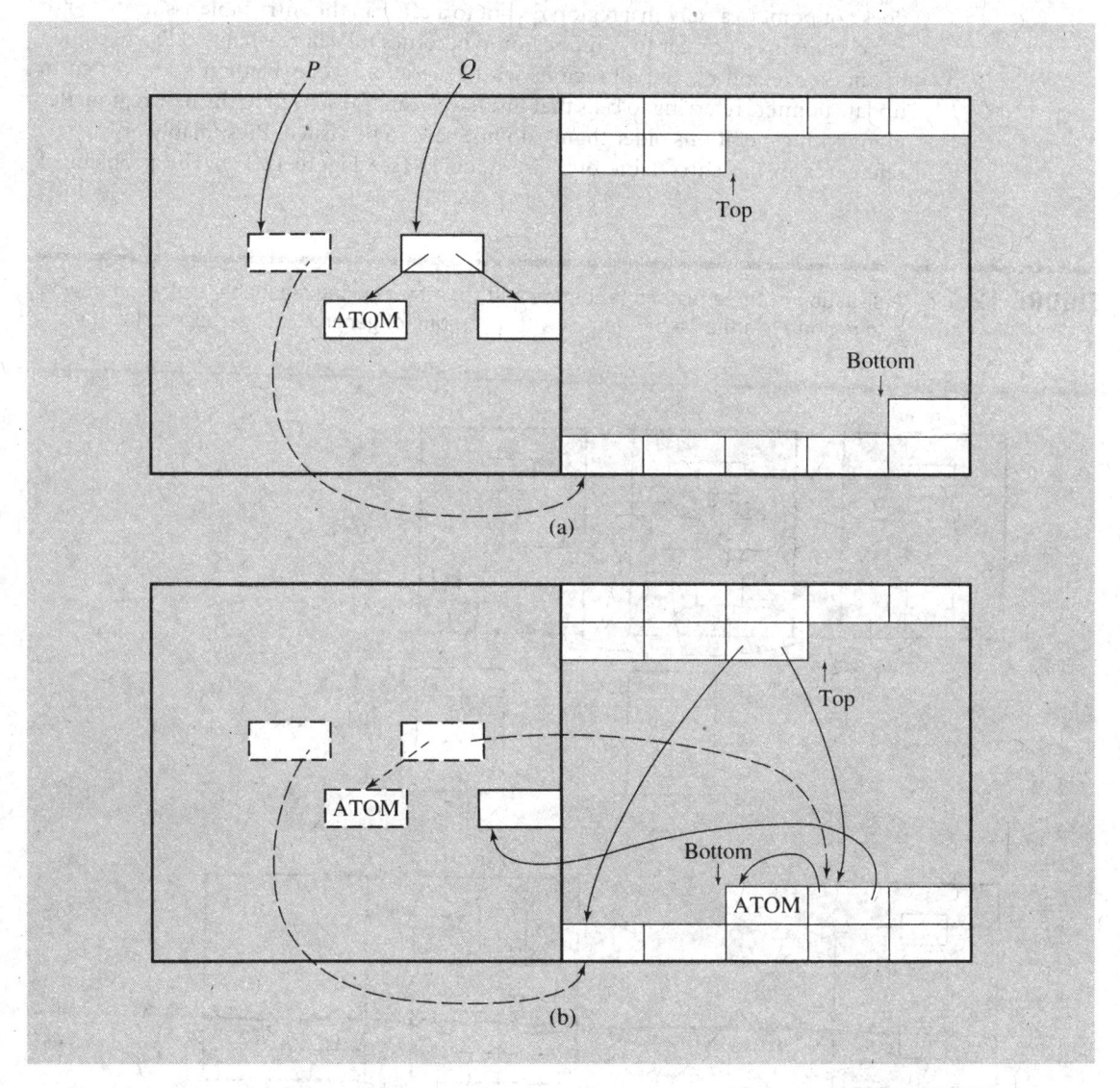

to be copied, saving the garbage collector some work. Moreover, the constant checking and copying of long-lived cells is unnecessarily wasteful, so testing garbage production among such cells is performed only infrequently.

In a classic version of a generational garbage collector, the address space is divided into several regions, $r_1, \ldots, r_n$, not just into tospace and fromspace; each of these regions holds cells of the same generation (Lieberman and Hewitt, 1983). Most pointers point to cells of an older generation. Some of them, however, can point forward in time (e.g., when Lisp's *rplaca* is used). In this method, such forward references are made indirectly through an *entry table* associated with each region. A pointer from a region r_i does not point to a cell c in a region r_{i+j} but to a cell c' in the entry table associated with r_{i+j}; c' contains a pointer to c. If a region r_i becomes full, all reachable cells are copied to another region r'_i, and all regions with generations younger than r_i are visited to update pointers referring to cells that just have been transferred to the new region. Regions with generations older than r_i do not have to be visited. Presumably only a few references in the entry table of r_i are updated (see Figure 12.12). The problem of

FIGURE 12.12 A situation in three regions (a) before and (b) after copying reachable cells from region r_i to region r'_i in the Lieberman-Hewitt technique of generational garbage collection.

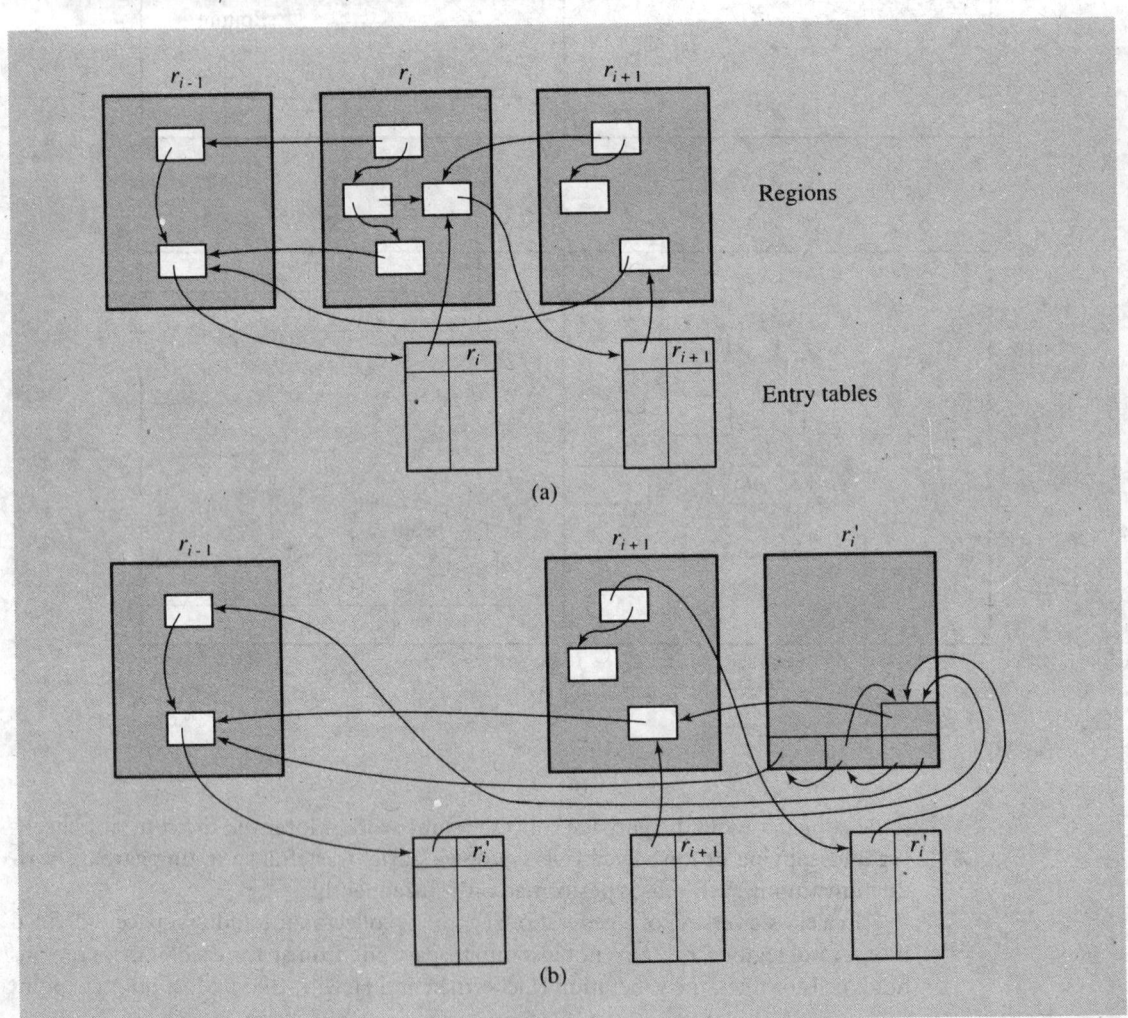

cleaning the entry tables can be solved by storing in each table, along with each pointer, a unique identifier for a region to which the pointer refers. The identifier is updated along with the pointer. Some pointers may be abandoned, as the pointer in the entry table for region r_{i+1} in Figure 12.12b, and they are ready to be cleaned up after the region itself is abandoned.

Noncopying Methods

In incremental methods based on copying, the problem is not so much with the content of the original cells and its copy, but with their positions or addresses in memory, which by necessity have to be different. The mutator must not treat these addresses on a par; otherwise, the program crashes. Therefore, some mechanisms are needed to maintain the integrity of addressing, and the read barrier serves this purpose. But we may need to avoid copying altogether; after all, the first garbage collection method, mark-and-sweep, did not use copies. However, because of the exhaustive and uninterrupted passes, the mark-and-sweep method was too costly, and in real-time systems, it is simply unacceptable. Yet, the simplicity of this method is very appealing, and an attempt was made by Taiichi Yuasa to adapt it to real-time constraints, with satisfactory results.

Yuasa's algorithm also has two phases; one for marking reachable (used) cells and one for sweeping the heap by including in *avail-list* all unused (unmarked) cells. The marking phase is similar to that used in the mark-and-sweep method except that it is incremental; each time the marking procedure is invoked, it marks only k_1 cells for some small constant k_1. After k_1 cells have been marked, the mutator resumes execution. The constant k_2 is used during the sweeping phase to decide how many cells have to be processed before execution is turned over to the mutator. The garbage collector remembers whether it is in the middle of marking or sweeping. The procedure for marking or sweeping is always invoked after one cell is requested from memory by a procedure that creates one new root pointer and initializes its *head* and *tail* fields, as in the following pseudocode:

```
createRootPtr(p,q,r)    // Lisp's cons
    if collector is in the marking phase
        mark up to k₁ cells;
    else if collector is in the sweeping phase
        sweep up to k₂ cells;
    else if the number of cells on availList is low
        push all root pointers onto collector's stack st;
    p = first cell on availList;
    head(p) = q;
    tail(p) = r;
    mark p if it is in the unswept portion of heap;
```

Remember that the mutator can scramble some graphs accessible from root pointers, which is particularly important if it happens during the marking phase, because it may cause certain cells to remain unmarked even though they are accessible. Figure 12.13 contains an example. After all of the roots have been pushed onto stack st (Figure 12.13a), and roots r_3 and r_2 have been processed and root r_1 is being

FIGURE 12.13 An inconsistency that results if, in Yuasa's noncopying incremental garbage collector, a stack is not used to record cells possibly unprocessed during the marking phase.

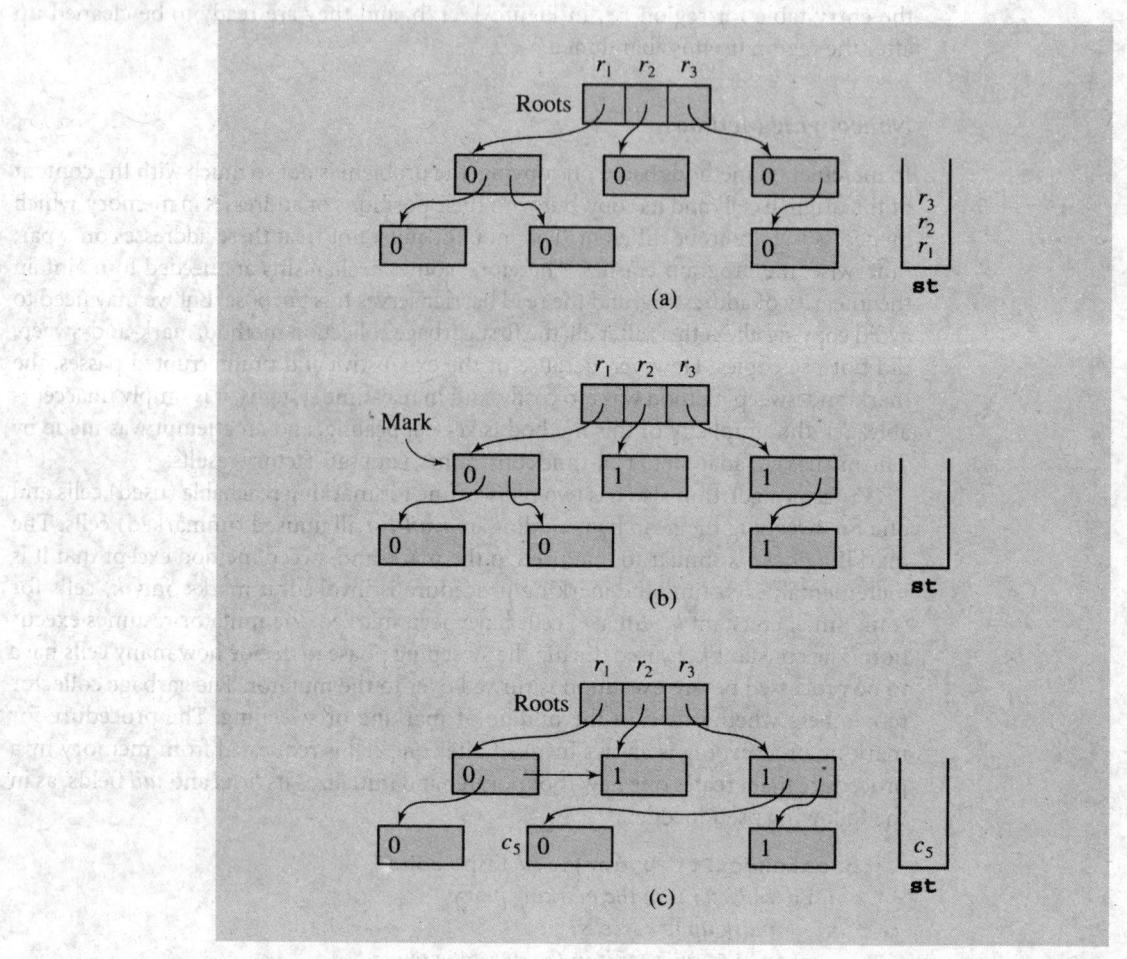

processed (Figure 12.13b), the mutator executes two assignments: $head(r_3)$ is changed to $tail(r_1)$, and $tail(r_1)$ is assigned r_2 (Figure 12.13c). If the marking process is now restarted, it has no chance to mark $head(r_3) = c_5$, because the entire graph r_3 is assumed to have been processed. This leads to including the cell $head(r_3)$ in the *avail-list* during the sweeping phase. To prevent that, the function that updates either *head* or *tail* of any cell pushes the old value of the field being updated onto the stack used by the garbage collector. For example,

```
updateTail(p,q)   // Lisp's rplacd
    if collector is in the marking phase
        mark tail(p);
        st.push(tail(p));
    tail(p) = q;
```

In the marking phase, the stack st is popped up k_1 times, and for each pointer p popped off, its *head* and *tail* are marked.

The sweeping phase incrementally goes through the heap, includes in *avail-list* all unmarked cells, and unmarks all marked cells. For the sake of consistency, if a new cell is allocated, it remains unmarked if a certain part of the heap has been already swept. Otherwise, the next round of marking could lead to distorted results. Figure 12.14 illustrates an example. The pointer *sweeper* has already reached cell c_3, and now the mutator requests a new cell by executing `createRootPtr(`r_2`, `c_5`, `c_3`)`, whereby the first cell is detached from *avail-list* and made a new root (Figure 12.14b). But the newly allocated cell is not marked because it precedes *sweeper* in the heap.

FIGURE 12.14 Memory changes during the sweeping phase using Yuasa's method.

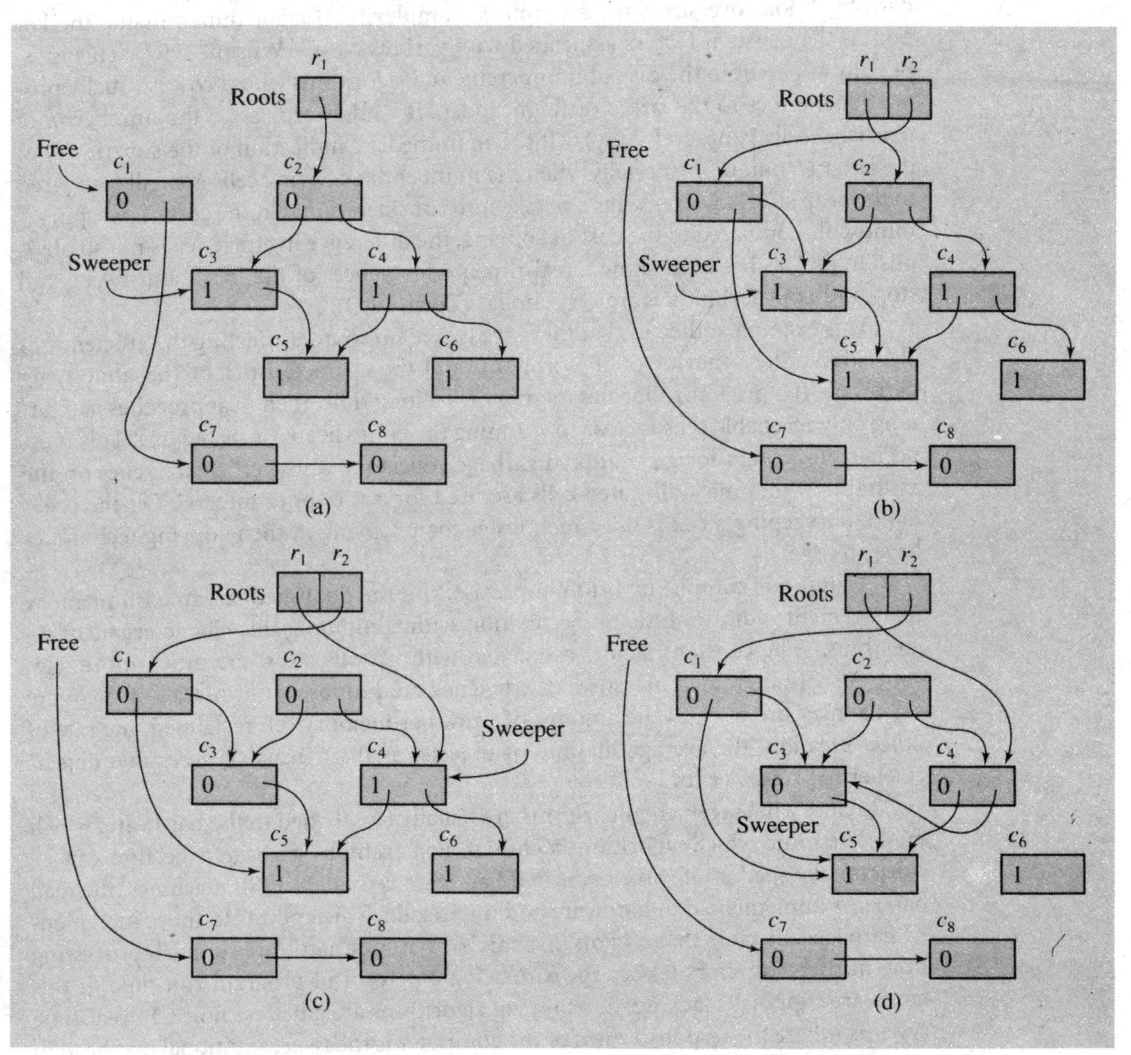

If some cell is released in the swept area, it becomes garbage, but it is not swept up until sweeping restarts from the beginning of memory. For example, after assigning $tail(r_1)$ to r_1, cell c_2 becomes unreachable, and yet it is not reclaimed now by adding it to *avail-list* (Figure 12.14c). The same thing happens to cells having higher addresses than the current value of *sweeper*, as is the case with cell c_6 after assigning $tail(r_2)$ to $tail(r_1)$ (Figure 12.14d). This is called *floating garbage,* and is collected in the next cycle.

12.4 Concluding Remarks

When assessing the efficiency of memory management algorithms, and especially garbage collectors, we have to be careful to avoid Paul Wilson's castigation that standard textbooks overstress the asymptotic complexity of algorithms missing the key point: "the constant factors associated with various costs" (Wilson, 1992). This is especially apparent in the case of nonincremental algorithms whose cost is usually proportional either to the size n of the heap (mark-and-sweep) or to the number m of reachable cells (stop-and-copy). This is an immediate indication of the superiority of the latter techniques, especially when the number of surviving cells is small compared to the heap size. However, when we take into consideration that the cost of sweeping is minuscule compared to the cost of copying, the difference in efficiency is not so obvious. In fact, as has been shown, real-time performances of the mark-and-sweep and stop-and-copy techniques are very similar (Zorn, 1990).

This example indicates that there are two main sources affecting the efficiency of algorithms: the behavior of the program and the characteristics of the underlying hardware. If a program allocates memory for a long time, then m approaches n; scanning only reachable cells is close to scanning the entire heap (or its region). This is especially important for generational garbage collectors, whose efficiency relies on the assumption that most allocated cells are used for a very brief interval. On the other hand, if sweeping a cell is not much faster than copying it, then copying techniques have an edge.

Asymptotic complexity is too imprecise, and the published research on memory management indicates little preoccupation with computing this characteristic of algorithms. "The constant factors associated with various costs" are much more relevant. Also, fine-grained measures of efficiency are proposed, but not all of them are easy to measure, such as the amount of work per memory cell reclaimed, the rate of object creation, the average lifetime of objects, or the density of accessible objects (Lieberman and Hewitt, 1983).

Memory management algorithms are usually closely tied to the hardware, which may determine which algorithm is chosen. For example, garbage collection can be substantially sped up if some dedicated hardware is used. In LISP machines, the read barrier is implemented in hardware and microcode, which points to those incremental garbage collectors that rely on this barrier. Without hardware support, processing time in such collectors takes approximately 50 percent of program run time. If this hardware support is lacking, noncopying algorithms are a better choice. In real-time systems, where the responsiveness of the computer is the issue, additional overhead of the garbage collector is added. It may be less noticeable than in the case of nonincremental collectors, because at no time does a program have to wait in a visible way for the collector to finish its task. However, the tuning of incremental methods should be proportional to real-time constraints.

12.5 CASE STUDY: AN IN-PLACE GARBAGE COLLECTOR

Copying algorithms for garbage collectors are efficient in that they do not require processing unused cells. Cells that are not processed are considered garbage at the end of collection. However, these algorithms are inefficient in copying reachable cells from one semispace to another. An in-place garbage collector attempts to retain the advantages of copying algorithms without producing copies of the reachable cells (Baker, 1992).

The in-place algorithm constantly maintains two doubly linked lists: `freeCells` and `nonFreeCells`. The list `freeCells` initially contains all cells of `heap[]`, and a cell is moved from `freeCells` to the other list if a request comes to construct a list or construct a new atom. After `freeCells` becomes empty, the function `collect()` is invoked. This function first transfers all root pointers from `nonFreeCells` to an intermediate list `markDescendants` and sets their `marked` field to true. Then, `collect()` detaches from `markDescendants` cell by cell to transfer each cell to another temporary list, `markedCells`. For each nonatom cell, `collect()` attaches to `markDescendants` unmarked `head` and `tail` pointers to be processed later. In the case study, they are attached to the beginning of `markDescendants`, thereby leading to depth-first traversal of list structures. For breadth-first traversal (as in Cheney's algorithm), they have to be attached to the end of `markDescendants`, which requires another pointer to the end of the list. Note that, although a cell is transferred to `markedCells`, it is also marked to prevent infinite loops in case of cyclic structures and redundant processing in case of interconnected noncycling structures.

After the list `markDescendants` becomes empty, all reachable cells of `heap[]` have been processed and `collect()` is almost done. Before returning from `collect()`, all cells left in `nonFreeCells` become members of `freeCells`, and all marked cells are put on `nonFreeCells` after setting their `marked` fields to false. The user `program()` can now resume.

To be sure, the garbage collector is part of the program's environment and is executed in the background almost unbeknownst to the user. To exemplify the workings of a garbage collector, some elements of the program background are simulated in the case study, in particular the heap and the symbol table.

The heap is implemented as an array of objects with two flag fields, atom/nonatom and marked/nonmarked, and two pointer fields, which are really integer fields indicating positions in `heap[]` of the previous and next cells (if any). In accordance with this implementation, both permanent lists, `freeCells` and `nonFreeCells`, and both temporary lists, `markDescendants` and `markedCells`, are simply integers indicating the index in `heap[]` of the first cell on a given list (if any).

The symbol table is implemented as an integer array `roots[]` of root pointers. No explicit variable names are used, only indexes to `heap[]` cells. For example, if `roots` is [3 2 4 0], then only four variables are currently in use by `program()`, `roots[0]` through `roots[3]`, and these variables are pointing to cells 3, 2, 4, and 0 in `heap[]`. The numbers 0–3 can be seen as subscripts to more palpable variable names, such as var_0, var_1, var_2 and var_3.

The user `program()` is just a coarse simulator that does nothing but require allocations and reallocations on `heap[]`. These requirements are generated randomly and classified by the type of requirement: 20 percent are atom (re)allocations, 20 percent are list (re)allocations, 20 percent are head updates, 20 percent are tail updates, and the remaining 20 percent are deallocations. Deallocations are simulated by the function `deallocate()`, which decides whether an existing root variable should be assigned `empty` (which represents the null pointer) or a local block is exited upon which all local variables are removed, which in turn means that memory assigned to them is free. The percentages can be assigned differently, and the distribution of assignments can be tuned to the number of assignments already made. This is just a matter of introducing changes in `program()`. Also, the size of `heap[]` and the size of `roots[]` can be modified.

The user `program()` randomly generates a number `rn` between 0 and 99 to indicate the operation to be performed. Then, variables are randomly chosen from `roots[]`. For example, if `rn` is 11, `roots[]` is [3 2 4 0], and p is 2, then the cell `roots[p] = 4` of `heap[]` indicated by variable 2 becomes an atom by storing the value of `val` in its `value` field and the `atom` field is set to true. If p is 4, this indicates that a new variable (variable 4, or var_4) has to be created in position 4 of `roots[]`, and position `roots[4]` is assigned the first value from `freeCells`.

To see that this program does something, a simple function `printList()` is supplied that prints elements on a given list, and the output operator is overloaded to print the contents of `heap[]` and of `roots[]`. Here is an example of an output generated by applying this operator to a heap of six cells:

```
roots: 1 5 3
(0: -1 2 0 0 0 0) (1: 5 4 0 0 1 4) (2: 0 -1 0 0 2 2)
(3: 4 -1 1 0 130 5) (4: 1 3 1 0 129 4) (5: -1 1 0 0 5 1)
freeCells: (0 0 0) (2 2 2)
nonFreeCells: (5 5 1) (1 1 4) (4 129 4) (3 130 5)
```

This output represents the situation illustrated in Figure 12.15. Figure 12.15a shows the contents of a heap with links `prev` and `next` used by lists `freeCells` and `nonFreeCells`, and head link `links[0]` and tail link `links[1]` of nonatom cells. For atom cells, some value is stored in `links` instead of links. Because of the number of crisscrossing links, the same situation is presented in Figure 12.15b, where the cells are organized by their connections rather than the positions in `heap`.

Figure 12.16 contains the listing of the program.

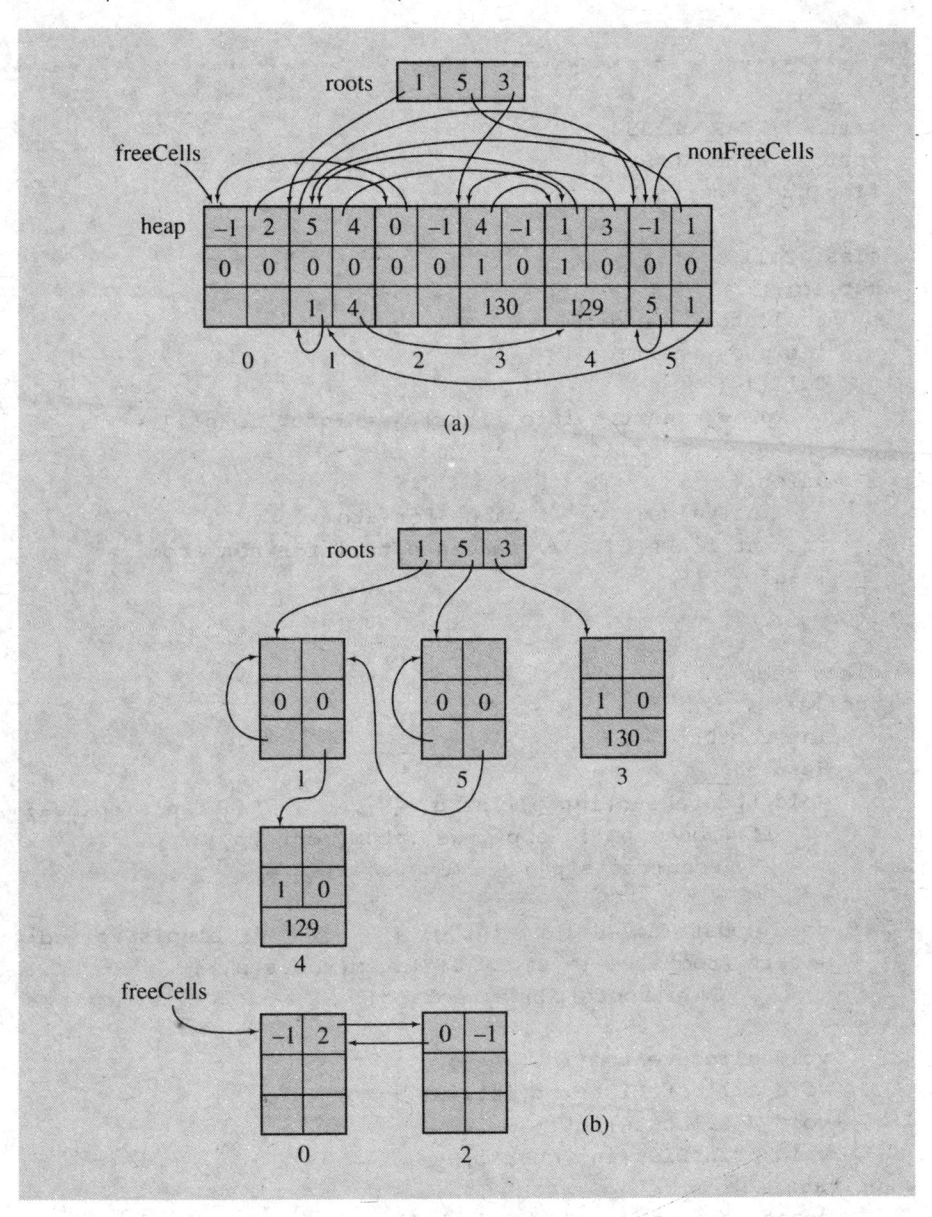

Section 12.5 Case Study: An In-Place Garbage Collector ■ 631

FIGURE 12.15 An example of a situation on the heap

FIGURE 12.16 Implementation of an in-place garbage collector.

```cpp
//********************     heap.h     ******************************
#ifndef HEAP_CLASS
#define HEAP_CLASS
#include <fstream>

class Cell {
public:
    bool atom, marked;
    int prev, next;
    Cell() {
        prev = next = info.links[0] = info.links[1] = -1;
    }
    union {
        int value;       // value for atom,
        int links[2];    // head and tail for non-atom;
    } info;
};

class Heap {
public:
    int rootCnt;
    Heap();
    void updateHead(int p, int q) {          // Lisp's rplaca;
        if (roots[p] != empty && !atom(roots[p]))
            Head(roots[p]) = roots[q];
    }
    void updateTail(int p, int q) {          // Lisp's rplacd;
        if (roots[p] != empty && !atom(roots[p]))
            Tail(roots[p]) = roots[q];
    }
    void allocateAtom(int,int);
    void allocateList(int,int,int);
    void deallocate(int);
    void printList(int,char*);
private:
    const int empty, OK, head, tail, maxHeap, maxRoot;
    Cell *heap;
    int *roots, freeCells, nonFreeCells;
    int& Head(int p) {
        return heap[p].info.links[head];
    }
```

FIGURE 12.16 (continued)

```cpp
    int& Tail(int p) {
        return heap[p].info.links[tail];
    }
    int& value(int p) {
        return heap[p].info.value;
    }
    int& prev(int p) {
        return heap[p].prev;
    }
    int& next(int p) {
        return heap[p].next;
    }
    bool& atom(int p) {
        return heap[p].atom;
    }
    bool& marked(int p) {
        return heap[p].marked;
    }
    void insert(int,int&);
    void detach(int,int&);
    void transfer(int cell, int& list1, int& list2) {
        detach(cell,list1); insert(cell,list2);
    }
    void collect();
    int allocateAux(int);
    friend ostream& operator<< (ostream&,Heap&);
};

Heap::Heap() : empty(-1), OK(1), head(0), tail(1), maxHeap(6),
maxRoot(50) {
    freeCells = nonFreeCells = empty;
    rootCnt = 0;
    heap = new Cell[maxHeap];
    roots = new int[maxRoot];
    for (int i = maxRoot-1; i >= 0; i--) {
        roots[i] = empty;
    for (i = maxHeap-1; i >= 0; i--) {
        insert(i,freeCells);
        marked(i) = false;
    }
}
```

Continues

FIGURE 12.16 *(continued)*

```cpp
void Heap::detach(int cell, int& list) {
    if (next(cell) != empty)
        prev(next(cell)) = prev(cell);
    if (prev(cell) != empty)
        next(prev(cell)) = next(cell);
    if (cell == list)                       // head of the list;
        list = next(cell);
}

void Heap::insert(int cell, int& list) {
    prev(cell) = empty;
    if (cell == list)        // don't create a circular list;
         next(cell) = empty;
    else next(cell) = list;
    if (list != empty)
        prev(list) = cell;
    list = cell;
}

void Heap::collect() {
    int markDescendants = empty, markedCells = empty;
    for (int p = 0; p < rootCnt; p++) {
        if (roots[p] != empty) {
            transfer(roots[p],nonFreeCells,markDescendants);
            marked(roots[p]) = true;
        }
    }
    printList(markDescendants,"markDescendants");
    for (p = markDescendants; p != empty; p = markDescendants) {
        transfer(p,markDescendants,markedCells);
        if (!atom(p) && !marked(Head(p))) {
            transfer(Head(p),nonFreeCells,markDescendants);
            marked(Head(p)) = true;
        }
        if (!atom(p) && !marked(Tail(p))) {
            transfer(Tail(p),nonFreeCells,markDescendants);
            marked(Tail(p)) = true;
        }
    }
    cout << *this;
```

FIGURE 12.16 *(continued)*

```cpp
        printList(markedCells,"markedCells");
        for (p = markedCells; p != empty; p = next(p))
            marked(p) = false;
        freeCells = nonFreeCells;
        nonFreeCells = markedCells;
    }

    int Heap::allocateAux(int p) {
        if (p == maxRoot) {
            cout << "No room for new roots\n";
            return !OK;
        }
        if (freeCells == empty)
            collect();
        if (freeCells == empty) {
            cout << "No room in heap for new cells\n";
            return !OK;
        }
        if (p == rootCnt)
            roots[rootCnt++] = p;
        roots[p] = freeCells;
        transfer(freeCells,freeCells,nonFreeCells);
        return OK;
    }

    void Heap::allocateAtom (int p, int val) {// an instance of Lisp's setf;
        if (allocateAux(p) == OK) {
            atom(roots[p]) = true;
            value(roots[p]) = val;
        }
    }

    void Heap::allocateList(int p, int q, int r) { // Lisp's cons;
        if (allocateAux(p) == OK) {
            atom(roots[p]) = false;
            Head(roots[p]) = roots[q];
            Tail(roots[p]) = roots[r];
        }
    }
```

Continues

FIGURE 12.16 *(continued)*

```cpp
void Heap::deallocate(int p) {
    if (rootCnt > 0)
        if (rand() % 2 == 0)
             roots[p] = roots[--rootCnt]; // remove variable when exiting
                                          // a block;
        else roots[p] = empty; // set variable to null;
}

void Heap::printList(int list, char *name) {
    cout << name << ": ";
    for (int i = list; i != empty; i = next(i))
        cout << "(" << i << " " << Head(i) << " " << Tail(i) << ")";
    cout << endl;
}

ostream& operator<< (ostream& out, Heap& h) {
    cout << "roots: ";
    for (int i = 0; i < h.rootCnt; i++)
        cout << h.roots[i] << " ";
    cout << endl;
    for (i = 0; i < h.maxHeap; i++)
        cout << "(" << i << ": " << h.prev(i) << " " << h.next(i)
             << h.atom(i) << " " << h.marked(i) << " "
             << " " << h.Head(i) << " " << h.Tail(i) << ") ";
    cout << endl;
    h.printList(h.freeCells,"freeCells");
    h.printList(h.nonFreeCells,"nonFreeCells");
    return out;
}

#endif

//************************* collector.cpp *************************

#include <iostream>
#include <cstdlib>
using namespace std;
#include "heap.h"

Heap heap;
```

FIGURE 12.16 *(continued)*

```
void program() {
    static int val = 123;
    int rn, p, q, r;
    if (heap.rootCnt == 0) { // call heap.allocateAtom(0,val++);
        p = 0;
        rn = 1;
    }
    else {
        rn = rand() % 100 + 1;
        p  = rand() % heap.rootCnt+1; // possibly new root;
        q  = rand() % heap.rootCnt;
        r  = rand() % heap.rootCnt;
    }
    if (rn <= 20)
        heap.allocateAtom(p,val++);
    else if (rn <= 40)
        heap.allocateList(p,q,r);
    else if (rn <= 60)
        heap.updateHead(q,r);
    else if (rn <= 80)
        heap.updateTail(q,r);
    else heap.deallocate(p);
    cout << heap;
}

int main() {
    for (int i = 0; i < 50; i++)
        program();
    return 0;
}
```

12.6 EXERCISES

1. What happens to the first-fit method if it is applied to a list ordered by block sizes?

2. How does the effort leading to coalescing blocks in sequential-fit methods depend on the order of blocks on the list? How can possible problems caused by these orders be solved?

3. The *optimal-fit* method determines which block to allocate after examining a sample of blocks to find the closest match to the request, and then finds the first block exceeding this match (Campbell, 1971). What does the efficiency of this method depend on? How does this algorithm compare to the efficiency of other sequential-fit methods?

4. In what circumstances can the size-list in the adaptive exact-fit method be empty (except at the beginning)? What is its maximal size, and when can it be this size?

5. Why in the buddy system are doubly linked, not singly linked, lists of blocks used?

6. Give an algorithm for returning blocks to the memory pool using the Fibonacci buddy system.

7. Apply `markingWithStack()` to the left degenerate and right degenerate list structures in Figure 12.17. How many calls to `pop()` and `push()` are executed for each case? Are all of them necessary? How would you optimize the code to avoid unnecessary operations?

FIGURE 12.17 (a) Left degenerate and (b) right degenerate list structures.

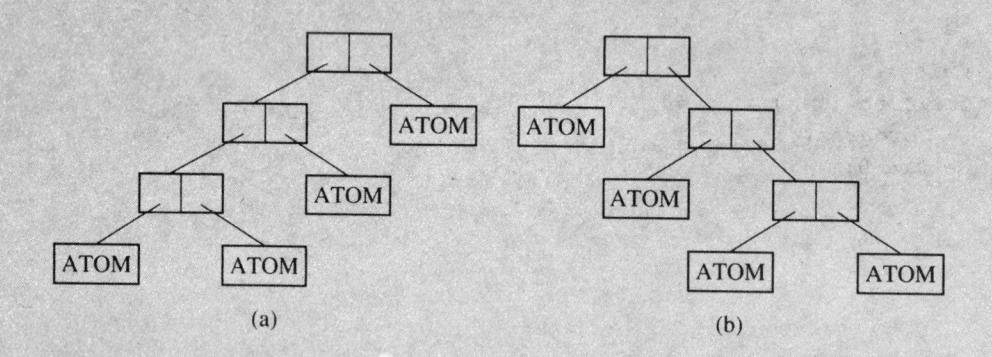

8. In a *reference count method* of garbage collection, each cell c has a counter field whose value indicates how many other cells refer (point) to it. The counter is incremented every time another cell refers to c and decremented if a reference is deleted. The garbage collector uses this counter when sweeping the heap: If a cell's count is zero, the cell can be reclaimed because it is not pointed to by any other cell. Discuss the advantages and disadvantages of this garbage collection method.

9. In Baker's algorithm, the scanning performed by the collector should be finished before *bottom* reaches *top* in tospace to flip spaces. What should the value of *k* be to ensure this? Assume that *n* is the maximum number of cells required by a program, and $2m$ is the number of cells in fromspace and tospace. What is the impact of doubling the value of *k* when it is an integer and when it is a fraction (for example, if it is .5, then one copy is made per two requests)?

10. In a modification of Baker's algorithm that requires updating heap pages in the case when the mutator's access is trapped (Ellis, Li, and Appel, 1988), there is a problem with objects that may cross the page boundary. Suggest a solution to this problem.

12.7 PROGRAMMING ASSIGNMENTS

1. Implement the following memory allocation method developed by W. A. Wulf, C. B. Weinstock, and C. B. Johnsson (Standish, 1980) called the *quick-fit* method. For an experimentally found number *n* of the most frequently requested sizes of blocks, this method uses an array *avail* of $n + 1$ cells, each cell *i* pointing to a linked list of blocks of size *i*. The last cell $(n + 1)$ refers to a block of other less frequently needed sizes. It may also be a pointer to a linked list, but because of possibly a large number of such blocks, another organization is recommended, such as a binary search tree. Write functions to allocate and deallocate blocks. If a block is returned, coalesce it with its neighbors. To test your program, randomly generate sizes of blocks to be allocated from memory simulated by an array whose size is a power of 2.

2. In the dual buddy system, two parts of memory are managed by the binary buddy method. But the number of such areas can be larger (Page and Hagins, 1986). Write a program to operate on three such areas with block sizes of the form 2^i, $3 \cdot 2^j$, and $5 \cdot 2^k$. For a requested block size *s*, round *s* to the nearest block size that can be generated by this method. For example, size 11 is rounded up to 12, that is the number from the second area. If this request cannot be accommodated in this area, 12 is rounded up to the next possibly available number, which is 15, a number from the third area. If there is no available block of this or greater size in this area either, the first area is tried. In case of failure, keep requests on a list and process them as soon as a block of sufficient size is coalesced. Run your program changing three parameters: the intervals for which blocks are reserved, the number of incoming requests, and the overall size of memory.

3. Implement a simple version of a generational garbage collector that uses only two regions (Appel, 1989). The heap is divided into two even parts. The upper part holds cells that have been copied from the lower part as cells reachable from the root pointers. The lower part is used for memory allocation and holds only newer cells (see Figure 12.18a). After this part becomes full, the garbage collector cleans it by copying all reachable cells to the upper part (Figure 12.18b), after which allocations are made starting from the beginning of the lower part. After several turns, the upper part becomes full too, and the cells being copied from the lower part are in reality copied to the lower part (Figure 12.18c). In this case, the cleanup process of the upper part is begun by copying all reachable cells from the upper part to the lower part (Figure 12.18d), and then all reachable cells are copied to the beginning of the upper part (Figure 12.18e).

FIGURE 12.18 A heap with two regions for Appel's generational garbage collection.

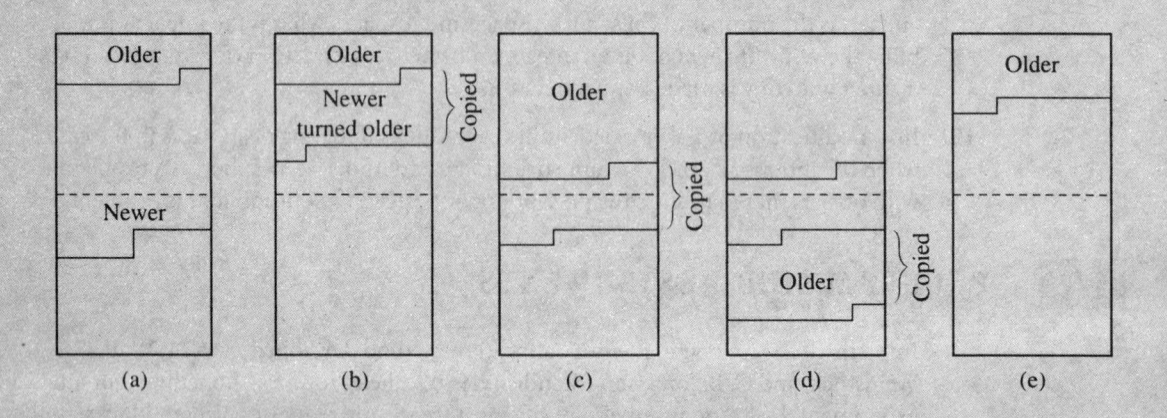

4. The case study presents an in-place nonincremental garbage collector. Modify and extend it to become an incremental collector. In this case, `program()` becomes `mutator()`, which allows the function `collect()` to process k cells for some value of k. To prevent `mutator()` from introducing inconsistencies in structures possibly not completely processed by `collect()`, `mutator()` should transfer any unmarked cells from `freeCells` to `markDescendants`.

Another very elegant modification is obtained by grouping all four lists in a circular list, creating what Henry Baker (1992) called a *treadmill* (Figure 12.19a). Pointer `free` is moved in a clockwise direction if a new cell is requested; pointer `toBeMarked` is moved k times when allowed by the mutator. For each nonatom cell currently scanned by `toBeMarked`, its *head* and *tail* are transferred in front of `toBeMarked` if they are not marked. After `toBeMarked` meets `endNonFree`, there are no cells to be marked, and after `free` meets `nonFree`, there are no free cells on the list of free cells (Figure 12.19b). In this case, what remains between `nonFree` and `endNonFree` (former `nonFreeCells`) is garbage, and hence, it can be utilized by the mutator. Therefore, the roles of `nonFree` and `endNonFree` are exchanged; it is as though `nonFreeCells` became `freeCells` (Figure 12.19c). All root pointers are transferred to a part of the treadmill between `toBeMarked` and `endFree` (to create a seed of the former `markDescendants`, and the mutator can resume execution.

FIGURE 12.19 Baker's treadmill.

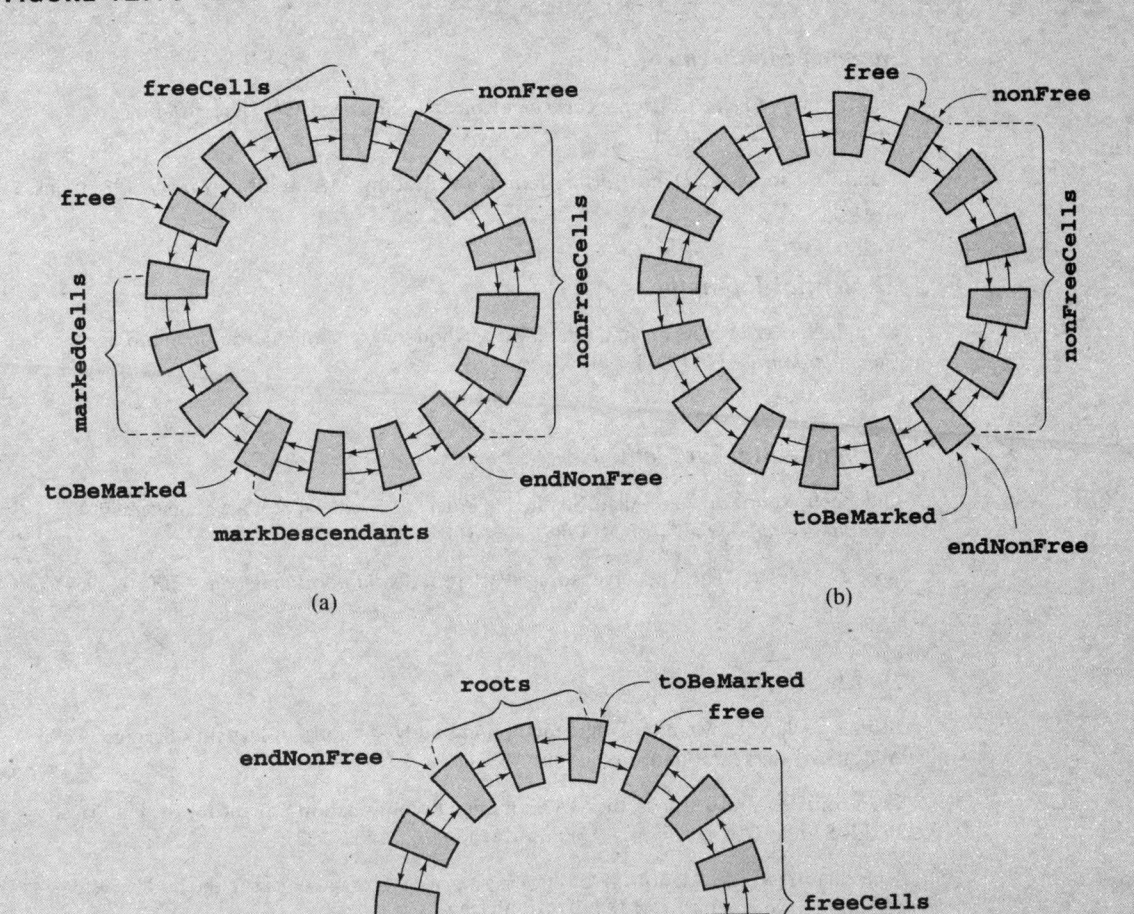

BIBLIOGRAPHY

Memory Management

Smith, Hary F., *Data Structures: Form and Function*, San Diego, CA: Harcourt-Brace-Jovanovich (1987), Ch. 11.

Standish, Thomas A., *Data Structure Techniques*, Reading, MA: Addison-Wesley (1980), Chs. 5 and 6.

Sequential-Fit Methods

Campbell, J. A., "A Note on an Optimal-Fit Method for Dynamic Allocation of Storage," *Computer Journal* 14 (1971), 7–9.

Nonsequential-Fit Methods

Oldehoeft, Rodney R., and Allan, Stephen J., "Adaptive Exact-Fit Storage Management," *Communications of the ACM* 28 (1985), 506–511.

Ross, Douglas T., "The AED Free Storage Package," *Communications of the ACM* 10 (1967), 481–492.

Buddy Systems

Bromley, Allan G., "Memory Fragmentation in Buddy Methods for Dynamic Storage Allocation," *Acta Informatica* 14 (1980), 107–117.

Cranston, Ben, and Thomas, Rick, "A Simplified Recombination Scheme for the Fibonacci Buddy System," *Communications of the ACM* 18 (1975), 331–332.

Hinds, James A., "An Algorithm for Locating Adjacent Storage Blocks in the Buddy System," *Communications of the ACM* 18 (1975), 221–222.

Hirschberg, Daniel S., "A Class of Dynamic Memory Allocation Algorithms," *Communications of the ACM* 16 (1973), 615–618.

Knowlton, Kenneth C., "A Fast Storage Allocator," *Communications of the ACM* 8 (1965), 623–625.

Page, Ivor P., and Hagins, Jeff, "Improving Performance of Buddy Systems," *IEEE Transactions on Computers* C-35 (1986), 441–447.

Shen, Kenneth K., and Peterson, James L., "A Weighted Buddy Method for Dynamic Storage Allocation," *Communications of the ACM* 17 (1974), 558–562.

Garbage Collection

Appel, Andrew W., "Garbage Collection Can Be Faster Than Stack Allocation," *Information Processing Letters* 25 (1987), 275–279.

Appel, Andrew W., "Simple Generational Garbage Collection and Fast Allocation," *Software—Practice and Experience* 19 (1989), 171–183.

Baker, Henry G., "List Processing in Real Time on a Serial Computer," *Communications of the ACM* 21 (1978), 280–294.

Baker, Henry G., "The Treadmill: Real-Time Garbage Collection Without Motion Sickness," *ACM SIGPLAN Notices* 27 (1992), No. 3, 66–70.

Brooks, Rodney A., "Trading Data Space for Reduced Time and Code Space in Real-Time Collection on Stock Hardware," *Conference Record of the 1984 ACM Symposium on Lisp and Functional Programming*, Austin, TX (1984), 108–113.

Cheney, C. J., "A Nonrecursive List Compacting Algorithm," *Communications of the ACM* 13 (1970), 677–678.

Cohen, Jacques, "Garbage Collection of Linked Data Structures," *Computing Surveys* 13 (1981), 341–367.

Ellis, John R., Li, Kai, and Appel, Andrew W., "Real-Time Concurrent Collection on Stock Multiprocessors," *SIGPLAN Notices* 23 (1988), No. 7, 11–20.

Fenichel, Robert R., and Yochelson, Jerome C., "A Lisp Garbage-Collector for Virtual-Memory Computer Systems," *Communications of the ACM* 12 (1969), 611–612.

Jones, Richard, and Lins, Rafael, *Garbage Collection: Algorithms for Automatic Dynamic Memory Management*, Chichester, United kingdom: Wiley, 1996.

Kurokawa, Toshiaki, "A New Fast and Safe Marking Algorithm," *Software—Practice and Experience* 11 (1981), 671–682.

Layer, D. Kevin, and Richardson, Chris, "Lisp Systems in the 1990s," *Communications of the ACM* 34 (1991), No. 9, 49–57.

Lieberman, Henry, and Hewitt, Carl, "A Real-Time Garbage Collector Based on the Lifetimes of Objects," *Communications of the ACM* 26 (1983), 419–429.

Schorr, H., and Waite, W. M., "An Efficient Machine-Independent Procedure for Garbage Collection in Various List Structures," *Communications of the ACM* 10 (1967), 501–506.

Wadler, Philip L., "Analysis of Algorithm for Real-Time Garbage Collection," *Communications of the ACM* 19 (1976), 491–500, 20 (1977), 120.

Wegbreit, Ben, "A Space-Efficient List Structure Tracing Algorithm," *IEEE Transactions on Computers* C-21 (1972), 1009–1010.

Wilson, Paul R., "Uniprocessor Garbage Collection Techniques," in Bekkers, Yves, and Cohen, Jacques (eds.), *Memory Management*, Berlin: Springer (1992), 1–42.

Yuasa, Taiichi, "Real-Time Garbage Collection on General-Purpose Machine," *Journal of Systems and Software* 11 (1990), 181–198.

Zorn, Benjamin, "Comparing Mark-and-Sweep and Stop-and-Copy Garbage Collection," *Proceedings of the 1990 ACM Conference on Lisp and Functional Programming* (1990), 87–98.

13 String Matching

String matching is important for virtually each computer user. When editing a text, the user processes it, organizes it into paragraphs and sections, reorders it, and, very often, searches for some subtext or pattern in the text to locate the pattern or replace it with something else. The larger the text that is being searched for, the more important is the efficiency of the searching algorithm. The algorithm cannot usually rely on, say, alphabetical ordering of words, as would be the case with a dictionary. For example, string searching algorithms are increasingly important in molecular biology, where they are used to extract information from DNA sequences by locating some pattern in them and comparing the sequences for common subsequences. Such processing has to be done frequently, under the assumption that an exact match cannot be expected. Problems of that type are addressed by what is often called *stringology*, whose major area of interest is *pattern matching*. Some stringological problems are discussed in this chapter.

This chapter uses the following notation: For a text T, which is a sequence of symbols, characters, or letters, $|T|$ signifies the length of T, T_j is the character at position j of T, and $T(i \ldots j)$ is a substring of T that begins at position i and ends at j. The first characters in pattern P and text T are in position 0. Also, a *regular expression* a^n stands for a string $a \ldots a$ of n as.

13.1 EXACT STRING MATCHING

Exact string matching consists of finding an exact copy of pattern P in text T. It is an all-or-nothing approach; if there is a very close similarity between P and a substring of T, the partial match is rejected.

13.1.1 Straightforward Algorithms

A simple approach to string matching is starting the comparison of P and T from the first letter of T and the first letter of P. If a mismatch occurs, the matching begins from the second character of T, and so on. Any information that can be useful in subsequent tries is not retained. The algorithm is given in this pseudocode:

```
bruteForceStringMatching(pattern P, text T)
    i = 0;
    while i ≤ |T| - |P|
        j = 0;
        while T_i == P_j and j < |P|
            i++;                        // try to match all characters in P;
            j++;
        if j == |P|
            return match at i - |P|;    // success if the end of P is reached;
                                        // if there is a mismatch,
        i = i - j + 1;                  // shift P to the right by one position;
    return no match;                    // failure if fewer characters left in T than |P|;
```

In the worst case the algorithm executes in $O(|T||P|)$ time. For example, if $P = a^{m-1}b$ and $T = a^n$, then the algorithm makes $(n - (m - 1))m = nm - m^2 + m$ comparisons, which is approximately nm for a large n and small m.

The average performance depends on the probability distribution of the characters in both the pattern and the text. As an example, assume that only two characters are used, and the probability of using any of the two characters equals 1/2. In this case, for a particular scan i, the probability equals 1/2 that only one comparison is made, the probability $1/2 \cdot 1/2 = 1/4$ that two comparisons are made, ..., and the probability $1/2 \cdot \ldots \cdot 1/2 = 2^{-|P|}$ that m comparisons are performed; that is, on average, for a given i, the number of comparisons equals

$$\sum_{k=1}^{|P|} \frac{k}{2^k} < 2$$

so that the average number of comparisons for all the scans equals $2(|T| - (|P| - 1)) < 2|T|$ for a large $|T|$. A far better estimate, $2^{|P|+1} - 2$, is found using the theory of absorbing Markov chains, and, more generally, for an alphabet A, the average number of comparisons is $(|A|^{|P|+1} - |A|) / (|A| - 1)$ (Barth, 1985).

Here is an example of execution of the brute force algorithm for $T = $ *ababcdabbabababad* and $P = $ *abababa*:

```
    ababcdabbabababad
 1  ababa ba
 2   ababa ba
 3    abababa
 4     abababa
 5      abababa
 6       abababa
 7        abababa
 8         abababa
 9          abababa
10           abababa
```
(13-1)

The corresponding characters in P and T are compared—which is marked by underlining characters in P—starting at the position where P is currently aligned with T.

After a mismatch is found, the scan through P and T is aborted and restarted after P is shifted to the right by one position. In the first iteration, the matching process begins at the first characters of P and T, and a mismatching occurs at the fifth character of $T(T_4 = c)$ and the fifth character of $P(P_4 = a)$. The next round begins from the first character of P, but this time, from the second character of T, which immediately leads to a mismatch. The third iteration reaches the third character of P, a, and the fifth character of T, c. The match of the entire pattern P is found in the tenth iteration.

Note that no actual shifting takes place; the shifting is accomplished by updating index i.

An improvement is accomplished by a not-so-naïve algorithm proposed by Hancart (1992). It begins comparisons from the second character of P, goes to the end, and ends comparisons with the first character. So the order of characters involved in comparisons is $P_1, P_2, \ldots, P_{|P|-1}, P_0$.

The information about equality of the first two characters of P is recorded and used in the matching process. Two cases are distinguished: $P_0 = P_1$ and $P_0 \neq P_1$. In the first case, if $P_1 \neq T_{i+1}$, text index i is incremented by 2, because $P_0 \neq T_{i+1}$; otherwise, i is incremented by 1. It is similar in the second case, if $P_1 = T_{i+1}$. In this way, a shift by two positions is possible. Here is the algorithm:

```
Hancart(pattern P, text T)
    if P₀ == P₁
        sEqual = 1;
        sDiff = 2;
    else sEqual = 2;
        sDiff = 1;
    i = 0;
    while i ≤ |T| - |P|
        if T_{i+1} ≠ P₁
            i = i + sDiff;
        else j = 1;
            while j < |P| and T_{i+j} == P_j
                j++;
            if j == |P| and P₀ == T_i
                return match at i;
            i = i + sEqual;
    return no match;
```

Matching begins from the second pattern character. If there is a mismatch between P_1 and T_{i+1}, then P can be shifted by two positions before beginning the next round, as long as the first two characters of P are the same, because this mismatch means that P_0 and T_{i+1} are also different:

```
    i
    ↓
    acaaca
1   aab
2     aab
```

However, after a mismatch occurs in the inner `while` loop, the pattern is shifted by only one position:

```
       i
       ↓
    acaaca
2    aab
3     aab
```

On the other hand, if the first two characters of P are different, then after noticing in the `if` statement that P_1 and T_{i+1} are different, P is shifted by one position only:

```
       i
       ↓
    aabaca
1    abb
2     abb
```

so that a possible occurrence of P is not missed. However, after a mismatch is found in any other position, P is shifted by two places:

```
       i
       ↓
    aabaca
2    abb
3      aab
```

This can be done safely, because P_1 and T_{i+1} have just been determined as equal and because P_0 and P_1 are different, P_0 and T_{i+1} must be also different, thus there is no need to check this in the third iteration. Here is another example:

```
    ababcdabbabababad
1   abababa
2    abababa
3     abababa
4      abababa
5       abababa
6        abababa
7         abababa
```

In the worst case, the algorithm executes in $O(|T||P|)$ time, but, as Hancart shows, it performs on average better than some of the more developed algorithms to be discussed in the next section.

13.1.2 The Knuth-Morris-Pratt Algorithm

The brute force algorithm is inefficient in that it shifts pattern P by one position after a mismatch is found. To speed up the process, Hancart's algorithm allows for a shift by two characters. However, a method is needed to shift P by as many positions to the right as possible but so that no match is missed.

The source of inefficiency of the brute force algorithm lies in performing redundant comparisons. The redundancy can be avoided by observing that pattern P includes identical substrings at the beginning of P and before the mismatched character. This fact can be used to shift P to the right by more than one position before beginning the next scan. Consider line 1 of the following diagram. The mismatch occurs at the fifth character, but until that point, both the prefix ab of P and the substring $P(2\ldots3)$, which is also ab, have been successfully processed. P can now be moved to the right to align its substring ab with the substring $T(2\ldots3)$ and the matching process can start from character P_2 and from the mismatched character in T, T_4. Because characters in the substring $P(2\ldots3)$ have just been successfully matched with $T(2\ldots3)$, it is as though the characters in the prefix $P(0\ldots1)$ were matched with $T(2\ldots3)$. In this way, the two redundant comparisons in line 2 can be omitted. After the mismatch

```
          i
          ↓
     ababcdabbababababad
  1  abababa
          ↑
          j
```

the matching process continues as in

```
          i
          ↓
     ababcdabbababababad
  2    abababa
          ↑
          j
```

thereby skipping $ab = P(0\ldots1)$. The two identical parts relevant to this shift are the prefix of P and suffix of this part of P that is currently successfully matched, which is the prefix $P(0\ldots1)$ and the suffix $P(2\ldots3)$ of the matched part of P, $P(0\ldots3)$.

Generally, to perform a shift, we first need to match a prefix of P with a suffix of $P(0\ldots j)$, where P_{j+1} is a mismatched character. This matching prefix should be the longest possible so that no potential match is passed after shifting P; that is, if the match is of length len and the current scan starts at position k of T, then no occurrence of P should begin in any position between k and $k + len$, but it may begin at position $k + len$, so that shifting P by len positions is safe.

This information will be used many times during the matching process; therefore, P should be preprocessed. Importantly, in this approach only the information about P is used; the configuration of characters in T is irrelevant.

Define the table *next*:

$$next[j] = \begin{cases} -1 & \text{for } j = 0 \\ \max\{k : 0 < k < j \text{ and } P[0\ldots k-1] = P[j-k\ldots j-1]\} & \text{if such a } k \text{ exists} \\ 0 & \text{otherwise} \end{cases}$$

that is, the number *next[j]* indicates the length of the longest suffix of substring $P(0 \ldots j-1)$ equal to a prefix of the same substring:

$$
\begin{array}{cc}
j - next[j] & j - 1 \\
\downarrow & \downarrow
\end{array}
$$

<u>a</u>...<u>b</u>c...d<u>a</u>...<u>b</u>e...
↑ ↑
0 *next[j]*

The condition $k < j$ indicates that the prefix is also a proper suffix. Without this condition, *next*[2] for $P(0 \ldots 2) = aab$ would be 2, because *aa* is at the same time a prefix and suffix of *aa*, but with the additional condition, *next*[2] = 1, not 2.

For example, for $P = abababa$,

P	a b a b a b a
j	0 1 2 3 4 5 6
next[j]	-1 0 0 1 2 3 4

Note that because of the condition requiring that the matching suffix be the longest, *next*[5] = 3 for $P(1 \ldots 6) = ababab$, because *aba* is the longest suffix of *ababa* matching its prefix (they overlap), not 1, although *a* is also both a prefix and a suffix of *ababa*.

The Knuth-Morris-Pratt algorithm can be obtained relatively easily from `bruteForceStringMatching()`:

```
KnuthMorrisPratt(pattern P, text T)
    findNext(P,next);
    i = j = 0;
    while i ≤ |T| - |P|
        while j == -1 or j < |P| and T_i == P_j
            i++;              // increment i only for matched characters;
            j++;
        if j == |P|
            return a match at i - |P|;
        j = next[j]           // in the case of a mismatch, i does not change;
    return no match;
```

The algorithm `findNext()` to determine the table *next* will be defined shortly. For example, for $P = abababa$, *next* = [-1 0 0 1 2 3 4], and $T = ababcdabbababababad$, the algorithm executes as follows:

```
    ababcdabbababababad
1   abababa
2    abababa
3        abababa
4         abababa
5          abababa
6           abababa
7            abababa
```

The diagram indicates that −1 in *next* means that the entire pattern *P* should be shifted past the mismatched text character; see the shift from line 4 to 5 and from line 6 to 7. One major difference between `bruteForceStringMatching()` and `KnuthMorrisPratt()` is that *i* is never decremented in the latter algorithm. It is incremented in the case of a match; in the case of a mismatch *i* stays the same so that the mismatched character in *T* is compared to another character in *P* in the next iteration of the outer `while` loop. The only case when *i* is incremented in the case of mismatch is when the first character in *P* is a mismatched character; to this end the subcondition j == −1 is needed in the inner loop. After finding a mismatch at position j ≠ 0 of *P*, *P* is shifted by *j* − *next*[*j*] positions; when the mismatch occurs at the first position of *P*, the pattern is shifted by one position.

To assess the computational complexity of `KnuthMorrisPratt()`, note that the outer loop executes $O(|T|)$ times. The inner loop executes at most $|T| - |P|$ times, because *i* is incremented in each iteration of the loop, and by the condition on the outer loop, $|T| - |P|$ is the maximum value for *i*. But for a mismatched character T_i, *j* can be assigned a new value $k \leq |P|$ times. When this happens, the first character in *P*, for which the mismatch occurs, is aligned with the character T_{i+k}. Consider *P* = *aaab* and *T* = *aaacaaadaaab*. In this case, *next* = [−1 0 1 2], and the trace of the execution of the algorithm is as follows:

```
  aaacaaadaaab
1 aaab
2  aaab
3   aaab
4    aaab
5     aaab
6      aaab
7       aaab
8        aaab
9         aaab
```

The mismatched *c* in *T* is compared to four characters in *P* in lines 1 through 4 because *b* is the fourth character in *P* for which the mismatch occurs for the first time and because *b* is aligned with *c*; that is, all the preceding characters have already been matched successfully. The next time, such a situation occurs for *d* in *T* and again for *b* in *P* on line 5, and by this time all the preceding characters in *P* are successfully matched. This means, that for some *i*, |*P*| comparisons can be performed, but this can not happen for every *i*, but only for every |*P*|th *i*, so that the number of unsuccessful comparisons can be up to $|P|(|T|/|P|) = |T|$. Up to $|T| - |P|$ successful comparisons have to be added to this number to obtain the running time $O(|T|)$.

The table *next* still remains to be determined. We can use the brute force algorithm to that end, which is not necessarily inefficient for short patterns. But we can also adapt the Knuth-Morris-Pratt algorithm to improve the efficiency of determining *next*.

Remember that *next* contains the lengths of the longest suffixes matching prefixes of *P*; that is, parts of *P* are being matched with other parts of *P*. But the problem of matching is solved already by the Knuth-Morris-Pratt algorithm. In this case, *P* is matched against itself. However, the Knuth-Morris-Pratt algorithm uses *next*, which is still unknown. Therefore, the Knuth-Morris-Pratt algorithm has to be modified so that it determines the values of the table *next* by using values already found. Let $next[0] = -1$. Assuming that values $next[0], \ldots, next[i-1]$ have already been determined, we want to find the value $next[i]$. There are two cases to consider.

In the first case, the longest suffix matching a prefix is found by simply attaching the character P_{i-1} to the suffix corresponding to position $next[i-1]$, which is true when $P_{i-1} = P_{next[i-1]}$:

```
a...bc..........da...bc...
   ↑                  ↑
next[i-1]-1         i-1
```

$$\Downarrow \quad next[i] = next[i-1] + 1$$

```
a...bc..........da...bc...
   ↑                   ↑
next[i-1]              i
```

In this case, the current suffix is longer by one character than the previously found suffix so that $next[i] = next[i-1] + 1$.

In the second case, $P_{i-1} \neq P_{next[i-1]}$. But this is simply a mismatch, and a mismatch can be handled with the table *next*, which is why it is being determined. Because $P_{next[i-1]}$ is a mismatched character, we need to go to $next[next[i-1]]$ to check whether P_{i-1} matches $P_{next[next[i-1]]}$. If they match, $next[i]$ is assigned $next[next[i-1]] + 1$:

```
a...bc...da...be..........fa...bc...da...bc...
            ↑                             ↑
        next[i-1]                       i-1
```

$$\Downarrow \quad next[i] = next[next[i-1]] + 1$$

```
a...bc...da...be..........fa...bc...da...bc...
   ↑                                         ↑
next[next[i-1]]                              i
```

otherwise P_{i-1} is compared to $P_{next[next[next[i-1]]]}$ to have $next[i] = next[next[next[i-1]]] + 1$ if the characters match; otherwise, the search continues until a match is found or the beginning of P is reached.

Note that in the previous diagram, the first prefix $a \ldots bc \ldots da \ldots b$ of $P(0 \ldots i-1)$ has a prefix $a \ldots b$ identical to its suffix. This is not an accident. The reason for $a \ldots b$ being both prefix and suffix of $a \ldots bc \ldots da \ldots b$ when $a \ldots b$ is about to be found as the longest prefix and suffix of $P(0 \ldots i-1)$ is as follows. The prefix $P(0 \ldots j-1) = a \ldots bc \ldots da \ldots b$ of $P(0 \ldots i-1)$ indicated by $next[i-1]$ is, by definition, equal to the suffix $P(i-j-1 \ldots i-2)$, which means that the suffix $P(j - next[j] \ldots j-1) = a \ldots b$ is also a suffix of $P(i-j-1 \ldots i-2)$. Therefore, to determine the value of $next[i]$ we refer to the already determined value $next[j]$ that specifies the length of this shorter suffix of $P(0 \ldots j-1)$ matching a prefix of P, and thus the length of the suffix $a \ldots b$ of $P(0 \ldots i-1)$ matching the same prefix.

The algorithm to find the table *next* is as follows:

```
findNext(pattern P, table next)
    next[0] = -1;
    i = 0;
    j = -1;
    while i < |P|
        while j == 0 or i < |P| and P_i == P_j
            i++;
            j++;
            next[i] = j;
        j = next[j];
```

Here is an example of finding *next* for pattern $P = ababacdd$. The values of indices i and j and the table *next* before entering the inner `while` loop are indicated with an arrow (and by the fact that i does not change); the remaining lines show these values at the end of the inner loop and a comparison that follows. For example, in line 2, after incrementing i to 1 and j to 0, 0 is assigned to $next[1]$, and then the first and second characters of P are compared, which leads to exiting the loop.

	i	j		next[ ]							P
→	0	-1	-1								ababacdd
	1	0	-1	0							a̲babacdd
→	1	-1	-1	0							
	2	0	-1	0	0						ab̲abacdd
	3	1	-1	0	0	1					aba̲bacdd
	4	2	-1	0	0	1	2				abab̲acdd
	5	3	-1	0	0	1	2	3			abaca̲cdd
→	5	1	-1	0	0	1	2	3			abab̲acdd
→	5	0	-1	0	0	1	2	3			aba̲bacdd
→	5	-1	-1	0	0	1	2	3			
	6	0	-1	0	0	1	2	3	0		ababac̲dd
→	6	-1	-1	0	0	1	2	3	0		
	7	0	-1	0	0	1	2	3	0	0	ababacd̲d
→	7	-1	-1	0	0	1	2	3	0	0	
	8	0	-1	0	0	1	2	3	0	0	

Because of the similarity of this algorithm and the Knuth-Morris-Pratt algorithm, we conclude that *next* can be determined in $O(|P|)$ time.

The outer `while` loop in `KnuthMorrisPratt()` executes in $O(|T|)$ time, so the Knuth-Morris-Pratt algorithm, including `findNext()`, executes in $O(|T| + |P|)$ time. Note that in the analysis of the complexity of the algorithm, no mention was made about the alphabet underlying the text T and pattern P; that is, the complexity is independent of the number of different characters constituting P and T.

The algorithm requires no backtracking in text T; that is, the variable i is never decremented during execution of the algorithm. This means that T can be processed one character at a time, which is very convenient for online processing.

The Knuth-Morris-Pratt algorithm can be improved if we eliminate unpromising comparisons. If the mismatch occurs for characters T_i and P_j, then the next match is attempted for the same character T_i and character $P_{next[j]+1}$. But if $P_j = P_{next[j]+1}$ then the same mismatch takes place, which means a redundant comparison is made. Consider $P = abababa$ and $T = ababcdabbababababad$, analyzed earlier, for which $next = [-1\ 0\ 0\ 1\ 2\ 3\ 4]$ and the Knuth-Morris-Pratt algorithm begins with:

```
     ababcdabbababababad
1    ababa̱ba
2      aba̱baba
```

The first mismatch occurs for a at the fifth position of P and for c in T. The table *next* indicates that in the case of the mismatch of the fifth character of P, P should be shifted by two positions to the right, because $4 - next[4] = 2$; that is, the two-character prefix of P should be aligned with the two-character suffix of $P(0 : \ldots 3)$. The situation is illustrated on the second line of the diagram. However, this means that the next comparison is made between c that just caused a mismatch and a at the third position of P. But this is a comparison that has just been made on line 1 of the diagram, where a in the fifth position of P was also compared to c. Therefore, if we knew that the prefix ab of P is followed by a, which is also a character following suffix ab of $P(0 \ldots 3)$, then the situation of the second line of the diagram could be avoided. To accomplish it, the table *next* has to be redesigned to exclude such redundant comparisons. This is done by extending the definition of *next* by one more condition, which leads to the following definition of a stronger *next*:

$$nextS[j] = \begin{cases} -1 & \text{for } j = 0 \\ \max\{k : 0 < k < j \text{ and } P[0 \ldots k-1] = P[j-k \ldots j-1] \text{ and } P_{k+1} \neq P_j\} & \text{if such a } k \text{ exists} \\ 0 & \text{otherwise} \end{cases}$$

To compute *nextS*, the algorithm `findNext()` needs to be modified slightly to account for the additional condition, as in

```
findNextS(pattern P, table nextS)
    nextS[0] = -1;
    i = 0;
    j = -1;
    while i < |P|
        while j == -1 or i < |P| and P_i == P_j
            i++;
            j++;
            if P_i ≠ P_j
                nextS[i] = j;
            else nextS[i] = nextS[j];
        j = nextS[j];
```

The rationale is as follows. If $P_i \neq P_j$—that is, the new subcondition defining *nextS* is satisfied—then clearly $next[i]$ and $nextS[i]$ are equal and so $nextS[i]$ in `findNextS()` is assigned the same value as $next[i]$ in `findNext()`. If the characters P_i and P_j are equal,

```
a...bc...da...be...fa...bc...da...be...
         ↑      ↑              ↑
         j     i-j              i
```

then the subcondition is violated, thus $nextS[i] < next[i]$, and the situation is as follows

$$\underline{a\ldots bc\ldots da\ldots be}\ldots fa\ldots \underline{bc\ldots da\ldots be}\ldots$$
$$\uparrow \qquad\qquad \uparrow \qquad\qquad\qquad \uparrow \qquad\quad \uparrow$$
$$j - nextS[j] \qquad j \qquad\qquad\qquad i - nextS[i] \quad i$$

The underlined substrings are the proper prefix and suffix of $P(0\ldots i-1)$ indicated by $nextS[i]$ that are shorter than $next[i]$ (they can be empty). But the prefix $P(0\ldots j-1) = a\ldots bc\ldots da\ldots b$ of $P(0\ldots i-1)$ indicated by $next[i]$ is, by definition, equal to the suffix $P(i-j\ldots i-1)$, which means that the suffix $P(j - nextS[j]\ldots j-1) = a\ldots b$ shown in italic is also a suffix of $P(i-j\ldots i-1)$. Therefore, to determine the value of $nextS[i]$ we refer to the already determined value $nextS[j]$ that specifies the length of the italicized suffix of $P(0\ldots j-1)$ matching a prefix of P, and thus the length of the suffix of $P(0\ldots i-1)$ matching the same prefix. If the prefix is followed by the character P_i, then $nextS[j]$ contains the length of a shorter prefix determined by the same process. For example, when processing position 11 in the string

```
P     = abcabdabcabdfabcabdabcabd
nextS = .....2.....2.............
```

number 2 is copied to $nextS[11]$ from $nextS[5]$, and the same number from position 11—that is, indirectly, from position 5—to $nextS[24]$:

```
P     = abcabdabcabdfabcabdabcabd
nextS = .....2.....2............2
```

The Knuth-Morris-Pratt algorithm is modified by replacing `findNext()` with `findNextS()`. The execution of this algorithm for $P = ababab$ generates $nextS = [-1\ 0\ -1\ 0\ -1\ 0\ -1]$ and then continues with comparisons as summarized in this diagram:

```
  ababcdabbabababad
1 abababa
2     abababa
3      abababa
4        abababa
```

The Knuth-Morris-Pratt algorithm exhibits the worst case performance for Fibonacci words defined recursively as follows:

$$F_1 = b,\ F_2 = a,\ F_n = F_{n-1}F_{n-2} \text{ for } n > 2$$

The words are: $b, a, ab, aba, abaab, abaababa, \ldots$.

In the case of mismatch, a Fibonacci word F_n can be shifted $\log_\varphi |F_n|$ times, where $\varphi = (1 + \sqrt{5})/2$ is the golden ratio. If the pattern $P = F_7 = abaababaabaab$, the Knuth-Morris-Pratt algorithm executes as follows:

```
  abaababaabaca...
1 abaababaabaa
2    abaabab
3      abaa
4       ab
5        a
6        a...
```

13.1.3 The Boyer-Moore Algorithm

In the Knuth-Morris-Pratt algorithm, each of the first $|T| - |P| + 1$ characters is used at least once in a comparison for an unsuccessful search. The source of this algorithm's better efficiency over the brute force approach lies in not starting the matching process from the beginning of pattern P when a mismatch is detected if possible. So the Knuth-Morris-Pratt algorithm goes through almost all characters in T from left to right and tries to minimize the number of characters in P involved in matching. It is not possible to skip any characters in T itself to avoid unpromising comparisons. To accomplish such skipping, the Boyer-Moore algorithm tries to match P with T by comparing them from right to left, not from left to right. In the case of a mismatch, it shifts P to the right and always begins the next matching from the end of P, but it shifts P to the right so that many characters in T are not involved in the comparisons. Thus, the Boyer-Moore algorithm attempts to gain speed by skipping characters in T rather than, as the Knuth-Morris-Pratt algorithm does, skipping them in P, which is more prudent because the length of P is usually negligible in comparison to the length of T.

The basic idea is very simple. In the case of detecting a mismatch at character T_i, P is shifted to the right to align T_i with the first encountered character equal to T_i, if such a character exists. For example, for $T = aaaaebdaabadbda$ and $P = dabacbd$, first, characters $T_6 = d$ and $P_6 = d$, then characters $T_5 = b$ and $P_5 = b$ are compared, and then the first mismatch is found at $T_4 = e$ and $P_4 = c$. But there is no occurrence of e in P. This means that there is no character in P to be aligned with e in T; that is, no character can be successfully matched with e. Therefore, P can be shifted to the right past the mismatched character:

```
  aaaaebdaabadbda
1 dabacbd
2     dabacbd
```

In this way, the first four characters of the text are excluded from later comparisons. Now, matching starts from the end of P and from position $11 = 4 + 7 =$ (the position of the mismatched character T_4) + $|P|$. A mismatch is found at $T_{10} = a$ and $P_5 = b$, and then the mismatched a is aligned with the first a to the left of mismatched P_5:

```
  aaaaebdaabadbda
2     dabacbd
3        dabacbd
```

that is, the position in T from which the matching process starts in the third line is $13 = 10 + 3 =$ (the position of the mismatched character $T_{10} = a$) + ($|P|$ − position of the rightmost a in P). After matching characters T_{13} with P_6 and T_{12} with P_5, a mismatch is found at $T_{11} = d$ and $P_4 = c$. If we aligned the mismatched d in text with the rightmost d in P, P would be moved backwards. Therefore, if there is a character in P equal to the mismatched character in T to the left of the mismatched character in P, the pattern P is shifted to the right by one position only:

```
  aaaaebdaabadbda
3        dabacbd
4         dabacbd
```

To sum up, the three rules can be termed character occurrence rules:

1. *No occurrence rule.* If the mismatched character T_i appears nowhere in P, align P_0 with T_{i+1}.
2. *Right side occurrence rule.* If there is a mismatch at T_i and P_j, and if there is an occurrence of character ch equal to T_i to the *right* of P_j, shift P by one position.
3. *Left side occurrence rule.* If there is an occurrence of character ch equal to T_i only to the *left* of P_j, align T_i with $P_k = ch$ closest to P_j.

To implement the algorithm, a table *delta1* specifies, for each character in the alphabet, by how much to increment i after a mismatch is detected. The table is indexed with characters and is defined as follows:

$$delta1[ch] = \begin{cases} |P| & \text{if } ch \text{ is not in } P \\ \min\{|P| - i - 1 : P_i = ch\} & \text{otherwise} \end{cases}$$

For the pattern $P = dabacbd$, $delta1['a'] = 3$, $delta1['b'] = 1$, $delta1['c'] = 2$, $delta1['d'] = 0$, and for remaining characters ch, $delta1[ch] = 7$.

The algorithm itself can be summarized as follows:

```
BoyerMooreSimple(pattern P, text T)
    initialize all cells of delta1 to |P|;
    for j = 0 to |P| - 1
        delta1[P_j] = |P| - j - 1;
    i = |P| - 1;
    while i < |T|
        j = |P| - 1;
        while j ≥ 0 and P_j == T_i
            i--;
            j--;
        if j == -1
            return match at i+1;
        i = i + max(delta1[T_i], |P|-j);
    return no match;
```

In the algorithm, i is incremented by $delta1[T_i]$ if the character T_i that caused a mismatch has in P an equivalent to the left of character P_j that caused the same mismatch and none to its right, which means shifting P to the right by $delta1[T_i] - (|P| - j)$ positions; otherwise, i is incremented by $|P| - j$ which is tantamount to shifting P by one position to the right. Without the latter provision, P would be shifted backwards to align the two characters.

In the worst case the algorithm executes in $O(|T||P|)$ time; for example, if $P = ba^{m-1}$ and $T = a^n$. Note that in this case, the algorithm rechecks characters in T that have already been checked.

The algorithm can be improved if we take into account the entire substring that follows a mismatched character P_j. Consider the following shift

```
  aaabcabcbabbaecabcab
1 abdabcabcab
2   abdabcabcab
```

which shifts P by one position in accordance with the left side occurrence rule. But a longer shift can result from aligning the substring of T equal to the already matched suffix that *directly* follows the mismatched character $P_8 = b$ with an equal substring in P that begins to the left of P_8.

```
  aaabcabcbabbaecabcab
1 abdabcabcab
2    abdabcabcab
```

However, note that after the shift, the mismatched character b in T is again aligned with $c = P_5$, which just caused a mismatch. Therefore, if matching ever reaches c after restarting from the end of P, the mismatch is guaranteed to reoccur. To avoid this mismatch, it is better to align the suffix ab of P that directly follows $P_8 = c$ with an equal substring of P that is preceded by a different character than c. In our example, substring ab in P that follows the mismatched character P_8 should be aligned with ab preceded by d because it is different from c:

```
  aaabcabcbabbaecabcab
1 abdabcabcab
2       abdabcabcab
```

after which the matching process restarts from the end of P.

What should be the shift if no substring begins to the left of a mismatched character P_j and is equal to the suffix that directly follows P_j? For example, what should be the shift after a mismatch is found in line 2? In this case, we align the longest suffix of P that follows the mismatched character P_j with an equal prefix of P:

```
  aaabcabcbabbaecabcab...
2       abdabcabcab
3                abdabcabcab
```

To sum up, there are two cases to consider:

1. *A full suffix rule.* If a mismatched character P_j is directly followed by a suffix that is equal to a substring of P that begins anywhere to the left of P_j, align the suffix with the substring.

2. *A partial suffix rule.* If there is a prefix of P equal to the longest suffix anywhere to the right of the mismatched character P_j, align the suffix with the prefix.

To accomplish these shifts, a table *delta2* is created that for each position in P holds a number by which the index i that scans T has to be incremented to restart the matching process; that is, if a mismatched character is P_j, then i is incremented by $delta2[j]$ (and j is set to $|P| - 1$). Formally, *delta2* is defined as follows:

$$delta2[j] = \min\{s + |P| - j - 1 : 1 \leq s \text{ and } (j \leq s \text{ or } P_{j-s} \neq P_j) \text{ and for } j < k < |P|: (k \leq s \text{ or } P_{k-s} = P_k)\}$$

and $delta2[|P| - 1] = delta2[|P| - 2]$ if the last two characters in P are the same (if they are different, then $delta2[|P| - 1] = 1$, because the third subcondition in the definition, "for ...", is vacuously true).

There are, as already indicated, two cases. In the first case, the suffix directly following the mismatched character P_j has a matching substring in P, so the situation after detecting a mismatch

658 ■ Chapter 13 String Matching

```
                                                i
                                                ↓
                           .................xb...cy...................y...
         1       ...ab...cd.............eb...c
                   ↑      ↑                   ↑
          |P| − delta2[j]  2|P| − delta2[j] − j − 2   j
```

when suffix $P(j + 1 \ldots |P| - 1)$ equals to the substring $P(|P| - delta2[j] \ldots 2|P| - delta2[j] - j - 2)$ changes to:

```
                                                                    i
                                                                    ↓
                           .................xb...cy...................y...
         2                      ...ab...cd.............eb...c
                                                                   ↑
                                                                   j
```

before resuming the matching process.

In the second case, the situation

```
                                i
                                ↓
                    .........x...ea...b..................y...
         1    a...bc..........d...ea...b
              ↑               ↑         ↑
         2|P| − delta2[j] − j − 2   j   delta2[j] − |P| + j + 1
```

when suffix $P(delta2[j] - |P| + j + 1 \ldots |P| - 1)$ equals the prefix $P(0 \ldots 2|P| - delta2[j] - j - 2)$ changes to:

```
                                                              i
                                                              ↓
                    .........x...ea...b..................y...
         2               a...bc........d...ea...b
                                                      ↑
                                                      j
```

Note that the inequalities in the *or* clauses in the definition of *delta2* are indispensable for the second case.

To compute *delta2*, a brute force algorithm can be used, as follows:

```
computeDelta2ByBruteForce(pattern P, table delta2)
    for k = 0 to |P|-1
        delta2[k] = 2*|P|-k-1;
    // partial suffix phase:
    for k = 0 to |P|-2                          // k is a mismatch position;
        for (i = 0, s = j = k+1; j < |P|; s++, j = s, i = 0)
            while j < |P| and P_i == P_j
                i++;
                j++;
            if j == |P|                         // a suffix to the right of k is detected
```

```
            delta2[k] = |P|-(k+1) + |P|-i;        // that is equal to a prefix of P,
            break;                                 // P(0... i-1) equals P(|P|-i... |P|-1);
// full suffix phase:
for k = |P|-2 downto 0                             // k is a mismatch position;
    for (i = |P|-1, s = j = |P|-2; j ≥ |P|-k-2; s--, j = s, i = |P|-1)
        if P_i == P_j
            while i > k and P_i == P_j
                i--;
                j--;
            if j == -1 or i == k and P_i ≠ P_j    // a substring in P is detected
                delta2[k] = |P|-j-1;               // that is equal to the suffix directly following k,
                break;                             // P(j+1... j+|P|-k-1) equals P(k+1... |P|-1);
if P_{|P|-1} == P_{|P|-2}
    delta2[|P|-1] = delta2[|P|-2];
else delta2[|P|-1] = 1;
```

The algorithm has three phases: initialization, partial suffix phase, and full suffix phase. Initialization prepares the pattern for the longest shift; after a mismatch, the pattern is shifted all the way past the mismatched character. The only exception is the mismatch at the last character in P, after which P is shifted by only one position. The partial suffix phase looks for the longest suffixes after a mismatch point by matching prefixes. The full suffix phase updates those values in *delta2* that correspond to a mismatch being followed by a suffix that has a matching substring in P. For $P = abdabcabcab$, the values in *delta2* after each phase are as follows:

	a	b	d	a	b	c	a	b	c	a	b	
delta2 =	21	20	19	18	17	16	15	14	13	12	*	after initialization
delta2 =	19	18	17	16	15	14	13	12	11	12	*	after partial suffix phase
delta2 =	19	18	17	16	15	8	13	12	8	12	1	after full suffix phase

The algorithm can be applied to short patterns only, because it is quadratic in the best case and cubic in the worst case. For $P = a^m$, the full suffix phase executes in total,

$$\sum_{k=0}^{m-2} \sum_{j=m-k-2}^{m-2} (m-k) = \frac{(m-1)m(m+4)}{6}$$

comparisons, because in one iteration of the inner for loop, $m - k - 1$ comparisons are performed in the while loop and one in the if statement. $P = a^{m-1}b$ is an example of the worst case for the partial suffix phase. Clearly, for longer patterns a faster algorithm is needed.

The algorithm can be significantly improved by using an auxiliary table f that is a counterpart of *next* for the reverse of P. The table f is defined as follows:

$$f[j] = \begin{cases} |P| & \text{if } j = |P| - 1 \\ \min\{k : j < k < |P| - 1 \text{ and } P(j+1 \ldots j+|P|-k) = P(k+1 \ldots |P|-1)\} & \text{if } 0 \le j < |P| - 1 \end{cases}$$

That is, $f[j]$ is the position preceding the starting position of the longest suffix of P of length $|P|-f[j]$ that is equal to the substring of P that begins at position $j+1$:

$$\begin{array}{cc} f[j]+1 & |P|-1 \\ \downarrow & \downarrow \end{array}$$

$$\ldots \underline{a}\ldots \underline{b}c\ldots d\underline{a}\ldots \underline{b}$$

$$\begin{array}{cc} \uparrow & \uparrow \\ j+1 & j+|P|-f[j] \end{array}$$

For example, for $P = aaabaaaba$, $f[0] = 4$, because substring $P(1\ldots 4) = aaba$ is the same as the suffix $P(5\ldots 8)$; $f[1] = 5$, because substring $P(2\ldots 4)$ is equal to the suffix $P(6\ldots 8)$; that is, the underlined substrings in $P = a\underline{aaba}\underline{aaba}$ are equal. The entire table $f = [4\ 5\ 6\ 7\ 7\ 7\ 8\ 8\ 9]$. Note that a substring of P equal to a suffix of P can overlap, as in $P = baaabaaaba$.

The table f allows us to go from a certain substring of P to a matching suffix of P. But the matching process during execution of the Boyer-Moore algorithm proceeds from right to left, so that after a mismatch is found, a suffix is known and we need to know a matching substring of P to align the substring with the suffix. In other words, we need to go from the suffix to the matching substring, which is the opposite direction with respect to information provided by f. That is why delta2 is created to have direct access to the needed information. This can be accomplished with the following algorithm, which is obtained from computeDelta2ByBruteForce().

```
computeDelta2UsingNext(pattern P, table delta2)
    findNext2(reverse(P),next);
    for i = 0 to |P|-1
        f[i] = |P| - next[|P|-i-1] - 1;
        delta2[i] = 2*|P| - i - 1;
    // full suffix phase:
    for i = 0 to |P|-2
        j = f[i];
        while j < |P|-1 and P_i ≠ P_j
            delta2[j] = |P| - i - 1;
            j = f[j];
    // partial suffix phase:
    for (i = 0; i < |P|-1 and P_0 == P_f[i]; i = f[i])
        for j = i to f[i]-1
            if delta2[j] == 2*|P| - j - 1  // if not updated during full suffix phase,
                delta2[j] = delta2[j] - (|P| - f[i]);  // update it now;
    if P_{|P|-1} == P_{|P|-2}
        delta2[|P|-1] = delta2[|P|-2];
    else delta2[|P|-1] = 1;
```

First, table next is created and used to initialize table f. Also, table delta2 is initialized to values that indicate shifting P past the mismatched character in T. For $P = dabcabeeeabcab$ it is

P	=	a	b	c	a	b	d	a	b	c	a	b	e	e	e	a	b	c	a	b
f	=	14	15	16	17	18	13	14	15	16	17	18	18	18	16	17	18	18	18	19
delta2	=	37	36	35	34	33	32	31	30	29	28	27	26	25	24	23	22	21	20	19

Section 13.1 Exact String Matching

As in the brute force algorithm, the full suffix phase addresses the first case when a mismatched character P_j is directly followed by the suffix that is equal to a substring of P that begins anywhere to the left of P_j. In that phase, positions directly accessible from f are processed. For example, in the sixth iteration of the `for` loop, $i = 5$ and $f[5] = 13$, which means that there is a suffix beginning at position 14, *abcab*, for which there is a substring beginning at position $|P| - f[5] = 6$, which is equal to the suffix. Because also $P_5 \neq P_{13}$, *delta2*[13] can be assigned a proper value:

```
                0                    5                   13          18
P      =  a  b  c  a  b  d  a  b  c  a  b  e  e  e  a  b  c  a  b
f      = 14 15 16 17 18 13 14 15 16 17 18 18 18 16 17 18 18 18 19
delta2 = 37 36 35 34 33 32 31 30 29 28 27 26 25 13 23 22 21 20 19
```

But the substring $P(6 \ldots 10) = abcab$ has a prefix *ab* that is the same as a suffix of the suffix *abcab*, and both the prefix *ab* and suffix *ab* are preceded by different characters. The position right before this shorter suffix is directly accessible only from the position $f[13] = f[f[5]] = 16$ because i is still 5. Therefore, after the second iteration of the `while` loop—still during the sixth iteration of the `for` loop—the situation changes to

```
                0                    5                   13    16    18
P      =  a  b  c  a  b  d  a  b  c  a  b  e  e  e  a  b  c  a  b
f      = 14 15 16 17 18 13 14 15 16 17 18 18 18 16 17 18 18 18 19
delta2 = 37 36 35 34 33 32 31 30 29 28 27 26 25 13 23 22 13 20 19
```

Number 13 is put in cell *delta2*[16], which means that when the matching process stops at P_{16}, index i that scans T is incremented by 13; that is, when scanning stops after detecting a mismatch with P_{16}, the situation is as in

```
                            i
                            ↓
      ...abcabdabcabeeeabcab.......x...
         abcabdabcabeeeabcab
```

so that scanning is resumed after updating i as in

```
                                          i
                                          ↓
      ...abcabdabcabeeeabcab.......x...
                      abcabdabcabeeeabcab
```

Note that the increment is the same as for the mismatch with P_{14}. This is because in both cases character $T_{i+1} = a$ (before updating i) is aligned with P_6, because both suffixes *ab* and *abcab* have a match that starts at P_6. However, as shown in the previous diagram, a match would be missed if i were incremented by 13. But the algorithm continues and for $i = 13$ and $j = f[13] = 16$ the `while` loop is entered to modify *delta2*:

```
                0                                        13    16    18
P      =  a  b  c  a  b  d  a  b  c  a  b  e  e  e  a  b  c  a  b
f      = 14 15 16 17 18 13 14 15 16 17 18 18 18 16 17 18 18 18 19
delta2 = 37 36 35 34 33 32 31 30 29 28 27 26 25 13 23 22  5 20 19
```

which prevents i from missing a match.

After finishing the first outer for loop, the second outer for loop is executed to decrease, if possible, other values in *delta2*. For $i = 0$ the inner for loop is entered and executed for j from $i = 0$ to $f[0] - 1 = 13$, so that *delta2* becomes

```
           0   1   2   3   4   5   6   7   8   9  10  11  12  13                  18
P      =   a   b   c   a   b   d   a   b   c   a   b   e   e   e   a   b   c   a   b
f      =  14  15  16  17  18  13  14  15  16  17  18  18  18  16  17  18  18  18  19
delta2 =  32  31  30  29  28  27  26  25  24  23  22  21  20  13  23  22   5  20  19
```

In this way, the first 13 values in *delta2* are decremented by 5. This corresponds to the situation when a mismatch occurs for any of the first 13 characters in P. When this happens, the suffix *abcab* is aligned with the prefix *abcab* because the suffix is to the right of any of these positions.

In the second iteration of the outer loop, when $i = 14$ and $f[14] - 1 = 16$, the inner loop updates *delta2*[14] and *delta2*[15] by decrementing them by 2 because this is the length of suffix *ab* to the right of P_{14} and P_{15} that have a matching prefix:

```
           0                                                          14  15  16      18
P      =   a   b   c   a   b   d   a   b   c   a   b   e   e   e   a   b   c   a   b
f      =  14  15  16  17  18  13  14  15  16  17  18  18  18  16  17  18  18  18  19
delta2 =  32  31  30  29  28  27  26  25  24  23  22  21  20  13  21  20   5  20  19
```

After exiting the loops, the last value is changed so that finally the situation is

```
           0                                                                      18
P      =   a   b   c   a   b   d   a   b   c   a   b   e   e   e   a   b   c   a   b
f      =  14  15  16  17  18  13  14  15  16  17  18  18  18  16  17  18  18  18  19
delta2 =  32  31  30  29  28  27  26  25  24  23  22  21  20  13  21  20   5  20   1
```

To find the complexity of the algorithm we claim that the while loop is executed at most $|P| - 1$ times in total. This is because after entering the while loop for an i and executing k iterations, the while loop is not entered for the next $|P| - f[i] - 1$ iterations of the outer for loop, where $|P| - f[i] - 1$ is the length of a matching suffix that begins at $P_{f[i]+1}$ that matches the substring that begins at P_1 and for which $k \le |P| - f[i] - 1$. It is because for each character P_r in the substring $P(i + 2 \ldots i + |P| - f[i])$, P_r and the corresponding characters $P_{r+f[i]}$ are preceded by the same character so that the condition in the while loop is false. Consider $P = badaacadaa$ for which $f = [5\ 6\ 7\ 8\ 9\ 8\ 9\ 8\ 9\ 10]$. Iterations of the while loop can be activated either for P_0 or for an s for which $f[s-1] > f[s]$, (e.g., for $s = 5$, $f[4] = 9$ and $f[5] = 8$). Increasing numbers in f indicate substrings that are extensions of following substrings and corresponding suffixes of P. For example, for $f[1] = 6$ and $f[2] = 7$, numbers 1 and 6 indicate that the substring $P(2 \ldots 4)$ equals suffix $P(7 \ldots 9)$ and numbers 2 and 7 indicate that the substring $P(3 \ldots 4)$ equals suffix $P(8 \ldots 9)$; that is, $P(2 \ldots 4)$ is an extension of $P(3 \ldots 4)$ and so is suffix $P(7 \ldots 9)$ a forward extension of suffix $P(8 \ldots 9)$. This means that $P_2 = P_7$, and for $j = 2$ the while loop cannot be entered.

In the worst case, the first half of P can lead to $|P|/2$ iterations of the while loop followed by $|P|/2$ iterations of the outer for loop without entering the while loop. Next, the first half of the second half of P can lead to $|P|/4$ iterations of the while loop followed by the same number of iterations of the outer for loop without entering the while loop, and so on, which gives

$$\sum_{k=1}^{\lg|P|} \frac{|P|}{2^k} = |P| - 1$$

iterations of the `while` loop in total. Because the outer `for` loop can iterate $|P| - 1$ times, this gives $2(|P| - 1)$ as the maximum number of values assigned to j, and hence the complexity of the outer `for` loop.

The last nested `for` loop is executed at most $|P| - 1$ times: For each i, it is executed for j from i to $f[i] - 1$ and then i is updated to $f[i]$; therefore, j refers to one position at most once. We can conclude that the algorithm is linear in the length of P.

To use *delta2*, algorithm `BoyerMooreSimple()` is modified by replacing the line that updates i

```
else i = i + max(delta1[T_i],|P|-j);
```

with the line

```
else i = i + max(delta1[T_i],delta2[j]);
```

In an involved proof, Knuth shows that the Boyer-Moore algorithm that utilizes tables *delta1* and *delta2* performs at most $7|T|$ comparisons if the text does not contain any occurrence of the pattern (Knuth, Morris, and Pratt, 1977). Guibas and Odlyzko (1980) improved the bound to $4|T|$ and Cole (1994) improved it to $3|T|$.

The Sunday Algorithms

Daniel Sunday (1990) begins his analyses with an observation that in the case of a mismatch with a text character T_i, the pattern shifts to the right by at least one position so that the character $T_{i+|P|}$ is included in the next iteration. The Boyer-Moore algorithm shifts the pattern according to the value in the *delta1* table (we leave the table *delta2* aside, for now) and this table includes shifts with respect to the mismatched character T_i. It would be more advantageous, Sunday submits, to build *delta1* with respect to character $T_{i+|P|}$. In this way, *delta1*[*ch*] is the position of character *ch* in P counted from the left. This is closely related to Boyer-Moore's *delta1* because by incrementing by one the values in the latter, we obtain Sunday's *delta1*.

One advantage of this solution is that the set of three rules used in the Boyer-Moore algorithm can be simplified. The no occurrence rule is slightly modified: If the character $T_{i+|P|}$ appears nowhere in P, align P_0 with $T_{i+|P|+1}$. The right side occurrence rule is no longer needed, because all characters in P are to the left of $T_{i+|P|}$. Finally, the left side occurrence rule can be simplified to the occurrence rule: If there is an occurrence in P of character *ch* equal to $T_{i+|P|}$, align $T_{i+|P|}$ with the closest (the rightmost) *ch* in P.

Although the definition of *delta1* depends on right-to-left scan of pattern, the matching process can be performed in any order, not only left to right or right to left. Sunday's `quickSearch()` performs the scan left to right. Here is its pseudocode:

```
quickSearch(pattern P, text T)
    initialize all cells of delta1 to |P| + 1;
    for i = 0 to |P|-1
        delta1[P_i] = |P| + 1 - i;
```

```
            i = 0
            while i ≤ |T|-|P|
                j = 0;
                while j < |P| and i < |T| and P_j == T_i
                    i++;
                    j++;
                if j > |P|
                    return success at i-|P|;
                i = i + delta1[T_{i+|P|}];
            return failure;
```

For example, for $P = cababa$, $delta1['a'] = 1$, $delta1['b'] = 2$, $delta1['c'] = 6$, and for remaining characters ch, $delta1[ch] = 7$. Here is an example:

```
   ffffaabcfacababafa
1  cababa
2     cababa
3            cababa
4             cababa
```

Line 1 has a mismatch right at the beginning, so that in line 2, character $T_{i+|P|} = T_{0+6} = b$ is aligned with the rightmost b in P and $i = 0$ is incremented by $delta1[T_{i+|P|}] = delta1['b'] = 2$, so that $i = 2$. Again there is a mismatch at the beginning of P; $i = 2$ is incremented by $delta1[T_{i+|P|}] = delta1['f'] = 6$, so that $i = 9$; that is, in effect, P is shifted past letter $T_{i+|P|} = f$. The fourth iteration is successful. Compare this trace with the trace for `BoyerMooreSimple()`, for which $delta1['a'] = 0$, $delta1['b'] = 1$, $delta1['c'] = 5$, and for remaining characters ch, $delta1[ch] = 6$:

```
   ffffaabcfacababafa
1  cababa
2    cababa
3     cababa
4            cababa
5             cababa
6              cababa
```

Sunday introduces two more algorithms, both based on a generalized $delta2$ table. Sunday's $delta2$ may be the same as Knuth-Morris-Pratt's *next* table if the $delta2$ table is initialized by scanning P left to right. If it is scanned in the reverse order, then $delta2$ is the same as Boyer-Moore's $delta2$. However, the matching process can be done in any order. Sunday's second algorithm, the maximal shift algorithm, uses $delta2$ such that $delta2[0]$ is associated with a character in P whose next leftward occurrence in P is maximum; $delta[1]$ refers to a character in P for which the next leftward occurrence in P is not less than $delta2[0]$, and so on. In the third algorithm, the optimal mismatch algorithm, characters are ordered in ascending order of frequency of occurrence. This is motivated by the fact that in English, 20 percent of words end with the letter e and 10 percent of the letters used in English are also e. Thus, it is very likely to match first characters tested by using Boyer-Moore's backward scanning. The testing of the least probable characters first improves the likelihood of early mismatch. However, Sunday's own tests show that although his three algorithms fare much better on searching for short English words than the Boyer-Moore algorithm, there is little difference between the three algorithms, and for all practical purposes `quickSearch()` is sufficient. This is particularly true when the overhead to find *delta2* is taken into account (see Pirklbauer, 1992). To address the problem of frequency of character occurrence, an adaptive technique can be used, as in Smith (1991).

Sunday points out that his *delta1* table usually allows for shifts one position greater than shifts based on Boyer-Moore's *delta1*. However, after pointing out that this is not always the case, Smith (1991) indicates that the larger of the two values should be used.

13.1.4 Multiple Searches

The algorithms presented in the preceding sections are designed to find an occurrence of a pattern in a text. Even if there are many occurrences, the algorithms are discontinued after finding the first. Many times, however, we are interested in finding all occurrences in the text. One way to accomplish this is to continue the search after an occurrence is detected, after shifting the pattern by one position. For example, the Boyer-Moore algorithm can be quickly modified to accommodate multiple searches in the following fashion:

```
BoyerMooreAllOccurrences(pattern P, text T)
    initialize all cells of delta1 to |P|;
    for i = 0 to |P| - 1
        delta1[P_i] = |P| - i - 1;
    compute delta2;
    i = |P| - 1;
    while i < |T|
        j = |P| - 1;
        while j ≥ 0 and P_j == T_i
            i--;
            j--;
        if j == -1
            output: match at i+1;
            i = i + |P| + 1;  // shift P by one position to the right;
        else i = i + max(delta1[T_i],delta2[j]);
```

But consider the process of finding all occurrences of $P = ababab a$ in $T = ababababa \ldots$:

```
    ababababa...
1   ababab a
2   ababab a
3     ababab a
4      ababab a
```

In every second iteration, the entire pattern is compared to the text only after shifting by two positions. For this reason, the algorithm requires $|P|(|T| - |P| + 1)/2$, or more generally, $O(|T||P|)$ steps. To reduce the number of comparisons it should be recognized that the pattern includes consecutive repetitive substrings, called *periods*, which should not be reexamined after they were matched with substrings in the text with which they are about to be matched.

The Boyer-Moore-Galil algorithm works the same as the Boyer-Moore algorithm until the first occurrence of the pattern is detected (Galil, 1979). After that, the pattern is shifted by $p = |$the period of the pattern$|$, and only the last p characters of the pattern need to be compared to the corresponding characters in the text to know whether the entire pattern matches a substring in the text. In this way, the part overlapping a previous occurrence does not have to be rechecked. For example, for $P = ababab a$ with the period ab, the new algorithm is executed as follows:

```
    ababababab...
1   abababa
2       ababa_ba
3          ababa_ba
```

However, if a mismatch is found, then the Boyer-Moore-Galil algorithm resumes its executions in the same way as the Boyer-Moore algorithm. The algorithm is as follows:

```
BoyerMooreGalil(pattern P, text T)
    p = period(P);
    compute delta1 and delta2;
    skip = -1;
    i = |P|-1;
    while i < |T|
        j = |P|-1;
        while j > skip and P_j == T_i
            i--;
            j--;
        if j == skip
            output: a match at i-skip;
            if p == 0
                i = i + |P|+1;
            else if skip == -1
                i = i + |P|+p;
            else i = i + 2*p;
            skip = |P|-p-1;
        else skip = -1;
            i = i + max(delta1[T_i],delta2[j]);
```

It is clear that the algorithm achieves better performance only for patterns with periods and only if the text contains a high number of overlapping occurrences of the pattern. For patterns with no periods, the two algorithms work the same way. For a pattern with periods but with no overlapping occurrence, the Boyer-Moore-Galil algorithm performs better shifts than the Boyer-Moore algorithm, but only when an occurrence is found.

13.1.5 Bit-Oriented Approach

In this approach, each state of the search is represented as a number—that is, a string of bits—and a transition from one state to the next is the result of a small number of bitwise operations. A shift-and algorithm that uses a bit-oriented approach for string matching was proposed by Baeza-Yates and Gonnet (1992) (see also Wu and Manber, 1992).

Consider a $(|P| + 1) \times (|T| + 1)$ table of bits defined as follows:

$$state[j,i] = \begin{cases} 0 & \text{if } i = -1 \text{ and } j > -1 \\ 1 & \text{if } j = -1 \\ 1 & \text{if } state[j-1, i-1] = 1 \text{ and } P_j = T_i \\ 0 & \text{otherwise} \end{cases}$$

The table includes information concerning all matches between prefixes of P and substrings ending at particular positions of the text. Number 1s in a row j of the table indicate the ending positions in T of substrings $T(i-j \ldots i)$ matching prefix $P(0 \ldots j)$ and number 1s in a column i indicate the prefixes of P that match substrings ending at i in T. For $P = abaabac$ and $T = bbababacaaba$, $state[0,4] = state[2,4] = 1$ because position 4 of T is the ending position of matches for prefixes $P(0 \ldots 0) = a$ and $P(0 \ldots 2) = aba$. Number 1 in the last row indicates an occurrence of the entire pattern P in T.

The algorithm computes a new state from the previous state, however, which can be done very efficiently without maintaining the entire table *state*. To accomplish it, a two-dimensional bit table is used to indicate for each character of the alphabet the positions at which it occurs in the pattern:

$$charactersInP[j,ch] = \begin{cases} 1 & \text{if } ch = P_j \\ 0 & \text{otherwise} \end{cases}$$

For example, the letter a occurs at positions 0, 2, and 4 of the pattern $P = abaabac$; therefore, $charactersInP[0,\text{'a'}] = charactersInP[2,\text{'a'}] = charactersInP[4,\text{'a'}] = 1$ and $charactersInP[1,\text{'a'}] = charactersInP[3,\text{'a'}] = charactersInP[5,\text{'a'}] = 0$. In practice, *charactersInP* is a one-dimensional table of numbers where bit positions in the number are implicitly used for row indices. Also, the table can only include information about the characters that appear in P. The tables *state* and *charactersInP* for $P = abaabac$ and $T = bbababacaaba$ are as follows:

```
                5 6 7
      b b a b a b a c a a b a           a b c
      1 1 1 1 1 1 1 1 1 1 1 1
    a 0 0 0 1 0 1 0 1 0 1 1 0 0         1 0 0
    b 0 0 0 0 1 0 1 0 0 0 0 1 0         0 1 0
    a 0 0 0 0 0 1 0 1 0 0 0 0 1         1 0 0
    b 0 0 0 0 0 0 1 0 0 0 0 0 0         0 1 0
    a 0 0 0 0 0 0 0 1 0 0 0 0 0         1 0 0
    c 0 0 0 0 0 0 0 0 1 0 0 0 0         0 0 1
```

With *charactersInP* it is easy now to compute a state corresponding to the currently processed position i of the text from the state corresponding to the preceding position. This is accomplished by executing the operation *shiftBits*, which shifts down bits corresponding to state $i-1$ so that the bottom bit is shifted out and 1 is shifted in as a new top bit. The result of the operation *shiftBits* is subjected to the bitwise-and operation with bits in *charactersInP* corresponding to the character T_i:

$$state[i] = shiftBits(state[i-1])) \text{ bitwise-and } charactersInP[T_i]$$

To see it, consider the transition from state 5 to state 6; that is, processing character T_6 after processing of T_5 has finished, and then to state 7. By that time, prefixes *abab* and *ab* of P have been found to match substrings $T(2 \ldots 5)$ and $T(4 \ldots 5)$:

		i = 5	6	7	7
		T = bbababacaaba	abababacaaba	bbababacaaba	bbababacaaba
a	0		a		a
b	1	ab		ab	
a	2		aba		
b	3	abab		abac	
a	4		ababa		
c	5			ababac	ababac

Passing to state 6 means that we try to match P with $T(1 \ldots 6)$. But it is also an attempt to extend partial matches. Therefore, the already matched prefixes are not shifted, but are extended by one character and tested as to whether the added character is the same as T_6. This is done not by comparing T_6 with the added character of each partially matched prefix of P, but by using information in *charactersInP*, which allows for checking all these partial matches at the same time. This is done by first shifting bits in column 5 of the table *state* down by one position. This amounts to giving the existing partial matches a chance to be successfully matched after extending them by one character. By shifting in 1 at the top of the column, the shortest prefix of P is given a chance as well. Whether the extended prefixes are matching substrings ending at position 6 is tested with *charactersInP*. Number 1 in row j of *state* indicates the occurrence of P's matched prefixes of length j. After a downward shift, the extended partial matches are tested by testing only the newly included character, which is in position j. An extended prefix is a match if the new character is the same as the current character in T, which is now $T_6 = a$. Therefore, if for a particular row j, the last character of a prefix is also a, then bitwise-and between 1 from *state*[j,6] and 1 from *charactersInP*[6,'a'] gives 1 as a result, which means a successful match. For example, *aba* in row 2 means that the prefix *ab* was successfully matched with a substring ending at position 5 of T, and now we are about to check whether the longer prefix *aba* matches a substring of T ending at 6. Because the last letter of this prefix, a, is the same as T_6, the tentative match becomes permanent. But consider the processing of T_7. The prefixes *a*, *aba*, and *ababa* are extended to become *ab*, *abab*, and *ababac*, and then the last characters are indirectly (with bitwise-and) compared with $T_7 = c$. Out of three extended prefixes, only one is retained, and because this prefix is equal to the pattern itself, an occurrence of P is reported.

The transition from state 6 to 7 is summarized as follows: first, the shift operation is executed:

$$\text{shiftBits} \begin{array}{c} 6 \\ b \\ 1 \\ 0 \\ 1 \\ 0 \\ 1 \\ 0 \end{array} \Rightarrow \begin{array}{c} \\ \\ 1 \\ 1 \\ 0 \\ 1 \\ 0 \\ 1 \end{array}$$

and then the bitwise-and generates state 7:

$$\begin{array}{c} \\ 1 \\ 1 \\ 0 \text{ bitwise-and} \\ 1 \\ 0 \\ 1 \end{array} \begin{array}{c} 7 \\ c \\ 0 \\ 0 \\ 0 \\ 0 \\ 0 \\ 1 \end{array} \Rightarrow \begin{array}{c} \\ c \\ 0 \\ 0 \\ 0 \\ 0 \\ 0 \\ 1 \end{array}$$

A pseudocode of the algorithm is quite simple:

```
shiftAnd(pattern P, text T)
    state = 0;
    matchBit = 1;
    // initialization:
    for i = 1 to |P|-1
        matchBit <<= 1;
    for i = 0 to 255
        charactersInP[i] = 0;
    for (i = 0, j = 1; i < |P|; i++, j <<= 1)
        charactersInP[P_i] |= j;
    // matching process:
    for i = 0 to |T|-1
        state = ((state << 1) | 1) & charactersInP[T_i];
        if ((matchBit & state) != 0)
            output: a match at i-|P|+1;
```

The two-dimensional table *charactersInP* is implemented as a one-dimensional table of long integers, where the second dimension is indicated by positions of bits in the integers. The function *shiftBits* is implemented as left shift followed by bitwise-or of the result of the shift and number 1. In this way, number 1 is placed in the least significant (rightmost) bit position. The algorithm performs $|T|$ iterations in the matching phase, executing four bitwise operations in each iteration, an assignment, and a comparison.

The algorithm requires no buffering of text as is the case for the Boyer-Moore algorithm. It is also clear that the length of the pattern must not exceed the size of an integer, which for many practical purposes is adequate. However, this limitation can be lifted by using a dynamic version of the shift-and algorithm. Here is one possible implementation of such a version:

```
dynamicShiftAnd(pattern P, text T)
    cellLen = size of long integer in bits;
    lastBit = 1;
    matchBit = 1;
    cellNum = (|P| % 8 == 0) ? (|P|/8) : (|P|/8 + 1);
    matchBit = 1;
    // initialization:
    for i = 1 to |P| - cellLen*(cellNum-1)-1
        matchBit <<= 1;
    for k = 0 to cellNum-1
        for i = 0.255
            charactersInP[k,i] = 0;
        for (i = k*cellLen, j = 1; i < (k+1)*cellLen && i < |P|; i++, j <<= 1)
            charactersInP[k,P_i] |= j;
    // matching process:
    for j = cellNum-1 down to 0
        state[j] = 0;
    for i = 0 to |T|-1
        for (j = cellNum-1 down to 1
            firstBit = ((state[j-1] & lastBit) == 0) ? 0 : 1;
            state[j] = ((state[j] << 1) | firstBit) & charactersInP[j,T_i];
        state[0] = ((state[0] << 1) | 1) & charactersInP[0,T_i];
        if ((matchBit & state[cellNum-1]) != 0)
            output: a match at i-|P|+1;
```

The table *charactersInP*, as before, records the occurrences of characters in the pattern, but in a piecemeal fashion. For example, for a pattern P of length 80 and 64-bit long integers, *charactersInP*[0] records character occurrences for subpattern $P(0 \ldots 63)$ and *charactersInP*[1] for $P(64 \ldots 79)$. The table *state* still implements one state of the matching process. For example, *state*[0] represents the state for $P(0 \ldots 63)$ and *state*[1] represents the state for $P(64 \ldots 79)$. Shifting bits consists of shifting bits to the right neighbor cell and shifting in bits from left neighbors, if any. For example, the last bit shifted out from *state*[0] is shifted in as the first bit to *state*[1].

The dynamic algorithm is not linear for all the lengths, but it is linear in intervals marked by the long integer size. For example, for a 64-bit word, patterns of length 1 through 64 require $O(|T|)$ operations; patterns of length 65 through 128 require $O(2|T|)$ operations and, generally, $O(\lceil |P|/64 \rceil |T|)$ operations.

13.1.6 Matching Sets of Words

The problem of matching sets of words arises in the situation when for a set *keywords* $= \{s_0, \ldots, s_{k-1}\}$ of strings and a text T, we need to identify in T all substrings that match strings in *keywords*. It is possible that the substrings overlap one another. In a brute force approach, the matching process is executed for each word in the set *keywords* separately. The running time of such an approach is $O(|keywords||T|)$. However, it is possible to considerably improve the run time by considering all relevant words at the same time during the matching process. An algorithm performing such matching was proposed by Aho and Corasick (1975).

Aho and Corasick construct a string-matching automaton that is composed of a set of states represented by numbers, an initial state 0, an alphabet, and two functions: a goto function *g* that assigns to each pair (state, character) a state or a special label *fail*, and a failure function *f* that assigns a state to a state. Also, the algorithm uses an output function *output*, which associates a set of keywords with each state. The states, for which a set of keywords is not empty, are accepting states. After reaching an accepting state, a set of keywords associated with it is output.

Two types of transitions from one state to another are made during execution of the algorithm: goto transitions and failure transitions. The automaton makes a failure transition, when for the current *state* and character T_i, $g(state, T_i) = fail$, in which case, the failure function *f* is used to determine the next current $state = f(state)$. If $g(state, T_i) = state_1 \neq fail$, then $state_1$ becomes the current state, T_{i+1} becomes the current character, and the transition from $state_1$ to $g(state_1, T_{i+1})$ is tried.

For no character *ch*, $g(0, ch) = fail$; that is, no failure transitions occur in the initial state. In this way, one character from *T* can be processed in each iteration of the algorithm. The algorithm is summarized as follows:

```
AhoCorasick(set keywords, text T)
    computeGotoFunction(keywords,g,output);    // the output function is computed
    computeFailureFunction(g,output,f);        // in these two functions;
    state = 0;
    for i = 0 to |T| - 1
        while g(state,T_i) == fail
            state = f(state);
        state = g(state,T_i);
        if output(state) is not empty
            output: a match ending at i;
            output(state);
```

The goto function is constructed in the form of a trie with numbered nodes representing states and the root representing the start state, $state_0$. A trie, as discussed in Section 7.2, is a multiway tree in which consecutive characters of a string are used to navigate the search in the tree. To enable such a search, links in the trie can be labeled with characters. When descending down a particular path and concatenating the characters encountered along the way, a word corresponding to this path is constructed. To make it possible, the insertion process has to construct such paths. Consider the construction of a trie for the set *keywords* = {*inner, input, in, outer, output, out, put, outing, tint*} (Figure 13.1). After inserting the word *inner*, there is only one path in the trie (Figure 13.1a). When inserting the next word, *input*, a part of the existing path is used for the prefix *in*, and then a new path is branching out from node 2 for the suffix *put* (Figure 13.1b). For the word *in*, no new path is created.

The remaining steps are summarized in Figure 13.1c. For the word *outer*, a new path is created. A path for the word *output* overlaps partially with the beginning of the path for *outer*, because the two words have the same prefix *out*.

FIGURE 13.1 (a) A trie for the string *inner*, (b) for the strings *inner* and *input*, and (c) for the set keywords = {inner, input, in, outer, output, out, put, outing, tint}; (d) the trie (c) with failure links; (e) scanning the trie (d) for the text T = outinputting.

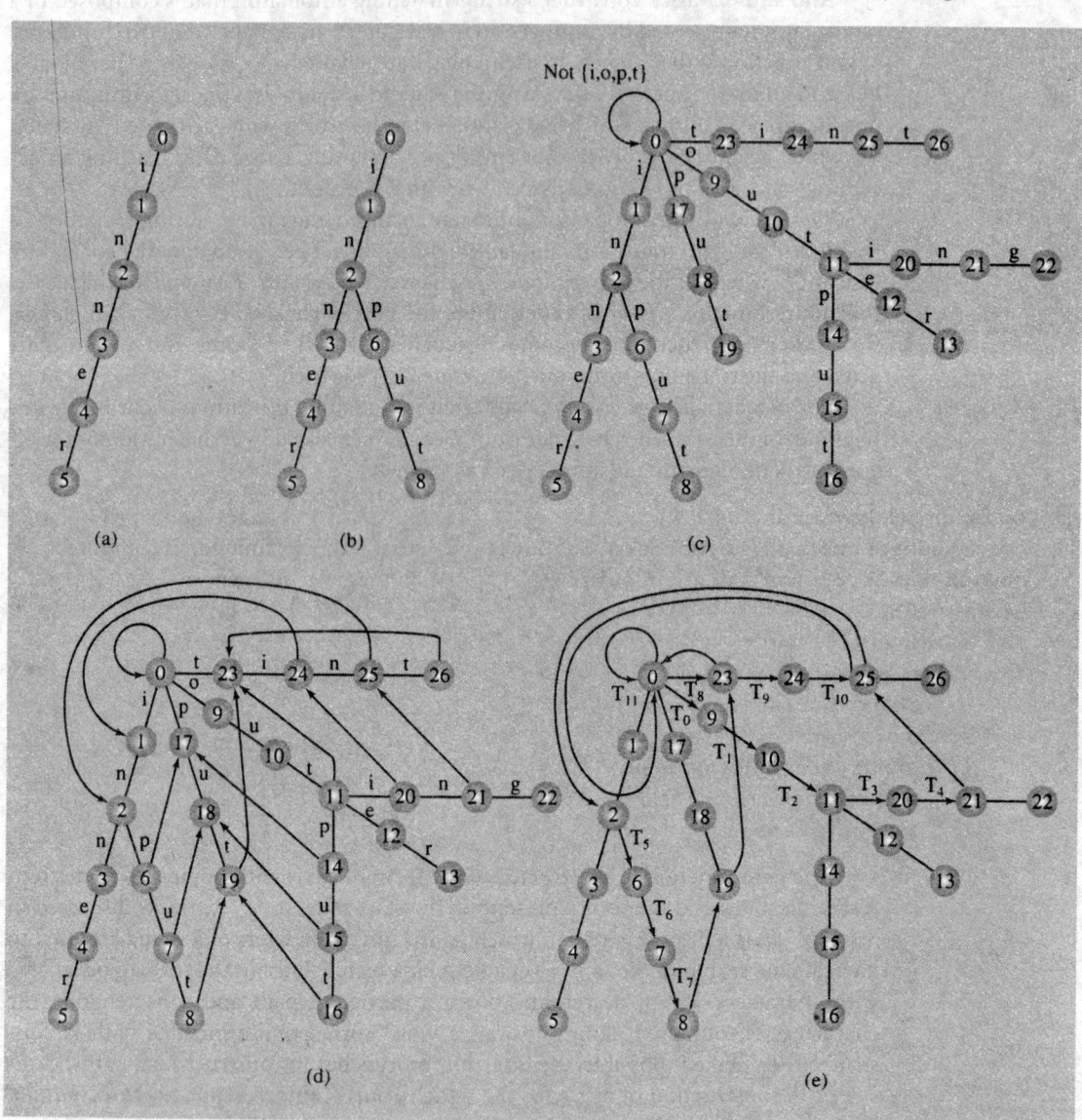

The last step is adding a loop in the root; that is, a one-branch-long path from the root to the root, for all characters other than the characters for which there are branches coming out of the root, in this case characters *i*, *o*, *p*, and *t*. An algorithm to construct the goto function is as follows:

```
computeGotoFunction(set keywords, function g, function output)
    newstate = 0;
    for i = 0 to |keywords| - 1
        state = 0;
        j = 0;
        P = keyword[i];
        while g(state,P_j) ≠ fail     // descend down an existing path;
            state = g(state,P_j);
            j++;
        for p = j to |P| - 1          // create a path for suffix P(j...|P|);
            newstate++;
            g(state,P_p) = newstate;
            state = newstate;
        add P to the set output(state);
    for all characters ch of the alphabet   // create loops in state 0;
        if g(0,ch) == fail
            g(0,ch) = 0;
```

Note that it is now impossible to tell which words are included in the trie. Is the word *inn* in the trie? There exists a path corresponding to it, but the word should not be counted as a part of the trie. To handle the problem of words whose paths are entirely included in other paths—that is, words, that are prefixes of other words—a special symbol can be used as a marker of the end of each word and included in the trie only when ambiguity may arise. Another solution is to include a flag as a part of each node to indicate the end of a word. The Aho-Corasick algorithm solves this problem with the output function, although this is not the only role played by this function.

The first stage of constructing the output function is included in `computeGotoFunction()`. For each state *s* (node of the trie), the output function tells whether there are paths that begin anywhere between the start state and *s* that correspond to words in the set *keywords*. After `computeGotoFunction()` is finished, the output function establishes the following mapping between states and keywords (states not indicated correspond to an empty set of words):

2 {*in*}	5 {*inner*}	8 {*input*}
11 {*out*}	13 {*outer*}	16 {*output*}
19 {*put*}	22 {*outing*}	26 {*tint*}

At this stage, the output function finds for each state *s* a word in *keywords* that corresponds to the path that begins at the start state and ends at *s*. At that moment, the output function plays the role of a flag, indicating for each node of the trie whether it corresponds to a keyword.

The goto function is now used to construct the failure function during a breadth-first trie traversal; that is, when traversing the trie level by level. For each node, the function records the fact that a suffix of the string ending in a particular node is also a prefix of a word that begins in the root, if any. In this, the failure function is a generalization of the table *next* used by the Knuth-Morris-Pratt algorithm, and it determines from which state the matching process should be resumed if a mismatch occurs. Consider the matching process executed as in the Knuth-Morris-Pratt algorithm with character comparison:

```
                 i
                 ↓
        ...outinput...
   1       outing
   2a         inner
   2b         input
   2c        · tint
```

After a mismatch in position i, the matching process should resume from the same character T_i, and a character in any of the keywords that follows a prefix equal to any of the suffixes in the substrings that end before the mismatched position in the pattern. In our example, the suffix *in* of the partially matched pattern *outin* is the same as a prefix of two keywords, *inner* and *input*, and the suffix *tin* of *outin* corresponds to a prefix of *tint*. As in the Knuth-Morris-Pratt algorithm, we do not want to repeat the comparisons already made, so the matching process is resumed from the mismatched character T_i, and the characters in the three keywords that follow the matched prefixes. But unlike in the Knuth-Morris-Pratt method, three patterns (keywords) have to be taken into account at the same time. How can the mismatched character T_i, be compared at the same time to three different characters, n, p, and t, in the three candidate keywords for a possible match? They are not compared at all. Instead of comparing characters between the text and keywords, the current character in the text is used to choose a transition in the trie, whereby some of the candidate keywords can be eliminated. After the matching process continues in round 2, three keywords are eliminated and the situation is as in

```
                 i
                 ↓
        ...outinput...
   2b         input
```

But how can candidate keywords be chosen for the next round? This is precisely the role of the failure function. The function adds to the trie the failure transitions. A failure transition is made if the matching process reaches a particular node (state) and there is no branch coming out of the node that corresponds to the current text character. In other words, the failure transition is made when a mismatch occurs between the current text character and each character accessible from the current node. Figure 13.1d shows the failure links corresponding to the failure function, which is:

state	1	2	3	4	5	6	7	8	9	10	11	12	13	14	15	16	17	18	19	20	21	22	23	24	25	26
f(state)	0	0	0	0	0	17	18	19	0	0	23	0	0	17	18	19	0	0	23	24	25	0	0	1	2	23

A failure transition exists for each state except for the start state. The transitions to the start state are not shown. With this function, a part of the matching process indicated in the previous two diagrams—an attempt to match *outing* in round 1 and then three keywords in round 2—corresponds to scanning the trie along the path shown in Figure 13.1e. Each nonfailure transition means descending the trie down a path. After reaching node 21, a mismatch occurs (the text character p mismatches the only character reachable from this node, g), and the failure transition leads to node 25 and, indirectly (through a failure transition from node 25), to node 2. This means that the words associated with paths on which nodes 25 and 2 are located are candidates for matches. The current text letter p mismatches the letter t accessible from node 25; therefore, the next failure transition is made, from node 25 to node 2, which eventually leads to a successful match.

Here is an algorithm to construct the failure function:

```
computeFailureFunction(function g, function output, function f)
    for each character ch of the alphabet
        if g(0,ch) ≠ 0
            enqueue(g(0,ch));
            f(g(0,ch)) = 0;
    while queue is not empty
        dequeue state r;
        for each character ch of the alphabet
            if g(r,ch) ≠ fail
                enqueue(g(r,ch));
                state = f(r);
                while g(state,ch) == fail     // follow failure links for
                    state = f(state);          // character ch;
                f(g(r,ch)) = g(state,ch);
                include in output(s) keywords from output(f(g(r,ch)));
```

For each *state* accessible through character *ch* from a dequeued state *r*, the algorithm adds a failure link. It follows failure links for *ch* until it finds a nonfailure (goto) transition for it. For example, when processing *state* 12 accessible through letter *e* from state $r = 11$, there is a failure transition for *e* from 11 to 0 and a nonfailure link for *e* from 0 to 0; therefore, $f(12) = 0$ (Figure 13.1d). For *state* 20 accessible from the same *state* 11 through letter *i*, there is a nonfailure transition from 11 for *i*; therefore, the inner `while` loop is not entered and $f(20) = 24$. The `while` loop iterates twice when determining the failure transition for *state* 22 accessible from state $r = 21$ through letter *g*; the first failure link for *g* leads from 21 to 25, then from 25 to 2, and finally from 2 to 0, where there is a nonfailure link for *g* to the same state 0.

The algorithm also completes the construction of the output function. For each node, the output function records words that end in this node, although they do not have to begin in the root. In the process of creating the output function during the breadth-first traversal, a list of words associated with the current node is expanded by adding words associated with a node reachable through the failure link. The first list expanded in this way is the empty list corresponding to node 25 on level 4 by including the word from the list {*in*} corresponding to node 2 and created when executing `computeGotoFunction()`. The next list expanded in this process is the list {*input*} associated with node 8 on level 6 by including the list {*put*} associated with node 19 accessible from node 8 through a failure link. In effect, the number of nonempty lists is expanded and some of the existing lists are extended by adding new keywords:

2 {in}	5 {inner}	8 {put, input}
11 {out}	13 {outer}	16 {put, output}
19 {put}	21 {in}	22 {outing}
25 {in}	26 {tint}	

In this way, every time a matching process reaches a node in the trie that is associated with a nonempty list of keywords, all the keywords can be output as matching a substring of text ending in the current text position *i*. For example, for text *outinputting* and *keywords* = {*inner, input, in, outer, output, out, put, outing, tint*}, the steps executed by AhoCorasick() are as follows:

```
outinputting    outinputting    outinputting    outinputting    outinputting
inner           out             out             outing          outing
input           outer           outer           outer           tint
in              outing          outing          output          in
outer           output          output          tint            inner
output                          tint            in              input
out                                             inner
put                                             input
outing
tint
0               9               10              11              20              21

outinputting    outinputting    outinputting    outinputting    outinputting
outing          input           input           tint            tint
  inner         put             put                             in
  input                         tint                            inner
  put                                                           input
       25 2 6              7              8        19 23 0 23        24

outinputting    outinputting
  tint            tint
  in
  inner
  input
  25                  2 0 0
```

The numbers indicate states. For example, the initial *o* leads from state 0 to state 9, and letter *p* leads from state 21 to state 25 through a failure link, then to state 2, also though a failure link, and then to state 6. Underlined are letters of the words that are on the path chosen by the algorithm or letters that are at the links branching out from the current state. Words not underlined are the ones that can be reached indirectly through the output function or through failure links.

The algorithm produces the following output:

```
a match ending at 2: out
A match ending at 4: in
a match ending at 7: put input
a match ending at 10: in
```

The goto function can be implemented as a two-dimensional array of size (number of states) · (number of characters). This implementation allows for an immediate access of the value corresponding to a pair (*state,ch*); however, the array would be very sparsely populated with nonfailure transitions. Therefore, a one-dimensional array of vectors or linked lists can be used instead: an array of linked lists (vectors) of characters indexed with the state numbers or an array indexed with characters of linked lists (vectors) of states (or state numbers) (see Section 3.6).

The failure function can be implemented as a one-dimensional array indexed with state numbers of states.

The output function can be implemented as an array of linked lists (or vectors) or words.

For the set *keywords* = $\{s_0, \ldots, s_{k-1}\}$ and the total length of keywords $m = |s_0| + \ldots + |s_{k-1}|$, the algorithm `computeGotoFunction()` is executed in linear time $O(m)$, and the algorithm `computeFailureFunction()` can be executed in the same time.

To determine the complexity of `AhoCorasick()`, note that in one iteration of the `for` loop for `state` on level l, the `while` loop is executed at most $l - 1$ times, which means that at most $l - 1$ failure transitions can be made for `state` corresponding to a node on level l, because these transitions always go up the trie by at least one level root. Therefore, the total number of failure transitions can be at most $|T| - 1$, and because the goto transitions go down the trie by exactly one level so that the number of goto transitions is exactly $|T|$, the number of state transitions during the entire matching process is $O(2|T|)$. Therefore, the complexity of the Aho-Corasick algorithm, including the creation of the failure and goto functions, is $O(|T| + m)$.

It is worth mentioning that the UNIX system's command *fgrep* is an implementation of Aho-Corasick.

13.1.7 Regular Expression Matching

In this section we address the problem of finding in text matches that are specified not by single or multiple patterns, but by regular expressions.

Regular expressions are defined as follows:

1. All letters of the alphabet are regular expressions.
2. If *r* and *s* are regular expressions then $r|s$, (r), r^*, and *rs* are regular expressions.
   a. Regular expression $r|s$ represents regular expression *r* or *s*.
   b. Regular expression r^* (where the star is called a Kleene closure) represents any finite sequence of *r*s: *r, rr, rrr, . . .*
   c. Regular expression *rs* represents a concatenation *rs*.
   d. (r) represents regular expression *r*.

An algorithm created by Ken Thompson constructs a nondeterministic finite automaton (NDFA) corresponding to a regular expression. An NDFA is a directed graph in which each node represents a state and each edge is labeled by a letter or a symbol ε that represents an empty string. The automaton has one initial state and may have multiple terminal or accepting states, but in the context of this section it has only one accepting state. An NDFA is used during the matching process. A match in the text is found if there is a path with letters on the edges in NDFA from the initial state to an accepting state that matches a substring of the text.

A construction of an NDFA is given in the form of the following recursive procedure:

1. An automaton representing one letter has one initial state i, one accepting state a, and an edge from the former to the latter labeled with the letter (Figure 13.2a).

FIGURE 13.2 (a) An automaton representing one letter c; an automaton a regular expression (b) $r|s$, (c) rs, (d) r^*.

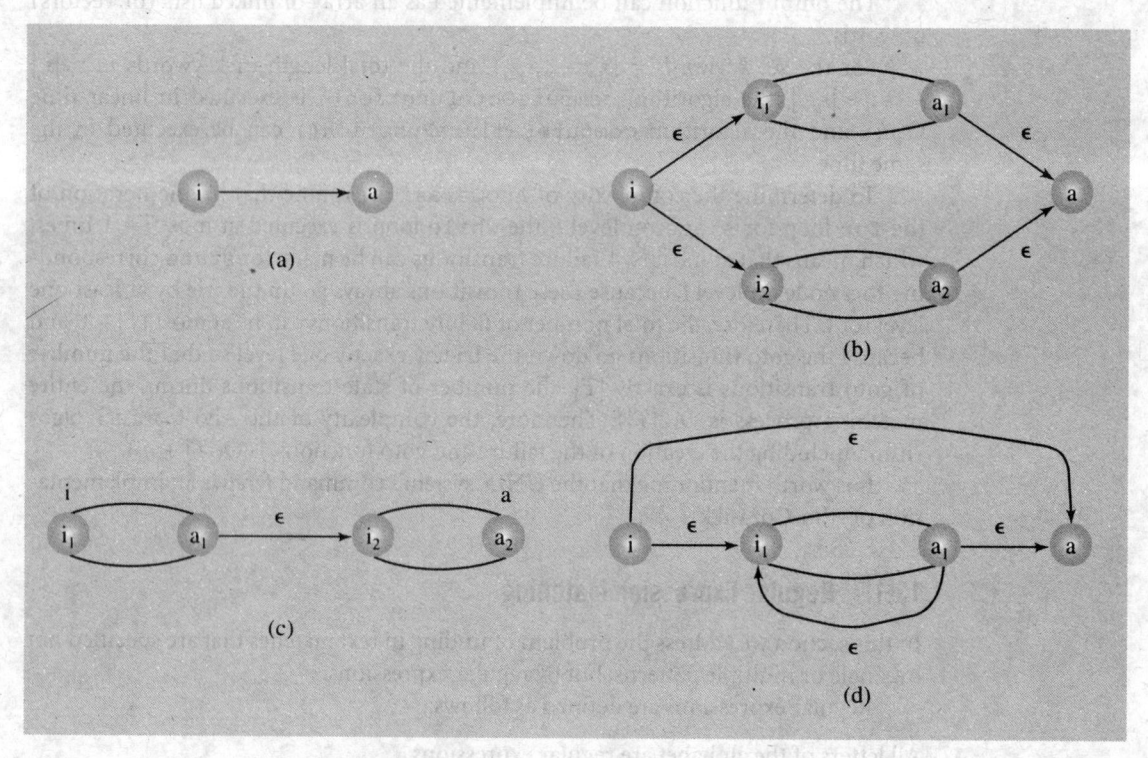

2. An automaton representing a regular expression $r|s$ is a union of the automata representing r and s. The union is constructed by
   a. creating an initial state i with two outcoming ε-edges, one to the initial state i_1 of the automaton representing r and one to the initial state i_2 of the automaton representing s;
   b. creating an accepting state with two incoming ε-edges from accepting states a_1 and a_2 of the two automata (Figure 13.2b).
3. An automaton representing a regular expression rs is a concatenation of the automata representing r and s. The concatenation is constructed by creating an ε-edge from the accepting state a_1 of the automaton representing r to the initial state i_2 of the automaton representing s; the initial state i_1 becomes the initial state of the concatenated automaton, and a_2 becomes its accepting state (Figure 13.2c).

4. An automaton representing a regular expression r^* is constructed as follows.
   a. A new initial state i is created with an ε-edge to the initial state i_1 of the automaton representing r;
   b. a new accepting state a is added with an ε-edge from the state a_1;
   c. an ε-edge is added from the initial state i to the accepting state a; and
   d. an ε-edge is added from the state a_1 to the state i_1 (Figure 13.2d).
5. An automaton representing a regular expression (r) is the same as the automaton representing r. The construction process indicates that an automaton corresponding to a regular expression
   a. has one initial state and one accepting state;
   b. each state has one outgoing edge labeled by a letter, one ε-edge, or two ε-edges;
   c. in each step two new nodes can be created (or none), so the number of states in the automaton is at most twice the length of the regular expression to which it corresponds, and the number of edges is at most four times that size.

An automaton can be created with the following routines:

```
component()
    if regExpr_i is a letter
        p = a character automaton as in Figure 13.2a;
        i++;
    else if regExpr_i == '('
        i++;
        p = regularExpr();
        if regExpr_i ≠ ')'
            failure;
        i++;
    if regExpr_i == '*'
        while regExpr_++i == '*';
        p = a star automaton as in Figure 13.2d;
    return p;

concatenation()
    p1 = component();
    while regExpr_i is a letter or '('
        p2 = component();
        p1 = concatenation of automata as in Figure 13.2c;
    return p1;

regularExpr() {
    p1 = concatenation();
    while i < |T| and regExpr_i == '|'
        i++;
        p2 = concatenation();
        p1 = union of automata as in Figure 13.2b;
    return p1;
```

and the processing begins by calling first `regularExpr()`. Note that the processing is done very much in the same spirit as the interpreter presented in Section 5.11, where `regularExpr()` corresponds to `expression()`, `concatenation()` corresponds to `term()`, and `component()` corresponds to `factor()`.

Two sets are needed to properly process regular expression. The set *epsilon(S)* is a set of states accessible from the states in *S* through ε-paths. The set *goto(S,ch)* is a set of states for which there is an edge labeled with character *ch* from a state in *S*. The sets can be created with the following algorithms:

```
gotoFunction(states, ch)
    for each state in states
        if there is a ch-transition from state to a state s
            include s in states2 if it is not already included;
    return states2;
```

```
epsilon(states)
    for each state in states
        remove state from states
        for each state s for which there is an ε-edge from state to s
            include s in states and in states2 if it is not already included;
    return states2;
```

The automaton and the sets so constructed are now used to process a text and detect the longest matching regular expressions in a text and print their positions in the text. The algorithm is as follows:

```
Thompson(regExpr, text T)
    initState = parse();
    from = 1;
    states = epsilon({initState});
    for i = 0 to |T|-1
        states = gotoFunction(states,T_i);
        if states is empty
            states = gotoFunction({initState});
            from = i;
        if accepting state is in states
            output: "match from " from " to " i;
        states = epsilon(states);
        if accepting state is in states
            output: "match from " from " to " i;
        if states is empty
            states = epsilon({initState});
            from = i+1;
```

The following table shows the steps in processing the string *T = aabbcdeffaefc* and regular expression *regExpr = a(b|cd)*ef* with initialization *states = epsilon({initState})* = {}. The automaton for the expression is given in Figure 13.3 with the number representing the order in which the states were generated. The substrings in the second column in boldface indicate the substrings detected by the program as matching *regExpr*; they are substrings $T(0 \ldots 6) = $ *abbcdef* and $T(8 \ldots 10) = $ *aef*.

i	ch	goto(states,ch)	states after `if`-stmt	epsilon(states,ch)	states after `if`-stmt
0	a	{}	{1}	{2 4 8 10 11 12}	{2 4 8 10 11 12}
1	a	{}	{1}	{2 4 8 10 11 12}	{2 4 8 10 11 12}
2	b	{3}	{3}	{2 4 8 9 11 12}	{2 4 8 9 11 12}
3	b	{3}	{3}	{2 4 8 9 11 12}	{2 4 8 9 11 12}
4	c	{5}	{5}	{6}	{6}
5	d	{7}	{7}	{2 4 8 9 11 12}	{2 4 8 9 11 12}
6	e	{13}	{13}	{14}	{14}
7	f	{15}	{15}	{}	{1}
8	f	{}	{1}	{2 4 8 10 11 12}	{2 4 8 10 11 12}
9	a	{}	{1}	{2 4 8 10 11 12}	{2 4 8 10 11 12}
10	e	{13}	{13}	{14}	{14}
11	f	{15}	{15}	{}	{1}
12	c	{}	{1}	{}	{1}

FIGURE 13.3 The Thompson automaton for the regular expression *a(b|cd)*ef*.

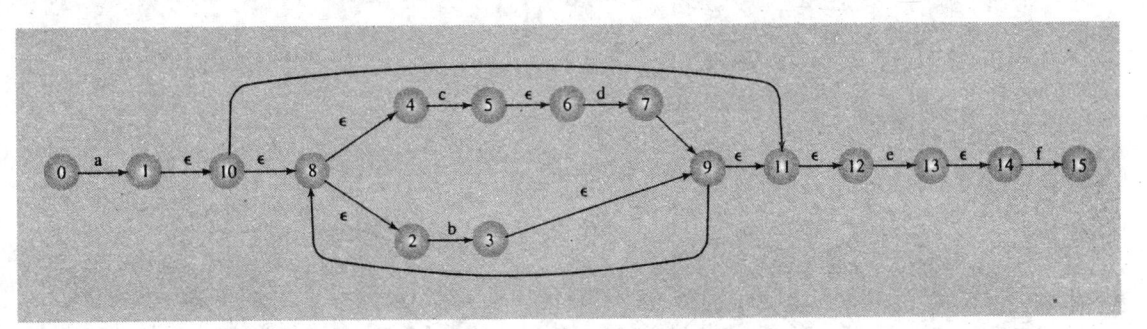

13.1.8 Suffix Tries and Trees

In many situations it is beneficial to preprocess a string or strings by creating a structure that allows further processing to be executed more efficiently than without using this structure. One such structure is a suffix trie and its generalization, a suffix tree.

A *suffix trie* for a text T is a tree structure in which each edge is labeled with one letter of T and each suffix of T is represented in the trie as a concatenation of edge labels from the root to some node of the trie. In a suffix trie, $head(i)$ is the longest prefix of string $T(i \ldots |T|-1)$ that matches a prefix of a suffix $T(j \ldots |T|-1)$ that is already in the tree. A trie for the word *caracas* is in Figure 13.4a. It should be clear that in the worst case, when all letters are different, a trie requires one node for the root and $|T|-i$ nodes for each suffix, where i is its starting position—that is, $(|T|+1)|T|/2$ in total—which means that the space requirement for a suffix trie is quadratic.

A simple algorithm for creating a suffix trie for a text T simply looks at one suffix at a time and extends paths corresponding to the suffix when necessary:

682 ■ Chapter 13 String Matching

FIGURE 13.4 (a) A suffix trie for the string *caracas*; (b) a suffix tree for the substring *caraca* and (c) for the string *caracas*.

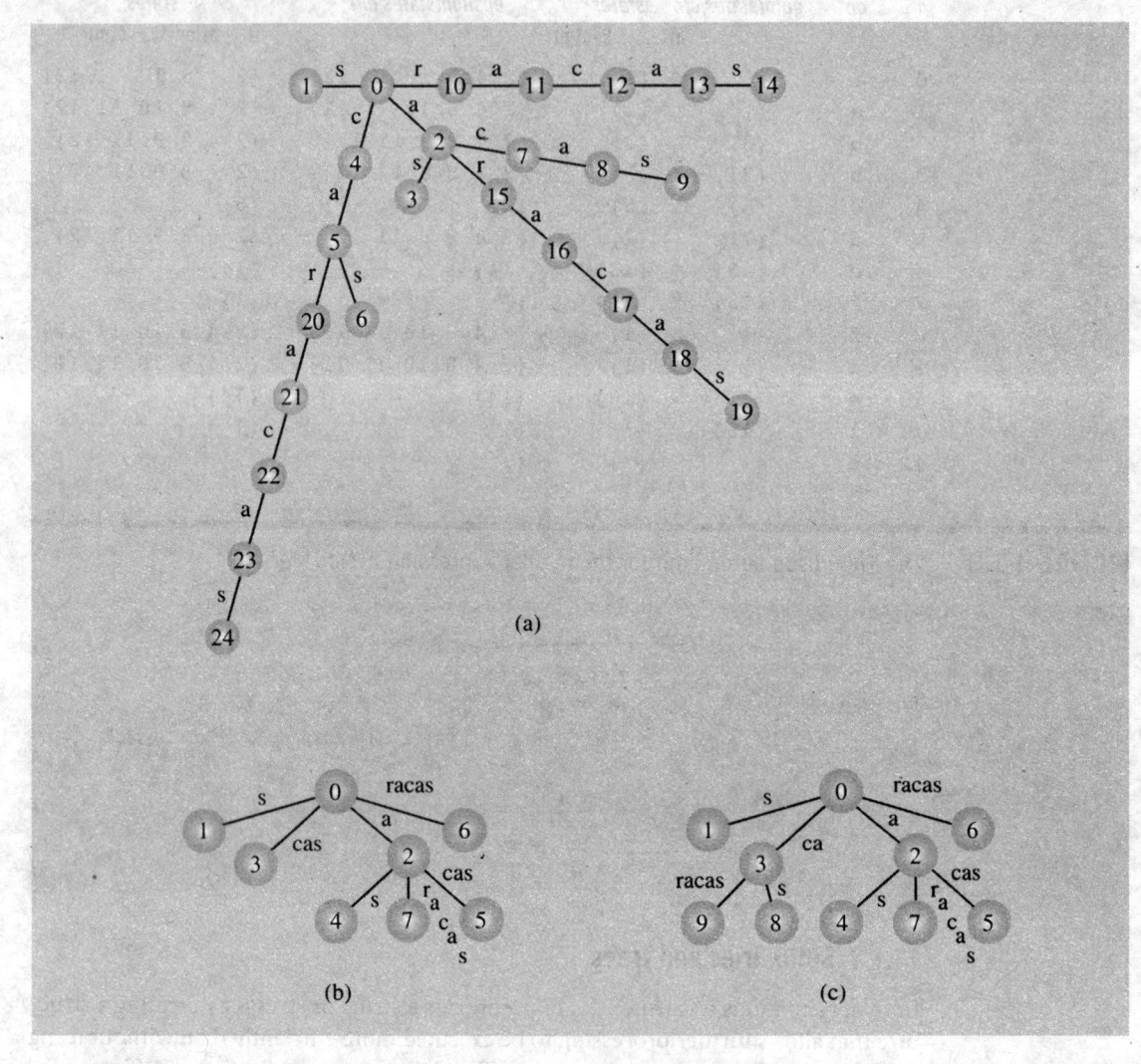

```
bruteForceSuffixTrie(text T) {
    // the order of processing suffixes does not matter; it can be
    // from the largest to the smallest:
//  for i = 0 to |T|-1
    // or from the smallest to the largest:
    for i = |T|-1 downto 0
        node = root;
        j = i; // to represent the suffix T(i...|T|-1);
        while T_j-edge from node exists
            node = the node accessible through T_j-edge from node;
```

```
        j++;
// store the suffix T(j...|T|-1) of the suffix T(i...|T|-1) in the trie:
for k = j to |T|-1
    create newNode;
    create edge(node,newNode,T_k);
    node = newNode;
```

The trie in Figure 13.4a has been created with this algorithm, which is reflected in the node numbers that indicate the order in which nodes were included in the trie.

The run time of the algorithm is quadratic because for each iteration i of the outer loop, the two inner loops combined perform $|T| - i + 1$ iterations.

A compact version of a suffix trie is a *suffix tree*, in which there are no internal nodes with only one descendant. A suffix tree can be obtained from a suffix trie by labeling each edge with the substring of T that corresponds to the concatenation of characters on the subpaths of the trie in which only nodes with one descendant are used. In other words, one-descendant nodes on such subpaths are merged into one single node, and the edges on such subpaths are merged into one edge. By converting the trie in Figure 13.4a, a suffix tree in Figure 13.4c is created. A suffix tree can have up to $|T|$ leaves and thus up to $|T| - 1$ nonleaves. It has exactly $|T| - 1$ nonleaves if it has $|T|$ leaves, each nonleaf has two descendants, and no suffix is represented by a path that ends in an nonleaf (see Section 6.1 and Figure 6.5). In this way, space requirements for a suffix tree are linear in the length of the text.

In a suffix tree, most of the nodes are implicit; these are the nodes that in a corresponding trie have one descendant. The problem with processing a suffix tree is determining when an implicit node has to be made explicit. The situation is illustrated by inserting the suffix *caracas* into the tree in Figure 13.4b, which requires splitting the edge labeled *cas* into two edges, labeled *ca* and *s*, and then inserting a new edge labeled *racas*, as shown in Figure 13.4c. For practical purposes, it is more convenient to label edges in a suffix tree with two indexes that represent starting and ending positions of a substring of T that is the word label of the edge.

Here is a summary of an algorithm:

```
bruteForceSuffixTree(text T)
    for each suffix s of T
        determine the head of s (find the longest match for the longest prefix of s and a path from the root);
        if the suffix of the head matches an entire label of an edge that ends in a leaf
            extend the label by the unmatched part of s;
        else if the suffix of the head matches an entire label of an edge that ends in a nonleaf
            create a leaf connected to the nonleaf with an edge labeled by the unmatched part of s;
        else split the edge(u,v) partially matched with the suffix of the head into
            edge(u,w) labeled with the part of the current label matching the longest suffix of the head and
            edge(w,v) labeled with the unmatched part of the current label and
            create edge(w,u) labeled with the unmatched part of s;
```

Processing can be done in at least two ways, left to right or right to left. The suffix tree in Figure 13.4c is created by processing suffixes of *caracas* right to left, from the smallest suffix to the largest, which is reflected in the order in which nodes have been created.

The algorithm executes in $O(|T|^2)$ time due to a straightforward strategy to determine the head: The search always starts at the root, which requires performing $|head(i)|$ steps in each iteration, which in the worst case is $|T| - i$, that is, the length of suffix $- 1$. To see that, consider string $a^k b$. The complexity can be improved by devising quicker ways of determining heads of suffixes, which is accomplished by maintaining additional links in the suffix tree. One such suffix tree is constructed by an algorithm devised by Esko Ukkonen.

Conceptually, a suffix tree is a compressed suffix trie, so the presentation of the Ukkonen algorithm for a suffix tree (Ukkonen, 1995) is better understood when started with the discussion of an algorithm for suffix tries (Ukkonen and Wood, 1993).

Ukkonen suffix tries and suffix trees use *suffix links* in the construction (these links are the same as failure transitions in the Aho-Corasick algorithm).

A new suffix trie is obtained from an existing suffix trie by extending paths corresponding to all suffixes of the subtext $T(0 \ldots i-1)$ by adding new transitions corresponding to the character T_i. In this way, a new trie has paths for all suffixes of the subtext $T(0 \ldots i)$. The states for which new transitions are added can be found using suffix links that form a path from the deepest node to the root. This path is called the *boundary path*. The path is traversed, and for each encountered node p a new leaf q is created with the $edge(p,q,ch)$ (i.e., an edge from p to q labeled ch) if there is no ch-edge coming out of p. The path traversal is aborted after the first node is encountered for which there is a ch-edge.

Also, new suffix links are created that form a path that joins together the newly added nodes. The new suffix links are a part of the boundary path of the updated trie.

The algorithm is as follows:

```
UkkonenSuffixTrie(text T)
    create newNode;
    root = deepestNode = oldNewNode = newNode;
    suffixLink(root) = null;
    for i = 0 to |T|-1
        node = deepestNode;
        while node is not null and T_i-edge from node does not exist
            create newNode;
            create edge(node,newNode,T_i);
            if node ≠ deepestNode
                suffixLink(oldNewNode) = newNode;
            oldNewNode = newNode;
            node = suffixLink(node);
        if node is null
            suffixLink(newNode) = root;
        else suffixLink(newNode) = child of node through T_i-edge;
        deepestNode = child of deepestNode through T_i-edge;
```

The algorithm is executed in the time proportional to the number of different substrings of the text T, which can be quadratic, as, for example, for a text with all different characters.

For an example, consider building a suffix trie for the word *pepper*. The trie is initialized to a one-node tree (Figure 13.5a). In the first iteration of the for loop, a new node is created with an edge corresponding to letter p between the root and the new node; a suffix link from the new node to the root also is created (Figure 13.5b). In the second

Section 13.1 Exact String Matching 685

FIGURE 13.5 Creating an Ukkonen suffix trie for the string *pepper*.

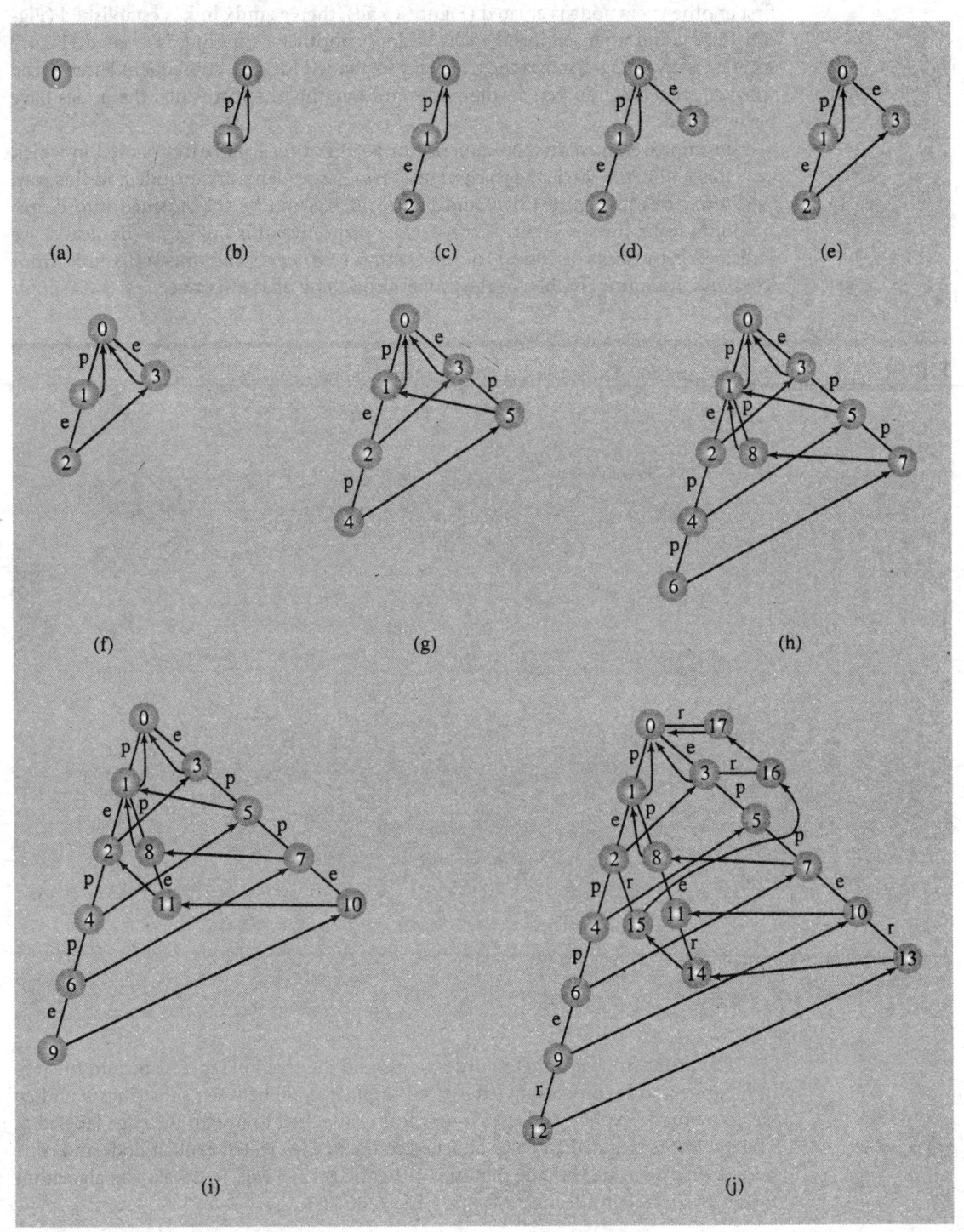

iteration of the `for` loop, the letter *e* is processed. A new node is created in the first iteration of the `while` loop (Figure 13.5c). In the second iteration of the `while` loop, first another new node is created (Figure 13.5d), then a suffix link is established (Figure 13.5e), and after exiting the `while` loop, another suffix link is created (Figure 13.5f). Figure 13.5g–f shows the trie being expanded for each subsequent letter of the word *pepper*. The numbers in the nodes indicate the order in which the nodes have been created.

To improve on space requirements for a suffix trie, a suffix tree is used in which only those nodes of a trie that have at least two descendants are included. In this way, the suffix tries in Figure 13.5b, and Figure 13.5f–j can be transformed into corresponding suffix trees in Figure 13.6a–f. Note that only suffix links from nonleaves are indicated because, as discussed later in Section 13.3, leaves and thus suffix links from leaves are not indispensable for the proper processing of a suffix tree.

FIGURE 13.6 Creating an Ukkonen suffix tree for the string *pepper*.

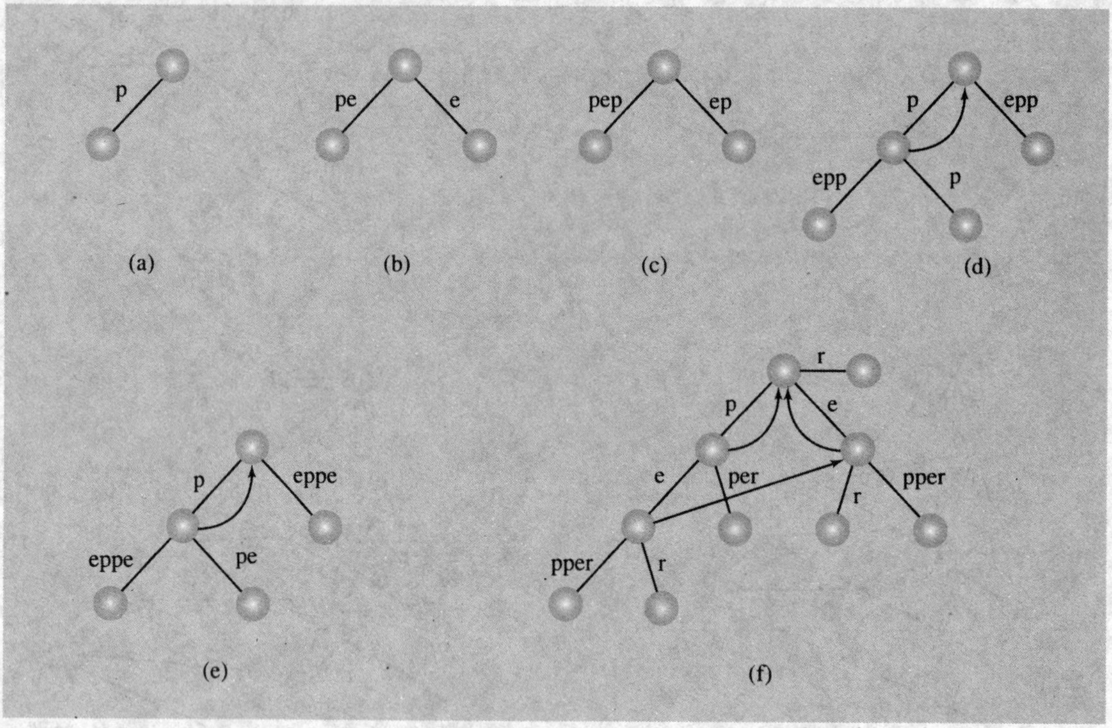

The problem now is that in order to expand the tree in Figure 13.6c into the tree in Figure 13.6d by processing letter *p*, an implicit node between substrings *p* and *ep* has to be made explicit and then a new leaf is attached to it with the edge labeled *p*. The old leaf is retained and also attached to the newly created explicit node and connected with the edge with the modified (extended) label *epp*. To develop an algorithm for a suffix tree, consider again the processing of a trie.

The first nonleaf on the boundary path is called an *active point*. When processing a trie, a new *ch*-edge and a new leaf are added for each leaf on the boundary path—that is, for each node that precedes the active point. Then, each node on the subpath between the active point and the so-called *endpoint* also receives a new *ch*-edge to a new leaf. Therefore, an endpoint is a node for which a *ch*-edge already exists, and thus such a *ch*-edge exists for all the nodes on the boundary path from the endpoint to the root. An endpoint can be a virtual parent of the root if the root also acquires a new *ch*-edge (it is assumed that between such a virtual parent and the root there is a *ch*-edge for each letter). For example, in Figure 13.5i, on the boundary path 9 10 11 2 3 0 −1, node 2 is the active point and node −1, the virtual parent of the root, is the endpoint with respect to character *r*, which is inserted into this trie. If *p* were inserted into this trie, then the active point would be the same, but node 2, the active point, would be the endpoint.

When processing a tree, an update of the equivalent of the beginning of the boundary path that includes all leaves requires only updating the labels of the edges that connect these leaves with their parents. However, each nonleaf on the boundary path up to the node before the endpoint may need to be made explicit in the tree so that a leaf can be attached to it. Here is a summary of the algorithm:

```
UkkonenSuffixTree(text T)
    initialize the root and active point to a new node corresponding to T;
    for i = 0 to |T|-1
        for each leaf on the boundary path (i.e., from the beginning to a node before the active point)
            update the label of the edge between the leaf and its parent;
        for each node on the boundary path from the active point to a node before the endpoint
            create newNode;
            if node is not explicit
                make node explicit by inserting it between its parent p and a node q;
                update the edge between node and p;
                create an edge between node and q;
            create edge(node, newNode, T_i);
```

An implementation of the algorithm is presented in the case study at the end of this chapter.

13.1.9 Suffix Arrays

Sometimes suffix trees may require too much space. A very simple alternative to suffix trees are suffix arrays (Manber and Myers, 1993).

Suffix array *pos* is the array of positions 0 through $|T| - 1$ of suffixes taken in lexicographic order. It is obvious that the suffix array requires $|T|$ cells. For example, suffixes of text *T = proposition* are ordered as follows:

    8 *ion*
    6 *ition*
   10 *n*
    9 *on*
    2 *oposition*
    4 *osition*

3 *position*
0 *proposition*
1 *roposition*
5 *sition*
7 *tion*

with their positions in the text indicated on the left-hand side; these positions form a suffix array *pos* = [8 6 10 9 2 4 3 0 1 5 7] that corresponds to the order of the suffixes.

The suffix array can be created in $O(|T| \lg |T|)$ time by sorting an array initialized as [0 1 ... |T| − 1]. The sorting routine compares suffixes but moves numbers in *pos* when the suffixes indicated by the positions are out of sequence.

The suffix array can be created from an existing suffix tree on which an ordered depth-first traversal is performed. For each node of the tree, the traversal traverses its subtrees according to the order of word labels of the outgoing edges, which is the order of the first letters in these labels. The traversal inserts in the suffix array the leaf numbers in the order of reaching the leaves. The edges can be sorted in $O(|A| \lg |A|)$ time so that traversal is executed in $O(|T||A| \lg |A|)$ time. The suffix tree routine can maintain the edges on a linked list in sorted order, which amounts to $O(|A|^2)$ time to maintain one such list and thus $O(|T||A|^2)$ time to maintain the tree. Only then can the traversal be done in linear time, $O(|T|)$.

With the suffix array, a pattern *P* can be found very quickly in text *T* by using binary search (see Section 2.7) and then all suffixes with prefix *P* are grouped together. To locate the beginning of the cluster of such suffixes, the following version of binary search can be used:

```
binarySearch(pattern P, text T, suffix array pos)
    left = 0;
    right = |T|;    // that is, |pos|;
    while left < right
        middle = (left+right)/2;
        if P ( T(pos[middle]...|T|-1)
            right = middle;
        else left = middle+1;
    if P is equal to T(pos[left]...pos[left]+|P|)
        return left;
    else return -1; // failure;
```

To locate the end of the cluster of such suffixes, the inequality ≤ in the `if` statement should be turned into ≥, and `left` should be replaced by `right` in the `if` statement following the loop. Pattern *P* is found in *T* at positions located in the cluster just determined. The cluster is determined in $O(|P| \lg |T|)$ time, where $\lg |T|$ refers to the number of iterations of the `while` loop of `binarySearch()`, and $|P|$ refers to the number of character comparisons between *P* and suffixes of *T* in one iteration of the loop.

13.2 APPROXIMATE STRING MATCHING

In preceding sections, algorithms for exact matching were analyzed, which was an all-or-nothing proposition: A search for a pattern P in a text T is considered successful if there is at least one substring in T that is equal to P. If there is at least one character difference, the substring is not considered a match for P. In many situations, however, the requirement for exact match can be relaxed by stating that only a certain level of similarity between P and T (or its substring) is needed to consider a match successful.

A popular measure of the similarity of two strings is the number of elementary edit operations that are needed to transform one string into another. Three elementary operations on strings are considered: insertion I, deletion D, and substitution S. The differences between two strings is sought in terms of these operations. These differences can be represented in at least three ways: trace, alignment (matching), and listing (derivation). For example,

```
alignment:
-app--le            source
capital-            target

listing:
apple               source
capple      (I)
capile      (S)
capitle     (I)
capitale    (I)
capital     (D)     target

trace:
apple               source
↓↓↓ ↘
capital             target
```

Lines in the trace cannot cross one other, and only one line can connect a source character with a target letter. A letter with no line in the source indicates a deletion; a letter with no line in the target indicates an insertion. Lines connect a source letter with the same letter of the target or a letter that is substituted for another letter.

Alignment is obtained by aligning two strings that may include null characters indicated by dashes. A dash in the source indicates an insertion; a dash in the target indicates a deletion.

Listings directly correspond to the way strings are processed by a particular algorithm; alignments and traces summarize the work more succinctly and legibly.

The most popular measure of distance between two strings is the *Levenshtein distance*. In fact, Levenshtein (1965) introduced two concepts of distance $d(Q,R)$ between two strings Q and R. One is the smallest number of insertions, deletions, and substitutions needed to convert Q into R. Another distance takes only deletions and insertions into account.

Mathematically, distance is a function d that satisfies the following conditions. For any Q, R, and U:

$d(Q,R) \geq 0$
$d(Q,R) = 0$ iff $Q = R$
$d(Q,R) = d(R,Q)$ (symmetry)
$d(Q,R) + d(Q,U) \geq d(Q,U)$ (triangle inequality)

Most distance functions used for string processing meet these requirements, including the Levenshtein distance, but exceptions are possible. Also, distances that include weights are possible. For example, in microbiological applications, a single deletion of two neighboring string elements may be much more likely than two separate deletions of single elements. In such a case, a larger weight is used for two consecutive deletions than for two separate deletions.

13.2.1 String Similarity

The string similarity problem arises when for two strings R and Q, the distance $d(Q,R)$ between these two strings needs to be determined.

Let $D(i,j) = d(Q(0 \ldots i-1), R(0 \ldots j-1))$ be the edit distance between prefixes $Q(0 \ldots i-1)$ and $R(0 \ldots j-1)$. The string similarity problem can be approached by reducing the problem of finding the minimum distance for a particular i and j to the problem of finding the minimum distance for values not larger than i and j. If the subproblems are solved, then the solution can be extended to i and j by observing which operation is needed to find correspondence between characters Q_i and R_j. There are four possibilities:

1. *Deletion.* When Q_i is deleted from $Q(0 \ldots i)$, then $D(i-1,j-1) = D(i-2,j-1) + 1$; that is, the minimum distance between $Q(0 \ldots i)$ and $R(0 \ldots j)$ equals the minimum distance between $Q(0 \ldots i-1)$ and $R(0 \ldots j)$ plus 1, where 1 signifies deletion of Q_i from the end of $Q(0 \ldots i)$.

2. *Insertion.* When R_j is inserted into $R(0 \ldots j-1)$, then $D(i-1,j-1) = D(i-1,j-2) + 1$; that is, the minimum distance between $Q(0 \ldots i)$ and $R(0 \ldots j)$ equals the minimum distance between $Q(0 \ldots i)$ and $R(0 \ldots j-1)$ plus 1, where 1 signifies insertion of R_j at the end of $R(0 \ldots j-1)$.

3. *Substitution.* When R_j is substituted in $R(0 \ldots j)$ for $Q_i \neq R_j$ in $Q(0 \ldots i)$, then $D(i-1,j-1) = D(i-2,j-2) + 1$; that is, the minimum distance between $Q(0 \ldots i)$ and $R(0 \ldots j)$ equals the minimum distance between $Q(0 \ldots i-1)$ and $R(0 \ldots j-1)$ plus 1, where 1 signifies the substitution operation.

4. *Match.* When $Q_i = R_j$, no additional operation is needed and then $D(i-1,j-1) = D(i-2,j-2)$; that is, the minimum distance between $Q(0 \ldots i)$ and $R(0 \ldots j)$ equals the minimum distance between $Q(0 \ldots i-1)$ and $R(0 \ldots j-1)$.

All these conditions can be combined into one recurrence relation:

$$D(i,j) = \min(D(i-1,j) + 1, D(i,j-1) + 1, D(i-1,j-1) + c(i,j))$$

where $c(i,j) = 0$ if $Q_i = R_j$ and 1 otherwise.

Moreover, to transform a nonempty string into an empty string, all its characters have to be deleted, so that

Section 13.2 Approximate String Matching

$$D(i,0) = i$$

and to transform an empty string into a nonempty string, all its characters have to be inserted, that is,

$$D(0,j) = j$$

The problem can be solved recursively by directly using the equations. However, in this way, the problem of large size is reduced to three problems that are only slightly smaller, which, in effect triples the effort needed at a particular level of recursion. The smaller problems have to be solved separately, which triples the effort for every one of them and means a nine-fold increase of effort for the previous level of recursion. Eventually, the original problem requires exponential effort to be solved. To avoid the use of excessive recursion, the problem is solved differently.

One solution is to use a 2D *edit table*, in which the results of iteratively solved subproblems, from smallest to largest, are recorded. We use an $(|R| + 1) \times (|Q| + 1)$ edit table $dist[0 \ldots |R|, 0 \ldots |Q|]$ for which $dist[i,j] = D(i,j)$; that is, its rows correspond to characters in R and columns to characters in Q. The first row corresponds to values $D(0,j)$ and thus is initialized with numbers $0, 1, \ldots, |R|$. Similarly, the first column corresponds to values $D(i,0)$ and so is initialized with numbers $0, 1, \ldots, |Q|$. Afterwards, for each cell of the table, the value stored in the cell is determined in accordance with the recurrence relation for $D(i,j)$, which means that it refers to three cells: one above, one to the left, and one positioned diagonally. Here is an algorithm devised by Wagner and Fischer (1974):

```
WagnerFischer(edit table dist, string Q, string R)
    for i = 0 to |Q|
        dist[i,0] = i;
    for j = 0 to |R|
        dist[0,j] = j;
    for i = 1 to |Q|
        for j = 1 to |R|
            x = dist[i-1,j]+1;  // upper
            y = dist[i,j-1]+1;  // left
            z = dist[i-1,j-1];  // diagonal
            if Q_{i-1} ≠ R_{j-1}
                z++;
            dist[i,j] = min(x,y,z);
```

It is clear from the use of nested `for` loops that the algorithm runs in $O(|Q||R|)$ time and space.

Consider strings Q = *capital* and R = *apple* and the edit table *dist* created for them:

		a	p	p	l	e
	0	1	2	3	4	5
c	1	1	2	3	4	5
a	2	1	2	3	4	5
p	3	2	1	2	3	4
i	4	3	2	2	3	4
t	5	4	3	3	3	4
a	6	5	4	4	4	4
	1	7	6	5	5	4	5

After initializing the first row and first column, the value for each cell is found by using values of the three mentioned neighboring cells. For example, to determine the value $D(6,3)$—that is, the value $dist[6,3] = 4$, which is shown in italics in the diagram—the three neighbors shown in bold type are consulted. There are two candidates for the minimum value, both with a value of 3: $dist[5,2]$ and $dist[5,3]$. If the first is chosen, then we choose the operation of substitution, because the sixth character in *capital*, character *a*, is different from the third character in *apple*, character *p*. In this way, $D[6,3] = d(capita,app) = d(capit,ap) + 1 = 4$. When the second candidate is chosen, then we choose the operation of insertion of *a* in the substring *capit* to obtain substring *capita*, so that $D[6,3] = d(capita,app) = d(capit,app) + 1 = 4$.

The number in the lower right corner of *dist*, number 5, is the minimum distance between strings *capital* and *apple*. The table can be used to generate an alignment so that not only $d(capital,apple)$ is known, but also a listing. If we are interested in only one possible listing, then it can be generated with the following algorithm:

```
WagnerFisherPrint(edit table dist, string Q, string R)
    i = |Q|;
    j = |R|;
    while i ≠ 0 or j ≠ 0
        output pair (i, j);
        if i > 0 and dist[i-1,j] < dist[i,j]              // up
            sQ.push(Q_{i-1});
            sR.push('-');
            i--;
        else if j > 0 and dist[i,j-1] < dist[i,j]   // left
            sQ.push('-');
            sR.push(R_{j-1});
            j--;
        else // if i > 0 and j > 0 and                    // diagonally
             //     (dist[i-1,j-1] == dist[i,j] and Q_{i-1} == R_{j-1} or
             //      dist[i-1,j-1] < dist[i,j] and Q_{i-1} ≠ R_{j-1})
            sQ.push(Q_{i-1});
            sR.push(R_{j-1});
            i--;
            j--;
    print stack sQ;
    print stack sR;
```

At least one of the indexes *i* and *j* is decremented in each iteration of the `while` loop, so the algorithm runs in $O(\max(|Q|,|R|))$ time.

Two stacks are used to generate the elements of the alignment. Stacks are very suitable here because the source and target of the alignment are generated backwards. Processing starts in the lower right corner, and for each cell *c* it goes to one of the three neighbors whose values were used to determine the value stored in *c*; depending on which neighbor it is, either a character from a string or a dash is stored—a dash in the target to indicate a deletion, and a dash in the source to indicate an insertion. The algorithm `WagnerFisherPrint()` generates this output:

```
path: [7 5] [6 5] [5 4] [4 3] [3 2] [2 1] [1 0]
capital
-apple-
```

As an extra, the path from the lower right corner to the upper left corner is also generated. The order of `if` statements in the algorithm determines which alignment is generated. If the first `if` statement is exchanged with the second, then the output is

```
path: [7 5] [7 4] [6 3] [5 3] [4 3] [3 2] [2 1] [1 0]
capital-
-app--le
```

It is also possible to generate all alignments and print them after removing duplicates.

The Wagner-Fisher algorithm can be improved in many ways, one way being the reduction of space from $O(|Q||R|)$ to $O(|R|)$ (Drozdek, 2002), another way is improving running time when sequences are far apart (Hunt and Szymanski, 1977).

For two strings Q and R, a common subsequence is the sequence of characters that occurs in both strings, not necessarily in consecutive order. For example, *es*, *ece*, and *ee* are common subsequences in *predecessor* and *descendant*. The longest common subsequence problem is the task of determining what is the longest subsequence found in two strings Q and R. A strong connection exists between the longest common subsequence and edit distance.

It is clear that the length of the longest common subsequence $lcs(Q, R)$ is the largest number of pairs (i, j) in any alignment in which equal characters Q_i and R_j are aligned. Let us consider such an alignment. Define a new edit distance d_2 in which the cost of insertion and deletion is 1, but the cost of substitution is 2 for unequal characters and 0 when they are equal. This amounts to the more restricted concept of Levenshtein edit distance, which includes only deletion and insertion, because substitution is, in effect, replaced by deletion followed by an insertion. In this case,

$$d_2(Q, R) = |Q| + |R| - 2lcs(Q, R)$$

because

$$d_2(Q, R) = \#deletions + \#insertions + 2 \cdot \#substitutions$$

$$d_2(Q, R) = |Q| - \#substitutions - lcs(Q, R) + |R| - \#substitutions - lcs(Q, R) + 2 \cdot \#substitutions$$

The longest common subsequence can be found with the following algorithm:

```
HuntSzymanski(Q, R)
    for i = 0 to |Q|-1
        matchlist[i] = list in descending order of positions j for which Q_i == R_j;
    for i = 1 to |Q|
        threshold[k] = |R|+1;
    threshold[0] = -1;
    for i = 0 to |Q|-1
        for each j in matchlist[i]
            find position k such that threshold[k] < j ≤ threshold[k+1];
            if j < threshold[k+1]
```

```
            threshold[k+1] = j;
            link[k] = new node(i,j,link[k-1]);
k = max{t: threshold[t] < |R|+1};
for (p = link[k]; p ≠ null; p = prev(p))  // print pairs in reverse order;
    output pair (i, j) in node p;
```

The *matchlist* is a table of lists of positions in descending order and a list *matchlist*[i] includes all positions j for which $Q_i = R_j$. The lists can be created in $O(|R|\lg|R| + |Q|\lg|R|)$ time by sorting in $O(|R|\lg|R|)$ time a copy of R while remembering original positions of its characters, and then for each position i extracting the list of positions corresponding to character Q_i using binary search to locate this list. For example, for strings $Q = rapidity$ and $R = paradox$, the lists are as follows:

```
r:  matchlist[0] = (2)
a:  matchlist[1] = (3 1)
p:  matchlist[2] = (0)
i:  matchlist[3] = ()
d:  matchlist[4] = (4)
i:  matchlist[5] = ()
t:  matchlist[6] = ()
y:  matchlist[7] = ()
```

that is, the list *matchlist*[0] that corresponds to character r in *rapidity* has one number, 2, which is the position of r in *paradox*; the list *matchlist*[1] that corresponds to character a has two numbers, 3 and 1, which are the positions of a in *paradox*, and so on. The lists represent the following table:

```
    p a r a d o x
    0 1 2 3 4 5 6
r 0     x
a 1   x   x
p 2 x
i 3
d 4         x
i 5
t 6
y 7
```

in which *x*s represent matching characters in the two strings.

To find the longest common sequence (marked with bold *x*s), the table *threshold* is used. Positions marked with *x*s are the positions indicated by the positions k in *threshold* and the numbers in this position, which can change during execution of the algorithm. Numbers in *threshold* are in ascending order so that the required position k can be found in $\lg|R|$ time with binary search. Denote by r the number of pairs (i, j) for which $Q_i = R_j$. Because there are r iterations of the inner for loop, one for each number in *matchlist*, this phase is executed in $O(r\lg|R|)$ time, during which up to r nodes in *link* can be created.

Here are the changes in *threshold*, *link*, and some variables during processing of strings $Q = rapidity$ and $R = paradox$:

```
    p a r a d o x    i j k        threshold              link
    0 1 2 3 4 5 7              0 1 2 3 4 5 6 7 8
r 0       x            0 2 0   -1 2 8 8 8 8 8 8 8    link[0] = (0,2)
a 1           x        1 3 1   -1 2 3 8 8 8 8 8 8    link[0] = (0,2)
                                                     link[1] = (1,3)↗
a 1   x                1 1 0   -1 1 3 8 8 8 8 8 8    link[0] = (1,1)
                                                     link[1] = (1,3)→(0,2)
p 2 x                  2 0 0   -1 0 3 8 8 8 8 8 8    link[0] = (2,0)
                                                     link[1] = (1,3)→(0,2)
i 3                    3       -1 0 3 8 8 8 8 8
d 4                x   4 4 2   -1 0 3 4 8 8 8 8 8    link[0] = (2,0)
                                                     link[1] = (1,3)→(0,2)
                                                     link[2] = (4,4)↗
i 5                    5       -1 0 3 4 8 8 8 8
t 6                    6       -1 0 3 4 8 8 8 8
y 7                    7       -1 0 3 4 8 8 8 8
```

Position 0 of *threshold* always equals -1. At the end of iteration i, $threshold[k+1]$ is the position j for which the length of common subsequence for $Q(0 \ldots i)$ and $R(0 \ldots j)$ equals $k+1$.

In the first iteration of the outer `for` loop, a match for the first character in *rapidity*, r, when $i = 0$, is retrieved from $matchlist[0]$, which is the position $j = 2$ of *paradox*, and 2 is assigned to $threshold[0]$, which indicates that there is a common subsequence of length 1 for $Q(0 \ldots 0) = r$ and $R(0 \ldots 2) = par$. The subsequences themselves are stored as a pair of indexes $(0,2)$ on $link[0]$ because list $link[k]$ defines a list of $k + 1$ pairs (i,j) that record a common subsequence of length $k + 1$.

In the second iteration of the outer `for` loop, when $i = 0$, the inner `for` loop is activated twice for the two numbers in $matchlist[i - 1] = (3\ 1)$. First, number $j = 3$ is stored in $threshold[k = 1]$ to indicate that substrings $Q(0 \ldots 1) = ra$ and $R(0 \ldots 3) = para$ have a common subsequence of length $k + 1 = 2$. In the second iteration of the inner `for` loop, number $j = 1$ overwrites 2 in $threshold[0]$ to indicate that substrings $Q(0 \ldots 1) = ra$ and $R(0 \ldots 1) = pa$ have a common subsequence of length 1. The subsequences are recorded in $link$. Note that $link[0]$ is updated but the second node on $link[1]$ remains the same. This node is now accessible only from the first node in list $link[1]$, whereas before the update of $link[0]$, it was also accessible from $link[0]$.

In its last phase, the algorithm prints the (i, j) pairs in reverse order:

[4 4] [1 3] [0 2]

The algorithm runs in $O((|Q| + |R| + r)\lg|R|)$ time and $O(r + |Q| + |R|)$ space. The algorithm is particularly efficient when sequences are far apart; that is, when most positions of one string match only a few positions in the other string, in which case r is small in comparison with the length of the two strings so that the algorithm runs in $O((|Q| + |R|)\lg|R|)$ time. However, in the worst case, for strings $aaa \ldots$ and $aaa \ldots, r = |Q||R|$ and the algorithm runs in unpromising $O(|Q||R|\lg|R|)$ time.

It is worth noticing that the Hunt-Szymanski algorithm is implemented as the *diff* command in UNIX.

13.2.2 String Matching with *k* Errors

Our task is now to determine all substrings of text T for which the Levenshtein distance does not exceed k; that is, we would like to perform string matching with at most k errors (or k differences).

In Section 13.1.5, we discussed the shiftAnd() algorithm for exact matching. The algorithm relied on bitwise operations, and Wu and Manber (1992) devised a generalization of shiftAnd() that can be used for approximate matching. The algorithm is implemented in UNIX as *agrep*, the approximate *grep* command.

Consider the case when the only edit operation is insertion and $k = 2$. For each prefix of P and a substring ending at character T_i, there may be now an exact match, a match with one insertion, and a match with two insertions. To handle all three possibilities, three tables are used—$state_0 = state$ as used in shiftAnd(), $state_1$, and $state_2$, where $state_k$ indicates all matches with up to k insertions. A value in $state_k$ is determined from the corresponding value in $state_{k-1}$ and the characters P_j and T_i being compared.

If there is an exact match between $P(0 \ldots j-1)$ and $T(i-j \ldots i-1)$ and $P_j = T_i$, then the exact match continues for $P(0 \ldots j)$ and $T(i-j \ldots i)$ and this fact has to be reflected in all three state tables, $state_0$, $state_1$, and $state_2$. If there is an exact match between $P(0 \ldots j-1)$ and $T(i-j \ldots i-1)$ and $P_j \neq T_i$, then an approximate match with one insertion of T_i is marked in $state_1$ and $state_2$. In effect, $P_0 P_1 \ldots P_{j-1}$– (note the dash) is approximately matched with $T_{i-j} T_{i-j+1} \ldots T_i$. Finally, if there is an approximate match with one insertion between $P(0 \ldots j-1)$ and $T(i-j \ldots i-1)$ and $P_j \neq T_i$, then an approximate match with two insertions, of T_{i-j+s} for some $0 \leq s \leq j-1$, and of T_i, is marked in $state_2$. This means that $P_0 P_1 \ldots P_s - P_{s+1} \ldots P_{j-1}$– (note two dashes) is approximately matched with $T_{i-j} T_{i-j+1} \ldots T_i$. However, if $P_j = T_i$, then the one-insertion match continues, which has to be reflected in both $state_1$ and $state_2$. Therefore, all the matches indicated in $state_e$ are also found in $state_s$ for $e < s \leq k$; that is, the amount of information grows with the increase of subscript e in $state_e$ because criteria for matching become more and more relaxed. Consider pattern $P = abc$ and text $T = abaccabc$. The situation in table $state_0$ changes as follows:

	i = 0	1	2	3	4	5	6	7
	T = abaccabc	abaccabc	abaccabc	abaccabc	abaccabc	abaccabc	abaccabc	abaccabc
a 0	a			a			a	
b 1		ab						ab
c 2								abc

For $i = 0$, substrings $P(0 \ldots 0)$ and $T(0 \ldots 0)$ are matched, for $i = 1$, a match is continued for substrings $P(0 \ldots 1)$ and $T(0 \ldots 1)$, but for $i = 2$, $P_2 \neq T_2$, so the matching for $P(0 \ldots 1)$ and $T(0 \ldots 1)$ has to be discontinued. But there is a match for $P(0 \ldots 0)$ and $T(2 \ldots 2)$. However, the latter match is also discontinued for $i = 3$. An exact match is found for $i = 7$.

Table $state_1$ is richer with information:

	i = 0	1	2	3	4	5	6	7
	T = abaccabc	abaccabc	abaccabc	abaccabc	abaccabc	abaccabc	abaccabc	abaccabc
a 0	a		a-	a	a-		a	a-
b 1		ab	ab-				ab	ab-
c 2				ab-c				abc

For $i = 0$, as before, substrings $P(0 \ldots 0)$ and $T(0 \ldots 0)$ are matched, for $i = 1$, a match is continued for substrings $P(0 \ldots 1)$ and $T(0 \ldots 1)$, but also a match with one insertion, for $P(0 \ldots 0)$– (one dash) and $T(0 \ldots 1)$. This approximate match cannot be continued for $i = 2$, when $P_2 \neq T_2$, but the approximate match $P(0 \ldots 1)$– (one dash) and $T(0 \ldots 2)$ can, which leads to a successful approximate match of the entire pattern with substring $T(0 \ldots 3)$ in step $i = 3$. Note that the substrings in $state_0$ appear in $state_1$, which leads to reflecting the exact match between P and $T(5 \ldots 7)$ also in $state_1$.

Finally, table $state_2$:

	i = 0	1	2	3	4	5	6	7
	T = abaccabc	abaccabc	abaccabc	abaccabc	abaccabc	abaccabc	abaccabc	abaccabc
a 0	a	a-	a	a-	a--	a	a-	a--
			a--					
b 1		ab	ab-	ab--			ab	ab-
c 2				ab-c	ab--c			abc

Substrings are matched and extended similarly to entries in $state_0$ and $state_1$. Note that for $i = 2$, there are two matches for $P(0 \ldots 0)$: with $T(0 \ldots 2)$, with two insertions, and an exact match with $T(2 \ldots 2)$. For $i = 4$, there is a match with two insertions between P and $T(0 \ldots 4)$, a match that did not appear in the previous tables.

In the tables used thus far, only exact matches and matches with insertions were analyzed. The situation is similar for two other edit operations, deletion and substitution. Moreover, we need to consider a general case for k errors in a match.

All the possibilities for matching $P(0 \ldots j)$ with a substring of T that ends at position i with $e \leq k$ errors can be summarized as follows.

1. *Match.* $P_j = T_i$, and there is a match with e errors between $P(0 \ldots j-1)$ and a substring ending at T_{j-1}.
2. *Substitution.* There is a match with $e - 1$ errors between $P(0 \ldots j-1)$ and a substring ending at T_{j-1}.
3. *Insertion.* There is a match with $e - 1$ errors between $P(0 \ldots j)$ and a substring ending at T_{j-1}.
4. *Deletion.* There is a match with $e - 1$ errors between $P(0 \ldots j-1)$ and a substring ending at T_j.

A $state_e$ can be derived from a preceding $state_{e-1}$ in a remarkably simple manner by generalizing the formula used in implementation of `shiftAnd()`:

$$state_{e,i+1} = 11 \ldots 1100 \ldots 00 \text{ with } e \text{ 1s.}$$

$$state_{e,i+1} = (shiftBits(state_{e,i}) \text{ AND } charactersInP[T_i]) \text{ OR } shiftBits(state_{e-1,i})$$
$$\text{OR } shiftBits(state_{e-1,i+1}) \text{ OR } state_{e-1,i}$$

where AND and OR are bitwise operations. AND is used to shift information out; OR is used to accumulate it. With bitwise-or, information included in the preceding state is also included in the current state. Here is an implementation of the algorithm:

```
WuManber(pattern P, text T, int k)
    matchBit = 1;
    for i = 1 to |P|-1
        matchBit <<= 1;
```

```
initialize charactersInP;
oldState[0] = 0;
for e = 1 to k
    oldState[e] = (oldState[e-1] << 1) | 1;
for i = 0 to |T|-1
    state[0] = ((state[0] << 1) | 1) & charactersInP[T_i];
    for e = 1 to k
        state[e] = ((oldState[e]    << 1) | 1) & charactersInP[T_i] |
                                                               // insertion
                   ((oldState[e-1] << 1) | 1) |   // substitution/match
                   ((state[e-1]    << 1) | 1) |   // deletion
                   oldState[e-1];                 // match
    for e = 0 to k
        oldState[e] = state[e];
    if matchBit & state[k] ≠ 0
        output "a match ending at " i;
```

Creating `charactersInP` takes $O(|P||A|)$ time; the state arrays require k space and k initialization steps. The matching process takes $O(|T|k)$ steps.

The algorithm can be accelerated using the *partition approach* in the case when k is small in comparison to $|P|$. In this case, the pattern P is divided into $k+1$ or $k+2$ blocks, each of the first $k+1$ blocks of size $r = |P|/(k+1)$. If there is a match with up to k errors for P in T, then at least one of the first $k+1$ blocks is matched without any error. Therefore, if one of the blocks is matched exactly, then an approximate match can be found in the neighborhood of size $|P|$ of the exact match.

To locate exact matches of the first $k+1$ blocks, an algorithm can be used that searches for all the blocks at the same time. To that end, the Aho-Corasick algorithm can be used (Baeza-Yates and Perleberg, 1992), but Wu and Manber propose a small modification of algorithm `shiftAnd()`. Consider the following example. For $P = abcdefghi$ and $k = 3$, the pattern is divided into five blocks *ab*, *cd*, *ef*, *gh*, and *i*, out of which only the first four are considered. The four blocks are interleaved to form a new pattern *acegbdfh* to which `WuManber()` is applied with one difference: Instead of shifting by one in each iteration of the main loop, *state* is shifted by four, in which step four 1s are also shifted in. A match is detected if any of the last four bits is 1. Consider text $T = aibcdiefgabb\ldots$ Modified `shiftAnd()` renders the following changes of bits in *state*:

	1	2	3	4	5	6	7	8	9	10		charactersInP							
	a	i	b	c	d	i	e	f	g	a b b		a	b	c	d	e	f	g	h
	1	1	1	1	1	1	1	1	1	1 1 1									
a	0	1	0	0	0	0	0	0	0	1 0 0		1	0	0	0	0	0	0	0
c	0	0	0	0	1	0	0	0	0	0 0 0		0	0	1	0	0	0	0	0
e	0	0	0	0	0	0	0	1	0	0 0 0		0	0	0	0	1	0	0	0
g	0	0	0	0	0	0	0	0	0	1 0 0		0	0	0	0	0	0	1	0
b	0	0	0	0	0	0	0	0	0	0 1 0		0	1	0	0	0	0	0	0
d	0	0	0	0	0	1	0	0	0	0 0 0		0	0	0	1	0	0	0	0
f	0	0	0	0	0	0	0	0	1	0 0 0		0	0	0	0	0	1	0	0
h	0	0	0	0	0	0	0	0	0	0 0 0		0	0	0	0	0	0	0	1

In step 4, when $T_3 = c$, P_1 is also c, which is reflected in setting the bit in row 2 (the third row) to 1. Afterwards, *state* is shifted by four and the result is matched with *charactersInP*$[T_4 =$ 'd'$]$ using bitwise-and, which gives a 1 in row 6, and row 6 is one of the last four rows. This signifies a detection of a block in T, which happens to be the block *cd*, and consequently to a search for an approximate match in its neighborhood. The latter search matches P with $T(0 \ldots 9) = $ *aibcdiefga*. Then, we return to finding another occurrence of a block in T, which occurs at T_7, for block *ef* and then at T_7, for block *ab*.

13.3 CASE STUDY: LONGEST COMMON SUBSTRING

Finding the longest common substring of two strings Q and R is a classical problem in string processing. It was once conjectured that it is impossible to solve the problem in linear time (Knuth, Morris, and Pratt 1977); however, using a suffix tree makes it possible. Therefore, before discussing the problem, an implementation of the Ukkonen algorithm to construct a suffix tree from Section 13.1.8 is introduced.

A node in the suffix tree is implemented as an object that includes an array of references to descendants; the array is indexed with letters of the alphabet from which a text T being processed is built. Also, the node includes arrays for right and left indexes of letters in T to indicate the label of the edge leading to a descendant. For example, for $T = $ *abaabaac* and node 1 in Figure 13.7i, `left['a'-offset] = right['a'-offset] = 3`, which amounts to the label *a*; `left['b'-offset] = 1, right['b'-offset] = 3`, which identifies the label *baa*; and finally, `left['c'-offset] = right['c'-offset] = 7`, which corresponds to the label *c*. Also, for node 1, `descendants['a'-offset] = node 4, descendants['b'-offset] = node 2`, and `descendants['c'-offset] = null`. Using edge labels and nodes from which the edges originate, each edge can be uniquely identified, which is important in a suffix tree, where some nodes may not be explicit. To that end, the notation *node*(explicit node, edge label) = *node*(explicit node, right, left) is used. The notation is called a *canonical reference*, and the explicit node used in it, when it is closest to the implicit node for which the reference is used, is called a *canonical node*. For example, in Figure 13.7h, *node*(node 1, *ba*) = *node*(node1, 1, 2) identifies an implicit node that falls between *ba* and *a* on the edge between node 1 and node 2; *node*(node 1, null string) = *node*(node1, 2, 1) identifies node 1 itself.

As discussed in Section 13.1.8, only the nodes of the equivalent of trie's boundary path have to be updated. The first part of the path includes only leaves; however, there is no need to process the leaves on the boundary path because after processing of a text T, all the leaves are connected to their parents with edges that are suffixes of T, so it can be assumed at the outset that each edge is labeled with such a suffix. Therefore, the edges leading to the leaves do not have to be updated, and thus the first `for` loop in `UkkonenSuffixTree()`, presented in Section 13.1.8, can be eliminated. Moreover, as a space saving device, the leaves can also be eliminated by retaining only the edges leading to them. In this way, the worst case for a trie for T with all different letters when $1 + (1 + 2 + \ldots + |T|)$ nodes are required, turns into a best case tree with one node only, the root.

FIGURE 13.7 (a–h) Creating an Ukkonen suffix tree for the string *abaabaac*; (i) a data structure used for implementation of the Ukkonen tree (h).

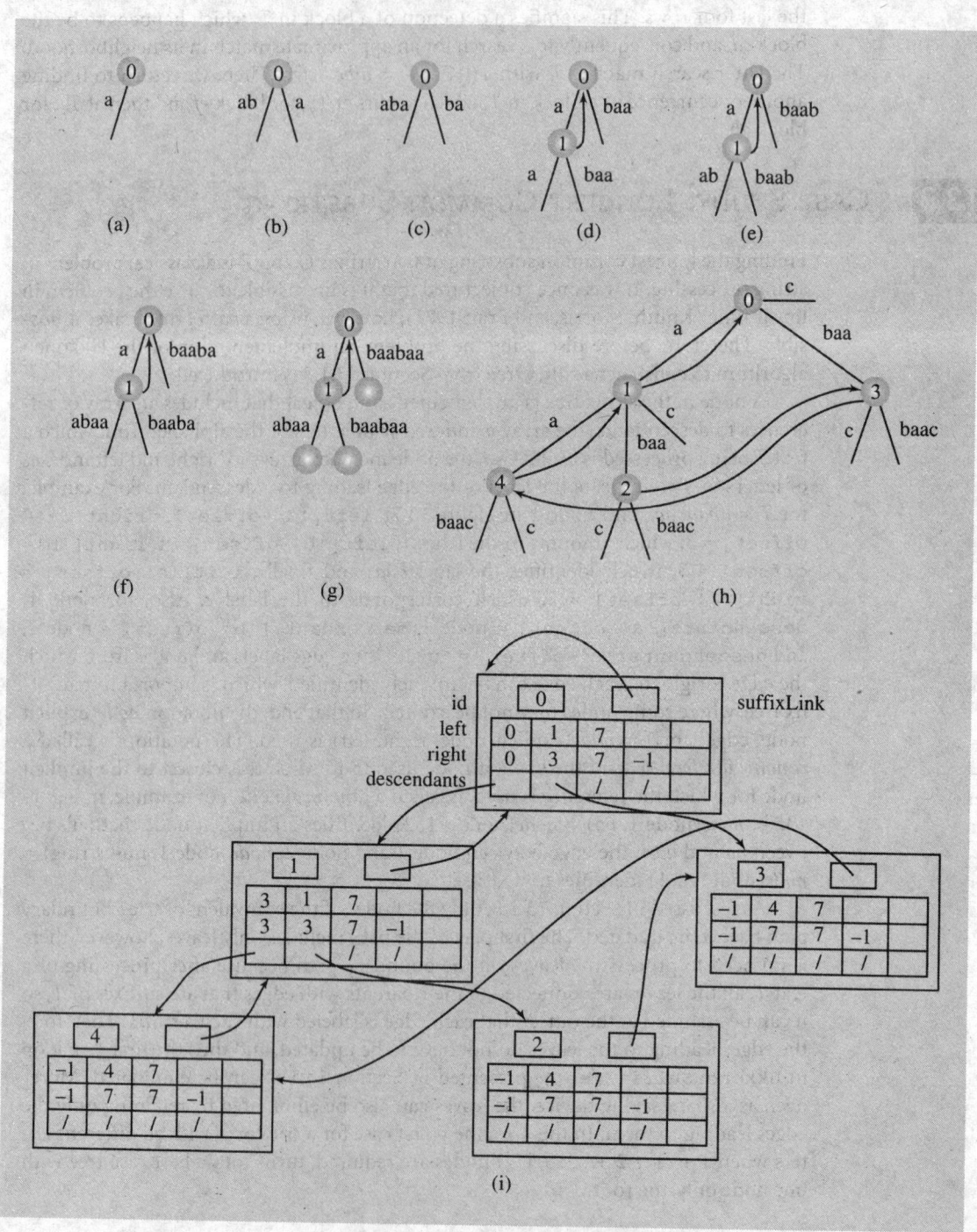

The second part of the boundary path begins with the first nonleaf, the active point, and ends right before the endpoint. Processing of the suffix tree concentrates on these nodes.

To create a new edge for a node, the node, if it is implicit, has to be made explicit first. To make it explicit, an edge from an explicit parent of the node to the node itself has to be split. To split it, the parent, a canonical node, has to be found first. This is the role of the function `findCanonicalNode()` that for an *implicitNode* = *node*(*explicitNode*, left, right) determines whether *explicitNode* is canonical. If it is, the search is finished; if not, the canonical node is found. For example, for the tree in Figure 13.7g and q = *node*(node 0, 5, 6) = *node*(node 0, *aa*), the canonical node is node 1 and q becomes explicit as node 4 and a descendant of node 1. After an implicit node r is made explicit, the edges between r and its parent and between r and its descendant are updated; afterwards, whether it was explicit or implicit, the node acquires a new T_i-edge.

The task of modifying the tree by processing nodes from the active point to the node before the endpoint is performed by the function `update()`. The task of determining whether the current node is an endpoint is performed by `testAndSplit()`.

The processing of a letter T_i begins from the active point. The point is easily determined because it is the endpoint reached after processing letter T_{i-1} is finished. To see this, consider processing letter T_{i-1} in a trie. In the trie's boundary path, each node acquires a new leaf reachable from its parent through a T_{i-1}-edge. Processing ends at the endpoint that already has a T_{i-1}-edge. Therefore, all the added leaves are linked in a new boundary path that includes the endpoint. This endpoint is the first nonleaf in the path, and thus it becomes the active point before processing of the letter T_i begins, and the processing starts from this active point.

By default, each node carries three arrays of 128 cells each, so each text letter would be an index for the arrays. If a range of characters is known, the first and the last characters of the range can be given as arguments to a constructor, whereby the variable `offset` is set to be the first character, and for each text character, the index is found by subtracting the offset.

The code for the suffix tree, which closely follows pseudocode given by Ukkonen (1995), is given in Figure 13.8.

With the suffix tree, a solution of the longest common substring for strings Q and R is now rather simple. First we need to create a suffix tree for the string $T = Q\$R\#$ where $ and # represent characters not used in the two strings. In this tree, no suffix ends in an internal (implicit or explicit) node. A leaf corresponding to Q is a leaf that is connected to its parent with an edge whose label includes $ (and #); a leaf corresponding to R is connected to its parent with an edge whose label includes # (but not $). Now the tree is traversed to find a node that meets two conditions. The node is the root of a subtree with edge labels corresponding to both strings. Moreover, the node should correspond to the longest string obtained by concatenating labels from the root to this node; this concatenated string is the sought longest substring for Q and R.

In the implementation provided in Figure 13.8, the symbols $ and # are the characters that directly follow the range specified by the user, and are automatically attached to the two strings. For example, if the range is from *a* to *z*, and the strings are *abccab* and *daababca*, then the suffix tree is built for the string T = *abccab{daababca|*, because in the ASCII character set, characters '{' and '|' directly follow 'z'.

FIGURE 13.8 Listing of the program to find longest common substring.

```cpp
#include <iostream>
#include <string>

using namespace std;

class SuffixTreeNode {
public:
    SuffixTreeNode **descendants;
    int *left, *right;
    SuffixTreeNode *suffixLink;
    int id;          // for printing only;
    SuffixTreeNode() {
        SuffixTreeNode(128);
    }
    SuffixTreeNode(int sz) {
        id = cnt++;
        descendants = new SuffixTreeNode*[sz];
        suffixLink = 0;
        left  = new int[sz];
        right = new int[sz];
        for (int i = 0; i < sz; i++) {
            descendants[i] = 0;
            left[i] = -1;
        }
    }
private:
    static int cnt; // for printing only;
};

int SuffixTreeNode::cnt;

class UkkonenSuffixTree {
public:
    UkkonenSuffixTree() {
        UkkonenSuffixTree(0,127);
    }
    UkkonenSuffixTree(int from, int to) {
        size = to - from + 1;
        offset = from;
        root = new SuffixTreeNode(size);
        root->suffixLink = root;
    }
```

FIGURE 13.8 *(continued)*

```
    void printTree(int pos) {
        cout << endl;
        printTree(root,0,0,0,pos);
    }
    void createTree(string text) {
        T = text;
        int Lt = 1;
        bool endPoint;
        const int n = T.length(), pos = T[0]-offset;
        SuffixTreeNode *canonicalNodeAP = root, *canonicalNodeEP;
        root->left [pos] = 0;
        root->right[pos] = n-1;
        for (int i = 1; i < n; i++) {
            canonicalNodeEP = update(canonicalNodeAP,i,Lt);
            // and thus, endpoint = node(canonicalNodeEP,Lt,i);
            canonicalNodeAP = findCanonicalNode(canonicalNodeEP,i,Lt);
            // and so, active point = node(canonicalNodeAP,Lt,i);
            printTree(i);
        }
    }
protected:
    SuffixTreeNode *root;
    int size, offset;
    string T;
private:
    void printTree(SuffixTreeNode *p, int lvl, int lt, int rt, int pos) {
        for (int i = 1; i <= lvl; i++)
            cout << "   ";
        if (p != 0) {          // if a nonleaf;
            if (p == root)
                cout << p->id << endl;
            else if (p->suffixLink != 0)    // to print in the middle
                cout << T.substr(lt,lt-rt+1)      // of update;
                     << " " << p->id << " " << p->suffixLink->id
                     << " [" << lt << " " << rt << "]\n";
            else cout << T.substr(lt,pos-lt+1) << " " << p->id;
            for (char i = 0; i < size; i++)
                if (p->left[i] != -1)        // if a tree node;
                    printTree(p->descendants[i],lvl+1,p->left[i],p->right[i],pos);
        }
        else cout << T.substr(lt,pos-lt+1) <<" [" << lt << " " << rt << "]\n";
```

Continues

FIGURE 13.8 (continued)

```
    }
    SuffixTreeNode* testAndSplit(SuffixTreeNode *p, int i, int& Lt, bool&
endPoint) {
        int Rt = i-1;
        if (Lt <= Rt) {
            int pos = T[Lt]-offset;
            SuffixTreeNode *pp = p->descendants[pos];
            int lt = p->left[pos];
            int rt = p->right[pos];
            if (T[i] == T[lt+Rt-Lt+1]) {         // if T(lt...rt) is
                endPoint = true;                 // and extension of
                return p;                        // T(Lt...i);
            }
            else{// insert a new node r between s and ss by splitting
                // edge(p,pp) = T(lt...rt) into
                // edge(p,r)  = T(lt...lt+Rt-Lt) and
                // edge(r,pp) = T(lt+Rt-Lt+1...rt);
                pos = T[lt]-offset;
                SuffixTreeNode *r = p->descendants[pos] = new SuffixTreeNode(size);
                p->right[pos] = lt+Rt-Lt;
                pos = T[lt+Rt-Lt+1]-offset;
                r->descendants[pos] = pp;
                r->left [pos] = lt+Rt-Lt+1;
                r->right[pos] = rt;
                endPoint = false;
                return r;
            }
        }
        else if (p->left[T[i]-offset] == -1)
            endPoint = false;
        else endPoint = true;
        return p;
    }
    SuffixTreeNode* findCanonicalNode(SuffixTreeNode *p, int Rt, int& Lt) {
        if (Rt >= Lt) {
            int pos = T[Lt]-offset;
            SuffixTreeNode *pp = p->descendants[pos];
            int lt = p->left[pos];
            int rt = p->right[pos];
            while (rt - lt <= Rt - Lt) {
                Lt = Lt + rt - lt + 1;
                p = pp;
```

FIGURE 13.8 (continued)

```
                if (Lt <= Rt) {
                    pos = T[Lt]-offset;
                    pp = p->descendants[pos];
                    lt = p->left[pos];
                    rt = p->right[pos];
                        if (p == root)
                          pp = root;
                }
            }
        }
        return p;
    }
    SuffixTreeNode* update(SuffixTreeNode *p, int i, int& Lt) {
      bool endPoint;
      SuffixTreeNode *prev = 0, *r = testAndSplit(p,i,Lt,endPoint);
      while (!endPoint) {
          int pos = T[i]-offset;
          r->left [pos] = i;   // add a T(i)-edge to r;
          r->right[pos] = T.length()-1;
          if (prev != 0)
             prev->suffixLink = r;
          prev = r;
          if (p == root)
             Lt++;
          else p = p->suffixLink;
          p = findCanonicalNode(p,i-1,Lt);
          r = testAndSplit(p,i,Lt,endPoint); // check if not the endpoint;
      }
      if (prev != 0)
          prev->suffixLink = p;
      return p;
      }
};

class LongestCommonSubstring : public UkkonenSuffixTree {
public:
    LongestCommonSubstring(int from, int to) : UkkonenSuffixTree(from,to+2) {
    }
    void run(string s1, string s2) {
         createTree(s1 + char(size+offset-2) + s2 + char(size+offset-1));
         findLongest(s1,s2);
    }
```

Continues

FIGURE 13.8 (continued)

```
private:
    int sllength, position, length;
    void findLongest(string s1, string s2) {
      bool dummy[] = {false, false};
      position = length = 0;
      sllength = s1.length();
      traverseTree(root,0,0,dummy);
      if (length == 0)
           cout << "Strings \"" << s1 << "\" and \"" << s2
                << "\" have no common substring\n";
      else cout << "A longest common substring for \""
                << s1 << "\" and \"" << s2 << "\" is " << "\""
                << T.substr(position-length,length) << "\" of length "
                << length << endl;
    }
    void traverseTree(SuffixTreeNode *p, int lt, int len, bool *whichEdges) {
      bool edges[] = {false, false};
      for (char i = 0; i < size; i++)
          if (p->left[i] != -1) {
              if (p->descendants[i] == 0)   // if it is an edge to
                  if (p->left[i] <= sllength)      // a leaf corresponding
                      whichEdges[0] = edges[0] = true; // to s1
                  else whichEdges[1] = edges[1] = true; // to s2
              else {
                  traverseTree(p->descendants[i],p->left[i],
                               len+(p->right[i]-p->left[i]+1),edges);
                  if (edges[0])
                      whichEdges[0] = true;
                    if (edges[1])
                       whichEdges[1] = true;
              }
              if (edges[0] && edges[1] && len > length) {
                  position = p->left[i];
                  length = len;
              }
          }
    }
};
```

FIGURE 13.8 (continued)

```
int main(int argc, string argv[]) {
    string s1 = "abcabc";
    string s2 = "cabaca";
    if (argc == 3) {
        s1 = argv[1];
        s2 = argv[2];
    }
    (new LongestCommonSubstring('a','z'))->run(s1,s2);
    return 0;
}
```

To learn during a tree traversal whether a particular subtree contains edges corresponding to both strings (only edges that lead to leaves can provide this information), a two-cell Boolean array is associated with each node. When an edge is detected that leads to a leaf (leaves are implicit nodes), then the left index of its label is tested. If the index is not greater than the length of Q, then the leaf corresponds to a suffix of Q; otherwise, to a suffix of R. The program maintains the maximum length of the common substring, and when it detects a node with a longer common substring and with suffixes of both strings in its subtree, the maximum length is updated, as is the position at which the substring ends.

It is clear that because the suffix tree can be created in linear time and because tree traversal can also be done in linear time, the problem of finding the longest common substrings for strings Q and R is solved in linear time $O(|Q| + |R|)$.

13.4 EXERCISES

1. Apply the Knuth-Morris-Pratt algorithm first with *next* and then with *nextS* to $P = $ *bacbaaa* and $T = $ *bacbacabcbacbbbacabacbabcbbba*.

2. Determine all three positions *i, j, k*, such that for the string *abcabdabcabdfabcabdab-cabd* `findNextS()` executes first $nextS[i] = nextS[j]$ and then $nextS[j] = nextS[k]$.

3. As mentioned in Section 13.3, $P = a^{m-1}b$ is an example of the worst case for the partial suffix phase for `computeDelta2ByBruteForce()`. What exactly is the total number of comparisons for this phase?

4. Consider the search for a pattern in a text when the pattern is not in the text. What is the smallest number of character comparisons in this case executed by

    a. Knuth-Morris-Pratt?

    b. Boyer-Moore?

5. An example of the worst case for `bruteForceStringMatching()` are strings $P = a^{m-1}b$ and $T = a^n$, and for `BoyerMooreSimple()`, strings $P = ba^{m-1}$ and $T = a^n$. Explain this symmetry.

6. As it stands, `BoyerMooreSimple()` shifts *P* by one position if a mismatched text character occurs also in *P* to the right of the mismatched pattern characters, for example,

    ```
         abbaabac...
    1    aabbcbac
    2     aabbcbac
    ```

    where the mismatched text character *a* occurs also in *P* to the right of the mismatched pattern character *c*. However, it is clear that it would be more efficient to align the mismatched text character with the same character in *P* that is closest to the left of the mismatched pattern character, as in

    ```
         abbaabac...
    1    aabbcbac
    2        aabbcbac
    ```

    where mismatched text character *a* is aligned with an *a* to the right of the mismatched pattern character *c* and closest to *c*. Generalize this rule and propose an implementation of *delta1* for the new rule.

7. Horspool gives a version of the Boyer-Moore algorithm that uses only one table, *delta12*, which is just like *delta1* except that for the last character of *P*, the entry in *delta12* is $|P|$, not a value $< |P|$ as in *delta1*:

    ```
    BoyerMooreHorspool(pattern P, text T)
        initialize all cells of delta12 to |P|;
        for j = 0 to |P|-2  // |P|-2, not |P|-1 as for delta1;
            delta12[P_j] = |P| - j - 1;
        i = |P| - 1;
    ```

```
while i < |T|
    j = |T| - 1;
    if T_i == P_{|P|-1}
        if T(i-|P|+1...i) is equal to P
            return match at i+|P|+1;
    i = i + delta12[T_i];
return no match;
```

Apply `BoyerMooreHorspool()` and `BoyerMooreSimple()` to $T = abababab\text{-}bababba$ and $P = aacaab$.

8. Implement the function `period()` to be used by the `BoyerMooreGalil()` algorithm.

9. `BoyerMooreGalil()` may be less efficient than `BoyerMoore()` because for patterns with no periods it checks the condition in an `if` statement in each iteration of the outer `while` loop. Change the algorithm so that a driver function preprocesses the pattern to check for a period, and if a period is found, it calls `BoyerMoore()`; otherwise, it calls `BoyerMooreGalil()` without the `if` statement.

10. Adopt `quickSearch()` so that it performs matching right to left.

11. Show an example for the case when `BoyerMooreSimple()` provides better shift than Sunday's `quickSearch()`.

12. The `shiftAnd()` algorithm performs four bitwise operations in each iteration of the last `for` loop. The number can be reduced to three if the roles of bits are reversed, as it is done originally by Baeza-Yates and Gonnet (1992); for example, in *charactersInP*, 0 represents the positions at which a character occurs in the pattern. With this bit role reversal, the assignment

    ```
    state = ((state << 1) | 1) & charactersInP[T[i]];
    ```

    can be changed to

    ```
    state = (state << 1) | charactersInP[T[i]];
    ```

    Write an algorithm `shiftOr()` by making all the necessary changes in `shiftAnd()` including the indicated modifications.

13. What is the maximum number of keywords in a set *output*(*state*) for some *state*?

14. Draw Ukkonen suffix tries for

    a. *aaaa*
    b. *aabb*
    c. *abba*
    d. *abcd*
    e. *baaa*
    f. *abaa*
    g. *aaba*
    h. *aaab*

15. How can the number of occurrences of pattern P in text T be determined using a suffix tree for T?

16. How can a suffix tree be used to determine all substrings of Q that are not substrings of R?

17. The problem of string matching with k differences can be solved by simple modification of the Wagner-Fischer algorithm. This is accomplished by having the entries in the edit distance table represent the minimum distance between prefix $Q(0 \ldots i)$ and any suffix of $R(0 \ldots j)$ (Sellers, 1980). This is done by defining the entries in the matrix with the same recurrence relation as $D(i,j) = \min(D(i-1,j) + 1, D(i,j-1) + 1, D(i-1,j-1) + c(i,j))$, the same boundary condition for column 0: $D(i,0) = i$, but a different condition for row 0: $D(0,j) = 0$ to indicate that an occurrence can start at any position of R. In the resulting edit table, any number in the last row that is not greater than k indicates a position in R of the end of a substring of R that has at most k differences with Q. Build such an edit table for strings $Q = abcabb$ and $R = acbdcbbcdd$.

13.5 PROGRAMMING ASSIGNMENTS

1. Write a program that implements and tests the brute force algorithm to create a suffix tree.

2. Extend the program from the case study so that it can find all substrings of length greater than k common to both strings.

3. Write a program that uses a suffix tree to find the longest repeated substring in a string s. After creating the tree, perform a tree traversal and find a node that has only leaf descendants and a longest substring determined by the path from the root to this node. Consider extending your program in one of three ways:
   a. Find only nonoverlapping substrings.
   b. Find the longest substring repeated at least k times.
   c. Find all repeated substrings longer than m characters.

4. One constructor of the UkkonenSuffixTree class allows for using a range of characters to save space for the three arrays used in each node. Sometimes, however, only a few nonconsecutive characters are used in the text, for example, letters A, C, G, and T in DNA sequences. In this situation, the constructor UkkonenSuffixTree('A','T') can be used, which would create three arrays of 'T'-'A' +1 = 20 cells each, although only four cells would be used in each array. Modify the program so that it accepts a set of characters to be used—as with the string "ACGT" in our example—and operates on arrays of the size equal to the number of characters used in the set.

BIBLIOGRAPHY

Aho, Alfred V., and Corasick, Margaret J., "Efficient String Matching: An Aid to Bibliographic Search," *Communications of the ACM* 18 (1975), 333–340.

Baeza-Yates, Ricardo A., and Perleberg, Chris H., "Fast and Practical Approximate String Matching," in A. Apostolico, M. Crochemore, Z. Galil, U. Manber (eds.), *Combinatorial Pattern Matching*, Berlin: Springer (1992), 185–192.

Baeza-Yates, Ricardo, and Gonnet, Gaston H., "A New Approach to Text Searching," *Communications of the ACM* 35 (1992), No. 10, 74–82.

Barth, Gerhard, "Relating the Average-Case Costs of the Brute-Force and Knuth-Morris-Pratt Matching Algorithm," in A. Apostolico and Z. Galil (eds.), *Combinatorial Algorithms on Words*, Berlin: Springer (1985), 45–58.

Boyer, Robert S., and Moore, J. Strother, "A Fast Searching Algorithm," *Communications of the ACM* 20 (1977), 762–772.

Cole, Richard, "Right Bounds on the Complexity of the Boyer-Moore String Matching Algorithm," *SIAM Journal on Computing* 23 (1994), 1075–1091.

Drozdek, Adam, "Hirschberg's Algorithm for Approximate Matching," *Computer Science* 4 (2002), 91–100.

Galil, Zvi, "On Improving the Worst Case Running Time of the Boyer-Moore String Matching," *Communications of the ACM* 22 (1979), 505–508.

Guibas, Leo J., and Odlyzko, Andrew M., "A New Proof of the Linearity of the Boyer-Moore String Searching Algorithm," *SIAM Journal on Computing* 9 (1980), 672–682.

Hancart, Christophe, "Un Analyse en Moyenne de l'Algorithm de Morris et Pratt et ses Raffinements," in D. Krob (ed.), *Actes des Deuxièmes Journées Franco-Belges*, Rouen: Université de Rouen (1992), 99–110.

Horspool, R. Nigel, "Practical Fast Searching in Strings," *Software—Practice and Experience* 10 (1980), 501–506.

Hunt, James W., and Szymanski, Thomas G., "A Fast Algorithm for Computing Longest Common Subsequences," *Communications of the ACM* 20 (1977), 350–353.

Knuth, Donald E., Morris, James H., and Pratt, Vaughan R., "Fast Pattern Matching in Strings," *SIAM Journal on Computing* 6 (1977), 323–350.

Levenshtein, V. I., "Binary Codes Capable of Correcting Deletions, Insertions, and Reversals," *Cybernetics and Control Theory* 10 (1966), 707–710, translation of a paper from *Doklady Akademii Nauk SSSR* 163 (1965), 845–848.

Manber, Udi, and Myers, Gene, "Suffix Arrays: A New Method for On-line String Searches," *SIAM Journal on Computing* 22 (1993), 935–948.

Pirklbauer, Klaus, "A Study of Pattern-Matching Algorithms," *Structured Programming* 13 (1992), 89–98.

Sellers, Peter H., "The Theory and Computation of Evolutionary Distances: Pattern Recognition," *Journal of Algorithms* 1 (1980), 359–373.

Smith, P. D., "Experiments with a Very Fast Substring Search Algorithm," *Software—Practice and Experience* 21 (1991), 1065–1074.

Sunday, Daniel M., "A Very Fast Substring Searching Algorithm," *Communications of the ACM* 33 (1990), 132–142.

Thompson, Ken, "Regular Expression Search Algorithm," *Communications of the ACM* 6 (1968), 419–422.

Ukkonen, Esko, "On-line Construction of Suffix Trees," *Algoritmica* 14 (1995), 249–260.

Ukkonen, Esko, and Wood, Derick, "Approximate String Matching with Suffix Automata," *Algoritmica* 10 (1993), 353–364.

Wagner, Richard A., and Fischer, M. J., "The String-to-String Correction Problem," *Journal of the ACM* 21 (1974), 168–173.

Wu, Sun, and Manber, Udi, "Fast Text Searching Allowing Errors," *Communications of the ACM* 35 (1992), No. 10, 83–91.

Computing Big-O

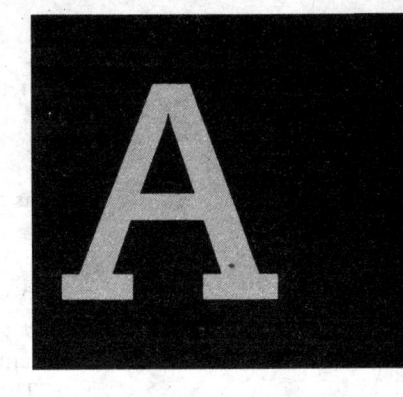

A.1 Harmonic Series

In some computations in this book, the convention H_n is used for harmonic numbers. The *harmonic numbers* H_n are defined as the sums of the *harmonic series*, a series of the form $\sum_{i=1}^{n} \frac{1}{i}$. This is a very important series for the analysis of searching and sorting algorithms. It is proved that

$$H_n = \ln n + \gamma + \frac{1}{2n} - \frac{1}{12n^2} + \frac{1}{120n^4} - \epsilon$$

where $n \geq 1$, $0 < \epsilon < \frac{1}{256n^6}$, and *Euler's constant* $\gamma \approx 0.5772$. This approximation, however, is very unwieldy and, in the context of our analyses, not necessary in this form. H_n's largest term is almost always $\ln n$, the only increasing term in H_n. Thus, H_n can be referred to as $O(\ln n)$.

A.2 Approximation of the Function lg(n!)

The roughest approximation of $\lg(n!)$ can be obtained by observing that each number in the product $n! = 1 \cdot 2 \cdot \cdots \cdot (n-1) \cdot n$ is less than or equal to n. Thus, $n! \leq n^n$ (only for $n = 1, n = n^n$), which implies that $\lg(n!) < \lg(n^n) = n \lg n$—that is, $n \lg n$ is an upper bound of $\lg(n!)$—and that $\lg(n!)$ is $O(n \lg n)$.

Let us also find a lower bound for $\lg(n!)$. If the elements of the product $n!$ are grouped appropriately, as in

$$P_{n!} = (1 \cdot n)(2 \cdot (n-1))(3 \cdot (n-2)) \cdots (i \cdot (n-i+1)) \ldots, \text{ for } 1 \leq i \leq \frac{n}{2}$$

then it can be noted that there are $\frac{n}{2}$ such terms and $n! = P_{n!}$ for even ns or $\frac{n+1}{2}$ terms and $n! = P_{n!} \frac{n+1}{2}$ for odd ns. We claim that each term of $P_{n!}$ is not less than n, or

$$1 \leq i \leq \frac{n}{2} \Rightarrow i(n-i+1) \geq n$$

In fact, this holds because

$$\frac{n}{2} \geq i = \frac{i(i-1)}{i-1} \Rightarrow i(n-2i+2) \geq n$$

and, as can easily be checked,

$$i \geq 1 \Rightarrow (n-2i+2) \leq (n-i+1)$$

We have shown that $n! = P_{n!} \geq n^{\frac{n}{2}}$, which means that $\lg(n!) \geq \frac{n}{2} \lg n$. This assumes that n is even. If n is odd, it has to be raised to the power of $\frac{n+1}{2}$, which introduces no substantial change.

The number $\lg(n!)$ has been estimated using the lower and upper bounds of this function, and the result is $\frac{n}{2} \lg n \leq \lg(n!) \leq n \lg n$. To approximate $\lg(n!)$, lower and upper bounds have been used that both grow at the rate of $n \lg n$. This implies that $\lg(n!)$ grows at the same rate as $n \lg n$ or that $\lg(n!)$ is not only $O(n \lg n)$, but also $\Theta(n \lg n)$. In other words, any sorting algorithm using comparisons on an array of size n must make at least $O(n \lg n)$ comparisons in the worst case. Thus, the function $n \lg n$ approximates the optimal number of comparisons in the worst case.

However, this result seems unsatisfactory because it refers only to the worst case, and such a case occurs only occasionally. Most of the time, average cases with random orderings of data occur. Is the number of comparisons really better in such cases, and is it a reasonable assumption that the number of comparisons in the average case can be better than $O(n \lg n)$? Unfortunately, this conjecture has to be rejected, and the following computations prove it false.

Our conjecture is that, in any binary tree with m leaves and two children for each nonterminal node, the average number of arcs leading from the root to a leaf is greater than or equal to $\lg m$.

For $m = 2$, $\lg m = 1$, if there is just a root with two leaves, then there is only one arc to every one of them. Assume that the proposition holds for a certain $m \geq 2$ and that

$$\text{Ave}_m = \frac{p_1 + \cdots + p_m}{m} \geq \lg m$$

where each p_i is a path (the number of arcs) from the root to node i. Now consider a randomly chosen leaf with two children about to be attached. This leaf converted to a nonterminal node has an index m (this index is chosen to simplify the notation) and a path from the root to the node m is p_m. After adding two new leaves, the total number of leaves is incremented by one and the path for both these appended leaves is $p_{m+1} = p_m + 1$. Is it now true that

$$\text{Ave}_{m+1} = \frac{p_1 + \cdots + p_{m-1} + 2p_m + 2}{m+1} \geq \lg(m+1)$$

From the definition of Ave_m and Ave_{m+1} and from the fact that $p_m = \text{Ave}_m$ (because leaf m was chosen randomly),

$$(m+1)\text{Ave}_{m+1} = m\text{Ave}_m + p_m + 2 = (m+1)\text{Ave}_m + 2$$

Is it now true that

$$(m+1)\text{Ave}_{m+1} \geq (m+1)\lg(m+1)$$

or

$$(m+1)\text{Ave}_{m+1} = (m+1)\text{Ave}_m + 2 \geq (m+1)\lg m + 2 \geq (m+1)\lg(m+1)$$

This is transformed into

$$2 \geq \lg\left(\frac{m+1}{m}\right)^{m+1} = \lg\left(1+\frac{1}{m}\right) + \lg\left(1+\frac{1}{m}\right)^m \to \lg 1 + \lg e = \lg e \approx 1.44$$

which is true for any $m \geq 1$. This completes the proof of the conjecture.

This proves that for a randomly chosen leaf of an m-leaf decision tree, the reasonable expectation is that the path from the root to the leaf is not less than $\lg m$. The number of leaves in such a tree is not less than $n!$, which is the number of all possible orderings of an n-element array. If $m \geq n!$, then $\lg m \geq \lg(n!)$. That is the unfortunate result indicating that an average case also requires, like the worst case, $\lg(n!)$ comparisons (length of path = number of comparisons), and as already estimated, $\lg(n!)$ is $O(n \lg n)$. This is also the best that can be expected in average cases.

A.3 BIG-O FOR AVERAGE CASE OF QUICKSORT

Let $C(n)$ be the number of comparisons required to sort an array of n cells. Because the arrays of size 1 and 0 are not partitioned, $C(0) = C(1) = 0$. Assuming a random ordering of an n-element array, any element can be chosen as the bound; the probability that any element will become the bound is the same for all elements. With $C(i-1)$ and $C(n-i)$ denoting the numbers of the comparisons required to sort the two subarrays, there are

$$C(n) = n - 1 + \frac{1}{n}\sum_{i=1}^{n}(C(i-1) + C(n-i)), \text{ for } n \geq 2$$

comparisons, where $n-1$ is the number of comparisons in the partition of the array of size n. First, some simplification can be done:

$$C(n) = n - 1 + \frac{1}{n}\left(\sum_{i=1}^{n}C(i-1) + \sum_{i=1}^{n}C(n-i)\right)$$

$$= n - 1 + \frac{1}{n}\left(\sum_{i=1}^{n}C(i-1) + \sum_{j=1}^{n}C(j-1)\right)$$

$$= n - 1 + \frac{2}{n}\sum_{i=0}^{n-1}C(i)$$

or

$$nC(n) = n(n-1) = 2\sum_{i=0}^{n-1}C(i)$$

To solve the equation, the summation operator is removed first. To that end, the last equation is subtracted from an equation obtained from it,

$$(n+1)C(n+1) = (n+1)n + 2\sum_{i=0}^{n}C(i)$$

resulting in

$$(n+1)C(n+1) - nC(n) = (n+1)n - n(n-1) + 2\left(\sum_{i=0}^{n} C(i) - \sum_{i=0}^{n-1} C(i)\right) = 2C(n) + 2n$$

from which

$$\frac{C(n+1)}{n+2} = \frac{C(n)}{n+1} + \frac{2n}{(n+1)(n+2)} = \frac{C(n)}{n+1} + \frac{4}{n+2} - \frac{2}{n+1}$$

This equation can be expanded, which gives

$$\frac{C(2)}{3} = \frac{C(1)}{2} + \frac{4}{3} - \frac{2}{2} = \frac{4}{3} - \frac{2}{2}$$

$$\frac{C(3)}{4} = \frac{C(2)}{3} + \frac{4}{4} - \frac{2}{3}$$

$$\frac{C(4)}{5} = \frac{C(3)}{4} + \frac{4}{5} - \frac{2}{4}$$

$$\vdots$$

$$\frac{C(n)}{n+1} = \frac{C(n-1)}{n} + \frac{4}{n+1} - \frac{2}{n}$$

$$\frac{C(n+1)}{n+2} = \frac{C(n)}{n+1} + \frac{4}{n+2} - \frac{2}{n+1}$$

from which

$$\frac{C(n+1)}{n+2} = \left(\frac{4}{3} - \frac{2}{2}\right) + \left(\frac{4}{4} - \frac{2}{3}\right) + \left(\frac{4}{5} - \frac{2}{4}\right) + \cdots + \left(\frac{4}{n+1} - \frac{2}{n}\right)$$

$$+ \left(\frac{4}{n+2} - \frac{2}{n+1}\right)$$

$$= -\frac{2}{2} + \frac{2}{3} + \frac{2}{4} + \frac{2}{5} + \cdots + \frac{2}{n} + \frac{2}{n+1} + \frac{4}{n+2}$$

$$= -4 + 2H_{n+2} + \frac{2}{n+2}$$

Note that H_{n+2} is a harmonic number. Using an approximation for this number (see Appendix A.1)

$$C(n) = (n+1)\left(-4 + 2H_{n+1} + \frac{2}{n+1}\right)$$

$$= (n+1)\left(-4 + 2O(\ln n) + \frac{2}{n+1}\right)$$

$$= O(n \lg n)$$

A.4 AVERAGE PATH LENGTH IN A RANDOM BINARY TREE

In Chapter 6, an approximation is used for the average path length in a randomly created binary search tree. Assuming that

$$P_n(i) = \frac{(i-1)(P_{i-1}+1) + (n-i)(P_{n-i}+1)}{n}$$

this approximation is given by this recurrence relation

$$P_1 = 0$$

$$P_n = \frac{1}{n}\sum_{i=1}^{n} P_n(i) = \frac{1}{n^2}\sum_{i=1}^{n}((i-1)(P_{i-1}+1) + (n-i)(P_{n-i}+1))$$

$$P_n = \frac{2}{n^2}\sum_{i=1}^{n-1} i(P_i + 1) \tag{1}$$

From this, we also have

$$P_{n-1} = \frac{2}{(n-1)^2}\sum_{i=1}^{n-2} i(P_i + 1) \tag{2}$$

After multiplying this equation by $\frac{(n-1)^2}{n^2}$ and subtracting the resulting equation from (1), we have

$$P_n = P_{n-1}\frac{(n-1)^2}{n^2} + \frac{2(n-1)(P_{n-1}+1)}{n^2} = \frac{(n-1)}{n^2}((n+1)P_{n-1}+2)$$

After successive applications of this formula to each P_{n-i}, we have

$$P_n = \frac{n-1}{n^2}\left((n+1)\frac{(n-2)}{(n-1)^2}\left(n\frac{(n-3)}{(n-2)^2}\left((n-1)\frac{(n-4)}{(n-3)^2}\left(\cdots\frac{1}{2^2}(3P_1+2)\cdots\right)+2\right)+2\right)+2\right)$$

$$P_n = 2\left(\frac{n-1}{n^2} + \frac{(n+1)(n-2)}{(n-1)n^2} + \frac{(n+1)(n-3)}{n(n-1)(n-2)} + \frac{(n+1)(n-4)}{n(n-2)(n-3)} + \cdots + \frac{n+1}{2 \cdot 3n}\right)$$

$$P_n = 2\left(\frac{n+1}{n}\right)\sum_{i=1}^{n-1}\frac{n-i}{(n-i+1)(n-i+2)} = 2\left(\frac{n+1}{n}\right)\sum_{i=1}^{n-1}\left(\frac{2}{n-i+2} - \frac{1}{n-i+1}\right)$$

$$P_n = 2\left(\frac{n+1}{n}\right)\frac{2}{n+1} + 2\left(\frac{n+1}{n}\right)\left(\sum_{i=1}^{n}\frac{1}{i} - 2\right) = 2\left(\frac{n+1}{n}\right)H_n - 4$$

So, P_n is $O(2 \ln n)$.

A.5 The Number of Nodes in an AVL Tree

The minimum number of nodes in an AVL tree is determined by the recurrence equation

$$AVL_h = AVL_{h-1} + AVL_{h-2} + 1$$

with $AVL_0 = 0$ and $AVL_1 = 1$. A comparison of this equation with the definition of the Fibonacci sequence (Section 5.8) indicates that $AVL_h = F_{h+2} - 1$; that is, using the de Moivre formula

$$F_h = \frac{1}{\sqrt{5}}\left(\frac{1+\sqrt{5}}{2}\right)^h - \frac{1}{\sqrt{5}}\left(\frac{1-\sqrt{5}}{2}\right)^h$$

we obtain

$$AVL_h = \frac{1}{\sqrt{5}}\left(\frac{1+\sqrt{5}}{2}\right)^{h+2} - \frac{1}{\sqrt{5}}\left(\frac{1-\sqrt{5}}{2}\right)^{h+2} - 1$$

Because $\left|\frac{1}{2}(1-\sqrt{5})\right| \approx 0.618034$, the second term in this equation quickly decreases with the increase of h and has the maximum value 0.17082 for $h = 0$; therefore,

$$AVL_h \geq \frac{1}{\sqrt{5}}\left(\frac{1+\sqrt{5}}{2}\right)^{h+2} - 0.17082 - 1 \geq \frac{1}{\sqrt{5}}\left(\frac{1+\sqrt{5}}{2}\right)^{h+2} - 2$$

or

$$AVL_h + 2 \geq \frac{1}{\sqrt{5}}\left(\frac{1+\sqrt{5}}{2}\right)^{h+2}$$

Taking lg of both sides renders

$$\lg(AVL_h + 2) \geq \lg\frac{1}{\sqrt{5}} + 2\lg\left(\frac{1+\sqrt{5}}{2}\right) + h\lg\left(\frac{1+\sqrt{5}}{2}\right) \approx 0.22787 + 0.69424h$$

from which we obtain an upper bound on h

$$h \leq 1.44042 \lg(AVL_h + 2) - 0.32824 \leq 1.44042 \lg(AVL_h + 2)$$

and thus

$$\lg(AVL_h + 1) \leq h < 1.44042 \lg(AVL_h + 2) - 0.32824.$$

Algorithms in the Standard Template Library

B.1 Standard Algorithms

To access these algorithms, the program has to include the statement

```
#include <algorithm>
```

In the two tables that follow, a phrase "elements in the range [`first`, `last`)" is an abbreviation of "elements indicated by iterators in the range [`first`, `last`)" or "elements referenced by iterators from `first` up to, but not including, `last`."

Member Function	Operation
`iterator adjacent_find (first, last)`	Find the first pair of duplicates in the range [`first`, `last`) and return an iterator indicating the position of the first duplicate element; return `last` if no duplicate is found.
`iterator adjacent_find (first, last, f())`	Find the first pair of duplicates in the range [`first`, `last`) and return an iterator indicating the position of the first duplicate element; use a two-argument Boolean function `f()` to compare elements; return `last` if no duplicate is found.
`bool binary_search(first, last, value)`	Return true if binary search locates `value` in the range [`first`, `last`), and false otherwise.
`bool binary_search(first, last, value, f())`	Return true if binary search locates `value` in the range [`first`, `last`) using a two-argument Boolean function `f()` to compare elements, and false otherwise.
`iterator copy(first, last, result)`	Copy all the elements in the range [`first`, `last`) to `result` and return an iterator indicating the end of the range of copied elements.
`iterator copy_backward(first, last, result)`	Copy all the elements in the range [`first`, `last`) to the range whose end is indicated by `result` and return an iterator indicating the beginning of the range.

`size_type count(first, last, value)`	Return the number of elements equal to `value` in the range [`first, last`).
`size_type count_if(first, last, f())`	Return the number of elements that return true for a one-argument Boolean function `f()` in the range [`first, last`).
`bool equal(first1, last1, first2)`	Compare range [`first1, last1`) and the range of the same length that starts in the position indicated by the iterator `first2` and return true if the ranges contain the same elements and false otherwise.
`bool equal(first1, last1, first2, f())`	Compare range [`first1, last1`) and the range of the same length that starts in the position indicated by the iterator `first2` and return true if the ranges contain similar elements where similarity is determined by a two-argument Boolean function `f()`; return false otherwise.
`pair<iterator, iterator> equal_range(first, last, value)`	Return a pair of iterators indicating the subrange of the range [`first, last`) of elements in the ascending order in which all elements are equal to `value`; if no such subrange is found, the returned pair contains two iterators equal to `first`.
`pair<iterator, iterator> equal_range(first, last, value, f())`	Return a pair of iterators indicating the subrange of the range [`first, last`) in the order determined by a two-argument Boolean function `f()` in which all elements are equal to `value`; if no such subrange is found, the return pair contains two iterators equal to `first`.
`void fill(first, last, value)`	Assign `value` to all the elements in the range [`first, last`).
`void fill_n(first, n, value)`	Assign `value` to all the elements in the range [`first, first+n`).
`iterator find(first, last, value)`	Return an iterator to the first occurrence of `value` in the range [`first, last`); return `last` if such an occurrence is not found.
`iterator find_end(first1, last1, first2, last2)`	Return the last iterator in the range [`first1, last1`) that indicates the beginning of a subrange of elements equal to the elements in the range [`first2, last2`); if not found, return `last1`.
`iterator find_end(first1, last1, first2, last2, f())`	Return the last iterator in the range [`first1, last1`) that indicates the beginning of a subrange of elements that are in the relation `f()` to the elements in the range [`first2, last2`); if not found, return `last1`.
`iterator find_first_of(first1, last1, first2, last2)`	Return the position in the range [`first1, last1`) of an element that is also present in the range [`first2, last2`); if not found, return `last1`.
`iterator find_first_of(first1, last1, first2, last2, f())`	Return the position in the range [`first1, last1`) of an element that is in the relation `f()` to an element in the range [`first2, last2`); if not found, return `last1`.
`iterator find_if(first, last, f())`	Return an iterator to the first occurrence of an element in the range [`first, last`) for which a one-argument Boolean function `f()` returns true; return `last` if such an occurrence is not found.
`function for_each(first, last, f())`	Apply the function `f()` to all the elements in the range [`first, last`) and return this function.
`void generate(first, last, f())`	Fill the elements in the range [`first, last`) with the successive values generated by the function `f()` that takes no arguments.

`void generate_n(first, n, f())`	Fill the elements in the range [`first`, `first+n`) with the successive values generated by the function `f()` that takes no arguments.
`bool includes(first1, last1, first2, last2)`	Return true if the elements in the ordered range [`first1`, `last1`) are included in the ordered range [`first2`, `last2`); return false otherwise; both ranges are in ascending order.
`bool includes(first1, last1, first2, last2, f())`	Return true if the elements in the ordered range [`first1`, `last1`) are included in the ordered range [`first2`, `last2`); return false otherwise; both ranges are in order by the relation `f()`.
`void inplace_merge(first, middle, last)`	Put at `first` the result of merging the ranges [`first`, `middle`) and [`middle`, `last`); the ranges are in ascending order.
`void inplace_merge(first, middle, last, f())`	Put at `first` the result of merging the ranges [`first`, `middle`) and [`middle`, `last`); the ranges are ordered with a relation `f()`.
`void iter_swap(i1, i2)`	Swap elements `*i1` and `*i2`.
`bool lexicographical_compare (first1, last1, first2, last2)`	Return true only if the range [`first1`, `last1`) is lexicographically less than the range [`first2`, `last2`).
`bool lexicographical_compare (first1, last1, first2, last2, f())`	Return true only if the range [`first1`, `last1`) is lexicographically less than the range [`first2`, `last2`) with respect to the relation `f()`.
`iterator lower_bound(first, last, value)`	Return an iterator that references the lowest position in the range [`first`, `last`) that is in ascending order before which `value` can be inserted without violating the order; return `last` if `value` is larger than all elements.
`iterator lower_bound(first, last, value, f())`	Return an iterator that references the lowest position in the range [`first`, `last`) that is in order by the relation `f()` before which `value` can be inserted without violating the order; return `last` if `value` succeeds all elements.
`void make_heap(first, last)`	Rearrange the elements in the range [`first`, `last`) to form a heap.
`void make_heap(first, last, f())`	Rearrange the elements in the range [`first`, `last`) to form a heap; use the relation `f()` to compare elements.
`const T& max(x, y)`	Return the maximum of the elements `x` and `y`.
`const T& max(x, y, f())`	Return the maximum of the elements `x` and `y` determined by the relation `f()`.
`iterator max_element(first, last)`	Return an iterator indicating the position of the largest element in the range [`first`, `last`).
`iterator max_element(first, last, f())`	Return an iterator indicating the position of the largest element in the range [`first`, `last`) determined by the relation `f()`.
`void merge(first1, last1, first2, last2, result)`	Put at `result` the result of merging the ranges [`first1`, `last1`) and [`first2`, `last2`); the ranges are in ascending order.
`void merge(first1, last1, first2, last2, result, f())`	Put at `result` the result of merging the ranges [`first1`, `last1`) and [`first2`, `last2`); the ranges are ordered by the relation `f()`.
`const T& min(x, y)`	Return the minimum of the elements `x` and `y`.

`const T& min(x, y, f())`	Return the minimum of the elements `x` and `y` determined by the relation `f()`.
`iterator min_element(first, last)`	Return an iterator indicating the position of the smallest element in the range [`first, last`).
`iterator min_element(first, last, f())`	Return an iterator indicating the position of the smallest element in the range [`first, last`) determined by the relation `f()`.
`pair<iterator, iterator> mismatch(first1, last1, first2)`	Compare elements in the range [`first1, last1`) and [`first2, first2+(last1-first1)`) and return a pair of iterators that refer to the first positions in the ranges in which a mismatch occurs.
`pair<iterator, iterator> mismatch(first1, last1, first2, f())`	Compare elements in the range [`first1, last1`) and [`first2, first2+(last1-first1)`) and return a pair of iterators that refer to the first positions in the ranges in which a mismatch occurs; use the relation `f()` to compare the corresponding elements.
`bool next_permutation(first, last)`	Generate a range of elements that is a permutation lexicographically greater than the elements in the range [`first, last`); return true if such a permutation exists and false otherwise.
`bool next_permutation(first, last, f())`	Generate a range of elements that is a permutation lexicographically greater than the elements in the range [`first, last`); return true if such a permutation exists and false otherwise; use the relation `f()` for lexicographical comparison.
`void nth_element(first, nth, last)`	Place in (`nth - first`)th position the (`nth - first`)th element in the range [`first, last`) and rearrange the remaining elements so that elements before `*nth` are not larger than `*nth` and the elements following it are not smaller than `*nth`.
`void nth_element(first, nth, last, f())`	Place in (`nth - first`)th position the (`nth - first`)th element in the range [`first, last`) and rearrange the remaining elements so that elements before `*nth` are not in the relation `f()` to `*nth` and the elements following it are in the relation `f()` or equal to `*nth`.
`void partial_sort(first, middle, last)`	Put in the range [`first, middle`) the smallest `middle - first` elements from the range [`first, last`) in ascending order and the remaining elements in the range [`middle, last`) in any order.
`void partial_sort(first, middle, last, f())`	Put in the range [`first, middle`) the smallest `middle - first` elements from the range [`first, last`) in the order determined by the relation `f()` and the remaining elements in the range [`middle, last`) in any order.
`iterator partial_sort_copy (first1, last1, first2, last2)`	Copy to the range [`first2, last2`) the smallest `min(last1 - first1, last2 - first2)` elements from the range [`first1, last1`) and output them in ascending order; return an iterator referring to one position past the last copied element.
`iterator partial_sort_copy (first1, last1, first2, last2, f())`	Copy to the range [`first2, last2`) the smallest `min(last1 - first1, last2 - first2)` elements from the range [`first1, last1`) and output them in the order determined by the relation `f()`; return an iterator referring to one position past the last copied element.

`iterator partition(first, last, f())`	Rearrange the elements in the range [`first`, `last`) so that all elements that satisfy the condition `f()` are placed before elements that violate it; return an iterator indicating the beginning of the second range.
`void pop_heap(first, last)`	Swap the root of the heap [`first`, `last`) with the last element and restore the heap for the range [`first`, `last-1`).
`void pop_heap(first, last, f())`	Swap the root of the heap [`first`, `last`) with the last element and restore the heap for the range [`first`, `last-1`); use the relation `f()` in organizing the heap.
`bool prev_permutation(first, last)`	Generate a range of elements that is a permutation lexicographically smaller than the elements in the range [`first`, `last`); return true if such a permutation exists and false otherwise.
`bool prev_permutation(first, last, f())`	Generate a range of elements that is a permutation lexicographically smaller than the elements in the range [`first`, `last`); return true if such a permutation exists and false otherwise; use the relation `f()` for lexicographical comparison.
`void push_heap(first, last)`	Make a heap out of the heap [`first`, `last-1`) and an element `*(last-1)`.
`void push_heap(first, last, f())`	Make a heap out of the heap [`first`, `last-1`) and an element `*(last-1)`; use the relation `f()` in organizing the heap.
`void random_suffle(first, last)`	Randomly rearrange elements in the [`first`, `last`) using internal random number generator.
`void random_suffle(first, last, f())`	Randomly rearrange elements in the [`first`, `last`) using random number generator `f()`.
`iterator remove(first, last, value)`	Remove in the range [`first`, `last`) all the elements equal to `value`; return an iterator that indicates the end of the new range.
`iterator remove_copy(first, last, result, value)`	Copy from the range [`first`, `last`) all elements not equal to `value` to a range that begins at `result`; return an iterator that indicates the end of the copied range.
`iterator remove_copy_if(first, last, result, f())`	Copy from the range [`first`, `last`) all elements for which the one-argument Boolean function `f()` is false to a range that begins at `result`; return an iterator that indicates the end of the copied range.
`iterator remove_if(first, last, f())`	Remove in the range [`first`, `last`) all the elements for which the one-argument Boolean function `f()` is true; return an iterator that indicates the end of the new range.
`void replace(first, last, oldValue, newValue)`	Replace all occurrences of `oldValue` by `newValue` in the range [`first`, `last`).
`iterator replace_copy(first, last, result, oldValue, newValue)`	Copy elements from the range [`first`, `last`) to the range that starts at `result` and replace all occurrences of `oldValue` by `newValue` in the new range; return an iterator indicating the end of the new range.
`iterator replace_copy_if(first, last, result, f(), value)`	Copy elements from the range [`first`, `last`) to the range that starts at `result` and replace in the new range all elements for which the function `f()` is true by `value`; return an iterator indicating the end of the new range.

`void replace_if(first, last, f(), value)`	Replace all elements for which the function `f()` is true by `value` in the range [`first, last`).
`void reverse(first, last)`	Reverse the order of elements in the range [`first, last`).
`iterator reverse_copy(first, last, result)`	Copy in reverse order the elements from the range [`first, last`) to the range that starts at `result`; return an iterator indicating the end of the range of copied elements.
`void rotate(first, middle, last)`	Rotate to the left all elements in the range [`first, last`) by `middle - first` positions.
`iterator rotate_copy(first, middle, last, result)`	Create in the range that begins at `result` a rotated copy of all elements in the range [`first, last`) rotated to the left by `middle - first` positions.
`iterator search(first1, last1, first2, last2)`	Search for the subrange [`first2, last2`) in the range [`first, last`) and return an iterator indicating the beginning of the subrange or `last1` if the subrange is not located.
`iterator search(first1, last1, first2, last2, f())`	Search for the subrange [`first2, last2`) in the range [`first1, last1`) and return an iterator indicating the beginning of the subrange or `last1` if the subrange is not located; use a two-argument Boolean function `f()` to compare elements.
`iterator search_n(first, last, n, value)`	Search the range [`first, last`) for a subrange of n elements equal to `value` and return an iterator indicating the beginning of the subrange or `last1` if the subrange is not located.
`iterator search_n(first, last, n, value, f())`	Search the range [`first, last`) for a subrange of n elements equal to `value` and return an iterator indicating the beginning of the subrange or `last1` if the subrange is not located; compare elements with a two-argument Boolean function `f()`.
`iterator set_difference(first1, last1, first2, last2, result)`	Starting at position `result`, place in ascending order all the elements that occur in the range [`first1, last1`) but not in the range [`first2, last2`) that are in ascending order; return an iterator indicating the end of the resulting range.
`iterator set_difference(first1, last1, first2, last2, result, f())`	Starting at position `result`, place in order all the elements that occur in the range [`first1, last1`) but not in the range [`first2, last2`) that are in order, where the order is determined by the relation `f()`; return an iterator indicating the end of the resulting range.
`iterator set_intersection (first1, last1, first2, last2, result)`	Starting at position `result`, place in ascending order all the elements that occur in both ranges [`first1, last1`) and [`first2, last2`) that are in ascending order; return an iterator indicating the end of the resulting range.
`iterator set_intersection (first1, last1, first2, last2, result, f())`	Starting at position `result`, place in order all the elements that occur in both ranges [`first1, last1`) and [`first2, last2`) that are in order, where the order is determined by the relation `f()`; return an iterator indicating the end of the resulting range.

`iterator set_symmetric_difference(first1, last1, first2, last2, result)`	Starting at position `result`, place in ascending order all the elements that occur in one of the ranges [`first1, last1`) and [`first2, last2`) that are in ascending order, but not in both; return an iterator indicating the end of the resulting range.
`iterator set_symmetric_difference(first1, last1, first2, last2, result, f())`	Starting at position `result`, place in order all the elements that occur in one of the ranges [`first1, last1`) and [`first2, last2`) that are in order, but not in both; the order is determined by the relation `f()`; return an iterator indicating the end of the resulting range.
`iterator set_union(first1, last1, first2, last2, result)`	Starting at position `result`, place in ascending order all the elements that occur at least once in the ranges [`first1, last1`) and [`first2, last2`) that are in ascending order; return an iterator indicating the end of the resulting range.
`iterator set_union(first1, last1, first2, last2, result, f())`	Starting at position `result`, place in order all the elements that occur at least once in the ranges [`first1, last1`) and [`first2, last2`) that are in order, where the order is determined by the relation `f()`; return an iterator indicating the end of the resulting range.
`void sort(first, last)`	Sort in ascending order the elements in the range [`first, last1`).
`void sort(first, last, f())`	Order the elements in the range [`first, last`) using the relation `f()`.
`void sort_heap(first, last)`	Sort in ascending order the elements in the heap [`first, last`).
`void sort_heap(first, last, f())`	Order the elements in the heap [`first, last`) using the relation `f()`.
`iterator stable_partition(first, last, f())`	Rearrange the elements in the range [`first, last`) so that all elements for which the condition `f()` is true come before all elements that violate the condition.
`void stable_sort(first, last)`	Sort the elements in the range [`first, last`) in ascending order without changing the relative order of equal elements.
`void stable_sort(first, last, f())`	Sort the elements in the range [`first, last`) in the order determined by the relation `f()` without changing the relative order of equivalent elements.
`void swap(x, y)`	Swap the elements `x` and `y`.
`iterator swap_ranges(first1, last1, first2)`	Swap corresponding elements in the ranges [`first1, last1`) and [`first2, first2+(last1-first1)`) and return the iterator `first2+(last1-first1)`.
`iterator transform(first, last, result, f())`	Transform the elements in the range [`first, last`) by applying to them the function `f()` and place the transformed elements in the range that begins at `result`; return an iterator indicating the end of this range.
`iterator transform(first1, last1, first2, result, f())`	Apply the two-argument function `f()` to the corresponding elements in the ranges [`first1, last1`) and [`first2, first2+(last1-first1)`) and place the resulting elements in the range that begins at `result`; return an iterator indicating the end of this range.

`iterator unique(first, last)`	Remove all duplicates from the range [`first`, `last`) that is in ascending order and return an iterator indicating the end of a possibly shortened range.
`iterator unique(first, last, f())`	Remove all duplicates from the range [`first`, `last`) that is ordered by the relation `f()` and return an iterator indicating the end of a possibly shortened range.
`iterator unique_copy(first, last, result)`	Copy the range [`first`, `last`) that is in ascending order to the range that begins at `result` and remove all duplicates during copying; return an iterator indicating the end of the range with copied elements.
`iterator unique_copy(first, last, result, f())`	Copy the range [`first`, `last`) that is ordered by the relation `f()` and to the range that begins at `result` and remove all duplicates during copying; return an iterator indicating the end of the range with copied elements.
`iterator upper_bound(first, last, value)`	Return an iterator that references the highest position in the range [`first`, `last`) that is in ascending order before which `value` can be inserted without violating the order; return `last` if `value` is larger than all elements.
`iterator upper_bound(first, last, value, f())`	Return an iterator that references the highest position in the range [`first`, `last`) that is in order by the relation `f()` before which `value` can be inserted without violating the order; return `last` if `value` succeeds all elements.

To access these algorithms, the program has to include the statement
`#include <numeric>`

Member Function	Operation
`T accumulate(first, last, value)`	Return `value` plus the sum of all the values in the range [`first`, `last`).
`T accumulate(first, last, op(), value)`	Return the result of applying a two-argument operation `op()` to `value` and to all the values in the range [`first`, `last`).
`iterator adjacent_difference (first, last, result)`	Calculate the difference between each pair of elements in the range [`first`, `last`) and store the result in a container referenced by the iterator `result`; return `result + (last-first)`; because there is one less different value than the number of elements between `first` and `last`, 0 is added at the beginning of the resulting container so that it has the same number of elements as the range [`first`, `last`).
`iterator adjacent_difference (first, last, result, op())`	As before, but use a two-argument operator `op()` instead of subtraction.
`T inner_product(first1, last1, first2, value)`	Return `value` plus the sum of products of the corresponding elements from the ranges [`first1`, `last1`) and [`first1`, `first2+(last1-first1)`); that is, $value + \sum_i(x_i \cdot y_i)$.
`T inner_product(first1, last1, first2, value, op1(), op2())`	Replace `value` `last1-first1` times by applying `op1()` to `value` and the result of application of `op2()` to the corresponding elements from the ranges [`first1`, `last1`) and [`first1`, `first2+(last1-first1)`); return `value`.

`iterator partial_sum(first, last, result)`	Assign to the range [`result`, `result+(last-first)`) cumulative sums of the preceding corresponding elements in the sequence [`first`, `last`), that is, `*(result+i)` = `*(first+0)+*(first+1)+ ... +*(first+i)`; return an iterator `result+(last-first)`.
`iterator partial_sum(first, last, result, f())`	Assign to the range [`result`, `result+(last-first)`) the results of cumulative application of the function `f()` to the preceding corresponding elements in the sequence [`first`, `last`); that is, `*(result+i) = f(f(...,f(*first,*(first+1)),...), *(first+i))`.

C NP-Completeness

C.1 COOK'S THEOREM

A Turing machine is a device that reads and manipulates symbols in the cells of an infinite tape. It processes symbols by using a head that can move in either direction. More formally, a *Turing machine M* is defined as a tuple

$$M = (Q, \Sigma, \Gamma, \delta, q_0, F)$$

where

$Q = \{q_0, q_1, \ldots, q_n\}$ is a finite set of states.

$\Sigma \subset \Gamma - \{\#\}$ is a finite input alphabet.

$\Gamma = \{a_0, a_1, \ldots, a_m\}$, a_0 = blank #, is a finite tape alphabet.

$\delta: Q \times \Gamma \to Q \times \Gamma \times \{-1, +1\}$ is a transition function.

q_0 is a start state.

$F \subseteq Q$ is a set of final states.

The machine accepts or rejects any string of symbols built from alphabet Σ. In this way, a Turing machine defines a language that is a set of all strings acceptable by the machine.

The following is a Turing machine that computes the function sign for binary integers (possibly beginning with redundant 0s), that is

$$\text{sgn}(n) = \begin{cases} 0 \text{ if } n = 0, \\ 1 \text{ if } x > 0 \end{cases}$$

For this machine, $Q = \{q_0, q_1, q_2, q_3\}$, $\Sigma = \{0, 1, \#\}$, $\Gamma = \{0, 1\}$, $F = \{q_3\}$, and the transition function δ is given by the following table:

δ	0	1	#
q_0	$(q_0,0,+1)$	$(q_1,1,+1)$	$(q_2,0,-1)$
q_1	$(q_1,0,+1)$	$(q_1,1,+1)$	$(q_2,1,-1)$
q_2	$(q_3,\#,+1)$	$(q_3,\#,+1)$	

For the binary representation of number 2 with one redundant 0, 010, the changes of states are indicated by the following modifications of the tape.

Initial situation:

↓

| 0 | 1 | 0 | # | # | ... |

Step 0. Because $\delta(q_0,0) = (q_0,0,+1)$—which reads: From state q_0 and 0 in the current cell, leave 0 in this cell, go to the cell to its right and to state q_0—the next situation is

↓

| 0 | 1 | 0 | # | # | ... |

Step 1. Because $\delta(q_0,1) = (q_1,1,+1)$, then

↓

| 0 | 1 | 0 | # | # | ... |

Step 2. Because $\delta(q_1,0) = (q_1,0,+1)$, we have

↓

| 0 | 1 | 0 | # | # | ... |

Step 3. Form $\delta(q_1,\#) = (q_2,1,-1)$, the next configuration is

↓

| 0 | 1 | 0 | 1 | # | ... |

Step 4. On account of the transition $\delta(q_2,0) = (q_3,\#,+1)$,

↓

| 0 | 1 | # | 1 | # | ... |

In this example, the head goes to the right when it encounters 0 or 1 in a cell. When it encounters a blank, it writes 1 over it, goes to the left, writes a blank over it, and the machine finishes execution. When an input is all 0s, then the head ignores these zeros by moving to the right until it finds a blank. Then it writes 0 over it, moves to the left, writes a blank over it, and finishes execution.

A *nondeterministic Turing machine* has a finite number of choices for the next move and it is not determined which move should be made.

Theorem (Cook). Satisfiability problem is NP-complete.

Proof. A nondeterministic Turing machine M is assumed to be polynomially time bounded; that is, a valid computation sequence on input I takes $N = p(|I|)$ steps, where $p(|I|)$ is a polynomial in the length $|I|$ of input I, and can use at most N tape cells. M has an infinite tape extended in one direction only, with cells $0, 1, \ldots$.

To show that every NP problem is polynomially transformable to a satisfiability problem, a mapping is constructed from any input I of an arbitrary nondeterministic Turing machine M to an instance $r(I)$ of a satisfiability problem. The mapping r is so constructed that the Boolean formula $r(I)$ is satisfiable iff M accepts I. Thus, the reduction $r(I)$ models (simulates) a computation on M for an input I.

There are three types of logical (Boolean) variables (proposition symbols) used by function r to construct $r(I)$:

$P(i,s,t)$[1] is true iff at step t, tape cell s contains symbol a_i.
$Q(j,t)$ is true iff at step t, M is in state q_j.
$S(s,t)$ is true iff at step t, cell s is scanned by M's head.

where $0 \leq i \leq m, 0 \leq j \leq n, 0 \leq s,t \leq N$. With these definitions, computations of M on I can be represented as assignments of truth values to sequences of Boolean variables. With these variables, Boolean clauses can be written that describe various situations in M during computation on I. The conjunction of these clauses that represent the history of computation of M on I is the formula $r(I)$.

The statement $r(I)$ is a conjunction of eight groups of statements, each group used to enforce one requirement that $r(I)$ models a computation on M.

Let $\bigwedge_{0 \leq i \leq m} Q(i,t) = Q(0,t) \wedge Q(1,t) \wedge \ldots \wedge Q(m,t)$, and similarly for the alternative.

1. At each step t, each tape cell contains exactly one symbol

$$\bigwedge_{0 \leq s,t \leq N} (\bigvee_{0 \leq i \leq m} P(i,s,t) \wedge \bigwedge_{0 \leq i < i' \leq N} (\neg P(i,s,t) \vee \neg P(i',s,t)))$$

This condition states that at each step t and each cell s, s includes at least a symbol a_i, $P(i,s,t)$, but, at the same time, no more than one symbol; that is, $\neg(P(i,s',t) \wedge P(i,s,t))$, or, by de Morgan's law, $\neg P(i,s,t) \vee \neg P(i',s,t)$ for two different symbols a_i and $a_{i'}$.

2. At each step t, M is in exactly one state q_j

$$\bigwedge_{0 \leq t \leq N} (\bigvee_{0 \leq j \leq n} Q(j,t) \wedge \bigwedge_{0 \leq j < j' \leq n} (\neg Q(j,t) \vee \neg Q(j',t)))$$

3. At each step t, M is scanning exactly one tape cell s

$$\bigwedge_{0 \leq t \leq N} (\bigvee_{0 \leq s \leq N} S(s,t) \wedge \bigwedge_{0 \leq s < s' \leq N} (\neg S(s,t) \vee \neg S(s',t)))$$

4. The computation process begins in state q_0 with input symbols $I = a_{s_1} a_{s_2} \ldots a_{s_{|I|}}$ occupying $0, \ldots, |I| - 1$ leftmost cells of the tape, with the remaining cells filled with blank characters #. This initial situation is represented by this formula:

---

[1] The notation $P_{i,s,t}$ can also be used.

$$Q(0,0) \wedge S(0,0) \wedge P(a_{s_1},0,0) \wedge \ldots \wedge P(a_{s_{|I|}},|I|-1,0) \wedge P(\#,|I|,0) \wedge \ldots \wedge P(\#,N,0)$$

5. The permitted transitions during computation are given by the transition function $\delta(q_j, a_i) = (q_{j'}, a_{i'}, d \in \{-1,+1\})$. Using this function it has to be stated now that in each step the values of functions P, Q, and S are properly updated. For example, for each cell s and step t, if M is in state q_j scanning symbol a_i, then in the next step it is in state $q_{j'}$, as specified by the function δ:

$$\bigwedge_{0 \leq t < N} \bigwedge_{0 \leq s \leq N} (P(i,s,t) \wedge Q(j,t) \wedge S(s,t)) \Rightarrow Q(j',t+1)$$

that is, by the definition of implication ($\alpha \Rightarrow \beta = \neg \alpha \vee \beta$) and by de Morgan's law:

$$\bigwedge_{0 \leq t < N} \bigwedge_{0 \leq s \leq N} (\neg P(i,s,t) \vee \neg Q(j,t) \vee \neg S(s,t) \vee Q(j',t+1))$$

After we generalize it for all steps and all cells, we obtain the requirement

$$\bigwedge_{0 \leq t < N} \bigwedge_{0 \leq s \leq N} \bigwedge_{0 \leq i \leq m} \bigwedge_{0 \leq j \leq n} (\neg P(i,s,t) \vee \neg Q(j,t) \vee \neg S(s,t) \vee Q(j',t+1))$$

where "for all" should be qualified to range over i and j for which $\delta(q_j, a_i)$ is defined. If M is in a halting state, then in the next step it remains in that state, in the same cell, and the symbol in the cell remains the same. If M is in cell 0 and the next step would require going to the left, thereby sliding off the tape, then M halts; that is, it remains in the same state, in the same cell with the same symbol in the cell.

6. Similarly, we obtain permitted updates for P:

$$\bigwedge_{0 \leq t < N} \bigwedge_{0 \leq s \leq N} \bigwedge_{0 \leq i \leq m} \bigwedge_{0 \leq j \leq n} (\neg P(i,s,t) \vee \neg Q(j,t) \vee \neg S(s,t) \vee P(i',s,t+1))$$

7. and for S:

$$\bigwedge_{0 \leq t < N} \bigwedge_{0 \leq s \leq N} \bigwedge_{0 \leq i \leq m} \bigwedge_{0 \leq j \leq n} (\neg P(i,s,t) \vee \neg Q(j,t) \vee \neg S(s,t) \vee S(s+d,t+1))$$

8. The machine finally reaches an accepting state, which is reflected in the simple formula

$$\bigvee_{\{j : q_j \in F\}} Q(j,N) \qquad \blacksquare$$

If input I belongs to language for which M was constructed, then M reaches the accepting state when processing I. This processing imposes truth value assignments that satisfy all the clauses from groups 1 through 8. Also, any assignment of truth values to statements 1–8 that satisfies them describes a computation that ends in an accepting state. Therefore, $r(I)$ is satisfiable iff input I is an element of language recognizable by M.

The construction process of $r(I)$ indicates that $r(I)$ can be constructed in polynomial time.

For the Turing machine M defined earlier, with $N = 4$, $m = 2$, and $n = 3$, the statement $r(010)$ is the following conjunction of alternatives:

$(P(0,0,0) \vee P(1,0,0) \vee P(2,0,0)) \wedge (\neg P(0,0,0) \vee \neg P(1,0,0)) \wedge (\neg P(0,0,0) \vee \neg P(2,0,0)) \wedge (\neg P(1,0,0) \vee \neg P(2,0,0)) \wedge$
$(P(0,0,1) \vee P(1,0,1) \vee P(2,0,1)) \wedge (\neg P(0,0,1) \vee \neg P(1,0,1)) \wedge (\neg P(0,0,1) \vee \neg P(2,0,1)) \wedge (\neg P(1,0,1) \vee \neg P(2,0,1)) \wedge$
$(P(0,0,2) \vee P(1,0,2) \vee P(2,0,2)) \wedge (\neg P(0,0,2) \vee \neg P(1,0,2)) \wedge (\neg P(0,0,2) \vee \neg P(2,0,2)) \wedge (\neg P(1,0,2) \vee \neg P(2,0,2)) \wedge$
$(P(0,0,3) \vee P(1,0,3) \vee P(2,0,3)) \wedge (\neg P(0,0,3) \vee \neg P(1,0,3)) \wedge (\neg P(0,0,3) \vee \neg P(2,0,3)) \wedge (\neg P(1,0,3) \vee \neg P(2,0,3)) \wedge$
$(P(0,0,4) \vee P(1,0,4) \vee P(2,0,4)) \wedge (\neg P(0,0,4) \vee \neg P(1,0,4)) \wedge (\neg P(0,0,4) \vee \neg P(2,0,4)) \vee (\neg P(1,0,4) \vee (P(2,0,4)) \wedge$
$(P(0,1,0) \vee P(1,1,0) \vee P(2,1,0)) \wedge (\neg P(0,1,0) \vee \neg P(1,1,0)) \wedge (\neg P(0,1,0) \vee \neg P(2,1,0)) \wedge (\neg P(1,1,0) \vee \neg P(2,1,0)) \wedge$

$(P(0,1,1) \lor P(1,1,1) \lor P(2,1,1)) \land (\neg P(0,1,1) \lor \neg P(1,1,1)) \land (\neg P(0,1,1) \lor \neg P(2,1,1)) \land (\neg P(1,1,1) \lor \neg P(2,1,1)) \land$
$(P(0,1,2) \lor P(1,1,2) \lor P(2,1,2)) \land (\neg P(0,1,2) \lor \neg P(1,1,2)) \land (\neg P(0,1,2) \lor \neg P(2,1,2)) \land (\neg P(1,1,2) \lor \neg P(2,1,2)) \land$
$(P(0,1,3) \lor P(1,1,3) \lor P(2,1,3)) \land (\neg P(0,1,3) \lor \neg P(1,1,3)) \land (\neg P(0,1,3) \lor \neg P(2,1,3)) \land (\neg P(1,1,3) \lor \neg P(2,1,3)) \land$
$(P(0,1,4) \lor P(1,1,4) \lor P(2,1,4)) \land (\neg P(0,1,4) \lor \neg P(1,1,4)) \land (\neg P(0,1,4) \lor \neg P(2,1,4)) \land (\neg P(1,1,4) \lor \neg P(2,1,4)) \land$
$(P(0,2,0) \lor P(1,2,0) \lor P(2,2,0)) \land (\neg P(0,2,0) \lor \neg P(1,2,0)) \land (\neg P(0,2,0) \lor \neg P(2,2,0)) \land (\neg P(1,2,0) \lor \neg P(2,2,0)) \land$
$(P(0,2,1) \lor P(1,2,1) \lor P(2,2,1)) \land (\neg P(0,2,1) \lor \neg P(1,2,1)) \land (\neg P(0,2,1) \lor \neg P(2,2,1)) \land (\neg P(1,2,1) \lor \neg P(2,2,1)) \land$
$(P(0,2,2) \lor P(1,2,2) \lor P(2,2,2)) \land (\neg P(0,2,2) \lor \neg P(1,2,2)) \land (\neg P(0,2,2) \lor \neg P(2,2,2)) \land (\neg P(1,2,2) \lor \neg P(2,2,2)) \land$
$(P(0,2,3) \lor P(1,2,3) \lor P(2,2,3)) \land (\neg P(0,2,3) \lor \neg P(1,2,3)) \land (\neg P(0,2,3) \lor \neg P(2,2,3)) \land (\neg P(1,2,3) \lor \neg P(2,2,3)) \land$
$(P(0,2,4) \lor P(1,2,4) \lor P(2,2,4)) \land (\neg P(0,2,4) \lor \neg P(1,2,4)) \land (\neg P(0,2,4) \lor \neg P(2,2,4)) \land (\neg P(1,2,4) \lor \neg P(2,2,4)) \land$
$(P(0,3,0) \lor P(1,3,0) \lor P(2,3,0)) \land (\neg P(0,3,0) \lor \neg P(1,3,0)) \land (\neg P(0,3,0) \lor \neg P(2,3,0)) \land (\neg P(1,3,0) \lor \neg P(2,3,0)) \land$
$(P(0,3,1) \lor P(1,3,1) \lor P(2,3,1)) \land (\neg P(0,3,1) \lor \neg P(1,3,1)) \land (\neg P(0,3,1) \lor \neg P(2,3,1)) \land (\neg P(1,3,1) \lor \neg P(2,3,1)) \land$
$(P(0,3,2) \lor P(1,3,2) \lor P(2,3,2)) \land (\neg P(0,3,2) \lor \neg P(1,3,2)) \land (\neg P(0,3,2) \lor \neg P(2,3,2)) \land (\neg P(1,3,2) \lor \neg P(2,3,2)) \land$
$(P(0,3,3) \lor P(1,3,3) \lor P(2,3,3)) \land (\neg P(0,3,3) \lor \neg P(1,3,3)) \land (\neg P(0,3,3) \lor \neg P(2,3,3)) \land (\neg P(1,3,3) \lor \neg P(2,3,3)) \land$
$(P(0,3,4) \lor P(1,3,4) \lor P(2,3,4)) \land (\neg P(0,3,4) \lor \neg P(1,3,4)) \land (\neg P(0,3,4) \lor \neg P(2,3,4)) \land (\neg P(1,3,4) \lor \neg P(2,3,4)) \land$
$(P(0,4,0) \lor P(1,4,0) \lor P(2,4,0)) \land (\neg P(0,4,0) \lor \neg P(1,4,0)) \land (\neg P(0,4,0) \lor \neg P(2,4,0)) \land (\neg P(1,4,0) \lor \neg P(2,4,0)) \land$
$(P(0,4,1) \lor P(1,4,1) \lor P(2,4,1)) \land (\neg P(0,4,1) \lor \neg P(1,4,1)) \land (\neg P(0,4,1) \lor \neg P(2,4,1)) \land (\neg P(1,4,1) \lor \neg P(2,4,1)) \land$
$(P(0,4,2) \lor P(1,4,2) \lor P(2,4,2)) \land (\neg P(0,4,2) \lor \neg P(1,4,2)) \land (\neg P(0,4,2) \lor \neg P(2,4,2)) \land (\neg P(1,4,2) \lor \neg P(2,4,2)) \land$
$(P(0,4,3) \lor P(1,4,3) \lor P(2,4,3)) \land (\neg P(0,4,3) \lor \neg P(1,4,3)) \land (\neg P(0,4,3) \lor \neg P(2,4,3)) \land (\neg P(1,4,3) \lor \neg P(2,4,3)) \land$
$(P(0,4,4) \lor P(1,4,4) \lor P(2,4,4)) \land (\neg P(0,4,4) \lor \neg P(1,4,4)) \land (\neg P(0,4,4) \lor \neg P(2,4,4)) \land (\neg P(1,4,4) \lor \neg P(2,4,4)) \land$
$(Q(0,0) \lor Q(1,0) \lor Q(2,0) \lor Q(3,0)) \land$ // group 2
$(\neg Q(0,0) \lor \neg Q(1,0)) \land (\neg Q(0,0) \lor \neg Q(2,0)) \land (\neg Q(0,0) \lor \neg Q(3,0)) \land$
$(\neg Q(1,0) \lor \neg Q(2,0)) \land (\neg Q(1,0) \lor \neg Q(3,0)) \land$
$(\neg Q(2,0) \lor \neg Q(3,0)) \land$
$(Q(0,1) \lor Q(1,1) \lor Q(2,1) \lor Q(3,1)) \land$
$(\neg Q(0,1) \lor \neg Q(1,1)) \land (\neg Q(0,1) \lor \neg Q(2,1)) \land (\neg Q(0,1) \lor \neg Q(3,1)) \land$
$(\neg Q(1,1) \lor \neg Q(2,1)) \land (\neg Q(1,1) \lor \neg Q(3,1)) \land$
$(\neg Q(2,1) \lor \neg Q(3,1)) \land$
$(Q(0,2) \lor Q(1,2) \lor Q(2,2) \lor Q(3,2)) \land$
$(\neg Q(0,2) \lor \neg Q(1,2)) \land (\neg Q(0,2) \lor \neg Q(2,2)) \land (\neg Q(0,2) \lor \neg Q(3,2)) \land$
$(\neg Q(1,2) \lor \neg Q(2,2)) \land (\neg Q(1,2) \lor \neg Q(3,2)) \land$
$(\neg Q(2,2) \lor \neg Q(3,2)) \land$
$(Q(0,3) \lor Q(1,3) \lor Q(2,3) \lor Q(3,3)) \land$
$(\neg Q(0,3) \lor \neg Q(1,3)) \land (\neg Q(0,3) \lor \neg Q(2,3)) \land (\neg Q(0,3) \lor \neg Q(3,3)) \land$
$(\neg Q(1,3) \lor \neg Q(2,3)) \land (\neg Q(1,3) \lor \neg Q(3,3)) \land$
$(\neg Q(2,3) \lor \neg Q(3,3)) \land$
$(Q(0,4) \lor Q(1,4) \lor Q(2,4) \lor Q(3,4)) \land$
$(\neg Q(0,4) \lor \neg Q(1,4)) \land (\neg Q(0,4) \lor \neg Q(2,4)) \land (\neg Q(0,4) \lor \neg Q(3,4)) \land$
$(\neg Q(1,4) \lor \neg Q(2,4)) \land (\neg Q(1,4) \lor \neg Q(3,4)) \land$
$(\neg Q(2,4) \lor \neg Q(3,4)) \land$
$(S(0,0) \lor S(1,0) \lor S(2,0) \lor S(3,0) \lor S(4,0)) \land$ // group 3
$(\neg S(0,0) \lor \neg S(1,0)) \land (\neg S(0,0) \lor \neg S(2,0)) \land (\neg S(0,0) \lor \neg S(3,0)) \land (\neg S(0,0) \lor \neg S(4,0))$
$(\neg S(1,0) \lor \neg S(2,0)) \land (\neg S(1,0) \lor \neg S(3,0)) \land (\neg S(1,0) \lor \neg S(4,0)) \land$
$(\neg S(2,0) \lor \neg S(3,0)) \land (\neg S(2,0) \lor \neg S(4,0)) \land$
$(\neg S(3,0) \lor \neg S(4,0)) \land$
$(S(0,1) \lor S(1,1) \lor S(2,1) \lor S(3,1) \lor S(4,1)) \land$
$(\neg S(0,1) \lor \neg S(1,1)) \land (\neg S(0,1) \lor \neg S(2,1)) \land (\neg S(0,1) \lor \neg S(3,1)) \land (\neg S(0,1) \lor \neg S(4,1)) \land$

Section C.1 Cook's Theorem   ■   733

$(\neg S(1,1) \lor \neg S(2,1)) \land (\neg S(1,1) \lor \neg S(3,1)) \land (\neg S(1,1) \lor \neg S(4,1)) \land$
$(\neg S(2,1) \lor \neg S(3,1)) \land (\neg S(2,1) \lor \neg S(4,1)) \land$
$(\neg S(3,1) \lor \neg S(4,1)) \land$
$(S(0,2) \lor S(1,2) \lor S(2,2) \lor S(3,2) \lor S(4,2)) \land$
$(\neg S(0,2) \lor \neg S(1,2)) \land (\neg S(0,2) \lor \neg S(2,2)) \land (\neg S(0,2) \lor \neg S(3,2)) \land (\neg S(0,2) \lor \neg S(4,2)) \land$
$(\neg S(1,2) \lor \neg S(2,2)) \land (\neg S(1,2) \lor \neg S(3,2)) \land (\neg S(1,2) \lor \neg S(4,2)) \land$
$(\neg S(2,2) \lor \neg S(3,2)) \land (\neg S(2,2) \lor \neg S(4,2)) \land$
$(\neg S(3,2) \lor \neg S(4,2)) \land$
$(S(0,3) \lor S(1,3) \lor S(2,3) \lor S(3,3) \lor S(4,3)) \land$
$(\neg S(0,3) \lor \neg S(1,3)) \land (\neg S(0,3) \lor \neg S(2,3)) \land (\neg S(0,3) \lor \neg S(3,3)) \land (\neg S(0,3) \lor \neg S(4,3)) \land$
$(\neg S(1,3) \lor \neg S(2,3)) \land (\neg S(1,3) \lor \neg S(3,3)) \land (\neg S(1,3) \lor \neg S(4,3)) \land$
$(\neg S(2,3) \lor \neg S(3,3)) \land (\neg S(2,3) \lor \neg S(4,3)) \land$
$(\neg S(3,3) \lor \neg S(4,3)) \land$
$(S(0,4) \lor S(1,4) \lor S(2,4) \lor S(3,4) \lor S(4,4)) \land$
$(\neg S(0,4) \lor \neg S(1,4)) \land (\neg S(0,4) \lor \neg S(2,4)) \land (\neg S(0,4) \lor \neg S(3,4)) \land (\neg S(0,4) \lor \neg S(4,4)) \land$
$(\neg S(1,4) \lor \neg S(2,4)) \land (\neg S(1,4) \lor \neg S(3,4)) \land (\neg S(1,4) \lor \neg S(4,4)) \land$
$(\neg S(2,4) \lor \neg S(3,4)) \land (\neg S(2,4) \lor \neg S(4,4)) \land$
$(\neg S(3,4) \lor \neg S(4,4)) \land$
$Q(0,0) \land S(0,0) \land P(0,0,0) \land P(1,1,0) \land P(0,2,0) \land P(\#,3,0) \land P(\#,4,0) \land$ // group 4
$(\neg P(0,0,0) \lor \neg Q(0,0) \lor \neg S(0,0) \lor Q(0,1)) \land (\neg P(0,0,1) \lor \neg Q(0,1) \lor \neg S(0,1) \lor Q(0,2)) \land$ // group 5
$(\neg P(0,0,2) \lor \neg Q(0,2) \lor \neg S(0,2) \lor Q(0,3)) \land (\neg P(0,0,3) \lor \neg Q(0,3) \lor \neg S(0,3) \lor Q(0,4)) \land$
$(\neg P(1,0,0) \lor \neg Q(0,0) \lor \neg S(0,0) \lor Q(1,1)) \land (\neg P(1,0,1) \lor \neg Q(0,1) \lor \neg S(0,1) \lor Q(1,2)) \land$
$(\neg P(1,0,2) \lor \neg Q(0,2) \lor \neg S(0,2) \lor Q(1,3)) \land (\neg P(1,0,3) \lor \neg Q(0,3) \lor \neg S(0,3) \lor Q(1,4)) \land$
// A blank # in cell 0 in state 0 causes *M* to halt to prevent it from sliding off the tape:
$(\neg P(\#,0,0) \lor \neg Q(0,0) \lor \neg S(0,0) \lor Q(3,1)) \land (\neg P(\#,0,1) \lor \neg Q(0,1) \lor \neg S(0,1) \lor Q(3,2)) \land$
$(\neg P(\#,0,2) \lor \neg Q(0,2) \lor \neg S(0,2) \lor Q(3,3)) \land (\neg P(\#,0,3) \lor \neg Q(0,3) \lor \neg S(0,3) \lor Q(3,4)) \land$
$(\neg P(0,1,0) \lor \neg Q(0,0) \lor \neg S(1,0) \lor Q(0,1)) \land (\neg P(0,1,1) \lor \neg Q(0,1) \lor \neg S(1,1) \lor Q(0,2)) \land$
$(\neg P(0,1,2) \lor \neg Q(0,2) \lor \neg S(1,2) \lor Q(0,3)) \land (\neg P(0,1,3) \lor \neg Q(0,3) \lor \neg S(1,3) \lor Q(0,4)) \land$
$(\neg P(1,1,0) \lor \neg Q(0,0) \lor \neg S(1,0) \lor Q(1,1)) \land (\neg P(1,1,1) \lor \neg Q(0,1) \lor \neg S(1,1) \lor Q(1,2)) \land$
$(\neg P(1,1,2) \lor \neg Q(0,2) \lor \neg S(1,2) \lor Q(1,3)) \land (\neg P(1,1,3) \lor \neg Q(0,3) \lor \neg S(1,3) \lor Q(1,4)) \land$
$(\neg P(\#,1,0) \lor \neg Q(0,0) \lor \neg S(1,0) \lor Q(2,1)) \land (\neg P(\#,1,1) \lor \neg Q(0,1) \lor \neg S(1,1) \lor Q(2,2)) \land$
$(\neg P(\#,1,2) \lor \neg Q(0,2) \lor \neg S(1,2) \lor Q(2,3)) \land (\neg P(\#,1,3) \lor \neg Q(0,3) \lor \neg S(1,3) \lor Q(2,4)) \land$
$(\neg P(0,2,0) \lor \neg Q(0,0) \lor \neg S(2,0) \lor Q(0,1)) \land (\neg P(0,2,1) \lor \neg Q(0,1) \lor \neg S(2,1) \lor Q(0,2)) \land$
$(\neg P(0,2,2) \lor \neg Q(0,2) \lor \neg S(2,2) \lor Q(0,3)) \land (\neg P(0,2,3) \lor \neg Q(0,3) \lor \neg S(2,3) \lor Q(0,4)) \land$
$(\neg P(1,2,0) \lor \neg Q(0,0) \lor \neg S(2,0) \lor Q(1,1)) \land (\neg P(1,2,1) \lor \neg Q(0,1) \lor \neg S(2,1) \lor Q(1,2)) \land$
$(\neg P(1,2,2) \lor \neg Q(0,2) \lor \neg S(2,2) \lor Q(1,3)) \land (\neg P(1,2,3) \lor \neg Q(0,3) \lor \neg S(2,3) \lor Q(1,4)) \land$
$(\neg P(\#,2,0) \lor \neg Q(0,0) \lor \neg S(2,0) \lor Q(2,1)) \land (\neg P(\#,2,1) \lor \neg Q(0,1) \lor \neg S(2,1) \lor Q(2,2)) \land$
$(\neg P(\#,2,2) \lor \neg Q(0,2) \lor \neg S(2,2) \lor Q(2,3)) \land (\neg P(\#,2,3) \lor \neg Q(0,3) \lor \neg S(2,3) \lor Q(2,4)) \land$
$(\neg P(0,3,0) \lor \neg Q(0,0) \lor \neg S(3,0) \lor Q(0,1)) \land (\neg P(0,3,1) \lor \neg Q(0,1) \lor \neg S(3,1) \lor Q(0,2)) \land$
$(\neg P(0,3,2) \lor \neg Q(0,2) \lor \neg S(3,2) \lor Q(0,3)) \land (\neg P(0,3,3) \lor \neg Q(0,3) \lor \neg S(3,3) \lor Q(0,4)) \land$
$(\neg P(1,3,0) \lor \neg Q(0,0) \lor \neg S(3,0) \lor Q(1,1)) \land (\neg P(1,3,1) \lor \neg Q(0,1) \lor \neg S(3,1) \lor Q(1,2)) \land$
$(\neg P(1,3,2) \lor \neg Q(0,2) \lor \neg S(3,2) \lor Q(1,3)) \land (\neg P(1,3,3) \lor \neg Q(0,3) \lor \neg S(3,3) \lor Q(1,4)) \land$
$(\neg P(\#,3,0) \lor \neg Q(0,0) \lor \neg S(3,0) \lor Q(2,1)) \land (\neg P(\#,3,1) \lor \neg Q(0,1) \lor \neg S(3,1) \lor Q(2,2)) \land$
$(\neg P(\#,3,2) \lor \neg Q(0,2) \lor \neg S(3,2) \lor Q(2,3)) \land (\neg P(\#,3,3) \lor \neg Q(0,3) \lor \neg S(3,3) \lor Q(2,4)) \land$
$(\neg P(0,4,0) \lor \neg Q(0,0) \lor \neg S(4,0) \lor Q(0,1)) \land (\neg P(0,4,1) \lor \neg Q(0,1) \lor \neg S(4,1) \lor Q(0,2)) \land$
$(\neg P(0,4,2) \lor \neg Q(0,2) \lor \neg S(4,2) \lor Q(0,3)) \land (\neg P(0,4,3) \lor \neg Q(0,3) \lor \neg S(4,3) \lor Q(0,4)) \land$

$(\neg P(1,4,0) \lor \neg Q(0,0) \lor \neg S(4,0) \lor Q(1,1)) \land (\neg P(1,4,1) \lor \neg Q(0,1) \lor \neg S(4,1) \lor Q(1,2)) \land$
$(\neg P(1,4,2) \lor \neg Q(0,2) \lor \neg S(4,2) \lor Q(1,3)) \land (\neg P(1,4,3) \lor \neg Q(0,3) \lor \neg S(4,3) \lor Q(1,4)) \land$
$(\neg P(\#,4,0) \lor \neg Q(0,0) \lor \neg S(4,0) \lor Q(2,1)) \land (\neg P(\#,4,1) \lor \neg Q(0,1) \lor \neg S(4,1) \lor Q(2,2)) \land$
$(\neg P(\#,4,2) \lor \neg Q(0,2) \lor \neg S(4,2) \lor Q(2,3)) \land (\neg P(\#,4,3) \lor \neg Q(0,3) \lor \neg S(4,3) \lor Q(2,4)) \land$
$(\neg P(0,0,0) \lor \neg Q(1,0) \lor \neg S(0,0) \lor Q(1,1)) \land (\neg P(0,0,1) \lor \neg Q(1,1) \lor \neg S(0,1) \lor Q(1,2)) \land$ //can't be in state 1
$(\neg P(0,0,2) \lor \neg Q(1,2) \lor \neg S(0,2) \lor Q(1,3)) \land (\neg P(0,0,3) \lor \neg Q(1,3) \lor \neg S(0,3) \lor Q(1,4)) \land$ //in step 0
$(\neg P(1,0,0) \lor \neg Q(1,0) \lor \neg S(0,0) \lor Q(1,1)) \land (\neg P(1,0,1) \lor \neg Q(1,1) \lor \neg S(0,1) \lor Q(1,2)) \land$
$(\neg P(1,0,2) \lor \neg Q(1,2) \lor \neg S(0,2) \lor Q(1,3)) \land (\neg P(1,0,3) \lor \neg Q(1,3) \lor \neg S(0,3) \lor Q(1,4)) \land$
// A blank # in cell 0 in state 1 causes M to halt:
$(\neg P(\#,0,0) \lor \neg Q(1,0) \lor \neg S(0,0) \lor Q(3,1)) \land (\neg P(\#,0,1) \lor \neg Q(1,1) \lor \neg S(0,1) \lor Q(3,2)) \land$
$(\neg P(\#,0,2) \lor \neg Q(1,2) \lor \neg S(0,2) \lor Q(3,3)) \land (\neg P(\#,0,3) \lor \neg Q(1,3) \lor \neg S(0,3) \lor Q(3,4)) \land$
$(\neg P(0,1,0) \lor \neg Q(1,0) \lor \neg S(1,0) \lor Q(1,1)) \land (\neg P(0,1,1) \lor \neg Q(1,1) \lor \neg S(1,1) \lor Q(1,2)) \land$
$(\neg P(0,1,2) \lor \neg Q(1,2) \lor \neg S(1,2) \lor Q(1,3)) \land (\neg P(0,1,3) \lor \neg Q(1,3) \lor \neg S(1,3) \lor Q(1,4)) \land$
$(\neg P(1,1,0) \lor \neg Q(1,0) \lor \neg S(1,0) \lor Q(1,1)) \land (\neg P(1,1,1) \lor \neg Q(1,1) \lor \neg S(1,1) \lor Q(1,2)) \land$
$(\neg P(1,1,2) \lor \neg Q(1,2) \lor \neg S(1,2) \lor Q(1,3)) \land (\neg P(1,1,3) \lor \neg Q(1,3) \lor \neg S(1,3) \lor Q(1,4)) \land$
$(\neg P(\#,1,0) \lor \neg Q(1,0) \lor \neg S(1,0) \lor Q(2,1)) \land (\neg P(\#,1,1) \lor \neg Q(1,1) \lor \neg S(1,1) \lor Q(2,2)) \land$
$(\neg P(\#,1,2) \lor \neg Q(1,2) \lor \neg S(1,2) \lor Q(2,3)) \land (\neg P(\#,1,3) \lor \neg Q(1,3) \lor \neg S(1,3) \lor Q(2,4)) \land$
$(\neg P(0,2,0) \lor \neg Q(1,0) \lor \neg S(2,0) \lor Q(1,1)) \land (\neg P(0,2,1) \lor \neg Q(1,1) \lor \neg S(2,1) \lor Q(1,2)) \land$
$(\neg P(0,2,2) \lor \neg Q(1,2) \lor \neg S(2,2) \lor Q(1,3)) \land (\neg P(0,2,3) \lor \neg Q(1,3) \lor \neg S(2,3) \lor Q(1,4)) \land$
$(\neg P(1,2,0) \lor \neg Q(1,0) \lor \neg S(2,0) \lor Q(1,1)) \land (\neg P(1,2,1) \lor \neg Q(1,1) \lor \neg S(2,1) \lor Q(1,2)) \land$
$(\neg P(1,2,2) \lor \neg Q(1,2) \lor \neg S(2,2) \lor Q(1,3)) \land (\neg P(1,2,3) \lor \neg Q(1,3) \lor \neg S(2,3) \lor Q(1,4)) \land$
$(\neg P(\#,2,0) \lor \neg Q(1,0) \lor \neg S(2,0) \lor Q(2,1)) \land (\neg P(\#,2,1) \lor \neg Q(1,1) \lor \neg S(2,1) \lor Q(2,2)) \land$
$(\neg P(\#,2,2) \lor \neg Q(1,2) \lor \neg S(2,2) \lor Q(2,3)) \land (\neg P(\#,2,3) \lor \neg Q(1,3) \lor \neg S(2,3) \lor Q(2,4)) \land$
$(\neg P(0,3,0) \lor \neg Q(1,0) \lor \neg S(3,0) \lor Q(1,1)) \land (\neg P(0,3,1) \lor \neg Q(1,1) \lor \neg S(3,1) \lor Q(1,2)) \land$
$(\neg P(0,3,2) \lor \neg Q(1,2) \lor \neg S(3,2) \lor Q(1,3)) \land (\neg P(0,3,3) \lor \neg Q(1,3) \lor \neg S(3,3) \lor Q(1,4)) \land$
$(\neg P(1,3,0) \lor \neg Q(1,0) \lor \neg S(3,0) \lor Q(1,1)) \land (\neg P(1,3,1) \lor \neg Q(1,1) \lor \neg S(3,1) \lor Q(1,2)) \land$
$(\neg P(1,3,2) \lor \neg Q(1,2) \lor \neg S(3,2) \lor Q(1,3)) \land (\neg P(1,3,3) \lor \neg Q(1,3) \lor \neg S(3,3) \lor Q(1,4)) \land$
$(\neg P(\#,3,0) \lor \neg Q(1,0) \lor \neg S(3,0) \lor Q(2,1)) \land (\neg P(\#,3,1) \lor \neg Q(1,1) \lor \neg S(3,1) \lor Q(2,2)) \land$
$(\neg P(\#,3,2) \lor \neg Q(1,2) \lor \neg S(3,2) \lor Q(2,3)) \land (\neg P(\#,3,3) \lor \neg Q(1,3) \lor \neg S(3,3) \lor Q(2,4)) \land$
$(\neg P(0,4,0) \lor \neg Q(1,0) \lor \neg S(4,0) \lor Q(1,1)) \land (\neg P(0,4,1) \lor \neg Q(1,1) \lor \neg S(4,1) \lor Q(1,2)) \land$
$(\neg P(0,4,2) \lor \neg Q(1,2) \lor \neg S(4,2) \lor Q(1,3)) \land (\neg P(0,4,3) \lor \neg Q(1,3) \lor \neg S(4,3) \lor Q(1,4)) \land$
$(\neg P(1,4,0) \lor \neg Q(1,0) \lor \neg S(4,0) \lor Q(1,1)) \land (\neg P(1,4,1) \lor \neg Q(1,1) \lor \neg S(4,1) \lor Q(1,2)) \land$
$(\neg P(1,4,2) \lor \neg Q(1,2) \lor \neg S(4,2) \lor Q(1,3)) \land (\neg P(1,4,3) \lor \neg Q(1,3) \lor \neg S(4,3) \lor Q(1,4)) \land$
$(\neg P(\#,4,0) \lor \neg Q(1,0) \lor \neg S(4,0) \lor Q(2,1)) \land (\neg P(\#,4,1) \lor \neg Q(1,1) \lor \neg S(4,1) \lor Q(2,2)) \land$
$(\neg P(\#,4,2) \lor \neg Q(1,2) \lor \neg S(4,2) \lor Q(2,3)) \land (\neg P(\#,4,3) \lor \neg Q(1,3) \lor \neg S(4,3) \lor Q(2,4)) \land$
$(\neg P(0,0,0) \lor \neg Q(2,0) \lor \neg S(0,0) \lor Q(3,1)) \land (\neg P(0,0,1) \lor \neg Q(2,1) \lor \neg S(0,1) \lor Q(3,2)) \land$
$(\neg P(0,0,2) \lor \neg Q(2,2) \lor \neg S(0,2) \lor Q(3,3)) \land (\neg P(0,0,3) \lor \neg Q(2,3) \lor \neg S(0,3) \lor Q(3,4)) \land$
$(\neg P(1,0,0) \lor \neg Q(2,0) \lor \neg S(0,0) \lor Q(3,1)) \land (\neg P(1,0,1) \lor \neg Q(2,1) \lor \neg S(0,1) \lor Q(3,2)) \land$
$(\neg P(1,0,2) \lor \neg Q(2,2) \lor \neg S(0,2) \lor Q(3,3)) \land (\neg P(1,0,3) \lor \neg Q(2,3) \lor \neg S(0,3) \lor Q(3,4)) \land$
$(\neg P(0,1,0) \lor \neg Q(2,0) \lor \neg S(1,0) \lor Q(3,1)) \land (\neg P(0,1,1) \lor \neg Q(2,1) \lor \neg S(1,1) \lor Q(3,2)) \land$
$(\neg P(0,1,2) \lor \neg Q(2,2) \lor \neg S(1,2) \lor Q(3,3)) \land (\neg P(0,1,3) \lor \neg Q(2,3) \lor \neg S(1,3) \lor Q(3,4)) \land$
$(\neg P(1,1,0) \lor \neg Q(2,0) \lor \neg S(1,0) \lor Q(3,1)) \land (\neg P(1,1,1) \lor \neg Q(2,1) \lor \neg S(1,1) \lor Q(3,2)) \land$
$(\neg P(1,1,2) \lor \neg Q(2,2) \lor \neg S(1,2) \lor Q(3,3)) \land (\neg P(1,1,3) \lor \neg Q(2,3) \lor \neg S(1,3) \lor Q(3,4)) \land$
$(\neg P(0,2,0) \lor \neg Q(2,0) \lor \neg S(2,0) \lor Q(3,1)) \land (\neg P(0,2,1) \lor \neg Q(2,1) \lor \neg S(2,1) \lor Q(3,2)) \land$
$(\neg P(0,2,2) \lor \neg Q(2,2) \lor \neg S(2,2) \lor Q(3,3)) \land (\neg P(0,2,3) \lor \neg Q(2,3) \lor \neg S(2,3) \lor Q(3,4)) \land$
$(\neg P(1,2,0) \lor \neg Q(2,0) \lor \neg S(2,0) \lor Q(3,1)) \land (\neg P(1,2,1) \lor \neg Q(2,1) \lor \neg S(2,1) \lor Q(3,2)) \land$

$(\neg P(1,2,2) \lor \neg Q(2,2) \lor \neg S(2,2) \lor Q(3,3)) \land (\neg P(1,2,3) \lor \neg Q(2,3) \lor \neg S(2,3) \lor Q(3,4)) \land$
$(\neg P(0,3,0) \lor \neg Q(2,0) \lor \neg S(3,0) \lor Q(3,1)) \land (\neg P(0,3,1) \lor \neg Q(2,1) \lor \neg S(3,1) \lor Q(3,2)) \land$
$(\neg P(0,3,2) \lor \neg Q(2,2) \lor \neg S(3,2) \lor Q(3,3)) \land (\neg P(0,3,3) \lor \neg Q(2,3) \lor \neg S(3,3) \lor Q(3,4)) \land$
$(\neg P(1,3,0) \lor \neg Q(2,0) \lor \neg S(3,0) \lor Q(3,1)) \land (\neg P(1,3,1) \lor \neg Q(2,1) \lor \neg S(3,1) \lor Q(3,2)) \land$
$(\neg P(1,3,2) \lor \neg Q(2,2) \lor \neg S(3,2) \lor Q(3,3)) \land (\neg P(1,3,3) \lor \neg Q(2,3) \lor \neg S(3,3) \lor Q(3,4)) \land$
$(\neg P(0,4,0) \lor \neg Q(2,0) \lor \neg S(4,0) \lor Q(3,1)) \land (\neg P(0,4,1) \lor \neg Q(2,1) \lor \neg S(4,1) \lor Q(3,2)) \land$
$(\neg P(0,4,2) \lor \neg Q(2,2) \lor \neg S(4,2) \lor Q(3,3)) \land (\neg P(0,4,3) \lor \neg Q(2,3) \lor \neg S(4,3) \lor Q(3,4)) \land$
$(\neg P(1,4,0) \lor \neg Q(2,0) \lor \neg S(4,0) \lor Q(3,1)) \land (\neg P(1,4,1) \lor \neg Q(2,1) \lor \neg S(4,1) \lor Q(3,2)) \land$
$(\neg P(1,4,2) \lor \neg Q(2,2) \lor \neg S(4,2) \lor Q(3,3)) \land (\neg P(1,4,3) \lor \neg Q(2,3) \lor \neg S(4,3) \lor Q(3,4)) \land$
$(\neg P(0,0,0) \lor \neg Q(3,0) \lor \neg S(0,0) \lor Q(3,1)) \land (\neg P(0,0,1) \lor \neg Q(3,1) \lor \neg S(0,1) \lor Q(3,2)) \land$
$(\neg P(0,0,2) \lor \neg Q(3,2) \lor \neg S(0,2) \lor Q(3,3)) \land (\neg P(0,0,3) \lor \neg Q(3,3) \lor \neg S(0,3) \lor Q(3,4)) \land$
$(\neg P(1,0,0) \lor \neg Q(3,0) \lor \neg S(0,0) \lor Q(3,1)) \land (\neg P(1,0,1) \lor \neg Q(3,1) \lor \neg S(0,1) \lor Q(3,2)) \land$
$(\neg P(1,0,2) \lor \neg Q(3,2) \lor \neg S(0,2) \lor Q(3,3)) \land (\neg P(1,0,3) \lor \neg Q(3,3) \lor \neg S(0,3) \lor Q(3,4)) \land$
$(\neg P(\#,0,0) \lor \neg Q(3,0) \lor \neg S(0,0) \lor Q(3,1)) \land (\neg P(\#,0,1) \lor \neg Q(3,1) \lor \neg S(0,1) \lor Q(3,2)) \land$
$(\neg P(\#,0,2) \lor \neg Q(3,2) \lor \neg S(0,2) \lor Q(3,3)) \land (\neg P(\#,0,3) \lor \neg Q(3,3) \lor \neg S(0,3) \lor Q(3,4)) \land$
$(\neg P(0,1,0) \lor \neg Q(3,0) \lor \neg S(1,0) \lor Q(3,1)) \land (\neg P(0,1,1) \lor \neg Q(3,1) \lor \neg S(1,1) \lor Q(3,2)) \land$
$(\neg P(0,1,2) \lor \neg Q(3,2) \lor \neg S(1,2) \lor Q(3,3)) \land (\neg P(0,1,3) \lor \neg Q(3,3) \lor \neg S(1,3) \lor Q(3,4)) \land$
$(\neg P(1,1,0) \lor \neg Q(3,0) \lor \neg S(1,0) \lor Q(3,1)) \land (\neg P(1,1,1) \lor \neg Q(3,1) \lor \neg S(1,1) \lor Q(3,2)) \land$
$(\neg P(1,1,2) \lor \neg Q(3,2) \lor \neg S(1,2) \lor Q(3,3)) \land (\neg P(1,1,3) \lor \neg Q(3,3) \lor \neg S(1,3) \lor Q(3,4)) \land$
$(\neg P(\#,1,0) \lor \neg Q(3,0) \lor \neg S(1,0) \lor Q(3,1)) \land (\neg P(\#,1,1) \lor \neg Q(3,1) \lor \neg S(1,1) \lor Q(3,2)) \land$
$(\neg P(\#,1,2) \lor \neg Q(3,2) \lor \neg S(1,2) \lor Q(3,3)) \land (\neg P(\#,1,3) \lor \neg Q(3,3) \lor \neg S(1,3) \lor Q(3,4)) \land$
$(\neg P(0,2,0) \lor \neg Q(3,0) \lor \neg S(2,0) \lor Q(3,1)) \land (\neg P(0,2,1) \lor \neg Q(3,1) \lor \neg S(2,1) \lor Q(3,2)) \land$
$(\neg P(0,2,2) \lor \neg Q(3,2) \lor \neg S(2,2) \lor Q(3,3)) \land (\neg P(0,2,3) \lor \neg Q(3,3) \lor \neg S(2,3) \lor Q(3,4)) \land$
$(\neg P(1,2,0) \lor \neg Q(3,0) \lor \neg S(2,0) \lor Q(3,1)) \land (\neg P(1,2,1) \lor \neg Q(3,1) \lor \neg S(2,1) \lor Q(3,2)) \land$
$(\neg P(1,2,2) \lor \neg Q(3,2) \lor \neg S(2,2) \lor Q(3,3)) \land (\neg P(1,2,3) \lor \neg Q(3,3) \lor \neg S(2,3) \lor Q(3,4)) \land$
$(\neg P(\#,2,0) \lor \neg Q(3,0) \lor \neg S(2,0) \lor Q(3,1)) \land (\neg P(\#,2,1) \lor \neg Q(3,1) \lor \neg S(2,1) \lor Q(3,2)) \land$
$(\neg P(\#,2,2) \lor \neg Q(3,2) \lor \neg S(2,2) \lor Q(3,3)) \land (\neg P(\#,2,3) \lor \neg Q(3,3) \lor \neg S(2,3) \lor Q(3,4)) \land$
$(\neg P(0,3,0) \lor \neg Q(3,0) \lor \neg S(3,0) \lor Q(3,1)) \land (\neg P(0,3,1) \lor \neg Q(3,1) \lor \neg S(3,1) \lor Q(3,2)) \land$
$(\neg P(0,3,2) \lor \neg Q(3,2) \lor \neg S(3,2) \lor Q(3,3)) \land (\neg P(0,3,3) \lor \neg Q(3,3) \lor \neg S(3,3) \lor Q(3,4)) \land$
$(\neg P(1,3,0) \lor \neg Q(3,0) \lor \neg S(3,0) \lor Q(3,1)) \land (\neg P(1,3,1) \lor \neg Q(3,1) \lor \neg S(3,1) \lor Q(3,2)) \land$
$(\neg P(1,3,2) \lor \neg Q(3,2) \lor \neg S(3,2) \lor Q(3,3)) \land (\neg P(1,3,3) \lor \neg Q(3,3) \lor \neg S(3,3) \lor Q(3,4)) \land$
$(\neg P(\#,3,0) \lor \neg Q(3,0) \lor \neg S(3,0) \lor Q(3,1)) \land (\neg P(\#,3,1) \lor \neg Q(3,1) \lor \neg S(3,1) \lor Q(3,2)) \land$
$(\neg P(\#,3,2) \lor \neg Q(3,2) \lor \neg S(3,2) \lor Q(3,3)) \land (\neg P(\#,3,3) \lor \neg Q(3,3) \lor \neg S(3,3) \lor Q(3,4)) \land$
$(\neg P(0,4,0) \lor \neg Q(3,0) \lor \neg S(4,0) \lor Q(3,1)) \land (\neg P(0,4,1) \lor \neg Q(3,1) \lor \neg S(4,1) \lor Q(3,2)) \land$
$(\neg P(0,4,2) \lor \neg Q(3,2) \lor \neg S(4,2) \lor Q(3,3)) \land (\neg P(0,4,3) \lor \neg Q(3,3) \lor \neg S(4,3) \lor Q(3,4)) \land$
$(\neg P(1,4,0) \lor \neg Q(3,0) \lor \neg S(4,0) \lor Q(3,1)) \land (\neg P(1,4,1) \lor \neg Q(3,1) \lor \neg S(4,1) \lor Q(3,2)) \land$
$(\neg P(1,4,2) \lor \neg Q(3,2) \lor \neg S(4,2) \lor Q(3,3)) \land (\neg P(1,4,3) \lor \neg Q(3,3) \lor \neg S(4,3) \lor Q(3,4)) \land$
$(\neg P(\#,4,0) \lor \neg Q(3,0) \lor \neg S(4,0) \lor Q(3,1)) \land (\neg P(\#,4,1) \lor \neg Q(3,1) \lor \neg S(4,1) \lor Q(3,2)) \land$
$(\neg P(\#,4,2) \lor \neg Q(3,2) \lor \neg S(4,2) \lor Q(3,3)) \land (\neg P(\#,4,3) \lor \neg Q(3,3) \lor \neg S(4,3) \lor Q(3,4)) \land$
$(\neg P(0,0,0) \lor \neg Q(0,0) \lor \neg S(0,0) \lor P(0,0,1)) \land (\neg P(0,0,1) \lor \neg Q(0,1) \lor \neg S(0,1) \lor P(0,0,2)) \land$ // group 6
$(\neg P(0,0,2) \lor \neg Q(0,2) \lor \neg S(0,2) \lor P(0,0,3)) \land (\neg P(0,0,3) \lor \neg Q(0,3) \lor \neg S(0,3) \lor P(0,0,4)) \land$
$(\neg P(1,0,0) \lor \neg Q(0,0) \lor \neg S(0,0) \lor P(1,0,1)) \land (\neg P(1,0,1) \lor \neg Q(0,1) \lor \neg S(0,1) \lor P(1,0,2)) \land$
$(\neg P(1,0,2) \lor \neg Q(0,2) \lor \neg S(0,2) \lor P(1,0,3)) \land (\neg P(1,0,3) \lor \neg Q(0,3) \lor \neg S(0,3) \lor P(1,0,4)) \land$

// A blank # in cell 0 in state 0 causes M to halt and retain the blank in cell 0:
$(\neg P(\#,0,0) \lor \neg Q(0,0) \lor \neg S(0,0) \lor P(\#,0,1)) \land (\neg P(\#,0,1) \lor \neg Q(0,1) \lor \neg S(0,1) \lor P(\#,0,2)) \land$
$(\neg P(\#,0,2) \lor \neg Q(0,2) \lor \neg S(0,2) \lor P(\#,0,3)) \land (\neg P(\#,0,3) \lor \neg Q(0,3) \lor \neg S(0,3) \lor P(\#,0,4)) \land$

$(\neg P(0,1,0) \lor \neg Q(0,0) \lor \neg S(1,0) \lor P(0,1,1)) \land (\neg P(0,1,1) \lor \neg Q(0,1) \lor \neg S(1,1) \lor P(0,1,2)) \land$
$(\neg P(0,1,2) \lor \neg Q(0,2) \lor \neg S(1,2) \lor P(0,1,3)) \land (\neg P(0,1,3) \lor \neg Q(0,3) \lor \neg S(1,3) \lor P(0,1,4)) \land$
$(\neg P(1,1,0) \lor \neg Q(0,0) \lor \neg S(1,0) \lor P(1,1,1)) \land (\neg P(1,1,1) \lor \neg Q(0,1) \lor \neg S(1,1) \lor P(1,1,2)) \land$
$(\neg P(1,1,2) \lor \neg Q(0,2) \lor \neg S(1,2) \lor P(1,1,3)) \land (\neg P(1,1,3) \lor \neg Q(0,3) \lor \neg S(1,3) \lor P(1,1,4)) \land$
$(\neg P(\#,1,0) \lor \neg Q(0,0) \lor \neg S(1,0) \lor P(0,1,1)) \land (\neg P(\#,1,1) \lor \neg Q(0,1) \lor \neg S(1,1) \lor P(0,1,2)) \land$
$(\neg P(\#,1,2) \lor \neg Q(0,2) \lor \neg S(1,2) \lor P(0,1,3)) \land (\neg P(\#,1,3) \lor \neg Q(0,3) \lor \neg S(1,3) \lor P(0,1,4)) \land$
$(\neg P(0,2,0) \lor \neg Q(0,0) \lor \neg S(2,0) \lor P(0,2,1)) \land (\neg P(0,2,1) \lor \neg Q(0,1) \lor \neg S(2,1) \lor P(0,2,2)) \land$
$(\neg P(0,2,2) \lor \neg Q(0,2) \lor \neg S(2,2) \lor P(0,2,3)) \land (\neg P(0,2,3) \lor \neg Q(0,3) \lor \neg S(2,3) \lor P(0,2,4)) \land$
$(\neg P(1,2,0) \lor \neg Q(0,0) \lor \neg S(2,0) \lor P(1,2,1)) \land (\neg P(1,2,1) \lor \neg Q(0,1) \lor \neg S(2,1) \lor P(1,2,2)) \land$
$(\neg P(1,2,2) \lor \neg Q(0,2) \lor \neg S(2,2) \lor P(1,2,3)) \land (\neg P(1,2,3) \lor \neg Q(0,3) \lor \neg S(2,3) \lor P(1,2,4)) \land$
$(\neg P(\#,2,0) \lor \neg Q(0,0) \lor \neg S(2,0) \lor P(0,2,1)) \land (\neg P(\#,2,1) \lor \neg Q(0,1) \lor \neg S(2,1) \lor P(0,2,2)) \land$
$(\neg P(\#,2,2) \lor \neg Q(0,2) \lor \neg S(2,2) \lor P(0,2,3)) \land (\neg P(\#,2,3) \lor \neg Q(0,3) \lor \neg S(2,3) \lor P(0,2,4)) \land$
$(\neg P(0,3,0) \lor \neg Q(0,0) \lor \neg S(3,0) \lor P(0,3,1)) \land (\neg P(0,3,1) \lor \neg Q(0,1) \lor \neg S(3,1) \lor P(0,3,2)) \land$
$(\neg P(0,3,2) \lor \neg Q(0,2) \lor \neg S(3,2) \lor P(0,3,3)) \land (\neg P(0,3,3) \lor \neg Q(0,3) \lor \neg S(3,3) \lor P(0,3,4)) \land$
$(\neg P(1,3,0) \lor \neg Q(0,0) \lor \neg S(3,0) \lor P(1,3,1)) \land (\neg P(1,3,1) \lor \neg Q(0,1) \lor \neg S(3,1) \lor P(1,3,2)) \land$
$(\neg P(1,3,2) \lor \neg Q(0,2) \lor \neg S(3,2) \lor P(1,3,3)) \land (\neg P(1,3,3) \lor \neg Q(0,3) \lor \neg S(3,3) \lor P(1,3,4)) \land$
$(\neg P(\#,3,0) \lor \neg Q(0,0) \lor \neg S(3,0) \lor P(0,3,1)) \land (\neg P(\#,3,1) \lor \neg Q(0,1) \lor \neg S(3,1) \lor P(0,3,2)) \land$
$(\neg P(\#,3,2) \lor \neg Q(0,2) \lor \neg S(3,2) \lor P(0,3,3)) \land (\neg P(\#,3,3) \lor \neg Q(0,3) \lor \neg S(3,3) \lor P(0,3,4)) \land$
$(\neg P(0,4,0) \lor \neg Q(0,0) \lor \neg S(4,0) \lor P(0,4,1)) \land (\neg P(0,4,1) \lor \neg Q(0,1) \lor \neg S(4,1) \lor P(0,4,2)) \land$
$(\neg P(0,4,2) \lor \neg Q(0,2) \lor \neg S(4,2) \lor P(0,4,3)) \land (\neg P(0,4,3) \lor \neg Q(0,3) \lor \neg S(4,3) \lor P(0,4,4)) \land$
$(\neg P(1,4,0) \lor \neg Q(0,0) \lor \neg S(4,0) \lor P(1,4,1)) \land (\neg P(1,4,1) \lor \neg Q(0,1) \lor \neg S(4,1) \lor P(1,4,2)) \land$
$(\neg P(1,4,2) \lor \neg Q(0,2) \lor \neg S(4,2) \lor P(1,4,3)) \land (\neg P(1,4,3) \lor \neg Q(0,3) \lor \neg S(4,3) \lor P(1,4,4)) \land$
$(\neg P(\#,4,0) \lor \neg Q(0,0) \lor \neg S(4,0) \lor P(0,4,1)) \land (\neg P(\#,4,1) \lor \neg Q(0,1) \lor \neg S(4,1) \lor P(0,4,2)) \land$
$(\neg P(\#,4,2) \lor \neg Q(0,2) \lor \neg S(4,2) \lor P(0,4,3)) \land (\neg P(\#,4,3) \lor \neg Q(0,3) \lor \neg S(4,3) \lor P(0,4,4)) \land$
$(\neg P(0,0,0) \lor \neg Q(1,0) \lor \neg S(0,0) \lor P(0,0,1)) \land (\neg P(0,0,1) \lor \neg Q(1,1) \lor \neg S(0,1) \lor P(0,0,2)) \land$
$(\neg P(0,0,2) \lor \neg Q(1,2) \lor \neg S(0,2) \lor P(0,0,3)) \land (\neg P(0,0,3) \lor \neg Q(1,3) \lor \neg S(0,3) \lor P(0,0,4)) \land$
$(\neg P(1,0,0) \lor \neg Q(1,0) \lor \neg S(0,0) \lor P(1,0,1)) \land (\neg P(1,0,1) \lor \neg Q(1,1) \lor \neg S(0,1) \lor P(1,0,2)) \land$
$(\neg P(1,0,2) \lor \neg Q(1,2) \lor \neg S(0,2) \lor P(1,0,3)) \land (\neg P(1,0,3) \lor \neg Q(1,3) \lor \neg S(0,3) \lor P(1,0,4)) \land$

// A blank # in cell 0 in state 1 causes M to halt and retain the blank in cell 0:
$(\neg P(\#,0,0) \lor \neg Q(1,0) \lor \neg S(0,0) \lor P(\#,0,1)) \land (\neg P(\#,0,1) \lor \neg Q(1,1) \lor \neg S(0,1) \lor P(\#,0,2)) \land$
$(\neg P(\#,0,2) \lor \neg Q(1,2) \lor \neg S(0,2) \lor P(\#,0,3)) \land (\neg P(\#,0,3) \lor \neg Q(1,3) \lor \neg S(0,3) \lor P(\#,0,4)) \land$
$(\neg P(0,1,0) \lor \neg Q(1,0) \lor \neg S(1,0) \lor P(0,1,1)) \land (\neg P(0,1,1) \lor \neg Q(1,1) \lor \neg S(1,1) \lor P(0,1,2)) \land$
$(\neg P(0,1,2) \lor \neg Q(1,2) \lor \neg S(1,2) \lor P(0,1,3)) \land (\neg P(0,1,3) \lor \neg Q(1,3) \lor \neg S(1,3) \lor P(0,1,4)) \land$
$(\neg P(1,1,0) \lor \neg Q(1,0) \lor \neg S(1,0) \lor P(1,1,1)) \land (\neg P(1,1,1) \lor \neg Q(1,1) \lor \neg S(1,1) \lor P(1,1,2)) \land$
$(\neg P(1,1,2) \lor \neg Q(1,2) \lor \neg S(1,2) \lor P(1,1,3)) \land (\neg P(1,1,3) \lor \neg Q(1,3) \lor \neg S(1,3) \lor P(1,1,4)) \land$
$(\neg P(\#,1,0) \lor \neg Q(1,0) \lor \neg S(1,0) \lor P(1,1,1)) \land (\neg P(\#,1,1) \lor \neg Q(1,1) \lor \neg S(1,1) \lor P(1,1,2)) \land$
$(\neg P(\#,1,2) \lor \neg Q(1,2) \lor \neg S(1,2) \lor P(1,1,3)) \land (\neg P(\#,1,3) \lor \neg Q(1,3) \lor \neg S(1,3) \lor P(1,1,4)) \land$
$(\neg P(0,2,0) \lor \neg Q(1,0) \lor \neg S(2,0) \lor P(0,2,1)) \land (\neg P(0,2,1) \lor \neg Q(1,1) \lor \neg S(2,1) \lor P(0,2,2)) \land$
$(\neg P(0,2,2) \lor \neg Q(1,2) \lor \neg S(2,2) \lor P(0,2,3)) \land (\neg P(0,2,3) \lor \neg Q(1,3) \lor \neg S(2,3) \lor P(0,2,4)) \land$
$(\neg P(1,2,0) \lor \neg Q(1,0) \lor \neg S(2,0) \lor P(1,2,1)) \land (\neg P(1,2,1) \lor \neg Q(1,1) \lor \neg S(2,1) \lor P(1,2,2)) \land$
$(\neg P(1,2,2) \lor \neg Q(1,2) \lor \neg S(2,2) \lor P(1,2,3)) \land (\neg P(1,2,3) \lor \neg Q(1,3) \lor \neg S(2,3) \lor P(1,2,4)) \land$
$(\neg P(\#,2,0) \lor \neg Q(1,0) \lor \neg S(2,0) \lor P(1,2,1)) \land (\neg P(\#,2,1) \lor \neg Q(1,1) \lor \neg S(2,1) \lor P(1,2,2)) \land$
$(\neg P(\#,2,2) \lor \neg Q(1,2) \lor \neg S(2,2) \lor P(1,2,3)) \land (\neg P(\#,2,3) \lor \neg Q(1,3) \lor \neg S(2,3) \lor P(1,2,4)) \land$
$(\neg P(0,3,0) \lor \neg Q(1,0) \lor \neg S(3,0) \lor P(0,3,1)) \land (\neg P(0,3,1) \lor \neg Q(1,1) \lor \neg S(3,1) \lor P(0,3,2)) \land$
$(\neg P(0,3,2) \lor \neg Q(1,2) \lor \neg S(3,2) \lor P(0,3,3)) \land (\neg P(0,3,3) \lor \neg Q(1,3) \lor \neg S(3,3) \lor P(0,3,4)) \land$
$(\neg P(1,3,0) \lor \neg Q(1,0) \lor \neg S(3,0) \lor P(1,3,1)) \land (\neg P(1,3,1) \lor \neg Q(1,1) \lor \neg S(3,1) \lor P(1,3,2)) \land$

$(\neg P(1,3,2) \lor \neg Q(1,2) \lor \neg S(3,2) \lor P(1,3,3)) \land (\neg P(1,3,3) \lor \neg Q(1,3) \lor \neg S(3,3) \lor P(1,3,4)) \land$
$(\neg P(\#,3,0) \lor \neg Q(1,0) \lor \neg S(3,0) \lor P(1,3,1)) \land (\neg P(\#,3,1) \lor \neg Q(1,1) \lor \neg S(3,1) \lor P(1,3,2)) \land$
$(\neg P(\#,3,2) \lor \neg Q(1,2) \lor \neg S(3,2) \lor P(1,3,3)) \land (\neg P(\#,3,3) \lor \neg Q(1,3) \lor \neg S(3,3) \lor P(1,3,4)) \land$
$(\neg P(0,4,0) \lor \neg Q(1,0) \lor \neg S(4,0) \lor P(0,4,1)) \land (\neg P(0,4,1) \lor \neg Q(1,1) \lor \neg S(4,1) \lor P(0,4,2)) \land$
$(\neg P(0,4,2) \lor \neg Q(1,2) \lor \neg S(4,2) \lor P(0,4,3)) \land (\neg P(0,4,3) \lor \neg Q(1,3) \lor \neg S(4,3) \lor P(0,4,4)) \land$
$(\neg P(1,4,0) \lor \neg Q(1,0) \lor \neg S(4,0) \lor P(1,4,1)) \land (\neg P(1,4,1) \lor \neg Q(1,1) \lor \neg S(4,1) \lor P(1,4,2)) \land$
$(\neg P(1,4,2) \lor \neg Q(1,2) \lor \neg S(4,2) \lor P(1,4,3)) \land (\neg P(1,4,3) \lor \neg Q(1,3) \lor \neg S(4,3) \lor P(1,4,4)) \land$
$(\neg P(\#,4,0) \lor \neg Q(1,0) \lor \neg S(4,0) \lor P(1,4,1)) \land (\neg P(\#,4,1) \lor \neg Q(1,1) \lor \neg S(4,1) \lor P(1,4,2)) \land$
$(\neg P(\#,4,2) \lor \neg Q(1,2) \lor \neg S(4,2) \lor P(1,4,3)) \land (\neg P(\#,4,3) \lor \neg Q(1,3) \lor \neg S(4,3) \lor P(1,4,4)) \land$
$(\neg P(0,0,0) \lor \neg Q(2,0) \lor \neg S(0,0) \lor P(\#,0,1)) \land (\neg P(0,0,1) \lor \neg Q(2,1) \lor \neg S(0,1) \lor P(\#,0,2)) \land$
$(\neg P(0,0,2) \lor \neg Q(2,2) \lor \neg S(0,2) \lor P(\#,0,3)) \land (\neg P(0,0,3) \lor \neg Q(2,3) \lor \neg S(0,3) \lor P(\#,0,4)) \land$
$(\neg P(1,0,0) \lor \neg Q(2,0) \lor \neg S(0,0) \lor P(\#,0,1)) \land (\neg P(1,0,1) \lor \neg Q(2,1) \lor \neg S(0,1) \lor P(\#,0,2)) \land$
$(\neg P(1,0,2) \lor \neg Q(2,2) \lor \neg S(0,2) \lor P(\#,0,3)) \land (\neg P(1,0,3) \lor \neg Q(2,3) \lor \neg S(0,3) \lor P(\#,0,4)) \land$
$(\neg P(0,1,0) \lor \neg Q(2,0) \lor \neg S(1,0) \lor P(\#,1,1)) \land (\neg P(0,1,1) \lor \neg Q(2,1) \lor \neg S(1,1) \lor P(\#,1,2)) \land$
$(\neg P(0,1,2) \lor \neg Q(2,2) \lor \neg S(1,2) \lor P(\#,1,3)) \land (\neg P(0,1,3) \lor \neg Q(2,3) \lor \neg S(1,3) \lor P(\#,1,4)) \land$
$(\neg P(1,1,0) \lor \neg Q(2,0) \lor \neg S(1,0) \lor P(\#,1,1)) \land (\neg P(1,1,1) \lor \neg Q(2,1) \lor \neg S(1,1) \lor P(\#,1,2)) \land$
$(\neg P(1,1,2) \lor \neg Q(2,2) \lor \neg S(1,2) \lor P(\#,1,3)) \land (\neg P(1,1,3) \lor \neg Q(2,3) \lor \neg S(1,3) \lor P(\#,1,4)) \land$
$(\neg P(0,2,0) \lor \neg Q(2,0) \lor \neg S(2,0) \lor P(\#,2,1)) \land (\neg P(0,2,1) \lor \neg Q(2,1) \lor \neg S(2,1) \lor P(\#,2,2)) \land$
$(\neg P(0,2,2) \lor \neg Q(2,2) \lor \neg S(2,2) \lor P(\#,2,3)) \land (\neg P(0,2,3) \lor \neg Q(2,3) \lor \neg S(2,3) \lor P(\#,2,4)) \land$
$(\neg P(1,2,0) \lor \neg Q(2,0) \lor \neg S(2,0) \lor P(\#,2,1)) \land (\neg P(1,2,1) \lor \neg Q(2,1) \lor \neg S(2,1) \lor P(\#,2,2)) \land$
$(\neg P(1,2,2) \lor \neg Q(2,2) \lor \neg S(2,2) \lor P(\#,2,3)) \land (\neg P(1,2,3) \lor \neg Q(2,3) \lor \neg S(2,3) \lor P(\#,2,4)) \land$
$(\neg P(0,3,0) \lor \neg Q(2,0) \lor \neg S(3,0) \lor P(\#,3,1)) \land (\neg P(0,3,1) \lor \neg Q(2,1) \lor \neg S(3,1) \lor P(\#,3,2)) \land$
$(\neg P(0,3,2) \lor \neg Q(2,2) \lor \neg S(3,2) \lor P(\#,3,3)) \land (\neg P(0,3,3) \lor \neg Q(2,3) \lor \neg S(3,3) \lor P(\#,3,4)) \land$
$(\neg P(1,3,0) \lor \neg Q(2,0) \lor \neg S(3,0) \lor P(\#,3,1)) \land (\neg P(1,3,1) \lor \neg Q(2,1) \lor \neg S(3,1) \lor P(\#,3,2)) \land$
$(\neg P(1,3,2) \lor \neg Q(2,2) \lor \neg S(3,2) \lor P(\#,3,3)) \land (\neg P(1,3,3) \lor \neg Q(2,3) \lor \neg S(3,3) \lor P(\#,3,4)) \land$
$(\neg P(0,4,0) \lor \neg Q(2,0) \lor \neg S(4,0) \lor P(\#,4,1)) \land (\neg P(0,4,1) \lor \neg Q(2,1) \lor \neg S(4,1) \lor P(\#,4,2)) \land$
$(\neg P(0,4,2) \lor \neg Q(2,2) \lor \neg S(4,2) \lor P(\#,4,3)) \land (\neg P(0,4,3) \lor \neg Q(2,3) \lor \neg S(4,3) \lor P(\#,4,4)) \land$
$(\neg P(1,4,0) \lor \neg Q(2,0) \lor \neg S(4,0) \lor P(\#,4,1)) \land (\neg P(1,4,1) \lor \neg Q(2,1) \lor \neg S(4,1) \lor P(\#,4,2)) \land$
$(\neg P(1,4,2) \lor \neg Q(2,2) \lor \neg S(4,2) \lor P(\#,4,3)) \land (\neg P(1,4,3) \lor \neg Q(2,3) \lor \neg S(4,3) \lor P(\#,4,4)) \land$
$(\neg P(0,0,0) \lor \neg Q(3,0) \lor \neg S(0,0) \lor P(0,0,1)) \land (\neg P(0,0,1) \lor \neg Q(3,1) \lor \neg S(0,1) \lor P(0,0,2)) \land$
$(\neg P(0,0,2) \lor \neg Q(3,2) \lor \neg S(0,2) \lor P(0,0,3)) \land (\neg P(0,0,3) \lor \neg Q(3,3) \lor \neg S(0,3) \lor P(0,0,4)) \land$
$(\neg P(1,0,0) \lor \neg Q(3,0) \lor \neg S(0,0) \lor P(1,0,1)) \land (\neg P(1,0,1) \lor \neg Q(3,1) \lor \neg S(0,1) \lor P(1,0,2)) \land$
$(\neg P(1,0,2) \lor \neg Q(3,2) \lor \neg S(0,2) \lor P(1,0,3)) \land (\neg P(1,0,3) \lor \neg Q(3,3) \lor \neg S(0,3) \lor P(1,0,4)) \land$
$(\neg P(\#,0,0) \lor \neg Q(3,0) \lor \neg S(0,0) \lor P(\#,0,1)) \land (\neg P(\#,0,1) \lor \neg Q(3,1) \lor \neg S(0,1) \lor P(\#,0,2)) \land$
$(\neg P(\#,0,2) \lor \neg Q(3,2) \lor \neg S(0,2) \lor P(\#,0,3)) \land (\neg P(\#,0,3) \lor \neg Q(3,3) \lor \neg S(0,3) \lor P(\#,0,4)) \land$
$(\neg P(0,1,0) \lor \neg Q(3,0) \lor \neg S(1,0) \lor P(0,1,1)) \land (\neg P(0,1,1) \lor \neg Q(3,1) \lor \neg S(1,1) \lor P(0,1,2)) \land$
$(\neg P(0,1,2) \lor \neg Q(3,2) \lor \neg S(1,2) \lor P(0,1,3)) \land (\neg P(0,1,3) \lor \neg Q(3,3) \lor \neg S(1,3) \lor P(0,1,4)) \land$
$(\neg P(1,1,0) \lor \neg Q(3,0) \lor \neg S(1,0) \lor P(1,1,1)) \land (\neg P(1,1,1) \lor \neg Q(3,1) \lor \neg S(1,1) \lor P(1,1,2)) \land$
$(\neg P(1,1,2) \lor \neg Q(3,2) \lor \neg S(1,2) \lor P(1,1,3)) \land (\neg P(1,1,3) \lor \neg Q(3,3) \lor \neg S(1,3) \lor P(1,1,4)) \land$
$(\neg P(\#,1,0) \lor \neg Q(3,0) \lor \neg S(1,0) \lor P(\#,1,1)) \land (\neg P(\#,1,1) \lor \neg Q(3,1) \lor \neg S(1,1) \lor P(\#,1,2)) \land$
$(\neg P(\#,1,2) \lor \neg Q(3,2) \lor \neg S(1,2) \lor P(\#,1,3)) \land (\neg P(\#,1,3) \lor \neg Q(3,3) \lor \neg S(1,3) \lor P(\#,1,4)) \land$
$(\neg P(0,2,0) \lor \neg Q(3,0) \lor \neg S(2,0) \lor P(0,2,1)) \land (\neg P(0,2,1) \lor \neg Q(3,1) \lor \neg S(2,1) \lor P(0,2,2)) \land$
$(\neg P(0,2,2) \lor \neg Q(3,2) \lor \neg S(2,2) \lor P(0,2,3)) \land (\neg P(0,2,3) \lor \neg Q(3,3) \lor \neg S(2,3) \lor P(0,2,4)) \land$
$(\neg P(1,2,0) \lor \neg Q(3,0) \lor \neg S(2,0) \lor P(1,2,1)) \land (\neg P(1,2,1) \lor \neg Q(3,1) \lor \neg S(2,1) \lor P(1,2,2)) \land$
$(\neg P(1,2,2) \lor \neg Q(3,2) \lor \neg S(2,2) \lor P(1,2,3)) \land (\neg P(1,2,3) \lor \neg Q(3,3) \lor \neg S(2,3) \lor P(1,2,4)) \land$
$(\neg P(\#,2,0) \lor \neg Q(3,0) \lor \neg S(2,0) \lor P(\#,2,1)) \land (\neg P(\#,2,1) \lor \neg Q(3,1) \lor \neg S(2,1) \lor P(\#,2,2)) \land$

$(\neg P(\#,2,2) \lor \neg Q(3,2) \lor \neg S(2,2) \lor P(\#,2,3)) \land (\neg P(\#,2,3) \lor \neg Q(3,3) \lor \neg S(2,3) \lor P(\#,2,4)) \land$
$(\neg P(0,3,0) \lor \neg Q(3,0) \lor \neg S(3,0) \lor P(0,3,1)) \land (\neg P(0,3,1) \lor \neg Q(3,1) \lor \neg S(3,1) \lor P(0,3,2)) \land$
$(\neg P(0,3,2) \lor \neg Q(3,2) \lor \neg S(3,2) \lor P(0,3,3)) \land (\neg P(0,3,3) \lor \neg Q(3,3) \lor \neg S(3,3) \lor P(0,3,4)) \land$
$(\neg P(1,3,0) \lor \neg Q(3,0) \lor \neg S(3,0) \lor P(1,3,1)) \land (\neg P(1,3,1) \lor \neg Q(3,1) \lor \neg S(3,1) \lor P(1,3,2)) \land$
$(\neg P(1,3,2) \lor \neg Q(3,2) \lor \neg S(3,2) \lor P(1,3,3)) \land (\neg P(1,3,3) \lor \neg Q(3,3) \lor \neg S(3,3) \lor P(1,3,4)) \land$
$(\neg P(\#,3,0) \lor \neg Q(3,0) \lor \neg S(3,0) \lor P(\#,3,1)) \land (\neg P(\#,3,1) \lor \neg Q(3,1) \lor \neg S(3,1) \lor P(\#,3,2)) \land$
$(\neg P(\#,3,2) \lor \neg Q(3,2) \lor \neg S(3,2) \lor P(\#,3,3)) \land (\neg P(\#,3,3) \lor \neg Q(3,3) \lor \neg S(3,3) \lor P(\#,3,4)) \land$
$(\neg P(0,4,0) \lor \neg Q(3,0) \lor \neg S(4,0) \lor P(0,4,1)) \land (\neg P(0,4,1) \lor \neg Q(3,1) \lor \neg S(4,1) \lor P(0,4,2)) \land$
$(\neg P(0,4,2) \lor \neg Q(3,2) \lor \neg S(4,2) \lor P(0,4,3)) \land (\neg P(0,4,3) \lor \neg Q(3,3) \lor \neg S(4,3) \lor P(0,4,4)) \land$
$(\neg P(1,4,0) \lor \neg Q(3,0) \lor \neg S(4,0) \lor P(1,4,1)) \land (\neg P(1,4,1) \lor \neg Q(3,1) \lor \neg S(4,1) \lor P(1,4,2)) \land$
$(\neg P(1,4,2) \lor \neg Q(3,2) \lor \neg S(4,2) \lor P(1,4,3)) \land (\neg P(1,4,3) \lor \neg Q(3,3) \lor \neg S(4,3) \lor P(1,4,4)) \land$
$(\neg P(\#,4,0) \lor \neg Q(3,0) \lor \neg S(4,0) \lor P(\#,4,1)) \land (\neg P(\#,4,1) \lor \neg Q(3,1) \lor \neg S(4,1) \lor P(\#,4,2)) \land$
$(\neg P(\#,4,2) \lor \neg Q(3,2) \lor \neg S(4,2) \lor P(\#,4,3)) \land (\neg P(\#,4,3) \lor \neg Q(3,3) \lor \neg S(4,3) \lor P(\#,4,4)) \land$
$(\neg P(0,0,0) \lor \neg Q(0,0) \lor \neg S(0,0) \lor S(1,1)) \land (\neg P(0,0,1) \lor \neg Q(0,1) \lor \neg S(0,1) \lor S(1,2)) \land$ // group 7
$(\neg P(0,0,2) \lor \neg Q(0,2) \lor \neg S(0,2) \lor S(1,3)) \land (\neg P(0,0,3) \lor \neg Q(0,3) \lor \neg S(0,3) \lor S(1,4)) \land$
$(\neg P(1,0,0) \lor \neg Q(0,0) \lor \neg S(0,0) \lor S(1,1)) \land (\neg P(1,0,1) \lor \neg Q(0,1) \lor \neg S(0,1) \lor S(1,2)) \land$
$(\neg P(1,0,2) \lor \neg Q(0,2) \lor \neg S(0,2) \lor S(1,3)) \land (\neg P(1,0,3) \lor \neg Q(0,3) \lor \neg S(0,3) \lor S(1,4)) \land$
// passing from $S(0,0)$ to $S(-1,1)$ means sliding off the tape, that is, halting execution; M remains in cell 0:
$(\neg P(\#,0,0) \lor \neg Q(0,0) \lor \neg S(0,0) \lor S(0,1)) \land (\neg P(\#,0,1) \lor \neg Q(0,1) \lor \neg S(0,1) \lor S(0,2)) \land$
$(\neg P(\#,0,2) \lor \neg Q(0,2) \lor \neg S(0,2) \lor S(0,3)) \land (\neg P(\#,0,3) \lor \neg Q(0,3) \lor \neg S(0,3) \lor S(0,4)) \land$
$(\neg P(0,1,0) \lor \neg Q(0,0) \lor \neg S(1,0) \lor S(2,1)) \land (\neg P(0,1,1) \lor \neg Q(0,1) \lor \neg S(1,1) \lor S(2,2)) \land$
$(\neg P(0,1,2) \lor \neg Q(0,2) \lor \neg S(1,2) \lor S(2,3)) \land (\neg P(0,1,3) \lor \neg Q(0,3) \lor \neg S(1,3) \lor S(2,4)) \land$
$(\neg P(1,1,0) \lor \neg Q(0,0) \lor \neg S(1,0) \lor S(2,1)) \land (\neg P(1,1,1) \lor \neg Q(0,1) \lor \neg S(1,1) \lor S(2,2)) \land$
$(\neg P(1,1,2) \lor \neg Q(0,2) \lor \neg S(1,2) \lor S(2,3)) \land (\neg P(1,1,3) \lor \neg Q(0,3) \lor \neg S(1,3) \lor S(2,4)) \land$
$(\neg P(\#,1,0) \lor \neg Q(0,0) \lor \neg S(1,0) \lor S(0,1)) \land (\neg P(\#,1,1) \lor \neg Q(0,1) \lor \neg S(1,1) \lor S(0,2)) \land$
$(\neg P(\#,1,2) \lor \neg Q(0,2) \lor \neg S(1,2) \lor S(0,3)) \land (\neg P(\#,1,3) \lor \neg Q(0,3) \lor \neg S(1,3) \lor S(0,4)) \land$
$(\neg P(0,2,0) \lor \neg Q(0,0) \lor \neg S(2,0) \lor S(3,1)) \land (\neg P(0,2,1) \lor \neg Q(0,1) \lor \neg S(2,1) \lor S(3,2)) \land$
$(\neg P(0,2,2) \lor \neg Q(0,2) \lor \neg S(2,2) \lor S(3,3)) \land (\neg P(0,2,3) \lor \neg Q(0,3) \lor \neg S(2,3) \lor S(3,4)) \land$
$(\neg P(1,2,0) \lor \neg Q(0,0) \lor \neg S(2,0) \lor S(3,1)) \land (\neg P(1,2,1) \lor \neg Q(0,1) \lor \neg S(2,1) \lor S(3,2)) \land$
$(\neg P(1,2,2) \lor \neg Q(0,2) \lor \neg S(2,2) \lor S(3,3)) \land (\neg P(1,2,3) \lor \neg Q(0,3) \lor \neg S(2,3) \lor S(3,4)) \land$
$(\neg P(\#,2,0) \lor \neg Q(0,0) \lor \neg S(2,0) \lor S(1,1)) \land (\neg P(\#,2,1) \lor \neg Q(0,1) \lor \neg S(2,1) \lor S(1,2)) \land$
$(\neg P(\#,2,2) \lor \neg Q(0,2) \lor \neg S(2,2) \lor S(1,3)) \land (\neg P(\#,2,3) \lor \neg Q(0,3) \lor \neg S(2,3) \lor S(1,4)) \land$
$(\neg P(0,3,0) \lor \neg Q(0,0) \lor \neg S(3,0) \lor S(4,1)) \land (\neg P(0,3,1) \lor \neg Q(0,1) \lor \neg S(3,1) \lor S(4,2)) \land$
$(\neg P(0,3,2) \lor \neg Q(0,2) \lor \neg S(3,2) \lor S(4,3)) \land (\neg P(0,3,3) \lor \neg Q(0,3) \lor \neg S(3,3) \lor S(4,4)) \land$
$(\neg P(1,3,0) \lor \neg Q(0,0) \lor \neg S(3,0) \lor S(4,1)) \land (\neg P(1,3,1) \lor \neg Q(0,1) \lor \neg S(3,1) \lor S(4,2)) \land$
$(\neg P(1,3,2) \lor \neg Q(0,2) \lor \neg S(3,2) \lor S(4,3)) \land (\neg P(1,3,3) \lor \neg Q(0,3) \lor \neg S(3,3) \lor S(4,4)) \land$
$(\neg P(\#,3,0) \lor \neg Q(0,0) \lor \neg S(3,0) \lor S(2,1)) \land (\neg P(\#,3,1) \lor \neg Q(0,1) \lor \neg S(3,1) \lor S(2,2)) \land$
$(\neg P(\#,3,2) \lor \neg Q(0,2) \lor \neg S(3,2) \lor S(2,3)) \land (\neg P(\#,3,3) \lor \neg Q(0,3) \lor \neg S(3,3) \lor S(2,4)) \land$
$(\neg P(0,4,0) \lor \neg Q(0,0) \lor \neg S(4,0) \lor S(5,1)) \land (\neg P(0,4,1) \lor \neg Q(0,1) \lor \neg S(4,1) \lor S(5,2)) \land$
$(\neg P(0,4,2) \lor \neg Q(0,2) \lor \neg S(4,2) \lor S(5,3)) \land (\neg P(0,4,3) \lor \neg Q(0,3) \lor \neg S(4,3) \lor S(5,4)) \land$
$(\neg P(1,4,0) \lor \neg Q(0,0) \lor \neg S(4,0) \lor S(5,1)) \land (\neg P(1,4,1) \lor \neg Q(0,1) \lor \neg S(4,1) \lor S(5,2)) \land$
$(\neg P(1,4,2) \lor \neg Q(0,2) \lor \neg S(4,2) \lor S(5,3)) \land (\neg P(1,4,3) \lor \neg Q(0,3) \lor \neg S(4,3) \lor S(5,4)) \land$
$(\neg P(\#,4,0) \lor \neg Q(0,0) \lor \neg S(4,0) \lor S(3,1)) \land (\neg P(\#,4,1) \lor \neg Q(0,1) \lor \neg S(4,1) \lor S(3,2)) \land$
$(\neg P(\#,4,2) \lor \neg Q(0,2) \lor \neg S(4,2) \lor S(3,3)) \land (\neg P(\#,4,3) \lor \neg Q(0,3) \lor \neg S(4,3) \lor S(3,4)) \land$
$(\neg P(0,0,0) \lor \neg Q(1,0) \lor \neg S(0,0) \lor S(1,1)) \land (\neg P(0,0,1) \lor \neg Q(1,1) \lor \neg S(0,1) \lor S(1,2)) \land$
$(\neg P(0,0,2) \lor \neg Q(1,2) \lor \neg S(0,2) \lor S(1,3)) \land (\neg P(0,0,3) \lor \neg Q(1,3) \lor \neg S(0,3) \lor S(1,4)) \land$
$(\neg P(1,0,0) \lor \neg Q(1,0) \lor \neg S(0,0) \lor S(1,1)) \land (\neg P(1,0,1) \lor \neg Q(1,1) \lor \neg S(0,1) \lor S(1,2)) \land$

$(\neg P(1,0,2) \lor \neg Q(1,2) \lor \neg S(0,2) \lor S(1,3)) \land (\neg P(1,0,3) \lor \neg Q(1,3) \lor \neg S(0,3) \lor S(1,4)) \land$
// an attempt to slide off the tape; M halts and remains in cell 0:
$(\neg P(\#,0,0) \lor \neg Q(1,0) \lor \neg S(0,0) \lor S(0,1)) \land (\neg P(\#,0,1) \lor \neg Q(1,1) \lor \neg S(0,1) \lor S(0,2)) \land$
$(\neg P(\#,0,2) \lor \neg Q(1,2) \lor \neg S(0,2) \lor S(0,3)) \land (\neg P(\#,0,3) \lor \neg Q(1,3) \lor \neg S(0,3) \lor S(0,4)) \land$ //can't be in state 1
$(\neg P(0,1,0) \lor \neg Q(1,0) \lor \neg S(1,0) \lor S(2,1)) \land (\neg P(0,1,1) \lor \neg Q(1,1) \lor \neg S(1,1) \lor S(2,2)) \land$
$(\neg P(0,1,2) \lor \neg Q(1,2) \lor \neg S(1,2) \lor S(2,3)) \land (\neg P(0,1,3) \lor \neg Q(1,3) \lor \neg S(1,3) \lor S(2,4)) \land$
$(\neg P(1,1,0) \lor \neg Q(1,0) \lor \neg S(1,0) \lor S(2,1)) \land (\neg P(1,1,1) \lor \neg Q(1,1) \lor \neg S(1,1) \lor S(2,2)) \land$
$(\neg P(1,1,2) \lor \neg Q(1,2) \lor \neg S(1,2) \lor S(2,3)) \land (\neg P(1,1,3) \lor \neg Q(1,3) \lor \neg S(1,3) \lor S(2,4)) \land$
$(\neg P(\#,1,0) \lor \neg Q(1,0) \lor \neg S(1,0) \lor S(0,1)) \land (\neg P(\#,1,1) \lor \neg Q(1,1) \lor \neg S(1,1) \lor S(0,2)) \land$
$(\neg P(\#,1,2) \lor \neg Q(1,2) \lor \neg S(1,2) \lor S(0,3)) \land (\neg P(\#,1,3) \lor \neg Q(1,3) \lor \neg S(1,3) \lor S(0,4)) \land$
$(\neg P(0,2,0) \lor \neg Q(1,0) \lor \neg S(2,0) \lor S(3,1)) \land (\neg P(0,2,1) \lor \neg Q(1,1) \lor \neg S(2,1) \lor S(3,2)) \land$
$(\neg P(0,2,2) \lor \neg Q(1,2) \lor \neg S(2,2) \lor S(3,3)) \land (\neg P(0,2,3) \lor \neg Q(1,3) \lor \neg S(2,3) \lor S(3,4)) \land$
$(\neg P(1,2,0) \lor \neg Q(1,0) \lor \neg S(2,0) \lor S(3,1)) \land (\neg P(1,2,1) \lor \neg Q(1,1) \lor \neg S(2,1) \lor S(3,2)) \land$
$(\neg P(1,2,2) \lor \neg Q(1,2) \lor \neg S(2,2) \lor S(3,3)) \land (\neg P(1,2,3) \lor \neg Q(1,3) \lor \neg S(2,3) \lor S(3,4)) \land$
$(\neg P(\#,2,0) \lor \neg Q(1,0) \lor \neg S(2,0) \lor S(1,1)) \land (\neg P(\#,2,1) \lor \neg Q(1,1) \lor \neg S(2,1) \lor S(1,2)) \land$
$(\neg P(\#,2,2) \lor \neg Q(1,2) \lor \neg S(2,2) \lor S(1,3)) \land (\neg P(\#,2,3) \lor \neg Q(1,3) \lor \neg S(2,3) \lor S(1,4)) \land$
$(\neg P(0,3,0) \lor \neg Q(1,0) \lor \neg S(3,0) \lor S(4,1)) \land (\neg P(0,3,1) \lor \neg Q(1,1) \lor \neg S(3,1) \lor S(4,2)) \land$
$(\neg P(0,3,2) \lor \neg Q(1,2) \lor \neg S(3,2) \lor S(4,3)) \land (\neg P(0,3,3) \lor \neg Q(1,3) \lor \neg S(3,3) \lor S(4,4)) \land$
$(\neg P(1,3,0) \lor \neg Q(1,0) \lor \neg S(3,0) \lor S(4,1)) \land (\neg P(1,3,1) \lor \neg Q(1,1) \lor \neg S(3,1) \lor S(4,2)) \land$
$(\neg P(1,3,2) \lor \neg Q(1,2) \lor \neg S(3,2) \lor S(4,3)) \land (\neg P(1,3,3) \lor \neg Q(1,3) \lor \neg S(3,3) \lor S(4,4)) \land$
$(\neg P(\#,3,0) \lor \neg Q(1,0) \lor \neg S(3,0) \lor S(2,1)) \land (\neg P(\#,3,1) \lor \neg Q(1,1) \lor \neg S(3,1) \lor S(2,2)) \land$
$(\neg P(\#,3,2) \lor \neg Q(1,2) \lor \neg S(3,2) \lor S(2,3)) \land (\neg P(\#,3,3) \lor \neg Q(1,3) \lor \neg S(3,3) \lor S(2,4)) \land$
$(\neg P(0,4,0) \lor \neg Q(1,0) \lor \neg S(4,0) \lor S(5,1)) \land (\neg P(0,4,1) \lor \neg Q(1,1) \lor \neg S(4,1) \lor S(5,2)) \land$
$(\neg P(0,4,2) \lor \neg Q(1,2) \lor \neg S(4,2) \lor S(5,3)) \land (\neg P(0,4,3) \lor \neg Q(1,3) \lor \neg S(4,3) \lor S(5,4)) \land$
$(\neg P(1,4,0) \lor \neg Q(1,0) \lor \neg S(4,0) \lor S(5,1)) \land (\neg P(1,4,1) \lor \neg Q(1,1) \lor \neg S(4,1) \lor S(5,2)) \land$
$(\neg P(1,4,2) \lor \neg Q(1,2) \lor \neg S(4,2) \lor S(5,3)) \land (\neg P(1,4,3) \lor \neg Q(1,3) \lor \neg S(4,3) \lor S(5,4)) \land$
$(\neg P(\#,4,0) \lor \neg Q(1,0) \lor \neg S(4,0) \lor S(3,1)) \land (\neg P(\#,4,1) \lor \neg Q(1,1) \lor \neg S(4,1) \lor S(3,2)) \land$
$(\neg P(\#,4,2) \lor \neg Q(1,2) \lor \neg S(4,2) \lor S(3,3)) \land (\neg P(\#,4,3) \lor \neg Q(1,3) \lor \neg S(4,3) \lor S(3,4)) \land$
$(\neg P(0,0,0) \lor \neg Q(2,0) \lor \neg S(0,0) \lor S(1,1)) \land (\neg P(0,0,1) \lor \neg Q(2,1) \lor \neg S(0,1) \lor S(1,2)) \land$
$(\neg P(0,0,2) \lor \neg Q(2,2) \lor \neg S(0,2) \lor S(1,3)) \land (\neg P(0,0,3) \lor \neg Q(2,3) \lor \neg S(0,3) \lor S(1,4)) \land$
$(\neg P(1,0,0) \lor \neg Q(2,0) \lor \neg S(0,0) \lor S(1,1)) \land (\neg P(1,0,1) \lor \neg Q(2,1) \lor \neg S(0,1) \lor S(1,2)) \land$
$(\neg P(1,0,2) \lor \neg Q(2,2) \lor \neg S(0,2) \lor S(1,3)) \land (\neg P(1,0,3) \lor \neg Q(2,3) \lor \neg S(0,3) \lor S(1,4)) \land$
$(\neg P(\#,0,0) \lor \neg Q(2,0) \lor \neg S(0,0) \lor S(1,1)) \land (\neg P(\#,0,1) \lor \neg Q(2,1) \lor \neg S(0,1) \lor S(1,2)) \land$
$(\neg P(\#,0,2) \lor \neg Q(2,2) \lor \neg S(0,2) \lor S(1,3)) \land (\neg P(\#,0,3) \lor \neg Q(2,3) \lor \neg S(0,3) \lor S(1,4)) \land$
$(\neg P(0,1,0) \lor \neg Q(2,0) \lor \neg S(1,0) \lor S(2,1)) \land (\neg P(0,1,1) \lor \neg Q(2,1) \lor \neg S(1,1) \lor S(2,2)) \land$
$(\neg P(0,1,2) \lor \neg Q(2,2) \lor \neg S(1,2) \lor S(2,3)) \land (\neg P(0,1,3) \lor \neg Q(2,3) \lor \neg S(1,3) \lor S(2,4)) \land$
$(\neg P(1,1,0) \lor \neg Q(2,0) \lor \neg S(1,0) \lor S(2,1)) \land (\neg P(1,1,1) \lor \neg Q(2,1) \lor \neg S(1,1) \lor S(2,2)) \land$
$(\neg P(1,1,2) \lor \neg Q(2,2) \lor \neg S(1,2) \lor S(2,3)) \land (\neg P(1,1,3) \lor \neg Q(2,3) \lor \neg S(1,3) \lor S(2,4)) \land$
$(\neg P(\#,1,0) \lor \neg Q(2,0) \lor \neg S(1,0) \lor S(2,1)) \land (\neg P(\#,1,1) \lor \neg Q(2,1) \lor \neg S(1,1) \lor S(2,2)) \land$
$(\neg P(\#,1,2) \lor \neg Q(2,2) \lor \neg S(1,2) \lor S(2,3)) \land (\neg P(\#,1,3) \lor \neg Q(2,3) \lor \neg S(1,3) \lor S(2,4)) \land$
$(\neg P(0,2,0) \lor \neg Q(2,0) \lor \neg S(2,0) \lor S(3,1)) \land (\neg P(0,2,1) \lor \neg Q(2,1) \lor \neg S(2,1) \lor S(3,2)) \land$
$(\neg P(0,2,2) \lor \neg Q(2,2) \lor \neg S(2,2) \lor S(3,3)) \land (\neg P(0,2,3) \lor \neg Q(2,3) \lor \neg S(2,3) \lor S(3,4)) \land$
$(\neg P(1,2,0) \lor \neg Q(2,0) \lor \neg S(2,0) \lor S(3,1)) \land (\neg P(1,2,1) \lor \neg Q(2,1) \lor \neg S(2,1) \lor S(3,2)) \land$
$(\neg P(1,2,2) \lor \neg Q(2,2) \lor \neg S(2,2) \lor S(3,3)) \land (\neg P(1,2,3) \lor \neg Q(2,3) \lor \neg S(2,3) \lor S(3,4)) \land$
$(\neg P(\#,2,0) \lor \neg Q(2,0) \lor \neg S(2,0) \lor S(3,1)) \land (\neg P(\#,2,1) \lor \neg Q(2,1) \lor \neg S(2,1) \lor S(3,2)) \land$
$(\neg P(\#,2,2) \lor \neg Q(2,2) \lor \neg S(2,2) \lor S(3,3)) \land (\neg P(\#,2,3) \lor \neg Q(2,3) \lor \neg S(2,3) \lor S(3,4)) \land$
$(\neg P(0,3,0) \lor \neg Q(2,0) \lor \neg S(3,0) \lor S(4,1)) \land (\neg P(0,3,1) \lor \neg Q(2,1) \lor \neg S(3,1) \lor S(4,2)) \land$

$(\neg P(0,3,2) \lor \neg Q(2,2) \lor \neg S(3,2) \lor S(4,3)) \land (\neg P(0,3,3) \lor \neg Q(2,3) \lor \neg S(3,3) \lor S(4,4)) \land$
$(\neg P(1,3,0) \lor \neg Q(2,0) \lor \neg S(3,0) \lor S(4,1)) \land (\neg P(1,3,1) \lor \neg Q(2,1) \lor \neg S(3,1) \lor S(4,2)) \land$
$(\neg P(1,3,2) \lor \neg Q(2,2) \lor \neg S(3,2) \lor S(4,3)) \land (\neg P(1,3,3) \lor \neg Q(2,3) \lor \neg S(3,3) \lor S(4,4)) \land$
$(\neg P(\#,3,0) \lor \neg Q(2,0) \lor \neg S(3,0) \lor S(4,1)) \land (\neg P(\#,3,1) \lor \neg Q(2,1) \lor \neg S(3,1) \lor S(4,2)) \land$
$(\neg P(\#,3,2) \lor \neg Q(2,2) \lor \neg S(3,2) \lor S(4,3)) \land (\neg P(\#,3,3) \lor \neg Q(2,3) \lor \neg S(3,3) \lor S(4,4)) \land$
$(\neg P(0,4,0) \lor \neg Q(2,0) \lor \neg S(4,0) \lor S(5,1)) \land (\neg P(0,4,1) \lor \neg Q(2,1) \lor \neg S(4,1) \lor S(5,2)) \land$
$(\neg P(0,4,2) \lor \neg Q(2,2) \lor \neg S(4,2) \lor S(5,3)) \land (\neg P(0,4,3) \lor \neg Q(2,3) \lor \neg S(4,3) \lor S(5,4)) \land$
$(\neg P(1,4,0) \lor \neg Q(2,0) \lor \neg S(4,0) \lor S(5,1)) \land (\neg P(1,4,1) \lor \neg Q(2,1) \lor \neg S(4,1) \lor S(5,2)) \land$
$(\neg P(1,4,2) \lor \neg Q(2,2) \lor \neg S(4,2) \lor S(5,3)) \land (\neg P(1,4,3) \lor \neg Q(2,3) \lor \neg S(4,3) \lor S(5,4)) \land$
$(\neg P(\#,4,0) \lor \neg Q(2,0) \lor \neg S(4,0) \lor S(5,1)) \land (\neg P(\#,4,1) \lor \neg Q(2,1) \lor \neg S(4,1) \lor S(5,2)) \land$
$(\neg P(\#,4,2) \lor \neg Q(2,2) \lor \neg S(4,2) \lor S(5,3)) \land (\neg P(\#,4,3) \lor \neg Q(2,3) \lor \neg S(4,3) \lor S(5,4)) \land$
$(\neg P(0,0,0) \lor \neg Q(3,0) \lor \neg S(0,0) \lor S(0,1)) \land (\neg P(0,0,1) \lor \neg Q(3,1) \lor \neg S(0,1) \lor S(0,2)) \land$
$(\neg P(0,0,2) \lor \neg Q(3,2) \lor \neg S(0,2) \lor S(0,3)) \land (\neg P(0,0,3) \lor \neg Q(3,3) \lor \neg S(0,3) \lor S(0,4)) \land$
$(\neg P(1,0,0) \lor \neg Q(3,0) \lor \neg S(0,0) \lor S(0,1)) \land (\neg P(1,0,1) \lor \neg Q(3,1) \lor \neg S(0,1) \lor S(0,2)) \land$
$(\neg P(1,0,2) \lor \neg Q(3,2) \lor \neg S(0,2) \lor S(0,3)) \land (\neg P(1,0,3) \lor \neg Q(3,3) \lor \neg S(0,3) \lor S(0,4)) \land$
$(\neg P(\#,0,0) \lor \neg Q(3,0) \lor \neg S(0,0) \lor S(0,1)) \land (\neg P(\#,0,1) \lor \neg Q(3,1) \lor \neg S(0,1) \lor S(0,2)) \land$
$(\neg P(\#,0,2) \lor \neg Q(3,2) \lor \neg S(0,2) \lor S(0,3)) \land (\neg P(\#,0,3) \lor \neg Q(3,3) \lor \neg S(0,3) \lor S(0,4)) \land$
$(\neg P(0,1,0) \lor \neg Q(3,0) \lor \neg S(1,0) \lor S(1,1)) \land (\neg P(0,1,1) \lor \neg Q(3,1) \lor \neg S(1,1) \lor S(1,2)) \land$
$(\neg P(0,1,2) \lor \neg Q(3,2) \lor \neg S(1,2) \lor S(1,3)) \land (\neg P(0,1,3) \lor \neg Q(3,3) \lor \neg S(1,3) \lor S(1,4)) \land$
$(\neg P(1,1,0) \lor \neg Q(3,0) \lor \neg S(1,0) \lor S(1,1)) \land (\neg P(1,1,1) \lor \neg Q(3,1) \lor \neg S(1,1) \lor S(1,2)) \land$
$(\neg P(1,1,2) \lor \neg Q(3,2) \lor \neg S(1,2) \lor S(1,3)) \land (\neg P(1,1,3) \lor \neg Q(3,3) \lor \neg S(1,3) \lor S(1,4)) \land$
$(\neg P(\#,1,0) \lor \neg Q(3,0) \lor \neg S(1,0) \lor S(1,1)) \land (\neg P(\#,1,1) \lor \neg Q(3,1) \lor \neg S(1,1) \lor S(1,2)) \land$
$(\neg P(\#,1,2) \lor \neg Q(3,2) \lor \neg S(1,2) \lor S(1,3)) \land (\neg P(\#,1,3) \lor \neg Q(3,3) \lor \neg S(1,3) \lor S(1,4)) \land$
$(\neg P(0,2,0) \lor \neg Q(3,0) \lor \neg S(2,0) \lor S(2,1)) \land (\neg P(0,2,1) \lor \neg Q(3,1) \lor \neg S(2,1) \lor S(2,2)) \land$
$(\neg P(0,2,2) \lor \neg Q(3,2) \lor \neg S(2,2) \lor S(2,3)) \land (\neg P(0,2,3) \lor \neg Q(3,3) \lor \neg S(2,3) \lor S(2,4)) \land$
$(\neg P(1,2,0) \lor \neg Q(3,0) \lor \neg S(2,0) \lor S(2,1)) \land (\neg P(1,2,1) \lor \neg Q(3,1) \lor \neg S(2,1) \lor S(2,2)) \land$
$(\neg P(1,2,2) \lor \neg Q(3,2) \lor \neg S(2,2) \lor S(2,3)) \land (\neg P(1,2,3) \lor \neg Q(3,3) \lor \neg S(2,3) \lor S(2,4)) \land$
$(\neg P(\#,2,0) \lor \neg Q(3,0) \lor \neg S(2,0) \lor S(2,1)) \land (\neg P(\#,2,1) \lor \neg Q(3,1) \lor \neg S(2,1) \lor S(2,2)) \land$
$(\neg P(\#,2,2) \lor \neg Q(3,2) \lor \neg S(2,2) \lor S(2,3)) \land (\neg P(\#,2,3) \lor \neg Q(3,3) \lor \neg S(2,3) \lor S(2,4)) \land$
$(\neg P(0,3,0) \lor \neg Q(3,0) \lor \neg S(3,0) \lor S(3,1)) \land (\neg P(0,3,1) \lor \neg Q(3,1) \lor \neg S(3,1) \lor S(3,2)) \land$
$(\neg P(0,3,2) \lor \neg Q(3,2) \lor \neg S(3,2) \lor S(3,3)) \land (\neg P(0,3,3) \lor \neg Q(3,3) \lor \neg S(3,3) \lor S(3,4)) \land$
$(\neg P(1,3,0) \lor \neg Q(3,0) \lor \neg S(3,0) \lor S(3,1)) \land (\neg P(1,3,1) \lor \neg Q(3,1) \lor \neg S(3,1) \lor S(3,2)) \land$
$(\neg P(1,3,2) \lor \neg Q(3,2) \lor \neg S(3,2) \lor S(3,3)) \land (\neg P(1,3,3) \lor \neg Q(3,3) \lor \neg S(3,3) \lor S(3,4)) \land$
$(\neg P(\#,3,0) \lor \neg Q(3,0) \lor \neg S(3,0) \lor S(3,1)) \land (\neg P(\#,3,1) \lor \neg Q(3,1) \lor \neg S(3,1) \lor S(3,2)) \land$
$(\neg P(\#,3,2) \lor \neg Q(3,2) \lor \neg S(3,2) \lor S(3,3)) \land (\neg P(\#,3,3) \lor \neg Q(3,3) \lor \neg S(3,3) \lor S(3,4)) \land$
$(\neg P(0,4,0) \lor \neg Q(3,0) \lor \neg S(4,0) \lor S(4,1)) \land (\neg P(0,4,1) \lor \neg Q(3,1) \lor \neg S(4,1) \lor S(4,2)) \land$
$(\neg P(0,4,2) \lor \neg Q(3,2) \lor \neg S(4,2) \lor S(4,3)) \land (\neg P(0,4,3) \lor \neg Q(3,3) \lor \neg S(4,3) \lor S(4,4)) \land$
$(\neg P(1,4,0) \lor \neg Q(3,0) \lor \neg S(4,0) \lor S(4,1)) \land (\neg P(1,4,1) \lor \neg Q(3,1) \lor \neg S(4,1) \lor S(4,2)) \land$
$(\neg P(1,4,2) \lor \neg Q(3,2) \lor \neg S(4,2) \lor S(4,3)) \land (\neg P(1,4,3) \lor \neg Q(3,3) \lor \neg S(4,3) \lor S(4,4)) \land$
$(\neg P(\#,4,0) \lor \neg Q(3,0) \lor \neg S(4,0) \lor S(4,1)) \land (\neg P(\#,4,1) \lor \neg Q(3,1) \lor \neg S(4,1) \lor S(4,2)) \land$
$(\neg P(\#,4,2) \lor \neg Q(3,2) \lor \neg S(4,2) \lor S(4,3)) \land (\neg P(\#,4,3) \lor \neg Q(3,3) \lor \neg S(4,3) \lor S(4,4)) \land$
$Q(3,4)$

// group 8

Name Index

Ackermann, Wilhelm, 187
Adel'son-Vel'skii, G. M., 255, 257, 298
Aho, Alfred V., 671, 711
Ahuja, Ravindra K., 471
Allan, Stephen J., 602, 642
Allen, Brian, 263, 298
Appel, Andrew W., 620, 621, 639, 642, 643
Auslander, M. A., 214

Bachmann, Paul, 53
Baer, J. L., 298
Baeza-Yates, Ricardo A., 667, 698, 709, 711
Baker, Henry G., 620, 622, 629, 639, 640, 643
Barber, Angus, 523
Barron, David W., 214
Barth, Gerhard, 645, 711
Bayer, Rudolf, 302, 317, 318, 322, 324, 374
Bell, James R., 269, 298, 560, 562, 597
Bell, Timothy C., 597, 598
Bentley, Jon L., 136, 375
Berge, Claude, 426, 471
Berlioux, Pierre, 214
Bertsekas, Dimitri P., 390, 471
Berztiss, Alfs, 297
Bird, Richard S., 214
Bitner, James R., 263, 298
Bizard, Philippe, 214
Blackstone, John H., 154, 169
Bollobés, B., 298
Bondy, John A., 430, 438, 471
Boruvka, Otakar, 467
Bourne, Charles P., 353, 375
Boyer, Robert S., 711
Brassard, G., 75
Brélaz, Daniel, 446, 471
Bretholz, E., 132, 136
Breymann, Ulrich, 51
Briandais, Rene de la, 352, 375
Bromley, Allan G., 610, 642
Brooks, Rodney A., 621, 643
Budd, Timothy, 51
Burge, William H., 214
Burkhard, W. A., 297
Busacker, R. G., 419

Campbell, J. A., 638, 642
Cardelli, Luca, 51
Carlsson, Svente, 524
Carroll, Lewis, 146
Celis, P., 561, 562
Chang, Hsi, 298
Chen, Qi F., 562
Chen, Wen C., 563
Cheney, C. J., 619, 629, 643
Chvátal, V., 438, 471
Cichelli, Richard J., 539, 561, 562
Cleary, J. G., 597, 598
Cohen, Jacques, 643
Cole, Richard, 711
Comer, Douglas, 313, 353, 374, 375
Cook, Curtis R., 499, 524, 541, 562
Cook, Stephen A., 70, 71, 75, 499, 541
Copes, Wayne, 169
Corasick, Margaret J., 671, 711
Cormack, Gordon V., 579, 598
Cranston, Ben, 608, 642
Culberson, Joseph, 250, 297
Czech, Zbigniew J., 562

Daoud, Amjad M., 562
Day, A. Colin, 254, 298
Deitel, Harvey M., 51
Deitel, P. J., 51
deMaine, P. A. D., 353, 375
Deo, Narsingh, 388, 471
D'Esopo, D., 389
Dijkstra, Edsger W., 173, 214, 384, 386, 388, 398, 471, 472
Ding, Yuzheng, 493, 524
Dinic, Efim A., 415, 472
Doberkat, E. E., 277, 298
Dobosiewicz, W., 488, 523
Dood, M., 562
Dromey, R. G., 524
Drozdek, Adam, 598, 693, 711
Durian, B., 524
Dvorak, S., 524
Dy, H. C., 562
Dymes, Ruth, 146

Edmonds, Jack, 391, 413, 433, 436, 472
Ege, Raimund K., 51
Ellis, John R., 621, 639, 643
Enbody, R. J., 562
Eppinger, Jeffrey L., 250, 297
Euler, Leonhard, 434

Fagin, Ronald, 544, 562
Faller, Newton, 576, 598
Faloutsos, Christos, 322, 374
Fenichel, Robert R., 618, 643
Ferguson, David E., 318, 374
Finkel, R. A., 375
Fischer, M. J., 691, 712
Flaming, Bryan, 51
Fleury, 434-435, 472
Flores, Ivan, 523
Floyd, Robert W., 274, 277, 293, 298, 390–391, 466, 472, 490
Folk, Michael J., 374
Ford, Donald F., 353, 375
Ford, Lester R., 386, 408, 409, 419, 472
Foster, Caxton C., 262, 298
Foster, John M., 136
Fox, Edward A., 542, 562
Frazer, William D., 524
Fredkin, Edward, 349, 375
Frieze, A., 299
Fulkerson, D. Ray, 408, 409, 419, 472
Fuller, S. H., 298

Gale, David, 429, 472, 523
Galil, Zvi, 666, 711
Gallager, Robert G., 576, 598
Gallo, Giorgio, 384, 388, 389, 472
Garey, Michael R., 75
Gibbons, Alan, 436, 472
Glover, Fred, 390, 472
Glover, Randy, 390, 472
Gonnett, Gaston H., 299, 667, 709, 711
Gould, Ronald, 472
Gowen, P. J., 419
Graham, R. L., 397, 472
Guibas, Leo J., 324, 374, 663, 711
Gupta, Gopal, 269, 298
Guttman, Antonin, 321, 374

Haggard, G., 541, 562
Hagins, Jeff, 610, 639, 642
Hall, Philip, 453, 472
Hamilton, William, 49
Hancart, Christophe, 646, 647, 711
Hankamer, M., 596, 598

Hansen, Wilfred J., 136
Hanson, E., 322, 375
Hartmanis, Juris, 52, 75
Hayward, Ryan, 277, 299
Heaps, H. S., 597, 598
Heath, Lenwood S., 562
Heileman, Gregory L., 75
Hell, Pavol, 397, 472
Hendriksen, James O., 154, 169
Hester, James H., 136
Hewitt, Carl, 623, 628, 643
Hibbard, Thomas N., 247, 292, 297, 522
Hinds, James A., 608, 642
Hirschberg, Daniel S., 136, 598, 608, 642
Hoare, Charles A. R., 493, 524
Hogg, Gary L., 169
Hopcroft, John E., 75, 380
Horowitz, E., 355, 375
Horspool, R. Nigel, 579, 598, 708, 711
Huang, B. C., 524
Huffman, David A., 565, 598
Hunt, James W., 693, 711

Incerpi, Janet, 488, 523
Ingerman, P. Z., 390-391, 472
Iri, M., 419
Isaacson, Joel D., 473
Iyengar, S. Sitharama, 298

Jarník, Vojtech, 467
Jewell, W. S., 419
Johnson, David S., 69, 75, 373, 468, 473
Johnson, Donald B., 386, 472
Johnson, Elias L., 436, 472
Johnson, Theodore, 373, 374
Johnsonbaugh, Richard, 51
Johnsson, C. B., 639
Jonassen, Arne T., 250, 297
Jones, Douglas W., 154, 169
Jones, Richard, 643
Julstrom, A., 521, 524

Kaehler, E. B., 298
Kalaba, Robert, 398, 473
Kalin, Martin, 51
Kaman, Charles H., 560, 562
Karlton, P. L., 262, 298
Karp, Richard M., 75, 391, 413, 472, 523
Karplus, K., 541, 562
Kershenbaum, Aaron, 465, 473
Khoshafian, Setrag, 51
Kim, Do Jin, 499, 524
Klingman, Darwin, 390, 472

Knott, G. D., 544, 562
Knowlton, Kenneth C., 603, 642
Knuth, Donald E., 75, 247, 250, 258, 277, 292, 297, 298, 313, 488, 520, 522, 523, 524, 532, 533, 562, 576, 598, 663, 699, 711
Koch, Helge von, 182
Koffeman, K. L., 375
Kruskal, Joseph B., 397, 473
Kuhn, Harold W., 430, 473
Kurokawa, Toshiaki, 616, 643
Kwan, Mei-ko, 436, 473

Lajoie, Josée, 51
Landis, E. M., 257, 298
Langston, M. A., 524
Larson, Per A., 544, 562
Layer, D. Kevin, 643
Lee, Meng, 24
Lelever, Debra A., 598
Lempel, Abraham, 598
Leung, Clement H. C., 313, 374
Levenshtein, V. I., 689, 693, 711
Lewis, Philip M., 473
Lewis, Ted G., 541, 562
Li, Kai, 621, 639, 643
Lieberman, Henry, 623, 628, 643
Lins, Rafael, 643
Lippman, Stanley B., 51
Litwin, Witold, 544, 547, 562
Lomet, David B., 547, 562
Lorentz, Richard, 214
Lorin, Harold, 523
Lukasiewicz, Jan, 277, 292
Lum, V. Y., 526, 562

Magnanti, Thomas L., 471
Majewski, Bohdan S., 562
Maly, Kurt, 356, 375
Manber, Udi, 58, 75, 667, 687, 696, 698, 711, 712
Marble, George, 473
Martin, W. A., 298
Matthews, D., 132, 136
Matula, David W., 473
McCreight, Edward M., 302, 313, 374
McDiarmid, Colin, 277, 293, 299
McDonald, M. A., 292, 298
McGeoch, Catharine C., 136
McKellar, Archie C., 524
McLuckie, Keith, 523
Meertens, L., 75
Mehlhorn, Kurt, 523
Meyer, Bertrand, 51
Miller, Victor S., 582, 598

Moivre, A. de, 187
Moore, J. Strother, 711
Moret, B. M. E., 524
Morris, James H., 663, 699, 711
Morris, Joseph M., 238-241, 297
Morris, Robert, 531, 562
Morrison, Donald R., 372, 375
Motzkin, Dalia, 524
Mullin, James K., 550, 562
Munkres, James, 430, 473
Munro, J. Ian, 263, 298, 299, 561, 562, 563
Murty, U. S. R., 430, 471
Myers, Gene, 687, 711

Napier, John, 209
Ness, D. N., 298
Neumann, John von, 499
Ng, D. T. H., 132, 136
Nievergelt, Jurg, 562

Odlyzko, Andrew M., 663, 711
Oldehoeft, Rodney R., 602, 642
Oommen, B. J., 132, 136
Ore, Oystein, 438, 473
Orlin, James B., 471

Page, Ivor P., 610, 639, 642
Pagli, L., 534, 563
Pallottino, Stefano, 384, 388, 389, 472
Pang, Chi-yin, 471
Papadimitriou, Christos H., 69, 75, 468, 473
Pape, U., 389, 473
Papernov, A. A., 523
Perleberg, Chris H., 698, 711
Peter, Rozsa, 187
Peterson, James L., 642
Phillips, Don T., 169
Pippenger, Nicholas, 562
Pirklbauer, Klaus, 665, 711
Poblete, Patricio V., 561, 563
Pollack, Maurice, 389, 473
Poonen, Bjorn, 523
Pountain, Dick, 598
Powell, M. B., 445, 473
Pratt, Vaughan R., 523, 663, 699, 711
Preparata, Franco P., 75
Prim, Robert C., 467, 473
Pugh, William, 97, 136

Radke, Charles E., 530, 563
Ramakrishna, M. V., 561
Razmik, Abnous, 51
Reed, B. A., 277, 293, 299

Reynolds, Carl W., 523
Riccardi, Greg, 374
Rich, R., 523
Richardson, Chris, 643
Rivest, Ronald, 136
Roberts, Eric, 214
Rohl, Jeffrey S., 214
Rosenberg, Arnold L., 369, 374
Rosenkrantz, Daniel J., 468, 473
Ross, Douglas T., 601, 642
Rotem, D., 132, 136
Rotwitt, T., 353, 375
Roussopoulos, Nick, 322, 374
Rubin, Frank, 565, 598

Sacco, William, 169
Sager, Thomas J., 542, 563
Salomon, David, 598
Samet, Hanan, 370, 375
Schildt, Herbert, 51
Schorr, H., 612-616, 643
Schuegraf, E. J., 597, 598
Schwab, B., 298
Scroggs, R. E., 298
Sebesta, Robert W., 541, 563
Sedgewick, Robert, 324, 374, 488, 523, 524
Sellers, Peter H., 710, 711
Sellis, Timos, 322, 374, 375
Sethi, Ravi, 353, 375
Shannon, Claude E., 565
Shapley, L. S., 429, 472
Shasha, Dennis, 373, 374
Shell, Donald L., 486, 523
Shen, Kenneth K., 642
Simon, I., 298
Sleator, Daniel D., 136, 264, 267, 298
Sloyer, Clifford, 169
Smith, Harry F., 642
Smith, P. D., 665, 712
Snyder, Lawrence, 369, 374
Standish, Thomas A., 532, 639, 642
Starck, Robert, 169
Stasevich, G. V., 523
Stearns, Richard E., 52, 75, 473
Steiglitz, Kenneth, 473
Stepanov, Alexander, 24
Stonebraker, M., 322, 375
Stout, Quentin F., 254, 298
Strong, H. Raymond, 214, 562
Stroustrup, Bjarne, 51
Sunday, Daniel M., 663, 712

Al-Suwaiyel, M., 355, 375
Swamy, M. N. S., 430, 473
Szymanski, Thomas G., 693, 711

Tarjan, Robert E., 75, 136, 264, 267, 298, 380, 404, 473
Taylor, Mark A., 541, 563
Tharp, Alan L., 563
Thomas, Rick, 608, 642
Thompson, Ken, 677, 712
Thulasiraman, K., 430, 473

Ukkonen, Esko, 684, 701, 712
Unterauer, Karl, 317, 318, 374

Vitanyi, P. M. B., 75
Vitter, Jeffrey S., 563

Wadler, Philip L., 620, 643
Wagner, Richard A., 691, 712
Waite, W. M., 612-616, 643
Wang, Paul S., 51
Warren, Bette L., 254, 298
Warshall, Stephen, 390-391, 473
Wedekind, H., 314, 375
Wegbreit, Ben, 616, 643
Wegener, Ingo, 524
Wegman, Mark N., 582, 598
Wegner, Peter, 51
Weinstock, C. B., 639
Weiss, Mark A., 299, 493, 523, 524
Welch, Terry A., 582, 598
Welsh, D. J. A., 445, 473
Wiebenson, Walter, 389, 473
Wilkes, Maurice V., 136
Williams, J. W. J., 274, 277, 293, 299, 489, 524
Wilson, Paul R., 628, 643
Wirth, Niklaus, 214
Witten, Ian H., 597, 598
Wood, Derick, 684, 712
Wu, Sun, 667, 696, 698, 712
Wulf, W. A., 639

Yao, Andrew Chi-Chih, 312, 375
Yochelson, Jerome C., 618, 643
Yuasa, Taiichi, 625, 643
Yuen, P. S. T., 562

Ziv, Jacob, 598
Zoellick, Bill, 374
Zorn, Benjamin, 628, 643
Zweben, S. H., 292, 298

Subject Index

*page numbers in italic represent diagrams/figures/pseudocode.

abstract data types, 1, 137
access time, 301
Ackermann function, 187
acrostics, 146
activation record, 173–174
active point, 687
Ada, 23
  speed of, 52
adaptive exact-fit method, 602
adaptive Huffman coding, 575–580
adding polynomials: case study, 512–519
  implementation of program, *514–519*
adjacency list, 378
adjacency matrix, 378
adjacent vertices, 377
admissible trees. *See* AVL trees
Aho-Corasick algorithm, 671, 673, 698
algorithms
  acrostic, *146*
  adding large numbers (pseudocode), *141*
  Aho-Corasick algorithm, 671, 673
  backtracking (pseudocode), *192*
  Baker's algorithm, 620–622
  best, average, and worst cases of efficiency, 62–65
  binary search, 61
  Boruvka's algorithm, 467
  Boyer-Moore algorithm, 655–663, 666
  Boyer-Moore-Galil algorithm, 666
  for calculating Fibonacci numbers, *190*

Cichelli's algorithm, 539–542
comparing efficiency of, 52
comparison of run times for different sorting, *511*
complexities of, 58, 59, 60
constant, 58, 59
cubic, 59, 60, 393
for deleting by copying, implementation of, *249*
for deleting by merging, implementation of, *246–247*
delimiter matching, 138–139 *139*
depth-first search, 380–382
D'Esopo-Pape algorithm, 465
deterministic, 69–70
Dijkstra's algorithm, 384–386
Dinic's algorithm, 415–418
DSW algorithm, 254–257
efficient, 70
efficient sorting, 486–506
elementary sorting, 475–482
enqueuing, *271*
for escaping maze (pseudocode), *161*
evaluating efficiency of, 52–53, 55–57
exponential, 59
FHCD algorithm, 542–544
flagFlipping algorithm, 329–331
Fleury's algorithm, 435
Ford-Fulkerson algorithm, 409–415
Ford's algorithm, 387–388
Hancart's algorithm, 646–647
Huffman algorithm, 566–567
Hunt-Szymanski algorithm, 693–695

implementation of insertion, *242*
inefficient, 60
for inserting keys in B-trees, *307*
to insert node into threaded tree, implementation of, *243*
Jarník-Prim algorithm, 467, 470
Kershenbaum's algorithm, 465
Knuth-Morris-Pratt algorithm, 647–654, 673–674
Kruskal's algorithm, 397–400
Kuhn-Munkres algorithm, 430–431
linear, 59
logarithmic, 59
Morris algorithm, 238–241, *239, 240*
  with nested loops, 61
nondeterministic, 69–70
$O(n \lg n)$, 59
polynomial-time, 70
quadratic, 58, 59, 60
Schorr and Waite algorithm, 612–615
for searching key using bit-tree leaf, *320*
set-related, 343–344
sorting, 33–35
sources affecting efficiency of, 628
STL, 24, 25
stop-and-copy algorithm, 618, 628
straightforward string matching, 644–647
Sunday algorithms, 663–665
Tarjan's algorithm, 404

745

Ukkonen algorithm, 684–687, 699
Wagner-Fischer algorithm, 691–693
WFI algorithm, 391–393, 466
Yuasa's algorithm, 625–628
Ziv-Lempel algorithm, 582–583
alignments, 689
alternating path, 424
American Standard Code for Information Interchange (ASCII) code. *See* ASCII code
amortized complexity, 65–69, 105–107, 267–269
amortized cost, 68–69, 105–106
  of access in splaying, lemma specifying, 267
  estimating, *67*
analysis, amortized, 65–69
anchor, 170, 175, 186
approximate string matching, 689–699
  string matching with *k* errors, 696–699
  string similarity, 690–695
arcs, 215, 376
  in decision trees, 484
arrays. *See also* sorting
  advantages over linked lists of, 119–120
  of characters. *See* strings
  flexible. *See* vectors
  generalization of. *See* maps
  heaps implemented by, 269–270
  implementation of queues in, *147, 148–149*
  implementing trees as, 220
  indexes in, 344
  insertion in, 119
  limitations, 76, 119
  of numbers, 90
  ordered, 61
  organizing, as heaps, 272–277, *275*
  of pointers, 102, 117–118
  pointers and, 12–13
  of pointers to subtries, 349, 351
  pseudoflexible, 359, *360*

queue viewed as circular, *147*
and sparse table for storing student grades, *108*
transformed into heaps, 490, *491*
two-dimensional, 108, *109*
unordered, 63, 69
use of loops to calculate sum of numbers in, 60–62
articulation points, 401
artificial intelligence, 191
ASCII code, 45, 318, 573, 575, 580
  used in sorting, 474
  values in, 528
assignment operator, 18
  overloading, 5, 16, 24
  in vector, 32
assignment problem, 430–432
assignment statements, 60–62
asymptotic complexity, 53
  finding: examples, 60–62
  imprecision of, 628
  use of big-O notation for specifying, 53–55
augmenting path, 424, *425*
avail-list, 616
average case assessment, 62–65
average path length, 224
AVL trees, 102, *257*, 257–262, *260*
  rebalancing, after deleting node, *261*
  transforming, into vh-trees, *337, 338*. *See also* vh-trees

backbone, 254–257
  transforming binary search tree into, *255*
  transforming, into perfectly balanced tree, *256*
back edges, 381, 393
backtracking, 159, 191–198
  in graphs, 380
Backus-Naur form (BNF), 172
Baker's algorithm, 620–622
  modification to, 622
balanced trees, 251
balance factors, 257–262
Bank One, 150–153
  customer data example, *151*
  example: implementation code, *152–153*

base class, 7–9
BASIC
  compiler in, 599
  speed of, 52
Bessel functions, 250
best case assessment, 62–65
best-fit algorithm, 600, 601
biconnected components, 401
biconnected graphs, 400
big-endian systems, 558–559
big-omega notation, 57–58
big-O notation, 53–57, 475
  calculation according to definition of, *54*
  estimates, typical functions applied in, *59*
  inherent imprecision of, 55
  possible problems, 58
  properties of, 55–57
binary buddy system, 603–608, *606*
binary code, 564, 566
binary search, 119, 252, 525, 688
  algorithms, 61
  average case for, 63–65
  nondeterministic version of, 69–70
binary search trees, *219*, 251, 262, 266. *See also* 2–4 trees
  comparing tries with, 352
  creating, from ordered array, 253
  deleting nodes in, 244–250
  as an encumbrance, 302
  function for searching, *223*
  implementation of generic, *221–223*
  inserting nodes into, *242*
  property of, 218–219
  searching, 223–225
  transforming, into backbone, 255
binary trees, 215–289, *218*, 566. *See also* trees
  balanced, 251
  defined, 217
  heap properties of, 490
  heaps, 269–277
  height-balanced, 251
  implementing, 220–223

important characteristic of, 217–218
inserting nodes in, 241–244
ordered, 218
perfectly balanced, 251
red-black trees, 324
representing 2–4 trees as, 324
searching, 241
sorting algorithms expressed in terms of, 483
top-down or bottom-up creation of, 279
tries implemented as, *354*
vh-trees, 324–337
bipartite graphs, 423, 426, 542
bits, 35
  strings of, 667
bit-trees, 318, *320*, 320–321
bivariate integral equations, 250
black pointers, 324
blocks, 301
  algorithm for allocating, *602*
  in graphs, 401
  in Huffman trees, 576
  memory, 600
blossoms, 433
Bn-trees, 313
Boolean data members, 235
Boolean expressions, 449
  in conjunctive normal form (CNF), 70–71
  member functions, 90
  satisfiable, 71, 450
Boruvka's algorithm, 467
boundary folding, 526–527
boundary path, 684
bounds, 493–499
  finding best, 499
  selecting, 494
Boyer-Moore algorithm, 655–663, 666
Boyer-Moore-Galil algorithm, 666
breadth-first search, 382, *383*, 413, *414*, 415
breadth-first traversal, 226, 576, 620, 621, 629, 673
  copying lists using, 619
  implementation, *226*
Brelaz algorithm, 446–447
bridge, in graph, 401, 435
B**-trees, 313

B*-trees, 312–313
B+-trees, 314–316, *315*
B-trees, 302–337
  building, 322–323
  deleting keys from, 310–312, *311*
  family of, 301–337
  free-at-empty, 372–374
  important property of, 302–303
  inserting keys, 306–310
  merge-at-half, 372–374
  merging in, 310
  of order 4. *See* 2–4 trees
  of order m, properties of, 303
  searching, 305
  splitting, 306–308, 310
bubble sort, 480–482, 484, 488
bucket addressing, 536–537
buckets, 545–550. *See also* bucket addressing
  hashing with: case study, 550–559
  splitting, *547*
buddy systems, 603–610
bytes, 44
  files as collections of, 36
  used to specify separators, 318

C, 23
  speed of, 52
  unused memory in, 599
C++
  code, 576
  delimiters in, 138
  equivalent of factorial, 172
  features, 23
  implementation of selection sort, *479*
  and object-oriented programming (OOP), 23
  representing polynomials in, 513
  shifting responsibility to garbage collector in, 620
  unused memory in, 599
canonical node, 699
canonical reference, 699
capacity, of edge, 408
Cartesian plane, 321

case studies:
  adding polynomials, 512–519
  computing word frequencies, 282–289
  distinct representatives, 452–464
  exiting a maze, 159–165
  hashing with buckets, 550–559
  Huffman method with run-length encoding, 584–595
  in-place garbage collector, 629–631
  library, 120–130, *121*
  longest common substring, 699–707
  random access files, 36–46
  recursive descent interpreter, 199–207
  spell checker, 358–368
cellar, 535
  coalesced hashing using, *536*
chaining, *534*, 534–536
cheapest insertion algorithm, 469
children, on trees, 215, 300. *See also* nodes
  deleting nodes with, *245*
Chinese postman problem, 436
  solving, *437*
chromatic number, 445
Cichelli's method, 539–542
circuit, 377
circular lists, 95–97, 394, 506
  doubly linked, 96, *97*
  singly linked, *95*, *96*
class(es)
  building, 35–36
  declaration, *3*
  defined, 1–2
  definitions, 6–7, *34*
  functions defined in, 2
  initializing constants in, 44–46
clique, 447
clique problem, 447
clusters, in hashing, 529, 531, 532
coalesced hashing, 535, *535*
coalescing process, 600, 605, *607*, 607–608, *609*
COBOL, compiler in, 599
cocktail shaker sort, 488, 520
collision resolution, 528–537
  bucket addressing, 536–537

chaining, 534–536
open addressing, 528–532, 534
collisions, 526
compaction, 617–618
  data. See data compression
  heap, 611, *618*
compilation, 199
compilation errors, 20, 343
compilers, 199, 599
  generating code in, 280
  symbol tables in, 525
  use of Polish notation with, 278
complement graphs, 450
complete binary trees, 218
complete graphs, 377
complexity analysis, 52–71
  amortized complexity, 65–69
  best, average, and worst cases, 62–65
  big-O notation, 53–55
  computational and asymptotic complexity, 52–53
  estimating amortized cost, 67
  examples of complexities, 58, 60
  finding asymptotic complexity: examples, 60–62
  NP-completeness, 69
  omega and theta notations, 57–58
compression
  adaptive, 576
  of tries, *355*, 355–357
  using run-length encoding, 580–581
compression rate, 355, 566
computational complexity, 52–53
  for sequence of operations, 65
computing word frequencies: case study, 282–289
  implementation, *284–289*
  semisplay tree used for, *283*
conjunctive normal form (CNF), 70–71
connected graphs, 400
connectivity, 400–405
  in directed graphs, 403–404
  in undirected graphs, 400–401, 403
constant algorithm, 58, *59*
constants, initializing, 44–46

constructors
  default, 24, 113
  in nodes, 78
  in vector, 32
containers, 24
  deque, 114
  map, member functions of, *344–346*
  member functions in list, *111–113*
  set, member functions of, *339–340*
  STL, 24
  vector, 27–35, 113
copy constructors, 24
  necessity of using, for objects with pointer members, 15
  pointers and, 14–16
copying methods, of garbage collection, 618–625, 628
count method, of organization, *103*, 104–107
Cray computers, 52
C-tries, 356–357
cubic algorithm, *59*, 60, 393
cut-edge, 401
cut, in network, 409
cut-vertices, 401
cycle, 377
cycle detection, 393–394

dangling reference problem, 11, 94
data
  ordering of, 474–475
  on secondary storage, processing, 300
  spatial, 321
  transferral, 564
data compression, 564–587
  conditions for, 564–566
data encapsulation, 2
data members, 2
  in nodes, 77, 235
data structures. See also specific data structures; graphs
  composed of nodes, 76
  with contiguous blocks of memory, 27–28
  linear, 137. See also stacks
  need to define, 1

and object-oriented programming (OOP), 35–36
  sophisticated, 102. See also trees
  for spatial data, 321
  spell checker, 358
  subject to sequence of operations, 65–69
dBaseIII+, 23
D-bits, useful properties of, 318, 320
decision problems, 69
decision trees, 218, 482–485, *485*. See also sorting
declaration, for nodes, 78
default constructors, 24, 113
degree of vertex, 377
deleting by copying, 247–250, *250*
  implementation of algorithm for, *249*
  in vh-trees, 332–333
deleting by merging, 245–247, *248*
  implementation of algorithm for, *246–247*
  summary of, *246*
deletion
  AVL trees, 259–261, *260–262*
  binary search trees, 244–250
  B+-trees, 315
  B-trees, 310–312, *311*
  by copying, 247–250, 332–333
  elements in arrays, 119
  in hashing, 537–538
  linked lists, 84–90, *85*, *88*, *89*, *94*, 94–95, 96
  by merging, 245–247
  prefix B+-trees, 317, 318
  simple prefix B+-trees, 318
  2–4 trees, 328
  of vertices in graphs, 378
  vh-trees, 332–333, *334–337*
delimiters, matching, 138–139
dependency graph, 542
depth-first search, 380–382, 394, 397, 400, 403, 406, 411, 413, 415
  adapted to topological sort, 406–407
depth-first traversal, 227–234, 629, 688
  implementation, *228*
  stackless, 234–237
  tasks of interest in, 227

depth, of directory, 545, 546
deques, 388–389
    changes, in process of pushing new elements, 118
    inserting elements in, 118–119
    list of member functions in class, 115–116
    program demonstration operating of member functions, 117
    in the Standard Template Library (STL), 114–119
dequeuing, 146, 147, 150
    elements from heaps, 271–272, 273
dereferencing
    in deques, 119
    notation of iterators, 24
    pointer variables, 17, 18
derived classes, 7–9
D'Esopo-Pape algorithm, 465
destructors, 24
    defined, 16
    pointers and, 16
deterministic algorithms, 69–70
differentiation, 281
    of multiplication and division, tree transformations for, 281
digital trees, 370
digraphs, 376. See also directed graphs; networks
    depth-first search in, 381
    linearizing. See topological sort
Dijkstra's algorithm, 384–386
    label-correcting version of, 390
    modified, 420–421
Dijkstra's method, 398, 400, 470
diminishing increment sort. See Shell sort
Dinic's algorithm, 415–418, 416
directed graphs. See also digraphs
    connectivity in, 403–404
directories, 544–547
    depth of, 545, 546
disks, 301
distinction bits. See D-bits

distinct representatives: case study, 452–464
    problem, implementation, 457–464
division, in hashing, 526
double-ended queues. See deques
double hashing, 532, 533, 534
double-O (OO) notation, 58
doubly linked lists, 91, 91–95, 147, 569, 576–577, 600, 603, 629
    circular, 96, 97
    deleting nodes, 94
    implementation, 92–93
    inserting nodes, 93
    random access simulation, 116–117
DSW algorithm, 254–257
dual buddy system, 610
dynamic binding, 22
dynamic hashing, 544

echoprinting, 146
edges, 376
    weights assigned to, 383
edit table, 691
efficiency
    of algorithms, using big-O notation to estimate, 55–57
    best, average, and worst cases, 62–65
    comparing algorithm, 52, 59
    criteria, 52–53
    evaluating algorithm, 55–57
    of heap sort, 493
    of organization methods for linked lists, 105–107, 106
    of prefix B+-trees, 318
    of recursion, 198–199
    of searching in binary trees, 225
    of skip lists, 102, 107. See also skip lists
    of traversal procedures, 241
Eiffel, 23
    automatic storage reclamation in, 600
eight queens problem, 191, 191–198, 195, 196, 197–198

implementation, 193–194
encapsulation, 1–5
endpoint, 687
enqueuing, 146, 147
    algorithm, 271
    elements to heaps, 272, 272, 274
entropy, 564
entry table, 624–625
equality subgraph, 430
Eulerian cycle, 434, 435
Eulerian graphs, 434–437
    Chinese postman problem, 436
Eulerian trail, 434
exact string matching, 644–688
    bit-oriented approach, 667–670
    Boyer-Moore algorithm, 655–663
    Knuth-Morris-Pratt algorithm, 647–654
    matching sets of words, 670–677
    multiple searches, 665–666
    regular expression matching, 677–681
    straightforward algorithms, 644–647
    suffix arrays, 687–688
    suffix tries and trees, 681–687
    Sunday algorithms, 663–665
exceptions, throwing, 85–86
excessive recursion, 187–191
exiting a maze: case study, 159–165
    algorithm psuedocode for, 161
    example of processing maze, 161
    program for maze-processing, 162–164
    stacks used in, 160
expandable hashing, 544
expected value, 62
exponential algorithm, 59
expression trees, 278, 278–279, 281
    operations on, 279–282
Extended Binary Coded Decimal Interchange Code (EBCDIC), 474

extendible hashing, 117, 544, *545*, 545–547
external fragmentation, 600, 607, 610
extraction, in hashing, 527

factorial function, 171
farthest insertion algorithm, 469
fastmark algorithm, 616
fax images, 581
FHCD algorithm, 542–544
Fibonacci buddy system, 608
Fibonacci numbers, 188–191, 199
    algorithms for calculating, *190*
    calculating, *189*
Fibonacci trees, 292
Fibonacci words, 654
FIFO (first in/first out) structures, 145. *See also* queues
files, 36
fill factor, 313
first-fit algorithm, 600, 601
fixed-length records, 36
flag flipping, 329–331
Fleury's algorithm, 435
floating garbage, 628
float numbers, 4, 90
flow, 408–409
flow-augmenting path, 409
flower, on blossom, 433
folding, 526–527
Ford-Fulkerson algorithm, 409–415, *411–412, 414*, 418
Ford's algorithm, 387–388
forest, 381
Forth, use of postfix notation by, 279
FORTRAN, compiler in, 599
forward edges, 381
4-nodes, 324, *325*
fraction of data reduction, 566. *See also* compression rate
free-at-empty B-trees, 372–374
friend functions, 23
fromspace, 620–621
full suffix rule, 657
functional languages, 3
function calls, 18, 180
    and recursion implementation, 173–175
function members, 2
    virtual, 22

function objects, 26–27
    one- and two-argument, 33
functions
    complex, 52–53
    defined in a class, 2
    factorial, 171
    f, growth rate of all terms of, 53
    friend, 23
    generic, 27, 28, 33. *See also* algorithms
    pointers to, 19–20
    recursive, 175
    for searching binary search tree, *223*
    swap, 5

garbage collection, 610–628
    copying methods, 618–625, 628
    decreasing cost of, 620
    incremental garbage collection, 620–628
    mark-and-sweep, 611, 625, 628
    marking phase, 611–616
    noncopying methods of, 625–628
    reclamation phase, 616–617
garbage collector, 610–611
generalized Fibonacci systems, 608
generational garbage collection, 622–625, 628
    Lieberman-Hewitt technique, 624
generic classes, 4–5
generic entities, 24
generic functions, 27, 28, 33. *See also* algorithms
Godel numbers, 168
graphs, 376–464, *377*
    connectivity, 400–405
    cycle detection, 393–394
    defined, 376, 377
    Eulerian graphs, 434–437
    graph coloring, 444–447
    Hamiltonian graphs, 438–439
    matching, 423–433
    networks, 408–422
    NP-complete problems in graph theory, 445, 457–451
    representation, 378, *379*

    shortest paths, 383–393
    spanning trees, 381, 396–398, 400
    topological sort, 406–407
    traversing. *See* graph traversals
graph traversals, 380–382
ground case, 170

Hamiltonian cycle, 438–439, *440*
    problem, 451
Hamiltonian graphs, 438–439
    traveling salesman problem (TSP), 440–444
Hancart's algorithm, 646–647
hash functions, 525, 526–528, 584
    division, 526
    for extendible files, 544–550
    extendible hashing, 545–547
    extraction, 527
    folding, 526–527
    linear hashing, 547–550
    mid-square function, 527
    radix transformation, 528
hashing, 525–550
    with buckets: case study, 550–559
    collision resolution, 528–537
    deletion, 537–538
    hash functions, 526–528
    hash functions for extendible files, 544–550
    perfect hash functions, 538–544
    using buckets, implementation, *552–558*
heap, memory, 599, *603*
    compaction, 611, *618*. *See also* compaction
    division of, 600
heap property, 269
heaps, 269–277 386, 569–572. *See also* heap sort
    implemented by arrays, 269–270
    and nonheaps, *270*
    organizing arrays as, 272–277, *275*
    as priority queues, 271–272
    properties of, 269, 490
heap sort, 489–493
height
    -balanced trees, 251, 292

of empty and nonempty trees, 215
maximum number of nodes in binary trees of different, 252
tree, extension or reduction after deletion by merging, 248
of tries, determining, 351, 352
of vh-trees, 325–327, *327*
height-balanced trees, 251
one-sided, 292
hexadecimal notation, 558, 564
hiding, 2, 3, 8, 23
horizontal pointers, 324, *326*
Huffman algorithm, 566–567
implemented with heap, *572*
Huffman coding, 566–575
adaptive, 575–580
Huffman method
implementation of, *587–595*
with run-length encoding: case study, 584–595
Huffman trees, *568, 569*–575
Hungarian tree, 426, 433
Hunt-Szymanski algorithm, 693–695

incidence matrix, 378
incremental garbage collection, 620–628
copying methods in, 620–625
indexes, 344, 525
of buckets, 544
index set, 314, 315
of B+-tree, 317–318
indirect recursion, 185–186
inequalities, 65
infix notation, 279
information-hiding principle, 2–3, 8, 23, 220
compromising, 19
inheritance, 6–9, 35
public and protected, 8
initialization, 60, 61
of constants, 44–46
of nodes, 78
of sets, 342
of variables, 44–46
inorder tree traversal, 227–232, *229*, 314, 342–343
changes in run-time stack during, *231*

details of several first steps of, *230*
generation of infix notation by, 279
Morris algorithm for, 238–241, *239*
nonrecursive, 234
nonrecursive implementation of, *234*
of threaded tree, and implementation of generic threaded tree, *236–237*
in-place garbage collector: case study, 629–631
implementation, *632–637*
insertion
arrays, 119
AVL trees, 258–259
binary search trees, 242
binary trees, 241–244
B*-trees, 313
B+-trees, 314–315
B-trees, 306–310
deques, 118–119
linked lists, 82–84, *83, 84*, 91, *93, 94, 95–96, 96*
multisets, 338, 341
R-trees, 321–322
sets, 341
simple prefix B+-trees, 318
threaded trees, 243, *244*
tries, 352
2-4 trees, 328
of vertices in graphs, 378
vh-trees, 330–331
insertion sort, 475–478, 484, 486, 488, 499, 511
integers, 4, 35
sorting, 502–503
storage of, 90
internal fragmentation, 600, 610
internal path length (IPL), 224, 250
interpretation, 199
interpreter
implementation of simple language, *203–207*
recursive descent: case study, 199–207
sample, for limited programming language, 199–207
intervals, 20

intractable problems, 70
inversion, 105–106
ISBN code, 527
isolated vertex, 377
iteration
use in real-time systems, 198–199
use of, versus recursion, 198–199
iterators, 24–25
STL, 24
used by most STL algorithms, 25

Jarník-Prim algorithm, 442, 467, 470,
Java, automatic storage reclamation in, 600

k-colorable, 445
Kershenbaum's algorithm, 465
keys, 302
comparison of, 525
deletion from B-trees of, 310–312
finding, in vh-trees, 324
insertion into B-trees of, 306–310
separator, 317
in tries, 349
used in maps, 344
Kleene closure, 677
Knuth-Morris-Pratt algorithm, 647–654, 673–674
Kruskal's algorithm, 397–400, 470
Kuhn-Munkres algorithm, 430–431

label-correcting methods, 384–390
label, in network, 409, 421
label-setting methods, 383–390
largest first algorithm, 446
last-come-first-served hashing, 561
latency, 301
layered network, 415, 417
lazy deletion, 292
leaders, 576
leaves, 215, 218
adding, to trees, *219*
of bit-trees, 320

in B+-trees, 314
in decision trees, 484
deleting, 244
or nonleaves, merging in B-trees, 310
of tries, 349
left side occurrence rule, 656
length of path, 215
level network. *See* layered network
level, of nodes, 215, 218
Levenshtein distance, 689–690, 693
lg, 57
library case study, 120–130
   implementation, *123*
   library program, *124–130*
   linked lists indicating library status, *121*
LIFO (last in/first out) structures, 137. *See also* stacks
linear algorithm, 59
linear hashing, 544, 547–550
linear probing, 528–529, 532, *533*, 534
linked lists, 76–130, 567
   advantages of arrays over, 119–120
   circular lists, 95–97, 394
   creating three-node, 78–79
   deleting nodes, 84–90, 94–95, 96
   doubly linked lists, *91*, 91–95
   drawback of, 97
   implementation of queues, 149–150
   implementing stacks as, 143–144
   indicating library status, *121*
   inserting nodes, 82–84, *83*, *84*, 91, *93*, 94, 95–96, *96*
   library program, *124–130*
   of *n* elements, searching, 217
   processing stream of data, *104*
   scanning, 88
   searching operation, 90
   self-organizing lists, 102–107, *103*
   singly linked lists, 76–90, *77*
   skip lists, 97–102
   sparse tables, 107–110
   student grades implemented using, *110*
   transforming, into trees, *217*

linked structures, 76. *See also* linked lists
LISP, 3, 23
   automatic storage reclamation in, 599
   code, 576
   files, 587
   forward references in, 624
   garbage collection in, 611
   read barrier in, 628
   speed of, 52
   use of prefix notation by, 279
listings, 689
little-endian systems, 558–559
loading factor, *533*, 548
local depth, in hashing, 546
logarithmic algorithm, 59
logarithmic function, 56–57
LOGO, use of prefix notation by, 279
longest common substring: case study, 699–707
   listing of program to find, *702–707*
loops, 60–61
   to calculate sum of numbers in arrays, 60–62
   in functional languages, 20
   infinite, 380, 619, 629
   in linked lists, 96, 116–117
   nested, 60, 61
   relation of tail recursion to, 179
   simple, 61
   in singly linked lists, 91
LZ77, 581–582, *582*
LZW, 582–583, *583*

maps, 344–349. *See also* tables
mark-and-sweep, 611, 625, 628
marking, in garbage collection, 611–616
Markov chains, theory of absorbing, 645
marriage problem, 423
matching, 423–433, *425*
   assignment problem, 430–432
   in nonbipartite graphs, 432–433
   stable matching problem, 427–430
matching problem, 423

matchlist, 694
matrix, 378
max-flow min-cut theorem, 409
max-flow problem. *See* maximum-flow problem
max heap, 269
maximum-flow problem, 409
maximum flows, 408–418
   of minimum cost, 418–422, *422*
maximum matching, 423
member functions, 2, 4, 8
   algorithms defined as, 25
   application of set and multiset, 340–341
   Boolean, 90
   in class deque, *115–116*
   in class vector, *28–29*
   common to all containers, 24
   in list container, *111–113*
   of map container, 344–346
   overloaded, 32–33
   overloading operators as, 26
   for processing doubly linked lists, 91
   program demonstrating operation of list, *113–114*
   program demonstrating operation of vector, *30–31*
   of set container, 339–340
memory, 19
   allocation, dynamic, 599
   C-tries and, 357
   data structures with contiguous blocks of, 27–28
   dynamically allocating and deallocating, 10
   management. *See* memory management
   primary, 312
   secondary, 312
memory cells
   marking currently used, 611
   pool of free, 89, 611
memory management, 599–628
   garbage collection, 610–628
   nonsequential-fit methods, 601–610
   sequential-fit methods, 600–601
memory manager, 10, 600
merge-at-half B-trees, 372–374

mergesort, 499–502, *501*, 521
merging
    in B-trees, 310
        deleting by, 245–247
message passing, 4
methods, 2
mid-square hash function, 527
min heap, 269, 569–570
minimal vertex, 406
minimum spanning trees, 397–398, 400, 441
Modula-3, 600
Morris algorithm, 238–241, *239, 240*
Morse code, 564
move-to-end method, 132
move-to-front method, of organization, *103*, 103–107
multigraphs, 376, 378
multimaps, 349
multisets, 338, 341–344
multiway Patricia trees, 372
multiway search tree of order *m*, 168, 300
multiway tree of order *m*, 300
multiway trees, 300–357. *See also specific trees*
mutators, 620, 621, 625
m-way search trees, 300
m-way trees, 300

natural sort, 521
nearest addition algorithm, 443
nearest insertion algorithm, 469
nearest neighbor algorithm, 469
nearest merger algorithm, 470
n-connected graphs, 400
negative cycle, 384
nested loops, 60, 61
nested recursion, 187
networks, 408–422
    defined, 408
    maximum flows, 408–418
next-fit method, 601
Nicod's axiom, 292
nodes, 76
    on B-trees, 302–303
    definition of, 77–78
    deleting, in binary search trees, 244–250
    deleting, in linked lists, 84–90, *85, 88, 89, 94*, 94–95, *96*

implementing, 102
initialization, 78
inserting, in binary trees, 241–244
inserting, in linked lists, 82–84, *83, 84*, 91, *93*, 94, 95–96, *96*
inserting, in threaded tree, *244*
maximum number, in binary trees of different heights, *252*
in *m*-way search trees, 300
in skip list, *98*
in trees, 215
types of, in tries, 352
nonbipartite graphs, 432–433
noncopying methods, of garbage collection, 625–628
nondeterministic algorithms, 69–70
    polynomial, 69
nondeterministic finite automaton (NDFA), 677–681
nonsequential-fit methods, 601–610
nonsequential searching, 97
nontail recursion, 179–185
no occurrence rule, 656
normalization function, 528
notation, 644
    array, 12, 13
    dot, 348
    pointer, 12, 13
NP-completeness, 69–71
    clique problem, 447–448
    Hamiltonian cycle problem, 451
    problems in graph theory, 445, 447–451
    3-colorability problem, 448–450
    vertex cover problem, 450–451
null suppression, 581
numbers
    adding large, 140
    represented in binary form in computers, 564

$O(n \lg n)$ algorithm, *59*
object-oriented language (OOL), 1
object-oriented programming (OOP), 1–46
    abstract data types, 1, 137

C++ and, 23
data structures and, 35–36
encapsulation, 1–5
inheritance, 6–9
pointers, 9–20
polymorphism in context of, 21–23
random access files, 36–46
roots of, in simulation, 2
Standard Template Library (STL), 24–35
objects, 1–3
    instantiation of, 3
    invoking member functions, 4
    in OOL, 2
    organizing hierarchical representation of. *See* trees
    self-referential, 77
    structuring programs in terms of, 2
open addressing, 528–532, 534
operators, 8, 24, 26. *See also specific operators*
    built-in, 27
    overloading, 34, 45, 235
optimal assignment problem, 430
optimal code, 565–566
optimal-fit method, 638
optimal static ordering, 105
optimization problems, 69
ordered binary trees, 218
ordering method, of organization, *103*, 104–107
orderly trees, 217

parsing, 200, 202
partial suffix rule, 657
partition approach, 698
Pascal, 2, 23
    cross-reference program, 541
    -like languages, 3
    shifting responsibility to garbage collector in, 620
    unused memory in, 599
path, 215, 377
    all-to-all shortest, problem, 390–391, 393
    finding shortest, in graph theory, 383
Patricia trees, 372 n, *373*
pattern matching, 644
PCs, 52

perfect hash functions, 525, 538–544
  Cichelli's method, 539–542
  FHCD algorithm, 542–544
perfectly balanced trees, 251, 256
perfect matching, 423
periods, 666
pivots. See bounds
pointers, 9–20
  array of, 102, 117–118
  arrays and, 12–13
  attributes, 9
  to blocks or arrays of data, 117–118
  changes of values after assignments, using, 10–11
  in circular lists, 95
  and copy constructors, 14–16
  and destructors, 16
  in doubly linked lists, 91
  flexible implementation of linked structures using, 76. See also linked lists
  to functions, 19–20
  horizontal and vertical, 324
  inconvenience of using, 79
  to linked lists, 79, 82
  necessity of using copy constructor for objects with members that are, 15
  null, 76
  red and black, 324
  reference variables and, 17–19
  singly linked list of, 567
  in singly linked lists, 76, 78, 82–84, 87
  in skip lists, 97, 98, 101
  split, 547–548
  in trees, 235
  in tries, 351
  updating, 620
point quadtrees, 370
Polish notation, 277–279
polymorphism, 21–23
polynomials, 56
  adding: case study, 512–519
  nondeterministic algorithms considered, 69
postfix notation, 279
postorder tree traversal, 227, 233
  generation of postfix notation by, 279

nonrecursive implementation of, 233
nonrecursive version of, 233
threaded trees used for, 237
PostScript, use of postfix notation by, 279
potential, 68
prefix B+-trees, 317–318, 319
prefix notation, 279
prefix property, 565
preorder tree traversal, 227, 232–233
  generation of prefix notation by, 279
  nonrecursive implementation of, 232
  threaded trees used for, 237
presplitting, 307
primary memory, 312
priority queues, 154, 567, 569
  heaps as, 271–272
  member functions, 157
  program using member functions of container, 158
  in the Standard Template Library (STL), 157–159
private information, 23
probability distribution, 62–63
probability function, 62
probe, 528
probing function, 528, 529
problems
  intractable, 70
  NP-complete, 70
  tractable, 70
programs
  random access file management, 37–43
  structuring, in terms of objects, 2
Prolog
  automatic storage reclamation in, 600
  tail recursion in, 179
pseudographs, 376–377
pseudokeys, 545
public information, 23, 36

quadratic algorithm, 58, 59, 60
quadratic probing, 531, 533, 534
quaternions, 49

queues, 145–159, 382, 388, 389, 413, 503–506
  applications of, 146
  array implementation of, 147, 148–149
  double-ended. See deques
  in graphs, 382
  linked list implementation of, 149–150
  member functions, 156
  operations executed on abstract, 151
  operations needed to manage, 145
  piles organized as, 503
  priority, 154, 157–159
  series of operations executed on, 146
  in the Standard Template Library (STL), 155–157
  used in breadth-first traversal, 226
  used in simulations, 150
queuing theory, 150
quick-fit method, of memory allocation, 639
quicksort, 493–499, 511, 550
  implementation, 495

radix sort, 502–506, 504, 511
radix transformation, 528
random access, 27. See also random access files
  to deque positions, 114
  performed in constant time, 119
  of pointers stored in arrays, 353
  simulation in doubly linked lists, 116
random access files
  case study, 36–46
  program to manage, 37–43
read barrier, 621, 625
real-time systems, use of iteration in, 198–199
records, 44
  fixed-length, 36
recursion, 170–199, 228, 253, 494, 499
  anatomy of recursive call, 175–178

backtracking, 191–198
  double, 227
  efficiency of iteration versus, 198–199
  excessive, 187–191
  implementation, function calls and, 173–175
  indirect, 185–186
  as natural implementation of backtracking, 192
  nested, 187
  nontail, 179–185
  power of, 234
  recursive definitions, 170–173
  recursive descent interpreter: case study, 199–207
  replacing, with iteration in mergesort, 502
  tail, 178–179
  von Koch snowflake implementation, *183–185*
recursive definitions, 170–173
  parts of, 170
  purposes of, 171
  of sequences, undesirable feature of, 171–172
  uses of, 172
recursive descent, 200, 279
recursive descent interpreter
  case study, 199–207
  diagrams of functions used by, *201*
red-black trees, 324, *325*, *326*
  maps implemented as, 344
  sets implemented as, 342–343
  use in STL, 338
redistribution, in B-tree, 310
red pointers, 324
reducibility, 70, 71
reduction algorithm, 70
reduction function, 70
reference count method, of garbage collection, 638
reference variables, 17–19
  initializing, 17
relational databases, 580
right side occurrence rule, 656
Robin Hood hashing, 561
root, 20
  of SCC, 403
root pointers, 611
root set, 611

roots, tree, 215
rotation, 254–257, 265–266
  of child about parent, *254*
rotational delay, 301
R+-trees, 322
R-trees, 321–322
run-length encoding, 580–581
  candidates for compression using, 581
  Huffman method with: case study, 584–595
runs, 521
  defined, 580
run-time stacks, 599. *See also* stacks
  for eight queens problem, *196*
  during inorder traversal, *231*
  overflow of, 611–612
  in recursion, 172–173, *174*, *177*, *180*, 199
  reliance of recursive traversals on, 241
  in tree traversal, 227, 229–231

satisfiability problem, 70–71
saturation degree, of vertex, 446
scatter tables, 534
Schorr and Waite algorithm, 612–613
  example of execution of, *614–615*
scope operator, 8
searching
  binary search trees, 223–225
  binary trees, 241
  bit-tree, 320, 321
  B-trees, 305
  compressed tries, problem with, 356
  linked list of n elements, 217
  linked lists, 90, 97–98, 104–105
  nonsequential, 97
  process, tree operations that accelerate. *See* trees

  R-tree, 321–322
  sequential, 44, 63, 97, 102, 117, 525
  skip lists, 98
  2-4 tree, 324, 327
  vh-trees, 324–237
secondary clusters, 531, 532

secondary memory, 312
secondary storage
  decrease in speed related to, 301
  relation of B-trees to, 302–304
  use of trees to process data on, 300
seek time, 301
selection sort, 478–480, *479*, 484, 489–490, 511
self-adjusting trees, 102, 262–269
  self-restructuring trees, 263
  splaying, 264–269
self-organizing lists, 102–107
  count method, *103*, 104–107
  methods for organizing, 103–107
  move-to-front method, *103*, 103–107
  ordering method, *103*, 104–107
  transpose method, *103*, 103–107
self-referential objects, 77
self-restructuring trees, 263
semispaces, 618, 620–621
  fromspace and tospace, 620–621
semisplaying, *265*, 268–269, *269*, 283
separate chaining, 534
sequence set, 314
sequential coloring, 445
sequential-fit methods, 600–601
sequential scanning, 97, 117, 353
sequential searching, 44, 63, 97, 102, 117, 525
sets, 338–344
shakersort, 488
sharp sign (#), *350*, 351
Shell sort, 486–489, *487*
shift folding, 526, 527
sibling property, 576
side edges, 393
simple graphs, 376
simple prefix B+-trees, 317–318, *319*
Simula, 2
simulations
  queues used in, 150
  random access, in doubly linked lists, 116–117
  roots of OOP in, 2

singly linked lists, 76–90, *77*, 217, 314–315, 567
   circular, *95*, *96*
   deleting nodes, 84–90, *85*, *88*, *89*
   implementation, *79–81*
   implementation of circular, *95–96*
   inserting nodes, 82–84, *83*, *84*
   of integers, *82*
sink, 406, 408
skip lists, 97–102, *98*
   efficiency of, 102
   implementation, *99–101*
slack, 410
slow sorting, 521
small label first method, 390
Smalltalk, 23
   automatic storage reclamation in, 600
sorting, 65, 474–510. *See also* decision trees
   algorithms, 33–35
   algorithms, efficient, 486–506
   algorithms, elementary, 475–482
   bubble sort, 480–482
   criteria, 474
   heap sort, 489–493
   insertion sort, 475–478
   mergesort, 499–502
   quicksort, 493–499
   radix sort, 502–506
   selection sort, 478–480
   Shell sort, 486–489
   speed, estimating lower bound of, 482
   in Standard Template Library (STL), 506–510
   topological sort, 406–407
source, in network, 408
space
   complexities, algorithms classified by, 58, *59*, 60
   as efficiency criteria, 52
   wasted, 109, 324, 352
space reclamation, 616–617
spanning trees, 381, 396–398, *398–399*, 400
sparse tables, 107–110
   arrays and, for storing student grades, *108*

spell checker: case study, 358–368
   implementation, using tries, *361–368*
splaying, 264–269, *265*
   heterogeneous configuration, 264, *265*, 268
   homogeneous configuration, 264, *265*, 267
   restructuring tree with, *266*
splitting
   blocks into buddies, 609
   B-tree, 306–308, 310
   B*-trees, 313
   B+-trees, 315
   in hashing, 547–550
   prefix B+-trees, 318
   R-trees, 321–322
   2-4 trees, *326*, 328
   vh-trees, 328–330
stable marriage problem, 427–430
stable matching problem, 427–430
stable sorting, 510
stacked-node-checking algorithm, 616
stack frame, 173
stacks, 137–145, 229–232, 382, 403, 503, 612, 616. *See also* run-time stacks
   adding large numbers using, *141*
   applications of, 138, 140
   explicit, 611–612
   to implement recursion, 173
   linked list implementation, *143–144*, 144, *145*
   list of member functions, 155
   operations, 137
   overflow of, 616
   run-time, 172–173, *177*, *180*, *196*, 199
   series of operations executed on, *138*
   in the Standard Template Library (STL), 155
   vector implementation of, *142–143*, 144, *145*
Standard Template Library (STL), 24–27
   algorithms, 24, 25
   containers, 24
   deques, 114–119
   function objects, 26–27

   iterators, 24–25
   lists, 110–114
   maps, 344–349
   multimaps, 349
   multisets, 338, 341–344
   priority queues, 157–159
   queues, 155–157
   sets, 338–344
   sorting in, 506–510
   stacks, 155
   vectors, 27–35
star representation, 378
static binding, 22
stem, on blossom, 433
stop-and-copy algorithm, 618, 628
straight merging, 521
string matching, 644–699
   approximate string matching, 689–699
   exact string matching, 644–688
stringology, 644
strings, 13, 580. *See also* string matching
   inaccessible, 16
   in random access files, 44–45
strongly connected components (SCC), 403–404
strongly connected graphs, 403
structures
   FIFO (first in/first out), 145. *See also* queues
   LIFO (last in/first out), 137. *See also* stacks
   linked, 76. *See also* linked lists
subarrays, 61, 62
subclasses, 7–9
subgraphs, 377
suffix arrays, 687–688
suffix links, 684
suffix trees, 681–687, *682*
suffix tries, 681–687, *682*
Sunday algorithms, 663–665
superclass, 7–9
super-symbols, 585
swap functions, 5
swapping technique, 132
sweeping, 616–617
symmetric binary B-trees. *See* 2-4 trees
symmetric difference, 424
system of distinct representatives theorem, 453

tables, 107. *See also* maps
    adjacency list, represented by, 378
    sparse, 107–110
tag bit, 612
tail recursion, 178–179
Tarjan's algorithm, 404, *405*
templates, 1, 90
template structure, 34
temporary storage, 357
theta notation, 57–58
threaded trees, 234–237, *235*, 243
    implementation of algorithm to insert node into, *243*
    implementation of generic, and inorder traversal, *236–237*
threads, 235
3-colorability problem, 448–450
3-nodes, 324, *325*
three-satisfiability problem, 71, 447
threshold algorithm, 390
time
    algorithms, polynomial-, 70
    complexities, algorithms classified by, 58, *59*, 60
    constant, 119
    as efficiency criteria, 52
    polynomial, 450
topological sort, 406–407, *407*
tospace, 620–621
tournament, 468
traces, 689
tractable problems, 70
transfer time, 301
transitive tournament, 468
transpose method, of organization, *103*, 103–107
traveling salesman problem (TSP), 440–444
traversal. *See* tree traversal
treadmill, 640, *641*
tree edges, 381
trees, 69, 199, *216*. *See also specific trees*; tries
    AVL, 102, 257–262
    balancing, 251–262, *258*, *259*
    of calls for Fib (6), *189*
    complete binary, 218
    decision, 218
    defined, 215, 300

expression, 278–279
height of empty and non-empty, 215
limitations of, 376
multiway. *See* multiway trees
orderly, 217
representing hierarchical structures, 215, *216*
representing recursive calls as, *186*
restructuring, *263*, *264*
self-adjusting, 102, 262–269
threaded, 234–237
transforming linked lists into, *217*
tree traversal, 225–241
tree traversal, 225–241, 576
    breadth-first traversal, 226
    defined, 225
    depth-first traversal, 227–234
    with Morris method, *240*
    stackless depth-first traversal, 234–237
    through tree transformation, 237–241
trie *a tergo*, 353
tries, 349–357, *350*, 671
    compression of, *355*, 355–357
    defined, 349
    implementation of spell checker using, *361–368*
    implementation using pseudo-flexible arrays, *360*
    implemented as binary trees, *354*
    space problems of, 352
Turing machine, 70, 71
2-connected graphs, 400
2-4 trees, 322–337
    represented by red-black trees, *326*
two-pointer algorithm, 617
2-3-4 trees. *See* 2-4 trees

Ukkonen algorithm, 684–687, 699
uncolored degree, of vertex, 446
underflow, 310, 315
undirected graphs, connectivity in, 400–401, 403
uniform probability distribution, 63
union-find problem, 394–396

unions, 303
universal coding scheme, 581
UNIX, 584, 696
    Hunt-Szymanski algorithm implemented in, 695
    implementation of Aho-Corasick in, 677

variables
    auxiliary. *See* pointers
    Boolean, 71, 447, 449
    in common algebraic operations, 512
    distinguishing single, 3
    initializing, 44–46
    pointers and reference, 17–19. *See also* pointers
    temporary, 88
variant buddy system, 610
variant records, 303
vectors
    adding new elements to, 31–32
    containers, 113
    elements, accessing, 32
    implementing stacks as, *142–143*
    member functions list, *28–29*
    ordered, 338
    program demonstrating operation of member functions, *30–31*
    in the Standard Template Library (STL), 27–35
    unordered, 338
vertex, 376
    as articulation point, 401
    degree of, 377
    independent, 444
    isolated, 377
    saturation degree, 446
    uncolored degree, 446
vertex cover, 450
vertex cover problem, 450–451
vertical-horizontal trees. *See* vh-trees
vertical pointers, 324, *326*
vh-trees, 324–337
    building, *332*
    deleting nodes from, 332–333, *334–337*
    inserting numbers in, 330–331, *332*

properties of, 324
splitting, 328–333
vine. *See* backbone
virtual hashing, 544
von Koch snowflakes, 182–185
    construction of, 182
    drawing, *182*
    examples of, *182*
    recursive implementation of, 183–185

Wagner-Fischer algorithm, 691–693
weakly connected graphs, 403
weighted buddy system, 610
weighted graphs, 377
weighted path length, 567
WFI algorithm, 391–393, 466
    execution of, *392*
worst case assessment, 62–65
worst-fit method, 600–601

Yuasa's algorithm, 625–628

zero-order predictor, 597
Ziv-Lempel algorithm, 582–583
Ziv-Lempel code, 581–584